SCHAUM'S OUTLINE OF

Theory and Problems of

ELECTRIC CIRCUITS

Fourth Edition

MAHMOOD NAHVI, Ph.D.
Professor of Electrical Engineering
California Polytechnic State University

JOSEPH A. EDMINISTER
Professor Emeritus of Electrical Engineering
The University of Akron

Schaum's Outline Series
McGRAW-HILL
New York Chicago San Francisco Lisbon
London Madrid Mexico City Milan New Dehli
San Juan Seoul Singapore Sydney Toronto

The McGraw-Hill Companies

MAHMOOD NAHVI is Professor of Electrical Engineering at California Polytechnic State University in San Luis Obispo, California. He earned his B.Sc., M.Sc., and Ph.D., all in electrical engineering, and has 44 years of teaching and research in this field. Dr. Nahvi's areas of special interest and expertise include network theory, control theory, communications engineering, signal processing, neural networks, adaptive control and learning in synthetic and living systems, communication and control in the central nervous system, and engineering education. In the area of engineering education, he has developed computer modules for electric circuits, signals, and systems which improve teaching and learning of the fundamentals of electrical engineering.

JOSEPH A. EDMINISTER is Professor Emeritus of Electrical Engineering from the University of Akron in Akron, Ohio, where he also served as Assistant Dean and Acting Dean of Engineering. He was a member of the faculty from 1957 until his retirement in 1983. In 1984 he served on the staff of Congressman Dennis Eckart (D-11-OH) on an IEEE Congressional Fellowship. He then joined Cornell University as a patent attorney and later as Director of Corporate Relations for the College of Engineering until his retirement in 1995. He received his B.S.E.E. in 1957 and his M.S.E. in 1960 from the University of Akron. In 1974 he received his J.D., also from Akron. Professor Edminister is a registered Professional Engineer in Ohio, a member of the bar in Ohio, and a registered patent attorney. He is the author of *Schaum's Outline of Theory and Problems of Electromagnetics*.

Schaum's Outline of Theory and Problems of
ELECTRIC CIRCUITS

2 3 4 5 6 7 8 9 10 11 12 13 14 15 16 17 18 19 20 VLP VLP 0 9 8 7 6 5 4 3

ISBN 0-07-139307-2

PREFACE

This book is designed for use as a textbook for a first course in circuit analysis or as a supplement to standard texts and can be used by electrical engineering students as well as other engineereing and technology students. Emphasis is placed on the basic laws, theorems, and problem-solving techniques which are common to most courses.

The subject matter is divided into 17 chapters covering duly-recognized areas of theory and study. The chapters begin with statements of pertinent definitions, principles, and theorems together with illustrative examples. This is followed by sets of solved and supplementary problems. The problems cover a range of levels of difficulty. Some problems focus on fine points, which helps the student to better apply the basic principles correctly and confidently. The supplementary problems are generally more numerous and give the reader an opportunity to practice problem-solving skills. Answers are provided with each supplementary problem.

The book begins with fundamental definitions, circuit elements including dependent sources, circuit laws and theorems, and analysis techniques such as node voltage and mesh current methods. These theorems and methods are initially applied to DC-resistive circuits and then extended to RLC circuits by the use of impedance and complex frequency. Chapter 5 on amplifiers and op amp circuits is new. The op amp examples and problems are selected carefully to illustrate simple but practical cases which are of interest and importance in the student's future courses. The subject of waveforms and signals is also treated in a new chapter to increase the student's awareness of commonly used signal models.

Circuit behavior such as the steady state and transient response to steps, pulses, impulses, and exponential inputs is discussed for first-order circuits in Chapter 7 and then extended to circuits of higher order in Chapter 8, where the concept of complex frequency is introduced. Phasor analysis, sinuosidal steady state, power, power factor, and polyphase circuits are thoroughly covered. Network functions, frequency response, filters, series and parallel resonance, two-port networks, mutual inductance, and transformers are covered in detail. Application of Spice and PSpice in circuit analysis is introduced in Chapter 15. Circuit equations are solved using classical differential equations and the Laplace transform, which permits a convenient comparison. Fourier series and Fourier transforms and their use in circuit analysis are covered in Chapter 17. Finally, two appendixes provide a useful summary of the complex number system, and matrices and determinants.

This book is dedicated to our students from whom we have learned to teach well. To a large degree it is they who have made possible our satisfying and rewarding teaching careers. And finally, we wish to thank our wives, *Zahra Nahvi* and *Nina Edminister* for their continuing support, and for whom all these efforts were happily made.

MAHMOOD NAHVI
JOSEPH A. EDMINISTER

CONTENTS

CHAPTER 1

Introduction

1.1 ELECTRICAL QUANTITIES AND SI UNITS

The International System of Units (SI) will be used throughout this book. Four basic quantities and their SI units are listed in Table 1-1. The other three basic quantities and corresponding SI units, not shown in the table, are temperature in degrees kelvin (K), amount of substance in moles (mol), and luminous intensity in candelas (cd).

All other units may be derived from the seven basic units. The electrical quantities and their symbols commonly used in electrical circuit analysis are listed in Table 1-2.

Two supplementary quantities are plane angle (also called phase angle in electric circuit analysis) and solid angle. Their corresponding SI units are the radian (rad) and steradian (sr).

Degrees are almost universally used for the phase angles in sinusoidal functions, for instance, $\sin(\omega t + 30^\circ)$. Since ωt is in radians, this is a case of mixed units.

The decimal multiples or submultiples of SI units should be used whenever possible. The symbols given in Table 1-3 are prefixed to the unit symbols of Tables 1-1 and 1-2. For example, mV is used for millivolt, 10^{-3} V, and MW for megawatt, 10^6 W.

1.2 FORCE, WORK, AND POWER

The derived units follow the mathematical expressions which relate the quantities. From "force equals mass times acceleration," the *newton* (N) is defined as the unbalanced force that imparts an acceleration of 1 meter per second squared to a 1-kilogram mass. Thus, $1\,\text{N} = 1\,\text{kg}\cdot\text{m/s}^2$.

Work results when a force acts over a distance. A *joule* of work is equivalent to a *newton-meter*: $1\,\text{J} = 1\,\text{N}\cdot\text{m}$. Work and energy have the same units.

Power is the rate at which work is done or the rate at which energy is changed from one form to another. The unit of power, the *watt* (W), is one joule per second (J/s).

Table 1-1

Quantity	Symbol	SI Unit	Abbreviation
length	L, l	meter	m
mass	M, m	kilogram	kg
time	T, t	second	s
current	I, i	ampere	A

Table 1-2

Quantity	Symbol	SI Unit	Abbreviation
electric charge	Q, q	coulomb	C
electric potential	V, v	volt	V
resistance	R	ohm	Ω
conductance	G	siemens	S
inductance	L	henry	H
capacitance	C	farad	F
frequency	f	hertz	Hz
force	F, f	newton	N
energy, work	W, w	joule	J
power	P, p	watt	W
magnetic flux	ϕ	weber	Wb
magnetic flux density	$\mathbf{B}$	tesla	T

EXAMPLE 1.1. In simple rectilinear motion a 10-kg mass is given a constant acceleration of $2.0\,\mathrm{m/s^2}$. (*a*) Find the acting force F. (*b*) If the body was at rest at $t = 0$, $x = 0$, find the position, kinetic energy, and power for $t = 4\,\mathrm{s}$.

(*a*) $$F = ma = (10\,\mathrm{kg})(2.0\,\mathrm{m/s^2}) = 20.0\,\mathrm{kg\cdot m/s^2} = 20.0\,\mathrm{N}$$

(*b*) At $t = 4\,\mathrm{s}$,
$$x = \tfrac{1}{2}at^2 = \tfrac{1}{2}(2.0\,\mathrm{m/s^2})(4\,\mathrm{s})^2 = 16.0\,\mathrm{m}$$
$$\mathrm{KE} = Fx = (20.0\,\mathrm{N})(16.0\,\mathrm{m}) = 3200\,\mathrm{N\cdot m} = 3.2\,\mathrm{kJ}$$
$$P = \mathrm{KE}/t = 3.2\,\mathrm{kJ}/4\,\mathrm{s} = 0.8\,\mathrm{kJ/s} = 0.8\,\mathrm{kW}$$

1.3 ELECTRIC CHARGE AND CURRENT

The unit of current, the *ampere* (A), is defined as the constant current in two parallel conductors of infinite length and negligible cross section, 1 meter apart in vacuum, which produces a force between the conductors of 2.0×10^{-7} newtons per meter length. A more useful concept, however, is that current results from charges in motion, and 1 ampere is equivalent to 1 coulomb of charge moving across a fixed surface in 1 second. Thus, in time-variable functions, $i(\mathrm{A}) = dq/dt(\mathrm{C/s})$. The derived unit of charge, the *coulomb* (C), is equivalent to an ampere-second.

The moving charges may be positive or negative. Positive ions, moving to the left in a liquid or plasma suggested in Fig. 1-1(*a*), produce a current i, also directed to the left. If these ions cross the plane surface S at the rate of *one coulomb per second*, then the resulting current is 1 ampere. Negative ions moving to the right as shown in Fig. 1-1(*b*) also produce a current directed to the left.

Table 1-3

Prefix	Factor	Symbol
pico	10^{-12}	p
nano	10^{-9}	n
micro	10^{-6}	μ
milli	10^{-3}	m
centi	10^{-2}	c
deci	10^{-1}	d
kilo	10^{3}	k
mega	10^{6}	M
giga	10^{9}	G
tera	10^{12}	T

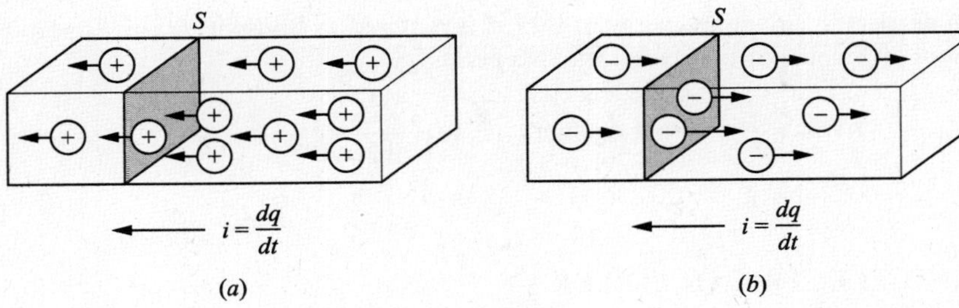

Fig. 1-1

Of more importance in electric circuit analysis is the current in metallic conductors which takes place through the motion of electrons that occupy the outermost shell of the atomic structure. In copper, for example, one electron in the outermost shell is only loosely bound to the central nucleus and moves freely from one atom to the next in the crystal structure. At normal temperatures there is constant, random motion of these electrons. A reasonably accurate picture of conduction in a copper conductor is that approximately 8.5×10^{28} *conduction* electrons per cubic meter are free to move. The electron charge is $-e = -1.602 \times 10^{-19}$ C, so that for a current of one ampere approximately 6.24×10^{18} electrons per second would have to pass a fixed cross section of the conductor.

EXAMPLE 1.2. A conductor has a constant current of five amperes. How many electrons pass a fixed point on the conductor in one minute?

$$5\,\text{A} = (5\,\text{C/s})(60\,\text{s/min}) = 300\,\text{C/min}$$

$$\frac{300\,\text{C/min}}{1.602 \times 10^{-19}\,\text{C/electron}} = 1.87 \times 10^{21}\ \text{electrons/min}$$

1.4 ELECTRIC POTENTIAL

An electric charge experiences a force in an electric field which, if unopposed, will accelerate the particle containing the charge. Of interest here is the work done to move the charge against the field as suggested in Fig. 1-2(*a*). Thus, if *1 joule* of work is required to move the charge Q, *1 coulomb* from position 0 to position 1, then position 1 is at a potential of *1 volt* with respect to position 0; 1 V = 1 J/C. This electric potential is capable of doing work just as the mass in Fig. 1-2(*b*), which was raised against the gravitational force g to a height h above the ground plane. The potential energy mgh represents an ability to do work when the mass m is released. As the mass falls, it accelerates and this potential energy is converted to kinetic energy.

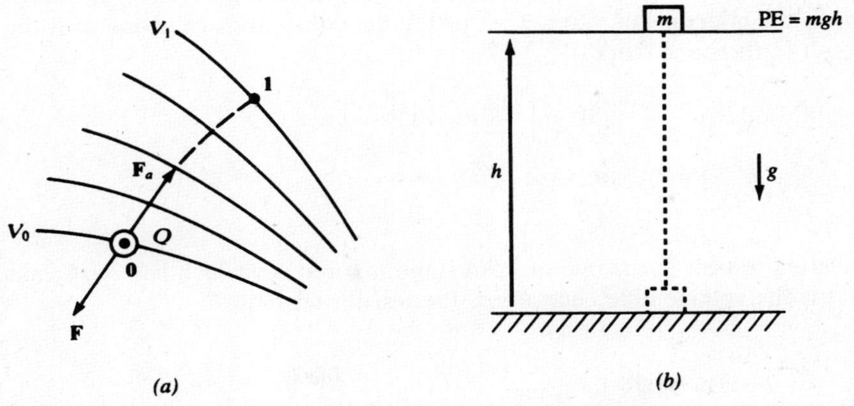

Fig. 1-2

EXAMPLE 1.3. In an electric circuit an energy of 9.25 μJ is required to transport 0.5 μC from point *a* to point *b*. What electric potential difference exists between the two points?

$$1 \text{ volt} = 1 \text{ joule per coulomb} \qquad V = \frac{9.25 \times 10^{-6}\,\text{J}}{0.5 \times 10^{-6}\,\text{C}} = 18.5\,\text{V}$$

1.5 ENERGY AND ELECTRICAL POWER

Electric energy in joules will be encountered in later chapters dealing with capacitance and inductance whose respective electric and magnetic fields are capable of storing energy. The rate, in *joules per second*, at which energy is transferred is electric power in *watts*. Furthermore, the product of voltage and current yields the electric power, $p = vi$; $1\,\text{W} = 1\,\text{V} \cdot 1\,\text{A}$. Also, $\text{V} \cdot \text{A} = (\text{J/C}) \cdot (\text{C/s}) = \text{J/s} = \text{W}$. In a more fundamental sense power is the time derivative $p = dw/dt$, so that instantaneous power p is generally a function of time. In the following chapters time average power P_{avg} and a root-mean-square (RMS) value for the case where voltage and current are sinusoidal will be developed.

EXAMPLE 1.4. A resistor has a potential difference of 50.0 V across its terminals and 120.0 C of charge per minute passes a fixed point. Under these conditions at what rate is electric energy converted to heat?

$$(120.0\,\text{C/min})/(60\,\text{s/min}) = 2.0\,\text{A} \qquad P = (2.0\,\text{A})(50.0\,\text{V}) = 100.0\,\text{W}$$

Since $1\,\text{W} = 1\,\text{J/s}$, the rate of energy conversion is one hundred joules per second.

1.6 CONSTANT AND VARIABLE FUNCTIONS

To distinguish between constant and time-varying quantities, capital letters are employed for the constant quantity and lowercase for the variable quantity. For example, a constant current of 10 amperes is written $I = 10.0\,\text{A}$, while a 10-ampere time-variable current is written $i = 10.0 f(t)\,\text{A}$. Examples of common functions in circuit analysis are the sinusoidal function $i = 10.0 \sin \omega t\,(\text{A})$ and the exponential function $v = 15.0\, e^{-at}\,(\text{V})$.

Solved Problems

1.1 The force applied to an object moving in the x direction varies according to $F = 12/x^2$ (N). (*a*) Find the work done in the interval $1\,\text{m} \le x \le 3\,\text{m}$. (*b*) What constant force acting over the same interval would result in the same work?

(*a*)
$$dW = F\,dx \qquad \text{so} \qquad W = \int_1^3 \frac{12}{x^2}\,dx = 12\left[\frac{-1}{x}\right]_1^3 = 8\,\text{J}$$

(*b*)
$$8\,\text{J} = F_c(2\,\text{m}) \qquad \text{or} \qquad F_c = 4\,\text{N}$$

1.2 Electrical energy is converted to heat at the rate of 7.56kJ/min in a resistor which has 270 C/min passing through. What is the voltage difference across the resistor terminals?

From $P = VI$,

$$V = \frac{P}{I} = \frac{7.56 \times 10^3\,\text{J/min}}{270\,\text{C/min}} = 28\,\text{J/C} = 28\,\text{V}$$

1.3 A certain circuit element has a current $i = 2.5 \sin \omega t$ (mA), where ω is the angular frequency in rad/s, and a voltage difference $v = 45 \sin \omega t$ (V) between terminals. Find the average power P_{avg} and the energy W_T transferred in one period of the sine function.

Energy is the time-integral of instantaneous power:

$$W_T = \int_0^{2\pi/\omega} vi\,dt = 112.5 \int_0^{2\pi/\omega} \sin^2 \omega t\,dt = \frac{112.5\pi}{\omega} \text{ (mJ)}$$

The average power is then

$$P_{avg} = \frac{W_T}{2\pi/\omega} = 56.25 \text{ mW}$$

Note that P_{avg} is independent of ω.

1.4 The unit of energy commonly used by electric utility companies is the kilowatt-hour (kWh). (*a*) How many joules are in 1 kWh? (*b*) A color television set rated at 75 W is operated from 7:00 p.m. to 11:30 p.m. What total energy does this represent in kilowatt-hours and in mega-joules?

(*a*) $1 \text{ kWh} = (1000 \text{ J/s})(3600 \text{ s/h}) = 3.6 \text{ MJ}$

(*b*) $(75.0 \text{ W})(4.5 \text{ h}) = 337.5 \text{ Wh} = 0.3375 \text{ kWh}$
$(0.3375 \text{ kWh})(3.6 \text{ MJ/kWh}) = 1.215 \text{ MJ}$

1.5 An AWG #12 copper wire, a size in common use in residential wiring, contains approximately 2.77×10^{23} free electrons per meter length, assuming one free conduction electron per atom. What percentage of these electrons will pass a fixed cross section if the conductor carries a constant current of 25.0 A?

$$\frac{25.0 \text{ C/s}}{1.602 \times 10^{-19} \text{ C/electron}} = 1.56 \times 10^{20} \text{ electron/s}$$

$$(1.56 \times 10^{20} \text{ electrons/s})(60 \text{ s/min}) = 9.36 \times 10^{21} \text{ electrons/min}$$

$$\frac{9.36 \times 10^{21}}{2.77 \times 10^{23}}(100) = 3.38\%$$

1.6 How many electrons pass a fixed point in a 100-watt light bulb in 1 hour if the applied constant voltage is 120 V?

$$100 \text{ W} = (120 \text{ V}) \times I(\text{A}) \qquad I = 5/6 \text{ A}$$

$$\frac{(5/6 \text{ C/s})(3600 \text{ s/h})}{1.602 \times 10^{-19} \text{ C/electron}} = 1.87 \times 10^{22} \text{ electrons per hour}$$

1.7 A typical 12 V auto battery is rated according to *ampere-hours*. A 70-A · h battery, for example, at a discharge rate of 3.5 A has a life of 20 h. (*a*) Assuming the voltage remains constant, obtain the energy and power delivered in a complete discharge of the preceding batttery. (*b*) Repeat for a discharge rate of 7.0 A.

(*a*) $(3.5 \text{ A})(12 \text{ V}) = 42.0 \text{ W}$ (or J/s)
$(42.0 \text{ J/s})(3600 \text{ s/h})(20 \text{ h}) = 3.02 \text{ MJ}$

(*b*) $(7.0 \text{ A})(12 \text{ V}) = 84.0 \text{ W}$
$(84.0 \text{ J/s})(3600 \text{ s/h})(10 \text{ h}) = 3.02 \text{ MJ}$

The ampere-hour rating is a measure of the energy the battery stores; consequently, the energy transferred for total discharge is the same whether it is transferred in 10 hours or 20 hours. Since power is the rate of energy transfer, the power for a 10-hour discharge is twice that in a 20-hour discharge.

Supplementary Problems

1.8 Obtain the work and power associated with a force of 7.5×10^{-4} N acting over a distance of 2 meters in an elapsed time of 14 seconds. *Ans.* 1.5 mJ, 0.107 mW

1.9 Obtain the work and power required to move a 5.0-kg mass up a frictionless plane inclined at an angle of 30° with the horizontal for a distance of 2.0 m along the plane in a time of 3.5 s. *Ans.* 49.0 J, 14.0 W

1.10 Work equal to 136.0 joules is expended in moving 8.5×10^{18} electrons between two points in an electric circuit. What potential difference does this establish between the two points? *Ans.* 100 V

1.11 A pulse of electricity measures 305 V, 0.15 A, and lasts 500 μs. What power and energy does this represent? *Ans.* 45.75 W, 22.9 mJ

1.12 A unit of power used for electric motors is the *horsepower* (hp), equal to 746 watts. How much energy does a 5-hp motor deliver in 2 hours? Express the answer in MJ. *Ans.* 26.9 MJ

1.13 For $t \geq 0$, $q = (4.0 \times 10^{-4})(1 - e^{-250t})$ (C). Obtain the current at $t = 3$ ms. *Ans.* 47.2 mA

1.14 A certain circuit element has the current and voltage

$$i = 10e^{-5000t} \text{ (A)} \qquad v = 50(1 - e^{-5000t}) \text{ (V)}$$

Find the total energy transferred during $t \geq 0$. *Ans.* 50 mJ

1.15 The *capacitance* of a circuit element is defined as Q/V, where Q is the magnitude of charge stored in the element and V is the magnitude of the voltage difference across the element. The SI derived unit of capacitance is the *farad* (F). Express the farad in terms of the basic units.
Ans. $1\text{ F} = 1\text{ A}^2 \cdot \text{s}^4/\text{kg} \cdot \text{m}^2$

CHAPTER 2

Circuit Concepts

2.1 PASSIVE AND ACTIVE ELEMENTS

An electrical device is represented by a *circuit diagram* or *network* constructed from series and parallel arrangements of two-terminal elements. The analysis of the circuit diagram predicts the performance of the actual device. A two-terminal element in general form is shown in Fig. 2-1, with a single device represented by the rectangular symbol and two perfectly conducting leads ending at connecting points *A* and *B*. *Active* elements are voltage or current sources which are able to supply energy to the network. Resistors, inductors, and capacitors are *passive* elements which take energy from the sources and either convert it to another form or store it in an electric or magnetic field.

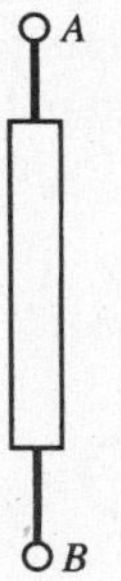

Fig. 2-1

Figure 2-2 illustrates seven basic circuit elements. Elements (*a*) and (*b*) are voltage sources and (*c*) and (*d*) are current sources. A voltage source that is not affected by changes in the connected circuit is an *independent* source, illustrated by the circle in Fig. 2-2(*a*). A *dependent* voltage source which changes in some described manner with the conditions on the connected circuit is shown by the diamond-shaped symbol in Fig. 2-2(*b*). Current sources may also be either independent or dependent and the corresponding symbols are shown in (*c*) and (*d*). The three passive circuit elements are shown in Fig. 2-2(*e*), (*f*), and (*g*).

The circuit diagrams presented here are termed *lumped-parameter* circuits, since a single element in one location is used to represent a distributed resistance, inductance, or capacitance. For example, a coil consisting of a large number of turns of insulated wire has resistance throughout the entire length of the wire. Nevertheless, a single resistance *lumped* at one place as in Fig. 2-3(*b*) or (*c*) represents the distributed resistance. The inductance is likewise lumped at one place, either in series with the resistance as in (*b*) or in parallel as in (*c*).

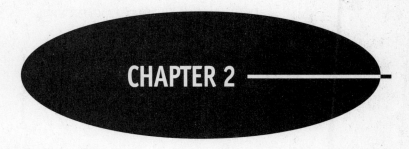

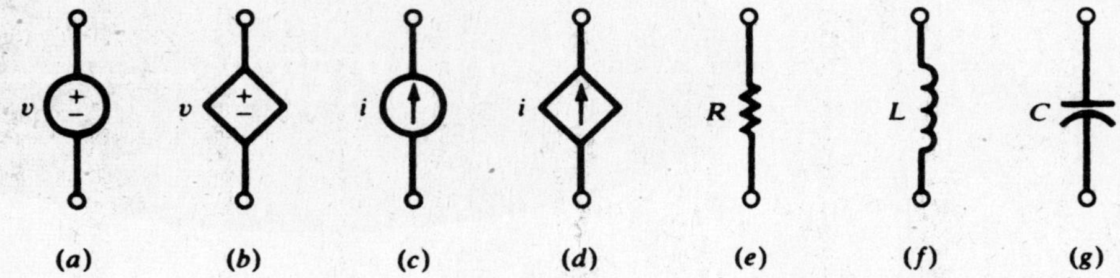

Fig. 2-2

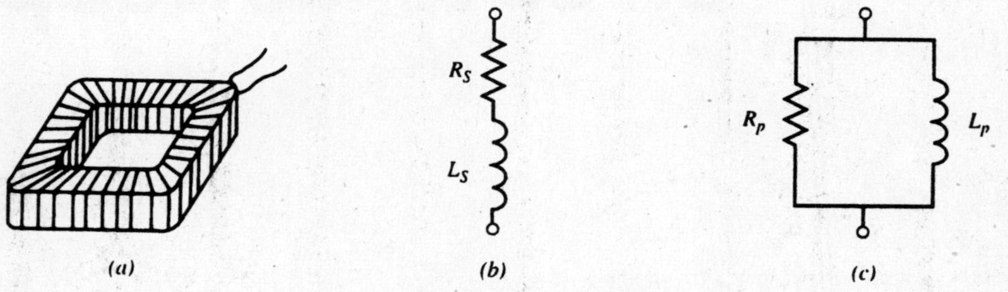

Fig. 2-3

In general, a coil can be represented by either a series or a parallel arrangement of circuit elements. The frequency of the applied voltage may require that one or the other be used to represent the device.

2.2 SIGN CONVENTIONS

A voltage function and a polarity must be specified to completely describe a voltage source. The polarity marks, + and −, are placed near the conductors of the symbol that identifies the voltage source. If, for example, $v = 10.0 \sin \omega t$ in Fig. 2-4(*a*), terminal A is positive with respect to B for $0 > \omega t > \pi$, and B is positive with respect to A for $\pi > \omega t > 2\pi$ for the first cycle of the sine function.

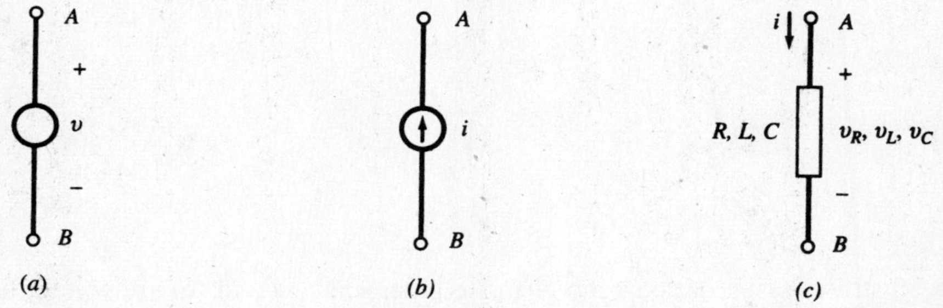

Fig. 2-4

Similarly, a current source requires that a direction be indicated, as well as the function, as shown in Fig. 2-4(*b*). For passive circuit elements R, L, and C, shown in Fig. 2-4(*c*), the terminal where the current enters is generally treated as positive with respect to the terminal where the current leaves.

The sign on power is illustrated by the dc circuit of Fig. 2-5(*a*) with constant voltage sources $V_A = 20.0$ V and $V_B = 5.0$ V and a single 5-Ω resistor. The resulting current of 3.0 A is in the clockwise direction. Considering now Fig. 2-5(*b*), power is absorbed by an element when the current enters the element at the positive terminal. Power, computed by VI or I^2R, is therefore absorbed by both the resistor and the V_B source, 45.0 W and 15 W respectively. Since the current enters V_A at the negative terminal, this element is the power source for the circuit. $P = VI = 60.0$ W confirms that the power absorbed by the resistor and the source V_B is provided by the source V_A.

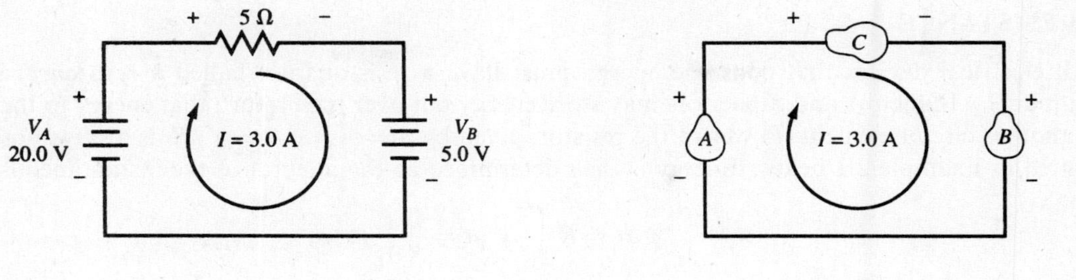

Fig. 2-5

2.3 VOLTAGE-CURRENT RELATIONS

The passive circuit elements resistance R, inductance L, and capacitance C are defined by the manner in which the voltage and current are related for the individual element. For example, if the voltage v and current i for a single element are related by a constant, then the element is a resistance, R is the constant of proportionality, and $v = Ri$. Similarly, if the voltage is the time derivative of the current, then the element is an inductance, L is the constant of proportionality, and $v = L\,di/dt$. Finally, if the current in the element is the time derivative of the voltage, then the element is a capacitance, C is the constant of proportionality, and $i = C\,dv/dt$. Table 2-1 summarizes these relationships for the three passive circuit elements. Note the current directions and the corresponding polarity of the voltages.

Table 2-1

Circuit element	Units	Voltage	Current	Power
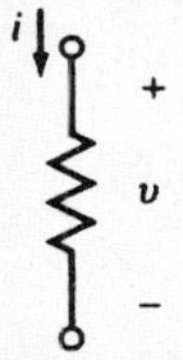 **Resistance**	ohms (Ω)	$v = Ri$ **(Ohms's law)**	$i = \dfrac{v}{R}$	$p = vi = i^2R$
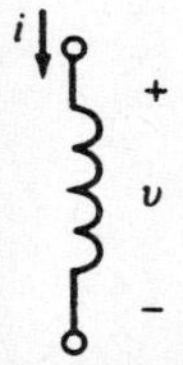 **Inductance**	henries (H)	$v = L\dfrac{di}{dt}$	$i = \dfrac{1}{L}\int v\,dt + k_1$	$p = vi = Li\dfrac{di}{dt}$
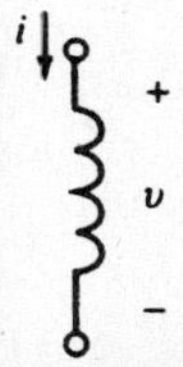 **Capacitance**	farads (F)	$v = \dfrac{1}{C}\int i\,dt + k_2$	$i = C\dfrac{dv}{dt}$	$p = vi = Cv\dfrac{dv}{dt}$

2.4 RESISTANCE

All electrical devices that consume energy must have a resistor (also called a *resistance*) in their circuit model. Inductors and capacitors may store energy but over time return that energy to the source or to another circuit element. Power in the resistor, given by $p = vi = i^2R = v^2/R$, is always positive as illustrated in Example 2.1 below. Energy is then determined as the integral of the instantaneous power

$$w = \int_{t_1}^{t_2} p\,dt = R\int_{t_1}^{t_2} i^2\,dt = \frac{1}{R}\int_{t_1}^{t_2} v^2\,dt$$

EXAMPLE 2.1. A 4.0-Ω resistor has a current $i = 2.5\sin\omega t$ (A). Find the voltage, power, and energy over one cycle. $\omega = 500$ rad/s.

$$v = Ri = 10.0\sin\omega t \text{ (V)}$$
$$p = vi = i^2R = 25.0\sin^2\omega t \text{ (W)}$$
$$w = \int_0^t p\,dt = 25.0\left[\frac{t}{2} - \frac{\sin 2\omega t}{4\omega}\right] \text{ (J)}$$

The plots of i, p, and w shown in Fig. 2-6 illustrate that p is always positive and that the energy w, although a function of time, is always increasing. This is the energy absorbed by the resistor.

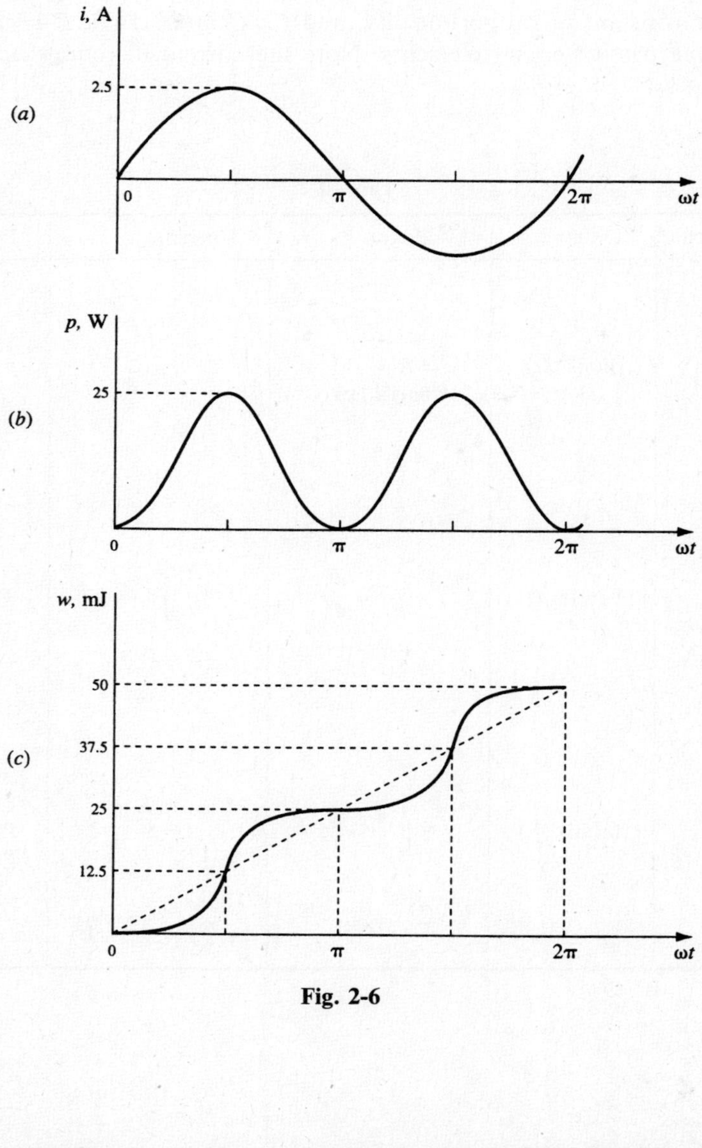

Fig. 2-6

2.5 INDUCTANCE

The circuit element that stores energy in a magnetic field is an inductor (also called an *inductance*). With time-variable current, the energy is generally stored during some parts of the cycle and then returned to the source during others. When the inductance is removed from the source, the magnetic field will collapse; in other words, no energy is stored without a connected source. Coils found in electric motors, transformers, and similar devices can be expected to have inductances in their circuit models. Even a set of parallel conductors exhibits inductance that must be considered at most frequencies. The power and energy relationships are as follows.

$$p = vi = L\frac{di}{dt}i = \frac{d}{dt}\left[\frac{1}{2}Li^2\right]$$

$$w_L = \int_{t_1}^{t_2} p\,dt = \int_{t_1}^{t_2} Li\,dt = \frac{1}{2}L[i_2^2 - i_1^2]$$

Energy stored in the magnetic field of an inductance is $w_L = \frac{1}{2}Li^2$.

EXAMPLE 2.2. In the interval $0 > t > (\pi/50)$ s a 30-mH inductance has a current $i = 10.0 \sin 50t$ (A). Obtain the voltage, power, and energy for the inductance.

$$v = L\frac{di}{dt} = 15.0 \cos 50t \text{ (V)} \qquad p = vi = 75.0 \sin 100t \text{ (W)} \qquad w_L = \int_0^t p\,dt = 0.75(1 - \cos 100t) \text{ (J)}$$

As shown in Fig. 2-7, the energy is zero at $t = 0$ and $t = (\pi/50)$ s. Thus, while energy transfer did occur over the interval, this energy was first stored and later returned to the source.

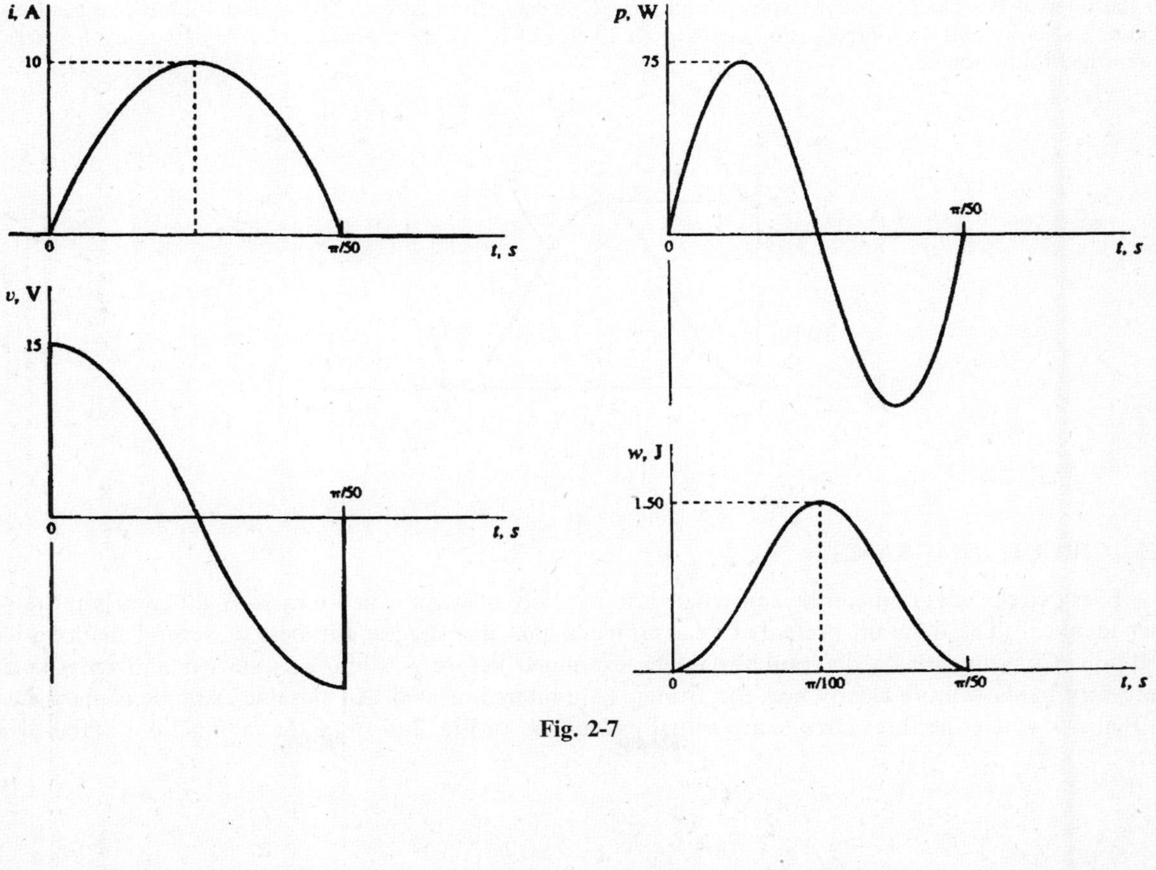

Fig. 2-7

2.6 CAPACITANCE

The circuit element that stores energy in an electric field is a *capacitor* (also called *capacitance*). When the voltage is variable over a cycle, energy will be stored during one part of the cycle and returned in the next. While an inductance cannot retain energy after removal of the source because the magnetic field collapses, the capacitor retains the charge and the electric field can remain after the source is removed. This charged condition can remain until a discharge path is provided, at which time the energy is released. The charge, $q = Cv$, on a capacitor results in an electric field in the dielectric which is the mechanism of the energy storage. In the simple parallel-plate capacitor there is an excess of charge on one plate and a deficiency on the other. It is the equalization of these charges that takes place when the capacitor is discharged. The power and energy relationships for the capacitance are as follows.

$$p = vi = Cv\frac{dv}{dt} = \frac{d}{dt}\left[\frac{1}{2}Cv^2\right]$$

$$w_C = \int_{t_1}^{t_2} p\,dt = \int_{t_1}^{t_2} Cv\,dv = \frac{1}{2}C[v_2^2 - v_1^2]$$

The energy stored in the electric field of capacitance is $w_C = \frac{1}{2}Cv^2$.

EXAMPLE 2.3. In the interval $0 > t > 5\pi$ ms, a 20-μF capacitance has a voltage $v = 50.0 \sin 200t$ (V). Obtain the charge, power, and energy. Plot w_C assuming $w = 0$ at $t = 0$.

$$q = Cv = 1000 \sin 200t \ (\mu\text{C})$$

$$i = C\frac{dv}{dt} = 0.20 \cos 200t \ (\text{A})$$

$$p = vi = 5.0 \sin 400t \ (\text{W})$$

$$w_C = \int_{t_1}^{t_2} p\,dt = 12.5[1 - \cos 400t] \ (\text{mJ})$$

In the interval $0 > t > 2.5\pi$ ms the voltage and charge increase from zero to 50.0 V and 1000 μC, respectively. Figure 2-8 shows that the stored energy increases to a value of 25 mJ, after which it returns to zero as the energy is returned to the source.

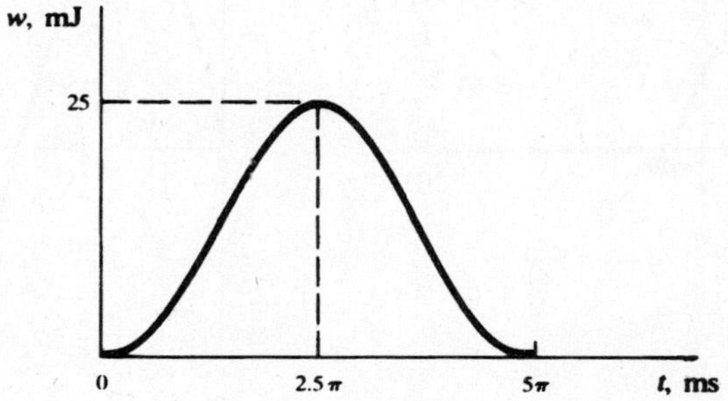

Fig. 2-8

2.7 CIRCUIT DIAGRAMS

Every circuit diagram can be constructed in a variety of ways which may look different but are in fact identical. The diagram presented in a problem may not suggest the best of several methods of solution. Consequently, a diagram should be examined before a solution is started and redrawn if necessary to show more clearly how the elements are interconnected. An extreme example is illustrated in Fig. 2-9, where the three circuits are actually identical. In Fig. 2-9(a) the three "junctions" labeled A

are shown as two "junctions" in (*b*). However, resistor R_4 is bypassed by a short circuit and may be removed for purposes of analysis. Then, in Fig. 2-9(*c*) the single junction A is shown with its three meeting branches.

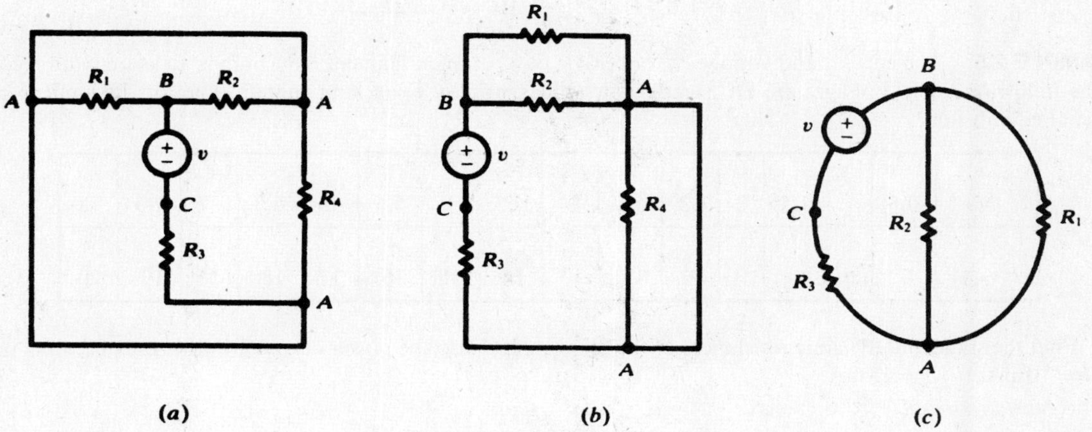

Fig. 2-9

2.8 NONLINEAR RESISTORS

The current-voltage relationship in an element may be instantaneous but not necessarily linear. The element is then modeled as a nonlinear resistor. An example is a filament lamp which at higher voltages draws proportionally less current. Another important electrical device modeled as a nonlinear resistor is a diode. A diode is a two-terminal device that, roughly speaking, conducts electric current in one direction (from anode to cathode, called forward-biased) much better than the opposite direction (reverse-biased). The circuit symbol for the diode and an example of its current-voltage characteristic are shown in Fig. 2-25. The arrow is from the anode to the cathode and indicates the forward direction ($i > 0$). A small positive voltage at the diode's terminal biases the diode in the forward direction and can produce a large current. A negative voltage biases the diode in the reverse direction and produces little current even at large voltage values. An ideal diode is a circuit model which works like a perfect switch. See Fig. 2-26. Its (i, v) characteristic is

$$\begin{cases} v = 0 & \text{when } i \geq 0 \\ i = 0 & \text{when } v \leq 0 \end{cases}$$

The *static resistance* of a nonlinear resistor operating at (I, V) is $R = V/I$. Its *dynamic resistance* is $r = \Delta V/\Delta I$ which is the inverse of the slope of the current plotted versus voltage. Static and dynamic resistances both depend on the operating point.

EXAMPLE 2.4. The current and voltage characteristic of a semiconductor diode in the forward direction is measured and recorded in the following table:

v (V)	0.5	0.6	0.65	0.66	0.67	0.68	0.69	0.70	0.71	0.72	0.73	0.74	0.75
i (mA)	2×10^{-4}	0.11	0.78	1.2	1.7	2.6	3.9	5.8	8.6	12.9	19.2	28.7	42.7

In the reverse direction (i.e., when $v < 0$), $i = 4 \times 10^{-15}$ A. Using the values given in the table, calculate the static and dynamic resistances (R and r) of the diode when it operates at 30 mA, and find its power consumption p.

From the table

$$R = \frac{V}{I} \approx \frac{0.74}{28.7 \times 10^{-3}} = 25.78\ \Omega$$
$$r = \frac{\Delta V}{\Delta I} \approx \frac{0.75 - 0.73}{(42.7 - 19.2) \times 10^{-3}} = 0.85\ \Omega$$
$$p = VI \approx 0.74 \times 28.7 \times 10^{-3}\text{W} = 21.238\ \text{mW}$$

EXAMPLE 2.5. The current and voltage characteristic of a tungsten filament light bulb is measured and recorded in the following table. Voltages are DC steady-state values, applied for a long enough time for the lamp to reach thermal equilibrium.

v (V)	0.5	1	1.5	2	3	3.5	4	4.5	5	5.5	6	6.5	7	7.5	8
i (mA)	4	6	8	9	11	12	13	14	15	16	17	18	18	19	20

Find the static and dynamic resistances of the filament and also the power consumption at the operating points (*a*) $i = 10$ mA; (*b*) $i = 15$ mA.

$$R = \frac{V}{I}, \qquad r = \frac{\Delta V}{\Delta I}, \qquad p = VI$$

(*a*) $R \approx \dfrac{2.5}{10 \times 10^{-3}} = 250\ \Omega,\ r \approx \dfrac{3 - 2}{(11 - 9) \times 10^{-3}} = 500\ \Omega,\ p \approx 2.5 \times 10 \times 10^{-3}\ \text{W} = 25\ \text{mW}$

(*b*) $R \approx \dfrac{5}{15 \times 10^{-3}} = 333\ \Omega,\ r \approx \dfrac{5.5 - 4.5}{(16 - 14) \times 10^{-3}} = 500\ \Omega,\ p \approx 5 \times 15 \times 10^{-3}\ \text{W} = 75\ \text{mW}$

Solved Problems

2.1 A 25.0-Ω resistance has a voltage $v = 150.0 \sin 377t$ (V). Find the corresponding current i and power p.

$$i = \frac{v}{R} = 6.0 \sin 377t\ \text{(A)} \qquad p = vi = 900.0 \sin^2 377t\ \text{(W)}$$

2.2 The current in a 5-Ω resistor increases linearly from zero to 10 A in 2 ms. At $t = 2^+$ ms the current is again zero, and it increases linearly to 10 A at $t = 4$ ms. This pattern repeats each 2 ms. Sketch the corresponding v.

Since $v = Ri$, the maximum voltage must be (5)(10) = 50 V. In Fig. 2-10 the plots of i and v are shown. The identical nature of the functions is evident.

2.3 An inductance of 2.0 mH has a current $i = 5.0(1 - e^{-5000t})$ (A). Find the corresponding voltage and the maximum stored energy.

$$v = L\frac{di}{dt} = 50.0e^{-5000t}\ \text{(V)}$$

In Fig. 2-11 the plots of i and v are given. Since the maximum current is 5.0 A, the maximum stored energy is

$$W_{\max} = \frac{1}{2}LI_{\max}^2 = 25.0\ \text{mJ}$$

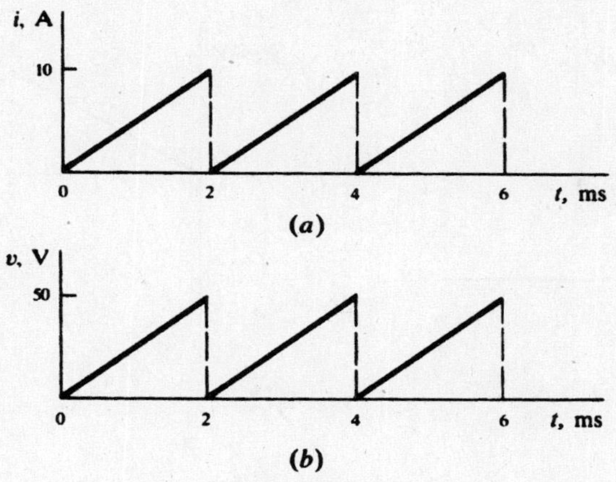

Fig. 2-10

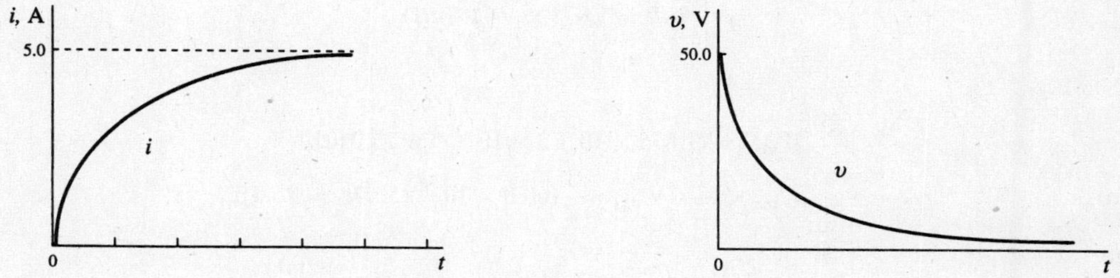

Fig. 2-11

2.4 An inductance of 3.0 mH has a voltage that is described as follows: for $0 > t > 2\,\text{ms}$, $V = 15.0\,\text{V}$ and, for $2 > t > 4\,\text{ms}$, $V = -30.0\,\text{V}$. Obtain the corresponding current and sketch v_L and i for the given intervals.

For $0 > t > 2\,\text{ms}$,

$$i = \frac{1}{L}\int_0^t v\,dt = \frac{1}{3\times 10^{-3}}\int_0^t 15.0\,dt = 5\times 10^3 t\ (\text{A})$$

For $t = 2\,\text{ms}$,

$$i = 10.0\ \text{A}$$

For $2 > t > 4\,\text{ms}$,

$$\begin{aligned} i &= \frac{1}{L}\int_{2\times 10^{-3}}^t v\,dt + 10.0 + \frac{1}{3\times 10^{-3}}\int_{2\times 10^{-3}}^t -30.0\,dt \\ &= 10.0 + \frac{1}{3\times 10^{-3}}[-30.0t + (60.0\times 10^{-3})]\ (\text{A}) \\ &= 30.0 - (10\times 10^3 t)\ (\text{A}) \end{aligned}$$

See Fig. 2-12.

2.5 A capacitance of 60.0 μF has a voltage described as follows: $0 > t > 2\,\text{ms}$, $v = 25.0\times 10^3 t$ (V). Sketch i, p, and w for the given interval and find W_{max}.

For $0 > t > 2\,\text{ms}$,

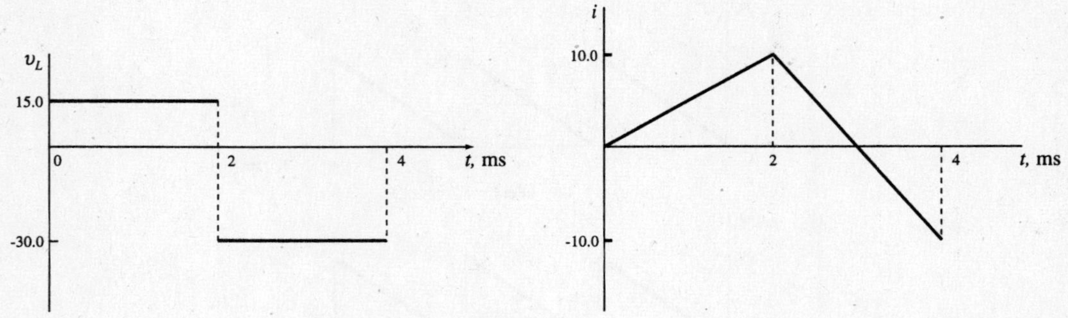

Fig. 2-12

$$i = C\frac{dv}{dt} = 60 \times 10^{-6}\frac{d}{dt}(25.0 \times 10^3 t) = 1.5\,\text{A}$$
$$p = vi = 37.5 \times 10^3 t\ (\text{W})$$
$$w_C = \int_0^t p\,dt = 1.875 \times 10^4 t^2\ (\text{mJ})$$

See Fig. 2-13.

$$W_{max} = (1.875 \times 10^4)(2 \times 10^{-3})^2 = 75.0\,\text{mJ}$$

or

$$W_{max} = \frac{1}{2}CV_{max}^2 = \frac{1}{2}(60.0 \times 10^{-6})(50.0)^2 = 75.0\,\text{mJ}$$

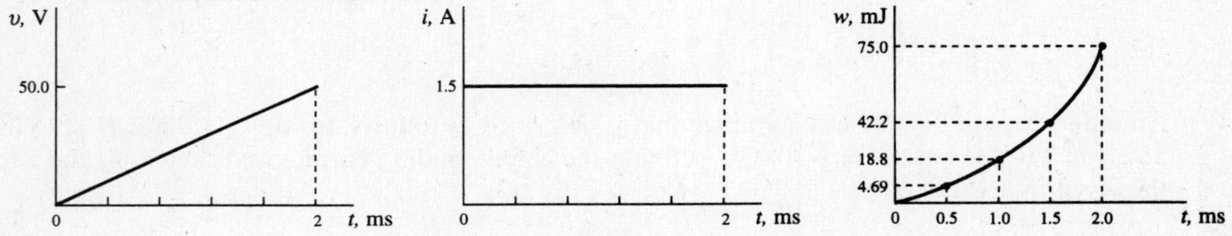

Fig. 2-13

2.6 A 20.0-μF capacitance is linearly charged from 0 to 400 μC in 5.0 ms. Find the voltage function and W_{max}.

$$q = \left(\frac{400 \times 10^{-6}}{5.0 \times 10^{-3}}\right)t = 8.0 \times 10^{-2} t\ (\text{C})$$
$$v = q/C = 4.0 \times 10^3 t\ (\text{V})$$
$$V_{max} = (4.0 \times 10^3)(5.0 \times 10^{-3}) = 20.0\,\text{V} \qquad W_{max} = \frac{1}{2}CV_{max}^2 = 4.0\,\text{mJ}$$

2.7 A series circuit with $R = 2\,\Omega$, $L = 2\,\text{mH}$, and $C = 500\,\mu\text{F}$ has a current which increases linearly from zero to 10 A in the interval $0 \le t \le 1$ ms, remains at 10 A for $1\,\text{ms} \le t \le 2\,\text{ms}$, and decreases linearly from 10 A at $t = 2$ ms to zero at $t = 3$ ms. Sketch v_R, v_L, and v_C.

v_R must be a time function identical to i, with $V_{max} = 2(10) = 20$ V.
For $0 < t < 1$ ms,

$$\frac{di}{dt} = 10 \times 10^3\,\text{A/s} \qquad \text{and} \qquad v_L = L\frac{di}{dt} = 20\,\text{V}$$

When $di/dt = 0$, for $1\,\text{ms} < t < 2\,\text{ms}$, $v_L = 0$.

Assuming zero initial charge on the capacitor,

$$v_C = \frac{1}{C}\int i\,dt$$

For $0 \le t \le 1$ ms,

$$v_C = \frac{1}{5 \times 10^{-4}}\int_0^t 10^4\,t\,dt = 10^7 t^2 \text{ (V)}$$

This voltage reaches a value of 10 V at 1 ms. For 1 ms $< t <$ 2 ms,

$$v_C = (20 \times 10^3)(t - 10^{-3}) + 10 \text{ (V)}$$

See Fig. 2-14.

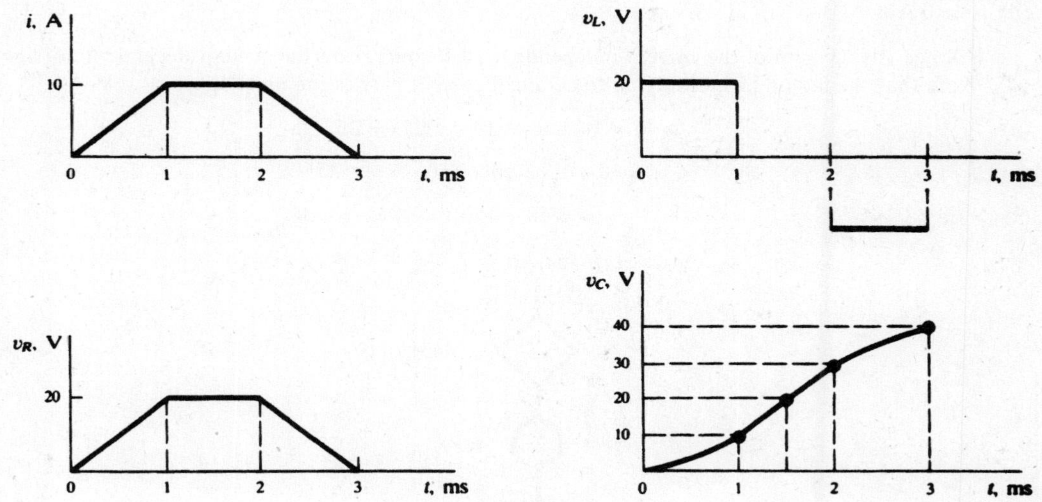

Fig. 2-14

2.8 A single circuit element has the current and voltage functions graphed in Fig. 2-15. Determine the element.

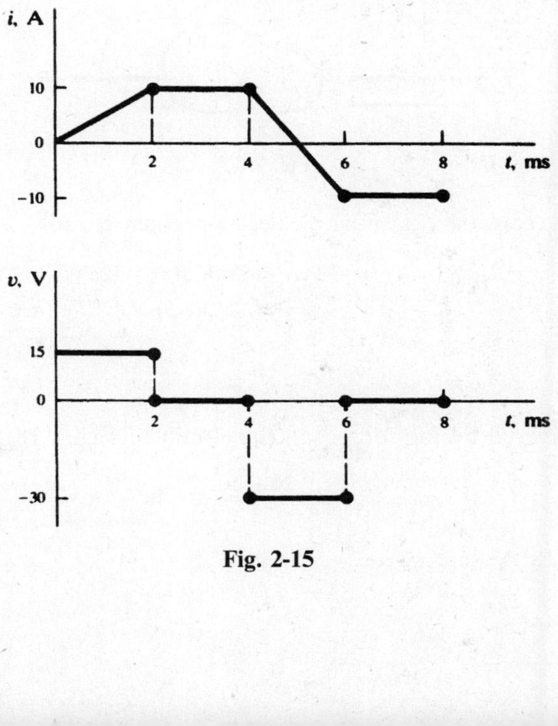

Fig. 2-15

The element cannot be a resistor since v and i are not proportional. v is an integral of i. For $2\,\text{ms} < t < 4\,\text{ms}$, $i \neq 0$ but v is constant (zero); hence the element cannot be a capacitor. For $0 < t < 2\,\text{ms}$,

$$\frac{di}{dt} = 5 \times 10^3\ \text{A/s} \qquad \text{and} \qquad v = 15\ \text{V}$$

Consequently,

$$L = v \bigg/ \frac{di}{dt} = 3\ \text{mH}$$

(Examine the interval $4\,\text{ms} < t < 6\,\text{ms}$; L must be the same.)

2.9 Obtain the voltage v in the branch shown in Fig. 2-16 for (*a*) $i_2 = 1\,\text{A}$, (*b*) $i_2 = -2\,\text{A}$, (*c*) $i_2 = 0\,\text{A}$.

Voltage v is the sum of the current-independent 10-V source and the current-dependent voltage source v_x. Note that the factor 15 multiplying the control current carries the units Ω.

(*a*) $$v = 10 + v_x = 10 + 15(1) = 25\ \text{V}$$

(*b*) $$v = 10 + v_x = 10 + 15(-2) = -20\ \text{V}$$

(*c*) $$v = 10 + 15(0) = 10\ \text{V}$$

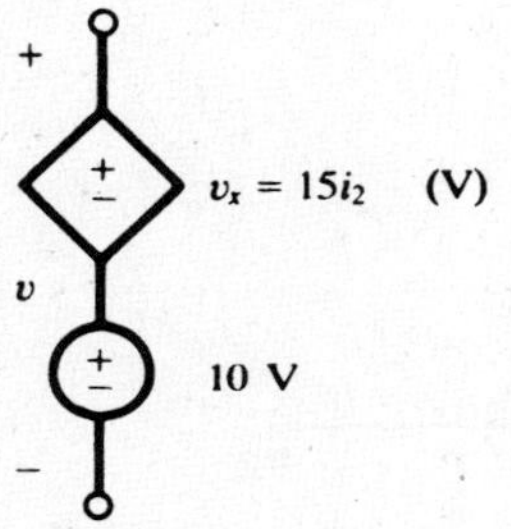

Fig. 2-16

2.10 Find the power absorbed by the generalized circuit element in Fig. 2-17, for (*a*) $v = 50\,\text{V}$, (*b*) $v = -50\,\text{V}$.

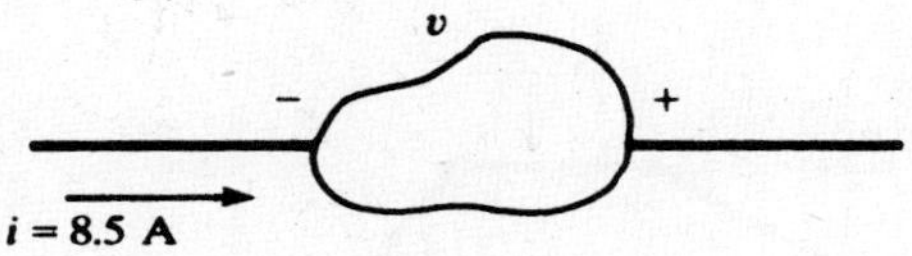

Fig. 2-17

Since the current enters the element at the negative terminal,

(*a*) $$p = -vi = -(50)(8.5) = -425\ \text{W}$$

(*b*) $$p = -vi = -(-50)(8.5) = 425\ \text{W}$$

2.11 Find the power delivered by the sources in the circuit of Fig. 2-18.

$$i = \frac{20 - 50}{3} = -10\ \text{A}$$

The powers *absorbed* by the sources are:

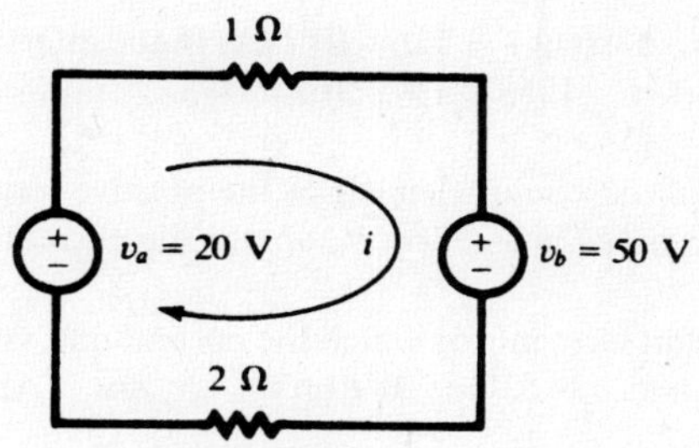

Fig. 2-18

$$p_a = -v_a i = -(20)(-10) = 200\ \text{W}$$
$$p_b = v_b i = (50)(-10) = -500\ \text{W}$$

Since power delivered is the negative of power absorbed, source v_b delivers 500 W and source v_a absorbs 200 W. The power in the two resistors is 300 W.

2.12 A 25.0-Ω resistance has a voltage $v = 150.0 \sin 377t$ (V). Find the power p and the average power p_{avg} over one cycle.

$$i = v/R = 6.0 \sin 377t\ \text{(A)}$$
$$p = vi = 900.0 \sin^2 377t\ \text{(W)}$$

The end of one period of the voltage and current functions occurs at $377t = 2\pi$. For P_{avg} the integration is taken over one-half cycle, $377t = \pi$. Thus,

$$P_{\text{avg}} = \frac{1}{\pi}\int_0^{\pi} 900.0 \sin^2(377t)d(377t) = 450.0\ \text{(W)}$$

2.13 Find the voltage across the 10.0-Ω resistor in Fig. 2-19 if the control current i_x in the dependent source is (*a*) 2 A and (*b*) −1 A.

$$i = 4i_x - 4.0; \qquad v_R = iR = 40.0i_x - 40.0\ \text{(V)}$$
$$i_x = 2; \qquad v_R = 40.0\ \text{V}$$
$$i_x = -1; \qquad v_R = -80.0\ \text{V}$$

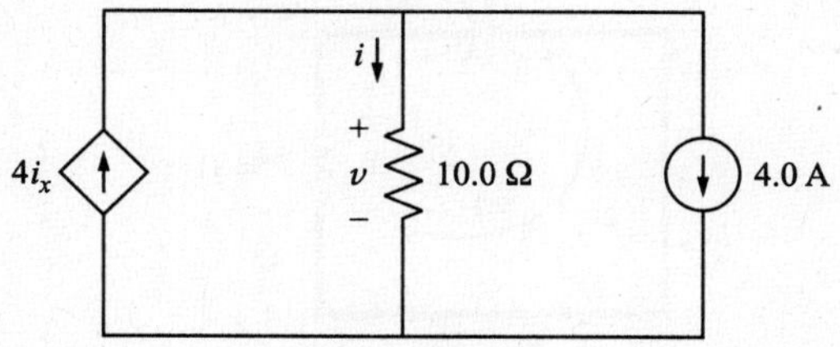

Fig. 2-19

Supplementary Problems

2.14 A resistor has a voltage of $V = 1.5$ mV. Obtain the current if the power absorbed is (*a*) 27.75 nW and (*b*) 1.20 μW. *Ans.* 18.5 μA, 0.8 mA

2.15 A resistance of 5.0 Ω has a current $i = 5.0 \times 10^3 t$ (A) in the interval $0 \geq t \geq 2$ ms. Obtain the instantaneous and average power. *Ans.* $125.0t^2$ (W), 167.0 (W)

2.16 Current i enters a generalized circuit element at the positive terminal and the voltage across the element is 3.91 V. If the power *absorbed* is −25.0 mW, obtain the current. *Ans.* −6.4 mA

2.17 Determine the single circuit element for which the current and voltage in the interval $0 \geq 10^3 t \geq \pi$ are given by $i = 2.0 \sin 10^3 t$ (mA) and $v = 5.0 \cos 10^3 t$ (mV). *Ans.* An inductance of 2.5 mH

2.18 An inductance of 4.0 mH has a voltage $v = 2.0e^{-10^3 t}$ (V). Obtain the maximum stored energy. At $t = 0$, the current is zero. *Ans.* 0.5 mW

2.19 A capacitance of 2.0 μF with an initial charge Q_0 is switched into a series circuit consisting of a 10.0-Ω resistance. Find Q_0 if the energy dissipated in the resistance is 3.6 mJ. *Ans.* 120.0 μC

2.20 Given that a capactance of C farads has a current $i = (V_m/R)e^{-t/(Rc)}$ (A), show that the maximum stored energy is $\frac{1}{2}CV_m^2$. Assume the initial charge is zero.

2.21 The current after $t = 0$ in a single circuit element is as shown in Fig. 2-20. Find the voltage across the element at $t = 6.5$ μs, if the element is (*a*) 10 kΩ, (*b*) 15 mH, (*c*) 0.3 nF with $Q(0) = 0$.
Ans. (*a*) 25 V; (*b*) −75 V; (*c*) 81.3 V

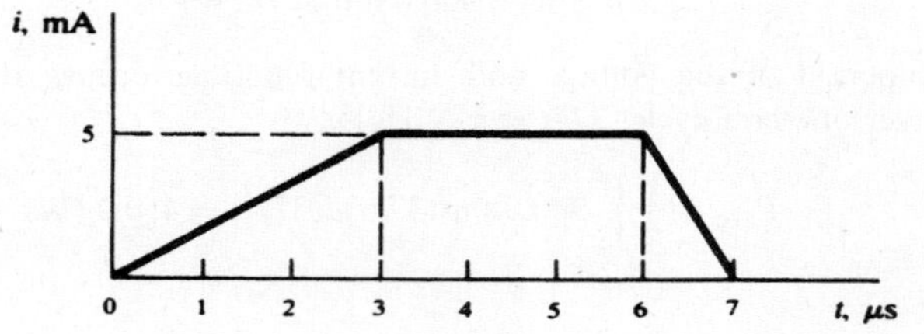

Fig. 2-20

2.22 The 20.0-μF capacitor in the circuit shown in Fig. 2-21 has a voltage for $t > 0$, $v = 100.0e^{-t/0.015}$ (V). Obtain the energy function that accompanies the discharge of the capacitor and compare the total energy to that which is absorbed by the 750-Ω resistor. *Ans.* $0.10\,(1 - e^{-t/0.0075})$ (J)

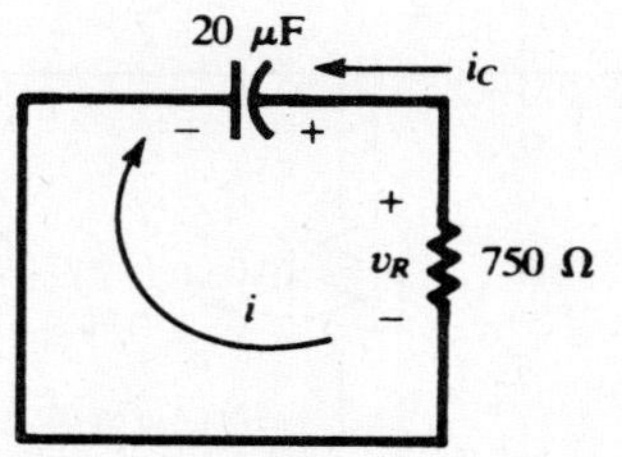

Fig. 2-21

2.23 Find the current i in the circuit shown in Fig. 2-22, if the control v_2 of the dependent voltage source has the value (*a*) 4 V, (*b*) 5 V, (*c*) 10 V. *Ans.* (*a*) 1 A; (*b*) 0 A; (*c*) −5 A

2.24 In the circuit shown in Fig. 2-23, find the current, i, given (*a*) $i_1 = 2$ A, $i_2 = 0$; (*b*) $i_1 = -1$ A, $i_2 = 4$ A; (*c*) $i_1 = i_2 = 1$ A. *Ans.* (*a*) 10 A; (*b*) 11 A; (*c*) 9A

2.25 A 1-μF capacitor with an initial charge of 10^{-4} C is connected to a resistor R at $t = 0$. Assume discharge current during $0 < t < 1$ ms is constant. Approximate the capacitor voltage drop at $t = 1$ ms for

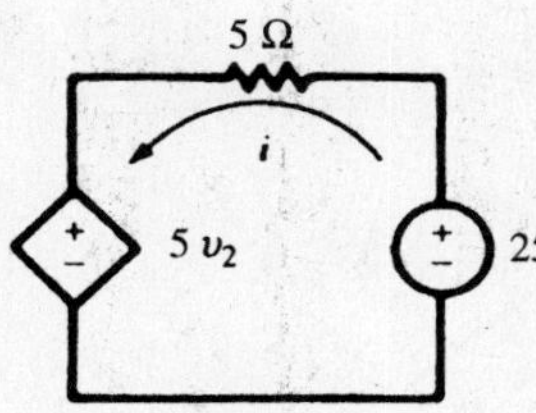

Fig. 2-22

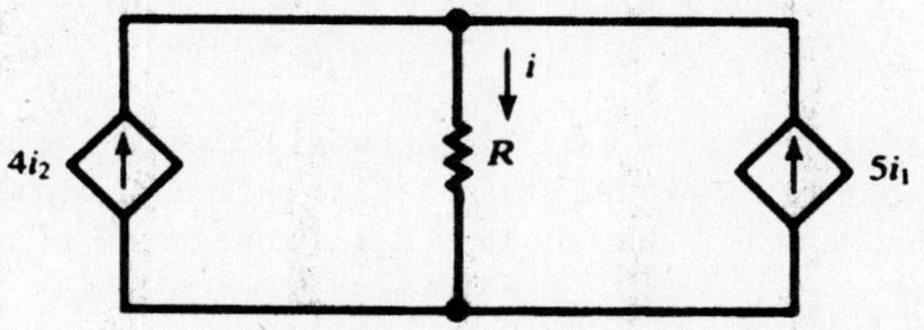

Fig. 2-23

(*a*) R = 1 $M\Omega$; (*b*) R = 100 $k\Omega$; (*c*) R = 10 $k\Omega$. Hint: Compute the charge lost during the 1-ms period.
Ans. (*a*) 0.1 V; (*b*) 1 V; (*b*) 10 V

2.26 The actual discharge current in Problem 2.25 is $i = (100/R)e^{-10^6 t/R}$ A. Find the capacitor voltage drop at 1 ms after connection to the resistor for (*a*) R = 1 $M\Omega$; (*b*) R = 100 $k\Omega$; (*c*) R = 10 $k\Omega$.
Ans. (*a*) 0.1 V; (*b*) 1 V; (*c*) 9.52 V

2.27 A 10-μF capacitor discharges in an element such that its voltage is $v = 2e^{-1000t}$. Find the current and power delivered by the capacitor as functions of time.
Ans. $i = 20e^{-1000t}$ mA, $p = vi = 40e^{-1000t}$ mJ

2.28 Find voltage v, current i, and energy W in the capacitor of Problem 2.27 at time t = 0, 1, 3, 5, and 10 ms. By integrating the power delivered by the capacitor, show that the energy dissipated in the element during the interval from 0 to t is equal to the energy lost by the capacitor.

Ans.

t	v	i	W
0	2 V	20 mA	20 μJ
1 ms	736 mV	7.36 mA	2.7 μJ
3 ms	100 mV	1 mA	0.05 μJ
5 ms	13.5 mV	135 μA	≈ 0.001 μJ
10 ms	91 μV	0.91 μA	≈ 0

2.29 The current delivered by a current source is increased linearly from zero to 10 A in 1-ms time and then is decreased linearly back to zero in 2 ms. The source feeds a 3-$k\Omega$ resistor in series with a 2-H inductor (see Fig. 2-24). (*a*) Find the energy dissipated in the resistor during the rise time (W_1) and the fall time (W_2). (*b*) Find the energy delivered to the inductor during the above two intervals. (*c*) Find the energy delivered by the current source to the series RL combination during the preceding two intervals. *Note*: Series elements have the same current. The voltage drop across their combination is the sum of their individual voltages.
Ans. (*a*) $W_1 = 100$, $W_2 = 200$; (*b*) $W_1 = 200$, $W_2 = -200$; (*c*) $W_1 = 300$, $W_2 = 0$, all in joules

2.30 The voltage of a 5-μF capacitor is increased linearly from zero to 10 V in 1 ms time and is then kept at that level. Find the current. Find the total energy delivered to the capacitor and verify that delivered energy is equal to the energy stored in the capacitor.
Ans. i = 50 mA during $0 < t < 1$ ms and is zero elsewhere, W = 250 μJ.

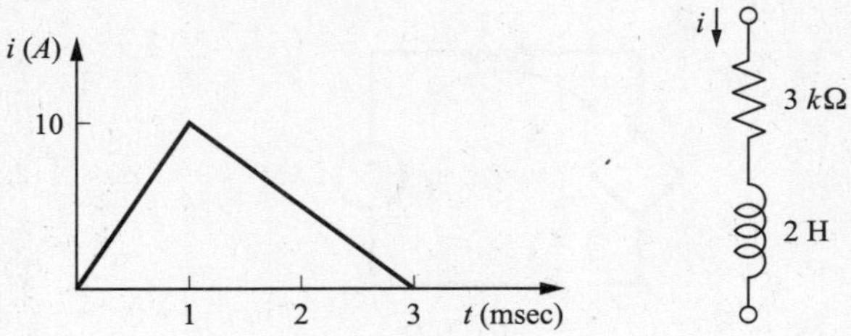

Fig. 2-24

2.31 A 10-μF capacitor is charged to 2 V. A path is established between its terminals which draws a constant current of I_0. (*a*) For $I_0 = 1$ mA, how long does it take to reduce the capacitor voltage to 5 percent of its initial value? (*b*) For what value of I_0 does the capacitor voltage remain above 90 percent of its initial value after passage of 24 hours?
Ans. (*a*) 19 ms, (*b*) 23.15pA

2.32 Energy gained (or lost) by an electric charge q traveling in an electric field is qv, where v is the electric potential gained (or lost). In a capacitor with charge Q and terminal voltage V, let all charges go from one plate to the other. By way of computation, show that the total energy W gained (or lost) is not QV but $QV/2$ and explain why. Also note that $QV/2$ is equal to the initial energy content of the capacitor.
Ans. $W = \int qvdt = Q[\frac{V-0}{2}] = QV/2 = \frac{1}{2}CV^2$. The apparent discrepancy is explained by the following. The starting voltage vetween the two plates is V. As the charges migrate from one plate of the capacitor to the other plate, the voltage between the two plates drops and becomes zero when all charges have moved. The average of the voltage during the migration process is $V/2$, and therefore, the total energy is $QV/2$.

2.33 **Lightning I**. The time profile of the discharge current in a typical cloud-to-ground lightning stroke is modeled by a triangle. The surge takes 1 μs to reach the peak value of 100 kA and then is reduced to zero in 99 μS. (*a*) Find the electric charge Q discharged. (*b*) If the cloud-to-ground voltage before the discharge is 400 MV, find the total energy W released and the average power P during the discharge. (*c*) If during the storm there is an average of 18 such lightning strokes per hour, find the average power released in 1 hour. *Ans.* (*a*) Q = 5 C; (*b*) $W = 10^9$ J, $P = 10^{13}$ W; (*c*) 5 MW

2.34 **Lightning II**. Find the cloud-to-ground capacitance in Problem 2.33 just before the lightning stroke.
Ans. 12.5 μF

2.35 **Lightning III**. The current in a cloud-to-ground lightning stroke starts at 200 kA and diminishes linearly to zero in 100 μs. Find the energy released W and the capacitance of the cloud to ground C if the voltage before the discharge is (*a*) 100 MV; (*b*) 500 MV.
Ans. (*a*) $W = 5 \times 10^8$ J, C = 0.1 μF; (*b*) $W = 25 \times 10^8$ J, C = 20 nF

2.36 The semiconductor diode of Example 2.4 is placed in the circuit of Fig. 2-25. Find the current for (*a*) $V_s = 1$ V, (*b*) $V_s = -1$ V. *Ans.* (*a*) 14 mA; (*b*) 0

2.37 The diode in the circuit of Fig. 2-26 is ideal. The inductor draws 100 mA from the voltage source. A 2-μF capacitor with zero initial charge is also connected in parallel with the inductor through an ideal diode such that the diode is reversed biased (i.e., it blocks charging of the capacitor). The switch *s* suddenly disconnects with the rest of the circuit, forcing the inductor current to pass through the diode and establishing 200 V at the capacitor's terminals. Find the value of the inductor. *Ans.* L = 8 H

2.38 Compute the static and dynamic resistances of the diode of Example 2.4 at the operating point $v = 0.66$ V.

Ans. $R \approx \dfrac{0.66}{1.2 \times 10^{-3}} = 550\ \Omega$ and $r \approx \dfrac{0.67 - 0.65}{(1.7 - 0.78) \times 10^{-3}} = 21.7\ \Omega$

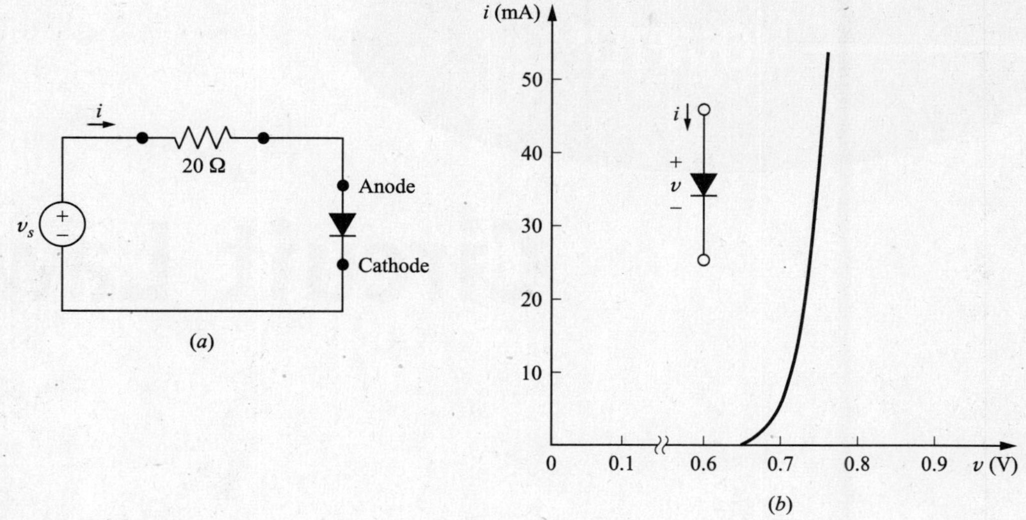

Fig. 2-25

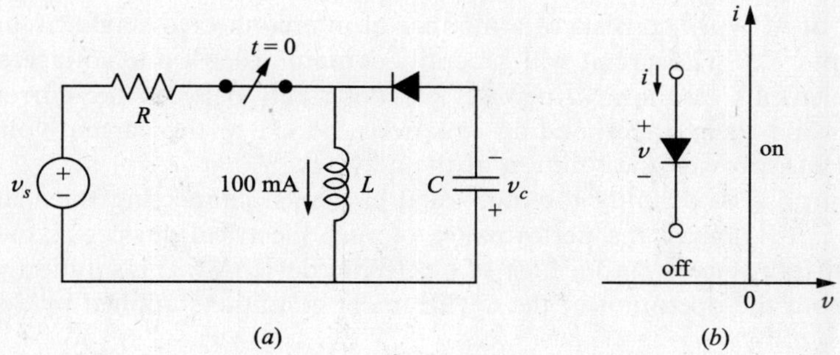

Fig. 2-26

2.39 The diode of Example 2.4 operates within the range $10 < i < 20$ mA. Within that range, approximate its terminal characteristic by a straight line $i = \alpha v + \beta$, by specifying α and β.
Ans. $i = 630\,v - 4407$ mA, where v is in V

2.40 The diode of Example 2.4 operates within the range of $20 < i < 40$ mA. Within that range, approximate its terminal characteristic by a straight line connecting the two operating limits.
Ans. $i = 993.33\,v - 702.3$ mA, where v is in V

2.41 Within the operating range of $20 < i < 40$ mA, model the diode of Example 2.4 by a resistor R in series with a voltage source V such that the model matches exactly with the diode performance at 0.72 and 0.75 V. Find R and V. *Ans.* $R = 1.007\,\Omega$, $V = 707$ mV

CHAPTER 3

Circuit Laws

3.1 INTRODUCTION

An electric circuit or network consists of a number of interconnected single circuit elements of the type described in Chapter 2. The circuit will generally contain at least one voltage or current source. The arrangement of elements results in a new set of constraints between the currents and voltages. These new constraints and their corresponding equations, added to the current-voltage relationships of the individual elements, provide the solution of the network.

The underlying purpose of defining the individual elements, connecting them in a network, and solving the equations is to analyze the performance of such electrical devices as motors, generators, transformers, electrical transducers, and a host of electronic devices. The solution generally answers necessary questions about the operation of the device under conditions applied by a source of energy.

3.2 KIRCHHOFF'S VOLTAGE LAW

For any closed path in a network, *Kirchhoff's voltage law* (KVL) states that the algebraic sum of the voltages is zero. Some of the voltages will be sosources, while others will result from current in passive elements creating a voltage, which is sometimes referred to as a *voltage drop*. The law applies equally well to circuits driven by constant sources, DC, time variable sources, $v(t)$ and $i(t)$, and to circuits driven by sources which will be introduced in Chapter 9. The mesh current method of circuit analysis introduced in Section 4.2 is based on Kirchhoff's voltage law.

EXAMPLE 3.1. Write the KVL equation for the circuit shown in Fig. 3-1.

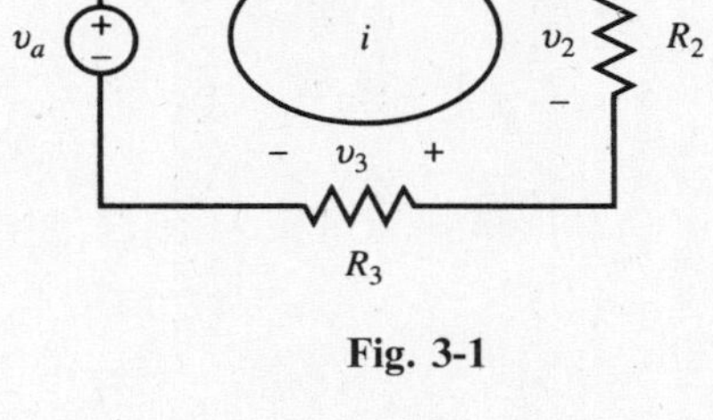

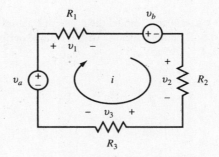

Fig. 3-1

Starting at the lower left corner of the circuit, for the current direction as shown, we have

$$-v_a + v_1 + v_b + v_2 + v_3 = 0$$
$$-v_a + iR_1 + v_b + iR_2 + iR_3 = 0$$
$$v_a - v_b = i(R_1 + R_2 + R_3)$$

3.3 KIRCHHOFF'S CURRENT LAW

The connection of two or more circuit elements creates a junction called a *node*. The junction between two elements is called a *simple node* and no division of current results. The junction of three or more elements is called a *principal node*, and here current division does take place. *Kirchhoff's current law* (KCL) states that the algebraic sum of the currents at a node is zero. It may be stated alternatively that the sum of the currents entering a node is equal to the sum of the currents leaving that node. The node voltage method of circuit analysis introduced in Section 4.3 is based on equations written at the principal nodes of a network by applying Kirchhoff's current law. The basis for the law is the conservation of electric charge.

EXAMPLE 3.2. Write the KCL equation for the principal node shown in Fig. 3-2.

$$i_1 - i_2 + i_3 - i_4 - i_5 = 0$$
$$i_1 + i_3 = i_2 + i_4 + i_5$$

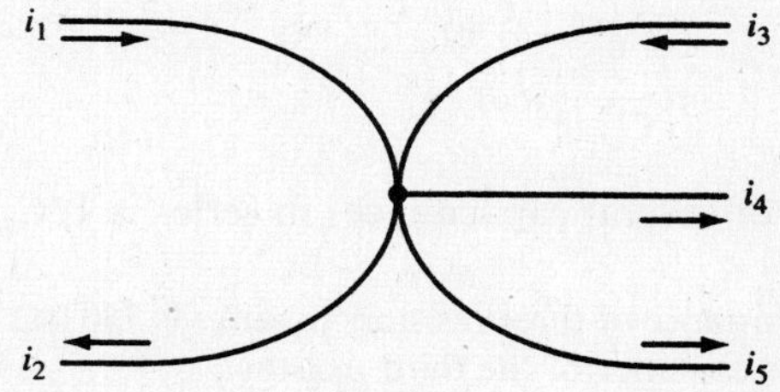

Fig. 3-2

3.4 CIRCUIT ELEMENTS IN SERIES

Three passive circuit elements in series connection as shown in Fig. 3-3 have the same current i. The voltages across the elements are v_1, v_2, and v_3. The total voltage v is the sum of the individual voltages; $v = v_1 + v_2 + v_3$.

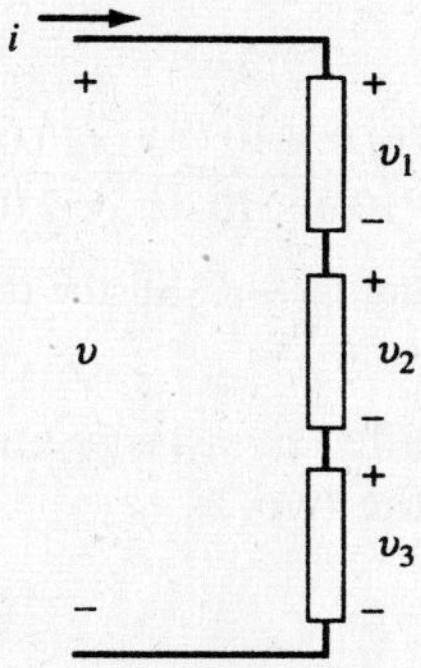

Fig. 3-3

If the elements are resistors,

$$\begin{aligned} v &= iR_1 + iR_2 + iR_3 \\ &= i(R_1 + R_2 + R_3) \\ &= iR_{eq} \end{aligned}$$

where a single equivalent resistance R_{eq} replaces the three series resistors. The same relationship between i and v will pertain.

For any number of resistors in series, we have $R_{eq} = R_1 + R_2 + \cdots$.

If the three passive elements are inductances,

$$\begin{aligned} v &= L_1 \frac{di}{dt} + L_2 \frac{di}{dt} + L_3 \frac{di}{dt} \\ &= (L_1 + L_2 + L_3) \frac{di}{dt} \\ &= L_{eq} \frac{di}{dt} \end{aligned}$$

Extending this to any number of inductances in series, we have $L_{eq} = L_1 + L_2 + \cdots$.

If the three circuit elements are capacitances, assuming zero initial charges so that the constants of integration are zero,

$$\begin{aligned} v &= \frac{1}{C_1} \int i\,dt + \frac{1}{C_2} \int i\,dt + \frac{1}{C_3} \int i\,dt \\ &= \left(\frac{1}{C_1} + \frac{1}{C_2} + \frac{1}{C_3}\right) \int i\,dt \\ &= \frac{1}{C_{eq}} \int i\,dt \end{aligned}$$

The equivalent capacitance of several capacitances in series is $1/C_{eq} = 1/C_1 + 1/C_2 + \cdots$.

EXAMPLE 3.3. The equivalent resistance of three resistors in series is 750.0 Ω. Two of the resistors are 40.0 and 410.0 Ω. What must be the ohmic resistance of the third resistor?

$$\begin{aligned} R_{eq} &= R_1 + R_2 + R_3 \\ 750.0 &= 40.0 + 410.0 + R_3 \qquad \text{and} \qquad R_3 = 300.0\ \Omega \end{aligned}$$

EXAMPLE 3.4. Two capacitors, $C_1 = 2.0\ \mu\text{F}$ and $C_2 = 10.0\ \mu\text{F}$, are connected in series. Find the equivalent capacitance. Repeat if C_2 is 10.0 pF.

$$C_{eq} = \frac{C_1 C_2}{C_1 + C_2} = \frac{(2.0 \times 10^{-6})(10.0 \times 10^{-6})}{2.0 \times 10^{-6} + 10.0 \times 10^{-6}} = 1.67\ \mu\text{F}$$

If $C_2 = 10.0$ pF,

$$C_{eq} = \frac{(2.0 \times 10^{-6})(10.0 \times 10^{-12})}{2.0 \times 10^{-6} + 10.0 \times 10^{-12}} = \frac{20.0 \times 10^{-18}}{2.0 \times 10^{-6}} = 10.0\ \text{pF}$$

where the contribution of 10.0×10^{-12} to the sum $C_1 + C_2$ in the denominator is negligible and therefore it can be omitted.

Note: When two capacitors in series differ by a large amount, the equivalent capacitance is essentially equal to the value of the smaller of the two.

3.5 CIRCUIT ELEMENTS IN PARALLEL

For three circuit elements connected in parallel as shown in Fig. 3-4, KCL states that the current i entering the principal node is the sum of the three currents leaving the node through the branches.

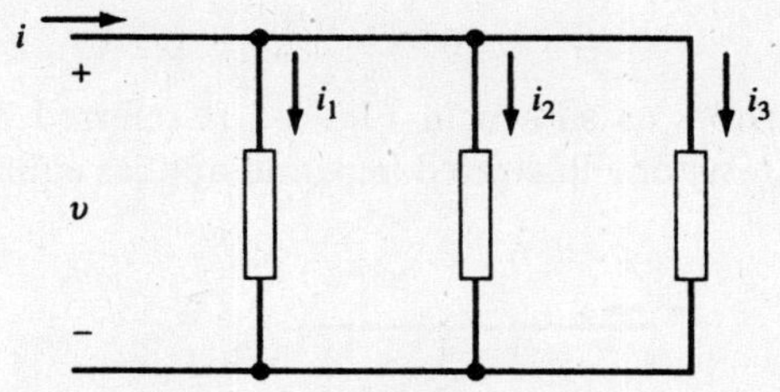

Fig. 3-4

$$i = i_1 + i_2 + i_3$$

If the three passive circuit elements are resistances,

$$i = \frac{v}{R_1} + \frac{v}{R_2} + \frac{v}{R_3} = \left(\frac{1}{R_1} + \frac{1}{R_2} + \frac{1}{R_3}\right)v = \frac{1}{R_{eq}}\,v$$

For several resistors in parallel,

$$\frac{1}{R_{eq}} = \frac{1}{R_1} + \frac{1}{R_2} + \cdots$$

The case of two resistors in parallel occurs frequently and deserves special mention. The equivalent resistance of two resistors in parallel is given by the *product over the sum.*

$$R_{eq} = \frac{R_1 R_2}{R_1 + R_2}$$

EXAMPLE 3.5. Obtain the equivalent resistance of (*a*) two 60.0-Ω resistors in parallel and (*b*) three 60.0-Ω resistors in parallel.

(*a*) $$R_{eq} = \frac{(60.0)^2}{120.0} = 30.0\,\Omega$$

(*b*) $$\frac{1}{R_{eq}} = \frac{1}{60.0} + \frac{1}{60.0} + \frac{1}{60.0} \qquad R_{eq} = 20.0\,\Omega$$

Note: For n identical resistors in parallel the equivalent resistance is given by R/n.

Combinations of inductances in parallel have similar expressions to those of resistors in parallel:

$$\frac{1}{L_{eq}} = \frac{1}{L_1} + \frac{1}{L_2} + \cdots \qquad \text{and, for two inductances,} \qquad L_{eq} = \frac{L_1 L_2}{L_1 + L_2}$$

EXAMPLE 3.6. Two inductances $L_1 = 3.0\,\text{mH}$ and $L_2 = 6.0\,\text{mH}$ are connected in parallel. Find L_{eq}.

$$\frac{1}{L_{eq}} = \frac{1}{3.0\,\text{mH}} + \frac{1}{6.0\,\text{mH}} \qquad \text{and} \qquad L_{eq} = 2.0\,\text{mH}$$

With three capacitances in parallel,

$$i = C_1\frac{dv}{dt} + C_2\frac{dv}{dt} + C_3\frac{dv}{dt} = (C_1 + C_2 + C_3)\frac{dv}{dt} = C_{eq}\frac{dv}{dt}$$

For several parallel capacitors, $C_{eq} = C_1 + C_2 + \cdots$, which is of the same form as resistors in series.

3.6 VOLTAGE DIVISION

A set of series-connected resistors as shown in Fig. 3-5 is referred to as a *voltage divider*. The concept extends beyond the set of resistors illustrated here and applies equally to impedances in series, as will be shown in Chapter 9.

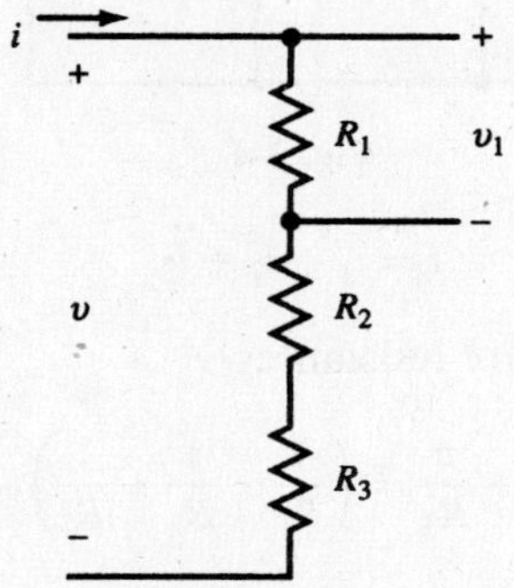

Fig. 3-5

Since $v_1 = iR_1$ and $v = i(R_1 + R_2 + R_3)$,

$$v_1 = v\left(\frac{R_1}{R_1 + R_2 + R_3}\right)$$

EXAMPLE 3.7. A voltage divider circuit of two resistors is designed with a total resistance of the two resistors equal to 50.0 Ω. If the output voltage is 10 percent of the input voltage, obtain the values of the two resistors in the circuit.

$$\frac{v_1}{v} = 0.10 \qquad 0.10 = \frac{R_1}{50.0 \times 10^3}$$

from which $R_1 = 5.0\,\Omega$ and $R_2 = 45.0\,\Omega$.

3.7 CURRENT DIVISION

A parallel arrangement of resistors as shown in Fig. 3-6 results in a *current divider*. The ratio of the branch current i_1 to the total current i illustrates the operation of the divider.

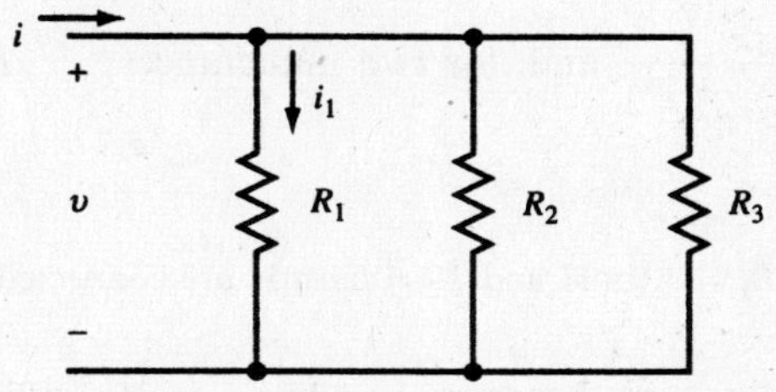

Fig. 3-6

$$i = \frac{v}{R_1} + \frac{v}{R_2} + \frac{v}{R_3} \qquad \text{and} \qquad i_1 = \frac{v}{R_1}$$

Then
$$\frac{i_1}{i} = \frac{1/R_1}{1/R_1 + 1/R_2 + 1/R_3} = \frac{R_2R_3}{R_1R_2 + R_1R_3 + R_2R_3}$$

For a two-branch current divider we have

$$\frac{i_1}{i} = \frac{R_2}{R_1 + R_2}$$

This may be expressed as follows: The ratio of the current in one branch of a two-branch parallel circuit to the total current is equal to the ratio of the resistance of the *other* branch resistance to the sum of the two resistances.

EXAMPLE 3.8. A current of 30.0 mA is to be divided into two branch currents of 20.0 mA and 10.0 mA by a network with an equivalent resistance equal to or greater than 10.0 Ω. Obtain the branch resistances.

$$\frac{20\,\text{mA}}{30\,\text{mA}} = \frac{R_2}{R_1 + R_2} \qquad \frac{10\,\text{mA}}{30\,\text{mA}} = \frac{R_1}{R_1 + R_2} \qquad \frac{R_1 R_2}{R_1 + R_2} \geq 10.0\,\Omega$$

Solving these equations yields $R_1 \geq 15.0\,\Omega$ and $R_2 \geq 30.0\,\Omega$.

Solved Problems

3.1 Find V_3 and its polarity if the current I in the circuit of Fig. 3-7 is 0.40 A.

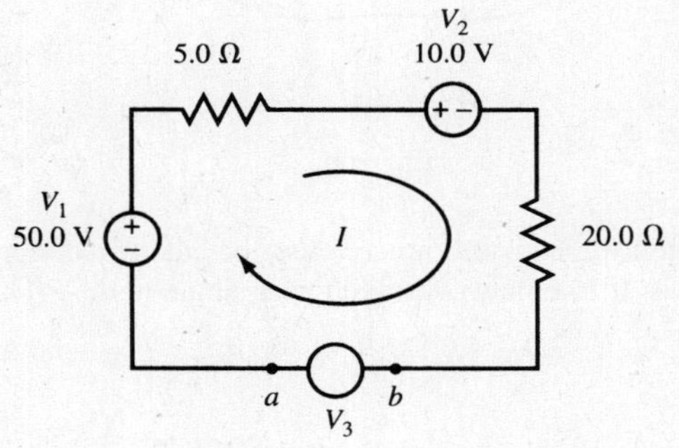

Fig. 3-7

Assume that V_3 has the same polarity as V_1. Applying KVL and starting from the lower left corner,

$$V_1 - I(5.0) - V_2 - I(20.0) + V_3 = 0$$
$$50.0 - 2.0 - 10.0 - 8.0 + V_3 = 0$$
$$V_3 = -30.0\,\text{V}$$

Terminal b is positive with respect to terminal a.

3.2 Obtain the currents I_1 and I_2 for the network shown in Fig. 3-8.

a and *b* comprise one node. Applying KCL,

$$2.0 + 7.0 + I_1 = 3.0 \qquad \text{or} \qquad I_1 = -6.0\,\text{A}$$

Also, *c* and *d* comprise a single node. Thus,

$$4.0 + 6.0 = I_2 + 1.0 \qquad \text{or} \qquad I_2 = 9.0\,\text{A}$$

3.3 Find the current I for the circuit shown in Fig. 3-9.

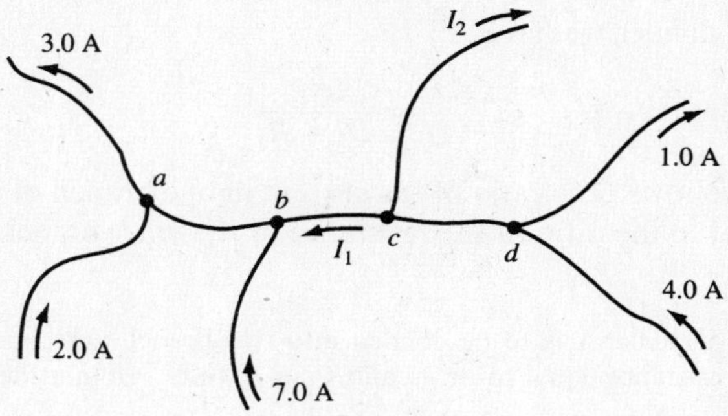

Fig. 3-8

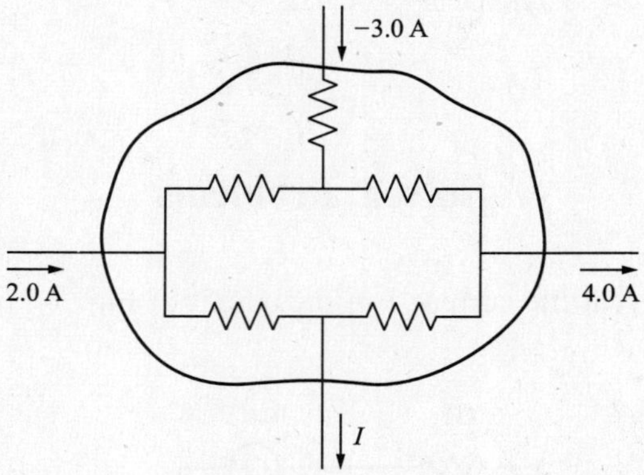

Fig. 3-9

The branch currents within the enclosed area cannot be calculated since no values of the resistors are given. However, KCL applies to the network taken as a single node. Thus,

$$2.0 - 3.0 - 4.0 - I = 0 \qquad \text{or} \qquad I = -5.0\ \text{A}$$

3.4 Find the equivalent resistance for the circuit shown in Fig. 3-10.

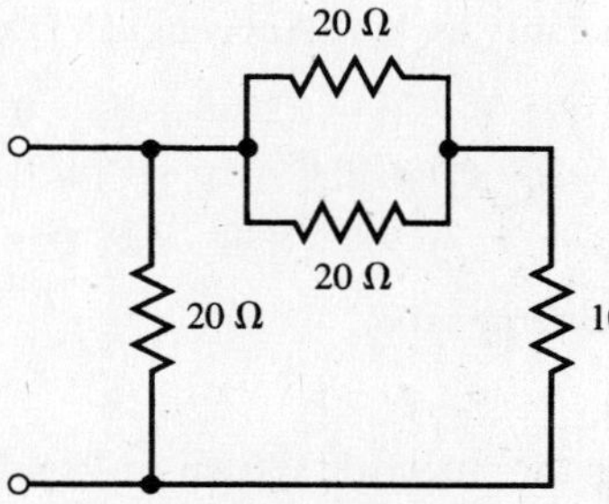

Fig. 3-10

The two 20-Ω resistors in parallel have an equivalent resistance $R_{eq} = [(20)(20)/(20 + 20)] = 10\ \Omega$. This is in series with the 10-Ω resistor so that their sum is 20 Ω. This in turn is in parallel with the other 20-Ω resistor so that the overall equivalent resistance is 10 Ω.

3.5 Determine the equivalent inductance of the three parallel inductances shown in Fig. 3-11.

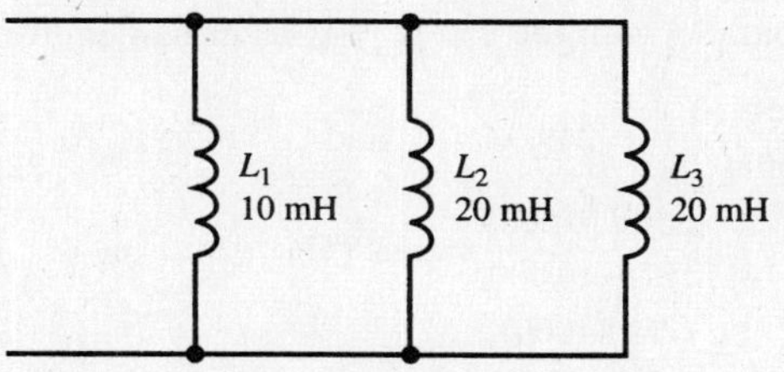

Fig. 3-11

The two 20-mH inductances have an equivalent inductance of 10 mH. Since this is in parallel with the 10-mH inductance, the overall equivalent inductance is 5 mH. Alternatively,

$$\frac{1}{L_{eq}} = \frac{1}{L_1} + \frac{1}{L_2} + \frac{1}{L_3} = \frac{1}{10\,\text{mH}} + \frac{1}{20\,\text{mH}} + \frac{1}{20\,\text{mH}} = \frac{4}{20\,\text{mH}} \qquad \text{or} \qquad L_{eq} = 5\,\text{mH}$$

3.6 Express the total capacitance of the three capacitors in Fig. 3-12.

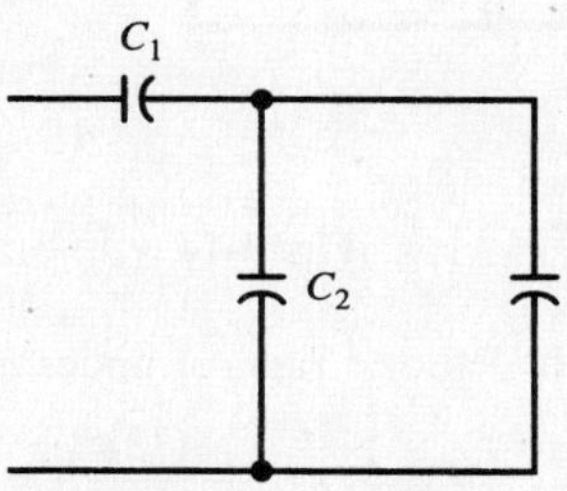

Fig. 3-12

For C_2 and C_3 in parallel, $C_{eq} = C_2 + C_3$. Then for C_1 and C_{eq} in series,

$$C_T = \frac{C_1 C_{eq}}{C_1 + C_{eq}} = \frac{C_1(C_2 + C_3)}{C_1 + C_2 + C_3}$$

3.7 The circuit shown in Fig. 3-13 is a voltage divider, also called an *attenuator*. When it is a single resistor with an adjustable tap, it is called a *potentiometer*, or *pot*. To discover the effect of loading, which is caused by the resistance R of the voltmeter VM, calculate the ratio V_{out}/V_{in} for (*a*) $R = \infty$, (*b*) 1 MΩ, (*c*) 10 kΩ, (*d*) 1 kΩ.

(*a*)
$$V_{out}/V_{in} = \frac{250}{2250 + 250} = 0.100$$

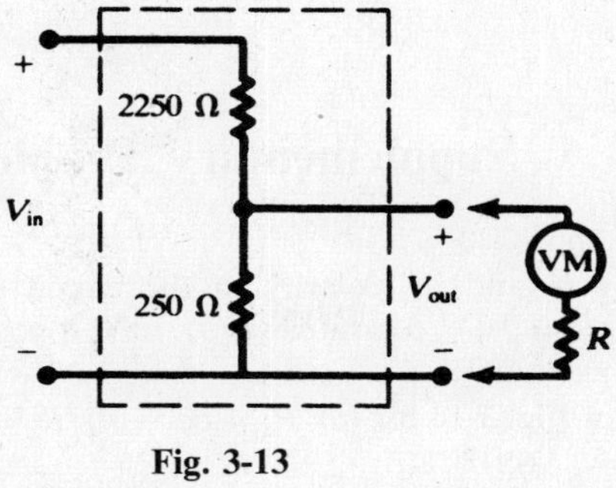

Fig. 3-13

(*b*) The resistance R in parallel with the 250-Ω resistor has an equivalent resistance

$$R_{eq} = \frac{250(10^6)}{250 + 10^6} = 249.9\ \Omega \qquad \text{and} \qquad V_{out}/V_{in} = \frac{249.9}{2250 + 249.9} = 0.100$$

(*c*) $$R_{eq} = \frac{(250)(10\,000)}{250 + 10\,000} = 243.9\ \Omega \qquad \text{and} \qquad V_{out}/V_{in} = 0.098$$

(*d*) $$R_{eq} = \frac{(250)(1000)}{250 + 1000} = 200.0\ \Omega \qquad \text{and} \qquad V_{out}/V_{in} = 0.082$$

3.8 Find all branch currents in the network shown in Fig. 3-14(*a*).

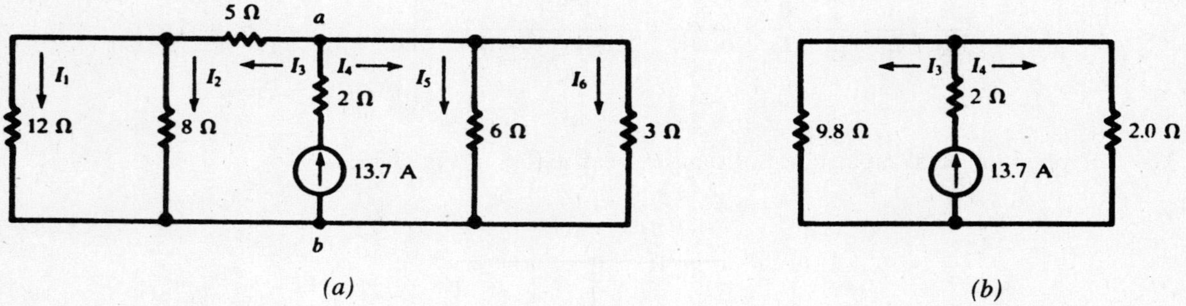

Fig. 3-14

The equivalent resistances to the left and right of nodes *a* and *b* are

$$R_{eq(left)} = 5 + \frac{(12)(8)}{20} = 9.8\ \Omega$$

$$R_{eq(right)} = \frac{(6)(3)}{9} = 2.0\ \Omega$$

Now referring to the reduced network of Fig. 3-14(*b*),

$$I_3 = \frac{2.0}{11.8}(13.7) = 2.32\ \text{A}$$

$$I_4 = \frac{9.8}{11.8}(13.7) = 11.38\ \text{A}$$

Then referring to the original network,

$$I_1 = \frac{8}{20}(2.32) = 0.93\ \text{A} \qquad I_2 = 2.32 - 0.93 = 1.39\ \text{A}$$

$$I_5 = \frac{3}{9}(11.38) = 3.79\ \text{A} \qquad I_6 = 11.38 - 3.79 = 7.59\ \text{A}$$

Supplementary Problems

3.9 Find the source voltage V and its polarity in the circuit shown in Fig. 3-15 if (*a*) $I = 2.0$ A and (*b*) $I = -2.0$ A. *Ans.* (*a*) 50 V, *b* positive; (*b*) 10 V, *a* positive.

3.10 Find R_{eq} for the circuit of Fig. 3-16 for (*a*) $R_x = \infty$, (*b*) $R_x = 0$, (*c*) $R_x = 5\ \Omega$.
Ans. (*a*) 36 Ω; (*b*) 16 Ω; (*c*) 20 Ω

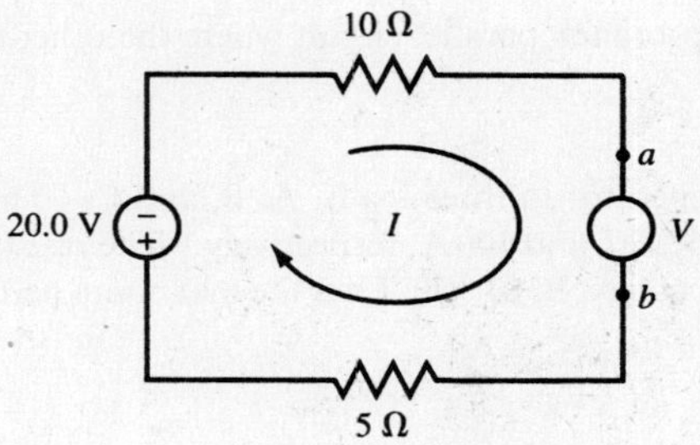

Fig. 3-15

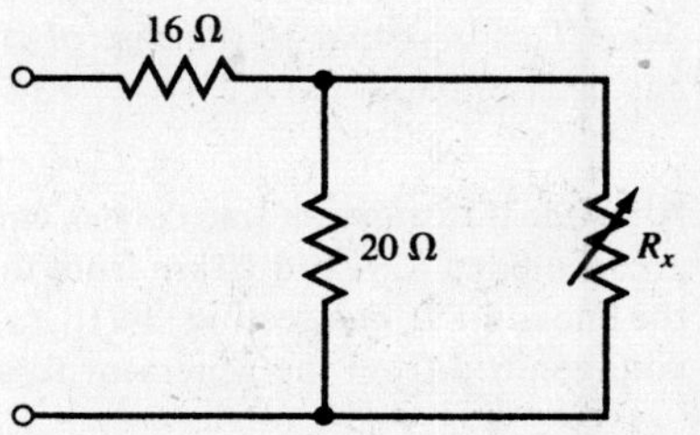

Fig. 3-16

3.11 An inductance of 8.0 mH is in series with two inductances in parallel, one of 3.0 mH and the other 6.0 mH. Find L_{eq}. *Ans.* 10.0 mH

3.12 Show that for the three capacitances of equal value shown in Fig. 3-17 $C_{eq} = 1.5\,C$.

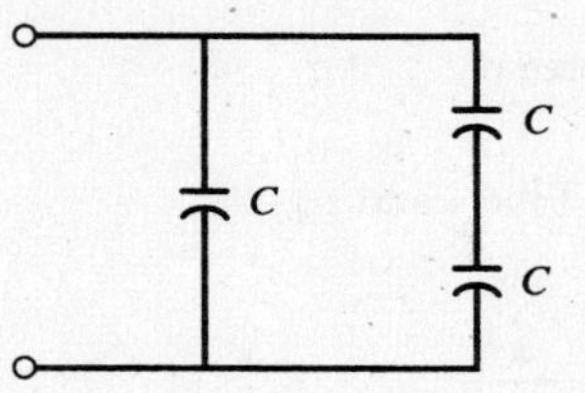

Fig. 3-17

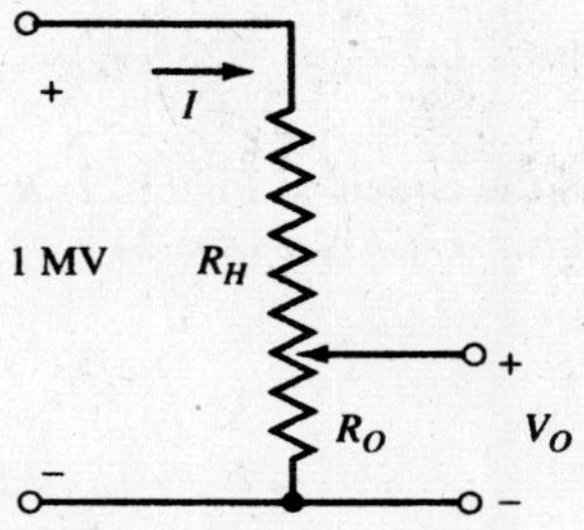

Fig. 3-18

3.13 Find R_H and R_O for the voltage divider in Fig. 3-18 so that the current I is limited to 0.5 A when $V_O = 100$ V. *Ans.* $R_H = 2\,\text{M}\Omega$, $R_O = 200\,\Omega$

3.14 Using voltage division, calculate V_1 and V_2 in the network shown in Fig. 3-19. *Ans.* 11.4 V, 73.1 V

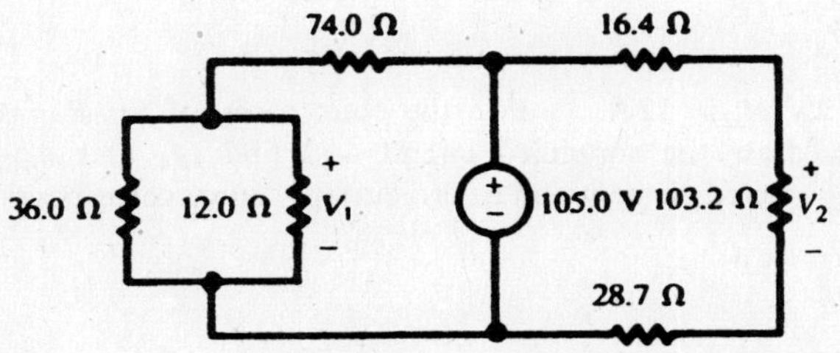

Fig. 3-19

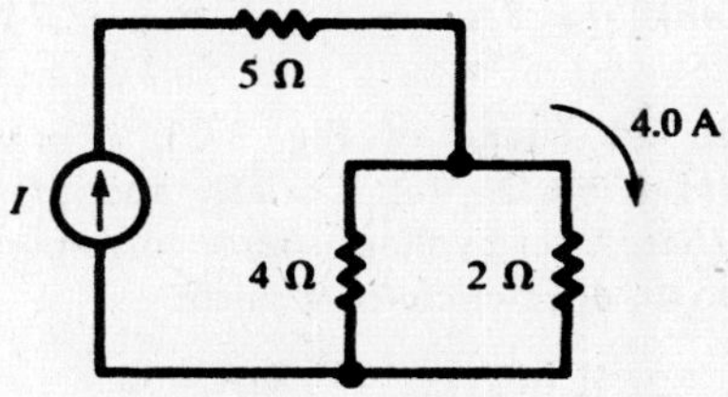

Fig. 3-20

3.15 Obtain the source current I and the total power delivered to the circuit in Fig. 3-20. *Ans.* 6.0 A, 228 W

3.16 Show that for four resistors in parallel the current in one branch, for example the branch of R_4, is related to the total current by

$$I_4 = I_T\left(\frac{R'}{R_4 + R'}\right) \qquad \text{where } R' = \frac{R_1R_2R_3}{R_1R_2 + R_1R_3 + R_2R_3}$$

Note: This is similar to the case of current division in a two-branch parallel circuit where the other resistor has been replaced by R'.

3.17 A power transmission line carries current from a 6000-V generator to three loads, A, B, and C. The loads are located at 4, 7, and 10 km from the generator and draw 50, 20, and 100 A, respectively. The resistance of the line is 0.1 Ω/km; see Fig. 3-21. (*a*) Find the voltage at loads A, B, C. (*b*) Find the maximum percentage voltage drop from the generator to a load.

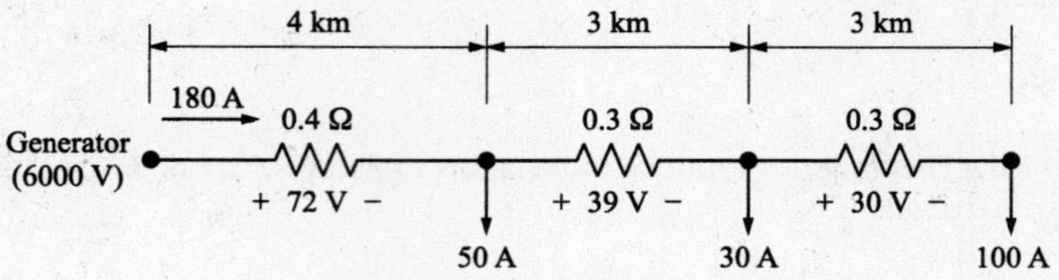

Fig. 3-21

Ans. (*a*) $v_A = 5928\text{ V}, v_B = 5889\text{ V}, v_C = 5859\text{ V}$; (*b*) 2.35 percent

3.18 In the circuit of Fig. 3-22, $R = 0$ and i_1 and i_2 are unknown. Find i and v_{AC}.
Ans. $i = 4\text{ A}, v_{AC} = 24\text{ V}$

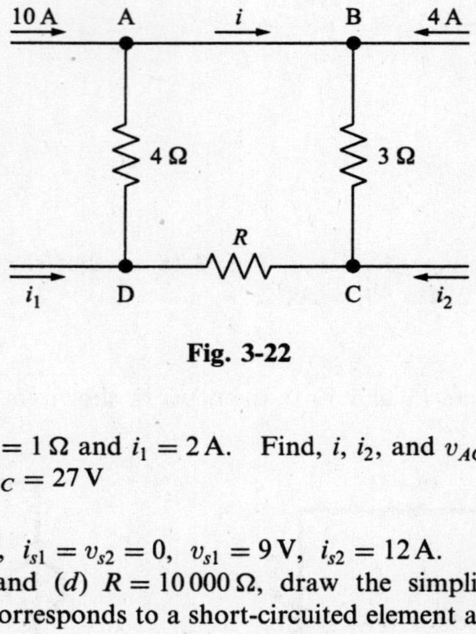

Fig. 3-22

3.19 In the circuit of Fig. 3-22, $R = 1\,\Omega$ and $i_1 = 2\text{ A}$. Find, i, i_2, and v_{AC}.
Ans. $i = 5\text{ A}, i_2 = -16\text{ A}, v_{AC} = 27\text{ V}$

3.20 In the circuit of Fig. 3-23, $i_{s1} = v_{s2} = 0$, $v_{s1} = 9\text{ V}$, $i_{s2} = 12\text{ A}$. For the four cases of (*a*) $R = 0$, (*b*) $R = 6\,\Omega$, (*c*) $R = 9\,\Omega$, and (*d*) $R = 10\,000\,\Omega$, draw the simplified circuit and find i_{BA} and v_{AC}. *Hint*: A zero voltage source corresponds to a short-circuited element and a zero current source corresponds to an open-circuited element.

Ans.
$$\begin{cases} (a)\quad i_{BA} = 7, v_{AC} = 30 \\ (b)\quad i_{BA} = 4.2, v_{AC} = 21.6 \\ (c)\quad i_{BA} = 3.5, v_{AC} = 19.5 \\ (d)\quad i_{BA} = 0.006 \approx 0, v_{AC} = 9.02 \approx 9 \end{cases}$$
(All in A and V)

3.21 In the circuit of Fig. 3-23, $v_{s1} = v_{s2} = 0, i_{s1} = 6\text{ A}, i_{s2} = 12\text{ A}$. For the four cases of (*a*) $R = 0$, (*b*) $R = 6\,\Omega$, (*c*) $R = 9\,\Omega$, and (*d*) $R = 10\,000\,\Omega$, draw the simplified circuit and find i_{BA} and v_{AC}.

Ans.
$$\begin{cases} (a)\quad i_{BA} = 6, v_{AC} = 36 \\ (b)\quad i_{BA} = 3.6, v_{AC} = 28.8 \\ (c)\quad i_{BA} = 3, v_{AC} = 27 \\ (d)\quad i_{BA} = 0.005 \approx 0, v_{AC} \approx 18 \end{cases}$$
(All in A and V)

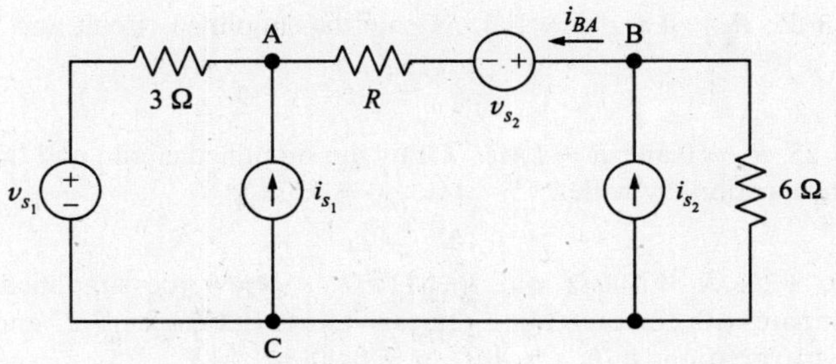

Fig. 3-23

3.22 In the circuit Fig. 3-23, $v_{s1} = 0$, $v_{s2} = 6\,\text{V}$, $i_{s1} = 6\,\text{A}$, $i_{s2} = 12\,\text{A}$. For the four cases of (*a*) $R = 0$, (*b*) $R = 6\,\Omega$, (*c*) $R = 9\,\Omega$, and (*d*) $R = 10\,000\,\Omega$, draw the simplified circuit and find i_{BA} and v_{AC}.

Ans.
$$\begin{cases} (a) & i_{BA} = 5.33, v_{AC} = 34 \\ (b) & i_{BA} = 3.2, v_{AC} = 27.6 \\ (c) & i_{BA} = 2.66, v_{AC} = 26 \\ (d) & i_{BA} = 0.005 \approx 0, v_{AC} = 18.01 \approx 18 \end{cases}$$
(All in A and V)

3.23 In the circuit of Fig. 3-24, (*a*) find the resistance seen by the voltage source, $R_{in} = v/i$, as a function of a, and (*b*) evaluate R_{in} for $a = 0, 1, 2$. *Ans.* (*a*) $R_{in} = R/(1 - a)$; (*b*) $R, \infty, -R$

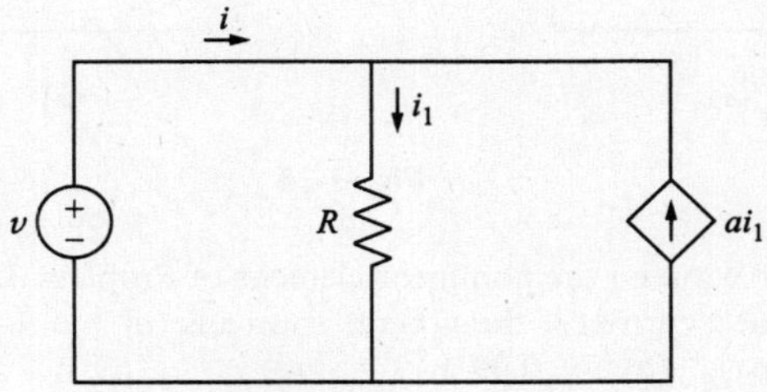

Fig. 3-24

3.24 In the circuit of Fig. 3-24, (*a*) find power P delivered by the voltage source as a function of a, and (*b*) evaluate P for $a = 0, 1, 2$. *Ans.* (*a*) $P = v^2(1 - a)/R$; (*b*) $v^2/R, 0, -v^2/R$

3.25 In the circuit of Fig. 3-24, let $a = 2$. Connect a resistor R_x in parallel with the voltage source and adjust it within the range $0 \le R_x \le 0.99R$ such that the voltage source delivers minimum power. Find (*a*) the value of R_x and (*b*) the power delivered by the voltage source.
Ans. (*a*) $R_x = 0.99R$, (*b*) $P = v^2/(99R)$

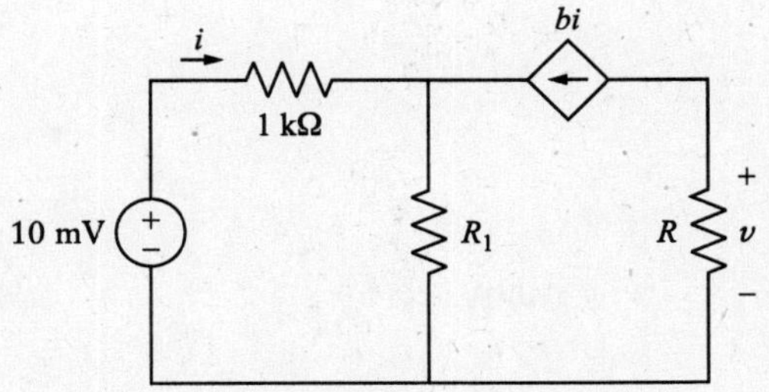

Fig. 3-25

3.26 In the circuit of Fig. 3-25, $R_1 = 0$ and $b = 100$. Draw the simplified circuit and find v for $R = 1\,k\Omega$ and $10\,k\Omega$. *Ans.* $v = 1, 10\,\text{V}$

3.27 In the circuit of Fig. 3-25, $R_1 = 0$ and $R = 1\,k\Omega$. Draw the simplified circuit and find v for $b = 50, 100, 200$. Note that v changes proportionally with b. *Ans.* $v = 0.5, 1, 2\,\text{V}$

3.28 In the circuit of Fig. 3-25, $R_1 = 100\,\Omega$ and $R = 11\,k\Omega$. Draw the simplified circuit and find v for $b = 50, 100, 200$. Compare with corresponding values obtained in Problem 3.27 and note that in the present case v is less sensitive to variations in b. *Ans.* $v = 0.90, 1, 1.04\,\text{V}$

3.29 A nonlinear element is modeled by the following terminal characteristic.

$$i = \begin{cases} 10v & \text{when } v \geq 0 \\ 0.1v & \text{when } v \leq 0 \end{cases}$$

Find the element's current if it is connected to a voltage source with (*a*) $v = 1 + \sin t$ and (*b*) $v = -1 + \sin t$. See Fig. 3-26(*a*). *Ans.* (*a*) $i = 10(1 + \sin t)$; (*b*) $i = 0.1(-1 + \sin t)$

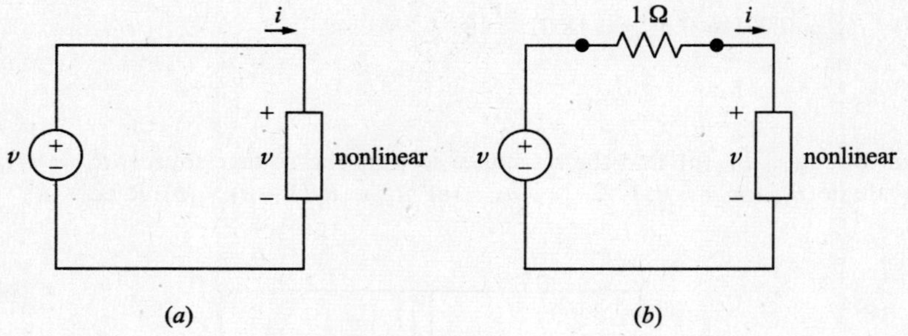

Fig. 3-26

3.30 Place a 1-Ω linear resistor between the nonlinear element of Problem 3.29 and the voltage source. See Fig. 3-26(*b*). Find the element's current if the voltage source is (*a*) $v = 1 + \sin t$ and (*b*) $v = -1 + \sin t$. *Ans.* (*a*) $i = 0.91(1 + \sin t)$; (*b*) $i = 0.091(-1 + \sin t)$

CHAPTER 4

Analysis Methods

4.1 THE BRANCH CURRENT METHOD

In the branch current method a current is assigned to each branch in an active network. Then Kirchhoff's current law is applied at the principal nodes and the voltages between the nodes employed to relate the currents. This produces a set of simultaneous equations which can be solved to obtain the currents.

EXAMPLE 4.1 Obtain the current in each branch of the network shown in Fig. 4-1 using the branch current method.

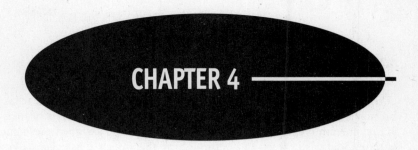

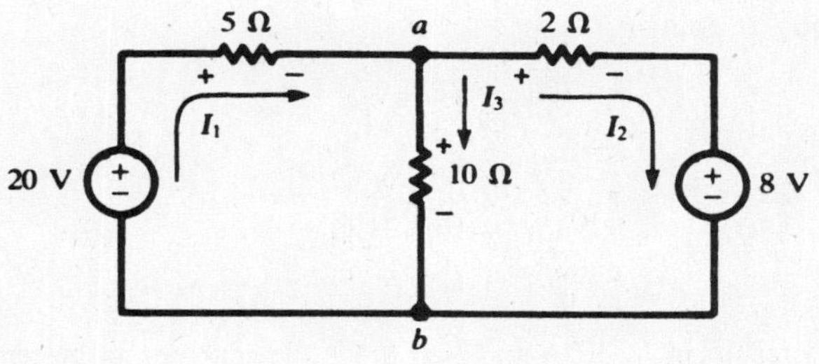

Fig. 4-1

Currents I_1, I_2, and I_3 are assigned to the branches as shown. Applying KCL at node a,

$$I_1 = I_2 + I_3 \tag{1}$$

The voltage V_{ab} can be written in terms of the elements in each of the branches; $V_{ab} = 20 - I_1(5)$, $V_{ab} = I_3(10)$ and $V_{ab} = I_2(2) + 8$. Then the following equations can be written

$$20 - I_1(5) = I_3(10) \tag{2}$$

$$20 - I_1(5) = I_2(2) + 8 \tag{3}$$

Solving the three equations (*1*), (*2*), and (*3*) simultaneously gives $I_1 = 2\text{ A}$, $I_2 = 1\text{ A}$, and $I_3 = 1\text{ A}$.

Other directions may be chosen for the branch currents and the answers will simply include the appropriate sign. In a more complex network, the branch current method is difficult to apply because it does not suggest either a starting point or a logical progression through the network to produce the necessary equations. It also results in more independent equations than either the mesh current or node voltage method requires.

4.2 THE MESH CURRENT METHOD

In the mesh current method a current is assigned to each *window* of the network such that the currents complete a closed loop. They are sometimes referred to as *loop currents*. Each element and branch therefore will have an independent current. When a branch has two of the mesh currents, the actual current is given by their algebraic sum. The assigned mesh currents may have either clockwise or counterclockwise directions, although at the outset it is wise to assign to all of the mesh currents a clockwise direction. Once the currents are assigned, Kirchhoff's voltage law is written for each loop to obtain the necessary simultaneous equations.

EXAMPLE 4.2 Obtain the current in each branch of the network shown in Fig. 4-2 (same as Fig. 4-1) using the mesh current method.

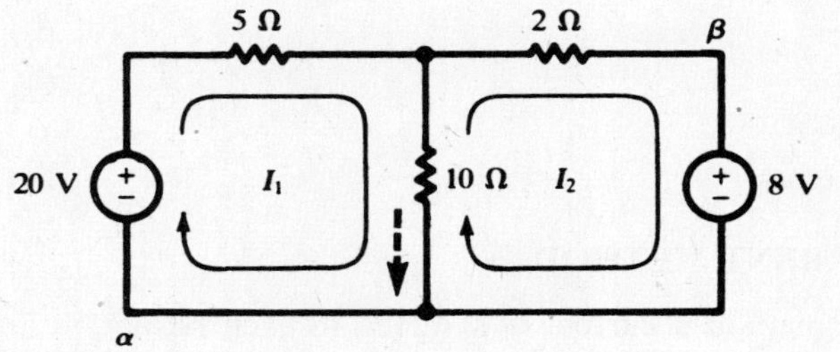

Fig. 4-2

The currents I_1 and I_2 are chosen as shown on the circuit diagram. Applying KVL around the left loop, starting at point α,

$$-20 + 5I_1 + 10(I_1 - I_2) = 0$$

and around the right loop, starting at point β,

$$8 + 10(I_2 - I_1) + 2I_2 = 0$$

Rearranging terms,

$$15I_1 - 10I_2 = 20 \quad (4)$$

$$-10I_1 + 12I_2 = -8 \quad (5)$$

Solving (*4*) and (*5*) simultaneously results in $I_1 = 2$ A and $I_2 = 1$ A. The current in the center branch, shown dotted, is $I_1 - I_2 = 1$ A. In Example 4.1 this was branch current I_3.

The currents do not have to be restricted to the *windows* in order to result in a valid set of simultaneous equations, although that is the usual case with the mesh current method. For example, see Problem 4.6, where each of the currents passes through the source. In that problem they are called loop currents. The applicable rule is that each element in the network must have a current or a combination of currents and no two elements in different branches can be assigned the same current or the same combination of currents.

4.3 MATRICES AND DETERMINANTS

The n simultaneous equations of an n-mesh network can be written in matrix form. (Refer to Appendix B for an introduction to matrices and determinants.)

EXAMPLE 4.3 When KVL is applied to the three-mesh network of Fig. 4-3, the following three equations are obtained:

$$\begin{aligned}(R_A+R_B)I_1 \quad - R_BI_2 \quad &= V_a\\ -R_BI_1+(R_B+R_C+R_D)I_2 \quad - R_DI_3 &= 0\\ -R_DI_2+(R_D+R_E)I_3 &= -V_b\end{aligned}$$

Placing the equations in matrix form,

$$\begin{bmatrix} R_A+R_B & -R_B & 0 \\ -R_B & R_B+R_C+R_D & -R_D \\ 0 & -R_D & R_D+R_E \end{bmatrix}\begin{bmatrix} I_1 \\ I_2 \\ I_3 \end{bmatrix} = \begin{bmatrix} V_a \\ 0 \\ -V_b \end{bmatrix}$$

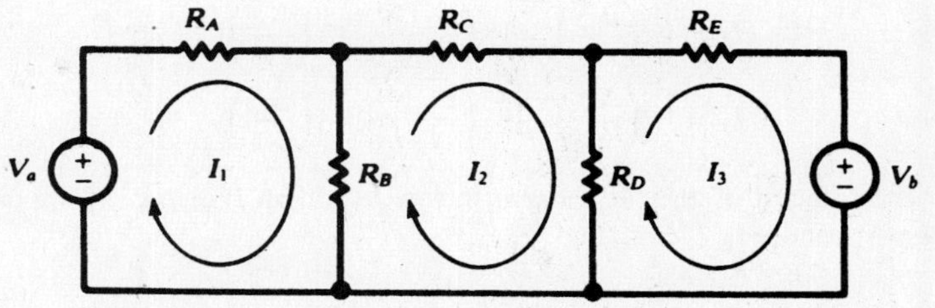

Fig. 4-3

The elements of the matrices can be indicated in general form as follows:

$$\begin{bmatrix} R_{11} & R_{12} & R_{13} \\ R_{21} & R_{22} & R_{23} \\ R_{31} & R_{32} & R_{33} \end{bmatrix}\begin{bmatrix} I_1 \\ I_2 \\ I_3 \end{bmatrix} = \begin{bmatrix} V_1 \\ V_2 \\ V_3 \end{bmatrix} \qquad (6)$$

Now element R_{11} (row 1, column 1) is the sum of all resistances through which mesh current I_1 passes. In Fig. 4-3, this is $R_A + R_B$. Similarly, elements R_{22} and R_{33} are the sums of all resistances through which I_2 and I_3, respectively, pass.

Element R_{12} (row 1, column 2) is the sum of all resistances through which mesh currents I_1 and I_2 pass. The sign of R_{12} is + if the two currents are in the same direction through each resistance, and − if they are in opposite directions. In Fig. 4-3, R_B is the only resistance common to I_1 and I_2; and the current directions are opposite in R_B, so that the sign is negative. Similarly, elements R_{21}, R_{23}, R_{13}, and R_{31} are the sums of the resistances common to the two mesh currents indicated by the subscripts, with the signs determined as described previously for R_{12}. It should be noted that for all i and j, $R_{ij} = R_{ji}$. As a result, the resistance matrix is symmetric about the principal diagonal.

The current matrix requires no explanation, since the elements are in a single column with subscripts 1, 2, 3, . . . to identify the current with the corresponding mesh. These are the unknowns in the mesh current method of network analysis.

Element V_1 in the voltage matrix is the sum of all source voltages driving mesh current I_1. A voltage is counted positive in the sum if I_1 passes from the − to the + terminal of the source; otherwise, it is counted negative. In other words, a voltage is positive if the source drives in the direction of the mesh current. In Fig. 4.3, mesh 1 has a source V_a driving in the direction of I_1; mesh 2 has no source; and mesh 3 has a source V_b driving opposite to the direction of I_3, making V_3 negative.

The matrix equation arising from the mesh current method may be solved by various techniques. One of these, the *method of determinants* (Cramer's rule), will be presented here. It should be stated, however, that other techniques are far more efficient for large networks.

EXAMPLE 4.4 Solve matrix equation (*6*) of Example 4.3 by the method of determinants.

The unknown current I_1 is obtained as the ratio of two determinants. The denominator determinant has the elements of resistance matrix. This may be referred to as the *determinant of the coefficients* and given the symbol Δ_R. The numerator determinant has the same elements as Δ_R except in the first column, where the elements of the voltage matrix replace those of the determinant of the coefficients. Thus,

$$I_1 = \begin{vmatrix} V_1 & R_{12} & R_{13} \\ V_2 & R_{22} & R_{23} \\ V_3 & R_{32} & R_{33} \end{vmatrix} \Bigg/ \begin{vmatrix} R_{11} & R_{12} & R_{13} \\ R_{21} & R_{22} & R_{23} \\ R_{31} & R_{32} & R_{33} \end{vmatrix} \equiv \frac{1}{\Delta_R}\begin{vmatrix} V_1 & R_{12} & R_{13} \\ V_2 & R_{22} & R_{23} \\ V_3 & R_{32} & R_{33} \end{vmatrix}$$

Similarly,

$$I_2 = \frac{1}{\Delta_R}\begin{vmatrix} R_{11} & V_1 & R_{13} \\ R_{21} & V_2 & R_{23} \\ R_{31} & V_3 & R_{33} \end{vmatrix} \qquad I_3 = \frac{1}{\Delta_R}\begin{vmatrix} R_{11} & R_{12} & V_1 \\ R_{21} & R_{22} & V_2 \\ R_{31} & R_{32} & V_3 \end{vmatrix}$$

An expansion of the numerator determinants by cofactors of the voltage terms results in a set of equations which can be helpful in understanding the network, particularly in terms of its driving-point and transfer resistances:

$$I_1 = V_1\left(\frac{\Delta_{11}}{\Delta_R}\right) + V_2\left(\frac{\Delta_{21}}{\Delta_R}\right) + V_3\left(\frac{\Delta_{31}}{\Delta_R}\right) \tag{7}$$

$$I_2 = V_1\left(\frac{\Delta_{12}}{\Delta_R}\right) + V_2\left(\frac{\Delta_{22}}{\Delta_R}\right) + V_3\left(\frac{\Delta_{32}}{\Delta_R}\right) \tag{8}$$

$$I_3 = V_1\left(\frac{\Delta_{13}}{\Delta_R}\right) + V_2\left(\frac{\Delta_{23}}{\Delta_R}\right) + V_3\left(\frac{\Delta_{33}}{\Delta_R}\right) \tag{9}$$

Here, Δ_{ij} stands for the cofactor of R_{ij} (the element in row i, column j) in Δ_R. Care must be taken with the signs of the cofactors—see Appendix B.

4.4 THE NODE VOLTAGE METHOD

The network shown in Fig. 4-4(*a*) contains five nodes, where *4* and *5* are simple nodes and *1*, *2*, and *3* are principal nodes. In the node voltage method, one of the principal nodes is selected as the reference and equations based on KCL are written at the other principal nodes. At each of these other principal nodes, a voltage is assigned, where it is understood that this is a voltage *with respect to the reference node*. These voltages are the unknowns and, when determined by a suitable method, result in the network solution.

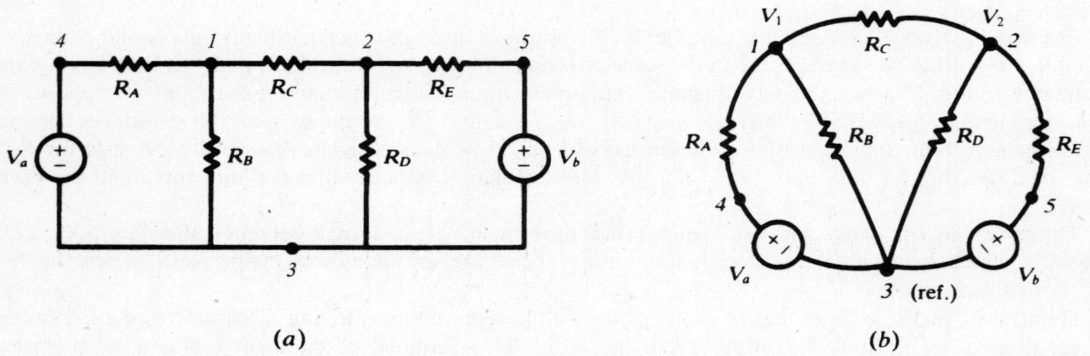

Fig. 4-4

The network is redrawn in Fig. 4-4(*b*) and node *3* selected as the reference for voltages V_1 and V_2. KCL requires that the total current out of node 1 be zero:

$$\frac{V_1 - V_a}{R_A} + \frac{V_1}{R_B} + \frac{V_1 - V_2}{R_C} = 0$$

Similarly, the total current out of node *2* must be zero:

$$\frac{V_2 - V_1}{R_C} + \frac{V_2}{R_D} + \frac{V_2 - V_b}{R_E} = 0$$

(Applying KCL in this form does not imply that the actual branch currents all are directed out of either node. Indeed, the current in branch *1–2* is necessarily directed *out of* one node and *into* the other.) Putting the two equations for V_1 and V_2 in matrix form,

$$\begin{bmatrix} \dfrac{1}{R_A}+\dfrac{1}{R_B}+\dfrac{1}{R_C} & -\dfrac{1}{R_C} \\ -\dfrac{1}{R_C} & \dfrac{1}{R_C}+\dfrac{1}{R_D}+\dfrac{1}{R_E} \end{bmatrix} \begin{bmatrix} V_1 \\ V_2 \end{bmatrix} = \begin{bmatrix} V_a/R_A \\ V_b/R_E \end{bmatrix}$$

Note the symmetry of the coefficient matrix. The 1,1-element contains the sum of the reciprocals of all resistances connected to note *1*; the 2,2-element contains the sum of the reciprocals of all resistances connected to node *2*. The 1,2- and 2,1-elements are each equal to the *negative* of the sum of the reciprocals of the resistances of all branches joining nodes *1* and *2*. (There is just one such branch in the present circuit.)

On the right-hand side, the current matrix contains V_a/R_A and V_b/R_E, the driving currents. Both these terms are taken positive because they both drive a current *into* a node. Further discussion of the elements in the matrix representation of the node voltage equations is given in Chapter 9, where the networks are treated in the sinusoidal steady state.

EXAMPLE 4.5 Solve the circuit of Example 4.2 using the node voltage method.

The circuit is redrawn in Fig. 4-5. With two principal nodes, only one equation is required. Assuming the currents are all directed out of the upper node and the bottom node is the reference,

$$\frac{V_1 - 20}{5} + \frac{V_1}{10} + \frac{V_1 - 8}{2} = 0$$

from which $V_1 = 10\,\text{V}$. Then, $I_1 = (10 - 20)/5 = -2\,\text{A}$ (the negative sign indicates that current I_1 flows into node *1*); $I_2 = (10 - 8)/2 = 1\,\text{A}$; $I_3 = 10/10 = 1\,\text{A}$. Current I_3 in Example 4.2 is shown dotted.

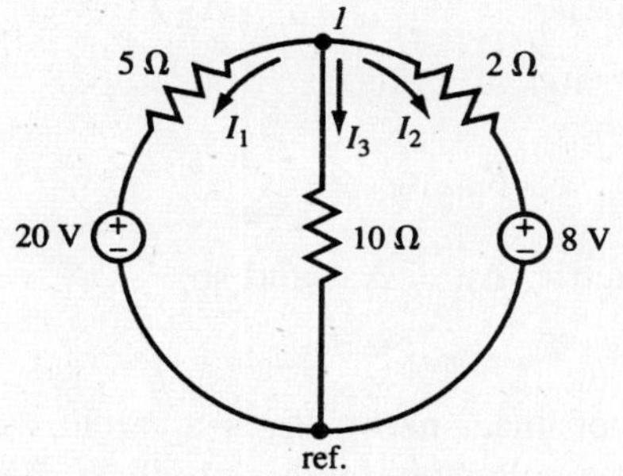

Fig. 4-5

4.5 INPUT AND OUTPUT RESISTANCES

In single-source networks, the *input* or *driving-point resistance* is often of interest. Such a network is suggested in Fig. 4-6, where the driving voltage has been designated as V_1 and the corresponding current as I_1. Since the only source is V_1, the equation for I_1 is [see (*7*) of Example 4.4]:

$$I_1 = V_1\left(\frac{\Delta_{11}}{\Delta_R}\right)$$

The input resistance is the ratio of V_1 to I_1:

$$R_{\text{input},1} = \frac{\Delta_R}{\Delta_{11}}$$

The reader should verify that Δ_R/Δ_{11} actually carries the units Ω.

A voltage source applied to a passive network results in voltages between all nodes of the network. An external resistor connected between two nodes will draw current from the network and in general will reduce the voltage between those nodes. This is due to the voltage across the output resistance (see

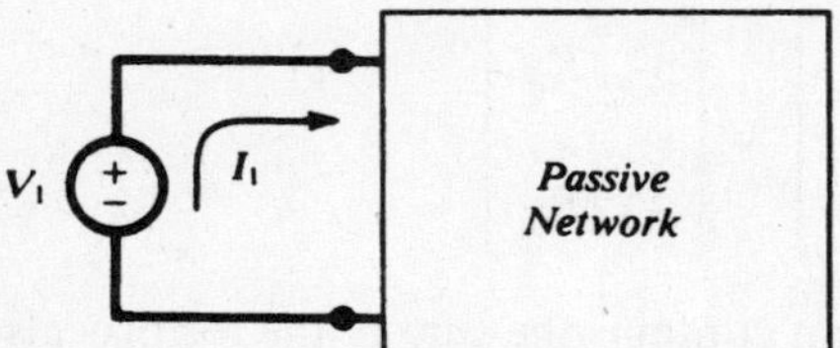

Fig. 4-6

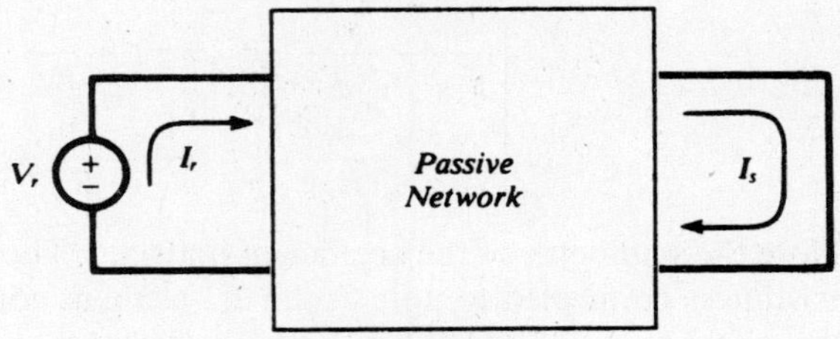

Fig. 4-7

Thévenin too). The output resistance is found by dividing the open-circuited voltage to the short-circuited current at the desired node. The short-circuited current is found in Section 4.6.

4.6 TRANSFER RESISTANCE

A driving voltage in one part of a network results in currents in all the network branches. For example, a voltage source applied to a passive network results in an output current in that part of the network where a load resistance has been connected. In such a case the network has an overall *transfer resistance.* Consider the passive network suggested in Fig. 4-7, where the voltage source has been designated as V_r and the output current as I_s. The mesh current equation for I_s contains only one term, the one resulting from V_r in the numerator determinant:

$$I_s = (0)\left(\frac{\Delta_{1s}}{\Delta_R}\right) + \cdots + 0 + V_r\left(\frac{\Delta_{rs}}{\Delta_R}\right) + 0 + \cdots$$

The network transfer resistance is the ratio of V_r to I_s:

$$R_{\text{transfer},rs} = \frac{\Delta_R}{\Delta_{rs}}$$

Because the resistance matrix is symmetric, $\Delta_{rs} = \Delta_{sr}$, and so

$$R_{\text{transfer},rs} = R_{\text{transfer},sr}$$

This expresses an important property of linear networks: If a certain voltage in mesh r gives rise to a certain current in mesh s, then the same voltage in mesh s produces the same current in mesh r.

Consider now the more general situation of an n-mesh network containing a number of voltage sources. The solution for the current in mesh k can be rewritten in terms of input and transfer resistances [refer to (*7*), (*8*), and (*9*) of Example 4.4]:

$$I_k = \frac{V_1}{R_{\text{transfer},1k}} + \cdots + \frac{V_{k-1}}{R_{\text{transfer},(k-1)k}} + \frac{V_k}{R_{\text{input},k}} + \frac{V_{k+1}}{R_{\text{transfer},(k+1)k}} + \cdots + \frac{V_n}{R_{\text{transfer},nk}}$$

There is nothing new here mathematically, but in this form the current equation does illustrate the superposition principle very clearly, showing how the resistances control the effects which the voltage sources have on a particular mesh current. A source far removed from mesh k will have a high transfer resistance into that mesh and will therefore contribute very little to I_k. Source V_k, and others in meshes adjacent to mesh k, will provide the greater part of I_k.

4.7 NETWORK REDUCTION

The mesh current and node voltage methods are the principal techniques of circuit analysis. However, the equivalent resistance of series and parallel branches (Sections 3.4 and 3.5), combined with the voltage and current division rules, provide another method of analyzing a network. This method is tedious and usually requires the drawing of several additional circuits. Even so, the process of reducing the network provides a very clear picture of the overall functioning of the network in terms of voltages, currents, and power. The reduction begins with a scan of the network to pick out series and parallel combinations of resistors.

EXAMPLE 4.6 Obtain the total power supplied by the 60-V source and the power absorbed in each resistor in the network of Fig. 4-8.

$$R_{ab} = 7 + 5 = 12\,\Omega$$

$$R_{cd} = \frac{(12)(6)}{12+6} = 4\,\Omega$$

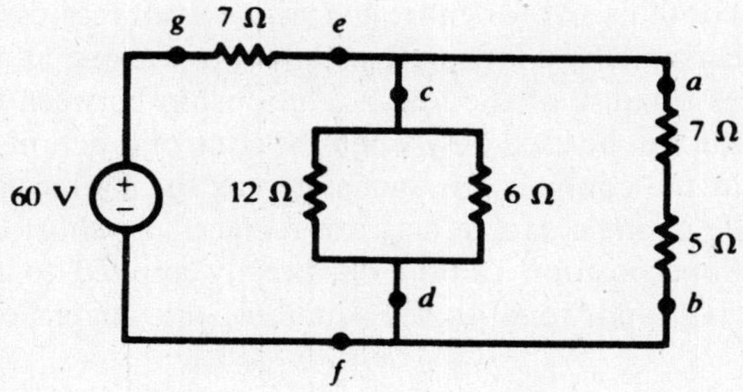

Fig. 4-8

These two equivalents are in parallel (Fig. 4-9), giving

$$R_{ef} = \frac{(4)(12)}{4+12} = 3\,\Omega$$

Then this 3-Ω equivalent is in series with the 7-Ω resistor (Fig. 4-10), so that for the entire circuit,

$$R_{\text{eq}} = 7 + 3 = 10\,\Omega$$

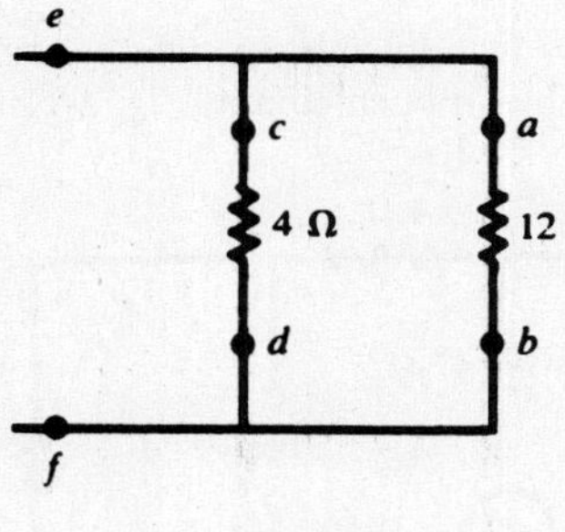

Fig. 4-9

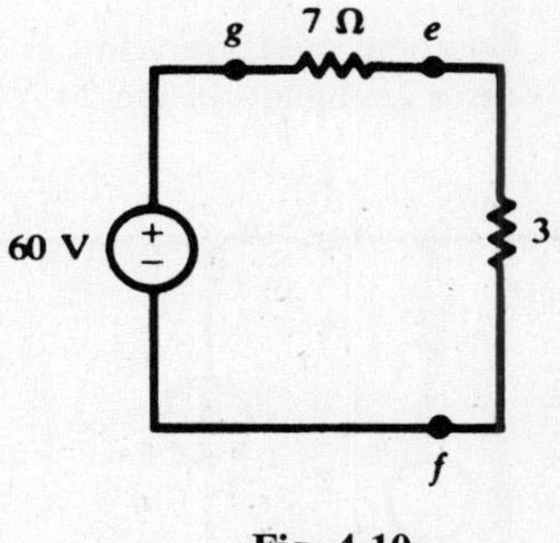

Fig. 4-10

The total power absorbed, which equals the total power supplied by the source, can now be calculated as

$$P_T = \frac{V^2}{R_{\text{eq}}} = \frac{(60)^2}{10} = 360\,\text{W}$$

This power is divided between R_{ge} and R_{ef} as follows:

$$P_{ge} = P_{7\Omega} = \frac{7}{7+3}(360) = 252\,\text{W} \qquad P_{ef} = \frac{3}{7+3}(360) = 108\,\text{W}$$

Power P_{ef} is further divided between R_{cd} and R_{ab} as follows:

$$P_{cd} = \frac{12}{4+12}(108) = 81\,\text{W} \qquad P_{ab} = \frac{4}{4+12}(108) = 27\,\text{W}$$

Finally, these powers are divided between the individual resistances as follows:

$$P_{12\Omega} = \frac{6}{12+6}(81) = 27\,\text{W} \qquad P_{7\Omega} = \frac{7}{7+5}(27) = 15.75\,\text{W}$$

$$P_{6\Omega} = \frac{12}{12+6}(81) = 54\,\text{W} \qquad P_{5\Omega} = \frac{5}{7+5}(27) = 11.25\,\text{W}$$

4.8 SUPERPOSITION

A linear network which contains two or more *independent* sources can be analyzed to obtain the various voltages and branch currents by allowing the sources to act one at a time, then superposing the results. This principle applies because of the linear relationship between current and voltage. With *dependent* sources, superposition can be used only when the control functions are external to the network containing the sources, so that the controls are unchanged as the sources act one at a time. Voltage sources to be suppressed while a single source acts are replaced by short circuits; current sources are replaced by open circuits. Superposition cannot be directly applied to the computation of power, because power in an element is proportional to the square of the current or the square of the voltage, which is nonlinear.

As a further illustration of superposition consider equation (7) of Example 4.4:

$$I_1 = V_1\left(\frac{\Delta_{11}}{\Delta_R}\right) + V_2\left(\frac{\Delta_{21}}{\Delta_R}\right) + V_3\left(\frac{\Delta_{31}}{\Delta_R}\right)$$

which contains the superposition principle implicitly. Note that the three terms on the right are added to result in current I_1. If there are sources in each of the three meshes, then each term contributes to the current I_1. Additionally, if only mesh 3 contains a source, V_1 and V_2 will be zero and I_1 is fully determined by the third term.

EXAMPLE 4.7 Compute the current in the 23-Ω resistor of Fig. 4-11(*a*) by applying the superposition principle. With the 200-V source acting alone, the 20-A current source is replaced by an open circuit, Fig. 4-11(*b*).

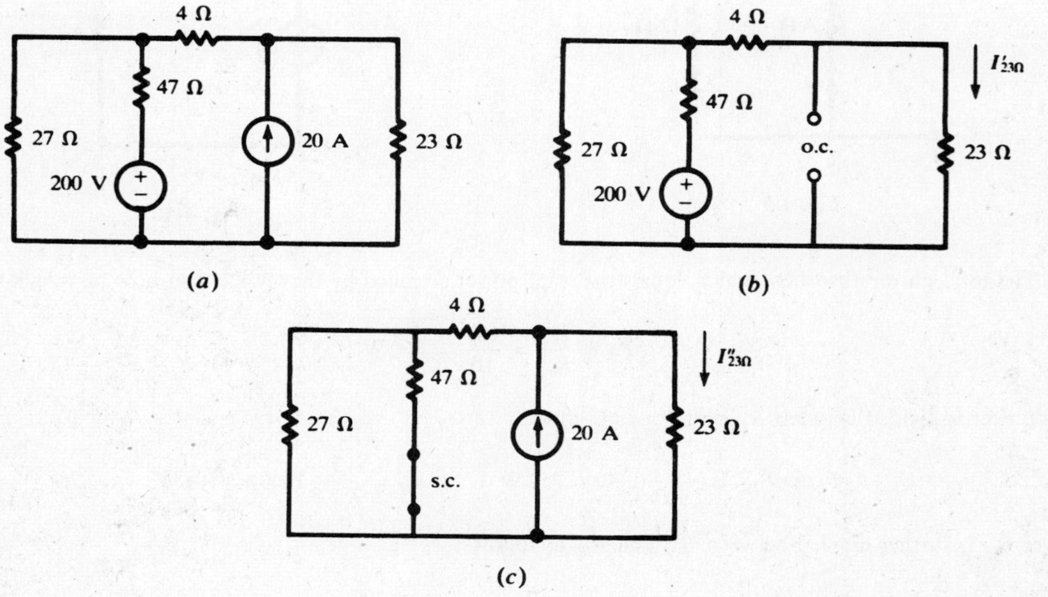

Fig. 4-11

$$R_{eq} = 47 + \frac{(27)(4+23)}{54} = 60.5\,\Omega$$

$$I_T = \frac{200}{60.5} = 3.31\,\text{A}$$

$$I'_{23\Omega} = \left(\frac{27}{54}\right)(3.31) = 1.65\,\text{A}$$

When the 20-A source acts alone, the 200-V source is replaced by a short circuit, Fig. 4-11(*c*). The equivalent resistance to the left of the source is

$$R_{eq} = 4 + \frac{(27)(47)}{74} = 21.15\,\Omega$$

Then

$$I''_{23\Omega} = \left(\frac{21.15}{21.15+23}\right)(20) = 9.58\,\text{A}$$

The total current in the 23-Ω resistor is

$$I_{23\Omega} = I'_{23\Omega} + I''_{23\Omega} = 11.23\,\text{A}$$

4.9 THÉVENIN'S AND NORTON'S THEOREMS

A linear, active, resistive network which contains one or more voltage or current sources can be replaced by a single voltage source and a series resistance (*Thévenin's theorem*), or by a single current source and a parallel resistance (*Norton's theorem*). The voltage is called the *Thévenin equivalent voltage*, V', and the current the *Norton equivalent current*, I'. The two resistances are the same, R'. When terminals *ab* in Fig. 4-12(*a*) are *open-circuited*, a voltage will appear between them.

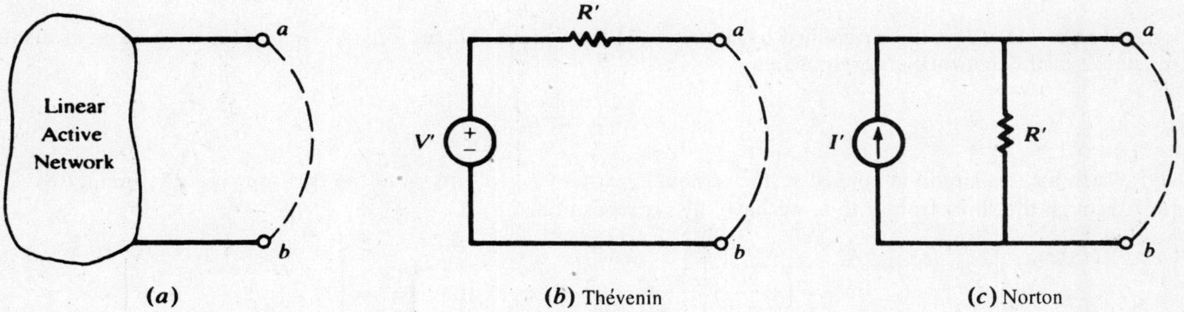

Fig. 4-12

From Fig. 4-12(*b*) it is evident that this must be the voltage V' of the Thévenin equivalent circuit. If a *short circuit* is applied to the terminals, as suggested by the dashed line in Fig. 4-12(*a*), a current will result. From Fig. 4-12(*c*) it is evident that this current must be I' of the Norton equivalent circuit. Now, if the circuits in (*b*) and (*c*) are equivalents of the same active network, they are equivalent to each other. It follows that $I' = V'/R'$. If both V' and I' have been determined from the active network, then $R' = V'/I'$.

EXAMPLE 4.8 Obtain the Thévenin and Norton equivalent circuits for the active network in Fig. 4-13(*a*).

With terminals *ab* open, the two sources drive a clockwise current through the 3-Ω and 6-Ω resistors [Fig. 4-13(*b*)].

$$I = \frac{20+10}{3+6} = \frac{30}{9}\,\text{A}$$

Since no current passes through the upper right 3-Ω resistor, the Thévenin voltage can be taken from either active branch:

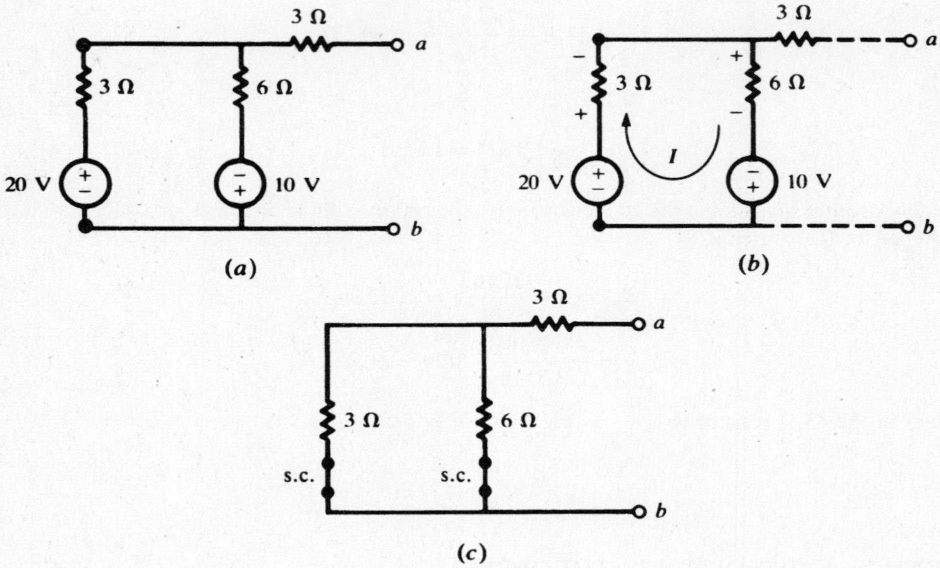

Fig. 4-13

$$V_{ab} = V' = 20 - \left(\frac{30}{9}\right)(3) = 10\,\text{V}$$

or

$$V_{ab} = V' = \left(\frac{30}{9}\right)6 - 10 = 10\,\text{V}$$

The resistance R' can be obtained by shorting out the voltage sources [Fig. 4.13(*c*)] and finding the equivalent resistance of this network at terminals *ab*:

$$R' = 3 + \frac{(3)(6)}{9} = 5\,\Omega$$

When a short circuit is applied to the terminals, current $I_{\text{s.c.}}$ results from the two sources. Assuming that it runs through the short from *a* to *b*, we have, by superposition,

$$I_{\text{s.c.}} = I' = \left(\frac{6}{6+3}\right)\left[\frac{20}{3+\dfrac{(3)(6)}{9}}\right] - \left(\frac{3}{3+3}\right)\left[\frac{10}{6+\dfrac{(3)(3)}{6}}\right] = 2\,\text{A}$$

Figure 4-14 shows the two equivalent circuits. In the present case, V', R', and I' were obtained independently. Since they are related by Ohm's law, any two may be used to obtain the third.

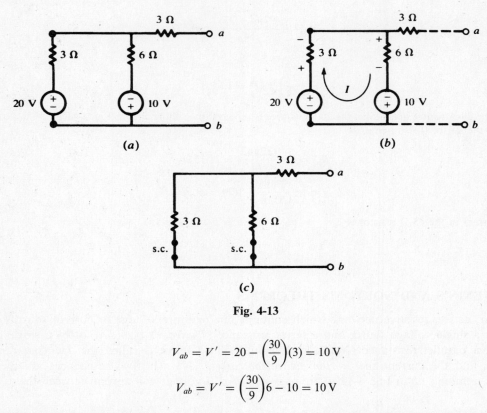

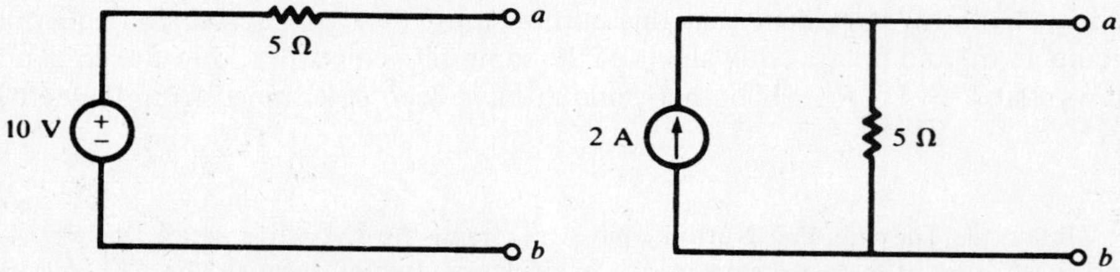

(*a*) **Thévenin Equivalent** (*b*) **Norton Equivalent**

Fig. 4-14

The usefulness of Thévenin and Norton equivalent circuits is clear when an active network is to be examined under a number of load conditions, each represented by a resistor. This is suggested in

Fig. 4-15, where it is evident that the resistors $R_1, R_2, \ldots, R_n$ can be connected one at a time, and the resulting current and power readily obtained. If this were attempted in the original circuit using, for example, network reduction, the task would be very tedious and time-consuming.

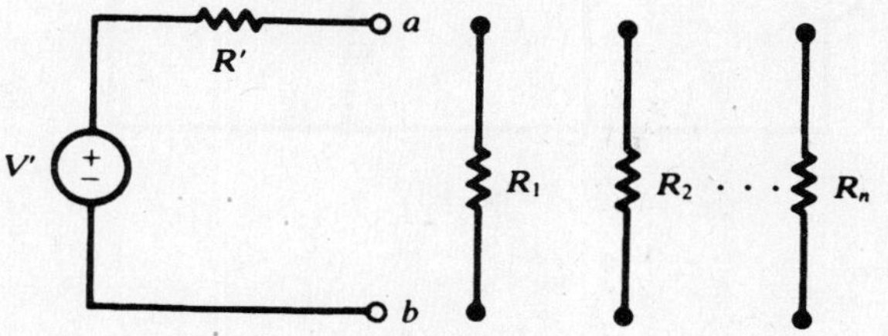

Fig. 4-15

4.10 MAXIMUM POWER TRANSFER THEOREM

At times it is desired to obtain the maximum power transfer from an active network to an external load resistor R_L. Assuming that the network is linear, it can be reduced to an equivalent circuit as in Fig. 4-16. Then

$$I = \frac{V'}{R' + R_L}$$

and so the power absorbed by the load is

$$P_L = \frac{V'^2 R_L}{(R' + R_L)^2} = \frac{V'^2}{4R'}\left[1 - \left(\frac{R' - R_L}{R' + R_L}\right)^2\right]$$

It is seen that P_L attains its maximum value, $V'^2/4R'$, when $R_L = R'$, in which case the power in R' is also $V'^2/4R'$. Consequently, when the power transferred is a maximum, the efficiency is 50 percent.

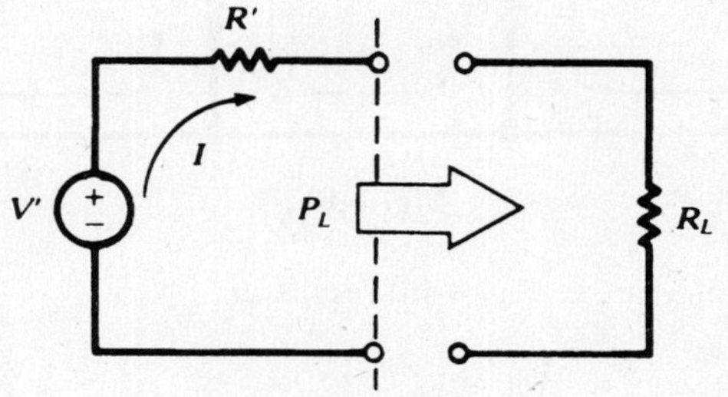

Fig. 4-16

It is noted that the condition for maximum power transfer to the load is not the same as the condition for maximum power delivered by the source. The latter happens when $R_L = 0$, in which case power delivered to the load is zero (i.e., at a minimum).

Solved Problems

4.1 Use branch currents in the network shown in Fig. 4-17 to find the current supplied by the 60-V source.

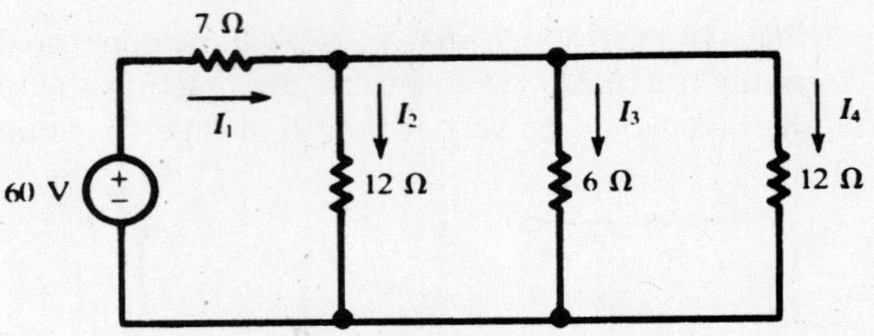

Fig. 4-17

KVL and KCL give:

$$I_2(12) = I_3(6) \tag{10}$$
$$I_2(12) = I_4(12) \tag{11}$$
$$60 = I_1(7) + I_2(12) \tag{12}$$
$$I_1 = I_2 + I_3 + I_4 \tag{13}$$

Substituting (*10*) and (*11*) in (*13*),

$$I_1 = I_2 + 2I_2 + I_2 = 4I_2 \tag{14}$$

Now (*14*) is substituted in (*12*):

$$60 = I_1(7) + \tfrac{1}{4}I_1(12) = 10I_1 \qquad \text{or} \qquad I_1 = 6\text{ A}$$

4.2 Solve Problem 4.1 by the mesh current method.

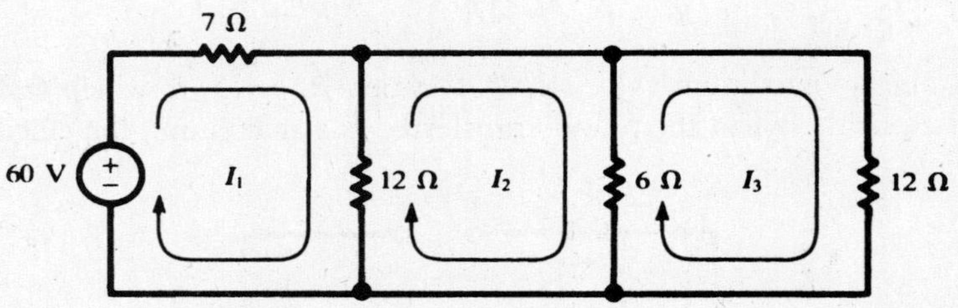

Fig. 4-18

Applying KVL to each mesh (see Fig. 4-18) results in

$$60 = 7I_1 + 12(I_1 - I_2)$$
$$0 = 12(I_2 - I_1) + 6(I_2 - I_3)$$
$$0 = 6(I_3 - I_2) + 12I_3$$

Rearranging terms and putting the equations in matrix form,

$$\begin{aligned} 19I_1 - 12I_2 \qquad &= 60 \\ -12I_1 + 18I_2 - 6I_3 &= 0 \\ -6I_2 + 18I_3 &= 0 \end{aligned} \qquad \text{or} \qquad \begin{bmatrix} 19 & -12 & 0 \\ -12 & 18 & -6 \\ 0 & -6 & 18 \end{bmatrix} \begin{bmatrix} I_1 \\ I_2 \\ I_3 \end{bmatrix} = \begin{bmatrix} 60 \\ 0 \\ 0 \end{bmatrix}$$

Using Cramer's rule to find I_1,

$$I_1 = \begin{vmatrix} 60 & -12 & 0 \\ 0 & 18 & -6 \\ 0 & -6 & 18 \end{vmatrix} \div \begin{vmatrix} 19 & -12 & 0 \\ -12 & 18 & -6 \\ 0 & -6 & 18 \end{vmatrix} = 17\,280 \div 2880 = 6\text{ A}$$

4.3 Solve the network of Problems 4.1 and 4.2 by the node voltage method. See Fig. 4-19.

With two principal nodes, only one equation is necessary.

$$\frac{V_1 - 60}{7} + \frac{V_1}{12} + \frac{V_1}{6} + \frac{V_1}{12} = 0$$

from which $V_1 = 18\,\text{V}$. Then,

$$I_1 = \frac{60 - V_1}{7} = 6\,\text{A}$$

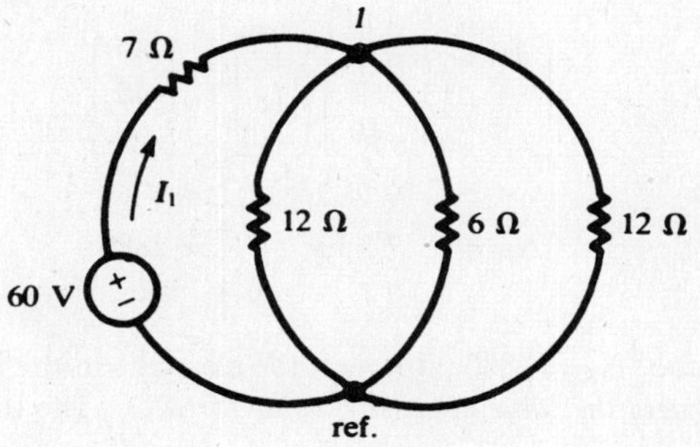

Fig. 4-19

4.4 In Problem 4.2, obtain $R_{\text{input},1}$ and use it to calculate I_1.

$$R_{\text{input},1} = \frac{\Delta_R}{\Delta_{11}} = \frac{2880}{\begin{vmatrix} 18 & -6 \\ -6 & 18 \end{vmatrix}} = \frac{2880}{288} = 10\,\Omega$$

Then
$$I_1 = \frac{60}{R_{\text{input},1}} = \frac{60}{10} = 6\,\text{A}$$

4.5 Obtain $R_{\text{transfer},12}$ and $R_{\text{transfer},13}$ for the network of Problem 4.2 and use them to calculate I_2 and I_3.

The cofactor of the 1,2-element in Δ_R must include a negative sign:

$$\Delta_{12} = (-1)^{1+2}\begin{vmatrix} -12 & -6 \\ 0 & 18 \end{vmatrix} = 216 \qquad R_{\text{transfer},12} = \frac{\Delta_R}{\Delta_{12}} = \frac{2880}{216} = 13.33\,\Omega$$

Then, $I_2 = 60/13.33 = 4.50\,\text{A}$.

$$\Delta_{13} = (-1)^{1+3}\begin{vmatrix} -12 & 18 \\ 0 & -6 \end{vmatrix} = 72 \qquad R_{\text{transfer},13} = \frac{\Delta_R}{\Delta_{13}} = \frac{2880}{72} = 40\,\Omega$$

Then, $I_3 = 60/40 = 1.50\,\text{A}$.

4.6 Solve Problem 4.1 by use of the loop currents indicated in Fig. 4-20.

The elements in the matrix form of the equations are obtained by inspection, following the rules of Section 4.2.

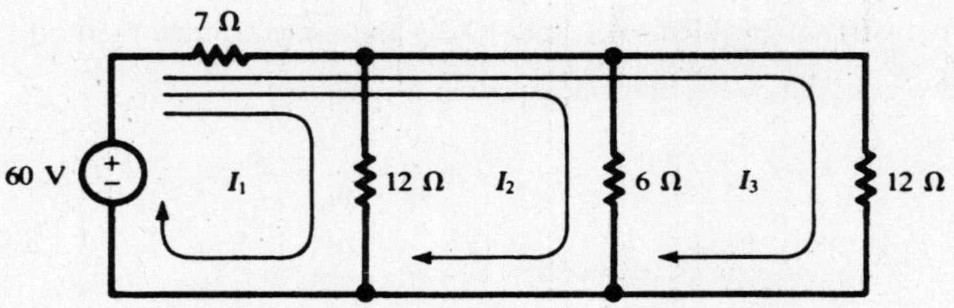

Fig. 4-20

$$\begin{bmatrix} 19 & 7 & 7 \\ 7 & 13 & 7 \\ 7 & 7 & 19 \end{bmatrix} \begin{bmatrix} I_1 \\ I_2 \\ I_3 \end{bmatrix} = \begin{bmatrix} 60 \\ 60 \\ 60 \end{bmatrix}$$

Thus,

$$\Delta_R = \begin{bmatrix} 19 & 7 & 7 \\ 7 & 13 & 7 \\ 7 & 7 & 19 \end{bmatrix} = 2880$$

Notice that in Problem 4.2, too, $\Delta_R = 2880$, although the elements in the determinant were different. *All valid sets of meshes or loops yield the same numerical value for* Δ_R. The three numerator determinants are

$$N_1 = \begin{vmatrix} 60 & 7 & 7 \\ 60 & 13 & 7 \\ 60 & 7 & 19 \end{vmatrix} = 4320 \qquad N_2 = 8642 \qquad N_3 = 4320$$

Consequently,

$$I_1 = \frac{N_1}{\Delta_R} = \frac{4320}{2880} = 1.5\,\text{A} \qquad I_2 = \frac{N_2}{\Delta_R} = 3\,\text{A} \qquad I_3 = \frac{N_3}{\Delta_R} = 1.5\,\text{A}$$

The current supplied by the 60-V source is the sum of the three loop currents, $I_1 + I_2 + I_3 = 6\,\text{A}$.

4.7 Write the mesh current matrix equation for the network of Fig. 4-21 by inspection, and solve for the currents.

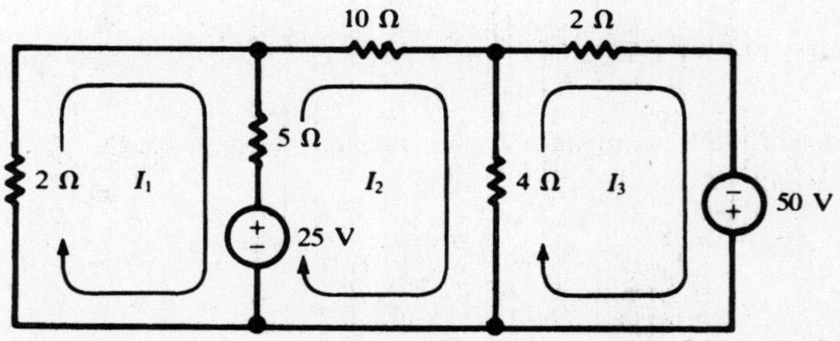

Fig. 4-21

$$\begin{bmatrix} 7 & -5 & 0 \\ -5 & 19 & -4 \\ 0 & -4 & 6 \end{bmatrix} \begin{bmatrix} I_1 \\ I_2 \\ I_3 \end{bmatrix} = \begin{bmatrix} -25 \\ 25 \\ 50 \end{bmatrix}$$

Solving,

$$I_1 = \begin{vmatrix} -25 & -5 & 0 \\ 25 & 19 & -4 \\ 50 & -4 & 6 \end{vmatrix} \div \begin{vmatrix} 7 & -5 & 0 \\ -5 & 19 & -4 \\ 0 & -4 & 6 \end{vmatrix} = (-700) \div 536 = -1.31\,\text{A}$$

Similarly,

$$I_2 = \frac{N_2}{\Delta_R} = \frac{1700}{536} = 3.17\,\text{A} \qquad I_3 = \frac{N_3}{\Delta_R} = \frac{5600}{536} = 10.45\,\text{A}$$

4.8 Solve Problem 4.7 by the node voltage method.

The circuit has been redrawn in Fig. 4-22, with two principal nodes numbered *1* and *2* and the third chosen as the reference node. By KCL, the net current out of node *1* must equal zero.

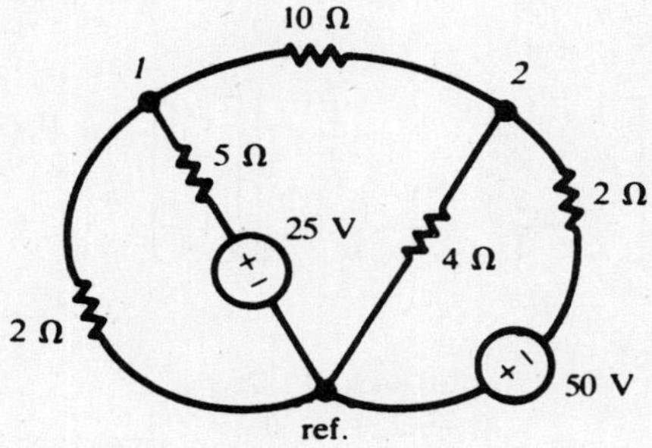

Fig. 4-22

$$\frac{V_1}{2} + \frac{V_1 - 25}{5} + \frac{V_1 - V_2}{10} = 0$$

Similarly, at node *2*,

$$\frac{V_2 - V_1}{10} + \frac{V_2}{4} + \frac{V_2 + 50}{2} = 0$$

Putting the two equations in matrix form,

$$\begin{bmatrix} \frac{1}{2} + \frac{1}{5} + \frac{1}{10} & -\frac{1}{10} \\ -\frac{1}{10} & \frac{1}{10} + \frac{1}{4} + \frac{1}{2} \end{bmatrix} \begin{bmatrix} V_1 \\ V_2 \end{bmatrix} = \begin{vmatrix} 5 \\ -25 \end{vmatrix}$$

The determinant of coefficients and the numerator determinants are

$$\Delta = \begin{vmatrix} 0.80 & -0.10 \\ -0.10 & 0.85 \end{vmatrix} = 0.670$$

$$N_1 = \begin{vmatrix} 5 & -0.10 \\ -25 & 0.85 \end{vmatrix} = 1.75 \qquad N_2 = \begin{vmatrix} 0.80 & 5 \\ -0.10 & -25 \end{vmatrix} = -19.5$$

From these,

$$V_1 = \frac{1.75}{0.670} = 2.61\,\text{V} \qquad V_2 = \frac{-19.5}{0.670} = -29.1\,\text{V}$$

In terms of these voltages, the currents in Fig. 4-21 are determined as follows:

$$I_1 = \frac{-V_1}{2} = -1.31\,\text{A} \qquad I_2 = \frac{V_1 - V_2}{10} = 3.17\,\text{A} \qquad I_3 = \frac{V_2 + 50}{2} = 10.45\,\text{A}$$

4.9 For the network shown in Fig. 4-23, find V_s which makes $I_0 = 7.5\,\text{mA}$.

The node voltage method will be used and the matrix form of the equations written by inspection.

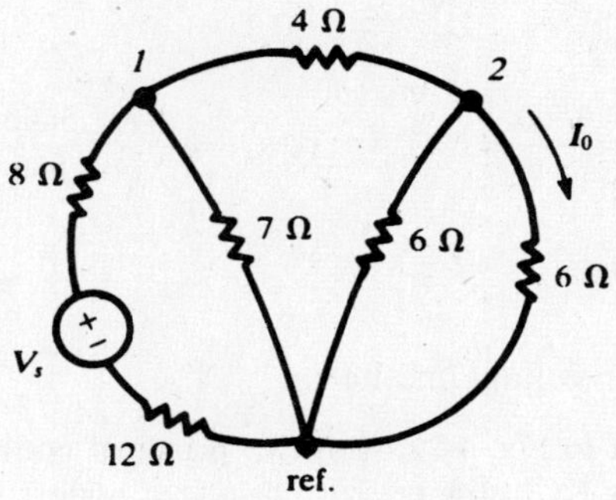

Fig. 4-23

$$\begin{bmatrix} \frac{1}{20}+\frac{1}{7}+\frac{1}{4} & -\frac{1}{4} \\ -\frac{1}{4} & \frac{1}{4}+\frac{1}{6}+\frac{1}{6} \end{bmatrix} \begin{bmatrix} V_1 \\ V_2 \end{bmatrix} = \begin{bmatrix} V_s/20 \\ 0 \end{bmatrix}$$

Solving for V_2,

$$V_2 = \frac{\begin{vmatrix} 0.443 & V_s/20 \\ -0.250 & 0 \end{vmatrix}}{\begin{vmatrix} 0.443 & -0.250 \\ -0.250 & 0.583 \end{vmatrix}} = 0.0638 V_s$$

Then $$7.5 \times 10^{-3} = I_0 = \frac{V_2}{6} = \frac{0.0638 V_s}{6}$$

from which $V_s = 0.705\,\text{V}$.

4.10 In the network shown in Fig. 4-24, find the current in the 10-Ω resistor.

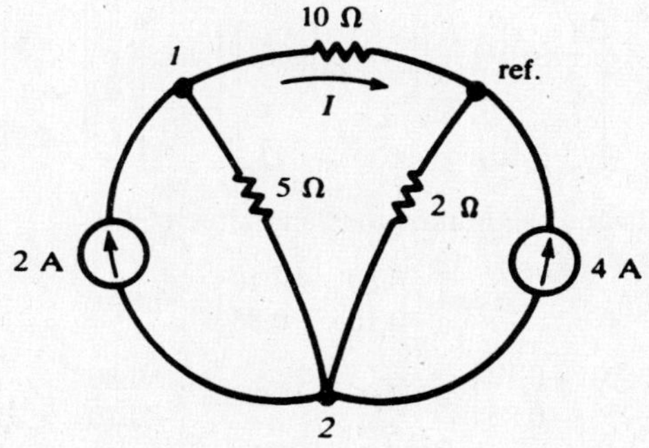

Fig. 4-24

The nodal equations in matrix form are written by inspection.

$$\begin{bmatrix} \frac{1}{5}+\frac{1}{10} & -\frac{1}{5} \\ -\frac{1}{5} & \frac{1}{5}+\frac{1}{2} \end{bmatrix} \begin{bmatrix} V_1 \\ V_2 \end{bmatrix} = \begin{bmatrix} 2 \\ -6 \end{bmatrix}$$

$$V_1 = \frac{\begin{vmatrix} 2 & -0.20 \\ -6 & 0.70 \end{vmatrix}}{\begin{vmatrix} 0.30 & -0.20 \\ -0.20 & 0.70 \end{vmatrix}} = 1.18\,\text{V}$$

Then, $I = V_1/10 = 0.118\,\mathrm{A}$.

4.11 Find the voltage V_{ab} in the network shown in Fig. 4-25.

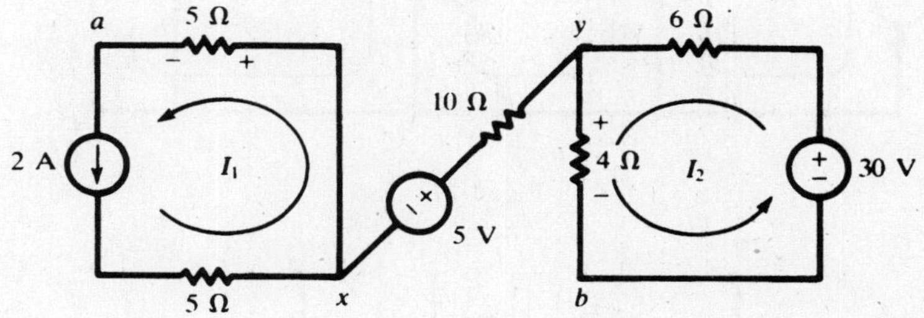

Fig. 4-25

The two closed loops are independent, and no current can pass through the connecting branch.

$$I_1 = 2\,\mathrm{A} \qquad I_2 = \frac{30}{10} = 3\,\mathrm{A}$$

$$V_{ab} = V_{ax} + V_{xy} + V_{yb} = -I_1(5) - 5 + I_2(4) = -3\,\mathrm{V}$$

4.12 For the ladder network of Fig. 4-26, obtain the transfer resistance as expressed by the ratio of V_{in} to I_4.

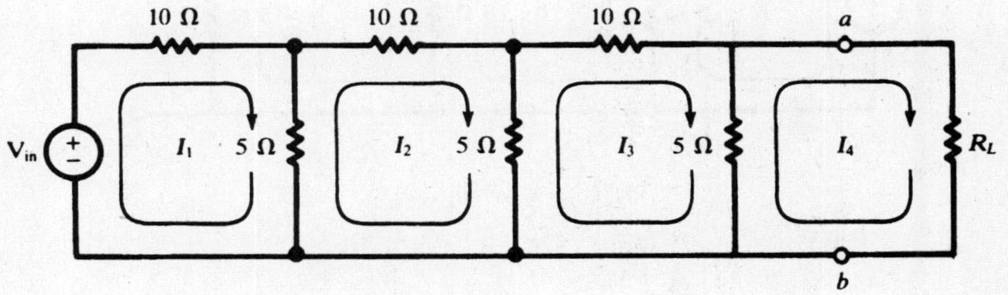

Fig. 4-26

By inspection, the network equation is

$$\begin{bmatrix} 15 & -5 & 0 & 0 \\ -5 & 20 & -5 & 0 \\ 0 & -5 & 20 & -5 \\ 0 & 0 & -5 & 5+R_L \end{bmatrix} \begin{bmatrix} I_1 \\ I_2 \\ I_3 \\ I_4 \end{bmatrix} = \begin{bmatrix} V_{in} \\ 0 \\ 0 \\ 0 \end{bmatrix}$$

$$\Delta R = 5125R_L + 18\,750 \qquad N_4 = 125V_{in}$$

$$I_4 = \frac{N_4}{\Delta_R} = \frac{V_{in}}{41R_L + 150}\ (\mathrm{A})$$

and

$$R_{\mathrm{transfer},14} = \frac{V_{in}}{I_4} = 41R_L + 150\ (\Omega)$$

4.13 Obtain a Thévenin equivalent for the circuit of Fig. 4-26 to the left of terminals *ab*.

The short-circuit current $I_{s.c.}$ is obtained from the three-mesh circuit shown in Fig. 4-27.

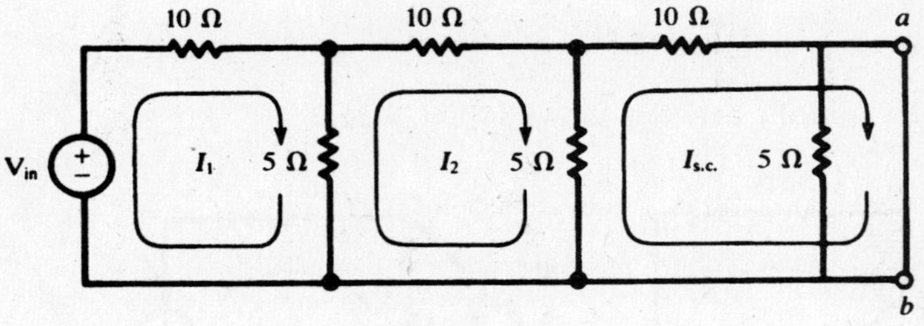

Fig. 4-27

$$\begin{bmatrix} 15 & -5 & 0 \\ -5 & 20 & -5 \\ 0 & -5 & 15 \end{bmatrix} \begin{bmatrix} I_1 \\ I_2 \\ I_{s.c.} \end{bmatrix} = \begin{bmatrix} V_{in} \\ 0 \\ 0 \end{bmatrix}$$

$$I_{s.c.} = \frac{V_{in}\begin{vmatrix} -5 & 20 \\ 0 & -5 \end{vmatrix}}{\Delta_R} = \frac{V_{in}}{150}$$

The open-circuit voltage $V_{o.c.}$ is the voltage across the 5-Ω resistor indicated in Fig. 4-28.

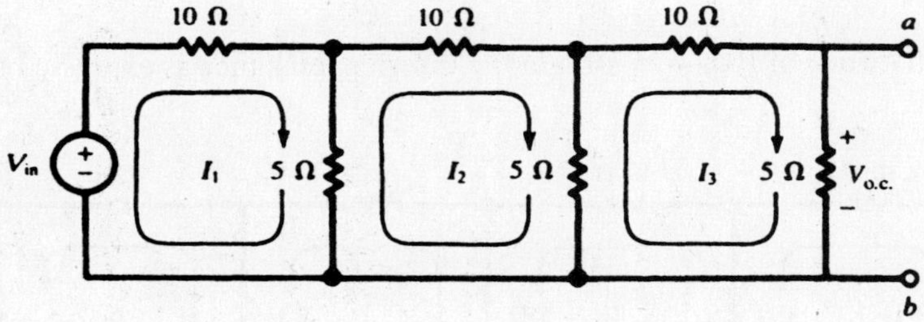

Fig. 4-28

$$\begin{bmatrix} 15 & -5 & 0 \\ -5 & 20 & -5 \\ 0 & -5 & 20 \end{bmatrix} \begin{bmatrix} I_1 \\ I_2 \\ I_3 \end{bmatrix} = \begin{bmatrix} V_{in} \\ 0 \\ 0 \end{bmatrix}$$

$$I_3 = \frac{25V_{in}}{5125} = \frac{V_{in}}{205} \text{ (A)}$$

Then, the Thévenin source $V' = V_{o.c.} = I_3(5) = V_{in}/41$, and

$$R_{Th} = \frac{V_{o.c.}}{I_{s.c.}} = \frac{150}{41}\ \Omega$$

The Thévenin equivalent circuit is shown in Fig. 4-29. With R_L connected to terminals ab, the output current is

$$I_4 = \frac{V_{in}/41}{(150/41) + R_L} = \frac{V_{in}}{41R_L + 150} \text{ (A)}$$

agreeing with Problem 4.12.

4.14 Use superposition to find the current I from each voltage source in the circuit shown in Fig. 4-30.

Loop currents are chosen such that each source contains only one current.

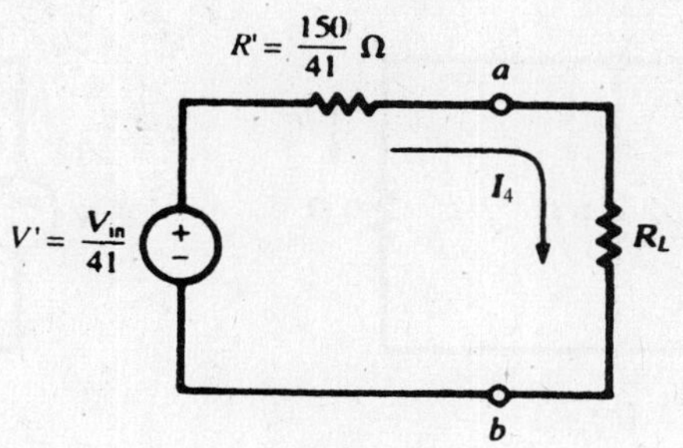

Fig. 4-29

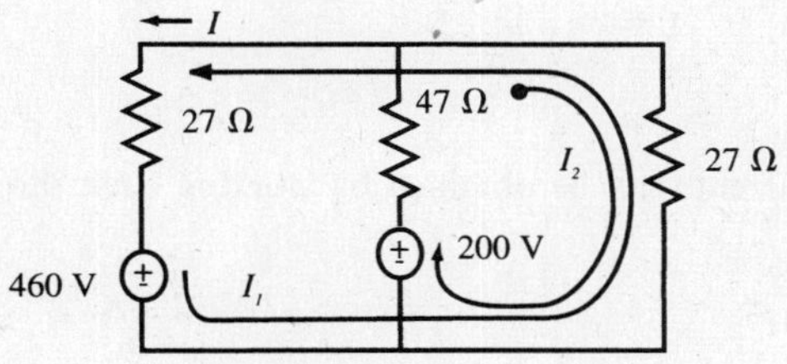

Fig. 4-30

$$\begin{bmatrix} 54 & -27 \\ -27 & 74 \end{bmatrix}\begin{bmatrix} I_1 \\ I_2 \end{bmatrix} = \begin{bmatrix} -460 \\ 200 \end{bmatrix}$$

From the 460-V source,

$$I_1' = I' = \frac{(-460)(74)}{3267} = -10.42\ \text{A}$$

and for the 200-V source

$$I_1'' = I'' = \frac{-(200)(-27)}{3267} = 1.65\ \text{A}$$

Then, $$I = I' + I'' = -10.42 + 1.65 = -8.77\ \text{A}$$

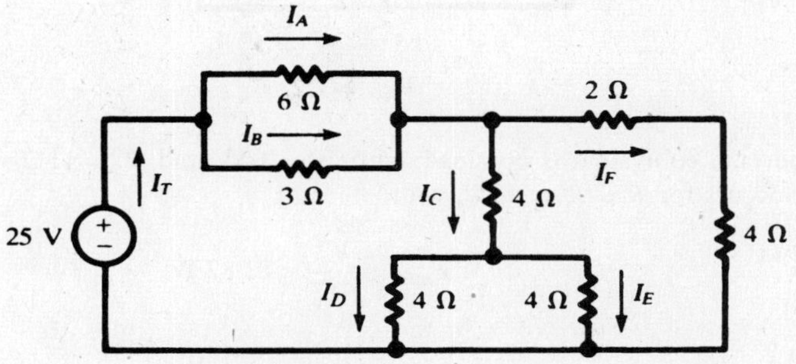

Fig. 4-31(*a*)

4.15 Obtain the current in each resistor in Fig. 4-31(*a*), using network reduction methods.

As a first step, two-resistor parallel combinations are converted to their equivalents. For the 6 Ω and 3 Ω, $R_{eq} = (6)(3)/(6+3) = 2\ \Omega$. For the two 4-Ω resistors, $R_{eq} = 2\ \Omega$. The circuit is redrawn with series resistors added [Fig. 4-31(*b*)]. Now the two 6-Ω resistors in parallel have the equivalent $R_{eq} = 3\ \Omega$, and this is in series with the 2 Ω. Hence, $R_T = 5\ \Omega$, as shown in Fig. 4-31(*c*). The resulting total current is

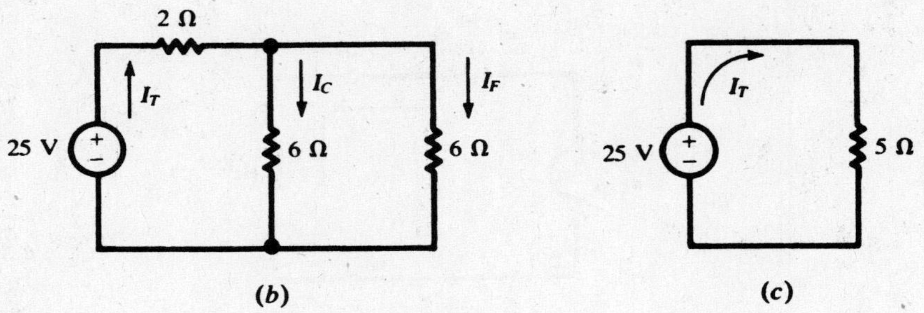

Fig. 4-31 *(cont.)*

$$I_T = \frac{25}{5} = 5\,\text{A}$$

Now the branch currents can be obtained by working back through the circuits of Fig. 4-31(*b*) and 4-31(*a*)

$$I_C = I_F = \tfrac{1}{2} I_T = 2.5\ \text{A}$$
$$I_D = I_E = \tfrac{1}{2} I_C = 1.25\ \text{A}$$
$$I_A = \frac{3}{6+3} I_T = \frac{5}{3}\ \text{A}$$
$$I_B = \frac{6}{6+3} I_T = \frac{10}{3}\ \text{A}$$

4.16 Find the value of the adjustable resistance R which results in maximum power transfer across the terminals *ab* of the circuit shown in Fig. 4-32.

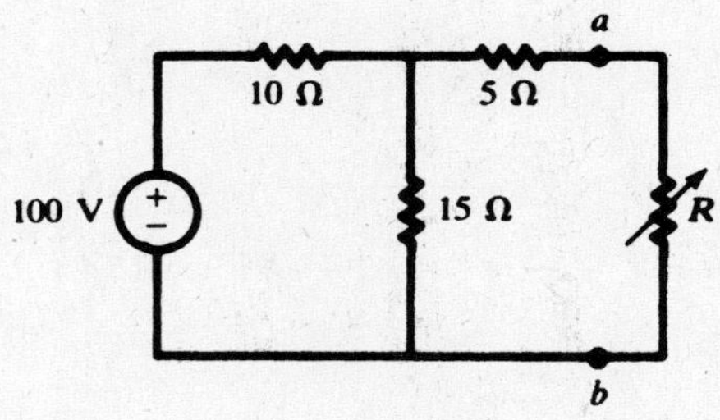

Fig. 4-32

First a Thévenin equivalent is obtained, with $V' = 60\,\text{V}$ and $R' = 11\,\Omega$. By Section 4.10, maximum power transfer occurs for $R = R' = 11\,\Omega$, with

$$P_{\text{max}} = \frac{V'^2}{4R'} = 81.82\ \text{W}$$

Supplementary Problems

4.17 Apply the mesh current method to the network of Fig. 4-33 and write the matrix equations by inspection. Obtain current I_1 by expanding the numerator determinant about the column containing the voltage sources to show that each source supplies a current of 2.13 A.

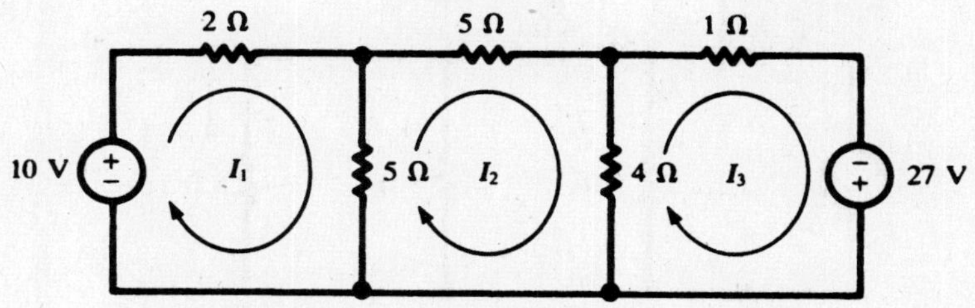

Fig. 4-33

4.18 Loop currents are shown in the network of Fig. 4-34. Write the matrix equation and solve for the three currents. *Ans.* 3.55 A, −1.98 A, −2.98 A

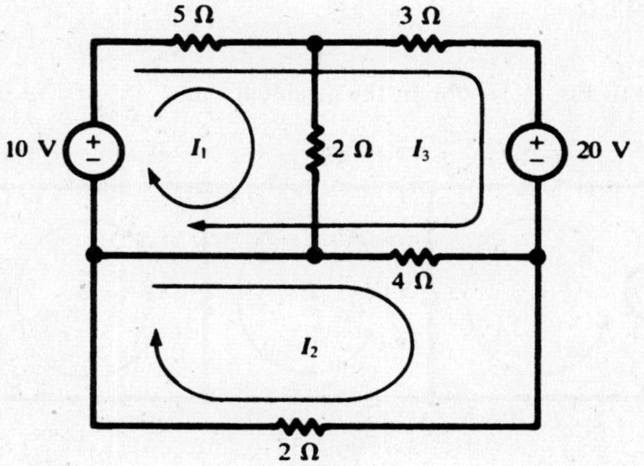

Fig. 4-34

4.19 The network of Problem 4.18 has been redrawn in Fig. 4-35 for solution by the node voltage method. Obtain node voltages V_1 and V_2 and verify the currents obtained in Problem 4.18.
Ans. 7.11 V, −3.96 V

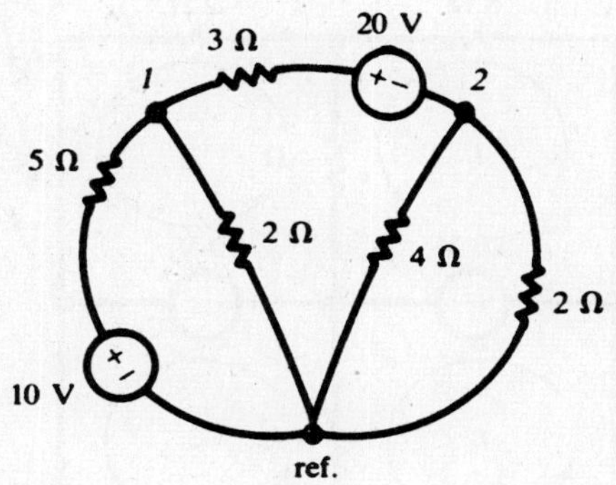

Fig. 4-35

4.20 In the network shown in Fig. 4-36 current $I_0 = 7.5\,\text{mA}$. Use mesh currents to find the required source voltage V_s. *Ans.* 0.705 V

4.21 Use appropriate determinants of Problem 4.20 to obtain the input resistance as seen by the source voltage V_s. Check the result by network reduction. *Ans.* 23.5 Ω

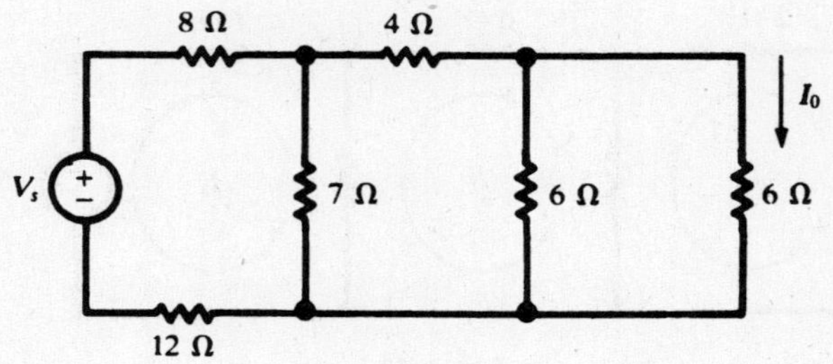

Fig. 4-36

4.22 For the network shown in Fig. 4-36, obtain the transfer resistance which relates the current I_0 to the source voltage V_s. *Ans.* 94.0 Ω

4.23 For the network shown in Fig. 4-37, obtain the mesh currents. *Ans.* 5.0 A, 1.0 A, 0.5 A

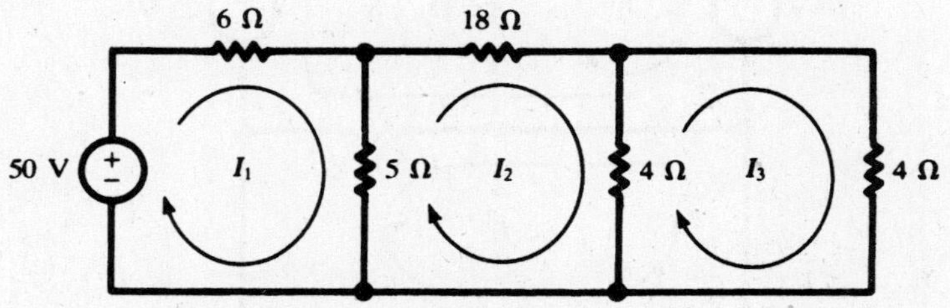

Fig. 4-37

4.24 Using the matrices from Problem 4.23 calculate $R_{\text{input},1}$, $R_{\text{transfer},12}$, and $R_{\text{transfer},13}$.
Ans. 10 Ω, 50 Ω, 100 Ω

4.25 In the network shown in Fig. 4-38, obtain the four mesh currents.
Ans. 2.11 A, −0.263 A, −2.34 A, 0.426 A

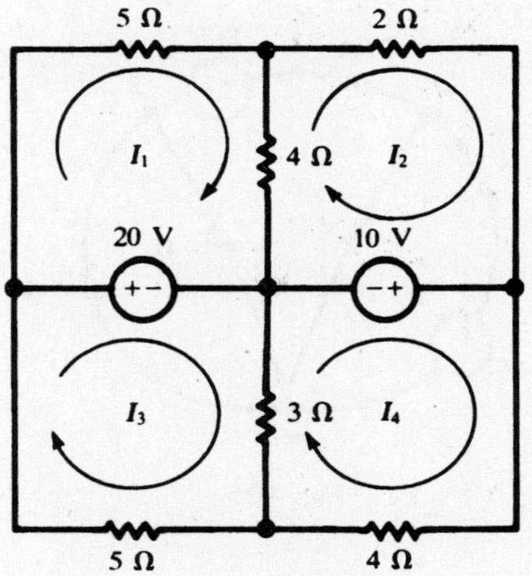

Fig. 4-38

4.26 For the circuit shown in Fig. 4-39, obtain $V_{\text{o.c.}}$, $I_{\text{s.c.}}$, and R' at the terminals *ab* using mesh current or node voltage methods. Consider terminal *a* positive with respect to *b*. *Ans.* − 6.29 V, −0.667 A, 9.44 Ω

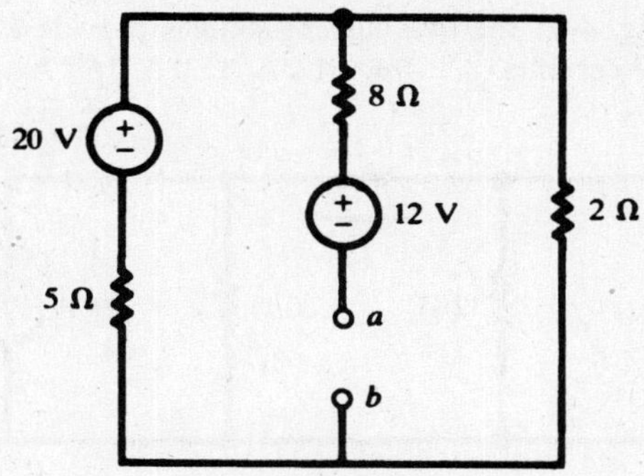

Fig. 4-39

4.27 Use the node voltage method to obtain $V_{o.c.}$ and $I_{s.c.}$ at the terminals *ab* of the network shown in Fig. 4-40. Consider *a* positive with respect to *b*. *Ans.* $-11.2\text{ V}, -7.37\text{ A}$

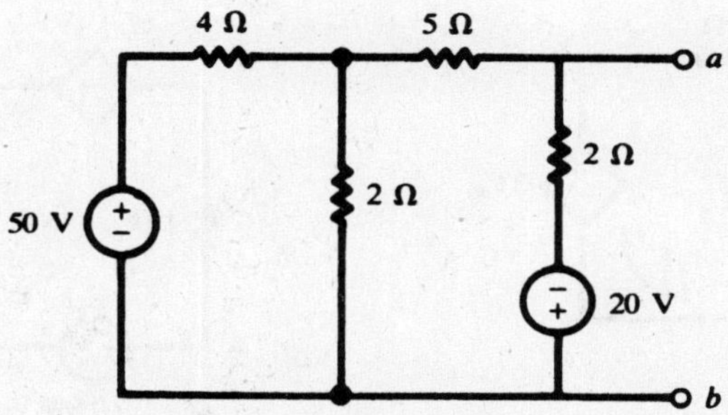

Fig. 4-40

4.28 Use network reduction to obtain the current in each of the resistors in the circuit shown in Fig. 4-41. *Ans.* In the 2.45-Ω resistor, 3.10 A; 6.7 Ω, 0.855 A; 10.0 Ω, 0.466 A; 12.0 Ω, 0.389 A; 17.47 Ω, 0.595 A; 6.30 Ω, 1.65 A

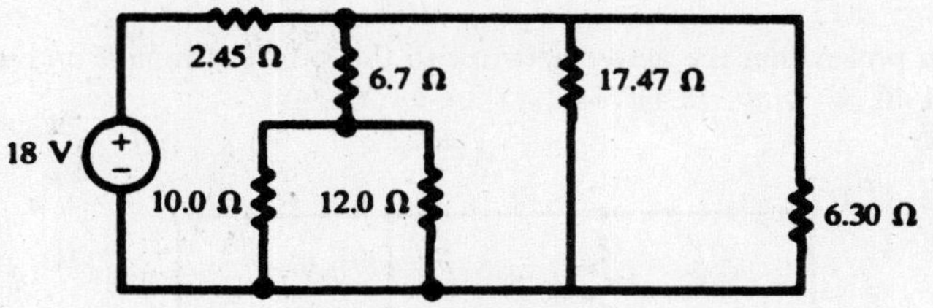

Fig. 4-41

4.29 Both ammeters in the circuit shown in Fig. 4-42 indicate 1.70 A. If the source supplies 300 W to the circuit, find R_1 and R_2. *Ans.* 23.9 Ω, 443.0 Ω

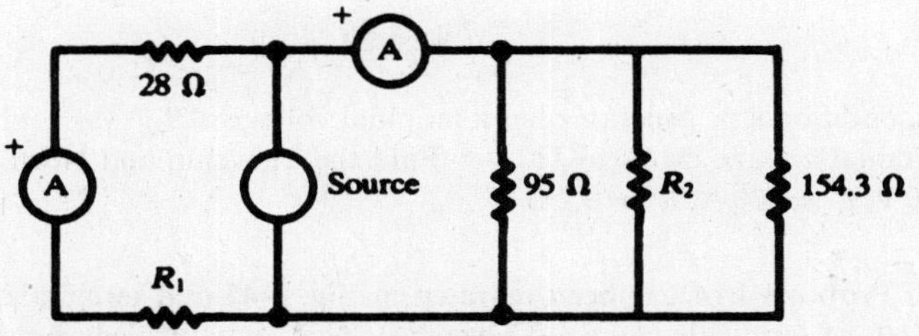

Fig. 4-42

4.30 In the network shown in Fig. 4-43 the two current sources provide I' and I'' where $I' + I'' = I$. Use superposition to obtain these currents. *Ans.* 1.2 A, 15.0 A, 16.2 A

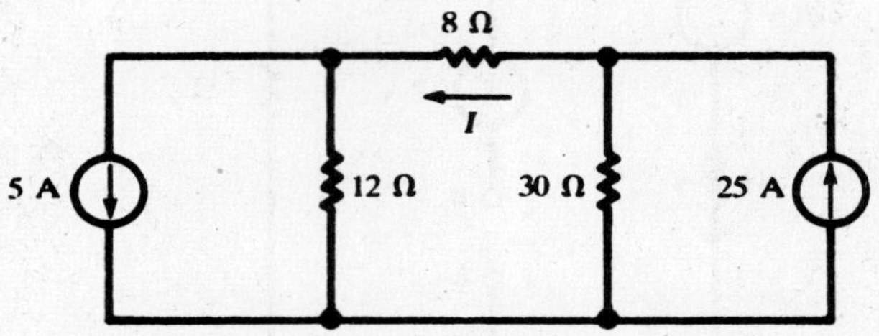

Fig. 4-43

4.31 Obtain the current I in the network shown in Fig. 4.44. *Ans.* -12 A

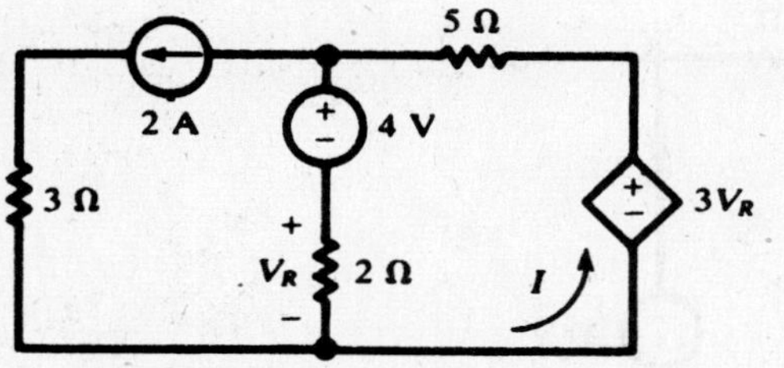

Fig. 4-44

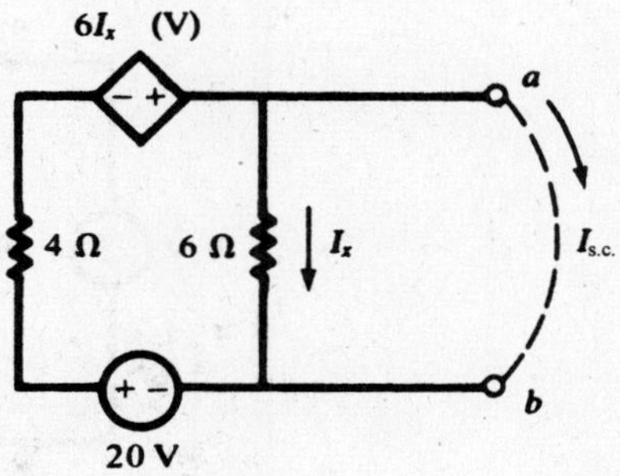

Fig. 4-45

4.32 Obtain the Thévenin and Norton equivalents for the network shown in Fig. 4.45.
Ans. $V' = 30$ V, $I' = 5$ A, $R' = 6\,\Omega$

4.33 Find the maximum power that the active network to the left of terminals *ab* can deliver to the adjustable resistor R in Fig. 4-46. *Ans.* 8.44 W

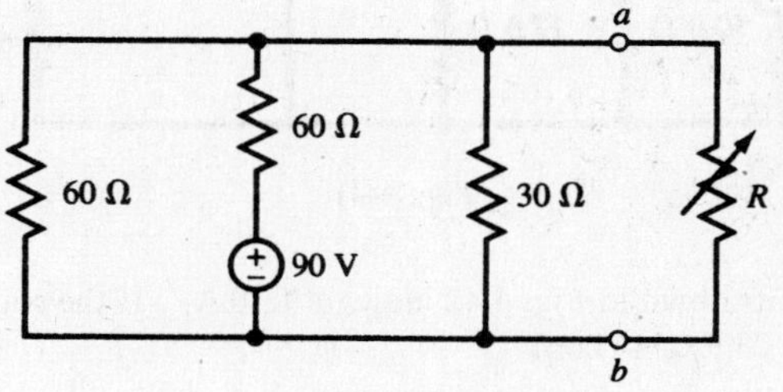

Fig. 4-46

4.34 Under no-load condition a dc generator has a terminal voltage of 120 V. When delivering its rated current of 40 A, the terminal voltage drops to 112 V. Find the Thévenin and Norton equivalents.
Ans. $V' = 120$ V, $I' = 600$ A, $R' = 0.2\,\Omega$

4.35 The network of Problem 4.14 has been redrawn in Fig. 4-47 and terminals a and b added. Reduce the network to the left of terminals *ab* by a Thévenin or Norton equivalent circuit and solve for the current I.
Ans. -8.77 A

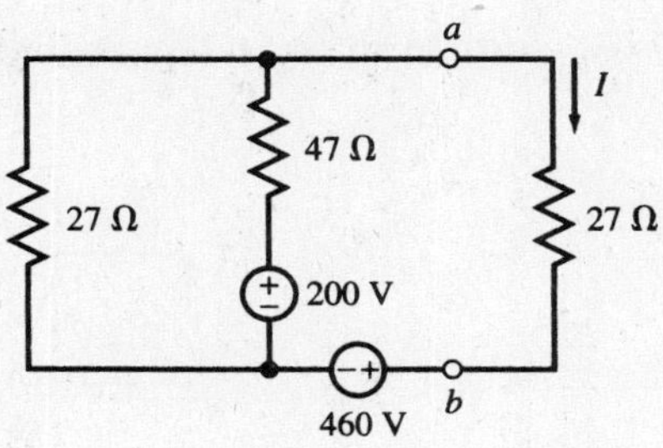

Fig. 4-47

4.36 **Node Voltage Method.** In the circuit of Fig. 4-48 write three node equations for nodes A, B, and C, with node D as the reference, and find the node voltages.

Ans.
$$\begin{cases} \text{Node A:} & 5V_A - 2V_B - 3V_C = 30 \\ \text{Node B:} & -V_A + 6V_B - 3V_C = 0 \\ \text{Node C:} & -V_A - 2V_B + 3V_C = 2 \end{cases}$$
from which $V_A = 17$, $V_B = 9$, $V_C = 12.33$ all in V

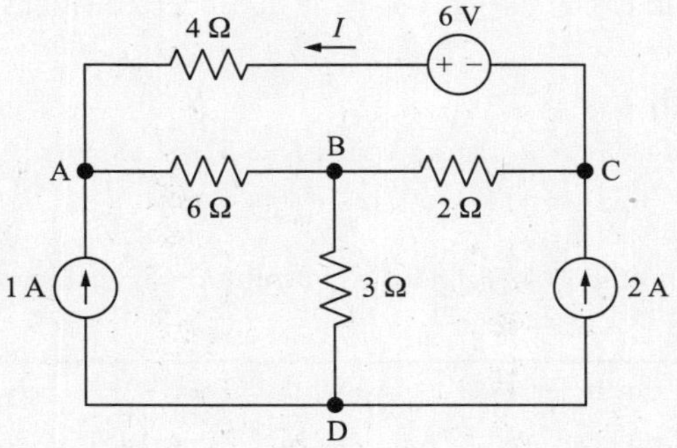

Fig. 4-48

4.37 In the circuit of Fig. 4-48 note that the current through the 3-Ω resistor is 3 A giving rise to $V_B = 9$ V. Apply KVL around the mesh on the upper part of the circuit to find current I coming out of the voltage source, then find V_A and V_C. *Ans.* $I = 1/3$ A, $V_A = 17$ V, $V_C = 37/3$ V

4.38 **Superposition.** In the circuit of Fig. 4-48 find contribution of each source to V_A, V_B, V_C, and show that they add up to values found in Problems 4.36 and 4.37.

Ans.

Contribution of the voltage source:	$V_A = 3$	$V_B = 0$	$V_C = -1$
Contribution of the 1 A current source:	$V_A = 6$	$V_B = 3$	$V_C = 4$
Contribution of the 2 A current source:	$V_A = 8$	$V_B = 6$	$V_C = 28/3$
Contribution of all sources:	$V_A = 17$	$V_B = 9$	$V_C = 37/3$

(All in V)

4.39 In the circuit of Fig. 4-48 remove the 2-A current source and then find the voltage $V_{o.c.}$ between the open-circuited nodes C and D. *Ans.* $V_{o.c.} = 3$ V

4.40 Use the values for V_C and $V_{o.c.}$ obtained in Problems 4.36 and 4.39 to find the Thévenin equivalent of the circuit of Fig. 4-48 seen by the 2-A current source. *Ans.* $V_{Th} = 3$ V, $R_{Th} = 14/3$ Ω

4.41 In the circuit of Fig. 4-48 remove the 2-A current source and set the other two sources to zero, reducing the circuit to a source-free resistive circuit. Find R, the equivalent resistance seen from terminals CD, and note that the answer is equal to the Thévenin resistance obtained in Problem 4.40. *Ans.* $R = 14/3$ Ω

4.42 Find Thévenin equivalent of the circuit of Fig. 4-49 seen from terminals AB.
ans. $V_{Th} = 12\,\text{V}, R_{Th} = 17\,\Omega$

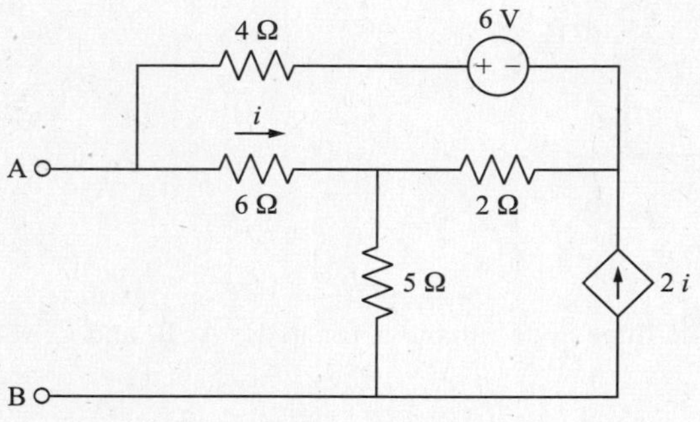

Fig. 4-49

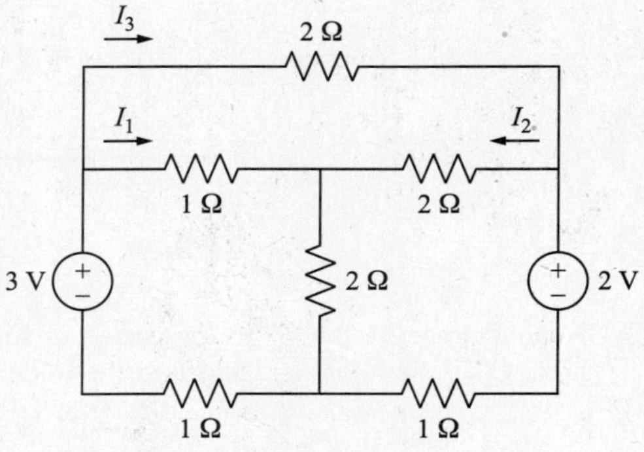

Fig. 4-50

4.43 **Loop Current Method.** In the circuit of Fig. 4-50 write three loop equations using I_1, I_2, and I_3. Then find the currents.

Ans. $\begin{cases} \text{Loop 1:} & 4I_1 + 2I_2 + I_3 = 3 \\ \text{Loop 2:} & 2I_1 + 5I_2 - I_3 = 2 \\ \text{Loop 3:} & -I_1 + 2I_2 + 2I_3 = 0 \end{cases}$ From which $I_1 = 32/51, I_2 = 9/51, I_3 = 7/51$ all in A

4.44 **Superposition.** In the circuit of Fig. 4-50 find the contribution of each source to I_1, I_2, I_3, and show that they add up to values found in Problem 4.43.

Ans.

From the source on the left:	$I_1 = 36/51$	$I_2 = -9/51$	$I_3 = 27/51$
From the source on the right:	$I_1 = -4/51$	$I_2 = 18/51$	$I_3 = -20/51$
From both sources:	$I_1 = 32/51$	$I_2 = 9/51$	$I_3 = 7/51$

(All in A)

4.45 **Node Voltage Method.** In the circuit of Fig. 4-51 write three node equations for nodes A, B, and C, with node D as the reference, and find the node voltages.

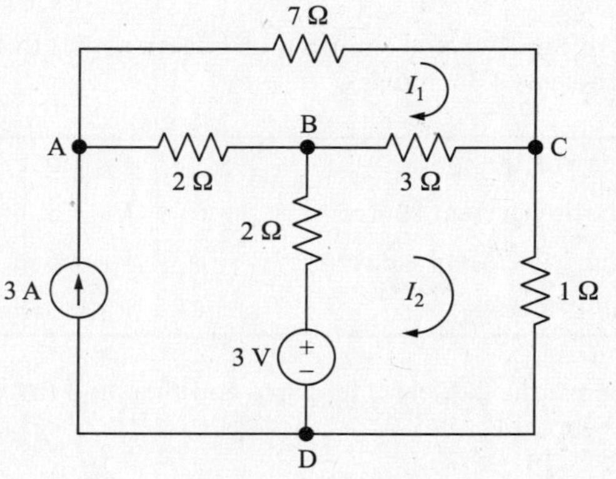

Fig. 4-51

Ans. $\begin{cases} \text{Node A:} & 9V_A - 7V_B - 2V_C = 42 \\ \text{Node B:} & -3V_A + 8V_B - 2V_C = 9 \\ \text{Node C:} & -3V_A - 7V_B + 31V_C = 0 \end{cases}$ From which $V_A = 9, V_B = 5, V_C = 2$ all in V

4.46 **Loop Current Method**. In the circuit of Fig. 4-51 write two loop equations using I_1 and I_2 as loop currents, then find the currents and node voltages.

Ans. $\begin{cases} \text{Loop 1:} & 4I_1 - I_2 = 2 \\ \text{Loop 2:} & -I_1 + 2I_2 = 3 \end{cases}$ from which, $\begin{cases} I_1 = 1\,\text{A}, & I_2 = 2\,\text{A} \\ V_A = 9\,\text{V}, & V_B = 5\,\text{V}, \; V_C = 2\,\text{V} \end{cases}$

4.47 **Superposition**. In the circuit of Fig. 4-51 find the contribution of each source to V_A, V_B, V_C, and show that they add up to values found in Problem 4.45.

Ans.

From the current source:	$V_A = 7.429$	$V_B = 3.143$	$V_C = 1.429$
From the voltage source:	$V_A = 1.571$	$V_B = 1.857$	$V_C = 0.571$
From both sources:	$V_A = 9$	$V_B = 5$	$V_C = 2$

(all in V)

4.48 Verify that the circuit of Fig. 4-52(*a*) is equivalent to the circuit of Fig. 4-51.
Ans. Move node B in Fig. 4-51 to the outside of the loop.

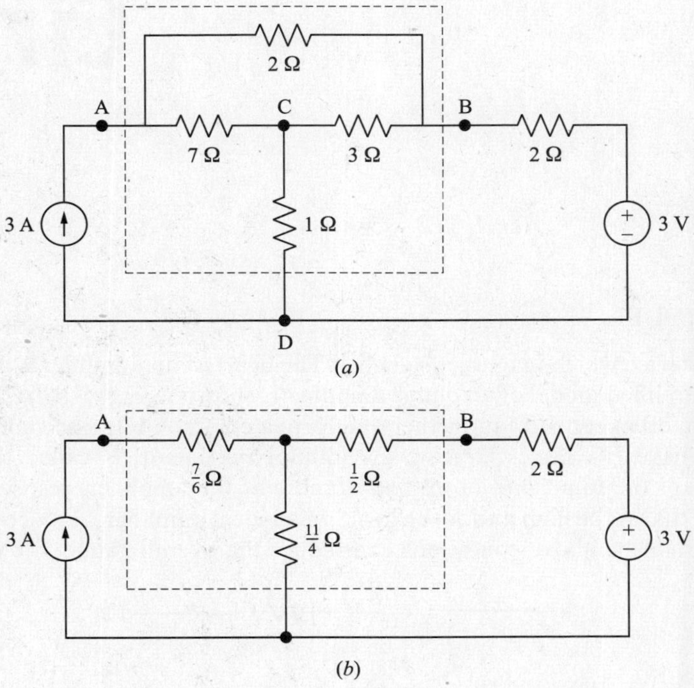

Fig. 4-52

4.49 Find V_A and V_B in the circuit of Fig. 4-52(*b*). *Ans.* $V_A = 9$, $V_B = 5$, both in V

4.50 Show that the three terminal circuits enclosed in the dashed boundaries of Fig. 4-52(*a*) and (*b*) are equivalent (i.e., in terms of their relation to other circuits). *Hint*: Use the linearity and superposition properties, along with the results of Problems 4.48 and 4.49.

CHAPTER 5

Amplifiers and Operational Amplifier Circuits

5.1 AMPLIFIER MODEL

An *amplifier* is a device which magnifies signals. The heart of an amplifier is a source controlled by an input signal. A simplified model of a voltage amplifier is shown in Fig. 5-1(*a*). The input and output reference terminals are often connected together and form a common reference node. When the output terminal is open we have $v_2 = kv_1$, where k, the multiplying factor, is called the open circuit *gain*. Resistors R_i and R_o are the input and output resistances of the amplifier, respectively. For a better operation it is desired that R_i be high and R_o be low. In an ideal amplifier, $R_i = \infty$ and $R_o = 0$ as in Fig. 5-1(*b*). Deviations from the above conditions can reduce the overall gain.

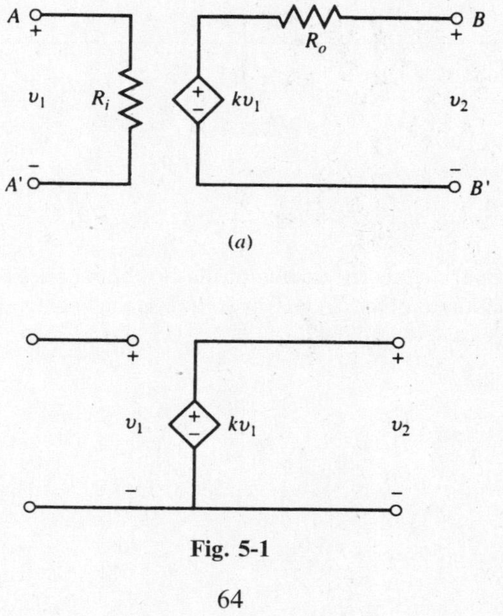

Fig. 5-1

EXAMPLE 5.1 A practical voltage source v_s with an internal resistance R_s is connected to the input of a voltage amplifier with input resistance R_i as in Fig. 5-2. Find v_2/v_s.

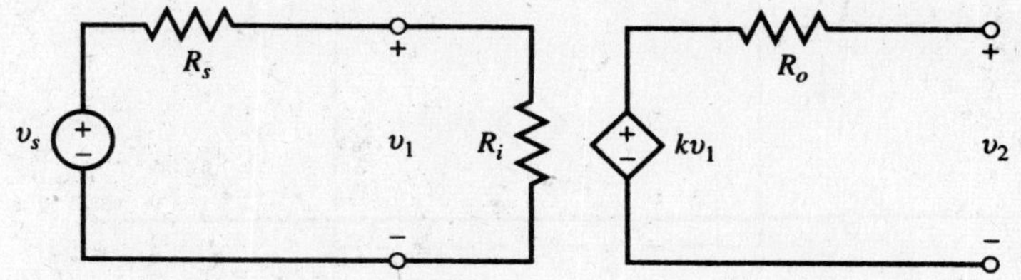

Fig. 5-2

The amplifier's input voltage, v_1, is obtained by dividing v_s between R_i and R_s.

$$v_1 = \frac{R_i}{R_i + R_s} v_s$$

The output voltage v_2 is

$$v_2 = kv_1 = \frac{kR_i}{R_i + R_s} v_s$$

from which

$$\frac{v_2}{v_s} = \frac{R_i}{R_i + R_s} k \tag{1}$$

The amplifier loads the voltage source. The open-loop gain is reduced by the factor $R_i/(R_i + R_s)$.

EXAMPLE 5.2 In Fig. 5-3 a practical voltage source v_s with internal resistance R_s feeds a load R_l through an amplifier with input and output resistances R_i and R_o, respectively. Find v_2/v_s.

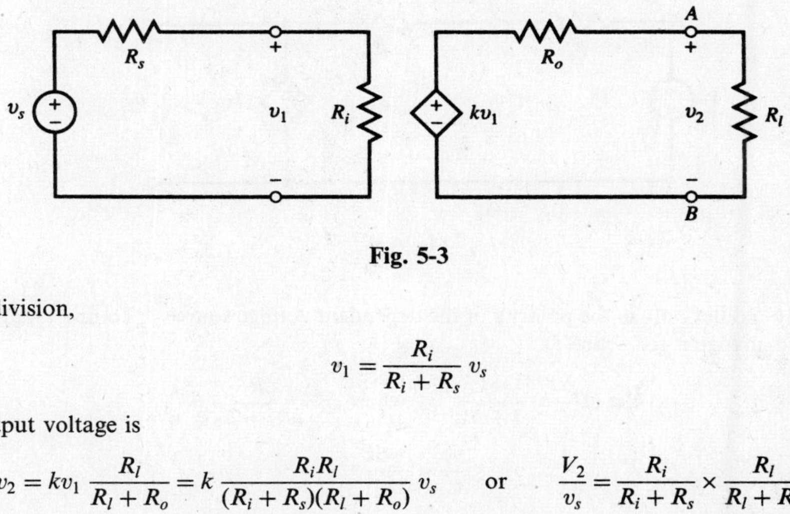

Fig. 5-3

By voltage division,

$$v_1 = \frac{R_i}{R_i + R_s} v_s$$

Similarly, the output voltage is

$$v_2 = kv_1 \frac{R_l}{R_l + R_o} = k \frac{R_i R_l}{(R_i + R_s)(R_l + R_o)} v_s \quad \text{or} \quad \frac{V_2}{v_s} = \frac{R_i}{R_i + R_s} \times \frac{R_l}{R_l + R_o} k \tag{2}$$

Note that the open-loop gain is further reduced by an additional factor of $R_l/(R_l + R_o)$, which also makes the output voltage dependent on the load.

5.2 FEEDBACK IN AMPLIFIER CIRCUITS

The gain of an amplifier may be controlled by feeding back a portion of its output to its input as done for the ideal amplifier in Fig. 5-4 through the feedback resistor R_2. The feedback ratio $R_1/(R_1 + R_2)$ affects the overall gain and makes the amplifier less sensitive to variations in k.

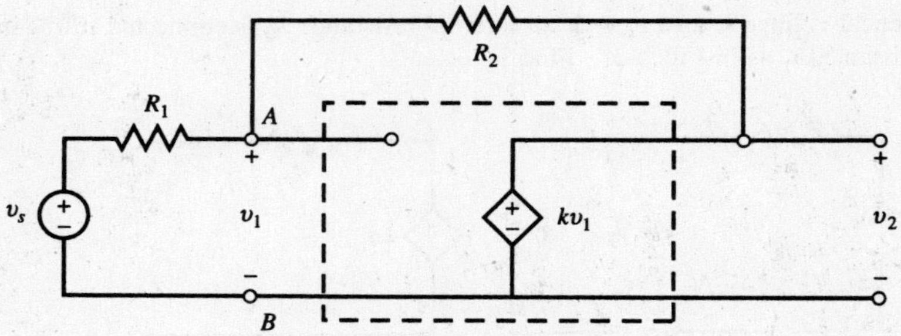

Fig. 5-4

EXAMPLE 5.3 Find v_2/v_s in Fig. 5-4 and express it as a function of the ratio $b = R_1/(R_1 + R_2)$.

From the amplifier we know that

$$v_2 = kv_1 \qquad \text{or} \qquad v_1 = v_2/k \tag{3}$$

Applying KCL at node A,

$$\frac{v_1 - v_s}{R_1} + \frac{v_1 - v_2}{R_2} = 0 \tag{4}$$

Substitute v_1 in (3) into (4) to obtain

$$\frac{v_2}{v_s} = \frac{R_2 k}{R_2 + R_1 - R_1 k} = (1 - b)\frac{k}{1 - bk} \qquad \text{where } b = \frac{R_1}{R_1 + R_2} \tag{5}$$

EXAMPLE 5.4 In Fig. 5-5, $R_1 = 1\,\text{k}\Omega$ and $R_2 = 5\,\text{k}\Omega$. (*a*) Find v_2/v_s as a function of the open-loop gain k. (*b*) Compute v_2/v_s for $k = 100$ and 1000 and discuss the results.

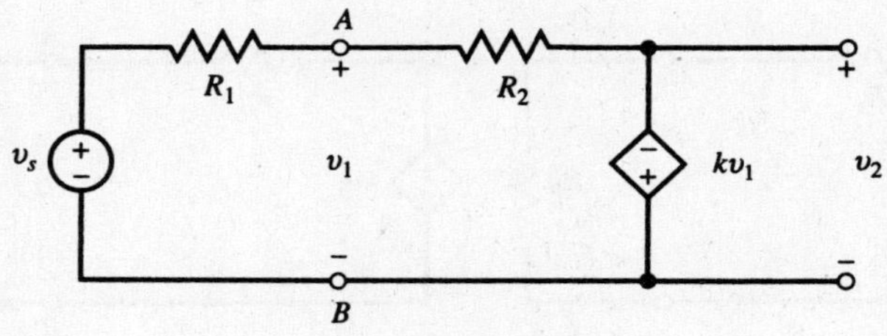

Fig. 5-5

(*a*) Figures 5-4 and 5-5 differ only in the polarity of the dependent voltage source. To find v_2/v_s, use the results of Example 5.3 and change k to $-k$ in (5).

$$\frac{v_2}{v_s} = (1 - b)\frac{-k}{1 + bk} \qquad \text{where } b = \frac{R_1}{R_1 + R_2} = \frac{1}{6}$$

$$\frac{v_2}{v_s} = \frac{-5k}{6 + k}$$

(*b*) At $k = 100$, $v_2/v_s = -4.72$; at $k = 1000$, $v_2/v_s = -4.97$. Thus, a tenfold increase in k produces only a 5.3 percent change in v_2/v_s; i.e., $(4.97 - 4.72)/4.72 = 5.3$ percent.
Note that for very large values of k, v_2/v_s approaches $-R_2/R_1$ which is independent of k.

5.3 OPERATIONAL AMPLIFIERS

The *operational amplifier* (op amp) is a device with two input terminals, labeled + and − or non-inverting and inverting, respectively. The device is also connected to dc power supplies ($+V_{cc}$ and

$-V_{cc}$). The common reference for inputs, output, and power supplies resides outside the op amp and is called the *ground* (Fig. 5-6).

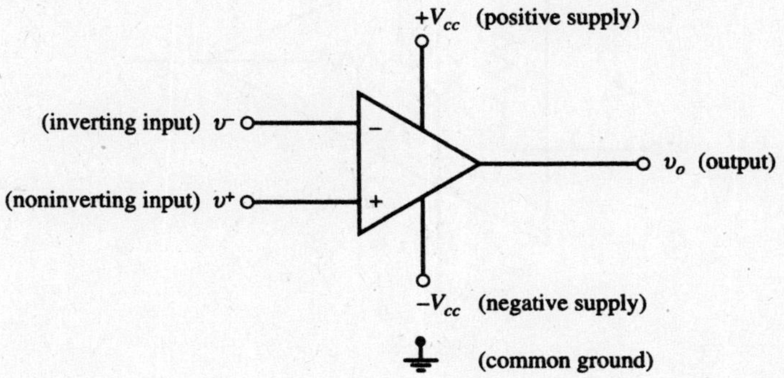

Fig. 5-6

The output voltage v_o depends on $v_d = v^+ - v^-$. Neglecting the capacitive effects, the transfer function is that shown in Fig. 5-7. In the linear range, $v_o = Av_d$. The open-loop gain A is generally very high. v_o saturates at the extremes of $+V_{cc}$ and $-V_{cc}$ when input v_d exceeds the linear range $|v_d| > V_{cc}/A$.

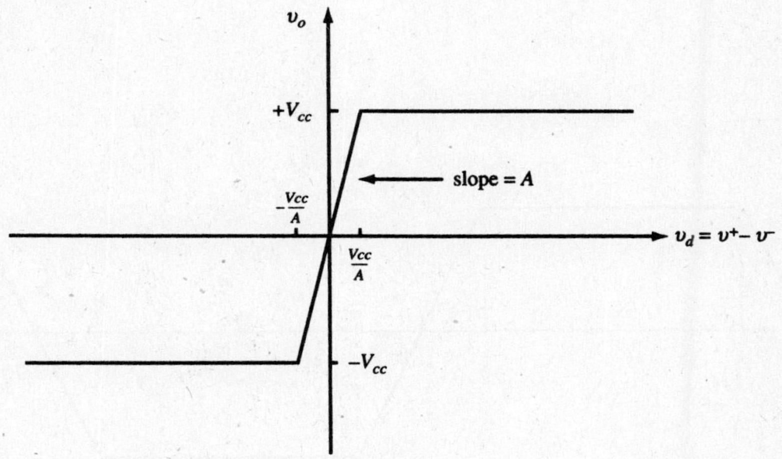

Fig. 5-7

Figure 5-8 shows the model of an op amp in the linear range with power supply connections omitted for simplicity. In practice, R_i is large, R_o is small, and A ranges from 10^5 to several millions. The model of Fig. 5-8 is valid as long as the output remains between $+V_{cc}$ and $-V_{cc}$. V_{cc} is generally from 5 to 18 V.

EXAMPLE 5.5 In the op amp of Fig. 5-8, $V_{cc} = 15$ V, $A = 10^5$, and $v^- = 0$. Find the upper limit on the magnitude of v^+ for linear operation.

$$|v_o| = |10^5 v^+| < 15 \text{ V} \qquad |v^+| < 15 \times 10^{-5} \text{ V} = 150 \,\mu\text{V}$$

EXAMPLE 5.6 In the op amp of Fig. 5-8, $V_{cc} = 5$ V, $A = 10^5$, $v^- = 0$ and $v^+ = 100 \sin 2\pi t$ (μV). Find and sketch the open-loop output v_o.

The input to the op amp is $v_d = v^+ - v^- = (100 \sin 2\pi t)10^{-6}$ (V). When the op amp operates in the linear range, $v_o = 10^5 v_d = 10 \sin 2\pi t$ (V). The output should remain between +5 and -5 V (Fig. 5-9). Saturation starts when $v_o = 10 \sin 2\pi t$ reaches the 5-V level. This occurs at $t = 1/12$ s. The op amp comes out of 5-V saturation at

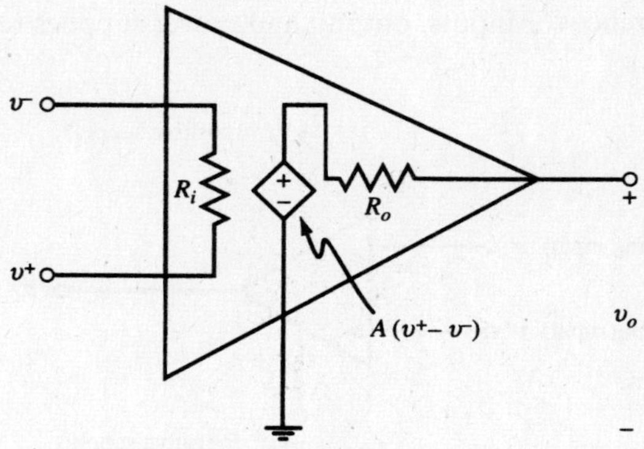

Fig. 5-8

$t = 5/12$. Similarly, the op amp is in −5-V saturation from $t = 7/12$ to $11/12$ s. One full cycle of the output, given in volts, from $t = 0$ to 1 s is

$$v_o = \begin{cases} 5 & 1/12 < t < 5/12 \\ -5 & 7/12 < t < 11/12 \\ 10 \sin 2\pi t & \text{otherwise} \end{cases}$$

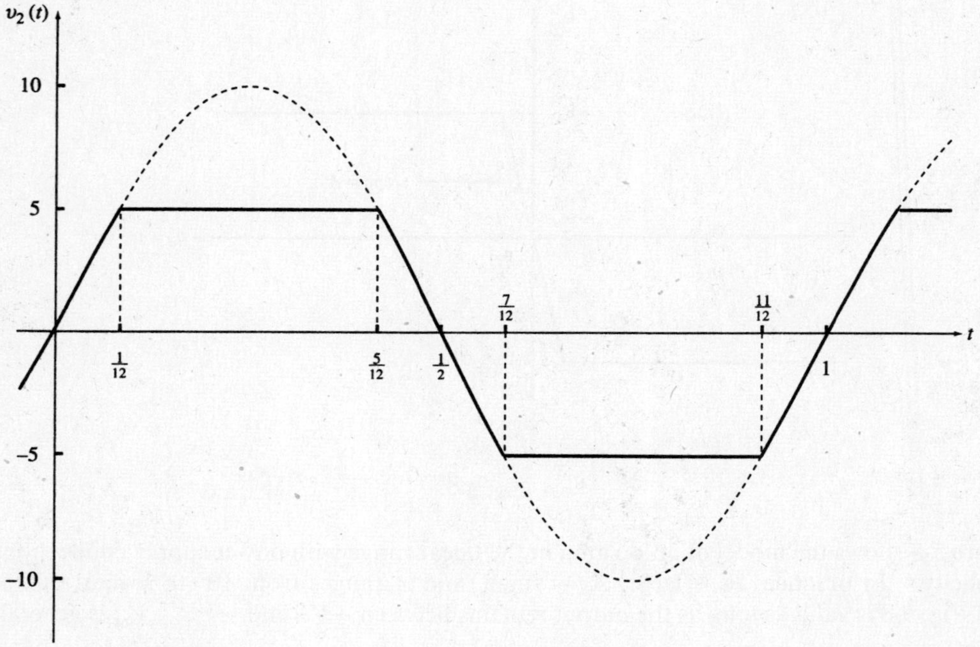

Fig. 5-9

EXAMPLE 5.7 Repeat Example 5.6 for $v^- = 25\,\mu\text{V}$ and $v^+ = 50 \sin 2\pi t$ (μV).

$$v_d = v^+ - v^- = (50 \sin 2\pi t)10^{-6} - 25 \times 10^{-6} = 50 \times 10^{-6}(\sin 2\pi t - 1/2) \text{ (V)}$$

When the op amp is within linear range, its output is

$$v_o = 10^5 v_d = 5(\sin 2\pi t - 1/2) \text{ (V)}$$

v_o saturates at the −5-V level when $5(\sin 2\pi t - 1/2) < -5$, $7/12 < t < 11/12$ (see Fig. 5-10). One cycle of v_o, in volts, from $t = 0$ to 1 s is

$$v_o = \begin{cases} -5 & 7/12 < t < 11/12 \\ 5(\sin 2\pi t - 1/2) & \text{otherwise} \end{cases}$$

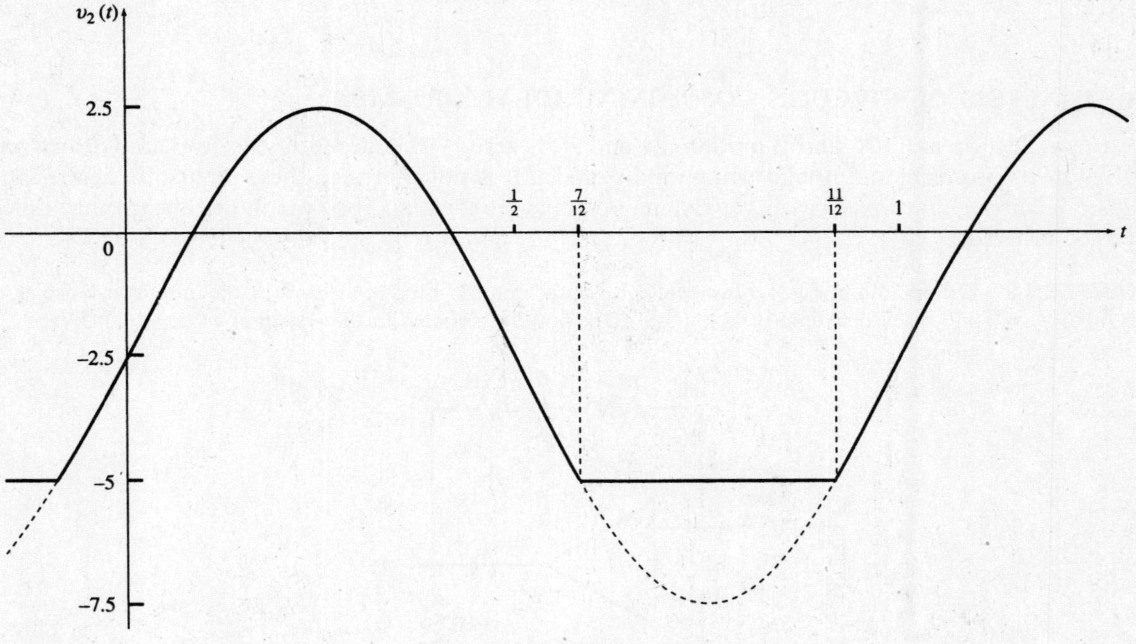

Fig. 5-10

EXAMPLE 5.8 In Fig. 5-11, $R_1 = 10\,\text{k}\Omega$, $R_2 = 50\,\text{k}\Omega$, $R_i = 500\,\text{k}\Omega$, $R_o = 0$, and $A = 10^5$. Find v_2/v_1. Assume the amplifier is not saturated.

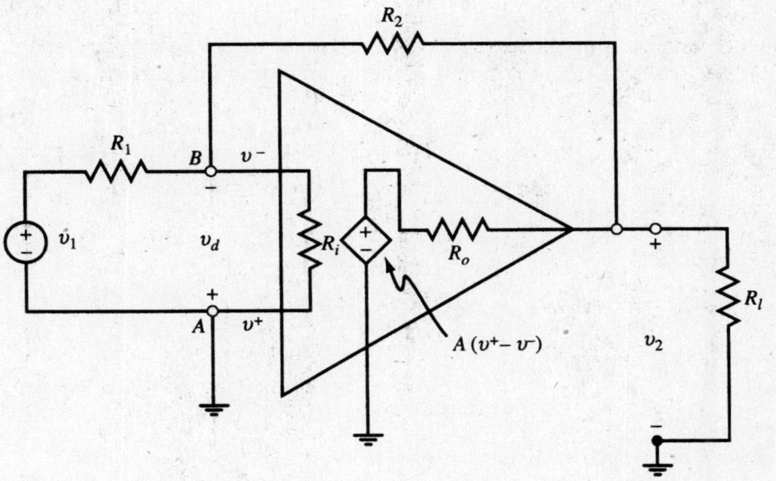

Fig. 5-11

The sum of currents arriving at node B is zero. Note that $v_A = 0$ and $v_B = -v_d$. Therefore,

$$\frac{v_1 + v_d}{10} + \frac{v_d}{500} + \frac{v_2 + v_d}{50} = 0 \tag{6}$$

Since $R_o = 0$, we have

$$v_2 = Av_d = 10^5 v_d \quad \text{or} \quad v_d = 10^{-5} v_2 \tag{7}$$

Substituting v_d in *(7)* into *(6)*, the ratio v_2/v_1 is found to be

$$\frac{v_2}{v_1} = \frac{-5}{1 + 10^{-5} + 5 \times 10^{-5} + 0.1 \times 10^{-5}} = -5$$

5.4 ANALYSIS OF CIRCUITS CONTAINING IDEAL OP AMPS

In an ideal op amp, R_i and A are infinite and R_o is zero. Therefore, the ideal op amp draws zero current at its inverting and noninverting inputs, and if it is not saturated these inputs are at the same voltage. Throughout this chapter we assume op amps are ideal and operate in the linear range unless specified otherwise.

EXAMPLE 5.9 The op amp in Fig. 5-12 is ideal and not saturated. Find *(a)* v_2/v_1; *(b)* the input resistance v_1/i_1; and *(c)* i_1, i_2, p_1 (the power delivered by v_1), and p_2 (the power dissipated in the resistors) given $v_1 = 0.5$ V.

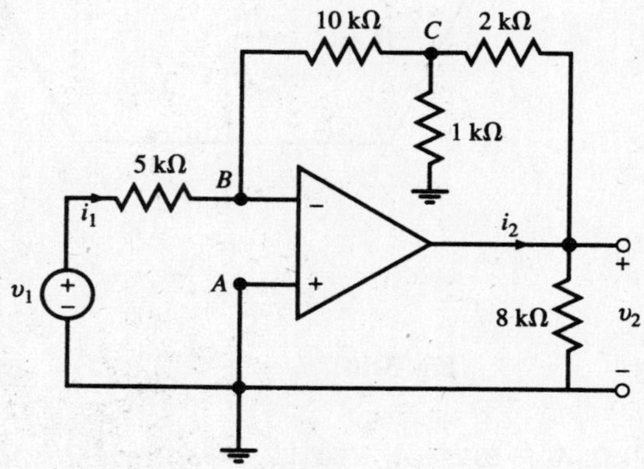

Fig. 5-12

(a) The noninverting terminal A is grounded and so $v_A = 0$. Since the op amp is ideal and not saturated, $v_B = 0$. Applying KCL at nodes B and C and noting that the op amp draws no current, we get

Node B: $$\frac{v_1}{5} + \frac{v_C}{10} = 0 \quad \text{or} \quad v_C = -2v_1 \tag{8}$$

Node C: $$\frac{v_C}{10} + \frac{v_C}{1} + \frac{v_C - v_2}{2} = 0 \quad \text{or} \quad v_2 = 3.2v_C \tag{9}$$

Substituting v_C in *(8)* into *(9)*,

$$v_2 = -6.4v_1 \quad \text{or} \quad v_2/v_1 = -6.4$$

(b) With $V_B = 0$, $i_1 = v_1/5000$ and so

$$\text{input resistance} = v_1/i_1 = 5\,\text{k}\Omega$$

(c) The input current is $i_1 = v_1/5000$. Given that $v_1 = 0.5$ V, $i_1 = 0.5/5000 = 0.1$ mA.

To find i_2, we apply KCL at the output of the op amp;

$$i_2 = \frac{v_2}{8000} + \frac{v_2 - v_C}{2000}$$

From part *(a)*, $v_2 = -3.2$ V and $v_C = -1$ V. Therefore, $i_2 = 1.5$ mA.

The power delivered by v_1 is

$$p_1 = v_1 i_1 = v_1^2/5000 = 50 \times 10^{-6}\,\text{W} = 50\,\mu\text{W}$$

Powers in the resistors are

1 kΩ: $p_{1\,\mathrm{k}\Omega} = v_C^2/1000 = 0.001\ \mathrm{W} = 1000\ \mu\mathrm{W}$

2 kΩ: $p_{2\,\mathrm{k}\Omega} = (v_2 - v_C)^2/2000 = 0.00242\ \mathrm{W} = 2420\ \mu\mathrm{W}$

5 kΩ: $p_{5\,\mathrm{k}\Omega} = v_1^2/5000 = 0.00005\ \mathrm{W} = 50\ \mu\mathrm{W}$

8 kΩ: $p_{8\,\mathrm{k}\Omega} = v_2^2/8000 = 0.00128\ \mathrm{W} = 1280\ \mu\mathrm{W}$

10 kΩ: $p_{10\,\mathrm{k}\Omega} = v_C^2/10\,000 = 0.0001\ \mathrm{W} = 100\ \mu\mathrm{W}$

The total power dissipated in the resistors is

$$p_2 = p_{1\,\mathrm{k}\Omega} + p_{2\,\mathrm{k}\Omega} + p_{5\,\mathrm{k}\Omega} + p_{8\,\mathrm{k}\Omega} + p_{10\,\mathrm{k}\Omega} = 1000 + 2420 + 50 + 1280 + 100 = 4850\ \mu\mathrm{W}$$

5.5 INVERTING CIRCUIT

In an *inverting circuit*, the input signal is connected through R_1 to the inverting terminal of the op amp and the output terminal is connected back through a feedback resistor R_2 to the inverting terminal. The noninverting terminal of the op amp is grounded (see Fig. 5-13).

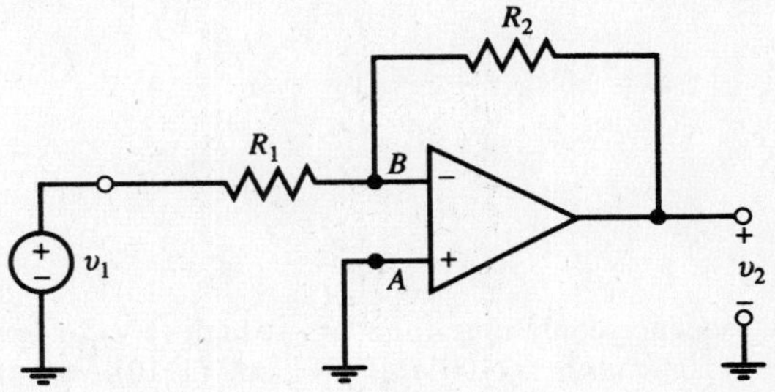

Fig. 5-13

To find the gain v_2/v_1, apply KCL to the currents arriving at node B:

$$\frac{v_1}{R_1} + \frac{v_2}{R_2} = 0 \quad \text{and} \quad \frac{v_2}{v_1} = -\frac{R_2}{R_1} \tag{10}$$

The gain is negative and is determined by the choice of resistors only. The input resistance of the circuit is R_1.

5.6 SUMMING CIRCUIT

The weighted sum of several voltages in a circuit can be obtained by using the circuit of Fig. 5-14. This circuit, called a *summing circuit*, is an extension of the inverting circuit.

To find the output, apply KCL to the inverting node:

$$\frac{v_1}{R_1} + \frac{v_2}{R_2} + \cdots + \frac{v_n}{R_n} + \frac{v_o}{R_f} = 0$$

from which

$$v_o = -\left(\frac{R_f}{R_1}v_1 + \frac{R_f}{R_2}v_2 + \cdots + \frac{R_f}{R_n}v_n\right) \tag{11}$$

EXAMPLE 5.10 Let the circuit of Fig. 5-14 have four input lines with $R_1 = 1$, $R_2 = \frac{1}{2}$, $R_3 = \frac{1}{4}$, $R_4 = \frac{1}{8}$, and $R_f = 1$, all values given in kΩ. The input lines are set either at 0 or 1 V. Find v_o in terms of v_4, v_3, v_2, v_1, given the following sets of inputs:

(*a*) $v_4 = 1\ \mathrm{V}$ $\quad v_3 = 0$ $\quad v_2 = 0$ $\quad v_1 = 1\ \mathrm{V}$

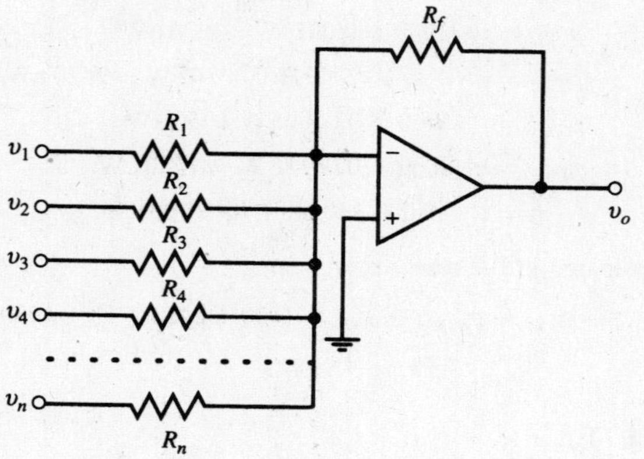

Fig. 5-14

(*b*) $v_4 = 1\text{ V}$ $\quad v_3 = 1\text{ V}$ $\quad v_2 = 1\text{ V}$ $\quad v_1 = 0$

From (*11*)

$$v_o = -(8v_4 + 4v_3 + 2v_2 + v_1)$$

Substituting for v_1 to v_4 we obtain

(*a*) $v_o = -9\text{ V}$

(*b*) $v_o = -14\text{ V}$

The set $\{v_4, v_3, v_2, v_1\}$ forms a binary sequence containing four bits at high (1 V) or low (0 V) values. Input sets given in (*a*) and (*b*) correspond to the binary numbers $(1001)_2 = (9)_{10}$ and $(1110)_2 = (14)_{10}$, respectively. With the inputs at 0 V (low) or 1 V (high), the circuit converts the binary number represented by the input set $\{v_4, v_3, v_2, v_1\}$ to a negative voltage which, when measured in V, is equal to the base 10 representation of the input set. The circuit is a digital-to-analog converter.

5.7 NONINVERTING CIRCUIT

In a *noninverting circuit* the input signal arrives at the noninverting terminal of the op amp. The inverting terminal is connected to the output through R_2 and also to the ground through R_1 (see Fig. 5-15).

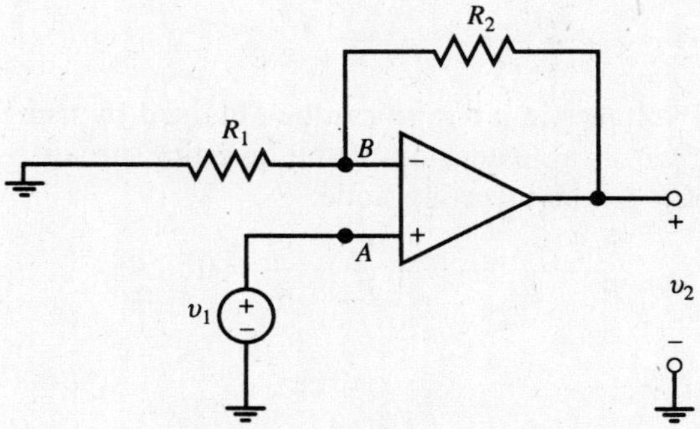

Fig. 5-15

To find the gain v_2/v_1, apply KCL at node B. Note that terminals A and B are both at v_1 and the op amp draws no current.

$$\frac{v_1}{R_1} + \frac{v_1 - v_2}{R_2} = 0 \quad \text{or} \quad \frac{v_2}{v_1} = 1 + \frac{R_2}{R_1} \tag{12}$$

The gain v_2/v_1 is positive and greater than or equal to one. The input resistance of the circuit is infinite as the op amp draws no current.

EXAMPLE 5.11 Find v_2/v_1 in the circuit shown in Fig. 5-16.

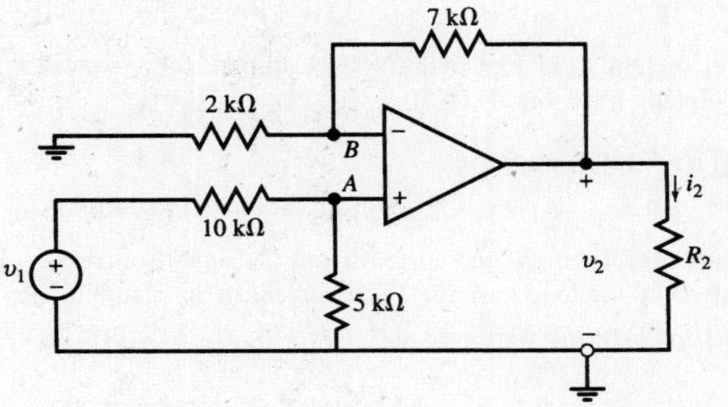

Fig. 5-16

First find v_A by dividing v_1 between the 10-kΩ and 5-kΩ resistors.

$$v_A = \frac{5}{5+10} v_1 = \frac{1}{3} v_1$$

From (12) we get

$$v_2 = \left(1 + \frac{7}{2}\right) v_A = \frac{9}{2} v_A = \frac{9}{2}\left(\frac{1}{3} v_1\right) = 1.5 v_1 \quad \text{and} \quad \frac{v_2}{v_1} = 1.5$$

Another Method

Find v_B by dividing v_2 between the 2-kΩ and 7-kΩ resistors and set $v_B = v_A$.

$$v_B = \frac{2}{2+7} v_2 = \frac{2}{9} v_2 = \frac{1}{3} v_1 \quad \text{and} \quad \frac{v_2}{v_1} = 1.5$$

EXAMPLE 5.12 Determine v_o in Fig. 5-17 in terms of v_1, v_2, v_3, and the circuit elements.

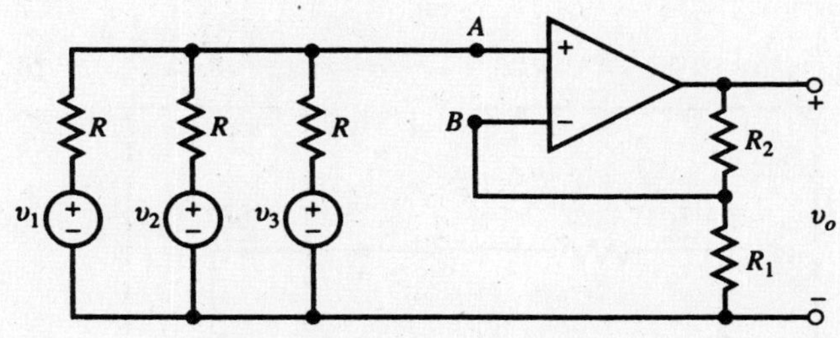

Fig. 5-17

First, v_A is found by applying KCL at node A.

$$\frac{v_1 - v_A}{R} + \frac{v_2 - v_A}{R} + \frac{v_3 - v_A}{R} = 0 \quad \text{or} \quad v_A = \frac{1}{3}(v_1 + v_2 + v_3) \tag{13}$$

From (12) and (13) we get

$$v_o = \left(1 + \frac{R_2}{R_1}\right) v_A = \frac{1}{3}\left(1 + \frac{R_2}{R_1}\right)(v_1 + v_2 + v_3) \tag{14}$$

5.8 VOLTAGE FOLLOWER

The op amp in the circuit of Fig. 5-18(*a*) provides a unity gain amplifier in which $v_2 = v_1$ since $v_1 = v^+$, $v_2 = v^-$ and $v^+ = v^-$. The output v_2 follows the input v_1. By supplying i_l to R_l, the op amp eliminates the loading effect of R_l on the voltage source. It therefore functions as a buffer.

EXAMPLE 5.13 (*a*) Find i_s, v_l, v_2, and i_l in Fig. 5-18(*a*). (*b*) Compare these results with those obtained when source and load are connected directly as in Fig. 5-18(*b*).

(*a*) With the op amp present [Fig. 5-18(*a*)], we have

$$i_s = 0 \qquad v_1 = v_s \qquad v_2 = v_1 = v_s \qquad i_l = v_s/R_l$$

The voltage follower op amp does not draw any current from the signal source v_s. Therefore, v_s reaches the load with no reduction caused by the load current. The current in R_l is supplied by the op amp.

(*b*) With the op amp removed [Fig. 5-18(*b*)], we have

$$i_s = i_l = \frac{v_s}{R_l + R_s} \qquad \text{and} \qquad v_1 = v_2 = \frac{R_l}{R_l + R_s} v_s$$

The current drawn by R_l goes through R_s and produces a drop in the voltage reaching it. The load voltage v_2 depends on R_l.

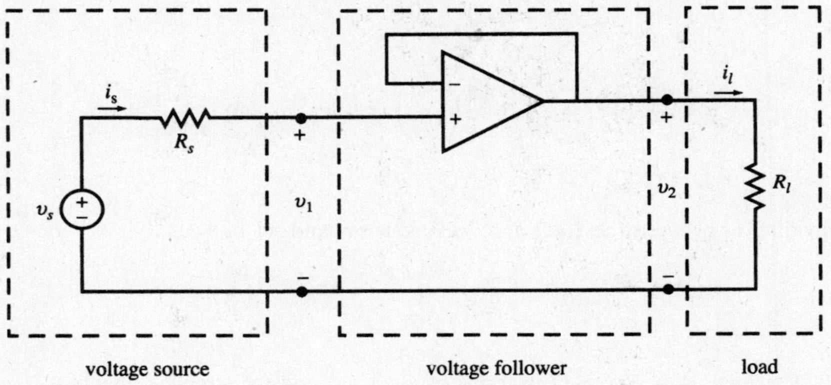

(*a*)

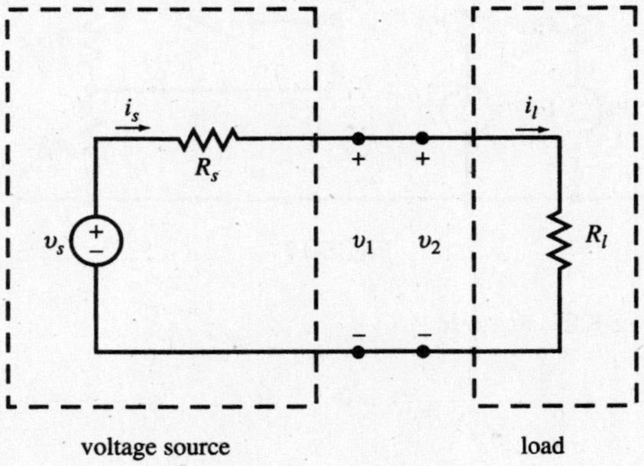

(*b*)

Fig. 5-18

5.9 DIFFERENTIAL AND DIFFERENCE AMPLIFIERS

A signal source v_f with no connection to ground is called a *floating source*. Such a signal may be amplified by the circuit of Fig. 5-19.

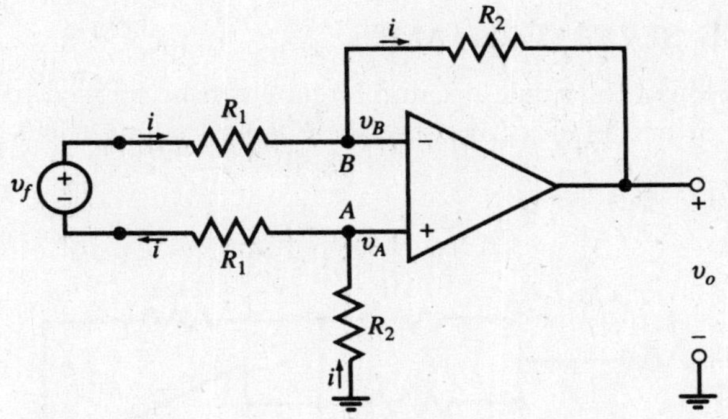

Fig. 5-19

Here the two input terminals A and B of the op amp are at the same voltage. Therefore, by writing KVL around the input loop we get

$$v_f = 2R_1 i \qquad \text{or} \qquad i = v_f/2R_1$$

The op amp inputs do not draw any current and so current i also flows through the R_2 resistors. Applying KVL around the op amp, we have

$$v_o + R_2 i + R_2 i = 0 \qquad v_o = -2R_2 i = -2R_2 v_f/2R_1 = -(R_2/R_1)v_f \tag{15}$$

In the special case when two voltage sources v_1 and v_2 with a common ground are connected to the inverting and noninverting inputs of the circuit, respectively (see Fig. 5-20), we have $v_f = v_1 - v_2$ and

$$v_o = (R_2/R_1)(v_2 - v_1) \tag{16}$$

EXAMPLE 5.14 Find v_o as a function of v_1 and v_2 in the circuit of Fig. 5-20.
Applying KCL at nodes A and B,

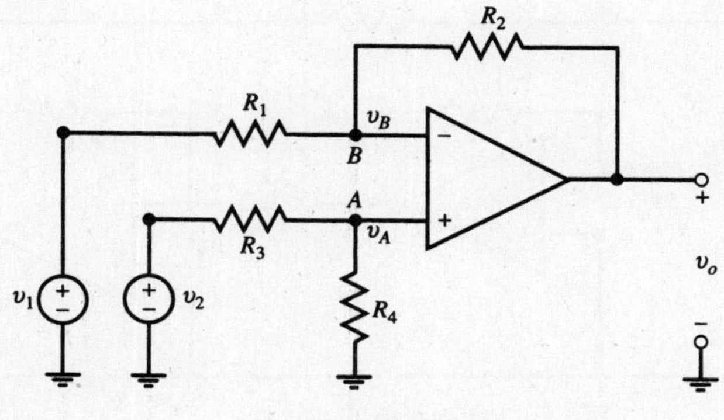

Fig. 5-20

Node A: $$\frac{v_A - v_2}{R_3} + \frac{v_A}{R_4} = 0$$

Node B: $$\frac{v_B - v_1}{R_1} + \frac{v_B - v_o}{R_2} = 0$$

Set $v_A = v_B$ and eliminate them from the preceding KCL equations to get

$$v_o = \frac{R_4(R_1 + R_2)}{R_1(R_3 + R_4)} v_2 - \frac{R_2}{R_1} v_1 \tag{17}$$

When $R_3 = R_1$ and $R_2 = R_4$, (17) is reduced to (16).

5.10 CIRCUITS CONTAINING SEVERAL OP AMPS

The analysis and results developed for single op amp circuits can be applied to circuits containing several ideal op amps in cascade or nested loops because there is no loading effect.

EXAMPLE 5.15 Find v_1 and v_2 in Fig. 5-21.

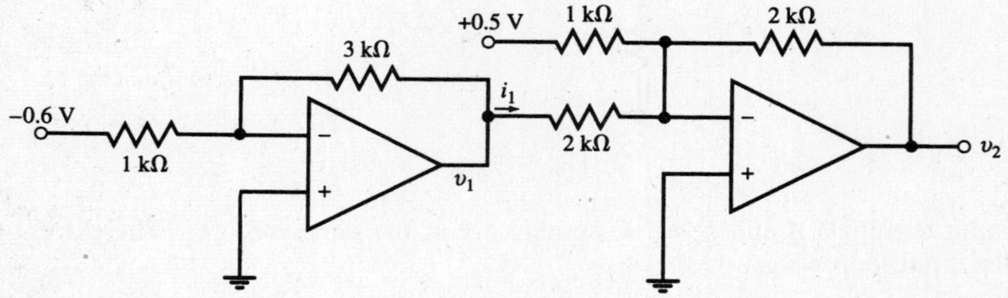

Fig. 5-21

The first op amp is an inverting circuit.

$$v_1 = -(3/1)(-0.6) = 1.8\text{ V}$$

The second op amp is a summing circuit.

$$v_2 = -(2/1)(0.5) - (2/2)(1.8) = -2.8\text{ V}$$

EXAMPLE 5.16 Let $R_s = 1\text{ k}\Omega$ in the circuit of Fig. 5-22, find v_1, v_2, v_o, i_s, i_1, and i_f as functions of v_s for (*a*) $R_f = \infty$ and (*b*) $R_f = 40\text{ k}\Omega$

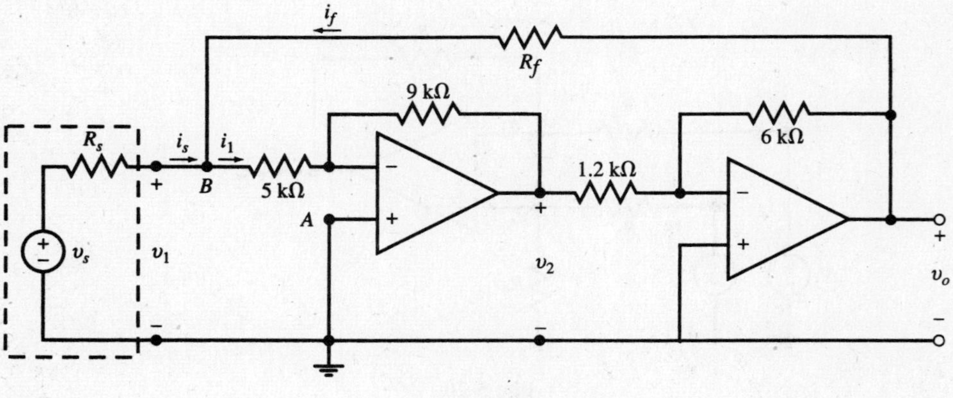

Fig. 5-22

(*a*) $R_f = \infty$. The two inverting op amps are cascaded, with $v^+ = 0$. By voltage division in the input loop we have

$$v_1 = \frac{5}{5+1} v_s = \frac{5}{6} v_s \tag{18}$$

From the inverting amplifiers we get

$$v_2 = -(9/5)v_1 = -(9/5)\left(\frac{5}{6}v_s\right) = -1.5v_s$$
$$v_o = -(6/1.2)v_2 = -5(-1.5v_s) = 7.5v_s$$
$$i_s = i_1 = \frac{v_s}{6000}\text{ (A)} = 0.166v_s\text{ (mA)}$$
$$i_f = 0$$

(*b*) $R_f = 40\,\text{k}\Omega$. From the inverting op amps we get $v_o = -5v_2$ and $v_2 = -(9/5)v_1$ so that $v_o = 9v_1$. Apply KCL to the currents leaving node *B*.

$$\frac{v_1 - v_s}{1} + \frac{v_1}{5} + \frac{v_1 - v_o}{40} = 0 \qquad (19)$$

Substitute $v_o = 9v_1$ in (*19*) and solve for v_1 to get

$$v_1 = v_s$$
$$v_2 = -(9/5)v_1 = -1.8v_s$$
$$v_o = -(6/1.2)v_2 = -5(-1.8v_s) = 9v_s$$
$$i_s = \frac{v_s - v_1}{1000} = 0$$

Apply KCL at node *B*.

$$i_f = i_1 = \frac{v_1}{5000}\text{ (A)} = \frac{v_s}{5000}\text{ (A)} = 0.2v_s\text{ (mA)}$$

The current i_1 in the 5-kΩ input resistor of the first op amp is provided by the output of the second op amp through the 40-kΩ feedback resistor. The current i_s drawn from v_s is, therefore, zero. The input resistance of the circuit is infinite.

5.11 INTEGRATOR AND DIFFERENTIATOR CIRCUITS

Integrator

By replacing the feedback resistor in the inverting amplifier of Fig. 5-13 with a capacitor, the basic *integrator circuit* shown in Fig. 5-23 will result.

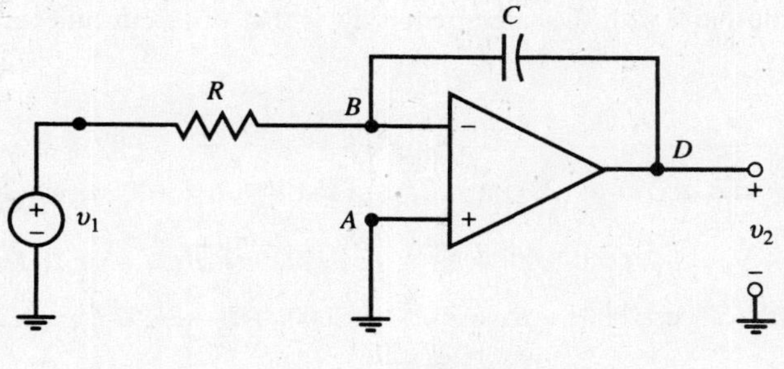

Fig. 5-23

To obtain the input-output relationship apply KCL at the inverting node:

$$\frac{v_1}{R} + C\frac{dv_2}{dt} = 0 \qquad \text{from which} \qquad \frac{dv_2}{dt} = -\frac{1}{RC}v_1$$

and

$$v_2 = -\frac{1}{RC}\int_{-\infty}^{t} v_1\,dt \qquad (20)$$

In other words, the output is equal to the integral of the input multiplied by a gain factor of $-1/RC$.

EXAMPLE 5.17 In Fig. 5-23 let $R = 1\,\text{k}\Omega$, $C = 1\,\mu\text{F}$, and $v_1 = \sin 2000t$. Assuming $v_2(0) = 0$, find v_2 for $t > 0$.

$$v_2 = -\frac{1}{10^3 \times 10^{-6}} \int_0^t \sin 2000t\, dt = 0.5(\cos 2000t - 1)$$

Leaky Integrator

The circuit of Fig. 5-24 is called a *leaky integrator*, as the capacitor voltage is continuously discharged through the feedback resistor R_f. This will result in a reduction in gain $|v_2/v_1|$ and a phase shift in v_2. For further discussion see Section 5.13.

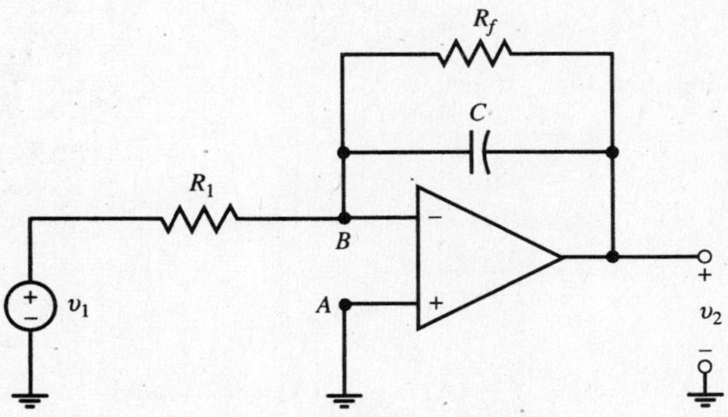

Fig. 5-24

EXAMPLE 5.18 In Fig. 5-24, $R_1 = R_f = 1\,\text{k}\Omega$, $C = 1\,\mu\text{F}$, and $v_1 = \sin 2000t$. Find v_2.

The inverting node is at zero voltage, and the sum of currents arriving at it is zero. Thus,

$$\frac{v_1}{R_1} + C\frac{dv_2}{dt} + \frac{v_2}{R_f} = 0 \qquad \text{or} \qquad v_1 + 10^{-3}\frac{dv_2}{dt} + v_2 = 0$$

$$10^{-3}\frac{dv_2}{dt} + v_2 = -\sin 2000t \tag{21}$$

The solution for v_2 in (*21*) is a sinusoidal with the same frequency as that of v_1 but different amplitude and phase angle, i.e.,

$$v_2 = A\cos(2000t + B) \tag{22}$$

To find A and B, we substitute v_2 and dv_2/dt in (*22*) into (*21*). First $dv/dt = -2000A\sin(2000t + B)$. Thus,

$$10^{-3}dv_2/dt + v_2 = -2A\sin(2000t + B) + A\cos(2000t + B) = -\sin 2000t$$

But

$$2A\sin(2000t + B) - A\cos(2000t + B) = A\sqrt{5}\sin(2000t + B - 26.57^\circ) = \sin 2000t$$

Therefore, $A = \sqrt{5}/5 = 0.447$, $B = 26.57^\circ$ and

$$v_2 = 0.447\cos(2000t + 26.57^\circ) \tag{23}$$

Integrator-Summer Amplifier

A single op amp in an inverting configuration with multiple input lines and a feedback capacitor as shown in Fig. 5-25 can produce the sum of integrals of several functions with desired gains.

EXAMPLE 5.19 Find the output v_o in the integrator-summer amplifier of Fig. 5-25, where the circuit has three inputs.

Apply KCL at the inverting input of the op amp to get

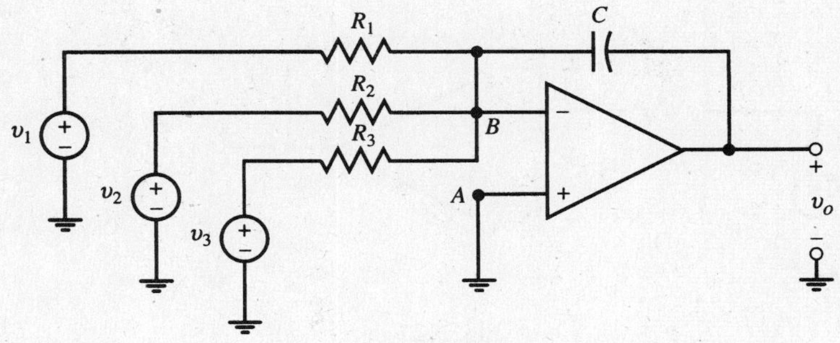

Fig. 5-25

$$\frac{v_1}{R_1} + \frac{v_2}{R_2} + \frac{v_3}{R_3} + C\frac{dv_o}{dt} = 0$$

$$v_o = -\int_{-\infty}^{t} \left(\frac{v_1}{R_1 C} + \frac{v_2}{R_2 C} + \frac{v_3}{R_3 C} \right) dt \tag{24}$$

Initial Condition of Integration

The desired initial condition, v_o, of the integration can be provided by a reset switch as shown in Fig. 5-26. By momentarily connecting the switch and then disconnecting it at $t = t_o$, an initial value of v_o is established across the capacitor and appears at the output v_2. For $t > t_o$, the weighted integral of input is added to the output.

$$v_2 = -\frac{1}{RC}\int_{t_o}^{t} v_1\,dt + v_o \tag{25}$$

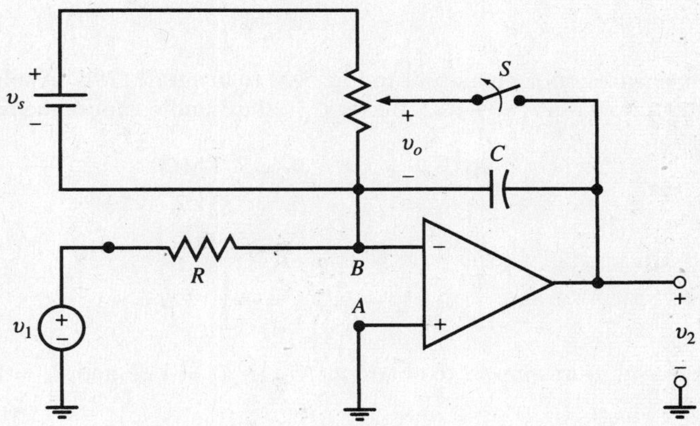

Fig. 5-26

Differentiator

By putting an inductor in place of the resistor in the feedback path of an inverting amplifier, the derivative of the input signal is produced at the output. Figure 5-27 shows the resulting *differentiator circuit*.

To obtain the input-output relationship, apply KCL to currents arriving at the inverting node:

$$\frac{v_1}{R} + \frac{1}{L}\int_{-\infty}^{t} v_2\,dt = 0 \qquad \text{or} \qquad v_2 = -\frac{L}{R}\frac{dv_1}{dt} \tag{26}$$

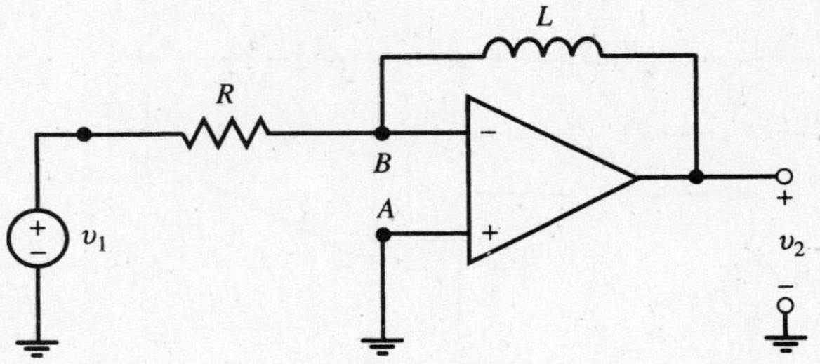

Fig. 5-27

5.12 ANALOG COMPUTERS

The inverting amplifiers, summing circuits, and integrators described in the previous sections are used as building blocks to form *analog computers* for solving linear differential equations. Differentiators are avoided because of considerable effect of noise despite its low level.

To design a computing circuit, first rearrange the differential equation such that the highest existing derivative of the desired variable is on one side of the equation. Add integrators and amplifiers in cascade and in nested loops as shown in the following examples. In this section we use the notations $x' = dx/dt$, $x'' = d^2x/dt^2$ and so on.

EXAMPLE 5.20 Design a circuit with $x(t)$ as input to generate output $y(t)$ which satisfies the following equation:

$$y''(t) + 2y'(t) + 3y(t) = x(t) \tag{27}$$

Step 1. Rearrange the differential equation (*27*) as follows:

$$y'' = x - 2y' - 3y \tag{28}$$

Step 2. Use the summer-integrator op amp #1 in Fig. 5-28 to integrate (*28*). Apply (*24*) to find R_1, R_2, R_3 and C_1 such that output of op amp #1 is $v_1 = -y'$. We let $C_1 = 1\ \mu\text{F}$ and compute the resistors accordingly:

$$\begin{aligned} R_1C_1 &= 1 & R_1 &= 1\ \text{M}\Omega \\ R_2C_1 &= 1/3 & R_2 &= 333\ \text{k}\Omega \\ R_3C_1 &= 1/2 & R_3 &= 500\ \text{k}\Omega \end{aligned}$$

$$v_1 = -\int (x - 3y - 2y')\,dt = -\int y''\,dt = -y' \tag{29}$$

Step 3. Integrate $v_1 = -y'$ by op amp #2 to obtain y. We let $C_2 = 1\ \mu\text{F}$ and $R_4 = 1\ \text{M}\Omega$ to obtain $v_2 = y$ at the output of op amp #2.

$$v_2 = -\frac{1}{R_4C_2}\int v_2\,dt = \int y'\,dt = y \tag{30}$$

Step 4. Supply inputs to op amp #1 through the following connections. Feed $v_1 = -y'$ directly back to the R_3 input of op amp #1. Pass $v_2 = y$ through the unity gain inverting op amp #3 to generate $-y$, and then feed it to the R_2 input of op amp #1. Connect the voltage source $x(t)$ to the R_1 input of op amp #1. The complete circuit is shown in Fig. 5-28.

EXAMPLE 5.21 Design an op amp circuit as an ideal voltage source $v(t)$ satisfying the equation $v' + v = 0$ for $t > 0$, with $v(0) = 1\ \text{V}$.

Following the steps used in Example 5.20, the circuit of Fig. 5-29 with $RC = 1\ \text{s}$ is assembled. The initial condition is entered when the switch is opened at $t = 0$. The solution $v(t) = e^{-t}$, $t > 0$, is observed at the output of the op amp.

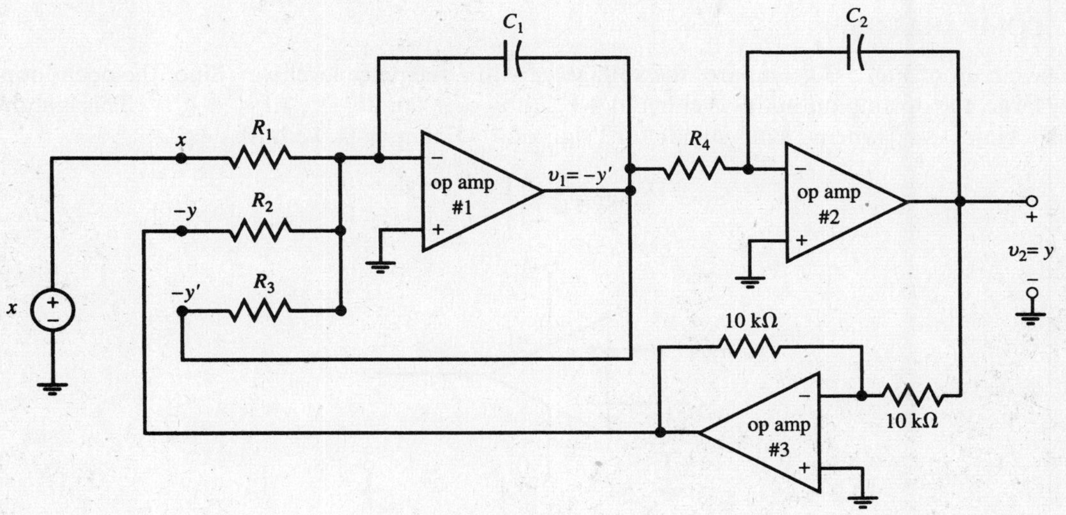

Fig. 5-28

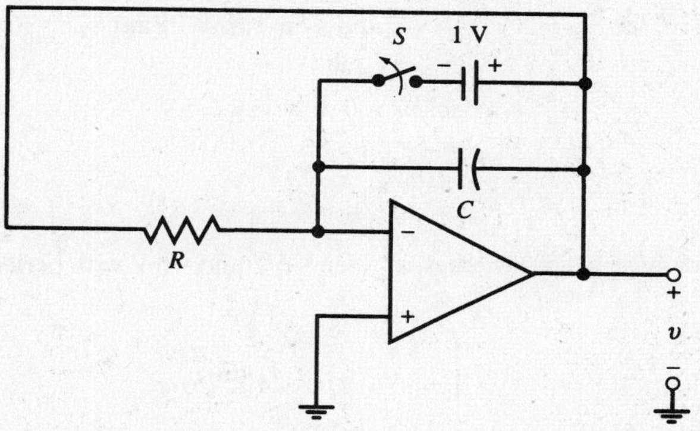

Fig. 5-29

5.13 LOW-PASS FILTER

A frequency-selective amplifier whose gain decreases from a finite value to zero as the frequency of the sinusoidal input increases from dc to infinity is called a *low-pass filter*. The plot of gain versus frequency is called a *frequency response*. An easy technique for finding the frequency response of filters will be developed in Chapter 13. The leaky integrator of Fig. 5-24 is a low-pass filter, as illustrated in the following example.

EXAMPLE 5.22 In Example 5.18 let $v_1 = \sin \omega t$. Find $|v_2|$ for $\omega = 0, 10, 100, 10^3, 10^4$, and 10^5 rad/s.

By repeating the procedure of Example 5.18, the frequency response is found and given in Table 5-1. The response amplitude decreases with frequency. The circuit is a low-pass filter.

Table 5-1. Frequency Response of the Low-pass Filter

ω, rad/s	0	10	100	10^3	10^4	10^5
f, Hz	0	1.59	15.9	159	1.59×10^3	15.9×10^3
$\lvert v_2/v_1 \rvert$	1	1	0.995	0.707	0.1	0.01

5.14 COMPARATOR

The circuit of Fig. 5-30 compares the voltage v_1 with a reference level v_o. Since the open-loop gain is very large, the op amp output v_2 is either at $+V_{cc}$ (if $v_1 > v_o$) or at $-V_{cc}$ (if $v_1 < v_o$). This is shown by $v_2 = V_{cc}\,\text{sgn}[v_1 - v_o]$ where "sgn" stands for "sign of." For $v_o = 0$, we have

$$v_2 = V_{cc}\,\text{sgn}[v_1] = \begin{cases} +V_{cc} & v_1 > 0 \\ -V_{cc} & v_1 < 0 \end{cases}$$

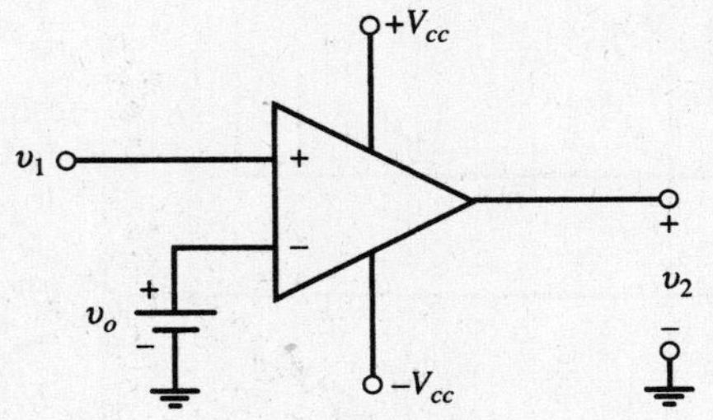

Fig. 5-30

EXAMPLE 5.23 In Fig. 5-30, let $V_{cc} = 5\text{ V}$, $v_o = 0$, and $v_1 = \sin \omega t$. Find v_2.

For $0 < t < \pi/\omega$,

$$v_1 = \sin \omega t > 0 \qquad v_2 = 5\text{ V}$$

For $\pi/\omega < t < 2\pi/\omega$,

$$v_1 = \sin \omega t < 0 \qquad v_2 = -5\text{ V}$$

The output v_2 is a square pulse which switches between $+5\text{ V}$ and -5 V with period of $2\pi/\omega$. One cycle of v_2 is given by

$$v_2 = \begin{cases} 5\text{ V} & 0 < t < \pi/\omega \\ -5\text{ V} & \pi/\omega < t < 2\pi/\omega \end{cases}$$

EXAMPLE 5.24 The circuit of Fig. 5-31 is a parallel analog-to-digital converter. The $+V_{cc}$ and $-V_{cc}$ connections are omitted for simplicity. Let $V_{cc} = 5\text{ V}$, $v_o = 4\text{ V}$, and $v_i = t$ (V) for $0 < t < 4\text{ s}$. Find outputs v_3, v_2, and v_1. Interpret the answer.

The op amps have no feedback, and they function as comparators. The outputs with values at +5 or −5 V are given in Table 5-2.

Table 5-2

time, s	input, V	outputs, V
$0 < t < 1$	$0 < v_i < 1$	$v_3 = -5 \quad v_2 = -5 \quad v_1 = -5$
$1 < t < 2$	$1 < v_i < 2$	$v_3 = -5 \quad v_2 = -5 \quad v_1 = +5$
$2 < t < 3$	$2 < v_i < 3$	$v_3 = -5 \quad v_2 = +5 \quad v_1 = +5$
$3 < t < 4$	$3 < v_i < 4$	$v_3 = +5 \quad v_2 = +5 \quad v_1 = +5$

The binary sequences $\{v_3, v_2, v_1\}$ in Table 5-2 uniquely specify the input voltage in discrete domain. However, in their present form they are not the binary numbers representing input amplitudes. Yet, by using a coder we could transform the above sequences into the binary numbers corresponding to the values of analog inputs.

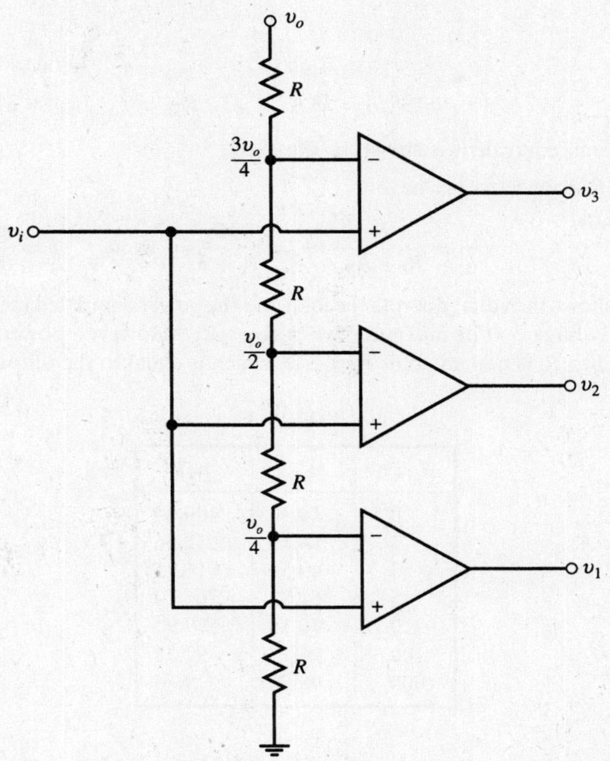

Fig. 5-31

Solved Problems

5.1 In Fig. 5-3, let $v_s = 20\,\text{V}$, $R_s = 10\,\Omega$, $R_i = 990\,\Omega$, $k = 5$, and $R_o = 3\,\Omega$. Find (*a*) the Thévenin equivalent of the circuit seen by R_l and (*b*) v_2 and the power dissipated in R_l for $R_l = 0.5$, 1, 3, 5, 10, 100, and $1000\,\Omega$.

(*a*) The open-circuit voltage and short-circuit current at *A–B* terminal are $v_{\text{o.c.}} = 5v_1$ and $i_{\text{s.c.}} = 5v_1/3$, respectively.

We find v_1 by dividing v_s between R_s and R_i. Thus,

$$v_1 = \frac{R_i}{R_s + R_i}\,v_s = \frac{990}{10 + 990}\,(20) = 19.8\,\text{V}$$

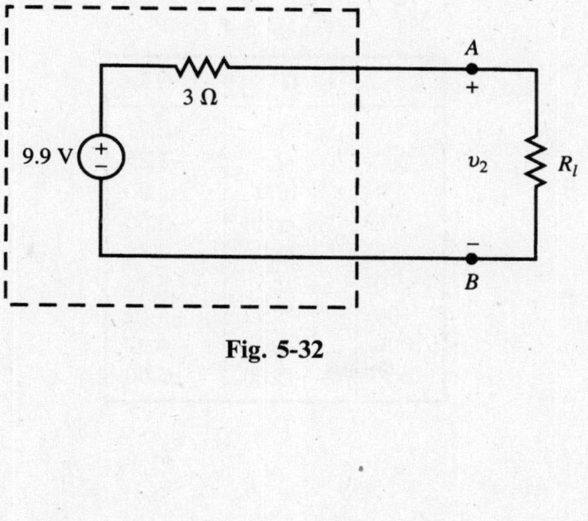

Fig. 5-32

Therefore,

$$v_{\text{o.c.}} = 5(19.8) = 99\,\text{V} \qquad v_{\text{Th}} = v_{\text{o.c.}} = 99\,\text{V}$$
$$i_{\text{s.c.}} = 99/3 = 33\,\text{A} \qquad R_{\text{Th}} = v_{\text{o.c.}}/i_{\text{s.c.}} = 3\,\Omega$$

The Thévenin equivalent is shown in Fig. 5-32.

(*b*) With the load R_l connected, we have

$$v_2 = \frac{R_l}{R_l + R_{\text{Th}}} v_{\text{Th}} = \frac{99 R_l}{R_l + 3} \quad \text{and} \quad p = \frac{v_2^2}{R_l}$$

Table 5-3 shows the voltage across the load and the power dissipated in it for the given seven values of R_l. The load voltage is at its maximum when $R_l = \infty$. However, power delivered to $R_l = \infty$ is zero. Power delivered to R_l is maximum at $R_l = 3\,\Omega$, which is equal to the output resistance of the amplifier.

Table 5-3

R_l, Ω	v_2, V	p, W
0.5	14.14	400.04
1	24.75	612.56
3	49.50	816.75
5	61.88	765.70
10	76.15	579.94
100	96.12	92.38
1000	98.70	9.74

5.2 In the circuits of Figs. 5-4 and 5-5 let $R_1 = 1\,\text{k}\Omega$ and $R_2 = 5\,\text{k}\Omega$. Find the gains $G^+ = v_2/v_s$ in Fig. 5-4 and $G^- = v_2/v_s$ in Fig. 5-5 for $k = 1, 2, 4, 6, 8, 10, 100, 1000$, and ∞. Compare the results.

From (*5*) in Example 5.3, at $R_1 = 1\,\text{k}\Omega$ and $R_2 = 5\,\text{k}\Omega$ we have

$$G^+ = \frac{v_2}{v_s} = \frac{5k}{6 - k} \qquad (31)$$

In Example 5.4 we found

$$G^- = \frac{v_2}{v_s} = -\frac{5k}{6 + k} \qquad (32)$$

The gains G^- and G^+ are calculated for nine values of k in Table 5-4. As k becomes very large, G^+ and G^- approach the limit gain of -5, which is the negative of the ratio R_2/R_1 and is independent of k. The circuit of Fig. 5-5 (with negative feedback) is always stable and its gain monotonically approaches the limit gain. However, the circuit of Fig. 5-4 (with positive feedback) is unstable. The gain G^+ becomes very large as k approaches six. At $k = 6$, $G^+ = \infty$.

Table 5-4

k	G^+	G^-
1	1.0	−0.71
2	2.5	−1.25
4	10.0	−2.00
6	∞	−2.50
8	−20.0	−2.86
10	−12.5	−3.12
100	−5.32	−4.72
1000	−5.03	−4.97
∞	−5.00	−5.00

5.3 Let $R_1 = 1\,\text{k}\Omega$, $R_2 = 5\,\text{k}\Omega$, and $R_i = 50\,\text{k}\Omega$ in the circuit of Fig. 5-33. Find v_2/v_s for $k = 1$, 10, 100, 1000, ∞ and compare the results with the values of G^- in Table 5-4.

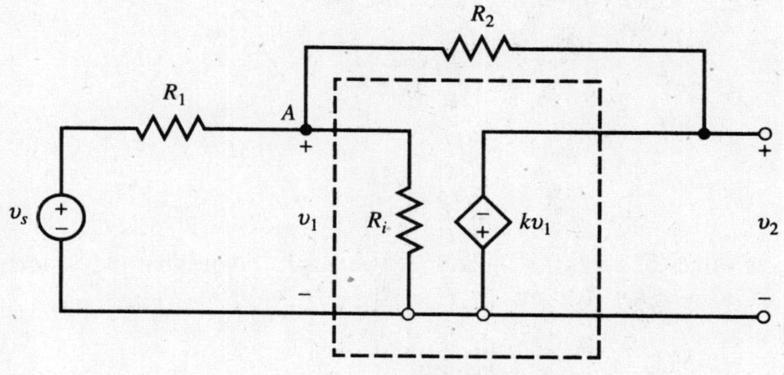

Fig. 5-33

This problem is solved by application of KCL at node A (another approach which uses the Thévenin equivalent is suggested in Problem 5.30). Thus,

$$\frac{v_1 - v_s}{1} + \frac{v_1 - v_2}{5} + \frac{v_1}{50} = 0 \qquad (33)$$

From the amplifier we obtain

$$v_2 = -kv_1 \qquad \text{or} \qquad v_1 = -v_2/k \qquad (34)$$

Replacing v_1 in (34) into (33) and rearranging terms, we obtain

$$\frac{v_2}{v_s} = \frac{-50k}{61 + 10k} = \frac{-5k}{6.1 + k} \qquad (35)$$

Values of v_2/v_s in (35) are shown in Table 5-5 as functions of k. The 50-kΩ input resistance of the amplifier reduces the overall gain very slightly, as seen by comparing Tables 5-4 and 5-5. The feedback has made the input resistance of the amplifier less effective in changing the overall gain.

Table 5-5

k	v_2/v_s
1	−0.704
10	−3.106
100	−4.713
1000	−4.97
∞	−5.00

5.4 Let again $R_1 = 1\,\text{k}\Omega$ and $R_2 = 5\,\text{k}\Omega$ in the circuit of Fig. 5-33.

(*a*) Find v_2/v_s as a function of k and R_i.
(*b*) Let $R_i = 1\,\text{k}\Omega$. Find v_2/v_1 for $k = 1, 10, 100, 1000, \infty$. Repeat for $R_i = \infty$.
(*c*) Discuss the effects of R_i and k on the overall gain. Show that, for $k = \infty$ and $R_i \neq 0$, the gain of the amplifier is independent of R_i and is equal to $-R_2/R_1$.

(*a*) Apply KCL to currents leaving node A to obtain

$$\frac{v_1 - v_s}{1} + \frac{v_1 - v_2}{5} + \frac{v_1}{R_i} = 0$$

From the amplifier we get $v_2 = -kv_1$ or $v_1 = -v_2/k$. Substituting for v_1 in the KCL equation and rearranging terms we get

$$\frac{v_2}{v_s} = -5\,\frac{ck}{1+ck} \qquad \text{where } c = \frac{R_i}{5+6R_i} \tag{36}$$

(*b*) For $R_i = 1\,\text{k}\Omega$, $c = 1/11$ which, substituted into (*36*), gives

$$\frac{v_2}{v_s} = \frac{-5k}{11+k} \tag{37}$$

For $R_i = \infty$ we get $c = 1/6$ and so

$$\frac{v_2}{v_s} = \frac{-5k}{6+k} \tag{38}$$

Table 5-6 gives values of v_2/v_s in (*37*) and (*38*) versus k. Note that (*38*) is identical with (*32*).

Table 5-6

	v_2/v_s	
k	$R_i = 1\,\text{k}\Omega$	$R_i = \infty$
1	−0.31	−0.71
10	−2.38	−3.12
100	−4.51	−4.72
1000	−4.95	−4.97
∞	−5.00	−5.00

(*c*) Comparing the two columns in Table 5-6 we see that the smaller R_i reduces the overall gain G^-. However, as the open-loop gain k increases, the effect of R_i is diminished. As k becomes very large, v_2/v_1 approaches -5 unless $R_i = 0$.

5.5 Let again $R_1 = 1\,k\Omega$ and $R_2 = 5\,k\Omega$ in the circuit of Fig. 5-33. Replace the circuit to the left of node A including v_s, R_1, and R_i by its Thévenin equivalent. Then use (*5*) to derive (*36*).

The Thévenin equivalent is given by

$$v_{\text{Th}} = \frac{R_i v_s}{R_1 + R_i} = \frac{R_i v_s}{1 + R_i}$$

$$R_{\text{Th}} = \frac{R_1 R_i}{R_1 + R_i} = \frac{R_i}{1 + R_i}$$

where the resistors are in kΩ.

From (*5*),

$$v_2 = (1-b)\,\frac{-k}{1+bk}\,v_{\text{Th}}$$

where
$$b = \frac{R_{\text{Th}}}{R_{\text{Th}} + R_2} = \frac{R_i}{6R_i + 5} \qquad \text{and} \qquad 1 - b = \frac{5(1+R_i)}{6R_i + 5}$$

Therefore,

$$v_2 = \frac{5(1+R_i)}{6R_i+5} \times \frac{-k}{1 + R_i k/(6R_i+5)} \times \frac{R_i}{1+R_i}\,v_s = \frac{-5R_i k}{6R_i + 5 + R_i k}\,v_s$$

which is identical with (*36*).

5.6 Find the output voltage of an op amp with $A = 10^5$ and $V_{cc} = 10\,\text{V}$ for $v^- = 0$ and $v^+ = \sin t$ (V). Refer to Figs. 5-7 and 5-8.

Because of high gain, saturation occurs quickly at

$$|v_2| = 10^5|v_d| = 10\text{ V} \quad \text{or} \quad |v_d| = 10^{-4}\text{ V}$$

We may ignore the linear interval and write

$$v_2 = \begin{cases} +10\text{ V} & v_d > 0 \\ -10\text{ V} & v_d < 0 \end{cases}$$

where $v_d = v^+ - v^- = \sin t$ (V). One cycle of the output is given by

$$v_2 = \begin{cases} +10\text{ V} & 0 < t < \pi \\ -10\text{ V} & \pi < t < 2\pi \end{cases}$$

For a more exact v_2, we use the transfer characteristic of the op amp in Fig. 5-7.

$$v_2 = \begin{cases} -10 & v_d < -10^{-4}\text{ V} \\ 10^5 v_d & -10^{-4} < v_d < 10^{-4}\text{ V} \\ +10 & v_d > 10^{-4}\text{ V} \end{cases}$$

Saturation begins at $|v_d| = |\sin t| = 10^{-4}$ V. Since this is a very small range, we may replace $\sin t$ by t. The output v_2 is then given by

$$\begin{aligned} v_2 &= 10^5 t & -10^{-4} &< t < 10^{-4}\text{ s} \\ v_2 &= 10 & 10^{-4} &< t < \pi - 10^{-4}\text{ s} \\ v_2 &= -10^5(t-\pi) & \pi - 10^{-4} &< t < \pi + 10^{-4}\text{ s} \\ v_2 &= -10 & \pi + 10^{-4} &< t < 2\pi - 10^{-4}\text{ s} \end{aligned}$$

To appreciate the insignificance of error in ignoring the linear range, note that during one period of 2π s the interval of linear operation is only 4×10^{-4} s, which gives a ratio of 64×10^{-6}.

5.7 Repeat Problem 5.6 for $v^+ = \sin 2\pi t$ (V) and $v^- = 0.5$ V.

The output voltage is

$$\begin{aligned} v_2 &= 10\text{ V} & &\text{when } v^+ > v^- \\ v_2 &= -10\text{ V} & &\text{when } v^+ < v^- \end{aligned}$$

Switching occurs when $\sin 2\pi t = 1/2$. This happens at $t = 1/12$, $5/12$, $13/12$, and so on. Therefore, one cycle of v_2 is given by

$$\begin{aligned} v_2 &= 10\text{ V} & 1/12 &< t < 5/12\text{ s} \\ v_2 &= -10\text{ V} & 5/12 &< t < 13/12\text{ s} \end{aligned}$$

Figure 5-34 shows the graphs of v^+, v^-, and v_2.

5.8 In the circuit of Fig. 5-35 $v_s = \sin 100t$. Find v_1 and v_2.

At nodes B and A, $v_B = v_A = 0$. Then,

$$v_1 = \frac{30}{20+30} v_s = 0.6 \sin 100t \text{ (V)}$$

$$v_2 = -\frac{100}{30} v_1 = -\frac{100}{30}(0.6 \sin 100t) = -2 \sin 100t \text{ (V)}$$

Alternatively, $$v_2 = -\frac{100}{20+30} v_s = -2 \sin 100t \text{ (V)}$$

5.9 Saturation levels for the op amps in Fig. 5-31 are $+V_{cc} = 5$ V and $-V_{cc} = -5$ V. The reference voltage is $v_o = 1$ V. Find the sequence of outputs corresponding to values of v_i from 0 to 1 V in steps of 0.25 V.

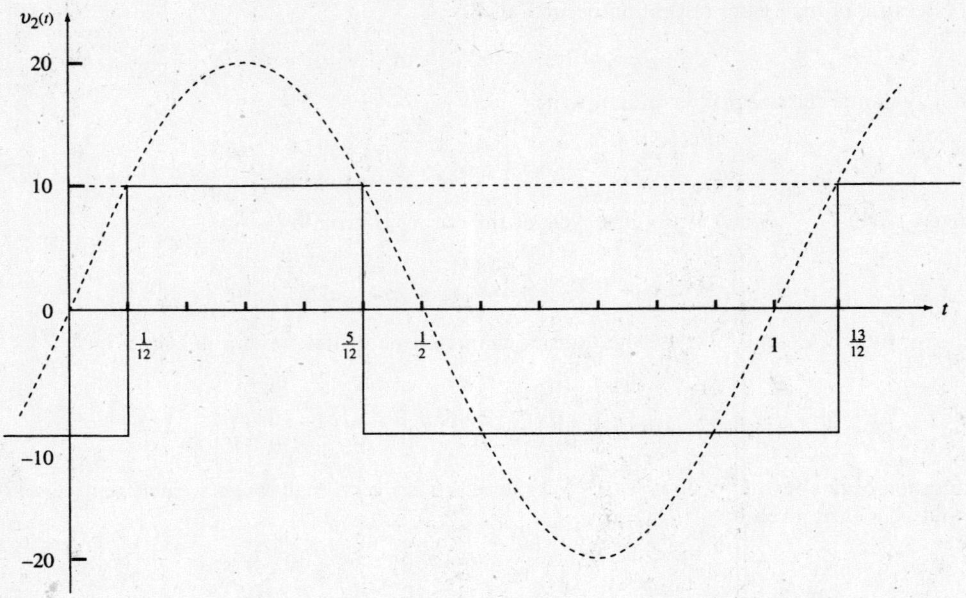

Fig. 5-34

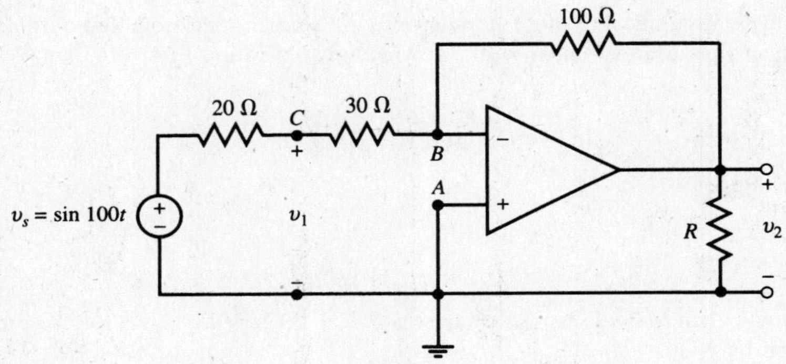

Fig. 5-35

See Table 5-7 where $L = -5$ V and $H = +5$ V.

Table 5-7

v_i, V	v_3	v_2	v_1
0 to 0.25^-	L	L	L
0.25^+ to 0.5^-	L	L	H
0.5^+ to 0.75^-	L	H	H
0.75^+ to 1	H	H	H

5.10 Find v in the circuit of Fig. 5-36.

Apply KCL at node A,

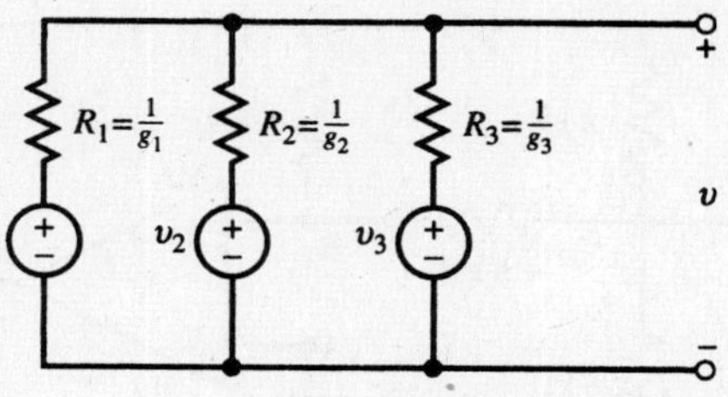

Fig. 5-36

$$(v - v_1)g_1 + (v - v_2)g_2 + (v - v_3)g_3 = 0$$

Then
$$v = \frac{v_1 g_1 + v_2 g_2 + v_3 g_3}{g_1 + g_2 + g_3} = \frac{v_1 R_2 R_3 + v_2 R_1 R_3 + v_3 R_2 R_1}{R_1 R_2 + R_2 R_3 + R_3 R_1}$$

5.11 In the circuit of Fig. 5-37 find v_C (the voltage at node C), i_1, R_{in} (the input resistance seen by the 9-V source), v_2, and i_2.

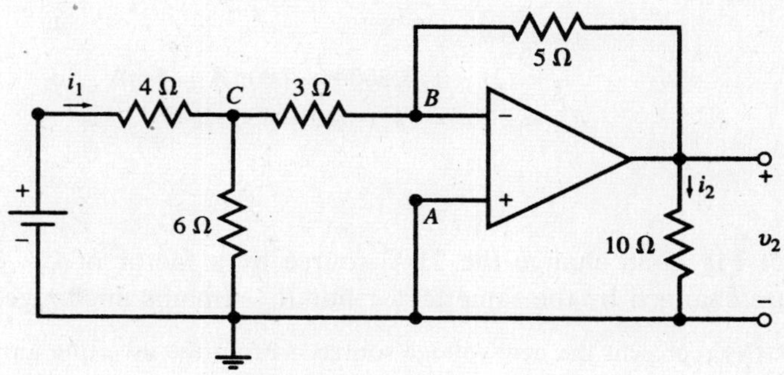

Fig. 5-37

At nodes B and A, $v_B = v_A = 0$. Applying KCL at node C, we get

$$(v_C - 9)/4 + v_C/6 + v_C/3 = 0 \qquad \text{from which} \qquad v_C = 3\text{ V}$$

Then
$$i_1 = (9 - v_C)/4 = 1.5\text{ A} \qquad \text{and} \qquad R_{in} = v_1/i_1 = 9/1.5 = 6\ \Omega$$

From the inverting amplifier circuit we have

$$v_2 = -(5/3)v_C = -5\text{ V} \qquad \text{and} \qquad i_2 = -5/10 = -0.5\text{ A}$$

5.12 Find v_2 in Problem 5.11 by replacing the circuit to the left of nodes A-B in Fig. 5-37 by its Thévenin equivalent.

$$R_{Th} = 3 + \frac{(6)(4)}{6+4} = 5.4\ \Omega \qquad \text{and} \qquad v_{Th} = \frac{6}{4+6}(9) = 5.4\text{ V}$$

Then $v_2 = -(5/5.4)(5.4) = -5$ V.

5.13 Find v_C, i_1, v_2, and R_{in}, the input resistance seen by the 21-V source in Fig. 5-38.

From the inverting amplifier we get

$$v_2 = -(5/3)v_C \tag{39}$$

Note that $v_B = v_A = 0$ and so KCL at node C results in

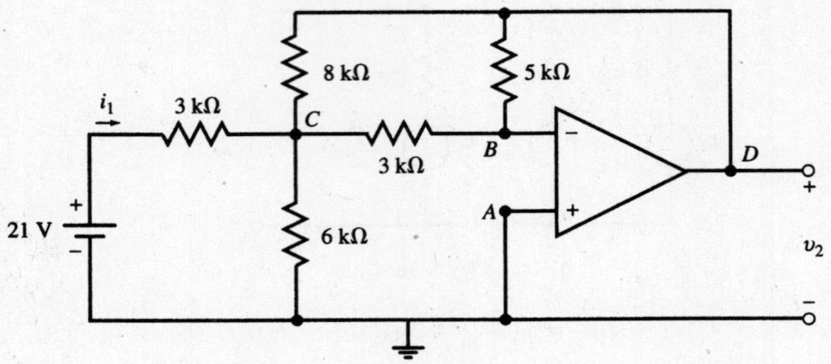

Fig. 5-38

$$\frac{v_C - 21}{3} + \frac{v_C}{6} + \frac{v_C}{3} + \frac{v_C - v_2}{8} = 0 \tag{40}$$

Substituting $v_C = -(3/5)v_2$ from (*39*) into (*40*) we get $v_2 = -10\,\text{V}$. Then

$$v_C = 6\,\text{V}$$
$$i_1 = (21 - v_C)/3000 = 0.005\,\text{A} = 5\,\text{mA}$$
$$R_{\text{in}} = 21/i_1 = 21/0.005 = 4200\,\Omega = 4.2\,\text{k}\Omega$$

5.14 In the circuit of Fig. 5-38 change the 21-V source by a factor of k. Show that v_C, i_1, v_2 in Problem 5.13 are changed by the same factor but R_{in} remains unchanged.

Let $v_s = 21k$ (V) represent the new voltage source. From the inverting amplifier we have [see (*39*)]

$$v_2 = -(5/3)v_C$$

Apply KCL at node C to obtain [see (*40*)]

$$\frac{v_C - v_s}{3} + \frac{v_C}{6} + \frac{v_C}{3} + \frac{v_C - v_2}{8} = 0$$

Solving for v_C and v_2, we have

$$v_C = (6/21)v_s = 6k\ (\text{V}) \qquad \text{and} \qquad v_2 = -(10/21)v_s = -10k\ (\text{V})$$
$$i_1 = (v_s - v_C)/3000 = (21 - 6)k/3000 = 0.005k\ \text{A}$$
$$R_{\text{in}} = v_s/i_1 = 21k/0.005k = 4200\,\Omega$$

These results are expected since the circuit is linear.

5.15 Find v_2 and v_C in Problem 5.13 by replacing the circuit to the left of node C in Fig. 5-38 (including the 21-V battery and the 3-kΩ and 6-kΩ resistors) by its Thévenin equivalent.

We first compute the Thévenin equivalent:

$$R_{\text{Th}} = \frac{(6)(3)}{6+3} = 2\,\text{k}\Omega \qquad \text{and} \qquad v_{\text{Th}} = \frac{6}{3+6}(21) = 14\,\text{V}$$

Replace the circuit to the left of node C by the above v_{Th} and R_{Th} and then apply KCL at C:

$$\frac{v_C - 14}{2} + \frac{v_C}{3} + \frac{v_C - v_2}{8} = 0 \tag{41}$$

For the inverting amplifier we have $v_2 = -(5/3)v_C$ or $v_C = -0.6\,v_2$, which results, after substitution in (*41*), in $v_2 = -10\,\text{V}$ and $v_C = 6\,\text{V}$.

5.16 (*a*) Find the Thévenin equivalent of the circuit to the left of nodes *A-B* in Fig. 5-39(*a*) and then find v_2 for $R_l = 1\,\text{k}\Omega$, $10\,\text{k}\Omega$, and ∞. (*b*) Repeat for Fig. 5-39(*c*) and compare with part (*a*).

(*a*) The Thévenin equivalent of the circuit in Fig. 5-39(*a*) is shown in Fig. 5-39(*b*).

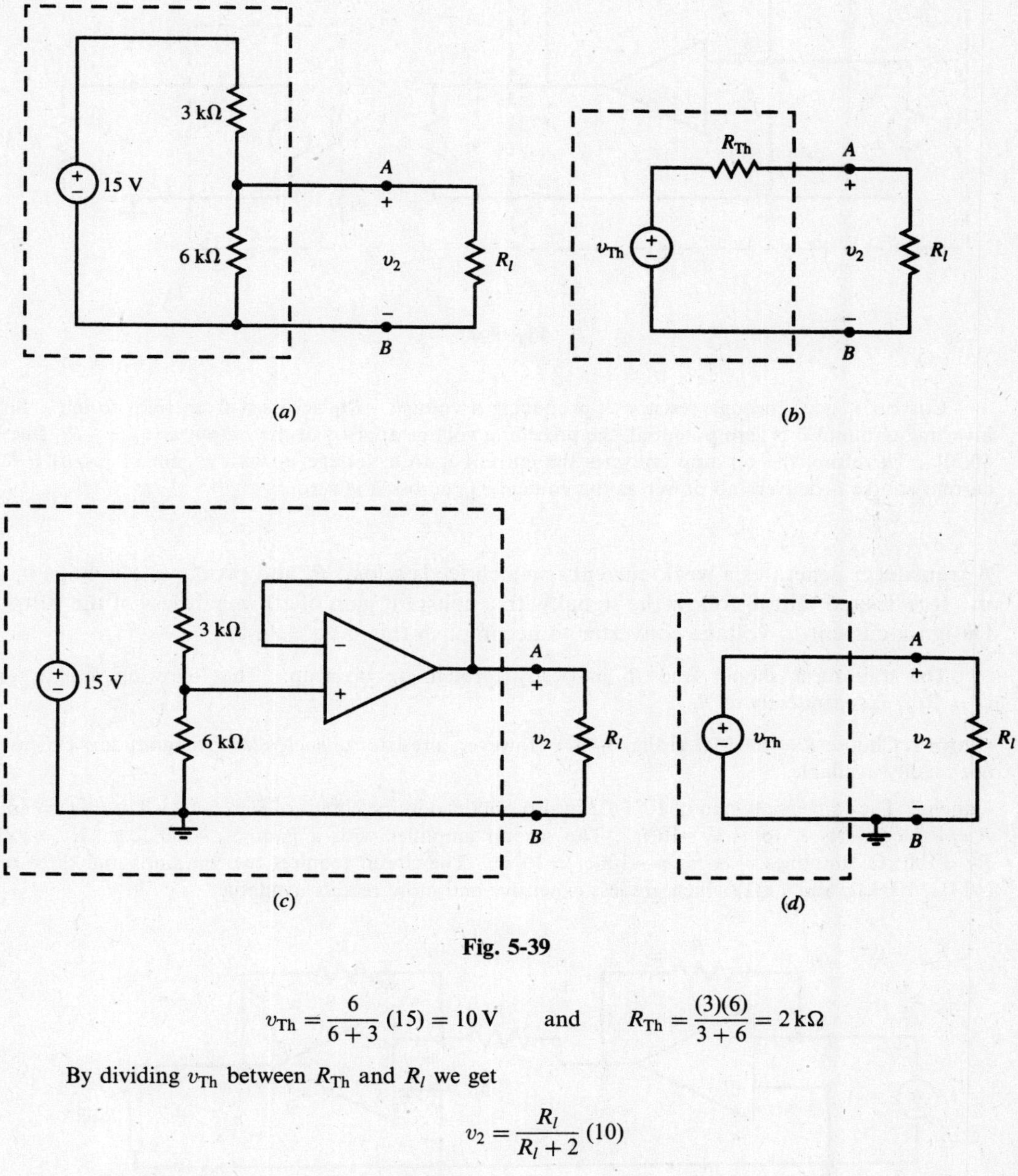

Fig. 5-39

$$v_{Th} = \frac{6}{6+3}(15) = 10\text{ V} \quad \text{and} \quad R_{Th} = \frac{(3)(6)}{3+6} = 2\,\text{k}\Omega$$

By dividing v_{Th} between R_{Th} and R_l we get

$$v_2 = \frac{R_l}{R_l + 2}(10)$$

For $R_l = 1\,\text{k}\Omega$, $v_2 = 3.33\text{ V}$
For $R_l = 10\,\text{k}\Omega$, $v_2 = 8.33\text{ V}$
For $R_l = \infty$ $v_2 = 10\text{ V}$

The output v_2 depends on R_l. The operation of the voltage divider is also affected by R_l.

(*b*) The Thévenin equivalent of the circuit in Fig. 5-39(*c*) is shown in Fig. 5-12(*d*). Here we have

$$v_{Th} = 10\text{ V} \quad \text{and} \quad R_{Th} = 0$$

and $v_2 = v_{Th} = 10\text{ V}$ for all values of R_l, that is, the output v_2 depends on R_1, R_2, and v_s only and is independent of R_l.

5.17 Find v_2 as a function of i_1 in the circuit of Fig. 5-40(a).

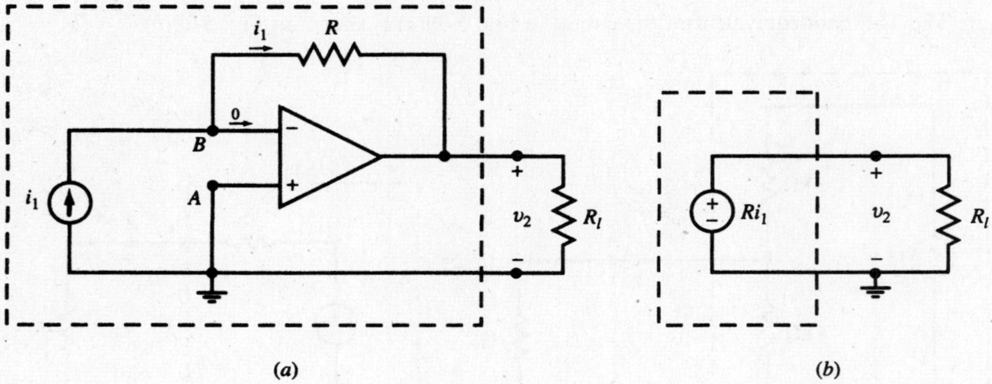

Fig. 5-40

Current i_1 goes through resistor R producing a voltage $-Ri_1$ across it from right to left. Since the inverting terminal B is zero potential, the preceding voltage appears at the output as $v_2 = -Ri_1$ [see Fig. 5-40(b)]. Therefore, the op amp converts the current i_1 to a voltage v_2 with a gain of $|v_2/i_1| = R$. The current source i_1 delivers no power as the voltage v_{AB} across it is zero.

5.18 A transducer generates a weak current i_1 which feeds a load R_l and produces a voltage v_1 across it. It is desired that v_1 follow the signal with a constant gain of 10^8 regardless of the value of R_l. Design a current-to-voltage converter to accomplish this task.

The transducer should feed R_l indirectly through an op amp. The following designs produce $v_1 = 10^8 i_1$ independently of R_l.

Design 1: Choose $R = 100\,\text{M}\Omega$ in Fig. 5-40. However, a resistor of such a large magnitude is expensive and not readily available.

Design 2: The conversion gain of $10^8\,V/A$ is also obtained in the circuit of Fig. 5-41. The first op amp with $R = 10^6$ converts i_1 to $v_1 = -10^6 i_1$. The second amplifier with a gain of -100 (e.g., $R_1 = 1\,\text{k}\Omega$ and $R_2 = 100\,\text{k}\Omega$) amplifies v_1 to $v_2 = -100v_1 = 10^8 i_1$. The circuit requires two op amps and three resistors ($1\,\text{M}\Omega$, $100\,\text{k}\Omega$, and $1\,\text{k}\Omega$) which are less expensive and more readily available.

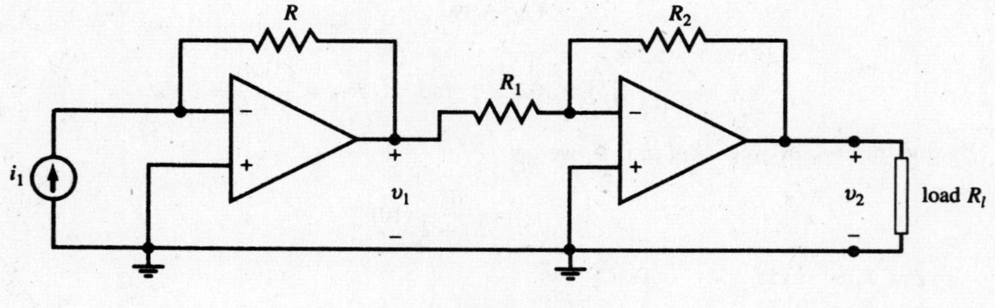

Fig. 5-41

Design 3: See Fig. 5-42 and Problem 5.19.

5.19 Determine the resistor values which would produce a current-to-voltage conversion gain of $v_2/i_1 = 10^8$ V/A in the circuit of Fig. 5-42.

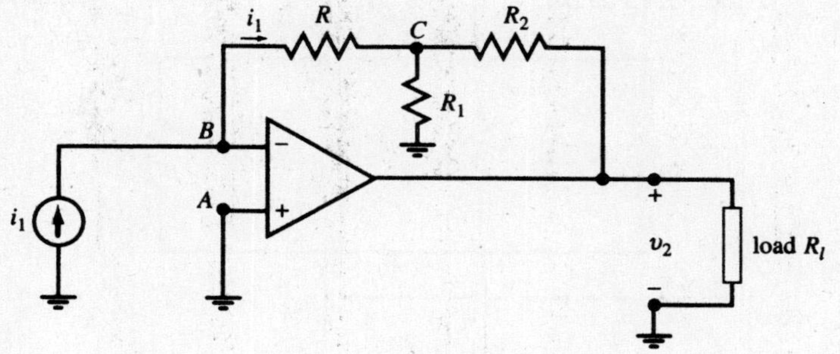

Fig. 5-42

Apply KCL at node C. Note that $v_B = v_A = 0$. Thus,

$$\frac{v_C}{R} + \frac{v_C}{R_1} + \frac{v_C - v_2}{R_2} = 0$$

Substituting $v_C = -Ri_1$ and solving for v_2 we get

$$v_2 = -R_{eq} i_1 \qquad \text{where } R_{eq} = R\left(1 + \frac{R_2}{R_1} + \frac{R_2}{R}\right)$$

For a conversion gain of $v_2/i_1 = R_{eq} = 10^8$ V/A $= 100$ MΩ, we need to find resistor values to satisfy the following equation:

$$R\left(1 + \frac{R_2}{R_1} + \frac{R_2}{R}\right) = 10^8\ \Omega$$

One solution is to choose $R = 1$ MΩ, $R_1 = 1$ kΩ, and $R_2 = 99$ kΩ. The design of Fig. 5-42 uses a single op amp and three resistors which are not expensive and are readily available.

5.20 Find i_2 as a function of v_1 in the circuit of Fig. 5-43.

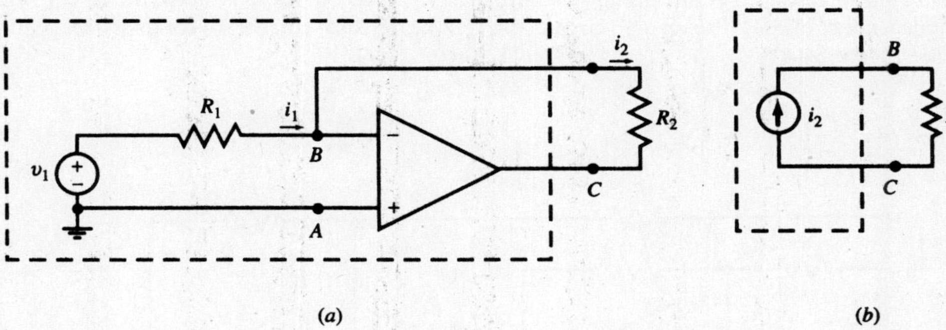

Fig. 5-43

We have

$$v_B = v_A = 0 \qquad i_1 = v_1/R_1 \qquad i_2 = i_1 = v_1/R_1$$

The op amp converts the voltage source to a floating current source. The voltage-to-current conversion ratio is R_1 and is independent of R_2.

5.21 A practical current source (i_s in parallel with internal resistance R_s) directly feeds a load R_l as in Fig. 5-44(a). (a) Find load current i_l. (b) Place an op amp between the source and the load as in Fig. 5-44(b). Find i_l and compare with part (a).

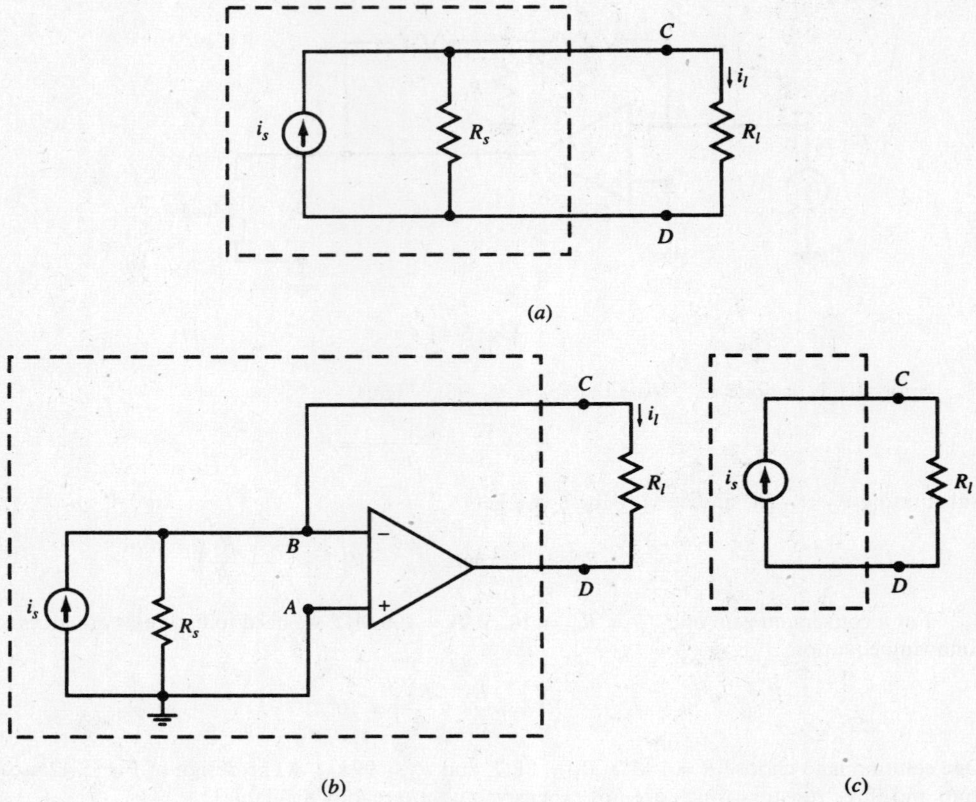

Fig. 5-44

(*a*) In the direct connection, Fig. 5-44(*a*), $i_l = i_s R_s/(R_s + R_l)$, which varies with R_l. (*b*) In Fig. 5-44(*b*), the op amp forces v_B to zero causing the current in R_s to become zero. Therefore, $i_l = i_s$ which is now independent of R_l. The op amp circuit converts the practical current source to an ideal current source. See Figure 5-44(*c*).

5.22 Find v_o in the circuit of Fig. 5-45.

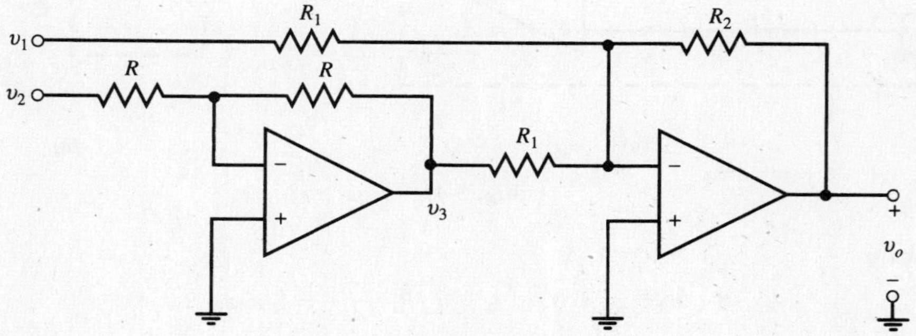

Fig. 5-45

The first op amp is a unity gain inverter with $v_3 = -v_2$. The second op amp is a summing circuit with a gain of $-R_2/R_1$ for both inputs v_1 and v_3. The output is

$$v_o = -\frac{R_2}{R_1}(v_1 + v_3) = \frac{R_2}{R_1}(v_2 - v_1)$$

The circuit is a difference amplifier.

5.23 Find v_o in the circuit of Fig. 5-46.

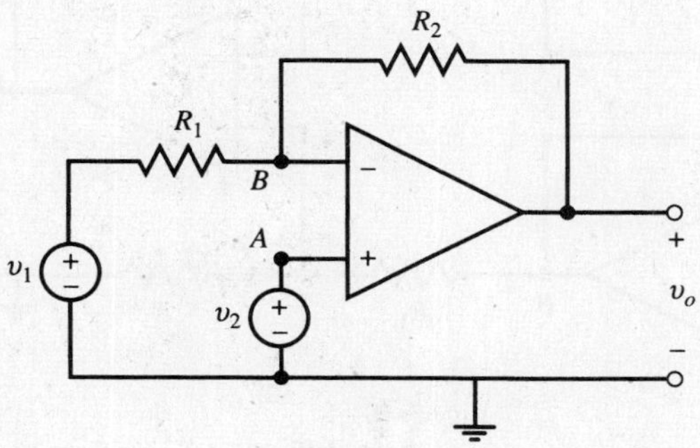

Fig. 5-46

Apply KCL at node B. Note that $v_B = v_A = v_2$. Thus,

$$\frac{v_2 - v_1}{R_1} + \frac{v_2 - v_o}{R_2} = 0$$

Solving for v_o, we get $v_o = v_2 + (R_2/R_1)(v_2 - v_1)$.

5.24 Find v_o in the circuit of Fig. 5-47.

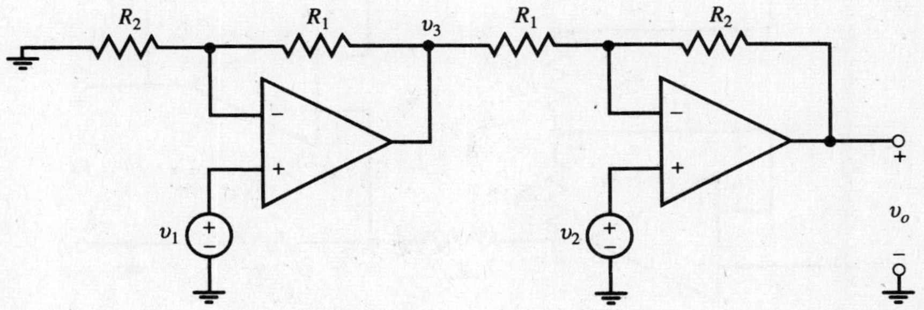

Fig. 5-47

The left part of the circuit has a gain of $(1 + R_1/R_2)$. Therefore, $v_3 = (1 + R_1/R_2)v_1$. Using results of Problem 5.23 and substituting for v_3 results in

$$v_o = v_2 + \frac{R_2}{R_1}(v_2 - v_3) = \left(1 + \frac{R_2}{R_1}\right)v_2 - \frac{R_2}{R_1}\left(1 + \frac{R_1}{R_2}\right)v_1 = \left(1 + \frac{R_2}{R_1}\right)(v_2 - v_1)$$

5.25 In Fig. 5-48 choose resistors for a differential gain of 10^6 so that $v_o = 10^6(v_2 - v_1)$.

The two frontal op amps are voltage followers.

$$v_A = v_1 \quad \text{and} \quad v_B = v_2$$

From *(16)*, Sec. 5.9, we have

$$v_o = \frac{R_2}{R_1}(v_B - v_A) = \frac{R_2}{R_1}(v_2 - v_1)$$

To obtain the required differential gain of $R_2/R_1 = 10^6$, choose $R_1 = 100\,\Omega$ and $R_2 = 100\,\text{M}\Omega$.

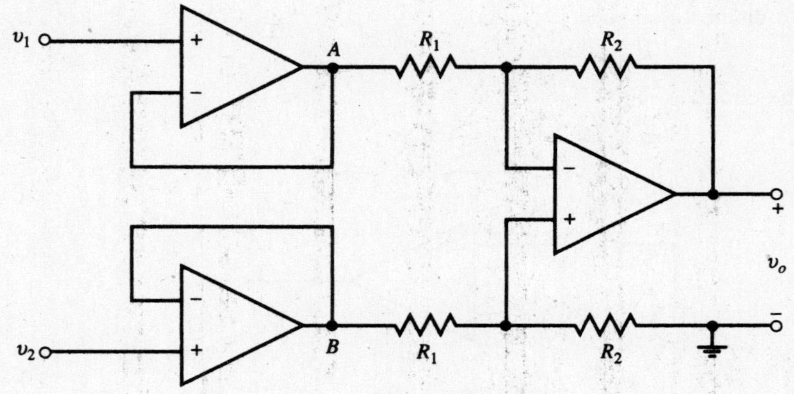

Fig. 5-48

The circuit of Fig. 5-48 can have the same gain as that of Fig. 5-45, but its input resistance is infinite. However, it employs two small and large resistors which are rather out of ordinary range.

5.26 Resistors having high magnitude and accuracy are expensive. Show that in the circuit of Fig. 5-49 we can choose resistors of ordinary range so that $v_o = 10^6(v_2 - v_1)$.

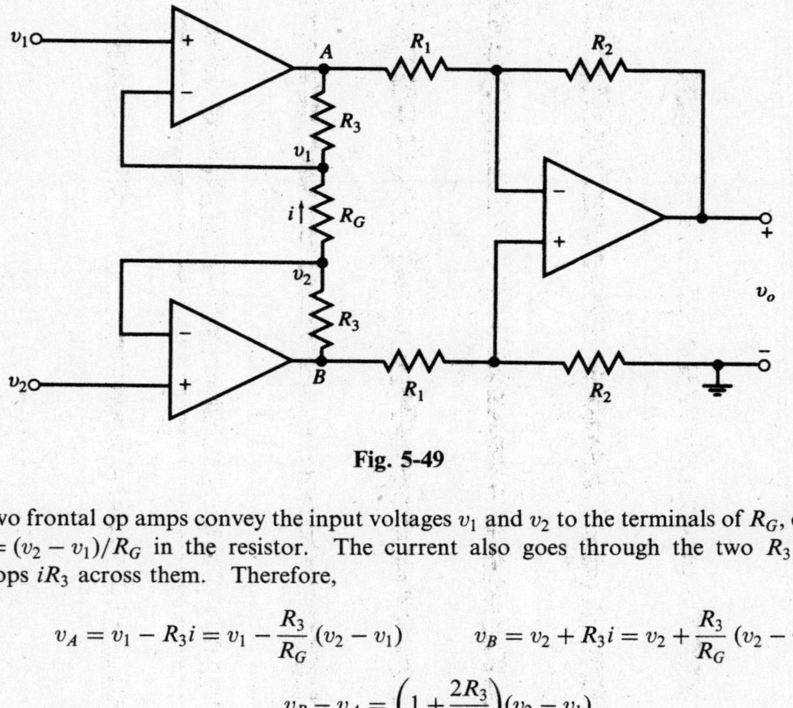

Fig. 5-49

The two frontal op amps convey the input voltages v_1 and v_2 to the terminals of R_G, creating an upward current $i = (v_2 - v_1)/R_G$ in the resistor. The current also goes through the two R_3 resistors, creating voltage drops iR_3 across them. Therefore,

$$v_A = v_1 - R_3 i = v_1 - \frac{R_3}{R_G}(v_2 - v_1) \qquad v_B = v_2 + R_3 i = v_2 + \frac{R_3}{R_G}(v_2 - v_1)$$

$$v_B - v_A = \left(1 + \frac{2R_3}{R_G}\right)(v_2 - v_1)$$

and

$$v_o = \frac{R_2}{R_1}(v_B - v_A) = \frac{R_2}{R_1}\left(1 + \frac{2R_3}{R_G}\right)(v_2 - v_1)$$

For a differential gain of 10^6 we must have

$$\frac{v_o}{v_2 - v_1} = \frac{R_2}{R_1}\left(1 + \frac{2R_3}{R_G}\right) = 10^6$$

Choose $R_1 = R_G = 1\,\text{k}\Omega$, $R_2 = 100\,\text{k}$, and $R_3 = 5\,\text{M}\Omega$.

The circuit of Fig. 5-49 has an infinite input resistance, employs resistors within ordinary range, and uses three op amps.

5.27 Show that in the circuit of Fig. 5-50 $i_1 = i_2$, regardless of the circuits of N_1 and N_2.

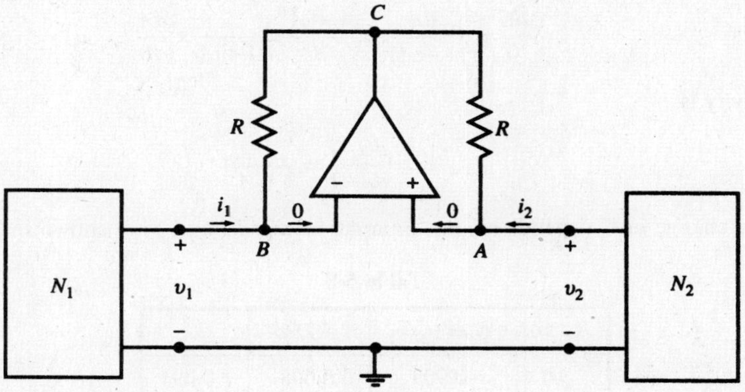

Fig. 5-50

Nodes A and B are at the same voltage $v_A = v_B$. Since the op amp draws no current, i_1 and i_2 flow through the two resistors and KVL around the op amp loop ABC gives $Ri_1 - Ri_2 = 0$. Therefore, $i_1 = i_2$.

5.28 Let N_1 be the voltage source v_1 and N_2 be the resistor R_2 in the circuit of Fig. 5-50. Find the input resistance $R_{in} = v_1/i_1$.

From the op amp we obtain $v_A = v_B$ and $i_1 = i_2$. From connections to N_1 and N_2 we obtain $v_1 = v_B = v_2 = v_A$ and $v_2 = -i_2R_2$, respectively. The input resistance is $v_1/i_1 = -i_2R_2/i_2 = -R_2$ which is the negative of the load. The op amp circuit is a negative impedance converter.

5.29 A voltage follower is constructed using an op amp with a finite open-loop gain A and $R_{in} = \infty$ (see Fig. 5-51). Find the gain $G = v_2/v_1$. Defining *sensitivity* s as the ratio of percentage change produced in G to the percentage change in A, find s.

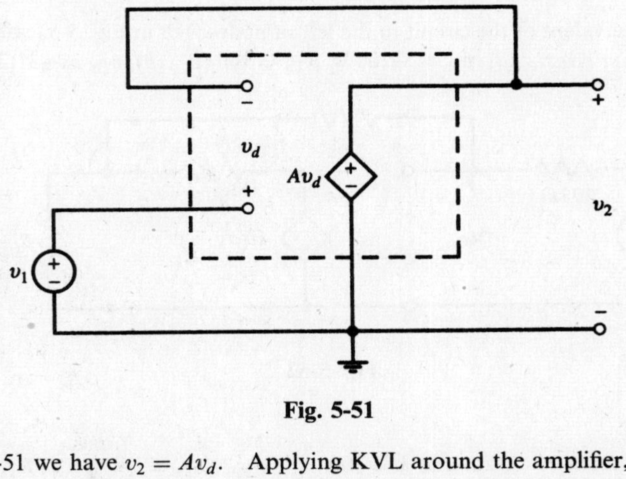

Fig. 5-51

From Fig. 5-51 we have $v_2 = Av_d$. Applying KVL around the amplifier, obtain

$$v_1 = v_d + v_2 = v_d + Av_d = v_d(1 + A) = v_2(1 + A)/A$$

$$G = \frac{v_2}{v_1} = \frac{A}{1 + A}$$

The rate of change of G with respect to A is

$$\frac{dG}{dA} = \frac{1}{(1+A)^2} \qquad \text{from which} \qquad dG = \frac{dA}{(1+A)^2}$$

The percentage change produced in G is $100(dG/G)$.

$$\frac{dG}{G} = \frac{dA}{(1+A)^2} \times \frac{1+A}{A} = \frac{1}{1+A} \times \frac{dA}{A}$$

and the sensitivity is

$$s = \frac{dG/G}{dA/A} = \frac{1}{1+A}$$

The percentage change in G depends on A. Samples of dG/dA and s are shown in Table 5-8.

Table 5-8

A	$G = v_2/v_1$	dG/dA	s
10	0.909	0.008	0.091
11	0.917	0.007	0.083
100	0.990	0.0001	0.01
1000	0.999	0	0

For high values of A, the gain G is not sensitive to changes in A.

Supplementary Problems

5.30 Repeat Problem 5.3 by replacing the circuit to the left of node B (including v_s, R_1, and R_i) by its Thévenin equivalent (see Fig. 5-33) Solve the problem by applying the results of Example 5.4.

5.31 Find the Thévenin equivalent of the circuit to the left of nodes A-B in Fig. 5-52 with $k = 10$ for (*a*) $R_2 = \infty$ and (*b*) $R_2 = 50\,\text{k}\Omega$. *Ans.* (*a*) $v_{\text{Th}} = -100\,\text{V}$, $R_{\text{Th}} = 100\,\Omega$; (*b*) $v_{\text{Th}} = -31.22\,\text{V}$, $R_{\text{Th}} = 37.48\,\Omega$

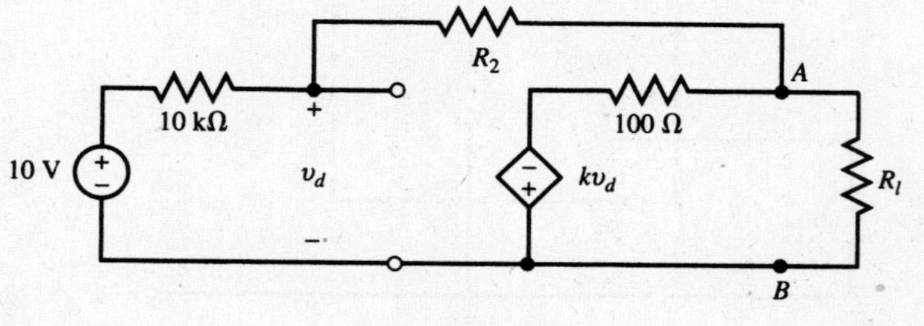

Fig. 5-52

5.32 Repeat Problem 5.31 for $R_2 = 50\,\text{k}\Omega$ and $k = 100$. *Ans.* $v_{\text{Th}} = -47.16\,\text{V}$, $R_{\text{Th}} = 5.66\,\Omega$

5.33 Determine the relationship between R, R_1, and R_2 in Fig. 5-41 such that the circuit has a gain of $v_2/i_1 = 10^6\,\text{V/A}$. *Ans.* $RR_2/R_1 = 10^6$

5.34 In the circuit of Fig. 5-13, $V_{cc} = 10\,\text{V}$, $R_1 = 2\,\text{k}\Omega$ and $v_1 = 1\,\text{V}$. Find the maximum value of R_2 before the op amp is saturated. *Ans.* $R_2 = 20\,\text{k}\Omega$

5.35 Let the summing circuit of Fig. 5-14 have two inputs with $v_1 = 1$ and $v_2 = \sin t$ (V). Let $R_1 = 3\,\text{k}\Omega$, $R_2 = 5\,\text{k}\Omega$, and $R_f = 8\,\text{k}\Omega$. Apply superposition to find v_o. *Ans.* $v_o = -(\frac{8}{3} + \frac{8}{5}\sin t)$

5.36 In Fig. 5-17 let $R_1 = 4\,\text{k}\Omega$ and $R_2 = 8\,\text{k}\Omega$. Apply superposition to find v_o in terms of the input voltages. *Ans.* $v_o = v_1 + v_2 + v_3$

5.37 Find the input resistance seen by v_f in Fig. 5-19. *Ans.* $R_{\text{in}} = 2R_1$

5.38 Use superposition to find v_o in Fig. 5-20 for $R_1 = 2$, $R_2 = 7$, $R_3 = 10$, $R_4 = 5$, all values in kΩ. *Ans.* $v_o = 1.5v_2 - 3.5v_1$

5.39 In the circuit of Fig. 5-20 find (*a*) v_0 for $R_1 = 1$, $R_2 = 3$, $R_3 = 2$, and $R_4 = 2$, all values in kΩ; (*b*) the input resistance $R_{2\text{in}}$ seen by v_2; (*c*) i_1 as a function of v_1 and v_2 and show that v_1 sees a variable load which depends on v_2. *Ans.* (*a*) $v_o = 2v_2 - 3v_1$, (*b*) $R_{2\text{in}} = 4\,\text{k}\Omega$, (*c*) $i_1 = v_1 - v_2/2$

5.40 Using a single op amp, design an amplifier with a gain of $v_2/v_1 = 3/4$, input resistance of $8\,\text{k}\Omega$, and zero output resistance. *Ans.* See Fig. 5-53.

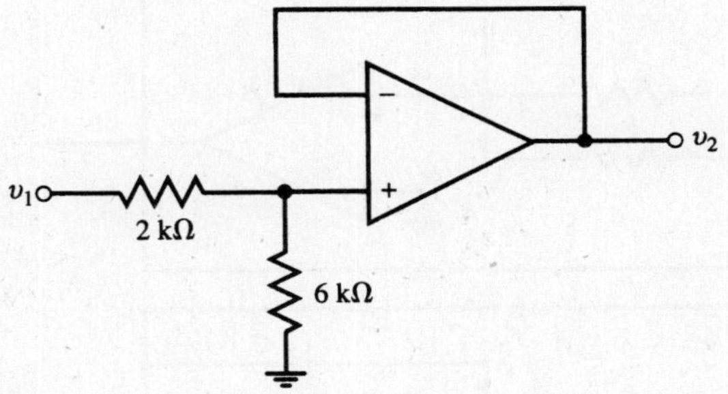

Fig. 5-53

5.41 Show that, given $R_1 = \infty$ and $R_2 = 0$, the noninverting op amp circuit of Fig. 5-15 and (*12*) is reduced to a voltage follower.

5.42 In the circuit of Fig. 5-22 let $R_s = 10\,\text{k}\Omega$. (*a*) Find R_f such that $i_s = 0$. (*b*) Is R_f independent of R_s? Discuss. *Ans.* (*a*) $40\,\text{k}\Omega$; (*b*) yes

5.43 The input to the circuit of Fig. 5-23 with $RC = 1$ is $v_1 = \sin \omega t$. Write KCL at node *B* and solve for v_2. *Ans.* $v_2 = -(1/\omega)\cos \omega t + C$

5.44 Show that the output v_2 in Fig. 5.54 is the same as the output of the integrator in Fig. 5-23.

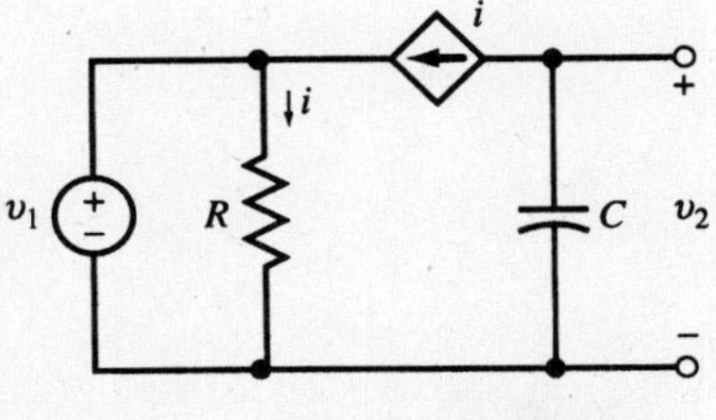

Fig. 5-54

5.45 Find v_2 in the leaky integrator of Fig. 5-24 with $R_1 = R_f = 1\,\text{k}\Omega$, $C = 1\,\mu\text{F}$, and $v_1 = \begin{cases} 1\text{ V} & t > 0 \\ 0 & t < 0 \end{cases}$.

Ans. $v_2(t) = \begin{cases} -1 + e^{-1000t}\ \text{(V)} & t > 0 \\ 0 & t < 0 \end{cases}$

5.46 Repeat Problem 5.45 for $v_1 = \begin{cases} 1\text{ V} & t < 0 \\ 0 & t > 0 \end{cases}$. *Ans.* $v_2(t) = \begin{cases} -e^{-1000t}\ \text{(V)} & t > 0 \\ -1\text{ V} & t < 0 \end{cases}$

5.47 In the differential equation $10^{-2} dv_2/dt + v_2 = v_s$, v_s is the forcing function and v_2 is the response. Design an op amp circuit to obtain v_2 from v_s. *Ans.* See Fig. 5-24, with $R_1 = R_f$, $RC = 10^{-2}$, and $v_1 = -v_s$.

5.48 Design a circuit containing op amps to solve the following set of equations:

$$y' + x = v_{s1}$$

$$2y + x' + 3x = -v_{s2}$$

Ans. See Fig. 5-55, with $R_1C = R_4C = 1$ s, $R_2C = \frac{1}{3}$ s, $R_3C = \frac{1}{2}$ s.

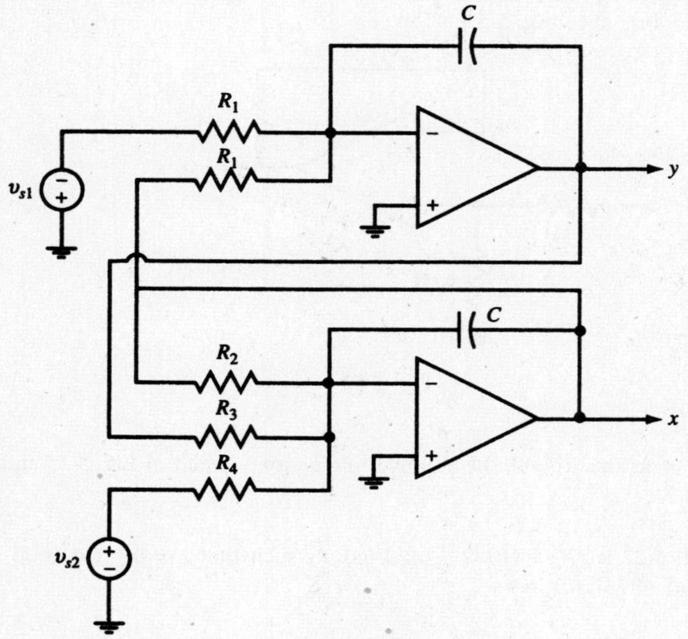

Fig. 5-55

CHAPTER 6

Waveforms and Signals

6.1 INTRODUCTION

The voltages and currents in electric circuits are described by three classes of time functions:

(i) Periodic functions
(ii) Nonperiodic functions
(iii) Random functions

In this chapter the time domain of all functions is $-\infty < t < \infty$ and the terms function, waveform, and signal are used interchangeably.

6.2 PERIODIC FUNCTIONS

A signal $v(t)$ is periodic with *period* T if

$$v(t) = v(t + T) \qquad \text{for all } t$$

Four types of periodic functions which are specified for one period T and corresponding graphs are as follows:

(*a*) Sine wave:

$$v_1(t) = V_0 \sin 2\pi t/T \tag{1}$$

See Fig. 6-1(*a*).

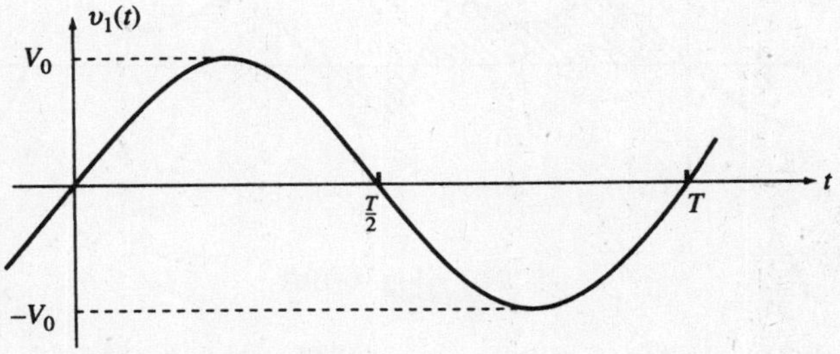

Fig. 6-1(*a*)

(*b*) Periodic pulse:

$$v_2(t) = \begin{cases} V_1 & \text{for } 0 < t < T_1 \\ -V_2 & \text{for } T_1 < t < T \end{cases} \tag{2}$$

See Fig. 6-1(*b*).

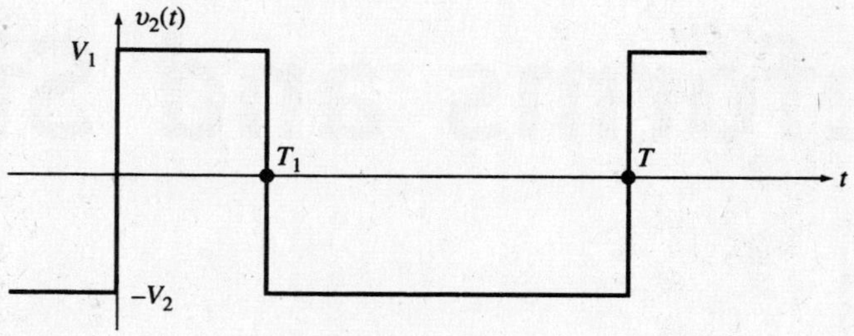

Fig. 6-1(*b*)

(*c*) Periodic tone burst:

$$v_3(t) = \begin{cases} V_0 \sin 2\pi t/\Lambda & \text{for } 0 < t < T_1 \\ 0 & \text{for } T_1 < t < T \end{cases} \tag{3}$$

where $T = k\Lambda$ and k is an integer. See Fig. 6-1(*c*).

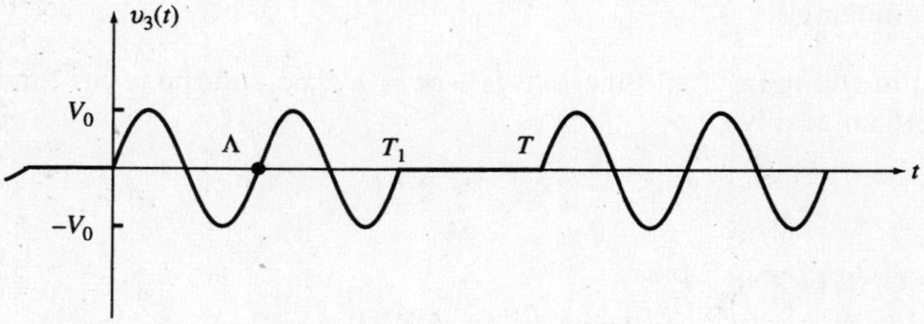

Fig. 6-1(*c*)

(*d*) Repetition of a recording every T seconds:

$$v_4(t) \tag{4}$$

See Fig. 6-1(*d*).

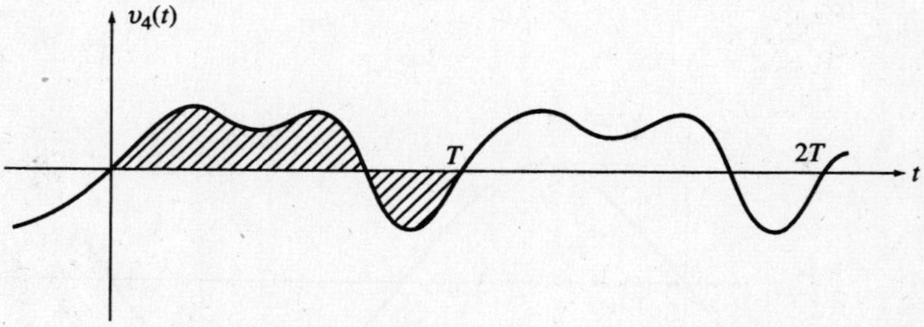

Fig. 6-1(*d*)

Periodic signals may be very complex. However, as will be seen in Chapter 17, they may be represented by a sum of sinusoids. This type of function will be developed in the following sections.

6.3 SINUSOIDAL FUNCTIONS

A sinusoidal voltage $v(t)$ is given by

$$v(t) = V_0 \cos(\omega t + \theta)$$

where V_0 is the amplitude, ω is the angular velocity, or angular frequency, and θ is the phase angle.

The angular velocity ω may be expressed in terms of the period T or the *frequency* f, where $f \equiv 1/T$. The frequency is given in *hertz*, Hz, or cycles/s. Since $\cos \omega t = \cos t(\omega t + 2\pi)$, ω and T are related by $\omega T = 2\pi$. And since it takes T seconds for $v(t)$ to return to its original value, it goes through $1/T$ cycles in one second.

In summary, for sinusoidal functions we have

$$\omega = 2\pi/T = 2\pi f \qquad f = 1/T = \omega/2\pi \qquad T = 1/f = 2\pi/\omega$$

EXAMPLE 6.1 Graph each of the following functions and specify period and frequency.

$$(a)\ v_1(t) = \cos t \qquad (b)\ v_2(t) = \sin t \qquad (c)\ v_3(t) = 2\cos 2\pi t$$

$$(d)\ v_4(t) = 2\cos(\pi t/4 - 45^\circ) = 2\cos(\pi t/4 - \pi/4) = 2\cos[\pi(t-1)/4]$$

$$(e)\ v_5(t) = 5\cos(10t + 60^\circ) = 5\cos(10t + \pi/3) = 5\cos 10(t + \pi/30)$$

(*a*) See Fig. 6-2(*a*). $T = 2\pi = 6.2832$ s and $f = 0.159$ Hz.

(*b*) See Fig. 6-2(*b*). $T = 2\pi = 6.2832$ s and $f = 0.159$ Hz.

(*c*) See Fig. 6-2(*c*). $T = 1$ s and $f = 1$ Hz.

(*d*) See Fig. 6-2(*d*). $T = 8$ s and $f = 0.125$ Hz.

(*e*) See Fig. 6-2(*e*). $T = 0.2\pi = 0.62832$ s and $f = 1.59$ Hz.

EXAMPLE 6.2 Plot $v(t) = 5\cos \omega t$ versus ωt.

See Fig. 6.3.

6.4 TIME SHIFT AND PHASE SHIFT

If the function $v(t) = \cos \omega t$ is delayed by τ seconds, we get $v(t - \tau) = \cos \omega(t - \tau) = \cos(\omega t - \theta)$, where $\theta = \omega\tau$. The delay shifts the graph of $v(t)$ to the right by an amount of τ seconds, which corresponds to a phase lag of $\theta = \omega\tau = 2\pi f\tau$. A *time shift* of τ seconds to the left on the graph produces $v(t + \tau)$, resulting in a leading phase angle called an *advance*.

Conversely, a *phase shift* of θ corresponds to a time shift of τ. Therefore, for a given phase shift the higher is the frequency, the smaller is the required time shift.

EXAMPLE 6.3 Plot $v(t) = 5\cos(\pi t/6 + 30^\circ)$ versus t and $\pi t/6$.

Rewrite the given as

$$v(t) = 5\cos(\pi t/6 + \pi/6) = 5\cos[\pi(t+1)/6]$$

This is a cosine function with period of 12 s, which is advanced in time by 1 s. In other words, the graph is shifted to the left by 1 s or 30° as shown in Fig. 6-4.

EXAMPLE 6.4 Consider a linear circuit with the following input-output pair valid for all ω and A:

$$\text{Input:}\quad v_i(t) = A\cos \omega t \qquad \text{Output:}\quad v_0(t) = A\cos(\omega t - \theta)$$

Given $v_i(t) = \cos \omega_1 t + \cos \omega_2 t$, find $v_0(t)$ when

(*a*) $\theta = 10^{-6}\omega$ [phase shift is proportional to frequency, Fig. 6-5(*a*)]

(*b*) $\theta = 10^{-6}$ [phase shift is constant, Fig. 6-5(*b*)]

The output is $v_0(t) = \cos(\omega_1 t - \theta_1) + \cos(\omega_2 t - \theta_2)$.

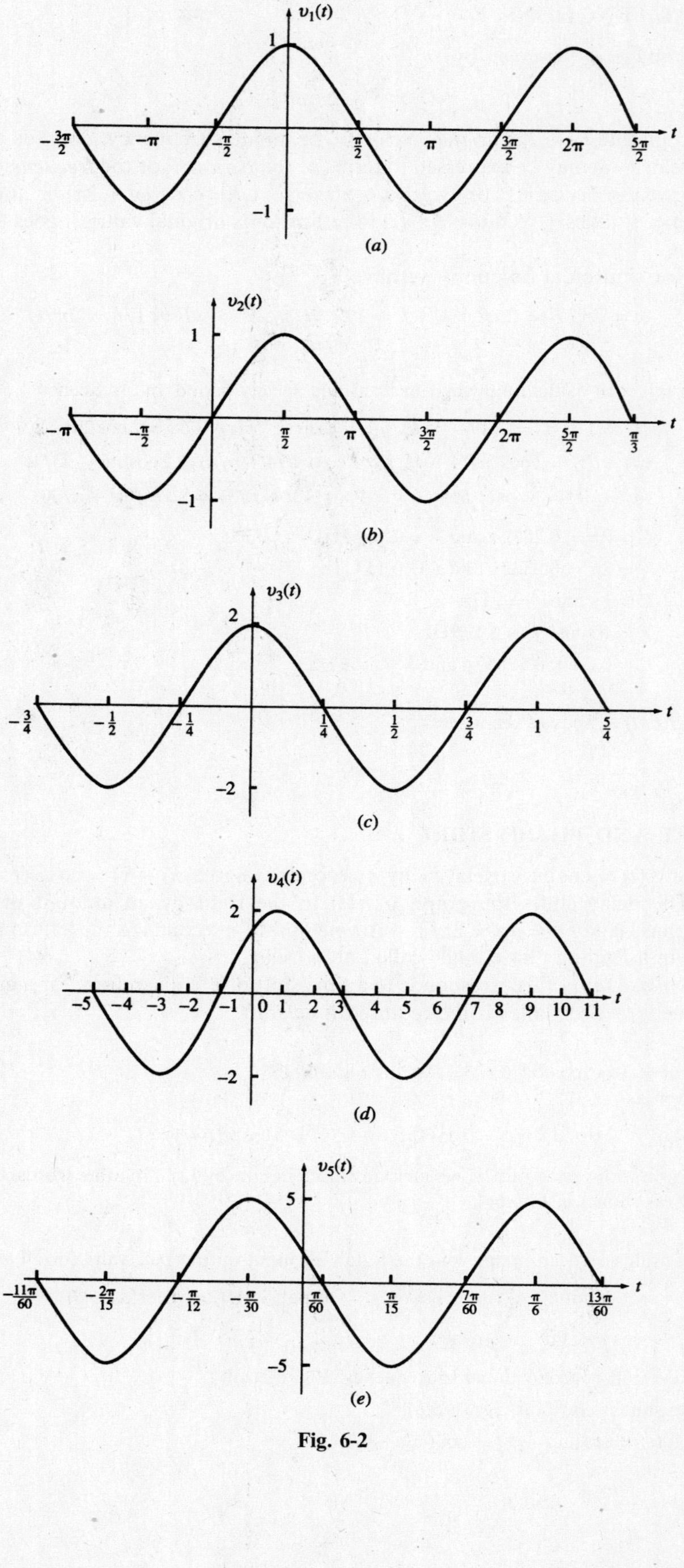

v1(t)
1
−1
−3π/2 −π −π/2 π/2 π 3π/2 2π 5π/2 t
(a)
v2(t)
1
−1
−π −π/2 π/2 π 3π/2 2π 5π/2 π/3 t
(b)
v3(t)
2
−2
−3/4 −1/2 −1/4 1/4 1/2 3/4 1 5/4 t
(c)
v4(t)
2
−2
−5 −4 −3 −2 −1 0 1 2 3 4 5 6 7 8 9 10 11 t
(d)
v5(t)
5
−5
−11π/60 −2π/15 −π/12 −π/30 π/60 π/15 7π/60 π/6 13π/60 t
(e)

Fig. 6-2

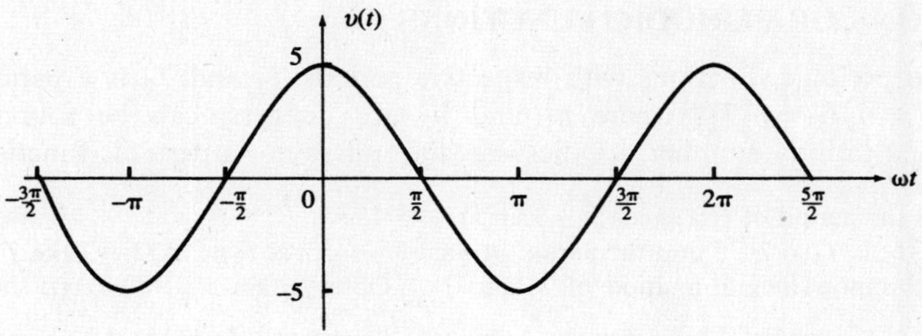

Fig. 6-3

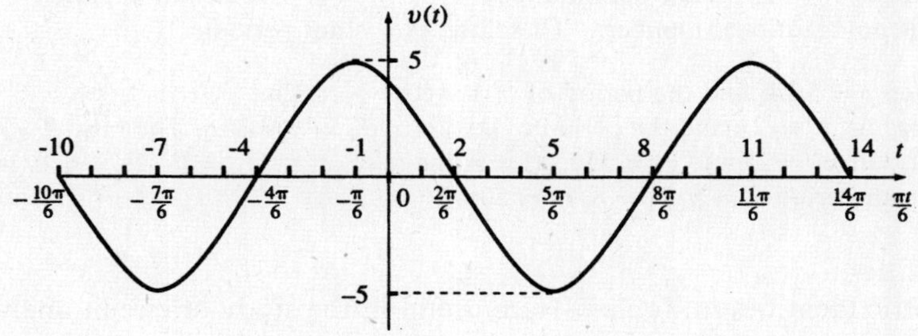

Fig. 6-4

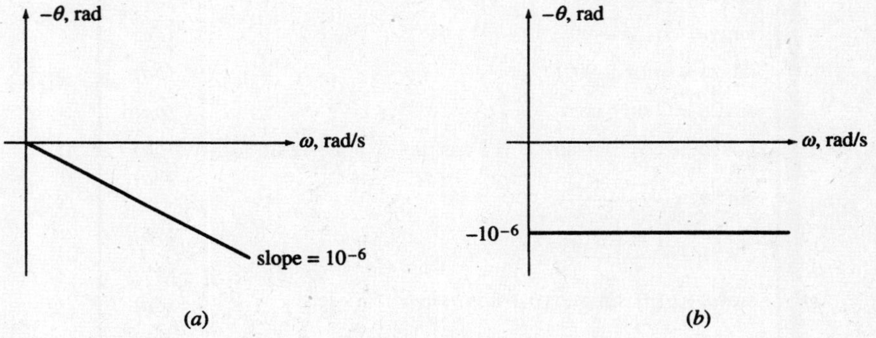

Fig. 6-5

(*a*) $\theta_1 = 10^{-6}\omega_1$, $\theta_2 = 10^{-6}\omega_2$. Then

$$v_0(t) = \cos(\omega_1 t - 10^{-6}\omega_1) + \cos(\omega_2 t - 10^{-6}\omega_2)$$
$$= \cos\omega_1(t - 10^{-6}) + \cos\omega_2(t - 10^{-6}) = v_i(t - 10^{-6}) = v_i(t - \tau)$$

where $\tau = 10^{-6}$ s $= 1\ \mu$s. Thus a phase shift proportional to ω [Fig. 6-5(*a*)] delays all frequency components of the input signal by 1 μs. The output follows the input with no distortion.

(*b*) $\theta_1 = \theta_2 = 10^{-6}$. Then

$$v_0(t) = \cos(\omega_1 t - 10^{-6}) + \cos(\omega_2 t - 10^{-6})$$
$$= \cos\omega_1(t - 10^{-6}/\omega_1) + \cos\omega_2(t - 10^{-6}/\omega_2)$$

A constant phase shift [Fig. 6-5(*b*)] delays different frequency components of the input signal by different amounts. The output is a distorted form of the input.

6.5 COMBINATIONS OF PERIODIC FUNCTIONS

The sum of two periodic functions with respective periods T_1 and T_2 is a periodic function if a common period $T = n_1 T_1 = n_2 T_2$, where n_1 and n_2 are integers, can be found. This requires $T_1/T_2 = n_2/n_1$ to be a rational number. Otherwise, the sum is not a periodic function.

EXAMPLE 6.5 Find the period of $v(t) = \cos 5t + 3\sin(3t + 45^\circ)$.

The period of $\cos 5t$ is $T_1 = 2\pi/5$ and the period of $3\sin(3t + 45^\circ)$ is $T_2 = 2\pi/3$. Take $T = 2\pi = 5T_1 = 3T_2$ which is the smallest common integral multiple of T_1 and T_2. Observe that $v(t + T) = v(t)$ since

$$v(t+T) = \cos 5(t + 2\pi) + 3\sin[3(t + 2\pi) + 45^\circ] = \cos 5t + 3\sin(3t + 45^\circ) = v(t)$$

Therefore, the period of $v(t)$ is 2π.

EXAMPLE 6.6 Is $v(t) = \cos t + \cos 2\pi t$ periodic? Discuss.

The period of $\cos t$ is $T_1 = 2\pi$. The period of $\cos 2\pi t$ is $T_2 = 1$. No common period $T = n_1 T_1 = n_2 T_2$ exists because $T_1/T_2 = 2\pi$ is not a rational number. Therefore, $v(t)$ is not periodic.

EXAMPLE 6.7 Given $p = 3.14$, find the period of $v(t) = \cos t + \cos 2pt$.

The period of $\cos t$ is $T_1 = 2\pi$ and the period of $\cos 2pt$ is $T_2 = \pi/3.14$. The ratio $T_1/T_2 = 6.28$ is a rational number. The integer pair $n_1 = 25$ and $n_2 = 157$ satisfies the relation $n_2/n_1 = T_1/T_2 = 628/100 = 157/25$. Therefore, $v(t)$ is periodic with period $T = n_1 T_1 = n_2 T_2 = 50\pi$ s.

Trigonometric Identities

The trigonometric identities in Table 6-1 are useful in the study of circuit analysis.

Table 6-1

$\sin a = -\sin(-a)$	(5a)
$\cos a = \cos(-a)$	(5b)
$\sin a = \cos(a - 90^\circ)$	(5c)
$\cos a = \sin(a + 90^\circ)$	(5d)
$\sin 2a = 2\sin a \cos a$	(6a)
$\cos 2a = \cos^2 a - \sin^2 a = 2\cos^2 a - 1 = 1 - 2\sin^2 a$	(6b)
$\sin^2 a = \dfrac{1 - \cos 2a}{2}$	(7a)
$\cos^2 a = \dfrac{1 + \cos 2a}{2}$	(7b)
$\sin(a + b) = \sin a \cos b + \cos a \sin b$	(8a)
$\cos(a + b) = \cos a \cos b - \sin a \sin b$	(8b)
$\sin a \sin b = \frac{1}{2}\cos(a - b) - \frac{1}{2}\cos(a + b)$	(9a)
$\sin a \cos b = \frac{1}{2}\sin(a + b) + \frac{1}{2}\sin(a - b)$	(9b)
$\cos a \cos b = \frac{1}{2}\cos(a + b) + \frac{1}{2}\cos(a - b)$	(9c)
$\sin a + \sin b = 2\sin\frac{1}{2}(a + b)\cos\frac{1}{2}(a - b)$	(10a)
$\cos a + \cos b = 2\cos\frac{1}{2}(a + b)\cos\frac{1}{2}(a - b)$	(10b)

EXAMPLE 6.8 Express $v(t) = \cos 5t \ \sin(3t + 45^\circ)$ as the sum of two cosine functions and find its period.

$$\begin{aligned} v(t) &= \cos 5t \ \sin(3t + 45^\circ) = [\sin(8t + 45^\circ) - \sin(2t - 45^\circ)]/2 \quad \text{[Eq. (9b)]} \\ &= [\cos(8t - 45^\circ) + \cos(2t + 45^\circ)]/2 \quad \text{[Eq. (5c)]} \end{aligned}$$

The period of $v(t)$ is π.

6.6 THE AVERAGE AND EFFECTIVE (RMS) VALUES

A periodic function $f(t)$, with a period T, has an average value F_{avg} given by

$$F_{avg} = \langle f(t) \rangle = \frac{1}{T} \int_0^T f(t)\,dt = \frac{1}{T} \int_{t_0}^{t_0+T} f(t)\,dt \tag{11}$$

The root-mean-square (rms) or effective value of $f(t)$ during the same period is defined by

$$F_{eff} = F_{rms} = \left[\frac{1}{T} \int_{t_0}^{t_0+T} f^2(t)\,dt \right]^{1/2} \tag{12}$$

It is seen that $F_{eff}^2 = \langle f^2(t) \rangle$.

Average and effective values of periodic functions are normally computed over one period.

EXAMPLE 6.9 Find the average and effective values of the cosine wave $v(t) = V_m \cos(\omega t + \theta)$.

Using (11),

$$V_{avg} = \frac{1}{T} \int_0^T V_m \cos(\omega t + \theta)\,dt = \frac{V_m}{\omega T} [\sin(\omega t + \theta)]_0^T = 0 \tag{13}$$

and using (12),

$$V_{eff}^2 = \frac{1}{T} \int_0^T V_m^2 \cos^2(\omega t + \theta)\,dt = \frac{1}{2T} \int_0^T V_m^2 [1 + \cos 2(\omega t + \theta)]\,dt = V_m^2/2$$

from which

$$V_{eff} = V_m/\sqrt{2} = 0.707 V_m \tag{14}$$

Equations (13) and (14) show that the results are independent of the frequency and phase angle θ. In other words, the average of a cosine wave and its rms value are always 0 and 0.707 V_m, respectively.

EXAMPLE 6.10 Find V_{avg} and V_{eff} of the half-rectified sine wave

$$v(t) = \begin{cases} V_m \sin \omega t & \text{when } \sin \omega t > 0 \\ 0 & \text{when } \sin \omega t < 0 \end{cases} \tag{15}$$

From (11),

$$V_{avg} = \frac{1}{T} \int_0^{T/2} V_m \sin \omega t\,dt = \frac{V_m}{\omega T} [-\cos \omega t]_0^{T/2} = V_m/\pi \tag{16}$$

and from (12),

$$V_{eff}^2 = \frac{1}{T} \int_0^{T/2} V_m^2 \sin^2 \omega t\,dt = \frac{1}{2T} \int_0^{T/2} V_m^2 (1 - \cos 2\omega t)\,dt = V_m^2/4$$

from which

$$V_{eff} = V_m/2 \tag{17}$$

EXAMPLE 6.11 Find V_{avg} and V_{eff} of the periodic function $v(t)$ where, for one period T,

$$v(t) = \begin{cases} V_0 & \text{for } 0 < t < T_1 \\ -V_0 & \text{for } T_1 < t < 3T_1 \end{cases} \qquad \text{Period } T = 3T_1 \tag{18}$$

We have

$$V_{avg} = \frac{V_0}{3T} (T_1 - 2T_1) = \frac{-V_0}{3} \tag{19}$$

and

$$V_{eff}^2 = \frac{V_0^2}{3T} (T_1 + 2T_1) = V_0^2$$

from which

$$V_{eff} = V_0 \tag{20}$$

The preceding result can be generalized as follows. If $|v(t)| = V_0$ then $V_{eff} = V_0$.

EXAMPLE 6.12 Compute the average power dissipated from 0 to T in a resistor connected to a voltage $v(t)$. Replace $v(t)$ by a constant voltage V_{dc}. Find V_{dc} such that the average power during the period remains the same.

$$p = vi = v^2/R$$

$$P_{avg} = \frac{1}{RT}\int_0^T v^2(t)\,dt = \frac{1}{R}V_{eff}^2 = \frac{V_{dc}^2}{R} \qquad \text{or} \qquad V_{dc} = V_{eff}$$

EXAMPLE 6.13 The current $i(t)$ shown in Fig. 6-6 passes through a 1-μF capacitor. Find (*a*) v_{ac} the voltage across the capacitor at $t = 5k$ ms ($k = 0, 1, 2, 3, \ldots$) and (*b*) the value of a constant current source I_{dc} which can produce the same voltage across the above capacitor at $t = 5k$ ms when applied at $t > 0$. Compare I_{dc} with $\langle i(t) \rangle$, the average of $i(t)$ in Fig. 6-6, for a period of 5 ms after $t > 0$.

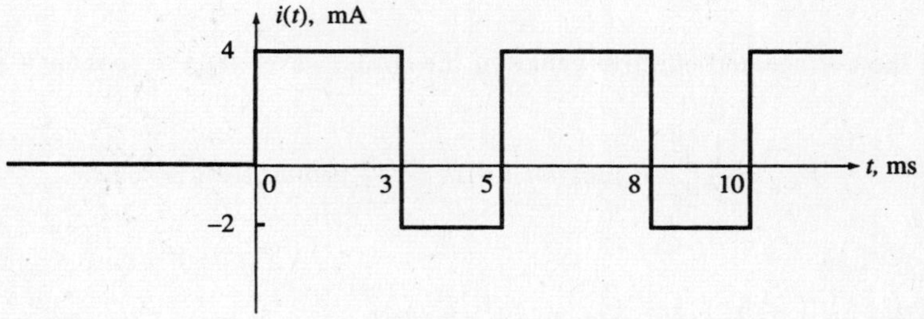

Fig. 6-6

(*a*) At $t = 5$ ms

$$v_{ac} = \frac{1}{C}\int_0^{5\times10^{-3}} i(t)\,dt = 10^6(10^{-3})\left[\int_0^{3\times10^{-3}} 4\,dt - \int_{3\times10^{-3}}^{5\times10^{-3}} 2\,dt\right] = 12 - 4 = 8\text{ V}$$

This is the net charging effect of $i(t)$ during each 5-ms interval. Every 5 ms the above amount is added to the capacitor voltage. Therefore, at $t = 5k$ ms, $v = 8k$ (V).

(*b*) With a constant current I_{dc}, the capacitor voltage v_{dc} at $t = 5k$ ms is

$$v_{dc} = \frac{1}{C}\int_0^{5k\times10^{-3}} I_{dc}\,dt = 10^6(I_{dc})(5k \times 10^{-3}) = 10^3(5k)(I_{dc}) \quad \text{(V)}$$

Since $v_{dc} = v_{ac}$ at $5k$ ms, we obtain

$$10^3(5k)(I_{dc}) = 8k \qquad \text{or} \qquad I_{dc} = 8k/(5k \times 10^3) = 1.6 \times 10^{-3}\text{ A} = 1.6\text{ mA}$$

Note that $I_{dc} = \langle i(t) \rangle$ of Fig. 6-6 for any period of 5 ms at $t > 0$.

6.7 NONPERIODIC FUNCTIONS

A nonperiodic function cannot be specified for all times by simply knowing a finite segment. Examples of nonperiodic functions are

(*a*)
$$v_1(t) = \begin{cases} 0 & \text{for } t < 0 \\ 1 & \text{for } t > 0 \end{cases} \tag{21}$$

(*b*)
$$v_2(t) = \begin{cases} 0 & \text{for } t < 0 \\ 1/T & \text{for } 0 < t < T \\ 0 & \text{for } t > T \end{cases} \tag{22}$$

(*c*)
$$v_3(t) = \begin{cases} 0 & \text{for } t < 0 \\ e^{-t/\tau} & \text{for } t > 0 \end{cases} \tag{23}$$

(*d*) $$v_4(t) = \begin{cases} 0 & \text{for } t < 0 \\ \sin \omega t & \text{for } t > 0 \end{cases} \tag{24}$$

(*e*) $$v_5(t) = \begin{cases} 0 & \text{for } t < 0 \\ e^{-t/\tau} \cos \omega t & \text{for } t > 0 \end{cases} \tag{25}$$

(*f*) $$v_6(t) = e^{-t/\tau} \quad \text{for all } t \tag{26}$$

(*g*) $$v_7(t) = e^{-a|t|} \quad \text{for all } t \tag{27}$$

(*h*) $$v_8(t) = e^{-a|t|} \cos \omega t \quad \text{for all } t \tag{28}$$

Several of these functions are used as mathematical models and building blocks for actual signals in analysis and design of circuits. Examples are discussed in the following sections.

6.8 THE UNIT STEP FUNCTION

The dimensionless *unit step function*, is defined by

$$u(t) = \begin{cases} 0 & \text{for } t < 0 \\ 1 & \text{for } t > 0 \end{cases} \tag{29}$$

The function is graphed in Fig. 6-7. Note that the function is undefined at $t = 0$.

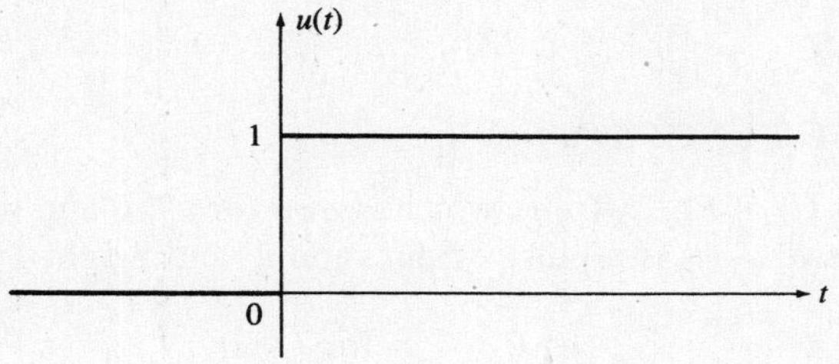

Fig. 6-7

To illustrate the use of $u(t)$, assume the switch S in the circuit of Fig. 6-8(*a*) has been in position *1* for $t < 0$ and is moved to position *2* at $t = 0$. The voltage across *A-B* may be expressed by $v_{AB} = V_0 u(t)$. The equivalent circuit for the voltage step is shown in Fig. 6-8(*b*).

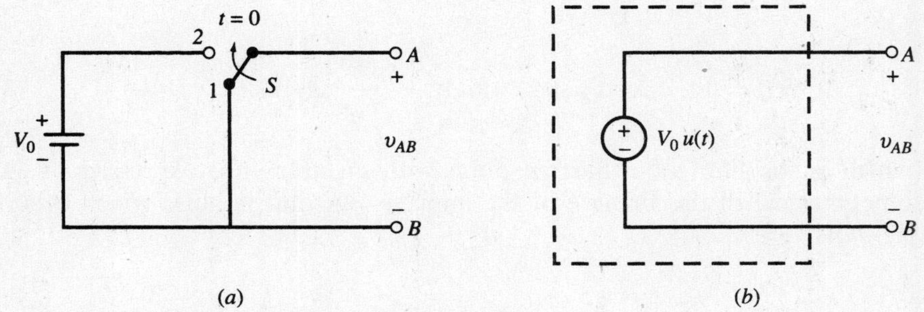

Fig. 6-8

EXAMPLE 6.14 The switch in the circuit of Fig. 6-8(*a*) is moved to position *2* at $t = t_0$. Express v_{AB} using the step function.

The appearance of V_0 across *A-B* is delayed until $t = t_0$. Replace the argument t in the step function by $t - t_0$ and so we have $v_{AB} = V_0 u(t - t_0)$.

EXAMPLE 6.15 If the switch in Fig. 6-8(*a*) is moved to position *2* at $t = 0$ and then moved back to position *1* at $t = 5$ s, express v_{AB} using the step function.

$$v_{AB} = V_0[u(t) - u(t-5)]$$

EXAMPLE 6.16 Express $v(t)$, graphed in Fig. 6-9, using the step function.

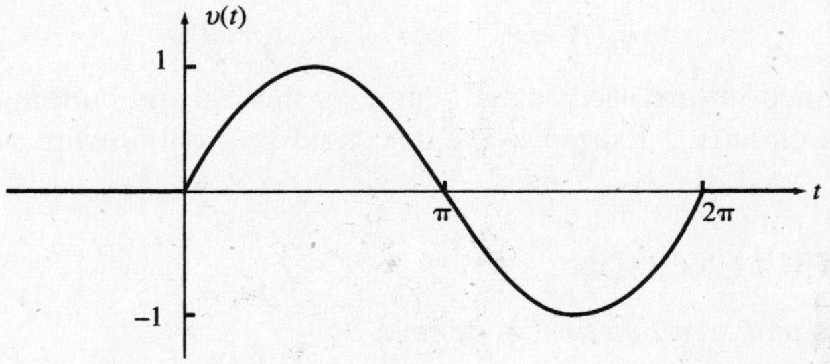

Fig. 6-9

$$v(t) = [u(t) - u(t-2\pi)]\sin t$$

6.9 THE UNIT IMPULSE FUNCTION

Consider the function $s_T(t)$ of Fig. 6-10(*a*), which is zero for $t < 0$ and increases uniformly from 0 to 1 in T seconds. Its derivative $d_T(t)$ is a pulse of duration T and height $1/T$, as seen in Fig. 6-10(*b*).

$$d_T(t) = \begin{cases} 0 & \text{for } t < 0 \\ 1/T & \text{for } 0 < t < T \\ 0 & \text{for } t > T \end{cases} \qquad (30)$$

If the transition time T is reduced, the pulse in Fig. 6-10(*b*) becomes narrower and taller, but the area under the pulse remains equal to 1. If we let T approach zero, in the limit function $s_T(t)$ becomes a unit step $u(t)$ and its derivative $d_T(t)$ becomes a unit pulse $\delta(t)$ with zero width and infinite height. The unit impulse $\delta(t)$ is shown in Fig. 6-10(*c*). The *unit impulse* or *unit delta function* is defined by

$$\delta(t) = 0 \quad \text{for } t \neq 0 \quad \text{and} \quad \int_{-\infty}^{\infty} \delta(t)\,dt = 1 \qquad (31)$$

An impulse which is the limit of a narrow pulse with an area A is expressed by $A\delta(t)$. The magnitude A is sometimes called the *strength* of the impulse. A unit impulse which occurs at $t = t_0$ is expressed by $\delta(t - t_0)$.

EXAMPLE 6.17 The voltage across the terminals of a 100-nF capacitor grows linearly, from 0 to 10 V, taking the shape of the function $s_T(t)$ in Fig. 6-10(*a*). Find (*a*) the charge across the capacitor at $t = T$ and (*b*) the current $i_C(t)$ in the capacitor for $T = 1$ s, $T = 1$ ms, and $T = 1$ μs.

(*a*) At $t = T$, $v_C = 10$ V. The charge across the capacitor is $Q = Cv_C = 10^{-7} \times 10 = 10^{-6}$.

(*b*)
$$i_c(t) = C\,\frac{dv_C}{dt}$$

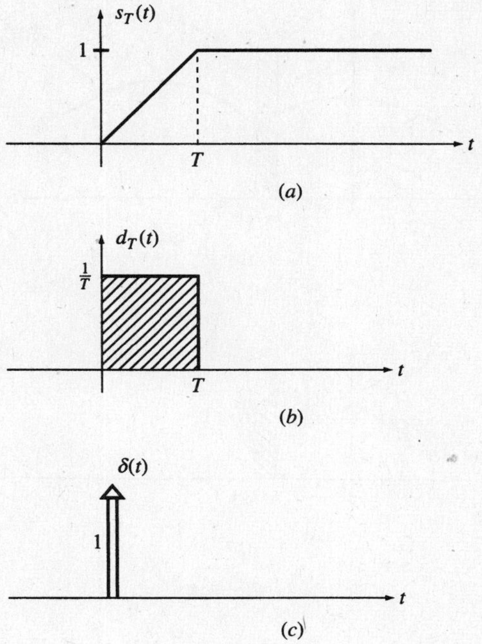

Fig. 6-10

From Fig. 6-10,

$$i_C(t) = \begin{cases} 0 & \text{for } t < 0 \\ I_0 = 10^{-6}/T \text{ (A)} & \text{for } 0 < t < T \\ 0 & \text{for } t > T \end{cases} \tag{32}$$

For $T = 1$ s, $I_0 = 10^{-6}$ A; for $T = 1$ ms, $I_0 = 10^{-3}$ A; and for $T = 1\ \mu$s, $I_0 = 1$ A.

In all the preceding cases, the charge accumulated across the capacitor at the end of the transition period is

$$Q = \int_0^T i_C(t)\, dt = I_0 T = 10^{-6}\text{ C}$$

The amount of charge at $t = T$ is independent of T. It generates a voltage $v_C = 10$ V across the capacitor.

EXAMPLE 6.18 Let $d_T(t - t_0)$ denote a narrow pulse of width T and height $1/T$, which starts at $t = t_0$. Consider a function $f(t)$ which is continuous between t_0 and $t_0 + T$ as shown in Fig. 6-11(*a*). Find the limit of integral I in (*33*) when T approaches zero.

$$I = \int_{-\infty}^{\infty} d_T(t - t_0) f(t)\, dt \tag{33}$$

$$d_T(t - t_0) = \begin{cases} 1/T & t_0 < t < t_0 + T \\ 0 & \text{elsewhere} \end{cases}$$

Substituting d_T in (*33*) we get

$$I = \frac{1}{T}\int_{t_0}^{t_0+T} f(t)\, dt = \frac{S}{T} \tag{34a}$$

where S is the hatched area under $f(t)$ between t_0 and $t_0 + T$ in Fig. 6.11(*b*). Assuming T to be small, the function $f(t)$ may be approximated by a line connecting A and B. S is the area of the resulting trapezoid.

$$S = \tfrac{1}{2}[f(t_0) + f(t_0 + T)]T \tag{34b}$$

$$I = \tfrac{1}{2}[f(t_0) + f(t_0 + T)] \tag{34c}$$

As $T \to 0$, $d_T(t - t_0) \to \delta(t - t_0)$ and $f(t_0 + T) \to f(t_0)$ and from (*34c*) we get

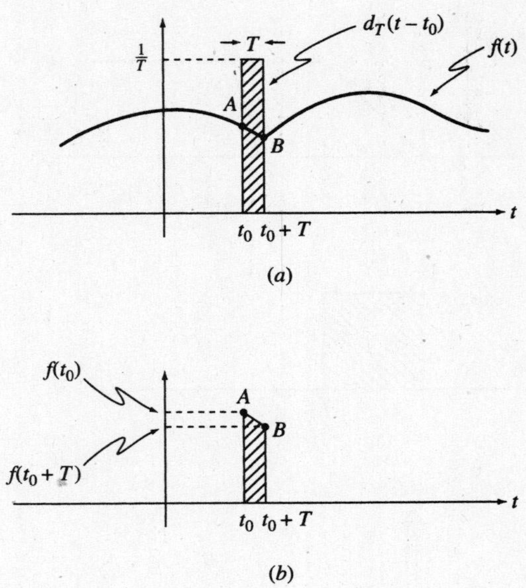

Fig. 6-11

$$\lim_{T\to 0} I = \lim_{T\to 0} \tfrac{1}{2}[f(t_0) + f(t_0+T)] \tag{34d}$$

We assumed $f(t)$ to be continuous between t_0 and $t_0 + T$. Therefore,

$$\lim_{T\to 0} I = f(t_0) \tag{34e}$$

But

$$\lim_{T\to 0} I = \int_{-\infty}^{\infty} \delta(t-t_0) f(t)\, dt \tag{34f}$$

and so

$$\int_{-\infty}^{\infty} \delta(t-t_0) f(t)\, dt = f(t_0) \tag{34g}$$

The identity (*34g*) is called the *sifting* property of the impulse function. It is also used as another definition for $\delta(t)$.

6.10 THE EXPONENTIAL FUNCTION

The function $f(t) = e^{st}$ with s a complex constant is called *exponential.* It decays with time if the real part of s is negative and grows if the real part of s is positive. We will discuss exponentials e^{at} in which the constant a is a real number.

The inverse of the constant a has the dimension of time and is called the *time constant* $\tau = 1/a$. A decaying exponential $e^{-t/\tau}$ is plotted versus t as shown in Fig. 6-12. The function decays from one at $t = 0$ to zero at $t = \infty$. After τ seconds the function $e^{-t/\tau}$ is reduced to $e^{-1} = 0.368$. For $\tau = 1$, the function e^{-t} is called a *normalized exponential* which is the same as $e^{-t/\tau}$ when plotted versus t/τ.

EXAMPLE 6.19 Show that the tangent to the graph of $e^{-t/\tau}$ at $t = 0$ intersects the t axis at $t = \tau$ as shown in Fig. 6-12.

The tangent line begins at point A ($v = 1, t = 0$) with a slope of $de^{-t/\tau}/dt|_{t=0} = -1/\tau$. The equation of the line is $v_{\tan}(t) = -t/\tau + 1$. The line intersects the t axis at point B where $t = \tau$. This observation provides a convenient approximate approach to plotting the exponential function as described in Example 6.20.

EXAMPLE 6.20 Draw an approximate plot of $v(t) = e^{-t/\tau}$ for $t > 0$.

Identify the initial point A ($t = 0, v = 1$) of the curve and the intersection B of its tangent with the t axis at $t = \tau$. Draw the tangent line AB. Two additional points C and D located at $t = \tau$ and $t = 2\tau$, with heights of 0.368 and

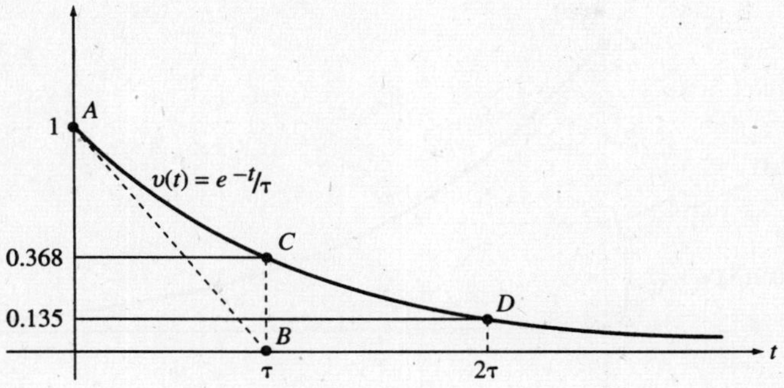

Fig. 6-12

$0.368^2 = 0.135$, respectively, also belong to the curve. Using the preceding indicators, the curve may be drawn with a rather good approximation (see Fig. 6-12).

EXAMPLE 6.21 (*a*) Show that the rate of change with respect to time of an exponential function $v = Ae^{st}$ is at any moment proportional to the value of the function at that moment. (*b*) Show that any linear combination of an exponential function and its n derivatives is proportional to the function itself. Find the coefficient of proportionality.

(*a*) The rate of change of a function is equal to the derivative of the function, which, for the given exponential function, is

$$\frac{dv}{dt} = sAe^{st} = sv$$

(*b*) Using the result of (*a*) we get

$$\frac{d^n v}{dt^n} = s^n A e^{st} = s^n v$$

$$a_0 v + a_1 \frac{dv}{dt} + \cdots + a_n \frac{d^n v}{dt^n} = (a_0 + a_1 s + \cdots + a_n s^n)v = Hv \tag{35}$$

where

$$H = a_0 + a_1 s + \cdots + a_n s^n \tag{36}$$

Specifying and Plotting $f(t) = Ae^{-at} + B$

We often encounter the function

$$f(t) = Ae^{-at} + B \tag{37}$$

This function is completely specified by the three numbers A, B, and a defined as

$A =$ initial value − final value $\qquad B =$ final value $\qquad a =$ inverse of the time constant

or, in another form,

Initial value $f(0) = A + B$ $\qquad$ Final value $f(\infty) = B$ $\qquad$ Time constant $= 1/a$

EXAMPLE 6.22 Find a function $v(t)$ which decays exponentially from 5 V at $t = 0$ to 1 V at $t = \infty$ with a time constant of 3 s. Plot $v(t)$ using the technique of Example 6.20.

From (*37*) we have $v(t) = Ae^{-t/\tau} + B$. Now $v(0) = A + B = 5$, $v(\infty) = B = 1$, $A = 4$, and $\tau = 3$. Thus

$$v(t) = 4e^{-t/3} + 1$$

The preceding result can be generalized in the following form:

$$v(t) = (\text{initial value} - \text{final value})e^{-t/\tau} + (\text{final value})$$

The plot is shown in Fig. 6-13.

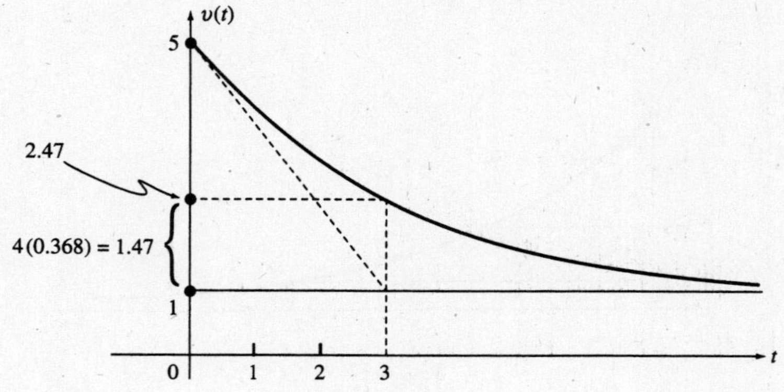

Fig. 6-13

EXAMPLE 6.23 The voltage $v = V_0 e^{-|t|/\tau}$, $\tau > 0$, is connected to a capacitor. Find the current i in the capacitor. Sketch v and i for $V_0 = 10\,\text{V}$, $C = 1\,\mu\text{F}$, and $\tau = 1\,\text{ms}$.

Using $i = C\,dv/dt$,

for $t < 0$, $\quad v = V_0 e^{t/\tau}$ and $i = I_0 e^{t/\tau}$

for $t > 0$, $\quad v = V_0 e^{-t/\tau}$ and $i = -I_0 e^{-t/\tau}$

where $I_0 = CV_0/\tau$.

For $V_0 = 10\,\text{V}$, $C = 1\,\mu\text{F}$, and $\tau = 10^{-3}\,\text{s}$, we get $I_0 = 10\,\text{mA}$. Graphs of v and i are shown in Figs. 6-14(*a*) and (*b*), respectively.

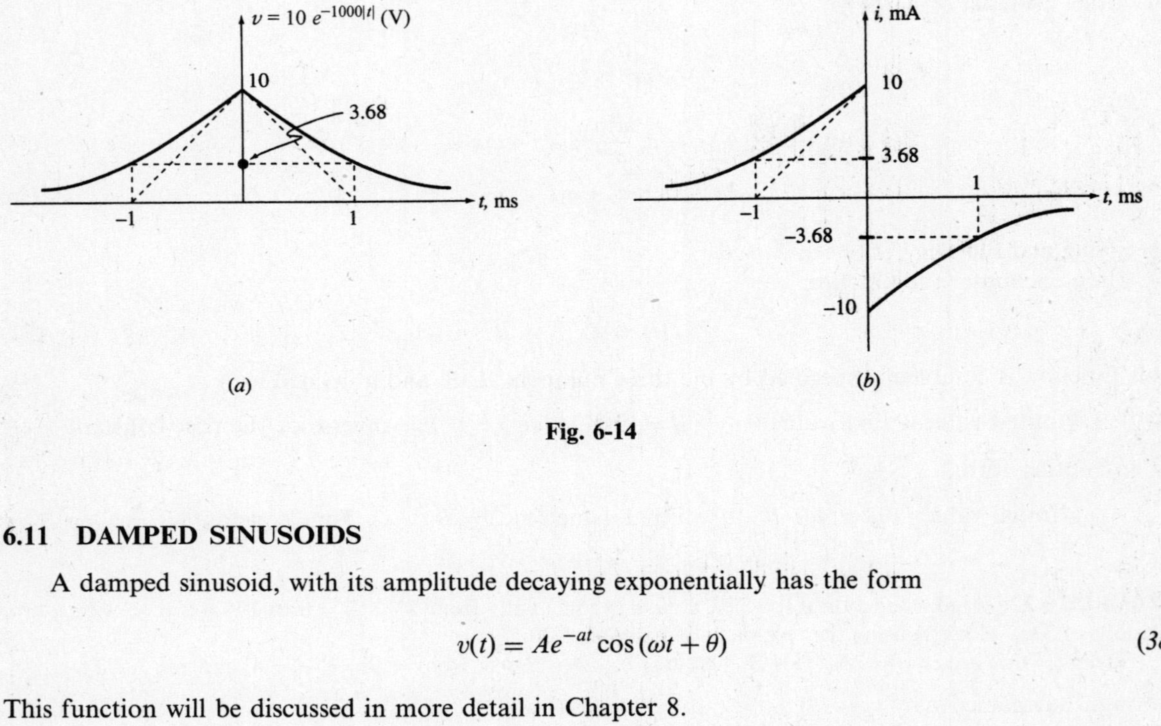

Fig. 6-14

6.11 DAMPED SINUSOIDS

A damped sinusoid, with its amplitude decaying exponentially has the form

$$v(t) = Ae^{-at}\cos(\omega t + \theta) \tag{38}$$

This function will be discussed in more detail in Chapter 8.

EXAMPLE 6.24 The current $i = I_0 e^{-at}\cos\omega t$ passes through a series RL circuit. (*a*) Find v_{RL}, the voltage across this combination. (*b*) Compute v_{RL} for $I_0 = 3\,\text{A}$, $a = 2$, $\omega = 40\,\text{rad/s}$, $R = 5\,\Omega$ and $L = 0.1\,\text{H}$. Sketch i as a function of time.

(*a*) We have

$$v_R = Ri = RI_0 e^{-at} \cos \omega t$$

$$v_L = L\frac{di}{dt} = -LI_0 e^{-at}(a \cos \omega t + \omega \sin \omega t)$$

$$v_{RL} = v_R + v_L = I_0 e^{-at}[(R - La)\cos \omega t - L\omega \sin \omega t] = V_0 e^{-at} \cos(\omega t + \theta)$$

where $$V_0 = I_0\sqrt{(R - La)^2 + L^2\omega^2} \quad \text{and} \quad \theta = \tan^{-1}[L\omega/(R - La)] \tag{39}$$

(*b*) Substituting the given data into (*39*), $V_0 = 18.75$ V and $\theta = 39.8°$. Current i and voltage v_{RL} are then given by

$$i = 3e^{-2t} \cos 40t \quad \text{and} \quad v_{RL} = 18.75e^{-2t} \cos(40t + 39.8°)$$

The current i is graphed in Fig. 6-15.

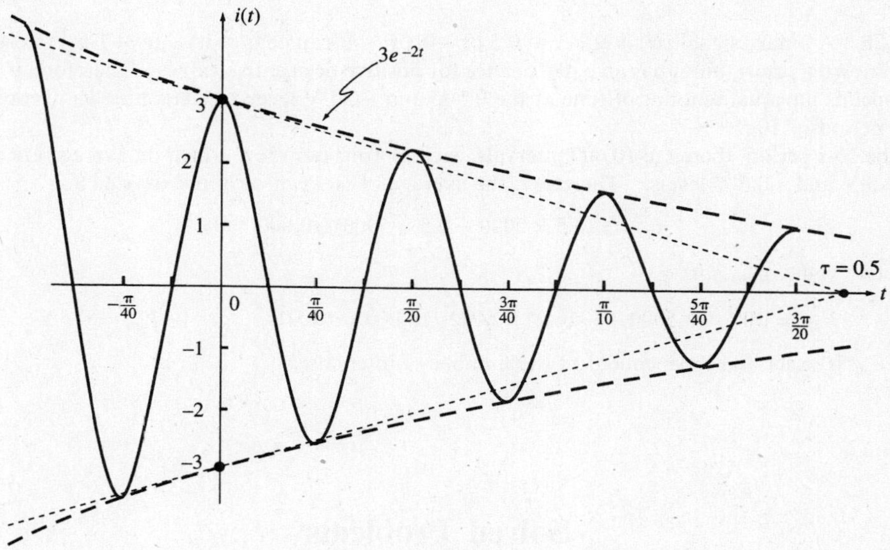

Fig. 6-15

6.12 RANDOM SIGNALS

So far we have dealt with signals which are completely specified. For example, the values of a sinusoidal waveform, such as the line voltage, can be determined for all times if its amplitude, frequency, and phase are known. Such signals are called *deterministic*.

There exists another class of signals which can be specified only partly through their time averages, such as their mean, rms value, and frequency range. These are called *random signals*. Random signals can carry information and should not be mistaken with noise, which normally corrupts the information contents of the signal.

The voltage recorded at the terminals of a microphone due to speech utterance and the signals picked up by an antenna tuned to a radio or TV station are examples of random signals. The future course and values of such signals can be predicted only in average and not precisely. Other examples of random signals are the binary waveforms in digital computers, image intensities over the area of a picture, and the speech or music which modulates the amplitude of carrier waves in an AM system.

It may not seem useful to discuss signals whose values are specified only in average. However, through harmonic analysis we can still find much about the average effect of such signals in electric circuits.

EXAMPLE 6.25 Samples from a random signal $x(t)$ are recorded every 1 ms and designated by $x(n)$. Approximate the mean and rms values of $x(t)$ from samples given in Table 6-2.

Table 6-2

n	0	1	2	3	4	5	6	7	8	9	10	11	12	13	14	15
$x(n)$	2	4	11	5	7	6	9	10	3	6	8	4	1	3	5	12

The time averages of $x(t)$ and $x^2(t)$ may be approximated from $x(n)$.

$$X_{\text{avg}} = (2+4+11+5+7+6+9+10+3+6+8+4+1+3+5+12)/16 = 6$$
$$X_{\text{eff}}^2 = (2^2+4^2+11^2+5^2+7^2+6^2+9^2+10^2+3^3+6^2+8^2+4^2+1^2+3^2+5^2+12^2)/16 = 46$$
$$X_{\text{eff}} = 6.78$$

EXAMPLE 6.26 A binary signal $v(t)$ is either at 0.5 or -0.5 V. It can change its sign at 1-ms intervals. The sign change is not known a priori, but it has an equal chance for positive or negative values. Therefore, if measured for a long time, it spends an equal amount of time at the 0.5-V and -0.5-V levels. Determine its average and effective values over a period of 10 s.

During the 10-s period, there are 10,000 intervals, each of 1-ms duration, which on average are equally divided between the 0.5-V and -0.5-V levels. Therefore, the average of $v(t)$ can be approximated as

$$v_{\text{avg}} = (0.5 \times 5000 - 0.5 \times 5000)/10{,}000 = 0$$

The effective value of $v(t)$ is

$$V_{\text{eff}}^2 = [(0.5)^2 \times 5000 + (-0.5)^2 \times 5000]/10{,}000 = (0.5)^2 \qquad \text{or} \qquad V_{\text{eff}} = 0.5\,\text{V}$$

The value of V_{eff} is exact and independent of the number of intervals.

Solved Problems

6.1 Find the maximum and minimum values of $v = 1 + 2\sin(\omega t + \theta)$, given $\omega = 1000$ rad/s and $\theta = 3$ rad. Determine if the function v is periodic, and find its frequency f and period T. Specify the phase angle in degrees.

$$V_{\max} = 1 + 2 = 3 \qquad V_{\min} = 1 - 2 = -1$$

The function v is periodic. To find the frequency and period, we note that $\omega = 2\pi f = 1000$ rad/s. Thus,

$$f = 1000/2\pi = 159.15\,\text{Hz} \qquad \text{and} \qquad T = 1/f = 2\pi/1000 = 0.00628\,\text{s} = 6.28\,\text{ms}$$
$$\text{Phase angle} = 3\,\text{rad} = 180^\circ \times 3/\pi = 171.9^\circ$$

6.2 In a microwave range measurement system the electromagnetic signal $v_1 = A\sin 2\pi ft$, with $f = 100$ MHz, is transmitted and its echo $v_2(t)$ from the target is recorded. The range is computed from τ, the time delay between the signal and its echo. (*a*) Write an expression for $v_2(t)$ and compute its phase angle for time delays $\tau_1 = 515$ ns and $\tau_2 = 555$ ns. (*b*) Can the distance be computed unambiguously from the phase angle in $v_2(t)$? If not, determine the additional needed information.

(*a*) Let $v_2(t) = B\sin 2\pi f(t - \tau) = B\sin(2\pi ft - \theta)$.
For $f = 100\,\text{MHz} = 10^8$ Hz, $\theta = 2\pi f\tau = 2 \times 10^8\pi\tau = 2\pi k + \phi$ where $0 < \phi < 2\pi$.
For $\tau_1 = 515 \times 10^{-9}$, $\theta_1 = 2\pi 10^8 \times 515 \times 10^{-9} = 103\pi = 51 \times 2\pi + \phi_1$ or $k_1 = 51$ and $\phi_1 = \pi$.
For $\tau_2 = 555 \times 10^{-9}$, $\theta_2 = 2\pi 10^8 \times 555 \times 10^{-9} = 111\pi = 55 \times 2\pi + \phi_2$ or $k_2 = 55$ and $\phi_2 = \pi$.

(*b*) Since phase angles ϕ_1 and ϕ_2 are equal, the time delays τ_1 and τ_2 may not be distinguished from each other based on the corresponding phase angles ϕ_1 and ϕ_2. For unambiguous determination of the distance, k and ϕ are both needed.

6.3 Show that if periods T_1 and T_2 of two periodic functions $v_1(t)$ and $v_2(t)$ have a common multiple, the sum of the two functions, $v(t) = v_1(t) + v_2(t)$, is periodic with a period equal to the smallest common multiple of T_1 and T_2. In such case show that $V_{avg} = V_{1,avg} + V_{2,avg}$.

If two integers n_1 and n_2 can be found such that $T = n_1 T_1 = n_2 T_2$, then $v_1(t) = v_1(t + n_1 T_1)$ and $v_2(t) = v_2(t + n_2 T_2)$. Consequently,

$$v(t+T) = v_1(t+T) + v_2(t+T) = v_1(t) + v_2(t) = v(t)$$

and $v(t)$ is periodic with period T.

The average is

$$V_{avg} = \frac{1}{T}\int_0^T [v_1(t) + v_2(t)]\,dt = \frac{1}{T}\int_0^T v_1(t)\,dt + \frac{1}{T}\int_0^T v_2(t)\,dt = V_{1,avg} + V_{2,avg}$$

6.4 Show that the average of $\cos^2(\omega t + \theta)$ is 1/2.

Using the identity $\cos^2(\omega t + \theta) = \frac{1}{2}[1 + \cos 2(\omega t + \theta)]$, the notation $\langle f \rangle = F_{avg}$, and the result of Problem 6.3, we have

$$\langle 1 + \cos 2(\omega t + \theta)\rangle = \langle 1 \rangle + \langle \cos 2(\omega t + \theta)\rangle$$

But $\langle \cos 2(\omega t + \theta)\rangle = 0$. Therefore, $\langle \cos^2(\omega t + \theta)\rangle = 1/2$.

6.5 Let $v(t) = V_{dc} + V_{ac}\cos(\omega t + \theta)$. Show that $V_{eff}^2 = V_{dc}^2 + \frac{1}{2}V_{ac}^2$.

$$\begin{aligned} V_{eff}^2 &= \frac{1}{T}\int_0^T [V_{dc} + V_{ac}\cos(\omega t + \theta)]^2\,dt \\ &= \frac{1}{T}\int_0^T [V_{dc}^2 + V_{ac}^2\cos^2(\omega t + \theta) + 2V_{dc}V_{ac}\cos(\omega t + \theta)]\,dt \\ &= V_{dc}^2 + \tfrac{1}{2}V_{ac}^2 \end{aligned}$$

Alternatively, we can write

$$\begin{aligned} V_{eff}^2 = \langle v^2(t)\rangle &= \langle [V_{dc} + V_{ac}\cos(\omega t + \theta)]^2\rangle \\ &= \langle V_{dc}^2 + V_{ac}^2\cos^2(\omega t + \theta) + 2V_{dc}V_{ac}\cos(\omega t + \theta)\rangle \\ &= V_{dc}^2 + V_{ac}^2\langle\cos^2(\omega t + \theta)\rangle + 2V_{dc}V_{ac}\langle\cos(\omega t + \theta)\rangle \\ &= V_{dc}^2 + \tfrac{1}{2}V_{ac}^2 \end{aligned}$$

6.6 Let f_1 and f_2 be two different harmonics of f_0. Show that the effective value of $v(t) = V_1\cos(2\pi f_1 t + \theta_1) + V_2\cos(2\pi f_2 t + \theta_2)$ is $\sqrt{\frac{1}{2}(V_1^2 + V_2^2)}$.

$$\begin{aligned} v^2(t) &= V_1^2\cos^2(2\pi f_1 t + \theta_1) + V_2^2\cos^2(2\pi f_2 t + \theta_2) \\ &\quad + 2V_1V_2\cos(2\pi f_1 t + \theta_1)\cos(2\pi f_2 t + \theta_2) \\ V_{eff}^2 = \langle v^2(t)\rangle &= V_1^2\langle\cos^2(2\pi f_1 t + \theta_1)\rangle + V_2^2\langle\cos^2(2\pi f_2 t + \theta_2)\rangle \\ &\quad + 2V_1V_2\langle\cos(2\pi f_1 t + \theta_1)\cos(2\pi f_2 t + \theta_2)\rangle \end{aligned}$$

But $\langle\cos^2(2\pi f_1 t + \theta_1)\rangle = \langle\cos^2(2\pi f_2 t + \theta_2)\rangle = 1/2$ (see Problem 6.4) and

$$\langle \cos(2\pi f_1 t + \theta_1)\cos(2\pi f_2 t + \theta_2)\rangle = \frac{1}{2}\langle \cos[2\pi(f_1 + f_2)t + (\theta_1 + \theta_2)]\rangle + \frac{1}{2}\langle \cos[2\pi(f_1 - f_2)t + (\theta_1 - \theta_2)]\rangle = 0$$

Therefore, $V_{\text{eff}}^2 = \frac{1}{2}(V_1^2 + V_2^2)$ and $V_{\text{eff}} = \sqrt{\frac{1}{2}(V_1^2 + V_2^2)}$.

6.7 The signal $v(t)$ in Fig. 6-16 is sinusoidal. Find its period and frequency. Express it in the form $v(t) = A + B\cos(\omega t + \theta)$ and find its average and rms values.

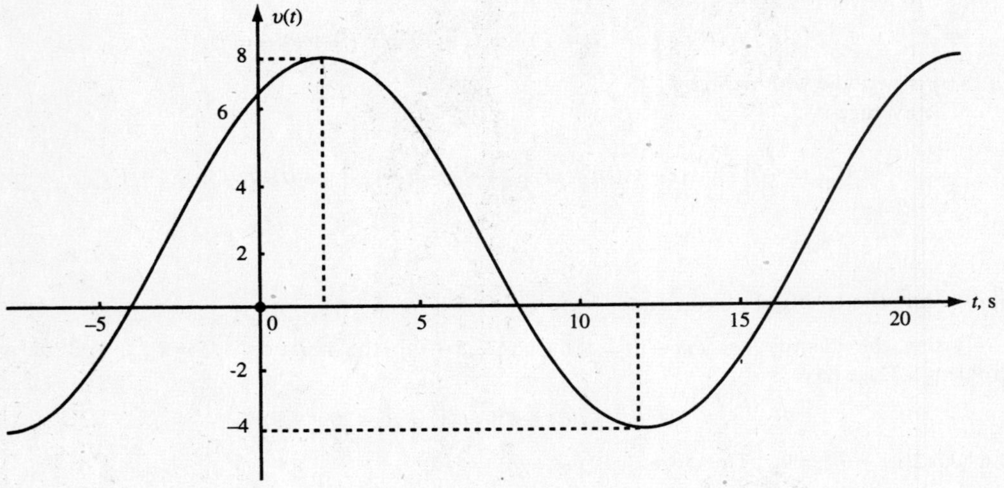

Fig. 6-16

The time between two positive peaks, $T = 20$ s, is one period corresponding to a frequency $f = 0.05$ Hz. The signal is a cosine function with amplitude B added to a constant value A.

$$B = \tfrac{1}{2}(V_{\text{max}} - V_{\text{min}}) = \tfrac{1}{2}(8 + 4) = 6 \qquad A = V_{\text{max}} - B = V_{\text{min}} + B = 2$$

The cosine is shifted by 2 s to the right, which corresponds to a phase lag of (2/20)360° = 36°. Therefore, the signal is expressed by

$$v(t) = 2 + 6\cos\left(\frac{\pi}{10}t - 36^\circ\right)$$

The average and effective values are found from A and B:

$$V_{\text{avg}} = A = 2, \qquad V_{\text{eff}}^2 = A^2 + B^2/2 = 2^2 + 6^2/2 = 22 \qquad \text{or} \qquad V_{\text{eff}} = \sqrt{22} = 4.69$$

6.8 Let $v_1 = \cos 200\pi t$ and $v_2 = \cos 202\pi t$. Show that $v = v_1 + v_2$ is periodic. Find its period, V_{max}, and the times when v attains its maximum value.

The periods of v_1 and v_2 are $T_1 = 1/100$ s and $T_2 = 1/101$ s, respectively. The period of $v = v_1 + v_2$ is the smallest common multiple of T_1 and T_2, which is $T = 100T_1 = 101T_2 = 1$ s. The maximum of v occurs at $t = k$ with k an integer when v_1 and v_2 are at their maxima and $V_{\text{max}} = 2$.

6.9 Convert $v(t) = 3\cos 100t + 4\sin 100t$ to $A\sin(100t + \theta)$.

Note that $3/\sqrt{3^2 + 4^2} = 3/5 = \sin 36.87^\circ$ and $4/\sqrt{3^2 + 4^2} = 4/5 = \cos 36.87^\circ$. Then,

$$v(t) = 3\cos 100t + 4\sin 100t = 5(0.6\cos 100t + 0.8\sin 100t)$$
$$= 5(\sin 36.87^\circ \cos 100t + \cos 36.87^\circ \sin 100t) = 5\sin(100t + 36.87^\circ)$$

6.10 Find the average and effective value of $v_2(t)$ in Fig. 6-1(*b*) for $V_1 = 2$, $V_2 = 1$, $T = 4T_1$.

$$V_{2,\text{avg}} = \frac{V_1T_1 - V_2(T - T_1)}{T} = \frac{V_1 - 3V_2}{4} = -0.25$$

$$V_{2,\text{eff}}^2 = \frac{V_1^2T_1 + V_2^2(T - T_1)}{T} = \frac{7}{4} \qquad \text{or} \qquad V_{2,\text{eff}} = \sqrt{7}/2 = 1.32$$

6.11 Find $V_{3,\text{avg}}$ and $V_{3,\text{eff}}$ in Fig. 6-1(*c*) for $T = 100T_1$.

From Fig. 6-1(*c*), $V_{3,\text{avg}} = 0$. To find $V_{3,\text{eff}}$, observe that the integral of v_3^2 over one period is $V_0^2T_1/2$. The average of v_3^2 over $T = 100T_1$ is therefore

$$\langle v_3^2(t) \rangle = V_{3,\text{eff}}^2 = V_0^2T_1/200T_1 = V_0^2/200 \qquad \text{or} \qquad V_{3,\text{eff}} = V_0\sqrt{2}/20 = 0.0707V_0$$

The effective value of the tone burst is reduced by the factor $\sqrt{T/T_1} = 10$.

6.12 Referring to Fig. 6-1(*d*), let $T = 6$ and let the areas under the positive and negative sections of $v_4(t)$ be $+5$ and -3, respectively. Find the average and effective values of $v_4(t)$.

$$V_{4,\text{avg}} = (5 - 3)/6 = 1/3$$

The effective value cannot be determined from the given data.

6.13 Find the average and effective value of the half-rectified cosine wave $v_1(t)$ shown in Fig. 6-17(*a*).

$$V_{1,\text{avg}} = \frac{V_m}{T}\int_{-T/4}^{T/4} \cos\frac{2\pi t}{T}\,dt = \frac{V_mT}{2\pi T}\left[\sin\frac{2\pi t}{T}\right]_{-T/4}^{T/4} = \frac{V_m}{\pi}$$

$$V_{1,\text{eff}}^2 = \frac{V_m^2}{T}\int_{-T/4}^{T/4} \cos^2\frac{2\pi t}{T}\,dt = \frac{V_m^2}{2T}\int_{-T/4}^{T/4}\left(1 + \cos\frac{4\pi t}{T}\right)dt$$

$$= \frac{V_m^2}{2T}\left[t + \frac{T}{4\pi}\sin\frac{4\pi t}{T}\right]_{-T/4}^{T/4} = \frac{V_m^2}{2T}\left(\frac{T}{4} + \frac{T}{4}\right) = \frac{V_m^2}{4}$$

from which $V_{1,\text{eff}} = V_m/2$.

6.14 Find the average and effective value of the full-rectified cosine wave $v_2(t) = V_m|\cos 2\pi t/T|$ shown in Fig. 6-17(*b*).

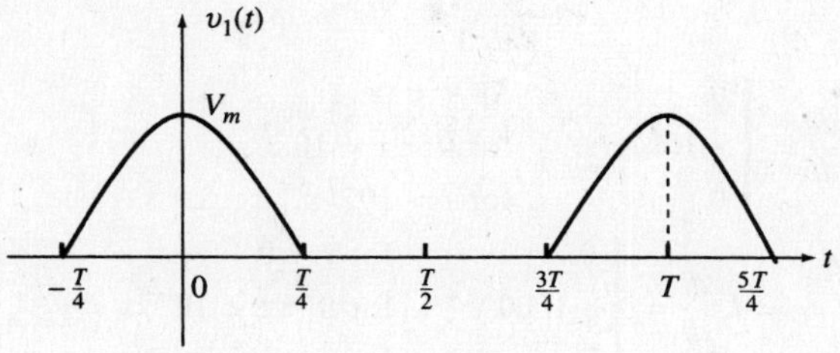

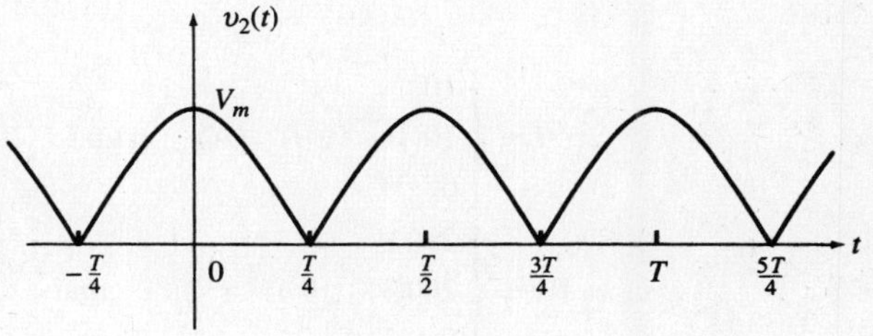

Fig. 6-17

Use the results of Problems 6.3 and 6.13 to find $V_{2,\text{avg}}$. Thus,

$$v_2(t) = v_1(t) + v_1(t - T/2) \qquad \text{and} \qquad V_{2,\text{avg}} = V_{1,\text{avg}} + V_{1,\text{avg}} = 2V_{1,\text{avg}} = 2V_m/\pi$$

Use the results of Problems 6.5 and 6.13 to find $V_{2,\text{eff}}$. And so,

$$V_{2,\text{eff}}^2 = V_{1,\text{eff}}^2 + V_{1,\text{eff}}^2 = 2V_{1,\text{eff}}^2 = V_m^2/2 \qquad \text{or} \qquad V_{2,\text{eff}} = V_m/\sqrt{2}$$

The rms value of $v_2(t)$ can also be derived directly. Because of the squaring operation, a full-rectified cosine function has the same rms value as the cosine function itself, which is $V_m/\sqrt{2}$.

6.15 A 100-mH inductor in series with 20-Ω resistor [Fig. 6-18(*a*)] carries a current i as shown in Fig. 6-18(*b*). Find and plot the voltages across R, L, and RL.

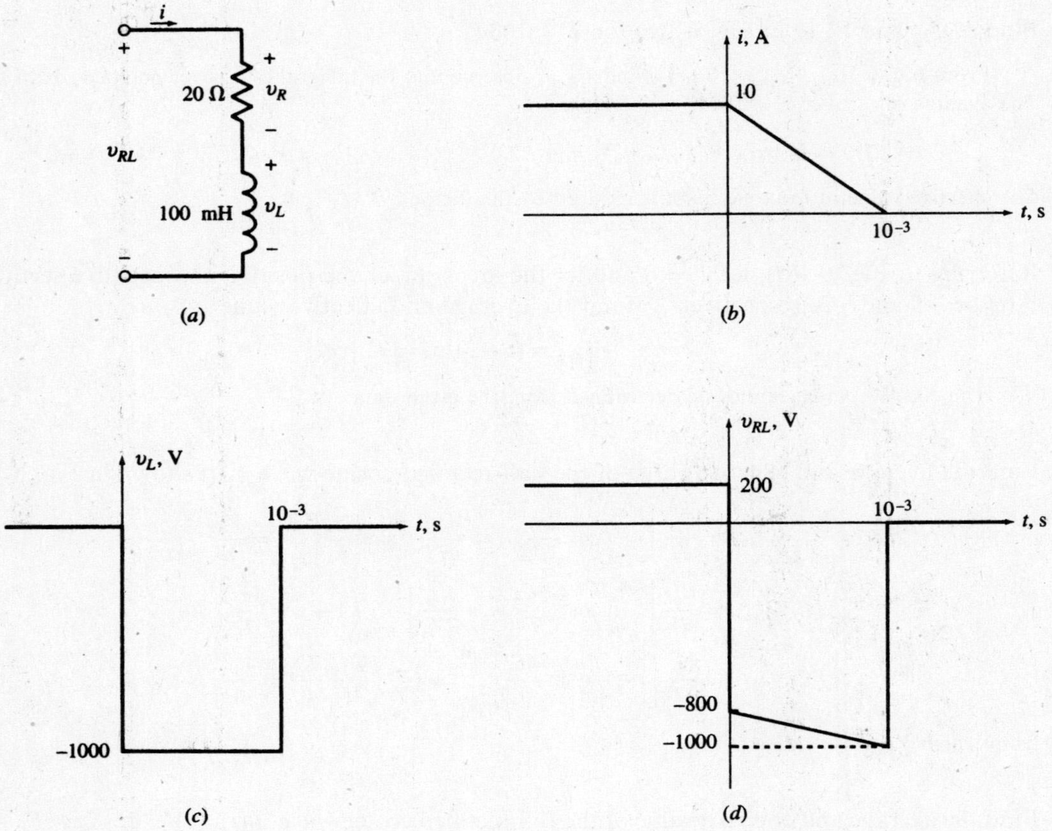

Fig. 6-18

$$i = \begin{cases} 10 \\ 10(1 - 10^3 t) \quad (\text{A}) \\ 0 \end{cases} \quad \text{and} \quad \frac{di}{dt} = \begin{cases} 0 & \text{for } t < 0 \\ -10^4 \text{ A/s} & \text{for } 0 < t < 10^{-3} \text{ s} \\ 0 & \text{for } t > 10^{-3} \text{ s} \end{cases}$$

$$v_R = \text{Ri} = \begin{cases} 200 \text{ V} \\ 200(1 - 10^3 t) \quad (\text{V}) \\ 0 \end{cases} \quad \text{and} \quad v_L = L\frac{di}{dt} = \begin{cases} 0 & \text{for } t < 0 \\ -1000 \text{ V} & \text{for } 0 < t < 10^{-3} \text{ s} \\ 0 & \text{for } t > 10^{-3} \text{ s} \end{cases}$$

Since the passive elements are in series, $v_{RL} = v_R + v_L$ and so

$$v_{RL} = \begin{cases} 200 \text{ V} & \text{for } t < 0 \\ -2(10^5 t) - 800 \quad (\text{V}) & \text{for } 0 < t < 10^{-3} \text{ s} \\ 0 & \text{for } t > 10^{-3} \text{ s} \end{cases}$$

The graphs of v_L and v_{RL} are given in Fig. 6-18(*c*) and (*d*), respectively. The plot of the resistor voltage v_R has the same shape as that of the current [see Fig. 6-18(*b*)], except for scaling by a factor of $+20\times$.

6.16 A radar signal $s(t)$, with amplitude $V_m = 100$ V, consists of repeated tone bursts. Each tone burst lasts $T_b = 50\ \mu s$. The bursts are repeated every $T_s = 10$ ms. Find S_{eff} and the average power in $s(t)$.

Let $V_{eff} = V_m\sqrt{2}$ be the effective value of the sinusoid within a burst. The energy contained in a single burst is $W_b = T_b V_{eff}^2$. The energy contained in one period of $s(t)$ is $W_s = T_s S_{eff}^2$. Since $W_b = W_s = W$, we obtain

$$T_b V_{eff}^2 = T_s S_{eff}^2 \qquad S_{eff}^2 = (T_b/T_s)V_{eff}^2 \qquad S_{eff} = \sqrt{T_b/T_s}\,V_{eff} \tag{40}$$

Substituting the values of T_b, T_s, and V_{eff} into (40), we obtain

$$S_{eff} = \sqrt{(50 \times 10^{-6})/(10 \times 10^{-3})}\,(100/\sqrt{2}) = 5\text{ V}$$

Then $W = 10^{-2}(25) = 0.25$ J. The average power in $s(t)$ is

$$P = W/T_s = T_s S_{eff}^2/T_s = S_{eff}^2 = 25\text{ W}$$

The average power of $s(t)$ is represented by S_{eff}^2 and its peak power by V_{eff}^2. The ratio of peak power to average power is $\sqrt{T_s/T_b}$. In this example the average power and the peak power are 25 W and 5000 W, respectively.

6.17 An appliance uses $V_{eff} = 120$ V at 60 Hz and draws $I_{eff} = 10$ A with a phase lag of 60°. Express v, i, and $p = vi$ as functions of time and show that power is periodic with a dc value. Find the frequency, and the average, maximum, and minimum values of p.

$$v = 120\sqrt{2}\cos\omega t \qquad i = 10\sqrt{2}\cos(\omega t - 60°)$$

$$p = vi = 2400\cos\omega t\cos(\omega t - 60°) = 1200\cos 60° + 1200\cos(2\omega t - 60°) = 600 + 1200\cos(2\omega t - 60°)$$

The power function is periodic. The frequency $f = 2 \times 60 = 120$ Hz and $P_{avg} = 600$ W, $p_{max} = 600 + 1200 = 1800$ W, $p_{min} = 600 - 1200 = -600$ W.

6.18 A narrow pulse i_s of 1-A amplitude and 1-μs duration enters a 1-μF capacitor at $t = 0$, as shown in Fig. 6-19. The capacitor is initially uncharged. Find the voltage across the capacitor.

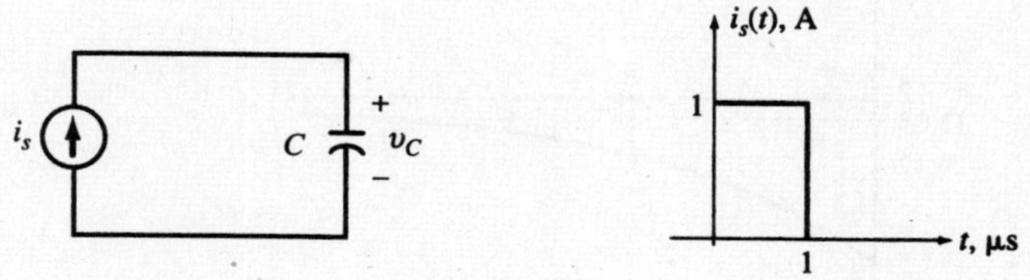

Fig. 6-19

The voltage across the capacitor is

$$V_C = \frac{1}{C}\int_{-\infty}^{t} i\,dt = \begin{cases} 0 & \text{for } t < 0 \\ 10^6 t \quad (\text{V}) & \text{for } 0 < t < 1\ \mu\text{s (charging period)} \\ 1\text{ V} & \text{for } t > 1\ \mu\text{s} \end{cases}$$

If the same amount of charge were deposited on the capacitor in zero time, then we would have $v = u(t)$ (V) and $i(t) = 10^{-6}\delta(t)$ (A).

6.19 The narrow pulse i_s of Problem 6.18 enters a parallel combination of a 1-μF capacitor and a 1-MΩ resistor (Fig. 6-20). Assume the pulse ends at $t = 0$ and that the capacitor is initially uncharged. Find the voltage across the parallel RC combination.

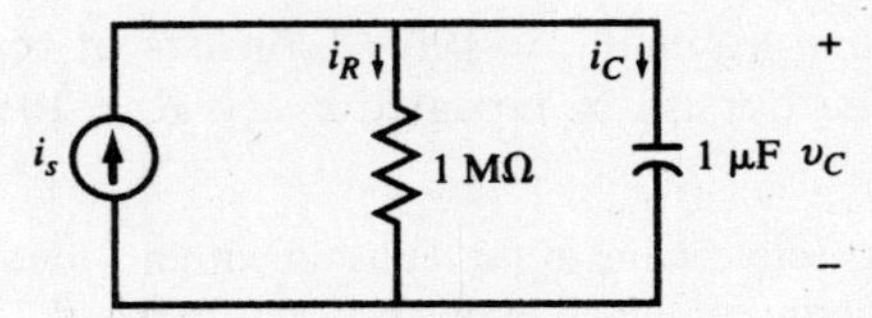

Fig. 6-20

Let v designate the voltage across the parallel RC combination. The current in R is $i_R = v/R = 10^{-6}v$. During the pulse, i_R remains negligible because v cannot exceed 1 V and i_R remains under 1 μA. Therefore, it is reasonable to assume that during the pulse, $i_C = 1$ A and consequently $v(0^+) = 1$ V. For $t > 0$, from application of KVL around the RC loop we get

$$v + \frac{dv}{dt} = 0, \qquad v(0^+) = 1\text{ V} \tag{41}$$

The only solution to (41) is $v = e^{-t}$ for $t > 0$ or $v(t) = e^{-t}u(t)$ for all t. For all practical purposes, i_s can be considered an impulse of size 10^{-6} A, and then $v = e^{-t}u(t)$ (V) is called the *response* of the RC combination to the current impulse.

6.20 Plot the function $v(t)$ which varies exponentially from 5 V at $t = 0$ to 12 V at $t = \infty$ with a time constant of 2 s. Write the equation for $v(t)$.

Identify the initial point A ($t = 0$ and $v = 5$) and the asymptote $v = 12$ in Fig. 6-21. The tangent at A intersects the asymptote at $t = 2$, which is point B on the line. Draw the tangent line AB. Identify point C belonging to the curve at $t = 2$. For a more accurate plot, identify point D at $t = 4$. Draw the curve as shown. The equation is $v(t) = Ae^{-t/2} + B$. From the initial and final conditions, we get $v(0) = A + B = 5$ and $v(\infty) = B = 12$ or $A = -7$, and $v(t) = -7e^{-t/2} + 12$.

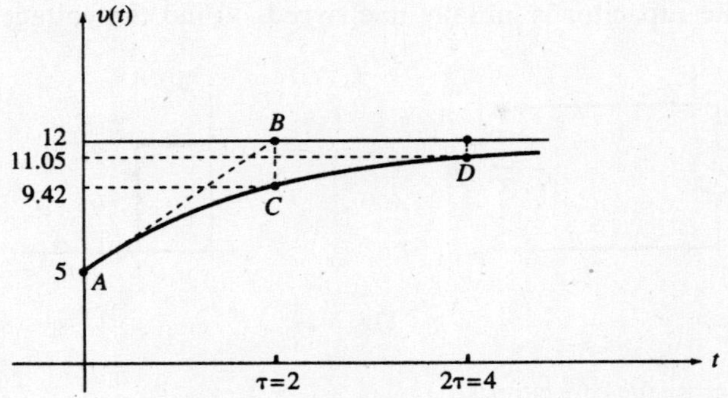

Fig. 6-21

6.21 The voltage $v = V_0e^{-a|t|}$ for $a > 0$ is connected across a parallel combination of a resistor and a capacitor as shown in Fig. 6-22(*a*). (*a*) Find currents i_C, i_R, and $i = i_C + i_R$. (*b*) Compute and graph v, i_C, i_R, and i for $V_0 = 10$ V, $C = 1$ μF, $R = 1$ MΩ, and $a = 1$.

(*a*) See (*a*) in Table 6-3 for the required currents.

(*b*) See (*b*) in Table 6-3. Figures 6-22(*b*)–(*e*) show the plots of v, i_C, i_R, and i, respectively, for the given data. During $t > 0$, $i = 0$, and the voltage source does not supply any current to the RC combination. The resistor current needed to sustain the exponential voltage across it is supplied by the capacitor.

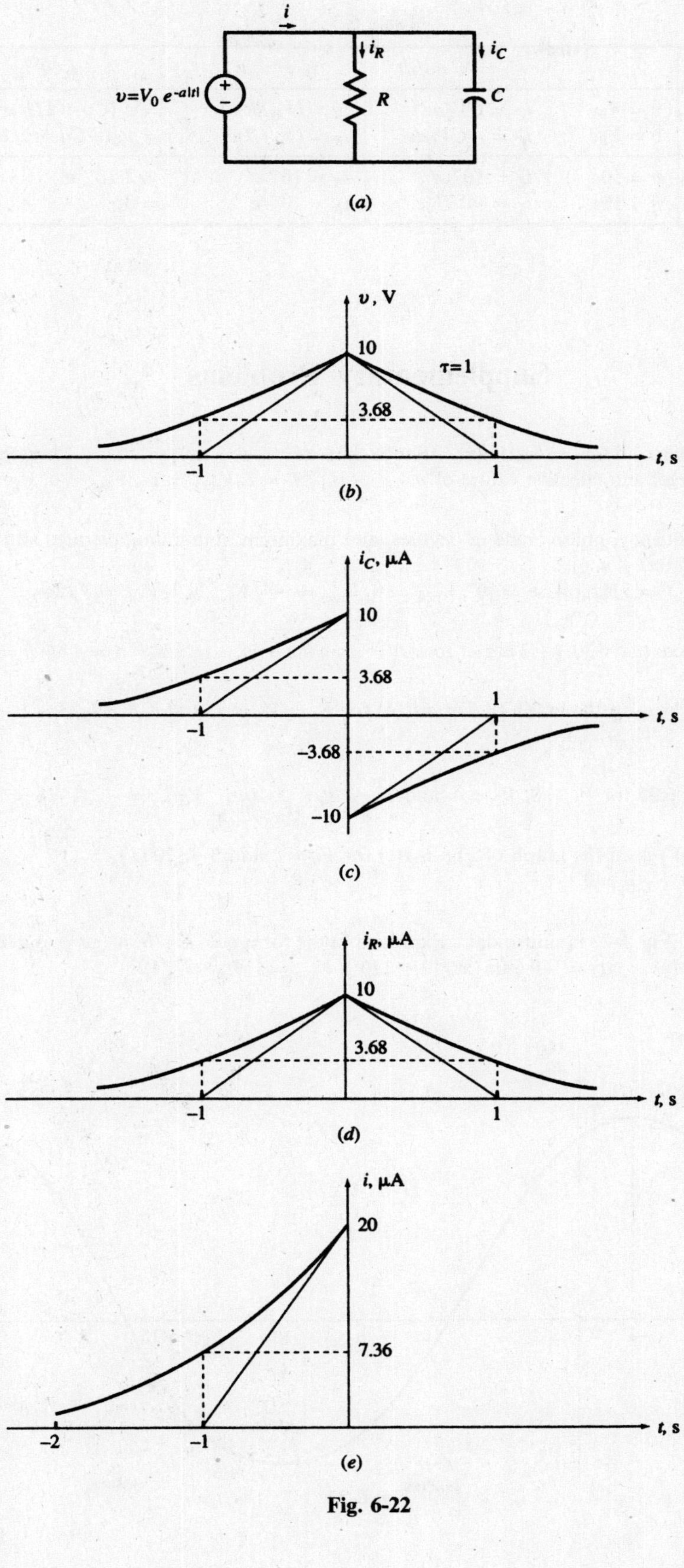

i
i_R
i_C
$v=V_0 e^{-a|t|}$
R
C
(a)
v, V
10
τ=1
3.68
−1
1
t, s
(b)
i_C, μA
10
3.68
1
t, s
−1
−3.68
−10
(c)
i_R, μA
10
3.68
t, s
−1
1
(d)
i, μA
20
7.36
t, s
−2
−1
(e)

Fig. 6-22

Table 6-3

	Time	v	$i_C = C\,dv/dt$	$i_R = v/R$	$i = i_C + i_R$
(*a*)	$t<0$ $t>0$	$v = V_0 e^{at}$ $v = V_0 e^{-at}$	$i_C = CV_0 a e^{at}$ $i_C = -CV_0 a e^{-at}$	$i_R = (V_0/R)e^{at}$ $i_R = (V_0/R)e^{-at}$	$i = v_0(Ca + 1/R)e^{at}$ $i = V_0(-Ca + 1/R)e^{-at}$
(*b*)	$t<0$ $t>0$	$v = 10e^{t}$ $v = 10e^{-t}$	$i_C = 10^{-5}e^{t}$ $i_C = -10^{-5}e^{-t}$	$i_R = 10^{-5}e^{t}$ $i_R = 10^{-5}e^{-t}$	$i = 2(10^{-5}e^{t})$ $i = 0$

Supplementary Problems

6.22 Let $v_1 = 8\sin 100\pi t$ and $v_2 = 6\sin 99\pi t$. Show that $v = v_1 + v_2$ is periodic. Find the period, and the maximum, average, and effective values of v. *Ans.* $T = 2$, $V_{\max} = 14$, $V_{\text{avg}} = 0$, $V_{\text{eff}} = 5\sqrt{2}$

6.23 Find period, frequency, phase angle in degrees, and maximum, minimum, average, and effective values of $v(t) = 2 + 6\cos(10\pi t + \pi/6)$.
Ans. $T = 0.2\text{ s}$, $f = 5\text{ Hz}$, phase $= 30°$, $V_{\max} = 8$, $V_{\min} = -4$, $V_{\text{avg}} = 2$, $V_{\text{eff}} = \sqrt{22}$

6.24 Reduce $v(t) = 2\cos(\omega t + 30°) + 3\cos\omega t$ to $v(t) = A\sin(\omega t + \theta)$. *Ans.* $A = 4.84$, $\theta = 102°$

6.25 Find $V_{2,\text{avg}}$ and $V_{2,\text{eff}}$ in the graph of Fig. 6-1(*b*) for $V_1 = V_2 = 3$, and $T = 4T_1/3$.
Ans. $V_{2,\text{avg}} = 1.5$, $V_{2,\text{eff}} = 3$

6.26 Repeat Problem 6.25 for $V_1 = 0$, $V_2 = 4$, and $T = 2T_1$. *Ans.* $V_{2,\text{avg}} = -2$, $V_{2,\text{eff}} = 2\sqrt{2}$

6.27 Find $V_{3,\text{avg}}$ and $V_{3,\text{eff}}$ in the graph of Fig. 6-1(*c*) for $V_0 = 2$ and $T = 200T_1$.
Ans. $V_{3,\text{avg}} = 0$, $V_{3,\text{eff}} = 0.1$

6.28 The waveform in Fig. 6-23 is sinusoidal. Express it in the form $v = A + B\sin(\omega t + \theta)$ and find its mean and rms values. *Ans.* $v(t) = 1 + 6\sin(\pi t/12 + 120°)$, $V_{\text{avg}} = 1$, $V_{\text{eff}} = \sqrt{19}$

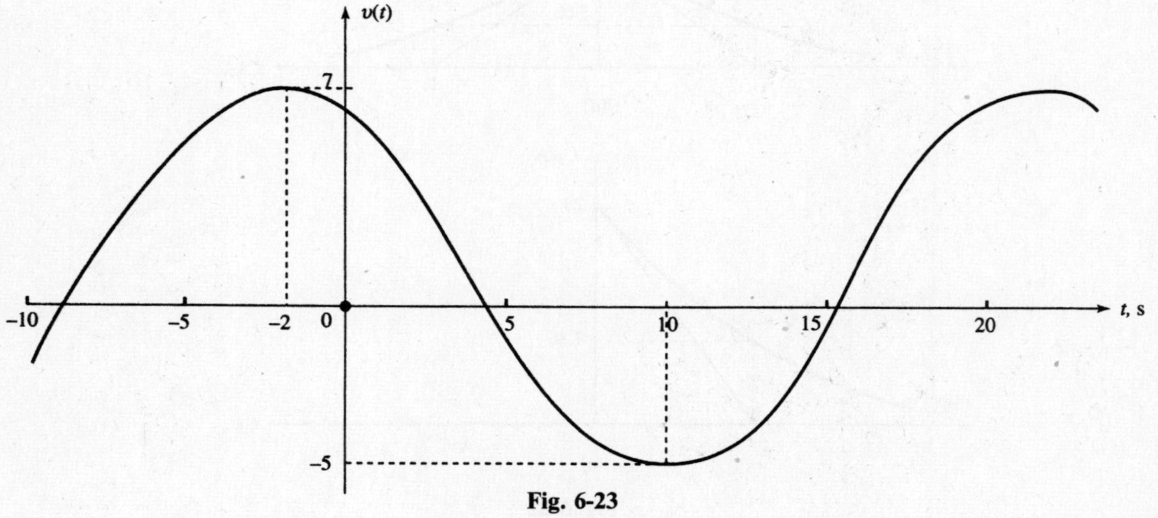

Fig. 6-23

6.29 Find the average and effective values of $v_1(t)$ in Fig. 6-24(a) and $v_2(t)$ in Fig. 6-24(b).

Ans. $V_{1,\mathrm{avg}} = -\dfrac{1}{3},\ V_{1,\mathrm{eff}} = \sqrt{\dfrac{17}{3}};\ V_{2,\mathrm{avg}} = -\dfrac{1}{2},\ V_{2,\mathrm{eff}} = \sqrt{\dfrac{13}{2}}$

6.30 The current through a series RL circuit with $R = 5\,\Omega$ and $L = 10\,\mathrm{H}$ is given in Fig. 6-10(a) where $T = 1$ s. Find the voltage across RL.

Ans. $$v = \begin{cases} 0 & \text{for } t < 0 \\ 10 + 5t & \text{for } 0 < t < 1 \\ 5 & \text{for } t > 1 \end{cases}$$

6.31 Find the capacitor current in Problem 6.19 (Fig. 6-20) for all t. *Ans.* $i_C = 10^{-6}[\delta(t) - e^{-t}u(t)]$

6.32 The voltage v across a 1-H inductor consists of one cycle of a sinusoidal waveform as shown in Fig. 6-25(a). (a) Write the equation for $v(t)$. (b) Find and plot the current through the inductor. (c) Find the amount and time of maximum energy in the inductor.

Ans. (a) $v = [u(t) - u(t-T)] \sin \dfrac{2\pi t}{T}$ (V)

(b) $i = (T/2\pi)[u(t) - u(t-T)]\left(1 - \cos \dfrac{2\pi t}{T}\right)$ (A). See Fig. 6-25(b).

(c) $W_{\max} = \dfrac{1}{2\pi^2} T^2$ (J) at $t = T/2$

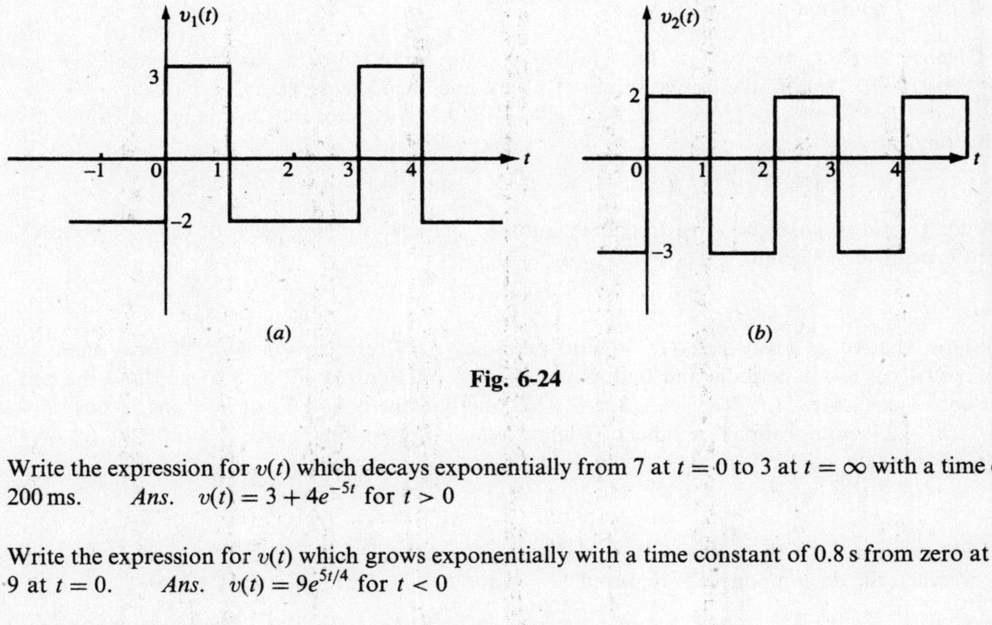

Fig. 6-24

6.33 Write the expression for $v(t)$ which decays exponentially from 7 at $t = 0$ to 3 at $t = \infty$ with a time constant of 200 ms. *Ans.* $v(t) = 3 + 4e^{-5t}$ for $t > 0$

6.34 Write the expression for $v(t)$ which grows exponentially with a time constant of 0.8 s from zero at $t = -\infty$ to 9 at $t = 0$. *Ans.* $v(t) = 9e^{5t/4}$ for $t < 0$

6.35 Express the current of Fig. 6-6 in terms of step functions.

Ans. $i(t) = 4u(t) + 6 \displaystyle\sum_{k=1}^{\infty} [u(t-5k) - u(t-5k+2)]$

6.36 In Fig. 6-10(a) let $T = 1$ s and call the waveform $s_1(t)$. Express $s_1(t)$ and its first two derivatives ds_1/dt and d^2s_1/dt^2, using step and impulse functions.
Ans. $s_1(t) = [u(t) - u(t-1)]t + u(t-1),\ ds_1/dt = u(t) - u(t-1),\ d^2s_1/dt^2 = \delta(t) - \delta(t-1)$

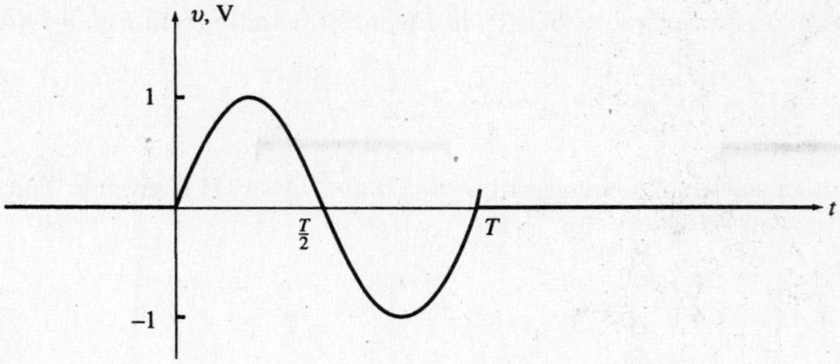

(a)

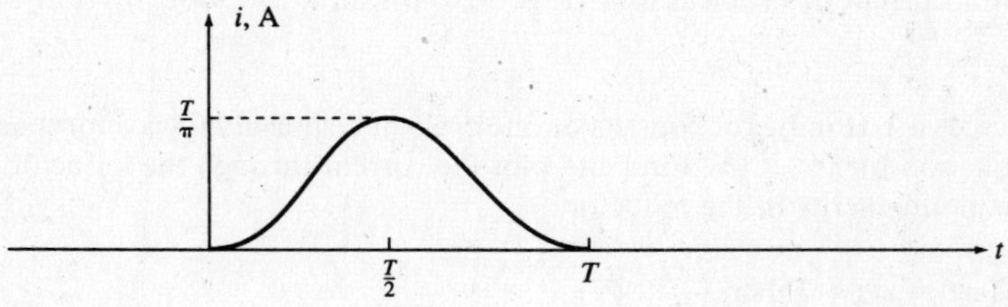

(b)

Fig. 6-25

6.37 Find an impulse voltage which creates a 1-A current jump at $t = 0$ when applied across a 10-mH inductor.
Ans. $v(t) = 10^{-2}\delta(t)$ (V)

6.38 (*a*) Given $v_1 = \cos t$, $v_2 = \cos(t + 30°)$ and $v = v_1 + v_2$, write v in the form of a single cosine function $v = A\cos(t + \theta)$. (*b*) Find effective values of v_1, v_2, and v. Discuss why $V_{\text{eff}}^2 > (V_{1,\text{eff}}^2 + V_{2,\text{eff}}^2)$.
Ans. (*a*) $v = 1.93\cos(t + 15°)$; (*b*) $V_{1,\text{eff}} = V_{2,\text{eff}} = 0.707$, $V_{\text{eff}} = 1.366$ V_{eff} is found from the following derivation

$$V_{\text{eff}}^2 = \langle v^2 \rangle = \langle (v_1 + v_2)^2 \rangle = \langle v_1^2 + v_2^2 + 2v_1v_2 \rangle = \langle v_1^2 \rangle + \langle v_2^2 \rangle + 2\langle v_1v_2 \rangle$$

Since v_1 and v_2 have the same frequency and are 30° out of phase, we get $\langle V_1V_2 \rangle = \frac{1}{2}\cos 30° = \sqrt{3}/4$, which is positive. Therefore, $V_{\text{eff}}^2 > (V_{1,\text{eff}}^2 + V_{2,\text{eff}}^2)$.

6.39 (*a*) Show that $v_1 = \cos t + \cos\sqrt{2}t$ is not periodic. (*b*) Replace $\sqrt{2}$ by 1.4 and then show that $v_2 = \cos t + \cos 1.4t$ is periodic and find its period T_2. (*c*) Replace $\sqrt{2}$ by 1.41 and find the period T_3 of $v_3 = \cos t + \cos 1.41t$. (*d*) Replace $\sqrt{2}$ by 1.4142 and find the period T_4 of $v_4 = \cos t + \cos 1.4142t$.
Ans. (*a*) $\sqrt{2}$ is not a rational number. Therefore, v_1 is not periodic. (*b*) $T_2 = 10\pi$ s. (*c*) $T_3 = 200\pi$ s. (*d*) $T_4 = 10\,000\pi$ s.

6.40 A random signal $s(t)$ with an rms value of 5 V has a dc value of 2 V. Find the rms value of $s_0(t) = s(t) - 2$, that is, when the dc component is removed. *Ans.* $S_{0,\text{eff}} = \sqrt{5^2 - 4} = \sqrt{21} = 4.58$ V

CHAPTER 7

First-Order Circuits

7.1 INTRODUCTION

Whenever a circuit is switched from one condition to another, either by a change in the applied source or a change in the circuit elements, there is a transitional period during which the branch currents and element voltages change from their former values to new ones. This period is called the *transient*. After the transient has passed, the circuit is said to be in the *steady state*. Now, the linear differential equation that describes the circuit will have two parts to its solution, the *complementary function* (or the *homogeneous solution*) and the *particular solution*. The complementary function corresponds to the transient, and the particular solution to the steady state.

In this chapter we will find the response of first-order circuits, given various initial conditions and sources. We will then develop an intuitive approach which can lead us to the same response without going through the formal solution of differential equations. We will also present and solve important issues relating to natural, force, step, and impulse responses, along with the dc steady state and the switching behavior of inductors and capacitors.

7.2 CAPACITOR DISCHARGE IN A RESISTOR

Assume a capacitor has a voltage difference V_0 between its plates. When a conducting path R is provided, the stored charge travels through the capacitor from one plate to the other, establishing a current i. Thus, the capacitor voltage v is gradually reduced to zero, at which time the current also becomes zero. In the RC circuit of Fig. 7-1(*a*), $Ri = v$ and $i = -C\,dv/dt$. Eliminating i in both equations gives

$$\frac{dv}{dt} + \frac{1}{RC}v = 0 \tag{1}$$

The only function whose linear combination with its derivative can be zero is an exponential function of the form Ae^{st}. Replacing v by Ae^{st} and dv/dt by sAe^{st} in (*1*), we get

$$sAe^{st} + \frac{1}{RC}Ae^{st} = A\left(s + \frac{1}{RC}\right)e^{st} = 0$$

from which $$s + \frac{1}{RC} = 0 \qquad \text{or} \qquad s = -\frac{1}{RC} \tag{2}$$

Given $v(0) = A = V_0$, $v(t)$ and $i(t)$ are found to be

$$v(t) = V_0 e^{-t/RC}, \qquad t > 0 \tag{3}$$

$$i(t) = -C\frac{dv}{dt} = \frac{V_0}{R}e^{-t/RC}, \qquad t > 0 \tag{4}$$

The voltage and current of the capacitor are exponentials with initial values of V_0 and V_0/R, respectively. As time increases, voltage and current decrease to zero with a time constant of $\tau = RC$. See Figs. 7-1(*b*) and (*c*).

EXAMPLE 7.1 The voltage across a 1-μF capacitor is 10 V for $t < 0$. At $t = 0$, a 1-MΩ resistor is connected across the capacitor terminals. Find the time constant τ, the voltage $v(t)$, and its value at $t = 5$ s.

$$\tau = RC = 10^6(10^{-6})\,\text{s} = 1\,\text{s} \qquad v(t) = 10e^{-t}\ (\text{V}),\ t > 0 \qquad v(5) = 10e^{-5} = 0.067\,\text{V}$$

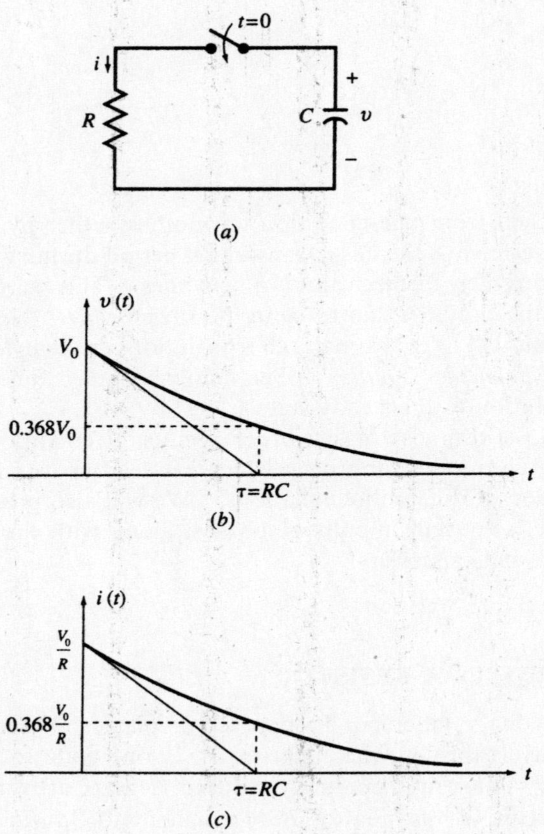

Fig. 7-1

EXAMPLE 7.2 A 5-μF capacitor with an initial voltage of 4 V is connected to a parallel combination of a 3-kΩ and a 6-kΩ resistor (Fig. 7-2). Find the current i in the 6-kΩ resistor.

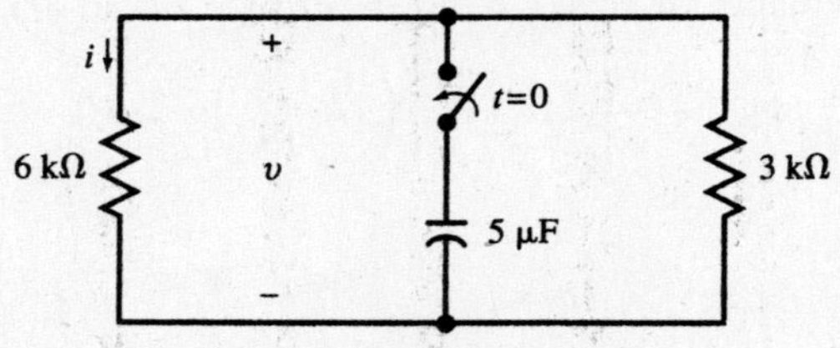

Fig. 7-2

The equivalent resistance of the two parallel resistors is $R = 2\,\text{k}\Omega$. The time constant of the circuit is $RC = 10^{-2}$ s. The voltage and current in the 6-kΩ resistor are, respectively,

$$v = 4e^{-100t}\ (\text{V}) \qquad \text{and} \qquad i = v/6000 = 0.67e^{-100t}\ (\text{mA})$$

7.3 ESTABLISHING A DC VOLTAGE ACROSS A CAPACITOR

Connect an initially uncharged capacitor to a battery with voltage V_0 through a resistor at $t = 0$. The circuit is shown in Fig. 7-3(*a*).

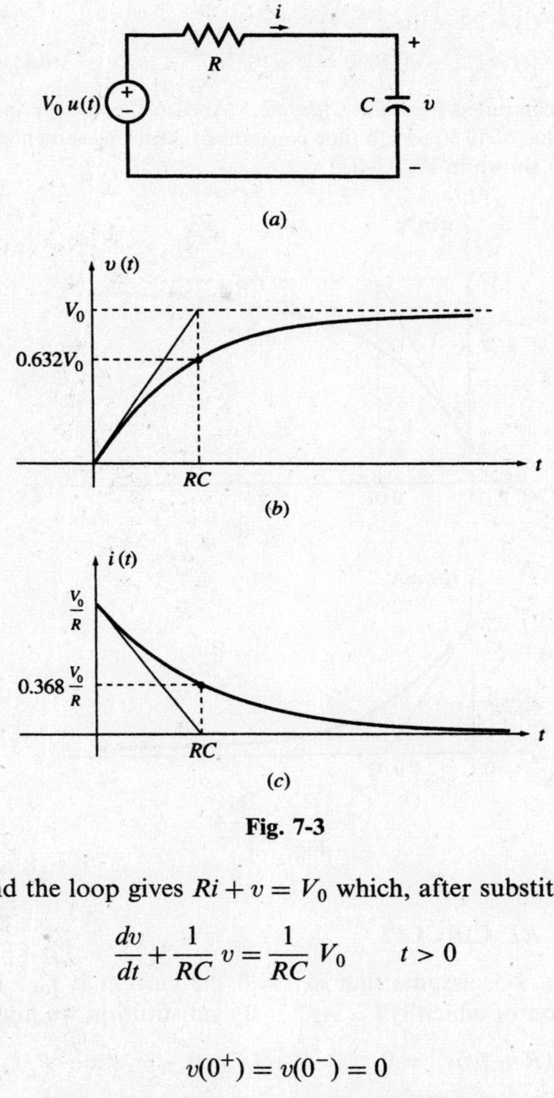

Fig. 7-3

For $t > 0$, KVL around the loop gives $Ri + v = V_0$ which, after substituting $i = C(dv/dt)$, becomes

$$\frac{dv}{dt} + \frac{1}{RC}v = \frac{1}{RC}V_0 \qquad t > 0 \tag{5a}$$

with the initial condition

$$v(0^+) = v(0^-) = 0 \tag{5b}$$

The solution should satisfy both (*5a*) and (*5b*). The *particular* solution (or *forced response*) $v_p(t) = V_0$ satisfies (*5a*) but not (*5b*). The *homogeneous* solution (or *natural response*) $v_h(t) = Ae^{-t/RC}$ can be added and its magnitude A can be adjusted so that the total solution (*6a*) satisfies both (*5a*) and (*5b*).

$$v(t) = v_p(t) + v_h(t) = V_0 + Ae^{-t/RC} \tag{6a}$$

From the initial condition, $v(0^+) = V_0 + A = 0$ or $A = -V_0$. Thus the total solution is

$$v(t) = V_0(1 - e^{-t/RC})u(t) \qquad \text{[see Fig. 7-3(b)]} \tag{6b}$$

$$i(t) = \frac{V_0}{R} e^{-t/RC} u(t) \qquad \text{[see Fig. 7-3(c)]} \tag{6c}$$

EXAMPLE 7.3 A 4-μF capacitor with an initial voltage of $v(0^-) = 2\,\text{V}$ is connected to a 12-V battery through a resistor $R = 5\,\text{k}\Omega$ at $t = 0$. Find the voltage across and current through the capacitor for $t > 0$.

The time constant of the circuit is $\tau = RC = 0.02\,\text{s}$. Following the analysis of Example 7.2, we get

$$v(t) = 12 + Ae^{-50t}$$

From the initial conditions, $v(0^-) = v(0^+) = 12 + A = 2$ or $A = -10$. Thus, for $t > 0$,

$$v(t) = 12 - 10e^{-50t} \text{ (V)}$$

$$i(t) = (12 - v)/5000 = 2 \times 10^{-3} e^{-50t} A = 2e^{-50t} \text{ (mA)}$$

The current may also be computed from $i = C(dv/dt)$. And so the voltage increases exponentially from an initial value of 2 V to a final value of 12 V, with a time constant of 20 ms, as shown in Fig. 7-4(*a*), while the current decreases from 2 mA to zero as shown in Fig. 7-4(*b*).

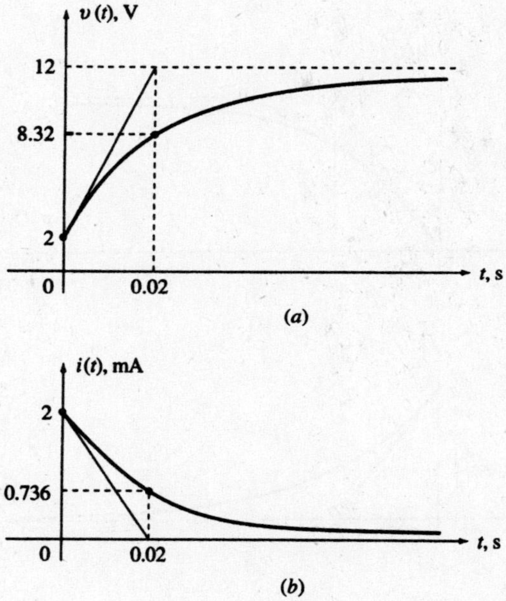

Fig. 7-4

7.4 THE SOURCE-FREE *RL* CIRCUIT

In the *RL* circuit of Fig. 7-5, assume that at $t = 0$ the current is I_0. For $t > 0$, i should satisfy $Ri + L(di/dt) = 0$, the solution of which is $i = Ae^{st}$. By substitution we find A and s:

$$A(R + Ls)e^{st} = 0, \qquad R + Ls = 0, \qquad s = -R/L$$

The initial condition $i(0) = A = I_0$. Then

$$i(t) = I_0 e^{-Rt/L} \qquad \text{for } t > 0 \tag{7}$$

The time constant of the circuit is L/R.

EXAMPLE 7.4 The 12-V battery in Fig. 7-6(*a*) is disconnected at $t = 0$. Find the inductor current and voltage v for all times.

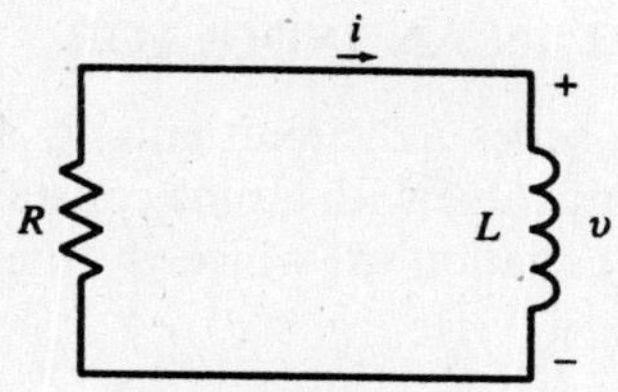

Fig. 7-5

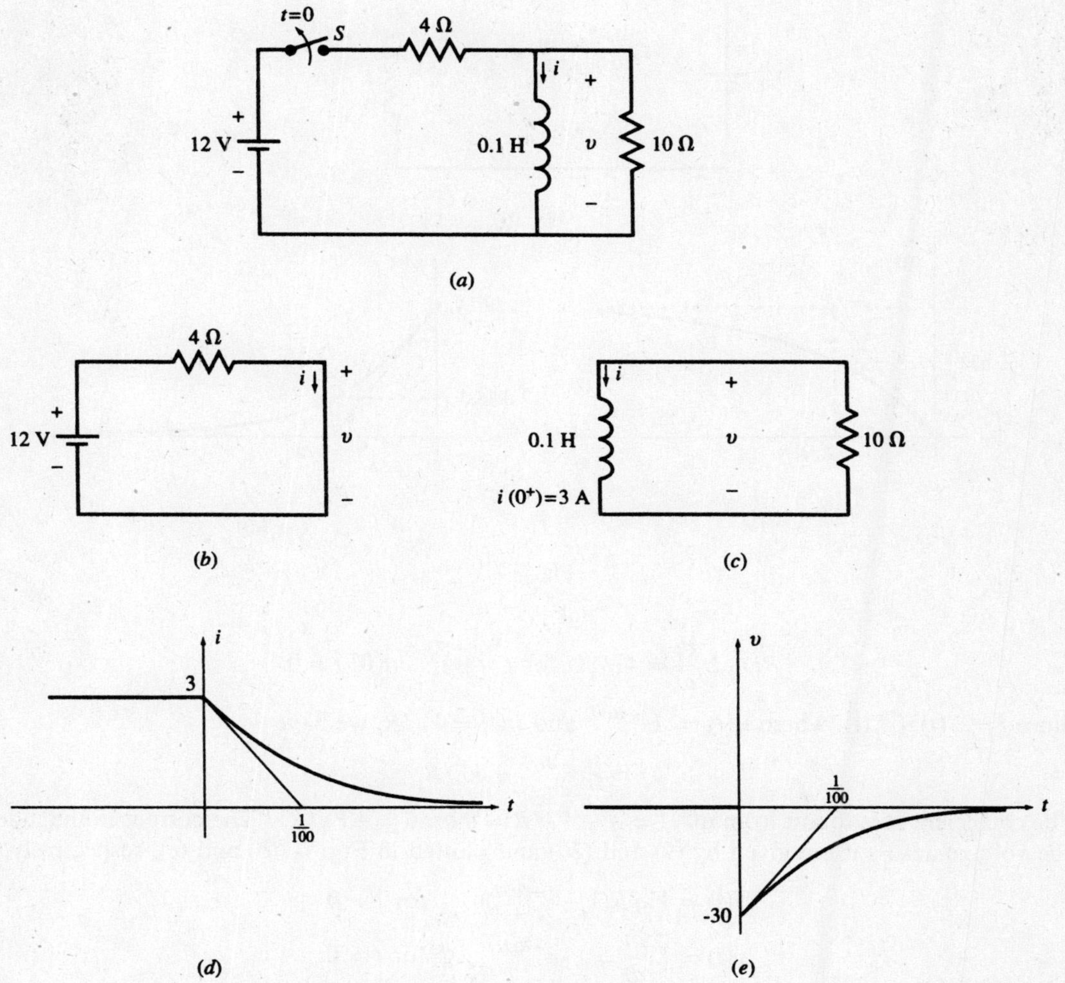

Fig. 7-6

Assume the switch S has been closed for a long time. The inductor current is then constant and its voltage is zero. The circuit at $t = 0^-$ is shown in Fig. 7-6(b) with $i(0^-) = 12/4 = 3$ A. After the battery is disconnected, at $t > 0$, the circuit will be as shown in Fig. 7-6(c). For $t > 0$, the current decreases exponentially from 3 A to zero. The time constant of the circuit is $L/R = (1/100)$ s. Using the results of Example 7.3, for $t > 0$, the inductor current and voltage are, respectively,

$$i(t) = 3e^{-100t}$$
$$v(t) = L(di/dt) = -30e^{-100t} \quad \text{(V)}$$

$i(t)$ and $v(t)$ are plotted in Figs. 7-6(d) and (e), respectively.

7.5 ESTABLISHING A DC CURRENT IN AN INDUCTOR

If a dc source is suddenly applied to a series *RL* circuit initially at rest, as in Fig. 7-7(*a*), the current grows exponentially from zero to a constant value with a time constant of L/R. The preceding result is the solution of the first-order differential equation (*8*) which is obtained by applying KVL around the loop. The solution follows.

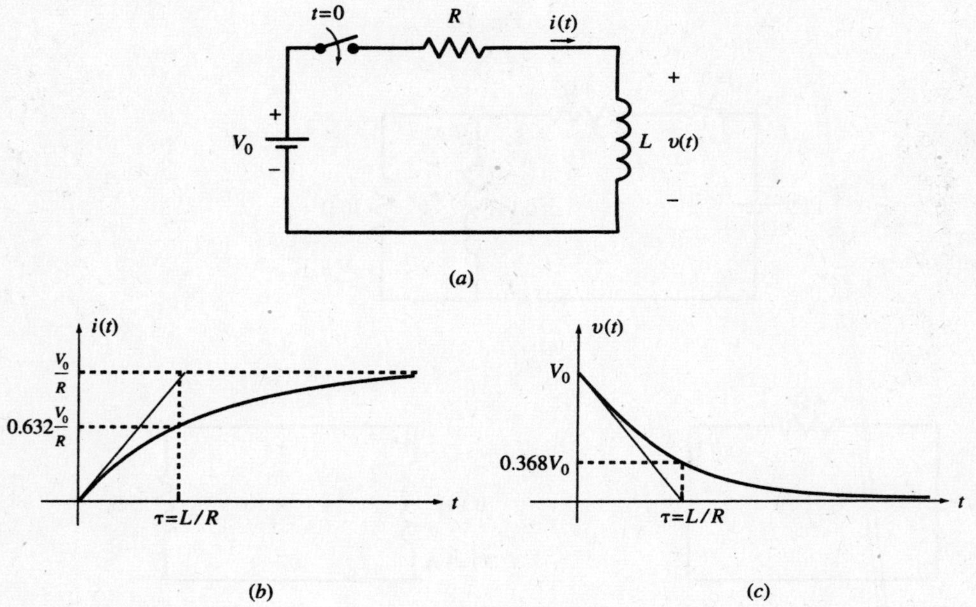

Fig. 7-7

$$Ri + L\frac{di}{dt} = V_0 \qquad \text{for } t > 0, \qquad i(0^+) = 0 \tag{8}$$

Since $i = i_h(t) + i_p(t)$, where $i_h(t) = Ae^{-Rt/L}$ and $i_p(t) = V_0/R$, we have

$$i = Ae^{-Rt/L} + V_0/R$$

The coefficient A is found from $i(0^+) = A + V_0/R = 0$ or $A = -V_0/R$. The current in the inductor and the voltage across it are given by (*9*) and (*10*) and plotted in Fig. 7-7(*b*) and (*c*), respectively.

$$i(t) = V_0/R(1 - e^{-Rt/L}) \qquad \text{for } t > 0 \tag{9}$$

$$v(t) = L\frac{di}{dt} = V_0 e^{-Rt/L} \qquad \text{for } t > 0 \tag{10}$$

7.6 THE EXPONENTIAL FUNCTION REVISITED

The exponential decay function may be written in the form $e^{-t/\tau}$, where τ is the *time constant* (in s). For the *RC* circuit of Section 7.2, $\tau = RC$; while for the *RL* circuit of Section 7.4, $\tau = L/R$. The general decay function

$$f(t) = Ae^{-t/\tau} \qquad (t > 0)$$

is plotted in Fig. 7-8, with time measured in multiples of τ. It is seen that

$$f(\tau) = Ae^{-1} = 0.368A$$

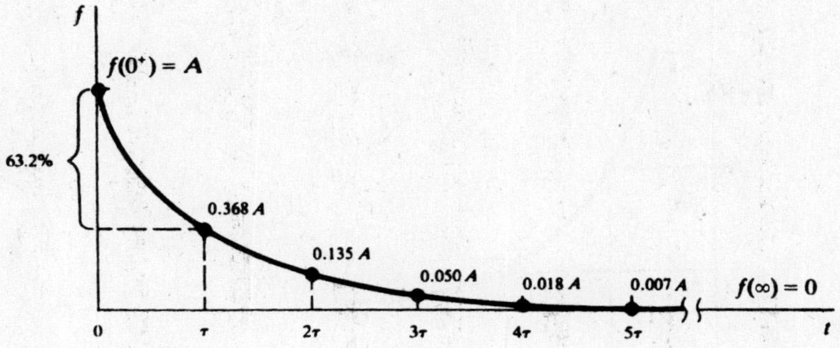

Fig. 7-8

that is, at $t = \tau$ the function is 36.8 percent of the initial value. It may also be said that the function has undergone 63.2 percent of the change from $f(0^+)$ to $f(\infty)$. At $t = 5\tau$, the function has the value $0.0067A$, which is less than 1 percent of the initial value. From a practical standpoint, the transient is often regarded as over after $t = 5\tau$.

The tangent to the exponential curve at $t = 0^+$ can be used to estimate the time constant. In fact, since

$$\text{slope} = f'(0^+) = -\frac{A}{\tau}$$

the tangent line must cut the horizontal axis at $t = \tau$ (see Fig. 7-9). More generally, the tangent at $t = t_0$ has horizontal intercept $t_0 + \tau$. Thus, if the two values $f(t_0)$ and $f'(t_0)$ are known, the entire curve can be constructed.

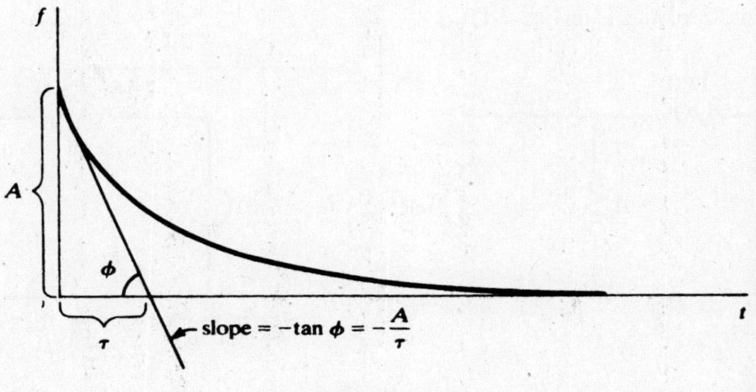

Fig. 7-9

At times a transient is only partially displayed (on chart paper or on the face of an oscilloscope), and the simultaneous values of function and slope needed in the preceding method are not available. In that case, any pair of data points, perhaps read from instruments, may be used to find the equation of the transient. Thus, referring to Fig. 7-10,

$$f_1 = Ae^{-t_1/\tau} \qquad f_2 = Ae^{-t_2/\tau}$$

which may be solved simultaneously to give

$$\tau = \frac{t_2 - t_1}{\ln f_1 - \ln f_2}$$

and then A in terms of τ and either f_1 or f_2.

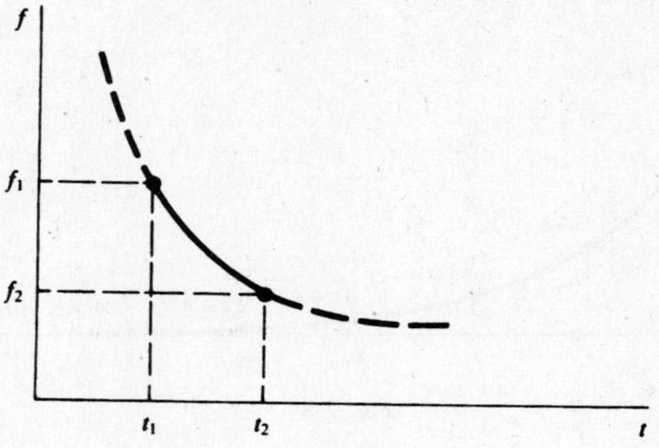

Fig. 7-10

7.7 COMPLEX FIRST-ORDER *RL* AND *RC* CIRCUITS

A more complex circuit containing resistors, sources, and a single energy storage element may be converted to a Thévenin or Norton equivalent as seen from the two terminals of the inductor or capacitor. This reduces the complex circuit to a simple *RC* or *RL* circuit which may be solved according to the methods described in the previous sections.

If a source in the circuit is suddently switched to a dc value, the resulting currents and voltages are exponentials, sharing the same time constant with possibly different initial and final values. The time constant of the circuit is either RC or L/R, where R is the resistance in the Thévenin equivalent of the circuit as seen by the capacitor or inductor.

EXAMPLE 7.5 Find i, v, and i_1 in Fig. 7-11(a).

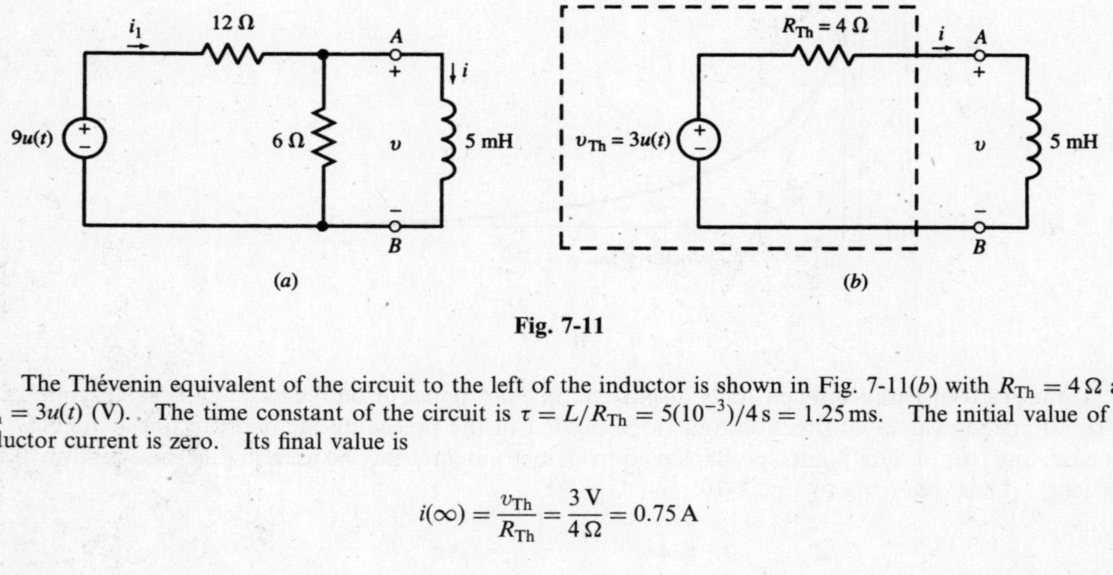

Fig. 7-11

The Thévenin equivalent of the circuit to the left of the inductor is shown in Fig. 7-11(b) with $R_{Th} = 4\,\Omega$ and $v_{Th} = 3u(t)$ (V).. The time constant of the circuit is $\tau = L/R_{Th} = 5(10^{-3})/4\,\text{s} = 1.25\,\text{ms}$. The initial value of the inductor current is zero. Its final value is

$$i(\infty) = \frac{v_{Th}}{R_{Th}} = \frac{3\,\text{V}}{4\,\Omega} = 0.75\,\text{A}$$

Therefore,

$$i = 0.75(1 - e^{-800t})u(t) \quad \text{(A)} \qquad v = L\frac{di}{dt} = 3e^{-800t}u(t) \quad \text{(V)} \qquad i_1 = \frac{9-v}{12} = \frac{1}{4}(3 - e^{-800t})u(t) \quad \text{(A)}$$

v can also be derived directly from its initial value $v(0^+) = (9 \times 6)/(12 + 6) = 3\,\text{V}$, its final value $v(\infty) = 0$ and the circuit's time constant.

EXAMPLE 7.6 In Fig. 7-12 the 9-μF capacitor is connected to the circuit at $t = 0$. At this time, capacitor voltage is $v_0 = 17$ V. Find v_A, v_B, v_C, i_{AB}, i_{AC}, and i_{BC} for $t > 0$.

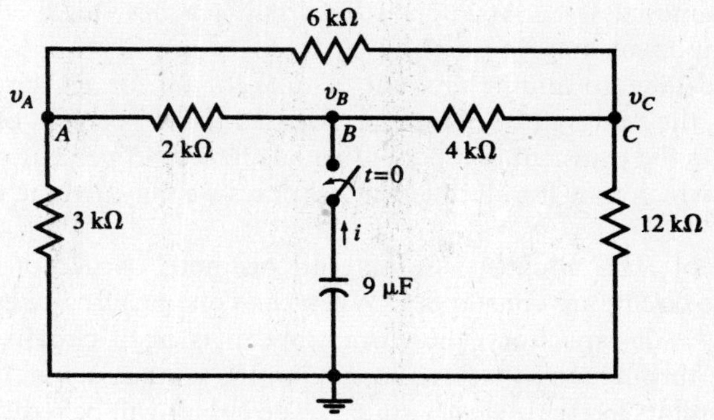

Fig. 7-12

Apply KCL at nodes A, B, and C for $t > 0$ to find voltages in term of i:

Node A: $$\left(\frac{1}{2}+\frac{1}{3}+\frac{1}{6}\right)v_A - \frac{1}{2}v_B - \frac{1}{6}v_C = 0 \quad \text{or} \quad 6v_A - 3v_B - v_C = 0 \tag{11}$$

Node B: $$-\frac{1}{2}v_A + \left(\frac{1}{2}+\frac{1}{4}\right)v_B - 10^3 i - \frac{1}{4}v_C = 0 \quad \text{or} \quad -2v_A + 3v_B - v_C = (4 \times 10^3)i \tag{12}$$

Node C: $$-\frac{1}{6}v_A - \frac{1}{4}v_B + \left(\frac{1}{4}+\frac{1}{6}+\frac{1}{12}\right)v_C = 0 \quad \text{or} \quad -2v_A - 3v_B + 6v_C = 0 \tag{13}$$

Solving (*11*), (*12*), and (*13*) simultaneously,

$$v_A = \tfrac{7}{3}(10^3)i \qquad v_B = \tfrac{34}{9}(10^3)i \qquad v_C = \tfrac{8}{3}(10^3)i$$

The circuit as seen by the capacitor is equivalent to a resistor $R = v_B/i = 34/9\,\text{k}\Omega$. The capacitor discharges its initial voltage V_0 in an exponential form with a time constant $\tau = RC = \frac{34}{9}(10^3)(9 \times 10^{-6}) = 0.034$ s. For $t > 0$, the voltages and currents are

$$v_B = V_0 e^{-t/\tau} = 17e^{-1000t/34} \quad \text{(V)}$$

$$i = -C\frac{dv_B}{dt} = (9 \times 17 \times 10^{-3}/34)e^{-1000t/34} = (4.5 \times 10^{-3})e^{-1000t/34} \quad \text{(A)}$$

$$v_A = \tfrac{7}{3}(10^3)i = 10.5e^{-1000t/34} \quad \text{(V)} \qquad v_C = \tfrac{8}{3}(10^3)i = 12e^{-1000t/34} \quad \text{(V)}$$

$$v_{AB} = v_A - v_B = -6.5e^{-1000t/34} \quad \text{(V)} \qquad i_{AB} = v_{AB}/2000 = (-3.25 \times 10^{-3})e^{-1000t/34} \quad \text{(A)}$$

$$v_{AC} = v_A - v_C = -1.5e^{-1000t/34} \quad \text{(V)} \qquad i_{AC} = v_{AC}/6000 = (-0.25 \times 10^{-3})e^{-1000t/34} \quad \text{(A)}$$

$$v_{BC} = v_B - v_C = 5e^{-1000t/34} \quad \text{(V)} \qquad i_{BC} = v_{BC}/4000 = (1.25 \times 10^{-3})e^{-1000t/34} \quad \text{(A)}$$

All voltages and currents are exponential functions and have the same time constant. For simplicity, it is customary to use units of V, mA, kΩ, and ms for voltage, current, resistance, and time, respectively, so that the multipliers 1000 and 10^{-3} can be omitted from the equations as summarized below.

$$v_A = 10.5e^{-t/34} \ \text{(V)} \qquad v_{AB} = -6.5e^{-t/34} \ \text{(V)} \qquad i_{AB} = -3.25e^{-t/34} \ \text{(mA)}$$

$$v_B = 17e^{-t/34} \ \text{(V)} \qquad v_{AC} = -1.5e^{-t/34} \ \text{(V)} \qquad i_{AC} = -0.25e^{-t/34} \ \text{(mA)}$$

$$v_C = 12e^{-t/34} \ \text{(V)} \qquad v_{BC} = 5e^{-t/34} \ \text{(V)} \qquad i_{BC} = 1.25e^{-t/34} \ \text{(mA)}$$

$$i = 4.5e^{-t/34} \ \text{(mA)}$$

7.8 DC STEADY STATE IN INDUCTORS AND CAPACITORS

As noted in Section 7.1, the natural exponential component of the response of *RL* and *RC* circuits to step inputs diminishes as time passes. At $t = \infty$, the circuit reaches steady state and the response is made of the forced dc component only.

Theoretically, it should take an infinite amount of time for *RL* or *RC* circuits to reach dc steady state. However, at $t = 5\tau$, the transient component is reduced to 0.67 percent of its initial value. After passage of 10 time constants the transient component equals to 0.0045 percent of its initial value, which is less than 5 in 100,000, at which time for all practical purposes we may assume the steady state has been reached.

At the dc steady state of *RLC* circuits, assuming no sustained oscillations exist in the circuit, all currents and voltages in the circuit are constants. When the voltage across a capacitor is constant, the current through it is zero. All capacitors, therefore, appear as open circuits in the dc steady state. Similarly, when the current through an inductor is constant, the voltage across it is zero. All inductors therefore appear as short circuits in the dc steady state. The circuit will be reduced to a dc-resistive case from which voltages across capacitors and currents through inductors can be easily found, as all the currents and voltages are constants and the analysis involves no differential equations.

The dc steady-state behavior presented in the preceding paragraph is valid for circuits containing any number of inductors, capacitors, and dc sources.

EXAMPLE 7.7 Find the steady-state values of i_L, v_{C1}, and v_{C2} in the circuit of Fig. 7-13(*a*).

When the steady state is reached, the circuit will be as shown in Fig. 7-13(*b*). The inductor current and capacitor voltages are obtained by applying KCL at nodes *A* and *B* in Fig. 7-13(*b*). Thus,

Node *A*: $$\frac{v_A}{3} + \frac{v_A - v_B}{6} + \frac{v_A + 18 - v_B}{6} = 3 \qquad \text{or} \qquad 2v_A - v_B = 0$$

Node *B*: $$\frac{v_B}{12} + \frac{v_B - v_A}{6} + \frac{v_B - 18 - v_A}{6} = 0 \qquad \text{or} \qquad -4v_A + 5v_B = 36$$

Solving for v_A and v_B we find $v_A = 6\text{ V}$ and $v_B = 12\text{ V}$. By inspection of Fig. 7-13(*b*), we have $i_L = 2\text{ mA}$, $v_{C1} = 8\text{ V}$, and $v_{C2} = 6\text{ V}$.

EXAMPLE 7.8 Find i and v in the circuit of Fig. 7-14.

At $t = 0$, the voltage across the capacitor is zero. Its final value is obtained from dc analysis to be -2 V. The time constant of the circuit of Fig. 7-14, as derived in Example 7.6, is 0.034 s. Therefore,

$$v = -2(1 - e^{-1000t/34})u(t) \quad \text{(V)}$$

$$i = C\frac{dv}{dt} = -\frac{(9 \times 10^{-6})(2 \times 10^{3})}{34} e^{-1000t/34}u(t) \quad \text{(A)} = -0.53e^{-1000t/34}u(t) \quad \text{(mA)}$$

7.9 TRANSITIONS AT SWITCHING TIME

A sudden switching of a source or a jump in its magnitude can translate into sudden jumps in voltages or currents in a circuit. A jump in the capacitor voltage requires an impulse current. Similarly, a jump in the inductor current requires an impulse voltage. If no such impulses can be present, the capacitor voltages and the inductor currents remain continuous. Therefore, the post-switching conditions of *L* and *C* can be derived from their pre-switching conditions.

EXAMPLE 7.9 In Fig. 7-15(*a*) the switch *S* is closed at $t = 0$. Find i and v for all times.

At $t = 0^-$, the circuit is at steady state and the inductor functions as a short with $v(0^-) = 0$ [see Fig. 7-15(*b*)]. The inductor current is then easily found to be $i(0^-) = 2\text{ A}$. After *S* is closed at $t = 0$, the circuit will be as shown in Fig. 7-15(*c*). For $t > 0$, the current is exponential with a time constant of $\tau = L/R = 1/30\text{ s}$, an initial value of $i(0^+) = i(0^-) = 2\text{ A}$, and a final value of $12/3 = 4\text{ A}$. The inductor's voltage and current are

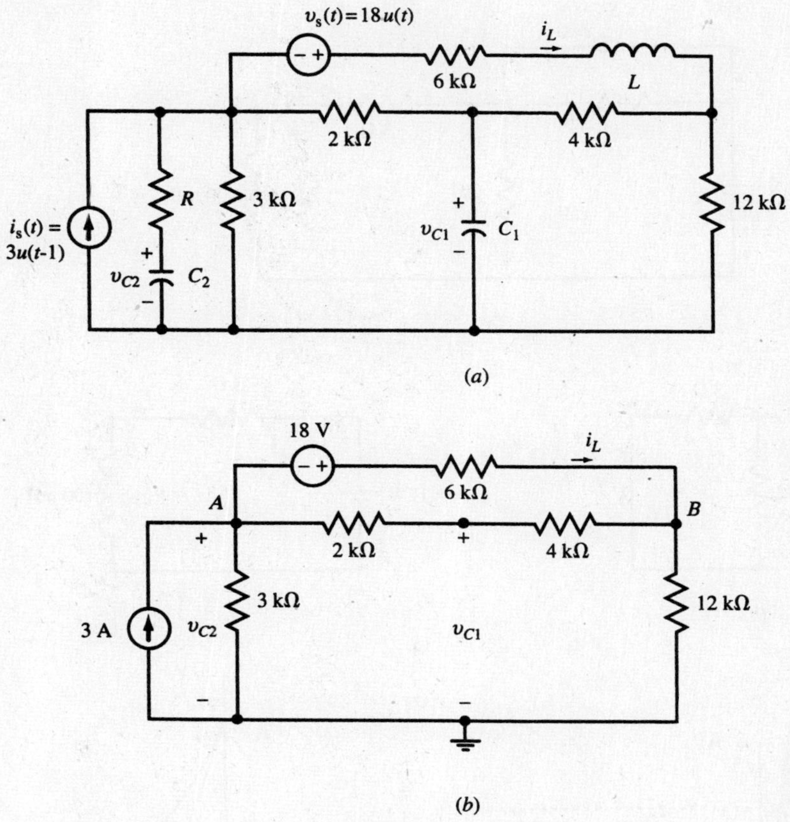

Fig. 7-13

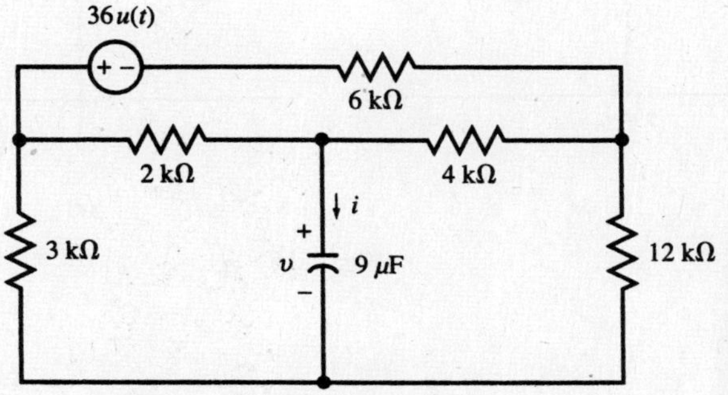

Fig. 7-14

For $t < 0$, $\quad i = 2\text{ A}$ and $v = 0$

For $t > 0$, $\quad i = 4 - 2e^{-30t}$ (A) and $v = L\dfrac{di}{dt} = 6e^{-30t}$ (V)

and plotted in Figs. 7-15(d) and (e).

EXAMPLE 7.10 Find i and v for $t = 0^-$ and $t = 0^+$ in the circuit of Fig. 7-16, given $R = 5\,\Omega$, $L = 10\,\text{mH}$, and

$$v_s = \begin{cases} 5\text{ V for } t < 0 \\ 5\sin\omega t\text{ (V) for } t > 0 \end{cases}$$

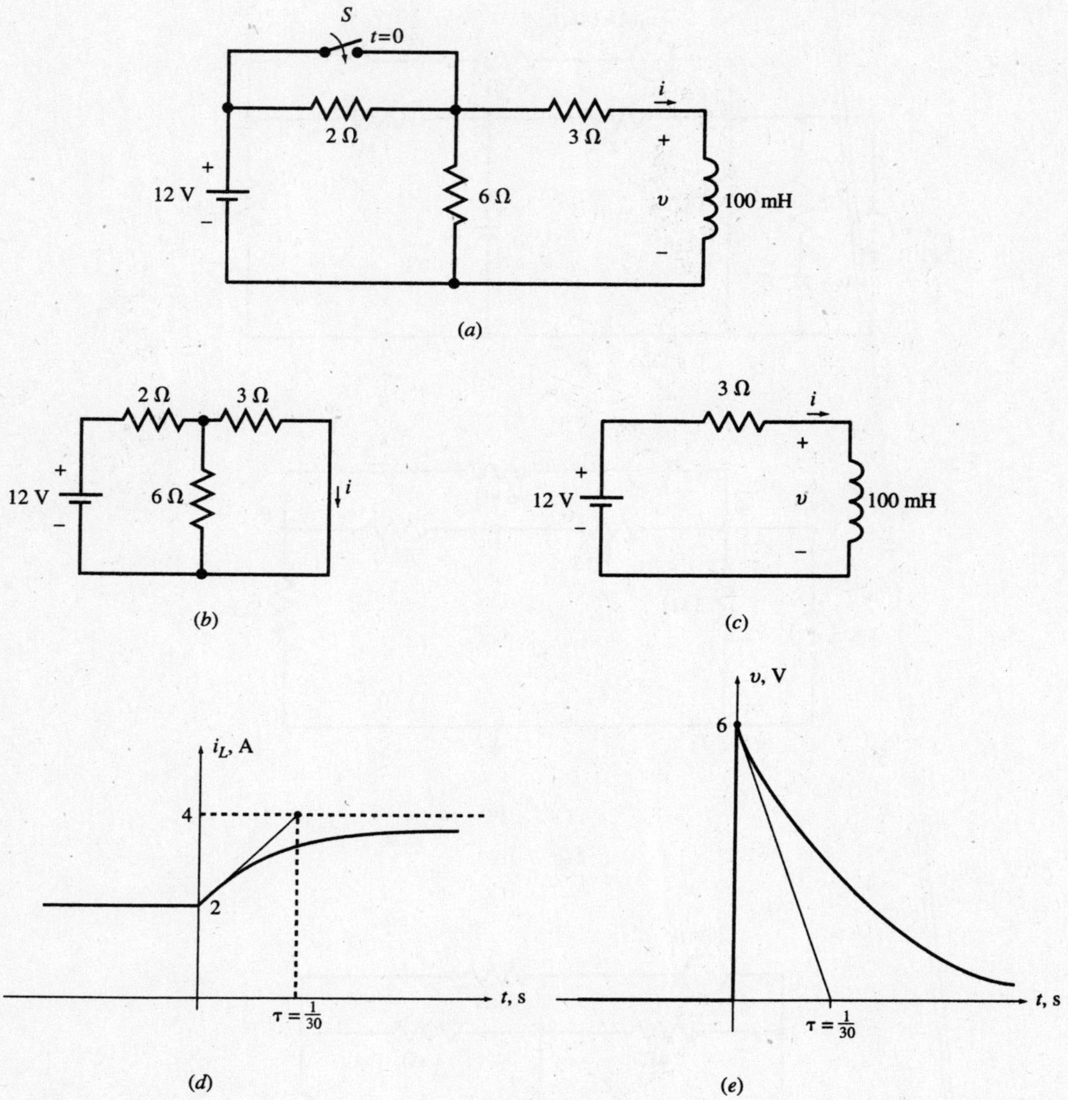

Fig. 7-15

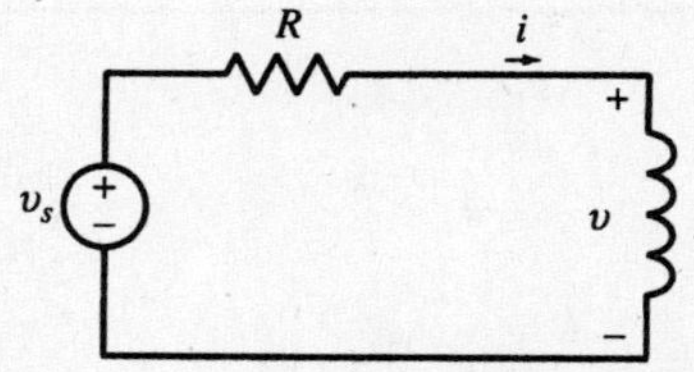

Fig. 7-16

At $t = 0^-$, $i(0^-) = 5/5 = 1\,\text{A}$ and $v(0^-) = 0$. During the transition time $t = 0^-$ to $t = 0^+$, the inductor current is continuous as there exists no voltage impulse to produce a discontinuity in it. Therefore, $i(0^+) = i(0^-) = 1\,\text{A}$. To find $v(0^+)$, write KVL at $t = 0^+ : v_s = RI + v$ and note that $v_s(0^+) = 0$. Therefore, $v(0^+) = v_s(0^+) - ri(0^+) = -5\,\text{V}$.

7.10 RESPONSE OF FIRST-ORDER CIRCUITS TO A PULSE

In this section we will derive the response of a first-order circuit to a rectangular pulse. The derivation applies to RC or RL circuits where the input can be a current or a voltage. As an example, we use the series RC circuit in Fig. 7-17(*a*) with the voltage source delivering a pulse of duration T and height V_0. For $t < 0$, v and i are zero. For the duration of the pulse, we use (*6b*) and (*6c*) in Section 7.3:

$$v = V_0(1 - e^{-t/RC}) \qquad (0 < t < T) \tag{14a}$$

$$i = \frac{V_0}{R} e^{-t/RC} \qquad (0 < t < T) \tag{14b}$$

When the pulse ceases, the circuit is source-free with the capacitor at an initial voltage V_T.

$$V_T = V_0(1 - e^{-T/RC}) \tag{14c}$$

Using (*3*) and (*4*) in Section 7.2, and taking into account the time shift T, we have

$$v = V_T e^{-(t-T)/RC} \qquad (t > T) \tag{15a}$$

$$i = -(V_T/R)e^{-(t-T)/RC} \qquad (t > T) \tag{15b}$$

The capacitor voltage and current are plotted in Figs. 7-17(*b*) and (*c*).

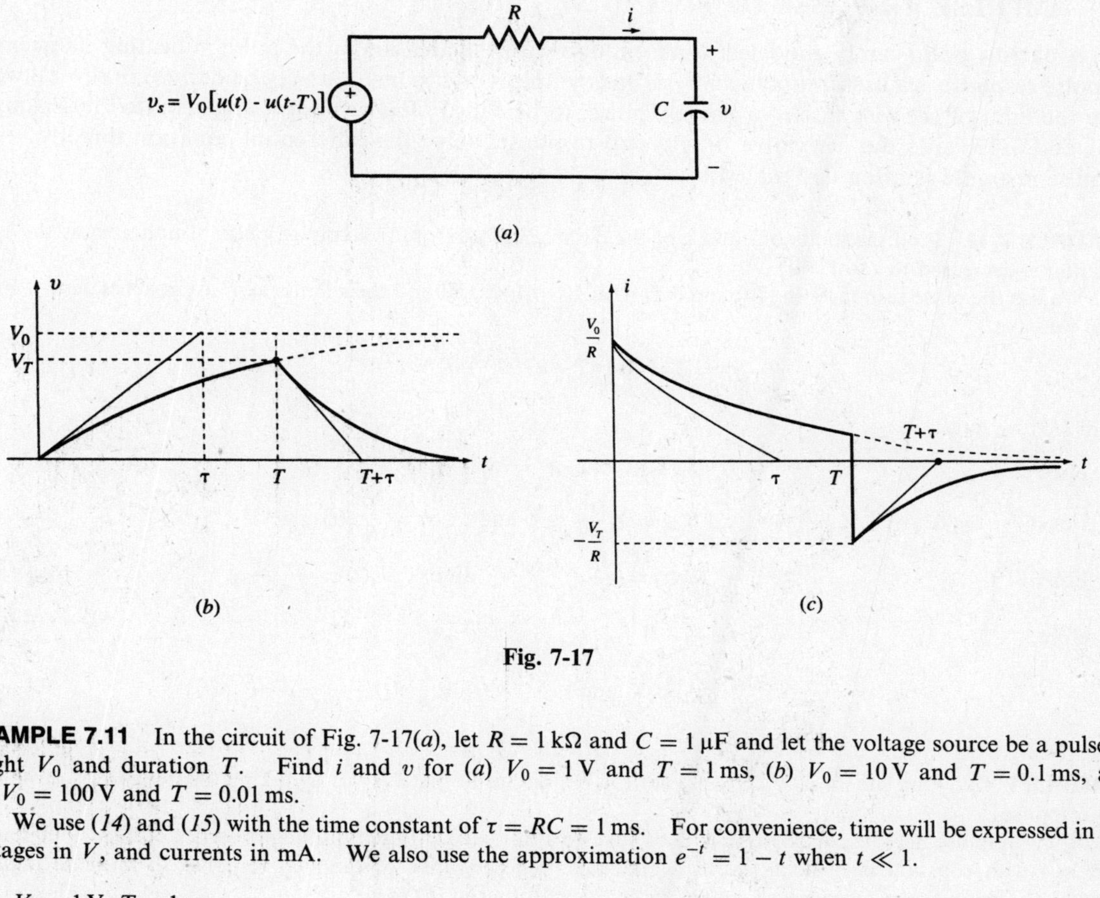

Fig. 7-17

EXAMPLE 7.11 In the circuit of Fig. 7-17(*a*), let $R = 1\,\text{k}\Omega$ and $C = 1\,\mu\text{F}$ and let the voltage source be a pulse of height V_0 and duration T. Find i and v for (*a*) $V_0 = 1\,\text{V}$ and $T = 1\,\text{ms}$, (*b*) $V_0 = 10\,\text{V}$ and $T = 0.1\,\text{ms}$, and (*c*) $V_0 = 100\,\text{V}$ and $T = 0.01\,\text{ms}$.

We use (*14*) and (*15*) with the time constant of $\tau = RC = 1\,\text{ms}$. For convenience, time will be expressed in ms, voltages in V, and currents in mA. We also use the approximation $e^{-t} = 1 - t$ when $t \ll 1$.

(*a*) $V_0 = 1\,\text{V}$, $T = 1\,\text{ms}$.
For $0 < t < 1\,\text{ms}$,

$$v = (1 - e^{-t}),\ i = e^{-t},\ \text{and}\ V_T = (1 - e^{-1}) = 0.632\,\text{V}$$

For $t > 1$ ms,

$$v = 0.632e^{-(t-1)} = 1.72e^{-t}, \text{ and } i = -1.72e^{-t}$$

(b) $V_0 = 10$ V, $T = 0.1$ ms.
For $0 < t < 0.1$ ms,

$$v = 10(1 - e^{-t}), i = 10e^{-t}, \text{ and } V_T = 10(1 - e^{-0.1}) = 0.95 \text{ V}$$

For $t > 0.1$ ms,

$$v = 0.95e^{-(t-0.1)} = 1.05e^{-t}, \text{ and } i = -1.05e^{-t}$$

(c) $V_0 = 100$ V, $T = 0.01$ ms.
For $0 < t < 0.01$ ms,

$$v = 100(1 - e^{-t}) \approx 100t, i = 100e^{-t} \approx 100(1 - t), \text{ and } V_T = 100(1 - e^{-0.01}) = 0.995 \text{ V}$$

For $t > 0.01$ ms,

$$v = 0.995e^{-(t-0.01)} = 1.01e^{-t} \text{ and } i = -1.01e^{-t}$$

As the input voltage pulse approaches an impulse, the capacitor voltage and current approach $v = e^{-t}u(t)$ (V) and $i = \delta(t) - e^{-t}u(t)$.

7.11 IMPULSE RESPONSE OF *RC* AND *RL* CIRCUITS

A narrow pulse can be modeled as an impulse with the area under the pulse indicating its strength. Impulse response is a useful tool in analysis and synthesis of circuits. It may be derived in several ways: take the limit of the response to a narrow pulse, to be called *limit approach*, as illustrated in Examples 7-11 and 7-12; take the derivative of the step response; solve the differential equation directly. The impulse response is often designated by $h(t)$.

EXAMPLE 7.12 Find the limits of i and v of the circuit Fig. 7-17(*a*) for a voltage pulse of unit area as the pulse duration is decreased to zero.

We use the pulse responses in (*14*) and (*15*) with $V_0 = 1/T$ and find their limits as T approaches zero. From (*14c*) we have

$$\lim_{T \to 0} V_T = \lim_{T \to 0} (1 - e^{-T/RC})/T = 1/RC$$

From (*15*) we have:

For $t < 0$, $\quad h_v = 0 \quad$ and $\quad h_i = 0$

For $0^- < t < 0^+$, $\quad 0 \le h_v \le \dfrac{1}{RC} \quad$ and $\quad h_i = \dfrac{1}{R}\,\delta(t)$

For $t > 0$, $\quad h_v(t) = \dfrac{1}{RC} e^{-t/RC} \quad$ and $\quad h_i(t) = -\dfrac{1}{R^2C}\, e^{-t/RC}$

Therefore,

$$h_v(t) = \frac{1}{RC}\, e^{-t/RC} u(t) \qquad \text{and} \qquad h_i(t) = \frac{1}{R}\,\delta(t) - \frac{1}{R^2C}\, e^{-t/RC} u(t)$$

EXAMPLE 7.13 Find the impulse responses of the *RC* circuit in Fig. 7-17(*a*) by taking the derivatives of its unit step responses.

A unit impulse may be considered the derivative of a unit step. Based on the properties of linear differential equations with constant coefficients, we can take the time derivative of the step response to find the impulse response. The unit step responses of an *RC* circuit were found in (*6*) to be

$$v(t) = (1 - e^{-t/RC})u(t) \qquad \text{and} \qquad i(t) = (1/R)e^{-t/RC} u(t)$$

We find the unit impulse responses by taking the derivatives of the step responses. Thus

$$h_v(t) = \frac{1}{RC}\, e^{-t/RC} u(t) \qquad \text{and} \qquad h_i(t) = \frac{1}{R}\,\delta(t) - \frac{1}{R^2 C}\, e^{-t/RC} u(t)$$

EXAMPLE 7.14 Find the impulse responses $h_i(t)$, $h_v(t)$, and $h_{i1}(t)$ of the RL circuit of Fig. 7-11(a) by taking the derivatives of its unit step responses.

The responses of the circuit to a step of amplitude 9 were already found in Example 7.5. Taking their derivatives and scaling them down by 1/9, we find the unit impulse responses to be

$$h_i(t) = \frac{1}{9}\frac{d}{dt}[0.75(1 - e^{-800t})u(t)] = \frac{200}{3}\, e^{-800t} u(t)$$
$$h_v(t) = \frac{1}{9}\frac{d}{dt}[3e^{-800t} u(t)] = -\frac{800}{3}\, e^{-800t} u(t) + \frac{1}{3}\,\delta(t)$$
$$h_{i1}(t) = \frac{1}{9}\frac{d}{dt}\left[\frac{1}{4}(3 - e^{-800t})u(t)\right] = \frac{200}{9}\, e^{-800t} u(t) + \frac{1}{18}\,\delta(t)$$

7.12 SUMMARY OF STEP AND IMPULSE RESPONSES IN RC AND RL CIRCUITS

Responses of RL and RC circuits to step and impulse inputs are summarized in Table 7-1. Some of the entries in this table have been derived in the previous sections. The remaining entries will be derived in the solved problems.

7.13 RESPONSE OF RC AND RL CIRCUITS TO SUDDEN EXPONENTIAL EXCITATIONS

Consider the first-order differential equation which is derived from an RL combination in series with a sudden exponential voltage source $v_s = V_0 e^{st} u(t)$ as in the circuit of Fig. 7-18. The circuit is at rest for $t < 0$. By applying KVL, we get

$$Ri + L\frac{di}{dt} = V_0 e^{st} u(t) \tag{16}$$

For $t > 0$, the solution is

$$i(t) = i_h(t) + i_p(t) \qquad \text{and} \qquad i(0^+) = 0 \tag{17a}$$

Table 7-1(*a*) Step and Impulse Responses in *RC* Circuits

RC circuit	Unit Step Response	Unit Impulse Response
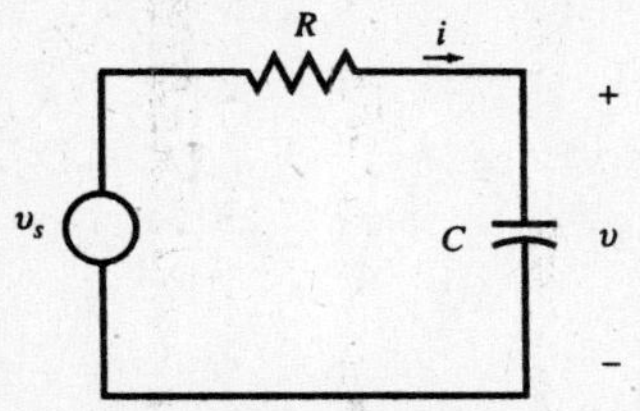	$v_s = u(t)$ $\begin{cases} v = (1 - e^{-t/Rc})u(t) \\ i = (1/R)e^{-t/Rc} u(t) \end{cases}$	$v_s = \delta(t)$ $\begin{cases} h_v = (1/RC)e^{-t/RC} u(t) \\ h_i = -(1/R^2 C)e^{-t/RC} u(t) + (1/R)\delta(t) \end{cases}$
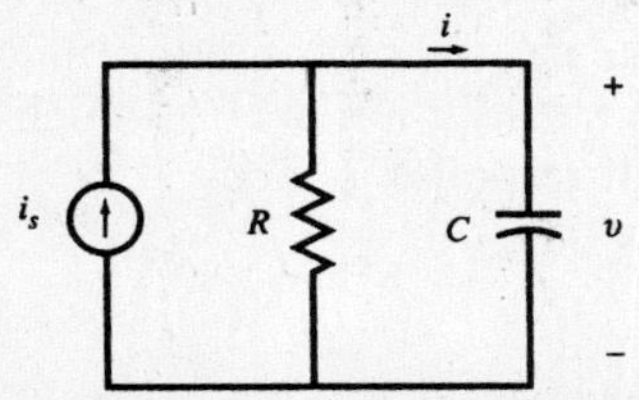	$i_s = u(t)$ $\begin{cases} v = R(1 - e^{-t/RC})u(t) \\ i = e^{-t/RC} u(t) \end{cases}$	$i_s = \delta(t)$ $\begin{cases} h_v = (1/C)e^{-t/RC} u(t) \\ h_i = -(1/RC)e^{-t/RC} u(t) + \delta(t) \end{cases}$

Table 7-1(*b*) Step and Impulse Responses in *RL* Circuits

RL circuit	Unit Step Response	Unit Impulse Response
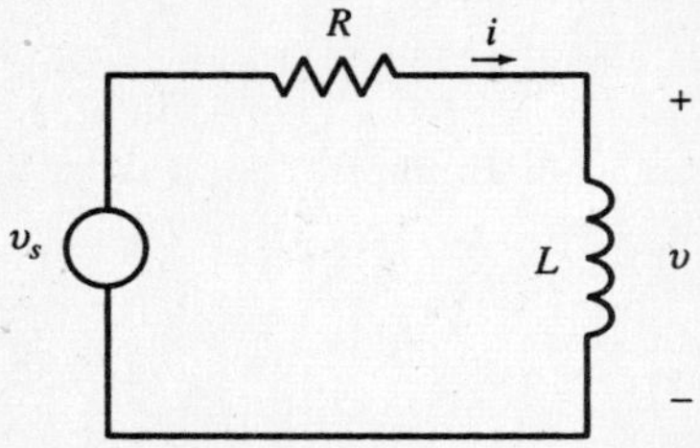	$v_s = u(t)$ $\begin{cases} v = e^{-Rt/L}u(t) \\ i = (1/R)(1 - e^{-Rt/L})u(t) \end{cases}$	$v_s = \delta(t)$ $\begin{cases} h_v = (R/L)e^{-Rt/L}u(t) + \delta(t) \\ h_i = -(1/L)e^{-Rt/L}u(t) \end{cases}$
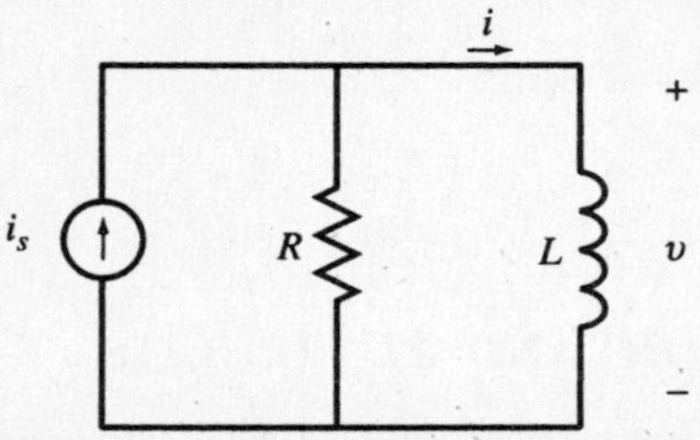	$i_s = u(t)$ $\begin{cases} v = Re^{-Rt/L}u(t) \\ i = (1 - e^{-Rt/L})u(t) \end{cases}$	$i_s = \delta(t)$ $\begin{cases} h_v = -(R^2/L)e^{-Rt/L}u(t) + R\delta(t) \\ h_i = (R/L)e^{-Rt/L}u(t) \end{cases}$

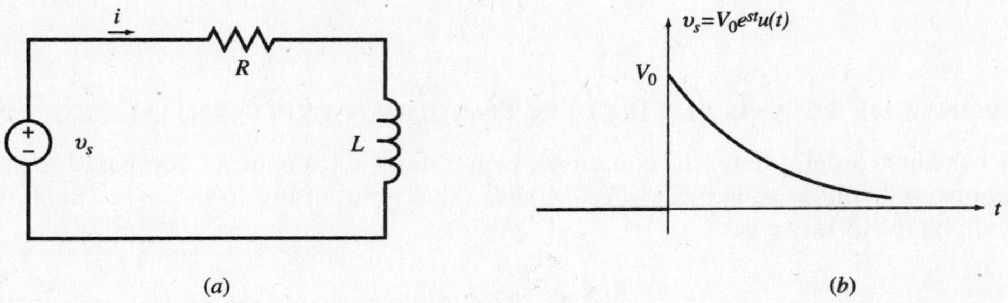

Fig. 7-18

The natural response $i_h(t)$ is the solution of $Ri + L(di/dt) = 0$; i.e., the case with a zero forcing function. Following an argument similar to that of Section 7.4 we obtain

$$i_h(t) = Ae^{-Rt/L} \tag{17b}$$

The forced response $i_p(t)$ is a function which satisfies (*16*) for $t > 0$. The only such function is

$$i_p(t) = I_0e^{st} \tag{17c}$$

After substituting i_p in (*16*), I_0 is found to be $I_0 = V_0/(R + Ls)$. By choosing $A = -V_0/\ (R + Ls)$, the boundary condition $i(0^+) = 0$ is also satisfied. Therefore,

$$i(t) = \frac{V_0}{R + Ls}(e^{st} - e^{-Rt/L})u(t) \tag{17d}$$

Special Case. If the forcing function has the same exponent as that of the natural response ($s = -R/L$), the forced response needs to be $i_p(t) = I_0te^{-Rt/L}$. This can be verified by substitution in (*16*), which also yields $I_0 = V_0/L$ The natural response is the same as (*17b*). The total response is then

$$i(t) = i_p(t) + i_h(t) = (I_0t + A)e^{-Rt/L}$$

From $i(0^-) = i(0^+) = 0$ we find $A = 0$, and so $i(t) = I_0te^{-Lt/R}u(t)$, where $I_0 = V_0/L$.

7.14 RESPONSE OF *RC* AND *RL* CIRCUITS TO SUDDEN SINUSOIDAL EXCITATIONS

When a series *RL* circuit is connected to a sudden ac voltage $v_s = V_0 \cos \omega t$ (Fig. 7-19), the equation of interest is

$$Ri + L\frac{di}{dt} = V_0(\cos \omega t)u(t) \tag{18}$$

The solution is

$$i(t) = i_h + i_p \qquad \text{where} \qquad i_h(t) = Ae^{-Rt/L} \qquad \text{and} \qquad i_p(t) = I_0 \cos(\omega t - \theta)$$

By inserting i_p in (*18*), we find I_0:

$$I_0 = \frac{V_0}{\sqrt{R^2 + L^2\omega^2}} \qquad \text{and} \qquad \theta = \tan^{-1}\frac{L\omega}{R}$$

Then
$$i(t) = Ae^{-Rt/L} + I_0 \cos(\omega t - \theta) \qquad t > 0$$

From $i(0^+) = 0$, we get $A = -I_0 \cos\theta$. Therefore,

$$i(t) = I_0[\cos(\omega t - \theta) - \cos\theta(e^{-Rt/L})]$$

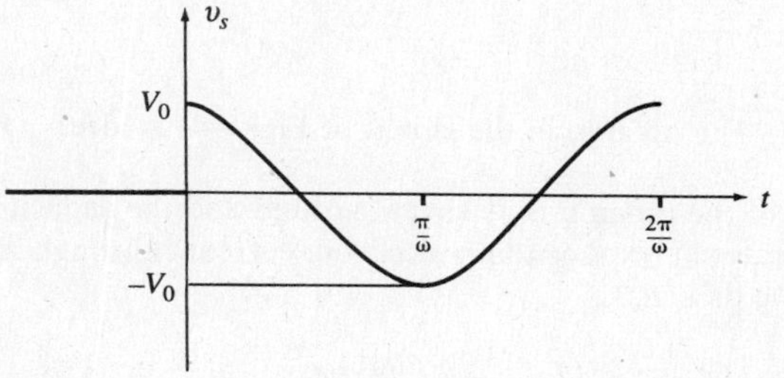

Fig. 7-19

7.15 SUMMARY OF FORCED RESPONSE IN FIRST-ORDER CIRCUITS

Consider the following differential equation:

$$\frac{dv}{dt}(t) + av(t) = f(t) \tag{19}$$

The forced response $v_p(t)$ depends on the forcing function $f(t)$. Several examples were given in the previous sections. Table 7-2 summarizes some useful pairs of the forcing function and what should be guessed for $v_p(t)$. The responses are obtained by substitution in the differential equation. By weighted linear combination of the entries in Table 7-2 and their time delay, the forced response to new functions may be deduced.

7.16 FIRST-ORDER ACTIVE CIRCUITS

Active circuits containing op amps are less susceptible to loading effects when interconnected with other circuits. In addition, they offer a wider range of capabilities with more ease of realization than passive circuits. In our present analysis of linear active circuits we assume ideal op amps; that is; (1) the current drawn by the op amp input terminals is zero and (2) the voltage difference between the inverting and noninverting terminals of the op amp is negligible (see Chapter 5). The usual methods of analysis are then applied to the circuit as illustrated in the following examples.

Table 7-2

$f(t)$	$v_p(t)$
1	$\dfrac{1}{a}$
t	$\dfrac{t}{a} - \dfrac{1}{a^2}$
$e^{st}, (s \neq -a)$	$\dfrac{e^{st}}{s+a}$
e^{-at}	te^{-at}
$\cos \omega t$	$A \cos(\omega t - \theta)$ where $A = \dfrac{1}{\sqrt{a^2 + \omega^2}}$ and $\tan \theta = \dfrac{\omega}{a}$
$e^{-bt} \cos \omega t$	$Ae^{-bt} \cos(\omega t - \theta)$ where $A = \dfrac{1}{\sqrt{(a-b)^2 + \omega^2}}$ and $\tan \theta = \dfrac{\omega}{a-b}$

EXAMPLE 7.15 **Highpass filter.** The op amp in the circuit of Fig. 7-44 is ideal. Find the unit-step response of the circuit; that is, v_2 for $v_1 = u(t)$.

The inverting input terminal of the op amp is at virtual ground and the capacitor has zero voltage at $t = 0^+$. The 1-V step input therefore generates an exponentially decaying current i through R_1C (from left to right, with a time constant R_1C and initial value of $1/R_1$).

$$i = \frac{1}{R_1} e^{-t/(R_1C)} u(t)$$

All of the preceding current passes through R_2 (the op amp draws no current), generating $v_2 = -R_2 i$ at the output terminal. The unit-step response is therefore

$$v_2 = -\frac{R_2}{R_1} e^{-t/(R_1C)} u(t)$$

EXAMPLE 7.16 In the circuit of Fig. 7-44 derive the differential equation relating v_2 to v_1. Find its unit-step response and compare with the answer in Example 7.15.

Since the inverting input terminal of the op amp is at virtual ground and doesn't draw any current, the current i passing through C, R_1, and R_2 from left to right is $-v_2/R_2$. Let v_A be the voltage of the node connecting R_1 and C. Then, the capacitor voltage is $v_1 - v_A$ (positive on the left side). The capacitor current and voltage are related by

$$-\frac{v_2}{R_2} = \frac{d(v_1 - v_A)}{dt}$$

To eliminate v_A, we note that the segment made of R_1, R_2, and the op amp form an inverting amplifier with $v_2 = -(R_2/R_1)v_A$, from which $v_A = -(R_1/R_2)v_2$. Substituting for v_A, we get

$$v_2 + R_1C \frac{dv_2}{dt} = -R_2C \frac{dv_1}{dt}$$

To find the unit-step response, we first solve the following equation:

$$v_2 + R_1C \frac{dv_2}{dt} = \begin{cases} -R_2C & t > 0 \\ 0 & t < 0 \end{cases}$$

The solution of the preceding equation is $-R_2C(1 - e^{-t/(R_1C)})u(t)$. The unit-step response of the circuit is the time-derivative of the preceding solution.

$$v_2(t) = -\frac{R_2}{R_1} e^{-t/(R_1C)} u(t)$$

Alternate Approach

The unit step response may also be found by the Laplace transform method (see Chapter 16).

EXAMPLE 7.17 Passive phase shifter. Find the relationship between v_2 and v_1 in the circuit of Fig. 7-45(a).

Let node D be the reference node. Apply KCL at nodes A and B to find

$$\text{KCL at node A:} \qquad C\frac{dv_A}{dt} + \frac{(v_A - v_1)}{R} = 0$$

$$\text{KCL at node B:} \qquad C\frac{d(v_B - v_1)}{dt} + \frac{v_B}{R} = 0$$

Subtracting the second equation from the first and noting that $v_2 = v_A - v_B$ we get

$$v_2 + RC\frac{dv_2}{dt} = v_1 - RC\frac{dv_1}{dt}$$

EXAMPLE 7.18 Active phase shifter. Show that the relationship between v_2 and v_1 in the circuit of Fig. 7-45(b) is the same as in Fig. 7-45(a).

Apply KCL at the inverting (node A) and non-inverting (node B) inputs of the op amp.

$$\text{KCL at node A:} \qquad \frac{(v_A - v_1)}{R_1} + \frac{(v_A - v_2)}{R_1} = 0$$

$$\text{KCL at node B:} \qquad \frac{(v_B - v_1)}{R} + C\frac{dv_B}{dt} = 0$$

From the op amp we have $v_A = v_B$ and from the KCL equation for node A, we have $v_A = (v_1 + v_2)/2$. Substituting the preceding values in the KCL at node B, we find

$$v_2 + RC\frac{dv_2}{dt} = v_1 - RC\frac{dv_1}{dt}$$

Solved Problems

7.1 At $t = 0^-$, just before the switch is closed in Fig. 7-20, $v_C = 100$ V. Obtain the current and charge transients.

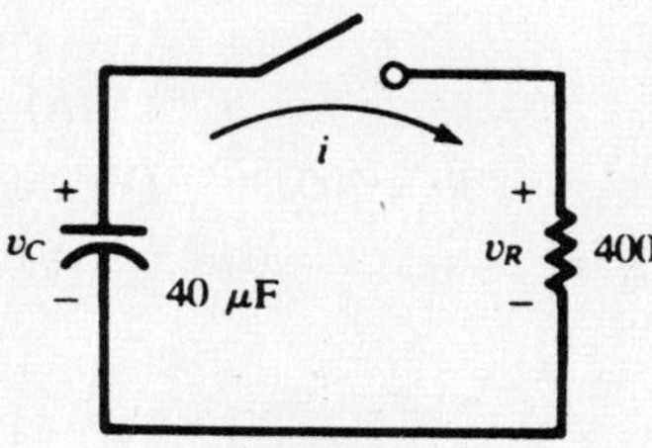

Fig. 7-20

With the polarities as indicated on the diagram, $v_R = v_C$ for $t > 0$, and $1/RC = 62.5\,\text{s}^{-1}$. Also, $v_C(0^+) = v_C(0^-) = 100$ V. Thus,

$$v_R = v_C = 100e^{-62.5t} \quad \text{(V)} \qquad i = \frac{v_R}{R} = 0.25e^{-62.5t} \quad \text{(A)} \qquad q = Cv_C = 4000e^{-62.5t} \quad (\mu\text{C})$$

7.2 In Problem 7.1, obtain the power and energy in the resistor, and compare the latter with the initial energy stored in the capacitor.

$$p_R = v_R i = 25e^{-125t} \quad \text{(W)}$$
$$w_R = \int_0^t p_R\,dt = \int_0^t 25e^{-125t}\,dt = 0.20(1 - e^{-125t}) \quad \text{(J)}$$

The initial stored energy is

$$W_0 = \tfrac{1}{2}CV_0^2 = \tfrac{1}{2}(40 \times 10^{-6})(100)^2\,\text{J} = 0.20 = w_R(\infty)$$

In other words, all the stored energy in the capacitor is eventually delivered to the resistor, where it is converted into heat.

7.3 An RC transient identical to that in Problems 7.1 and 7.2 has a power transient

$$p_R = 360e^{-t/0.00001} \quad \text{(W)}$$

Obtain the initial charge Q_0, if $R = 10\,\Omega$.

$$p_R = P_0 e^{-2t/RC} \quad \text{or} \quad \frac{2}{RC} = 10^5 \quad \text{or} \quad C = 2\,\mu\text{F}$$
$$w_R = \int_0^t p_R\,dt = 3.6(1 - e^{-t/0.00001}) \quad \text{(mJ)}$$

Then, $w_R(\infty) = 3.6\,\text{mJ} = Q_0^2/2C$, from which $Q_0 = 120\,\mu\text{C}$.

7.4 The switch in the RL circuit shown in Fig. 7-21 is moved from position *1* to position *2* at $t = 0$. Obtain v_R and v_L with polarities as indicated.

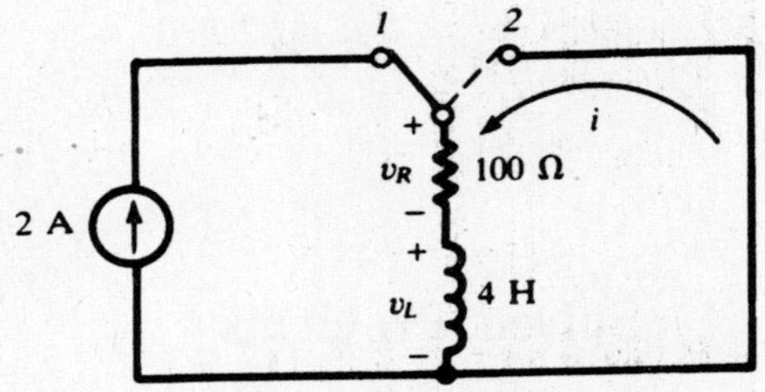

Fig. 7-21

The constant-current source drives a current through the inductance in the same direction as that of the transient current i. Then, for $t > 0$,

$$i = I_0 e^{-Rt/L} = 2e^{-25t} \quad \text{(A)}$$
$$v_R = Ri = 200e^{-25t} \quad \text{(V)}$$
$$v_L = -v_R = -200e^{-25t} \quad \text{(V)}$$

7.5 For the transient of Problem 7.4 obtain p_R and p_L.

$$p_R = v_R i = 400e^{-50t} \quad \text{(W)}$$
$$p_L = v_L i = -400e^{-50t} \quad \text{(W)}$$

Negative power for the inductance is consistent with the fact that energy is leaving the element. And, since this energy is being transferred to the resistance, p_R is positive.

7.6 A series RC circuit with $R = 5\,\text{k}\Omega$ and $C = 20\,\mu\text{F}$ has a constant-voltage source of 100 V applied at $t = 0$; there is no initial charge on the capacitor. Obtain i, v_R, v_C, and q, for $t > 0$.

The capacitor charge, and hence v_C, must be continuous at $t = 0$:

$$v_C(0^+) = v_C(0^-) = 0$$

As $t \to \infty$, $v_C \to 100$ V, the applied voltage. The time constant of the circuit is $\tau = RC = 10^{-1}$ s. Hence, from Section 6.10,

$$v_C = [v_C(0^+) - v_C(\infty)]e^{-t/\tau} + v_C(\infty) = -100e^{-10t} + 100 \quad \text{(V)}$$

The other functions follow from this. If the element voltages are both positive where the current enters, $v_R + v_C = 100\ V$, and so

$$v_R = 100e^{-10t} \quad \text{(V)}$$
$$i = \frac{v_R}{R} = 20e^{-10t} \quad \text{(mA)}$$
$$q = Cv_C = 2000(1 - e^{-10t}) \quad (\mu\text{C})$$

7.7 The switch in the circuit shown in Fig. 7-22(a) is closed at $t = 0$, at which moment the capacitor has charge $Q_0 = 500\,\mu\text{C}$, with the polarity indicated. Obtain i and q, for $t > 0$, and sketch the graph of q.

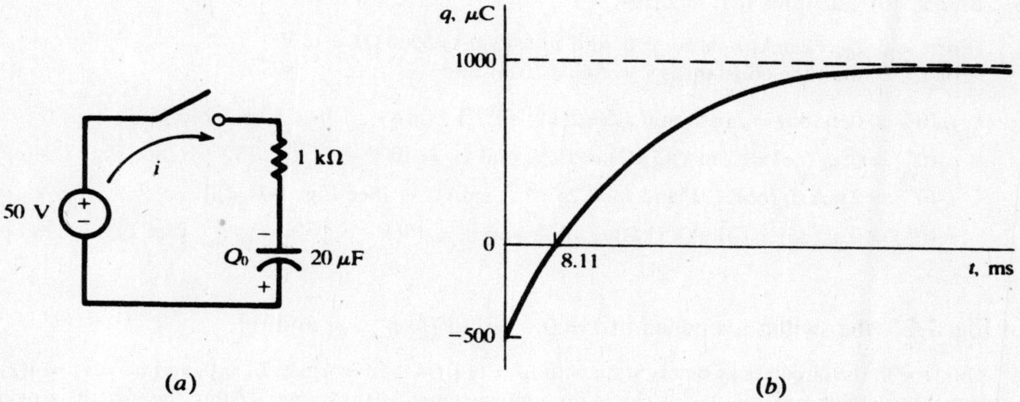

Fig. 7-22

The initial charge has a corresponding voltage $V_0 = Q_0/C = 25$ V, whence $v_C(0^+) = -25$ V. The sign is negative because the capacitor voltage, in agreement with the positive direction of the current, would be + on the top plate. Also $v_C(\infty) = +50$ V and $\tau = 0.02$ s. Thus, as in Problem 7.6,

$$v_C = -75e^{-50t} + 50 \quad \text{(V)}$$

from which

$$q = Cv_C = -1500e^{-50t} + 1000 \quad (\mu\text{C}) \qquad i = \frac{dq}{dt} = 75e^{-50t} \quad \text{(mA)}$$

The sketch in Fig. 7-22(b) shows that the charge changes from 500 μC of one polarity to 1000 μC of the opposite polarity.

7.8 Obtain the current i, for all values of t, in the circuit of Fig. 7-23.

For $t < 0$, the voltage source is a short circuit and the current source shares 2 A equally between the two 10-Ω resistors:

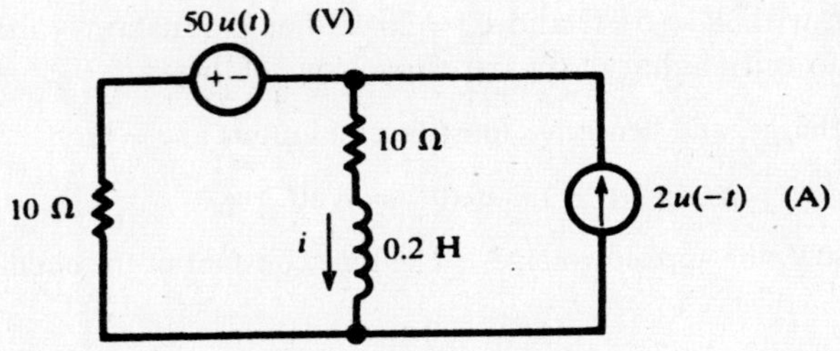

Fig. 7-23

$$i(t) = i(0^-) = i(0^+) = 1 \text{ A}$$

For $t > 0$, the current source is replaced by an open circuit and the 50-V source acts in the RL series circuit ($R = 20\,\Omega$). Consequently, as $t \to \infty$, $i \to -50/20 = -2.5$ A. Then, by Sections 6.10 and 7.3,

$$i(t) = [(i(0^+) - i(\infty)]e^{-Rt/L} + i(\infty) = 3.5e^{-100t} - 2.5 \quad \text{(A)}$$

By means of unit step functions, the two formulas may be combined into a single formula valid for all t:

$$i(t) = u(-t) + (3.5e^{-100t} - 2.5)u(t) \quad \text{(A)}$$

7.9 In Fig. 7-24(*a*), the switch is closed at $t = 0$. The capacitor has no charge for $t < 0$. Find i_R, i_C, v_C, and v_s for all times if $i_s = 2$ mA.

For $t < 0$, $i_R = 2$ mA, $i_C = v_C = 0$, and $v_s = (2\text{ mA})(5000\,\Omega) = 10$ V.
For $t > 0$, the time constant is $\tau = RC = 10$ ms and

$$i_R(0^+) = 0,\ i_R(\infty) = 2\text{ mA, and } i_R = 2(1 - e^{-100t}) \quad \text{(mA)} \qquad \text{[See Fig. 7-24(}b\text{).]}$$
$$v_C(0^+) = 0,\ v_C(\infty) = (2\text{ mA})(5\text{ k}\Omega) = 10\text{ V, and } v_C = 10(1 - e^{-100t}) \quad \text{(V)} \qquad \text{[See Fig. 7-24(}c\text{).]}$$
$$i_C(0^+) = 2\text{ mA},\ i_C(\infty) = 0\text{, and } i_C = 2e^{-100t} \quad \text{(mA)} \qquad \text{[See Fig. 7-24(}d\text{).]}$$
$$v_s(0^+) = 0,\ v_s(\infty) = (2\text{ mA})(5\text{ k}\Omega) = 10\text{ V, and } v_s = 10(1 - e^{-100t}) \quad \text{(V)} \qquad \text{[See Fig. 7-24(}e\text{).]}$$

7.10 In Fig. 7-25, the switch is opened at $t = 0$. Find i_R, i_C, v_C, and v_s.

For $t < 0$, the circuit is at steady state with $i_R = 6(4)/(4+2) = 4$ mA, $i_C = 0$, and $v_C = v_s = 4(2) = 8$ V. During the switching at $t = 0$, the capacitor voltage remains the same. After the switch is opened, at $t = 0^+$, the capacitor has the same voltage $v_C(0^+) = v_C(0^-) = 8$ V.

For $t > 0$, the capacitor discharges in the 5-kΩ resistor, produced from the series combination of the 3-kΩ and 2-kΩ resistors. The time constant of the circuit is $\tau = (2+3)(10^3)(2 \times 10^{-6}) = 0.01$ s. The currents and voltages are

$$v_C = 8e^{-100t} \quad \text{(V)}$$
$$i_R = -i_C = v_C/5000 = (8/5000)e^{-100t} = 1.6e^{-100t} \quad \text{(mA)}$$
$$v_s = (6\text{ mA})(4\text{ k}\Omega) = 24\text{ V}$$

since, for $t > 0$, all of the 6 mA goes through the 4-kΩ resistor.

7.11 The switch in the circuit of Fig. 7-26 is closed on position *1* at $t = 0$ and then moved to *2* after one time constant, at $t = \tau = 250\,\mu$s. Obtain the current for $t > 0$.

It is simplest first to find the charge on the capacitor, since it is known to be continuous (at $t = 0$ and at $t = \tau$), and then to differentiate it to obtain the current.

For $0 \le t \le \tau$, q must have the form

$$q = Ae^{-t/\tau} + B$$

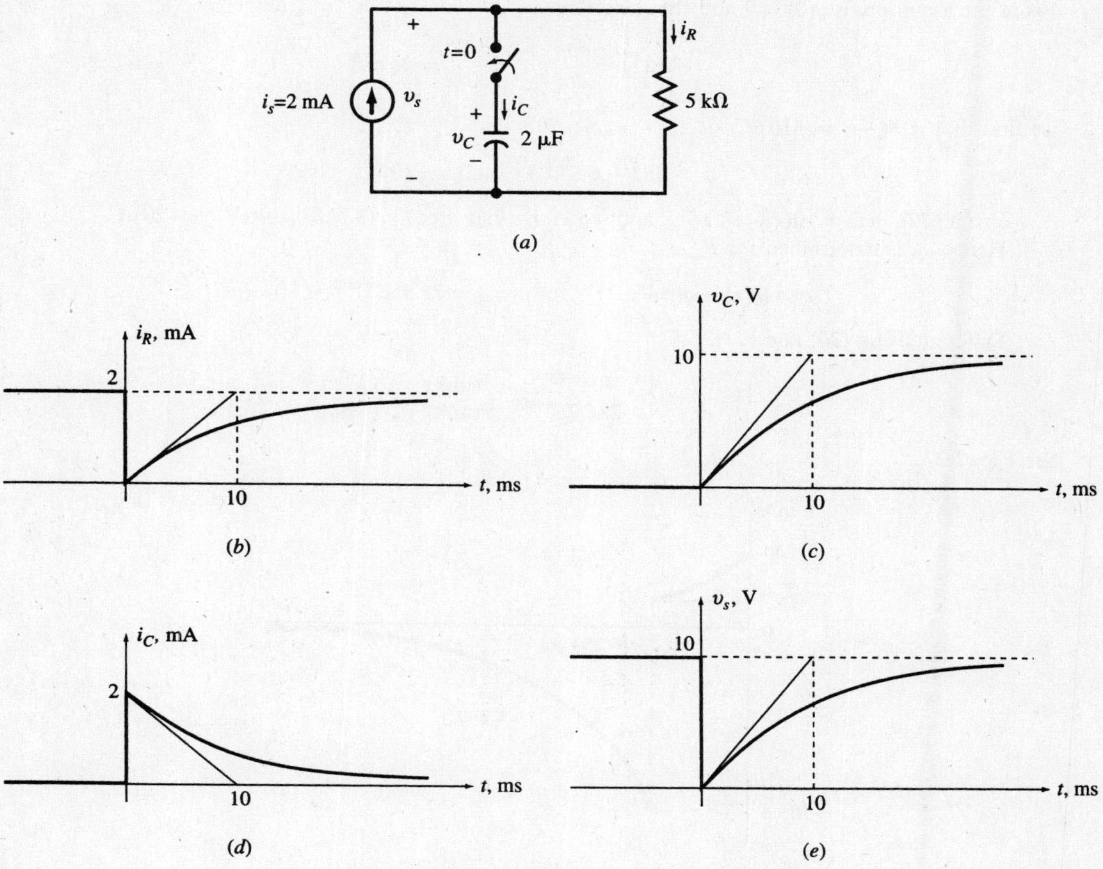

Fig. 7-24

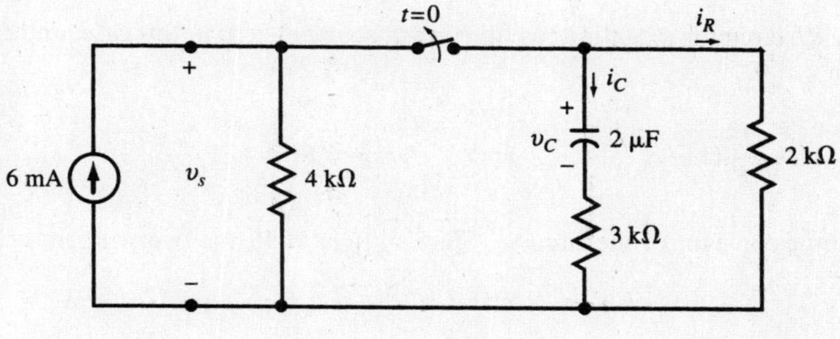

Fig. 7-25

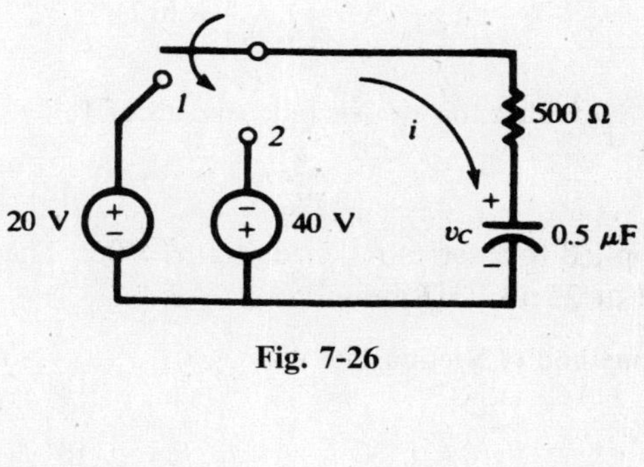

Fig. 7-26

From the assumption $q(0) = 0$ and the condition

$$i(0^+) = \frac{dq}{dt}\bigg|_{0^+} = \frac{20\text{ V}}{500\ \Omega} = 40\text{ mA}$$

we find that $A = -B = -10\ \mu\text{C}$, or

$$q = 10(1 - e^{-4000t})\quad (\mu\text{C}) \qquad (0 \le t \le \tau) \tag{20}$$

From (20), $q(\tau) = 10(1 - e^{-1})\ \mu\text{C}$; and we know that $q(\infty) = (0.5\ \mu\text{F})(-40\text{ V}) = -20\ \mu\text{C}$. Hence, q, is determined for $t \ge \tau$ as

$$q = [q(\tau) - q(\infty)]e^{-(t-\tau)/\tau} + q(\infty) = 71.55e^{-4000t} - 20\quad (\mu\text{C}) \tag{21}$$

Differentiating (20) and (21),

$$i = \frac{dq}{dt} = \begin{cases} 40e^{-4000t} & (\text{mA}) & (0 < t < \tau) \\ -286.2e^{-4000t} & (\text{mA}) & (t > \tau) \end{cases}$$

See Fig. 7-27.

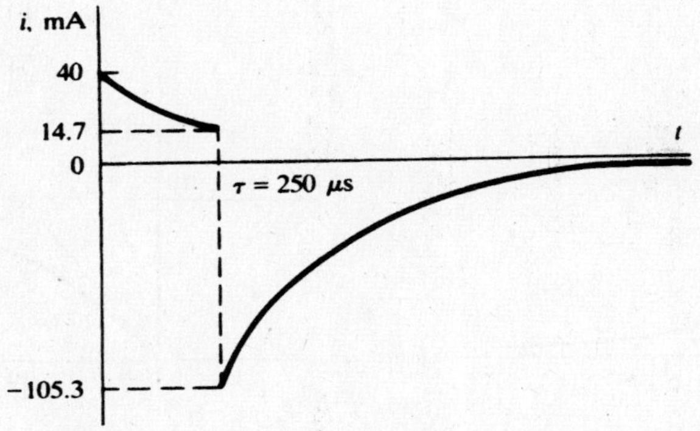

Fig. 7-27

7.12 A series RL circuit has a constant voltage V applied at $t = 0$. At what time does $v_R = v_L$?

The current in an RL circuit is a continuous function, starting at zero in this case, and reaching the final value V/R. Thus, for $t > 0$,

$$i = \frac{V}{R}(1 - e^{-t/\tau}) \qquad \text{and} \qquad v_R = Ri = V(1 - e^{-t/\tau})$$

where $\tau = L/R$ is the time constant of the circuit. Since $v_R + v_L = V$, the two voltages will be equal when

$$v_R = \tfrac{1}{2}V$$
$$V(1 - e^{-t/\tau}) = \tfrac{1}{2}V$$
$$e^{-t/\tau} = \tfrac{1}{2}$$
$$\frac{t}{\tau} = \ln 2$$

that is, when $t = 0.693\tau$. Note that this time is independent of V.

7.13 A constant voltage is applied to a series RL circuit at $t = 0$. The voltage across the inductance is 20 V at 3.46 ms and 5 V at 25 ms. Obtain R if $L = 2\ H$.

Using the two-point method of Section 7-6.

$$\tau = \frac{t_2 - t_1}{\ln v_1 - \ln v_2} = \frac{25 - 3.46}{\ln 20 - \ln 5} = 15.54\,\text{ms}$$

and so
$$R = \frac{L}{\tau} = \frac{2}{15.54 \times 10^{-3}} = 128.7\,\Omega$$

7.14 In Fig. 7-28, switch S_1 is closed at $t = 0$. Switch S_2 is opened at $t = 4\,\text{ms}$. Obtain i for $t > 0$.

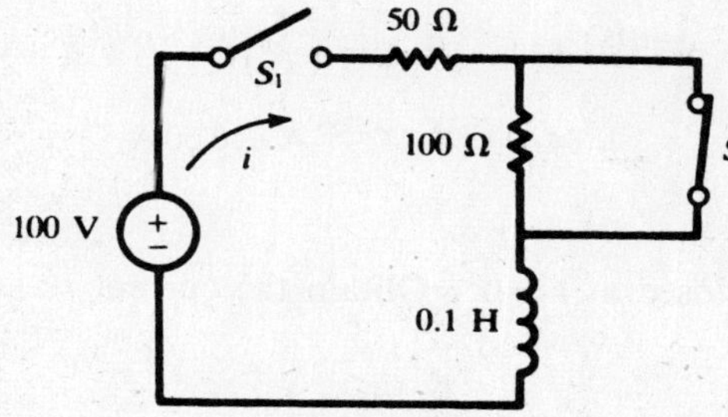

Fig. 7-28

As there is always inductance in the circuit, the current is a continuous function at all times. In the interval $0 \le t \le 4\,\text{ms}$, with the $100\,\Omega$ shorted out and a time constant $\tau = (0.1\,\text{H})/(50\,\Omega) = 2\,\text{ms}$, i starts at zero and builds toward

$$\frac{100\,\text{V}}{50\,\Omega} = 2\,\text{A}$$

even though it never gets close to that value. Hence, as in Problem 7.12

$$i = 2(1 - e^{-t/2}) \quad (\text{A}) \qquad (0 \le t \le 4) \tag{22}$$

wherein t is measured in ms. In particular,

$$i(4) = 2(1 - e^{-2}) = 1.729\,\text{A}$$

In the interval $t \ge 4\,\text{ms}$, i starts at 1.729 A and decays toward $100/150 = 0.667\,\text{A}$, with a time constant $0.1/150 = \frac{2}{3}\,\text{ms}$. Therefore, with t again in ms,

$$i = (1.729 - 0.667)e^{-(t-4)/(2/3)} + 0.667 = 428.4e^{-3t/2} + 0.667 \quad (\text{A}) \qquad (t \ge 4) \tag{23}$$

7.15 In the circuit of Fig. 7-29, the switch is closed at $t = 0$, when the 6-μF capacitor has charge $Q_0 = 300\,\mu\text{C}$. Obtain the expression for the transient voltage v_R.

The two parallel capacitors have an equivalent capacitance of 3 μF. Then this capacitance is in series with the 6 μF, so that the overall equivalent capacitance is 2 μF. Thus, $\tau = RC_{eq} = 40\,\mu\text{s}$.

At $t = 0^+$, KVL gives $v_R = 300/6 = 50\,\text{V}$; and, as $t \to \infty$, $v_R \to 0$ (since $i \to 0$). Therefore,

$$v_R = 50\, e^{-t/\tau} = 50e^{-t/40} \quad (\text{V})$$

in which t is measured in μs.

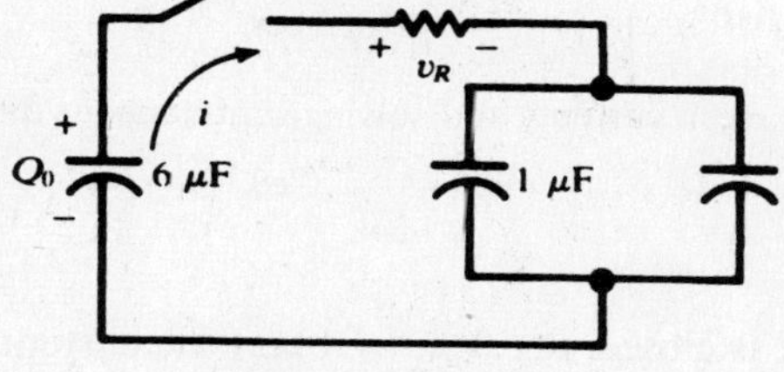

Fig. 7-29

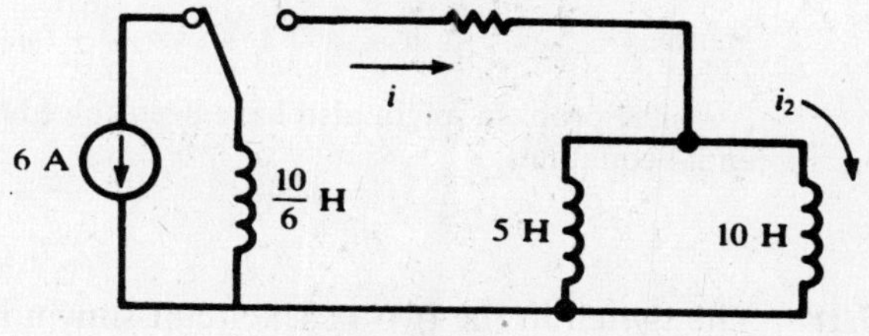

Fig. 7-30

7.16 In the circuit shown in Fig. 7-30, the switch is moved to position *2* at $t = 0$. Obtain the current i_2 at $t = 34.7\,\text{ms}$.

After the switching, the three inductances have the equivalent

$$L_{eq} = \frac{10}{6} + \frac{5(10)}{15} = 5\,\text{H}$$

Then $\tau = 5/200 = 25\,\text{ms}$, and so, with t in ms,

$$i = 6e^{-t/25} \quad (\text{A}) \qquad i_2 = \left(\frac{5}{15}\right) i = 2e^{-t/25} \quad (\text{A})$$

and

$$i_2(34.7) = 2e^{-34.7/25}\,\text{A} = 0.50\,\text{A}$$

7.17 In Fig. 7-31, the switch is closed at $t = 0$. Obtain the current i and capacitor voltage v_C, for $t > 0$.

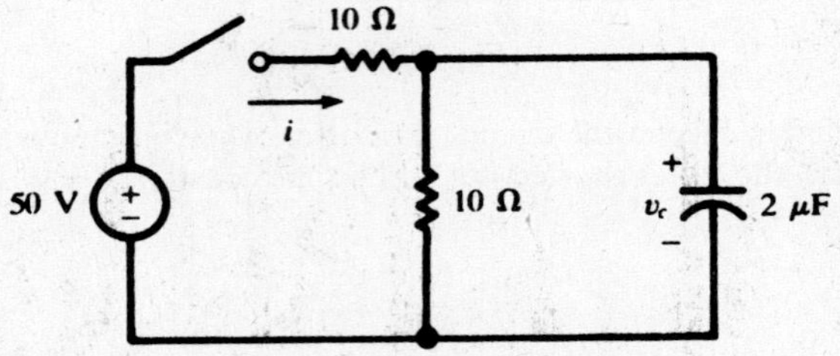

Fig. 7-31

As far as the natural response of the circuit is concerned, the two resistors are in parallel; hence,

$$\tau = R_{eq}C = (5\,\Omega)(2\,\mu\text{F}) = 10\,\mu\text{s}$$

By continuity, $v_C(0^+) = v_C(0^-) = 0$. Furthermore, as $t \to \infty$, the capacitor becomes an open circuit, leaving $20\,\Omega$ in series with the 50 V. That is,

$$i(\infty) = \frac{50}{20} = 2.5\,\text{A} \qquad v_C(\infty) = (2.5\,\text{A})(10\,\Omega) = 25\,\text{V}$$

Knowing the end conditions on v_C, we can write

$$v_C = [v_C(0^+) - v_C(\infty)]e^{-t/\tau} + v_C(\infty) = 25(1 - e^{-t/10}) \quad (\text{V})$$

wherein t is measured in μs.

The current in the capacitor is given by

$$i_C = C\frac{dv_C}{dt} = 5e^{-t/10} \quad (\text{A})$$

and the current in the parallel 10-Ω resistor is

$$i_{10\Omega} = \frac{v_C}{10\,\Omega} = 2.5(1 - e^{-t/10}) \quad (\text{A})$$

Hence,

$$i = i_C + i_{10\,\Omega} = 2.5(1 + e^{-t/10}) \quad (\text{A})$$

The problem might also have been solved by assigning mesh currents and solving simultaneous differential equations.

7.18 The switch in the two-mesh circuit shown in Fig. 7-32 is closed at $t = 0$. Obtain the currents i_1 and i_2, for $t > 0$.

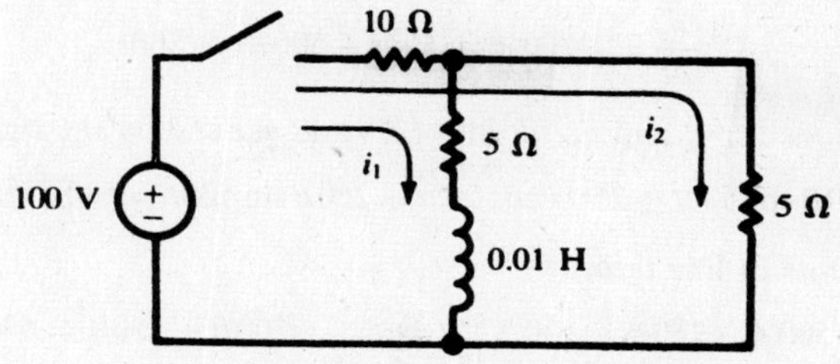

Fig. 7-32

$$10(i_1 + i_2) + 5i_1 + 0.01\frac{di_1}{dt} = 100 \qquad (24)$$

$$10(i_1 + i_2) + 5i_2 = 100 \qquad (25)$$

From (25), $i_2 = (100 - 10i_1)/15$. Substituting in (24),

$$\frac{di_1}{dt} + 833i_1 = 3333 \qquad (26)$$

The steady-state solution (particular solution) of (26) is $i_1(\infty) = 3333/833 = 4.0\,\text{A}$; hence

$$i_1 = Ae^{-833t} + 4.0 \quad \text{(A)}$$

The initial condition $i_1(0^-) = i_1(0^+) = 0$ now gives $A = -4.0\,\text{A}$, so that

$$i_1 = 4.0(1 - e^{-833t}) \quad \text{(A)} \qquad \text{and} \qquad i_2 = 4.0 + 2.67e^{-833t} \quad \text{(A)}$$

Alternate Method

When the rest of the circuit is viewed from the terminals of the inductance, there is equivalent resistance

$$R_{\text{eq}} = 5 + \frac{5(10)}{15} = 8.33\,\Omega$$

Then $1/\tau = R_{\text{eq}}/L = 833\,\text{s}^{-1}$. At $t = \infty$, the circuit resistance is

$$R_T = 10 + \frac{5(5)}{10} = 12.5\,\Omega$$

so that the total current is $i_T = 100/12.5 = 8\,\text{A}$. And, at $t = \infty$, this divides equally between the two 5-Ω resistors, yielding a final inductor current of 4 A. Consequently,

$$i_L = i_1 = 4(1 - e^{-833t}) \quad \text{(A)}$$

7.19 A series *RL* circuit, with $R = 50\,\Omega$ and $L = 0.2\,\text{H}$, has a sinusoidal voltage

$$v = 150 \sin(500t + 0.785) \quad \text{(V)}$$

applied at $t = 0$. Obtain the current for $t > 0$.

The circuit equation for $t > 0$ is

$$\frac{di}{dt} + 250i = 750 \sin(500t + 0.785) \qquad (27)$$

The solution is in two parts, the complementary function (i_c) and the particular solution (i_p), so that $i = i_c + i_p$. The complementary function is the general solution of (27) when the right-hand side is replaced by zero: $i_c = ke^{-250t}$. The *method of undetermined coefficients* for obtaining i_p consists in assuming that

$$i_p = A \cos 500t + B \sin 500t$$

since the right-hand side of (27) can also be expressed as a linear combination of these two functions. Then

$$\frac{di_p}{dt} = -500A \sin 500t + 500B \cos 500t$$

Substituting these expressions for i_p and di_p/dt into *(27)* and expanding the right-hand side,

$$-500A \sin 500t + 500B \cos 500t + 250A \cos 500t + 250B \sin 500t = 530.3 \cos 500t + 530.3 \sin 500t$$

Now equating the coefficients of like terms,

$$-500A + 250B = 530.3 \qquad \text{and} \qquad 500B + 250A = 530.3$$

Solving these simultaneous equations, $A = -0.4243\text{ A}$, $B = 1.273\text{ A}$.

$$i_p = -0.4243 \cos 500t + 1.273 \sin 500t = 1.342 \sin (500t - 0.322) \quad \text{(A)}$$

and

$$i = i_c + i_p = ke^{-250t} + 1.342 \sin (500t - 0.322) \quad \text{(A)}$$

At $t = 0$, $i = 0$. Applying this condition, $k = 0.425\text{ A}$, and, finally,

$$i = 0.425e^{-250t} + 1.342 \sin (500t - 0.322) \quad \text{(A)}$$

7.20 For the circuit of Fig. 7-33, obtain the current i_L, for all values of t.

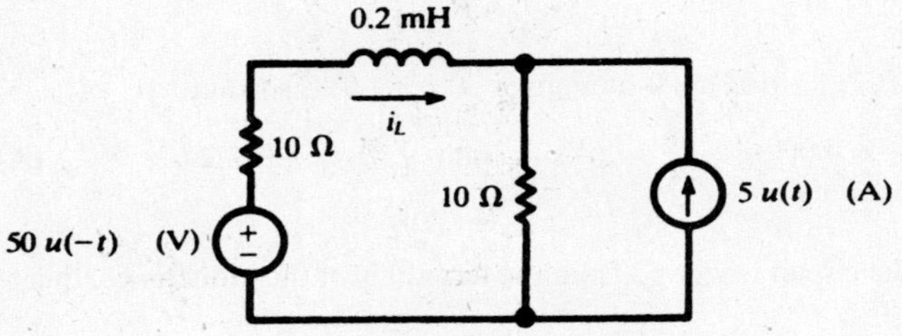

Fig. 7-33

For $t < 0$, the 50-V source results in inductor current $50/20 = 2.5\text{ A}$. The 5-A current source is applied for $t > 0$. As $t \to \infty$, this current divides equally between the two 10-Ω resistors, whence $i_L(\infty) = -2.5\text{ A}$. The time constant of the circuit is

$$\tau = \frac{0.2 \times 10^{-3}\text{ H}}{20\ \Omega} = \frac{1}{100}\text{ ms}$$

and so, with t in ms and using $i_L(0^+) = i_L(0^-) = 2.5\text{ A}$,

$$i_L = [i_L(0^+) - i_L(\infty)]e^{-t/\tau} + i_L(\infty) = 5.0e^{-100t} - 2.5 \quad \text{(A)}$$

Finally, using unit step functions to combine the expressions for $t < 0$ and $t > 0$,

$$i_L = 2.5u(-t) + (5.0e^{-100t} - 2.5)u(t) \quad \text{(A)}$$

7.21 The switch in Fig. 7-34 has been in position *1* for a long time; it is moved to *2* at $t = 0$. Obtain the expression for i, for $t > 0$.

With the switch on *1*, $i(0^-) = 50/40 = 1.25\text{ A}$. With an inductance in the circuit, $i(0^-) = i(0^+)$. Long after the switch has been moved to *2*, $i(\infty) = 10/40 = 0.25\text{ A}$. In the above notation,

$$B = i(\infty) = 0.25\text{ A} \qquad A = i(0^+) - B = 1.00\text{ A}$$

and the time constant is $\tau = L/R = (1/2000)\text{ s}$. Then, for $t > 0$,

$$i = 1.00e^{-2000t} + 0.25 \quad \text{(A)}$$

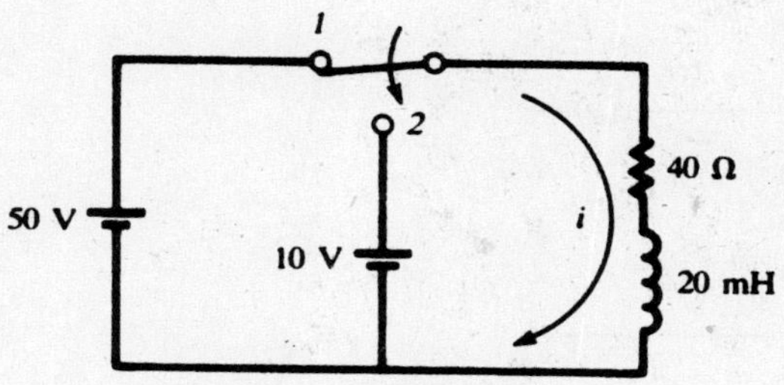

Fig. 7-34

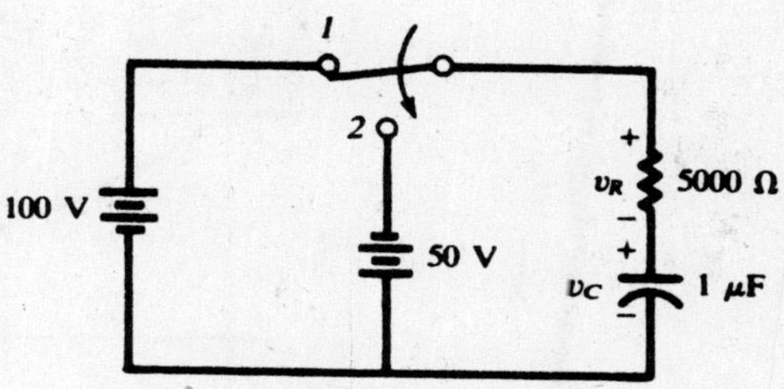

Fig. 7-35

7.22 The switch in the circuit shown in Fig. 7-35 is moved from *1* to *2* at $t = 0$. Find v_C and v_R, for $t > 0$.

With the switch on *1*, the 100-V source results in $v_C(0^-) = 100\text{ V}$; and, by continuity of charge, $v_C(0^+) = v_C(0^-)$. In position *2*, with the 50-V source of opposite polarity, $v_C(\infty) = -50\text{ V}$. Thus,

$$B = v_C(\infty) = -50\text{ V} \qquad A = v_C(0^+) - B = 150\text{ V}$$

$$\tau = RC = \frac{1}{200}\text{ s}$$

and

$$v_C = 150e^{-200t} - 50 \quad \text{(V)}$$

Finally, KVL gives $v_R + v_C + 50 = 0$, or

$$v_R = -150e^{-200t} \quad \text{(V)}$$

7.23 Obtain the energy functions for the circuit of Problem 7.22.

$$w_C = \tfrac{1}{2}Cv_C^2 = 1.25(3e^{-200t} - 1)^2 \quad \text{(mJ)}$$

$$w_R = \int_0^t \frac{v_R^2}{R}\,dt = 11.25(1 - e^{-400t}) \quad \text{(mJ)}$$

7.24 A series *RC* circuit, with $R = 5\text{ k}\Omega$ and $C = 20\,\mu\text{F}$, has two voltage sources in series,

$$v_1 = 25u(-t) \quad \text{(V)} \qquad v_2 = 25u(t - t') \quad \text{(V)}$$

Obtain the complete expression for the voltage across the capacitor and make a sketch, if t' is positive.

The capacitor voltage is continuous. For $t \le 0$, v_1 results in a capacitor voltage of 25 V.

For $0 \le t \le t'$, both sources are zero, so that v_C decays exponentially from 25 V towards zero:

$$v_C = 25e^{-t/RC} = 25e^{-10t} \quad \text{(V)} \qquad (0 \le t \le t')$$

In particular, $v_C(t') = 25e^{-10t'}$ (V).

For $t \ge t'$, v_C builds from $v_C(t')$ towards the final value 25 V established by v_2:

$$\begin{aligned} v_C &= [v_C(t') - v_C(\infty)]e^{-(t-t')/RC} + v_C(\infty) \\ &= 25[1 - (e^{10t'} - 1)e^{-10t}] \quad \text{(V)} \qquad (t \ge t') \end{aligned}$$

Thus, for all t,

$$v_C = 25u(-t) + 25e^{-10t}[u(t) - u(t - t')] + 25[1 - (e^{10t'} - 1)e^{-10t}]u(t - t') \quad \text{(V)}$$

See Fig. 7-36.

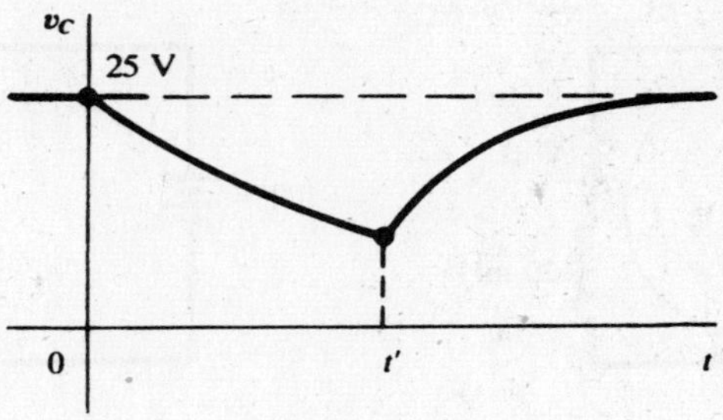

Fig. 7-36

Supplementary Problems

7.25 The capacitor in the circuit shown in Fig. 7-37 has initial charge $Q_0 = 800\ \mu C$, with polarity as indicated. If the switch is closed at $t = 0$, obtain the current and charge, for $t > 0$.
Ans. $i = -10e^{-25\,000t}$ (A), $q = 4 \times 10^{-4}(1 + e^{-25\,000t})$ (C)

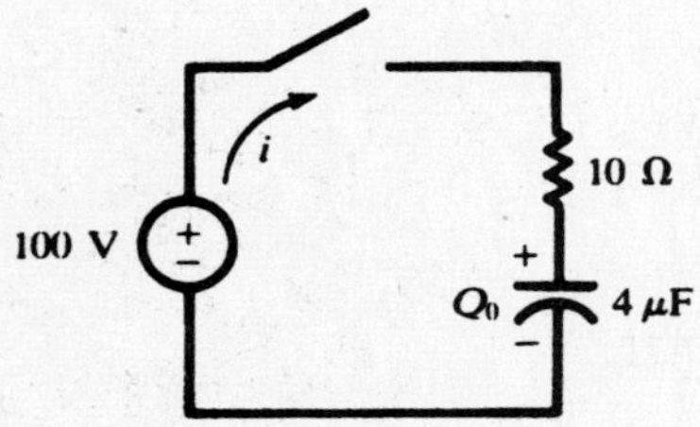

Fig. 7-37

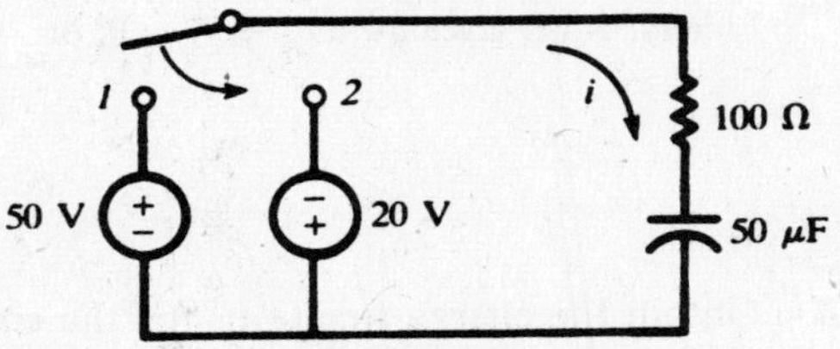

Fig. 7-38

7.26 A 2-μF capacitor, with initial charge $Q_0 = 100\ \mu C$, is connected across a 100-Ω resistor at $t = 0$. Calculate the time in which the transient voltage across the resistor drops from 40 to 10 volts. *Ans.* 0.277 ms

7.27 In the *RC* circuit shown in Fig. 7-38, the switch is closed on position *1* at $t = 0$ and then moved to *2* after the passage of one time constant. Obtain the current transient for (*a*) $0 < t < \tau$, (*b*) $t > \tau$.
Ans. (*a*) $0.5e^{-200t}$ (A); (*b*) $-0.516e^{-200(t-\tau)}$ (A)

7.28 A 10-μF capacitor, with initial charge Q_0, is connected across a resistor at $t = 0$. Given that the power transient for the capacitor is $800e^{-4000t}$ (W), find R, Q_0, and the initial stored energy in the capacitor.
Ans. 50 Ω, 2000 μC, 0.20 J

7.29 A series *RL* circuit, with $R = 10\ \Omega$ and $L = 1$ H, has a 100-V source applied at $t = 0$. Find the current for $t > 0$. *Ans.* $10(1 - e^{-10t})$ (A)

7.30 In Fig. 7-39, the switch is closed on position *1* at $t = 0$, then moved to *2* at $t = 1$ ms. Find the time at which the voltage across the resistor is zero, reversing polarity. *Ans.* 1.261 ms

7.31 A series *RL* circuit, with $R = 100\ \Omega$ and $L = 0.2$ H, has a 100-V source applied at $t = 0$; then a second source, of 50 V with the same polarity, is switched in at $t = t'$, replacing the first source. Find t' such that the current is constant at 0.5 A for $t > t'$. *Ans.* 1.39 ms

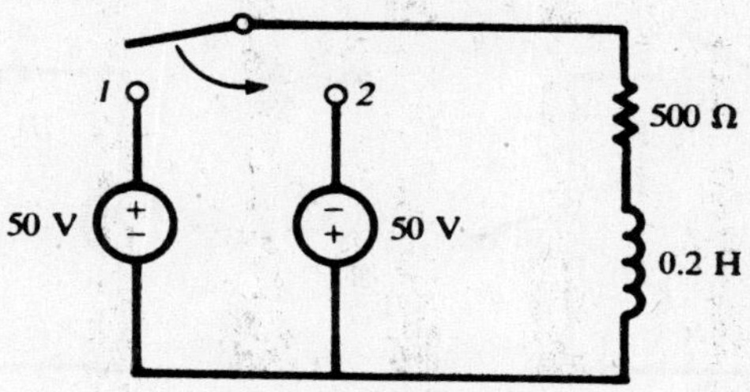

Fig. 7-39

7.32 The circuit of Problem 7.31 has a 50-V source of *opposite* polarity switched in at $t = 0.50$ ms, replacing the first source. Obtain the current for (*a*) $0 < t < 0.50$ ms, (*b*) $t > 0.50$ ms.
Ans. (*a*) $1 - e^{-500t}$ (A); (*b*) $0.721e^{-500(t-0.0005)} - 0.50$ (A)

7.33 A voltage transient, $35e^{-500t}$ (V), has the value 25 V at $t_1 = 6.73 \times 10^{-4}$ s. Show that at $t = t_1 + \tau$ the function has a value 36.8 percent of that at t_1.

7.34 A transient that increases from zero toward a positive steady-state magnitude is 49.5 at $t_1 = 5.0$ ms, and 120 at $t_2 = 20.0$ ms. Obtain the time constant τ. *Ans.* 12.4 ms

7.35 The circuit shown in Fig. 7-40 is switched to position *1* at $t = 0$, then to position *2* at $t = 3\tau$. Find the transient current i for (*a*) $0 < t < 3\tau$, (*b*) $t > 3\tau$.
Ans. (*a*) $2.5e^{-50\,000t}$ (A); (*b*) $-1.58e^{-66\,700(t-0.00006)}$ (A)

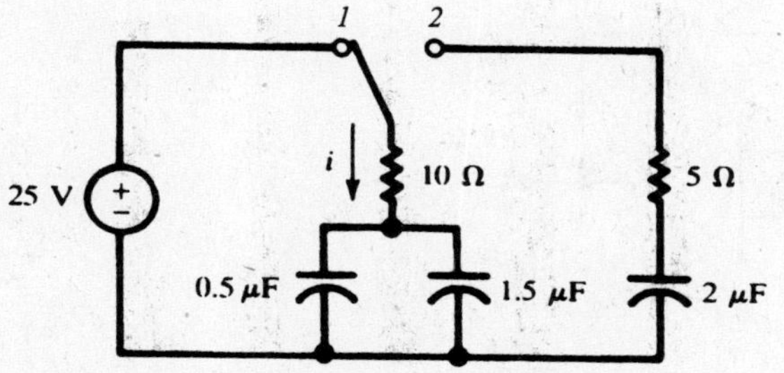

Fig. 7-40

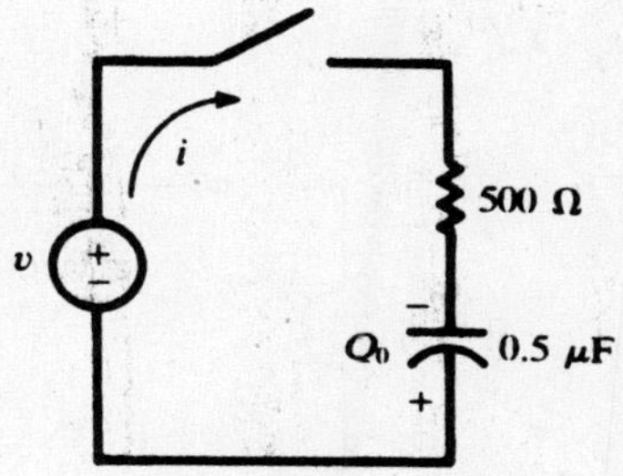

Fig. 7-41

7.36 An *RL* circuit, with $R = 300\ \Omega$ and $L = 1$ H, has voltage $v = 100\cos(100t + 45°)$ (V) applied by closing a switch at $t = 0$. [A convenient notation has been used for the phase of v, which, strictly, should be indicated as $100t + (\pi/4)$ (rad).] Obtain the resulting current for $t > 0$.
Ans. $-0.282e^{-300t} + 0.316\cos(100t + 26.6°)$ (A)

7.37 The *RC* circuit shown in Fig. 7-41 has an initial charge on the capacitor $Q_0 = 25\ \mu C$, with polarity as indicated. The switch is closed at $t = 0$, applying a voltage $v = 100\sin(1000t + 30°)$ (V). Obtain the current for $t > 0$. *Ans.* $153.5e^{-4000t} + 48.4\sin(1000t + 106°)$ (mA)

7.38 What initial charge on the capacitor in Problem 7.37 would cause the current to go directly into the steady state without a transient? *Ans.* 13.37 μC (+ on top plate)

7.39 Write simultaneous differential equations for the circuit shown in Fig. 7-42 and solve for i_1 and i_2. The switch is closed at $t = 0$ after having been open for an extended period of time. (This problem can also be solved by applying known initial and final conditions to general solutions, as in Problem 7-17.)
Ans. $i_1 = 1.67e^{6.67t} + 5$ (A), $i_2 = -0.555e^{-6.67t} + 5$ (A)

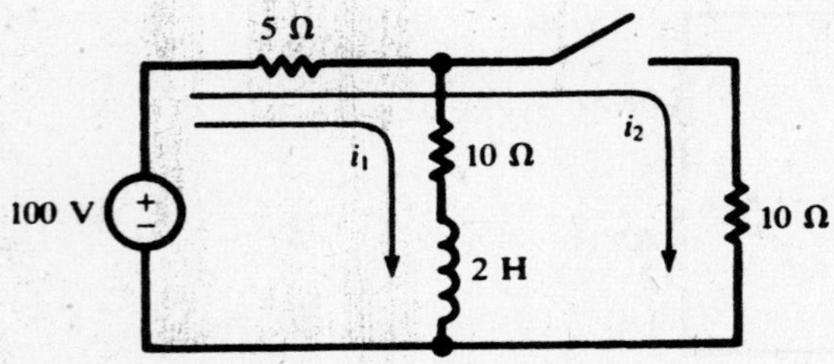

Fig. 7-42

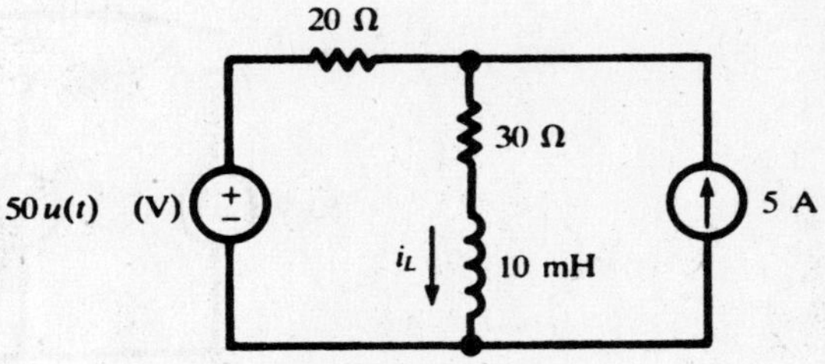

Fig. 7-43

7.40 For the *RL* circuit shown in Fig. 7-43, find the current i_L at the following times: (*a*) -1 ms, (*b*) 0^+, (*c*) 0.3 ms, (*d*) ∞. *Ans.* (*a*) 2.00 A; (*b*) 2.00 A; (*c*) 2.78 A; (*d*) 3.00 A

7.41 A series *RC* circuit, with $R = 2\,\text{k}\Omega$ and $C = 40\,\mu\text{F}$, has two voltage sources in series with each other, $v_1 = 50$ V and $v_2 = -100u(t)$ (V). Find (*a*) the capacitor voltage at $t = \tau$, (*b*) the time at which the capacitor voltage is zero and reversing polarity. *Ans.* (*a*) -13.2 V; (*b*) 55.5 ms

7.42 Find the unit-impulse response of the circuit of Fig. 7-44; i.e., v_2 for $v_1 = \delta(t)$ (a unit-area narrow voltage pulse).

Ans. $v_2 = -\dfrac{R_2}{R_1}\left[\delta(t) - \dfrac{1}{R_1 C}\, e^{-t/(R_1 C)} u(t)\right]$

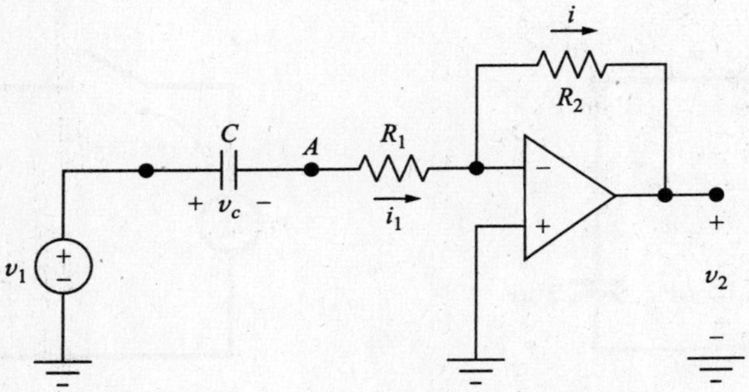

Fig. 7-44

7.43 In the circuits of Fig. 7-45, $RC = 5 \times 10^{-7}$ and $v_1(t) = 10 + \cos(1000t) + 3\cos(2000t)$. Find $v_2(t)$. Assume $\tan\theta \approx \theta$ when $\theta < 1°$. *Ans.* $v_2(t) \approx 10 + \cos[1000(t - 10^{-6})] + 3\cos[2000(t - 10^{-6})] = v_1(t - 10^{-6})$

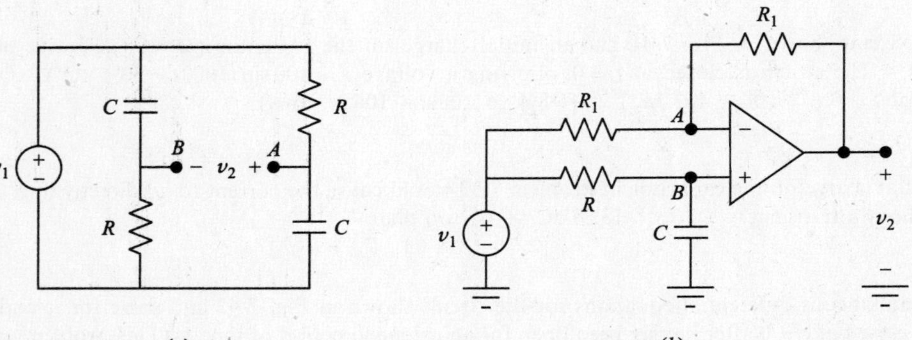

Fig. 7-45

7.44 The input voltage in the circuits of 7-45 is a weighted sum of sinusoids with the highest frequency f_0 Hz. Assuming that $RC < 1/(360 f_0)$, find $v_2(t)$. *Ans.* $v_2(t) \approx v_1(t - 2RC)$

7.45 Find the relationship between v_2 and v_1 in the circuit of Fig. 7-46.

Ans. $v_2 + RC\dfrac{dv_2}{dt} = 2v_1$

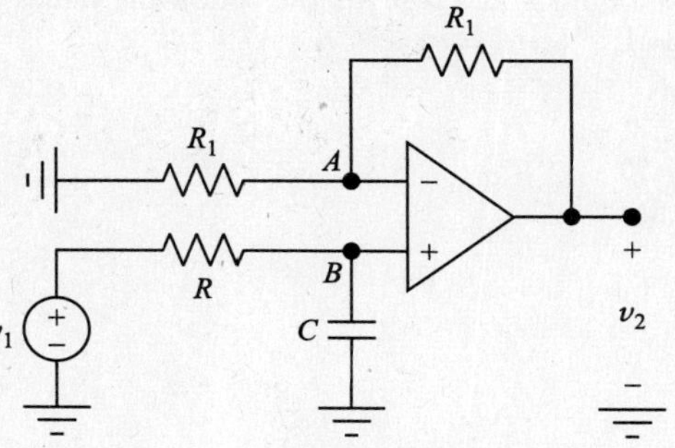

Fig. 7-46

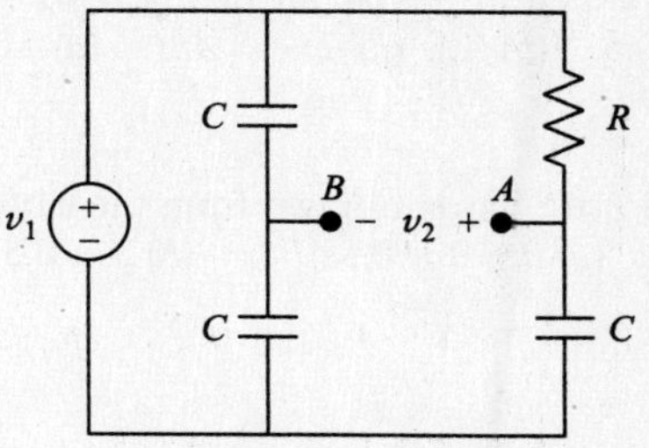

Fig. 7-47

7.46 In the circuit of Fig. 7-47, find the differential equation relating v_2 to v_1. Compare with the circuit of Fig. 7-45(*a*) of Example 7.17.

Ans. $v_2 + RC\dfrac{dv_2}{dt} = \dfrac{1}{2}\left(v_1 - RC\dfrac{dv_1}{dt}\right)$

7.47 In the circuit of Fig. 7-48, find the relationship between v_2 and v_1.

Ans. $v_2 + R_1C_1\dfrac{dv_2}{dt} = -\dfrac{C_1}{C_2}\left(v_1 - R_2C_2\dfrac{dv_1}{dt}\right)$

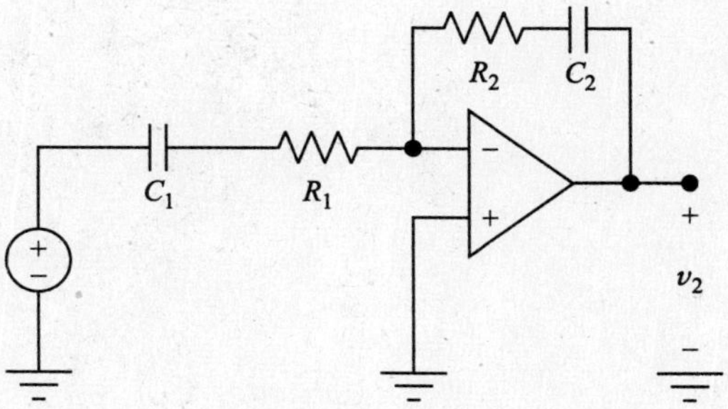

Fig. 7-48

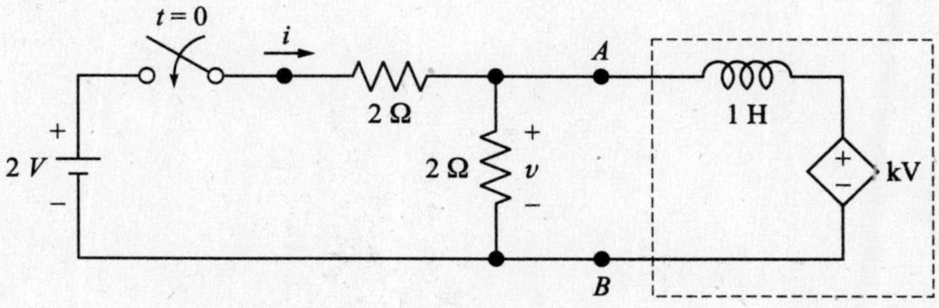

Fig. 7-49

7.48 In the circuit of Fig. 7-49, let $k = 0$. Find v and i after the switch is closed at $t = 0$.
Ans. $v = e^{-t}, i = 1 - 0.5e^{-t}$

7.49 Show that the segment of the circuit enclosed by the dashed box in the circuit of Fig. 7-49 is equivalent to an inductor with value $L = 1/(1 - k)$ H. *Hint*: Write KVL between terminals AB of the dashed box.

7.50 The switch in the circuit of Fig. 7-49 is closed at $t = 0$. Find v at $t > 0$ for the following values of k: (*a*) 0.5, (*b*) 1, (*c*) 2. *Ans.* (*a*) $v = e^{-t/2}$; (*b*) $v = 1$; (*c*) $v = e^{t}$

7.51 Find i, the current drawn from the battery, in Problem 7.50.
Ans. (*a*) $i = 1 - 0.5e^{-t/2}$; (*b*) $i = 0.5$; (*c*) $i = 1 - 0.5e^{t}$

CHAPTER 8

Higher-Order Circuits and Complex Frequency

8.1 INTRODUCTION

In Chapter 7, *RL* and *RC* circuits with initial currents or charge on the capacitor were examined and first-order differential equations were solved to obtain the transient voltages and currents. When two or more storage elements are present, the network equations will result in second-order differential equations. In this chapter, several examples of second-order circuits will be presented. This will then be followed by more direct methods of analysis, including complex frequency and pole-zero plots.

8.2 SERIES *RLC* CIRCUIT

The second-order differential equation, which will be examined shortly, has a solution that can take three different forms, each form depending on the circuit elements. In order to visualize the three possibilities, a second-order mechanical system is shown in Fig. 8-1. The mass M is suspended by a spring with a constant k. A damping device D is attached to the mass M. If the mass is displaced from its rest position and then released at $t = 0$, its resulting motion will be *overdamped*, *critically damped*, or *underdamped* (*oscillatory*). Figure 8-2 shows the graph of the resulting motions of the mass after its release from the displaced position z_1 (at $t = 0$).

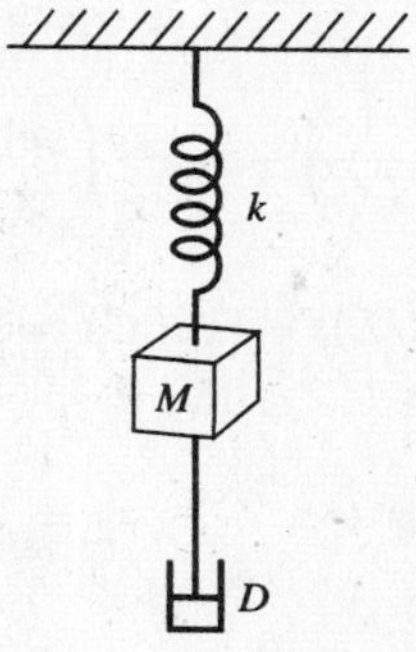

Fig. 8-1

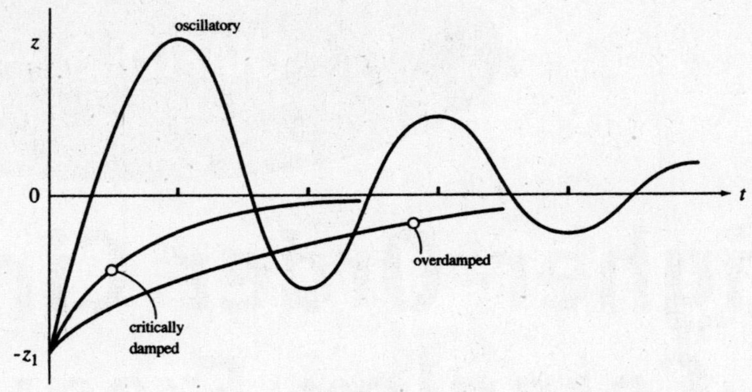

Fig. 8-2

The series *RLC* circuit shown in Fig. 8-3 contains no voltage source. Kirchhoff's voltage law for the closed loop after the switch is closed is

$$v_R + v_L + v_C = 0$$

or

$$Ri + L\frac{di}{dt} + \frac{1}{C}\int i\,dt = 0$$

Differentiating and dividing by L yields

$$\frac{d^2i}{dt^2} + \frac{R}{L}\frac{di}{dt} + \frac{1}{LC}i = 0$$

A solution of this second-order differential equation is of the form $i = A_1 e^{s_1 t} + A_2 e^{s_2 t}$. Substituting this solution in the differential equation obtains

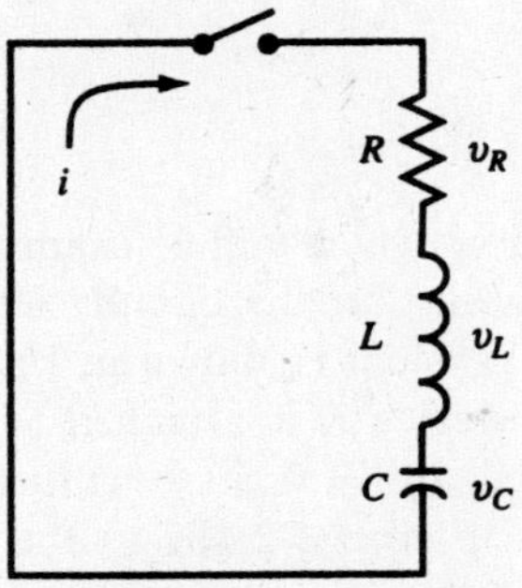

Fig. 8-3

$$A_1 e^{s_1 t}\left(s_1^2 + \frac{R}{L}s_1 + \frac{1}{LC}\right) + A_2 e^{s_2 t}\left(s_2^2 + \frac{R}{L}s_2 + \frac{1}{LC}\right) = 0$$

that is, if s_1 and s_2 are the roots of $s^2 + (R/L)s + (1/LC) = 0$,

$$s_1 = -\frac{R}{2L} + \sqrt{\left(\frac{R}{2L}\right)^2 - \frac{1}{LC}} \equiv -\alpha + \beta \qquad s_2 = -\frac{R}{2L} - \sqrt{\left(\frac{R}{2L}\right)^2 - \frac{1}{LC}} \equiv -\alpha - \beta$$

where $\alpha \equiv R/2L$, $\beta \equiv \sqrt{\alpha^2 - \omega_0^2}$, and $\omega_0 \equiv 1/\sqrt{LC}$.

Overdamped Case ($\alpha > \omega_0$)

In this case, both α and β are real positive numbers.

$$i = A_1 e^{(-\alpha+\beta)t} + A_2 e^{(-\alpha-\beta)t} = e^{-\alpha t}(A_1 e^{\beta t} + A_2 e^{-\beta t})$$

EXAMPLE 8.1 A series RLC circuit, with $R = 200\,\Omega$, $L = 0.10\,\text{H}$, and $C = 13.33\,\mu\text{F}$, has an initial charge on the capacitor of $Q_0 = 2.67 \times 10^{-3}\,\text{C}$. A switch is closed at $t = 0$, allowing the capacitor to discharge. Obtain the current transient. (See Fig. 8-4.)

For this circuit,

$$\alpha = \frac{R}{2L} = 10^3\,\text{s}^{-1}, \qquad \omega_0^2 = \frac{1}{LC} = 7.5 \times 10^5\,\text{s}^{-2}, \qquad \text{and} \qquad \beta = \sqrt{\alpha^2 - \omega_0^2} = 500\,\text{s}^{-1}$$

Then,

$$i = e^{-1000t}(A_1 e^{500t} + A_2 e^{-500t})$$

The values of the constants A_1 and A_2 are obtained from the initial conditions. The inductance requires that $i(0^+) = i(0^-)$. Also the charge and voltage on the capacitor at $t = 0^+$ must be the same as at $t = 0^-$, and $v_C(0^-) = Q_0/C = 200\,\text{V}$. Applying these two conditions,

$$0 = A_1 + A_2 \qquad \text{and} \qquad \pm 2000 = -500A_1 - 1500A_2$$

from which $A_1 = \pm 2$, $A_2 = \mp 2$, and, taking A_1 positive,

$$i = 2e^{-500t} - 2e^{-1500t} \quad \text{(A)}$$

If the negative value is taken for A_1, the function has simply flipped downward but it has the same shape. The signs of A_1 and A_2 are fixed by the polarity of the initial voltage on the capacitor and its relationship to the assumed positive direction for the current.

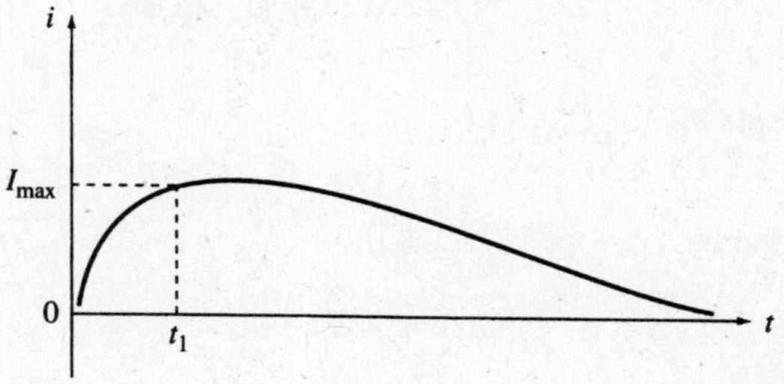

Fig. 8-4

Critically Damped Case ($\alpha = \omega_0$)

With $\alpha = \omega_0$, the differential equation takes on a different form and the two exponential terms suggested in the preceding will no longer provide a solution. The equation becomes

$$\frac{d^2 i}{dt^2} + 2\alpha \frac{di}{dt} + \alpha^2 i = 0$$

and the solution takes the form $i = e^{-\alpha t}(A_1 + A_2 t)$.

EXAMPLE 8.2 Repeat Example 8.1 for $C = 10\,\mu\text{F}$, which results in $\alpha = \omega_0$.

As in Example 8.1, the initial conditions are used to determine the constants. Since $i(0^-) = i(0^+)$, $0 = [A_1 + A_2(0)]$ and $A_1 = 0$. Then,

$$\frac{di}{dt} = \frac{d}{dt}(A_2 t e^{-\alpha t}) = A_2(-\alpha t e^{-\alpha t} + e^{-\alpha t})$$

from which $A_2 = (di/dt)|_{0^+} = \pm 2000$. Hence, $i = \pm 2000 t e^{-10^3 t}$ (A) (see Fig. 8-5).

Once again the polarity is a matter of the choice of direction for the current with respect to the polarity of the initial voltage on the capacitor.

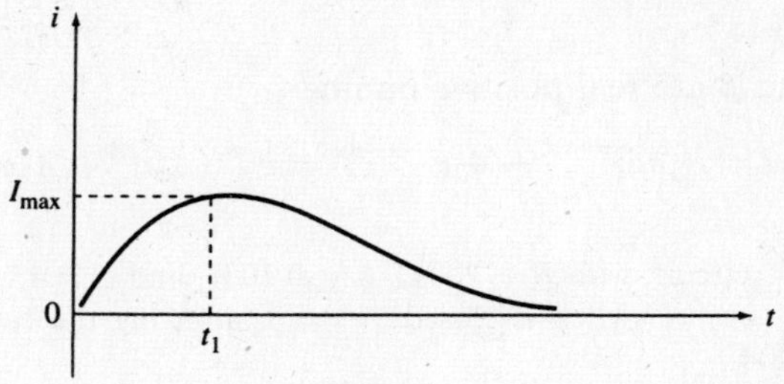

Fig. 8-5

The responses for the overdamped and critically damped cases plotted in Figs. 8-4 and 8-5, respectively, are quite similar. The reader is encouraged to examine the results, selecting several values for t, and comparing the currents. For example, find the time at which the current in each of the two cases reaches the values of 1.0 mA and 1.0 μA. Also, in each case, find t_1 for the maximum current.

Underdamped or Oscillatory Case ($\alpha < \omega_0$)

When $\alpha < \omega_0$, s_1 and s_2 in the solution to the differential equation suggested in the preceding are complex conjugates $s_1 = \alpha + j\beta$ and $s_2 = \alpha - j\beta$, where β is now given by $\sqrt{\omega_0^2 - \alpha^2}$. The solution can be written in the exponential form

$$i = e^{-\alpha t}(A_1 e^{j\beta t} + A_2 e^{-j\beta t})$$

or, in a readily derived sinusoidal form,

$$i = e^{-\alpha t}(A_3 \cos \beta t + A_4 \sin \beta t)$$

EXAMPLE 8.3 Repeat Example 8.1 for $C = 1\,\mu\text{F}$.

As before,

$$\alpha = \frac{R}{2L} = 1000\,\text{s}^{-1} \qquad \omega_0^2 = \frac{1}{LC} = 10^7\,\text{s}^{-2} \qquad \beta = \sqrt{10^7 - 10^6} = 3000\,\text{rad/s}$$

Then,

$$i = e^{-1000t}(A_3 \cos 3000t + A_4 \sin 3000t)$$

The constants A_3 and A_4 are obtained from the initial conditions as before, $i(0^+) = 0$ and $v_c(0^+) = 200\,\text{V}$. From this $A_3 = 0$ and $A_4 = \pm 0.667$. Thus,

$$i = \pm 0.667 e^{-1000t}(\sin 3000t) \quad \text{(A)}$$

See Fig. 8-6. The function $\pm 0.667e^{-1000t}$, shown dashed in the graph, provides an envelope within which the sine function is confined. The oscillatory current has a radian frequency of β (rad/s), but is *damped* by the exponential term $e^{-\alpha t}$.

8.3 PARALLEL *RLC* CIRCUIT

The response of the parallel *RLC* circuit shown in Fig. 8-7 will be similar to that of the series *RLC* circuit, since a second-order differential equation can be expected. The node voltage method gives

$$\frac{v}{R} + \frac{1}{L}\int_0^t v\,dt + C\frac{dv}{dt} = 0 \qquad (1)$$

Differentiating and dividing by C yields

$$\frac{d^2v}{dt^2} + \frac{1}{RC}\frac{dv}{dt} + \frac{v}{LC} = 0$$

A solution is of the form

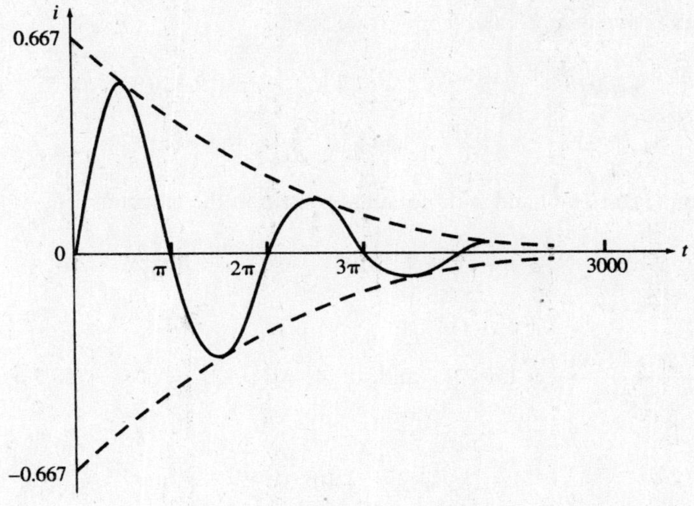

Fig. 8-6

$$v = A_1 e^{s_1 t} + A_2 e^{s_2 t} \tag{2}$$

where

$$s_1 = -\frac{1}{2RC} + \sqrt{\left(\frac{1}{2RC}\right)^2 - \frac{1}{LC}} = -\alpha + \sqrt{\alpha^2 - \omega_0^2}$$

$$s_2 = -\frac{1}{2RC} - \sqrt{\left(\frac{1}{2RC}\right)^2 - \frac{1}{LC}} = -\alpha - \sqrt{\alpha^2 - \omega_0^2}$$

where $\alpha = 1/2RC$ and $\omega_0 = 1/\sqrt{LC}$. Note that α, the *damping factor* of the transient, differs from α in the series *RLC* circuit.

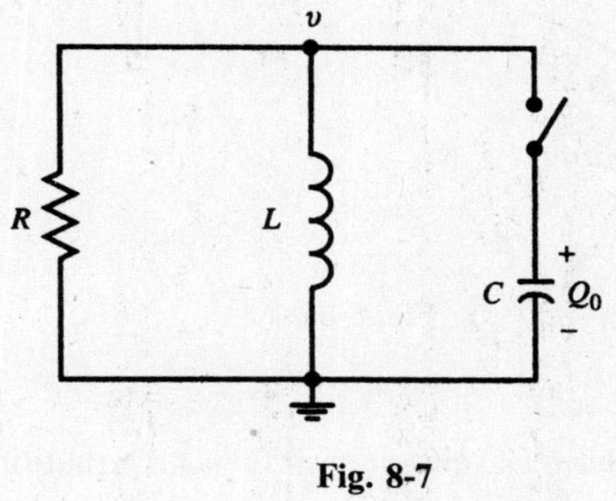

Fig. 8-7

The transient response is easiest to visualize by assuming an initial charge Q_0 on the capacitor and a switch that closes at $t = 0$. However, a step function voltage applied to the circuit will initiate the same transient response.

Overdamped Case ($\alpha^2 > \omega_0^2$)

In this case, the solution (*2*) applies.

EXAMPLE 8.4 A parallel *RLC* circuit, with $R = 1000\,\Omega$, $C = 0.167\,\mu\text{F}$, and $L = 1.0\,\text{H}$, has an initial voltage $V_0 = 50.0\,\text{V}$ on the capacitor. Obtain the voltage $v(t)$ when the switch is closed at $t = 0$.

We have

$$\alpha = \frac{1}{2RC} = 2994 \qquad \alpha^2 = 8.96 \times 10^6 \qquad \omega_0^2 = \frac{1}{LC} = 5.99 \times 10^6$$

Since $\alpha^2 > \omega_0^2$, the circuit is overdamped and from (2) we have

$$s_1 = -\alpha + \sqrt{\alpha^2 - \omega_0^2} = -1271 \qquad \text{and} \qquad s_2 = -\alpha - \sqrt{\alpha^2 - \omega_0^2} = -4717$$

At $t = 0$,

$$V_0 = A_1 + A_2 \qquad \text{and} \qquad \left.\frac{dv}{dt}\right|_{t=0} = s_1 A_1 + s_2 A_2$$

From the nodal equation (1), at $t = 0$ and with no initial current in the inductance L,

$$\frac{V_0}{R} + C\frac{dv}{dt} = 0 \qquad \text{or} \qquad \left.\frac{dv}{dt}\right|_{t=0} = -\frac{V_0}{RC}$$

Solving for A_1,

$$A_1 = \frac{V_0(s_2 + 1/RC)}{s_2 - s_1} = 155.3 \qquad \text{and} \qquad A_1 = V_0 - A_1 = 50.0 - 155.3 = -105.3$$

Substituting into (2)

$$v = 155.3e^{-1271t} - 105.3e^{-4717t} \quad \text{(V)}$$

See Fig. 8-8.

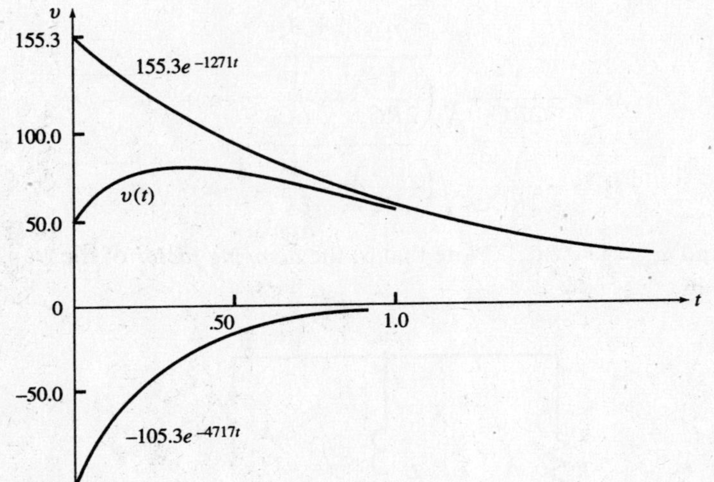

Fig. 8-8

Underdamped (Oscillatory) Case ($\omega_0^2 > \alpha^2$)

The oscillatory case for the parallel *RLC* circuit results in an equation of the same form as that of the underdamped series *RLC* circuit. Thus,

$$v = e^{-\alpha t}(A_1 \cos \omega_d t + A_2 \sin \omega_d t) \tag{3}$$

where $\alpha = 1/2RC$ and $\omega_d = \sqrt{\omega_0^2 - \alpha^2}$. ω_d is a radian frequency just as was the case with sinusoidal circuit analysis. Here it is the frequency of the damped oscillation. It is referred to as the *damped radian frequency*.

EXAMPLE 8.5 A parallel *RLC* circuit, with $R = 200\,\Omega$, $L = 0.28\,\text{H}$, and $C = 3.57\,\mu\text{F}$, has an initial voltage $V_0 = 50.0\,\text{V}$ on the capacitor. Obtain the voltage function when the switch is closed at $t = 0$.

$$\alpha = \frac{1}{2RC} = \frac{1}{2(200)(3.57 \times 10^{-6})} = 700 \qquad \alpha^2 = 4.9 \times 10^5 \qquad \omega_0^2 = \frac{1}{LC} = \frac{1}{(0.28)(3.57 \times 10^{-6})} = 10^6$$

Since $\omega_0^2 > \alpha^2$, the circuit parameters result in an oscillatory response.

$$\omega_d = \sqrt{\omega_0^2 - \alpha^2} = \sqrt{10^6 - (4.9 \times 10^5)} = 714$$

At $t = 0$, $V_0 = 50.0$; hence in (3) $A_1 = V_0 = 50.0$. From the nodal equation

$$\frac{V_0}{R} + \frac{1}{L}\int_0^t v\,dt + C\frac{dv}{dt} = 0$$

$$\left.\frac{dv}{dt}\right|_{t=0} = -\frac{V_0}{RC}$$

at $t = 0$,

Differentiating the expression for v and setting $t = 0$ yields

$$\left.\frac{dv}{dt}\right|_{t=0} = \omega_d A_2 - \alpha A_1 \qquad \text{or} \qquad \omega_d A_2 - \alpha A_1 = -\frac{V_0}{RC}$$

Since $A_1 = 50.0$,

$$A_2 = \frac{-(V_0/RC) + V_0\alpha}{\omega_d} = -49.0$$

and so

$$v = e^{-700t}(50.0\cos 714t - 49.0\sin 714t) \quad \text{(V)}$$

The critically damped case will not be examined for the parallel RLC circuit, since it has little or no real value in circuit design. In fact, it is merely a curiosity, since it is a set of circuit constants whose response, while damped, is on the verge of oscillation.

8.4 TWO-MESH CIRCUIT

The analysis of the response for a two-mesh circuit which contains two storage elements results in simultaneous differential equation as shown in the following.

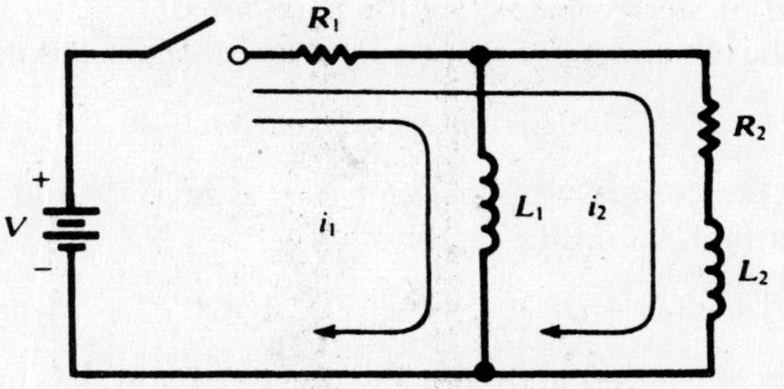

Fig. 8-9

For the circuit of Fig. 8-9, choose mesh currents i_1 and i_2, as indicated. KVL yields the two first-order differential equations

$$R_1 i_1 + L_1\frac{di_1}{dt} + R_1 i_2 = V \tag{4}$$

$$R_1 i_1 + (R_1 + R_2)i_2 + L_2\frac{di_2}{dt} = V \tag{5}$$

which must be solved simultaneously. To accomplish this, take the time derivative of (4),

$$R_1\frac{di_1}{dt} + L_1\frac{d^2 i_1}{dt^2} + R_1\frac{di_2}{dt} = 0 \tag{6}$$

and then eliminate i_2 and di_2/dt between (4), (5), and (6). The following result is a second-order equation for i_1, of the types treated in Sections 8.2 and 8.3, except for the constant term on the right:

$$\frac{d^2 i_1}{dt^2} + \frac{R_1 L_1 + R_2 L_1 + R_1 L_2}{L_1 L_2} \frac{di_1}{dt} + \frac{R_1 R_2}{L_1 L_2} i_1 = \frac{R_2 V}{L_1 L_2} \tag{7}$$

The steady-state solution of (7) is evidently $i_1(\infty) = V/R_1$; the transient solution will be determined by the roots s_1 and s_2 of

$$s^2 + \frac{R_1 L_1 + R_2 L_1 + R_1 L_2}{L_1 L_2} s + \frac{R_1 R_2}{L_1 L_2} = 0$$

together with the initial conditions

$$i_1(0^+) = 0 \qquad \left.\frac{di_1}{dt}\right|_{0^+} = \frac{V}{L_1}$$

(both i_1 and i_2 must be continuous at $t = 0$). Once the expression for i_1 is known, that for i_2 follows from (4).

There will be a damping factor that insures the transient will ultimately die out. Also, depending on the values of the four circuit constants, the transient can be overdamped or underdamped, which is oscillatory. In general, the current expression will be

$$i_1 = (\text{transient}) + \frac{V}{R_1}$$

The transient part will have a value of $-V/R_1$ at $t = 0$ and a value of zero as $t \to \infty$.

8.5 COMPLEX FREQUENCY

We have examined circuits where the driving function was a constant (e.g., $V = 50.0$ V), a sinusoidal function (e.g., $v = 100.0 \sin(500t + 30^\circ)$ (V), or an exponential function, e.g., $v = 10e^{-5t}$ (V). In this section, we introduce a *complex frequency*, **s**, which unifies the three functions and will simplify the analysis, whether the transient or steady-state response is required.

We begin by expressing the exponential function in the equivalent cosine and sine form:

$$e^{j(\omega t + \phi)} = \cos(\omega t + \phi) + j \sin(\omega t + \phi)$$

We will focus exclusively on the cosine term $\cos(\omega t + \phi) = \text{Re}\, e^{j(\omega t + \phi)}$ and for convenience drop the prefix Re. Introducing a constant A and the factor $e^{\sigma t}$,

$$Ae^{\sigma t} e^{j(\omega t + \phi)} \Rightarrow Ae^{\sigma t} \cos(\omega t + \phi) \qquad Ae^{j\phi} e^{(\sigma + j\omega)t} = Ae^{j\phi} e^{\mathbf{s}t} \qquad \text{where } \mathbf{s} = \sigma + j\omega$$

The complex frequency $\mathbf{s} = \sigma + j\omega$ has units s^{-1}, and ω, as we know, has units rad/s. Consequently, the units on σ must also be s^{-1}. This is the neper frequency with units Np/s. If both σ and ω are

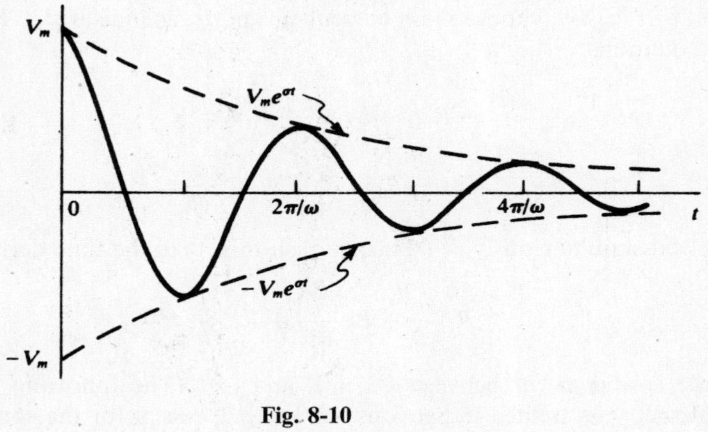

Fig. 8-10

nonzero, the function is a damped cosine. Only negative values of σ are considered. If σ and ω are zero, the result is a constant. And finally, with $\omega = 0$ and σ nonzero, the result is an exponential decay function. In Table 8-1, several functions are given with corresponding values of **s** for the expression $A e^{\mathbf{s}t}$.

Table 8-1

$f(t)$	$\mathbf{s}$	A
$10e^{-5t}$	$-5 + j0$	10
$5\cos(500t + 30°)$	$0 + j500$	5
$2e^{-3t}\cos(100t - 45°)$	$-3 + j100$	2
100.0	$0 + j0$	100.0

When Fig. 8-10 is examined for various values of **s**, the three cases are evident. If $\sigma = 0$, there is no damping and the result is a cosine function with maximum values of $\pm V_m$ (not shown). If $\omega = 0$, the function is an exponential decay with an initial value V_m. And finally, with both ω and σ nonzero, the damped cosine is the result.

8.6 GENERALIZED IMPEDANCE (R, L, C) IN s-DOMAIN

A driving voltage of the form $v = V_m e^{\mathbf{s}t}$ applied to a passive network will result in branch currents and voltages across the elements, each having the same time dependence $e^{\mathbf{s}t}$; e.g., $I_a e^{j\psi} e^{\mathbf{s}t}$, and $V_b e^{j\phi} e^{\mathbf{s}t}$. Consequently, only the magnitudes of currents and voltages and the phase angles need be determined (this will also be the case in sinusoidal circuit analysis in Chapter 9). We are thus led to consider the network in the s-domain (see Fig. 8-11).

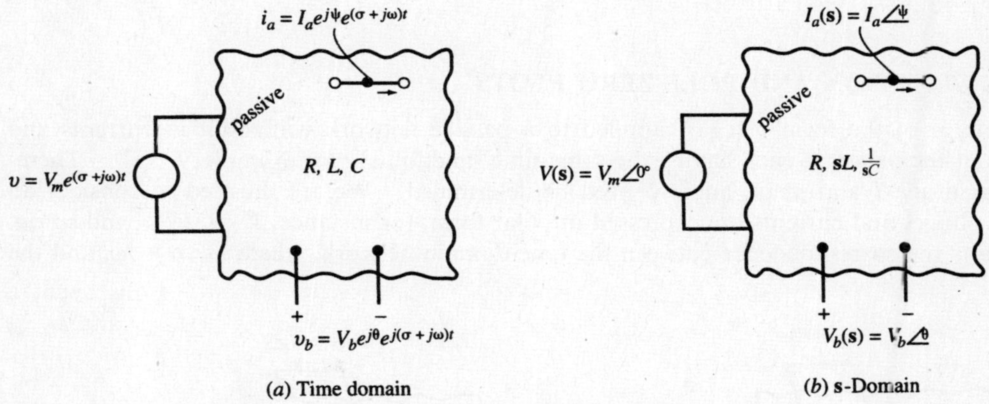

Fig. 8-11

A series RL circuit with an applied voltage $v = V_m e^{j\phi} e^{\mathbf{s}t}$ will result in a current $i = I_m e^{j\psi} e^{\mathbf{s}t} = \mathbf{I}_m e^{\mathbf{s}t}$, which, substituted in the nodal equation

$$Ri + L\frac{di}{dt} = V_m e^{j\phi} e^{\mathbf{s}t}$$

will result in

$$R\mathbf{I}_m e^{\mathbf{s}t} = \mathbf{s}L\mathbf{I}_m e^{\mathbf{s}t} = V_m e^{j\phi} e^{\mathbf{s}t} \qquad \text{from which} \qquad \mathbf{I}_m = \frac{V_m e^{j\phi}}{R + \mathbf{s}L}$$

Note that in the s-domain the *impedance* of the series RL circuit is $R + \mathbf{s}L$. The inductance therefore has an s-domain impedance $\mathbf{s}L$.

EXAMPLE 8.6 A series RL circuit, with $R = 10\,\Omega$ and $L = 2\,\text{H}$, has an applied voltage $v = 10\,e^{-2t}\cos(10t + 30°)$. Obtain the current i by an $\mathbf{s}$-domain analysis.

$$v = 10\,\underline{/30°}\,e^{\mathbf{s}t} = Ri + L\frac{di}{dt} = 10i + 2\frac{di}{dt}$$

Since $i = Ie^{\mathbf{s}t}$,

$$10\,\underline{/30°}\,e^{\mathbf{s}t} = 10Ie^{\mathbf{s}t} + 2\mathbf{s}Ie^{\mathbf{s}t} \qquad \text{or} \qquad I = \frac{10\,\underline{/30°}}{10 + 2\mathbf{s}}$$

Substituting $\mathbf{s} = -2 + j10$,

$$I = \frac{10\,\underline{/30°}}{10 + 2(-2 + j10)} = \frac{10\,\underline{/30°}}{6 + j20} = 0.48\,\underline{/-43.3°}$$

Then, $i = Ie^{\mathbf{s}t} = 0.48e^{-2t}\cos(10t - 43.3°)$ (A).

EXAMPLE 8.7 A series RC circuit, with $R = 10\,\Omega$ and $C = 0.2\,\text{F}$, has the same applied voltage as in Example 8.6. Obtain the current by an $\mathbf{s}$-domain analysis.

As in Example 8.6,

$$v = 10\,\underline{/30°}\,e^{\mathbf{s}t} = Ri + \frac{1}{C}\int i\,dt = 10i + 5\int i\,dt$$

Since $i = Ie^{\mathbf{s}t}$,

$$10\,\underline{/30°}\,e^{\mathbf{s}t} = 10Ie^{\mathbf{s}t} + \frac{5}{\mathbf{s}}Ie^{\mathbf{s}t} \qquad \text{from which} \qquad I = \frac{10\,\underline{/30°}}{10 + 5/\mathbf{s}} = 1.01\,\underline{/32.8°}$$

Then, $i = 1.01e^{-2t}\cos(10t + 32.8°)$ (A).

Note that the $\mathbf{s}$-domain impedance for the capacitance is $1/(\mathbf{s}C)$. Thus the $\mathbf{s}$-domain impedance of a series RLC circuit will be $\mathbf{Z}(\mathbf{s}) = R + \mathbf{s}L + 1/(\mathbf{s}C)$

8.7 NETWORK FUNCTION AND POLE-ZERO PLOTS

A driving voltage of the form $v = Ve^{\mathbf{s}t}$ applied to a passive network will result in currents and voltages throughout the network, each having the same time function $e^{\mathbf{s}t}$; for example, $Ie^{j\psi}e^{\mathbf{s}t}$. Therefore, only the magnitude I and phase angle ψ need be determined. We are thus led to consider an $\mathbf{s}$-domain where voltages and currents are expressed in polar form, for instance, $V\,\underline{/\theta}$, $I\,\underline{/\psi}$, and so on. Figure 8-12 suggests the correspondence between the time-domain network, where $\mathbf{s} = \sigma + j\omega$, and the

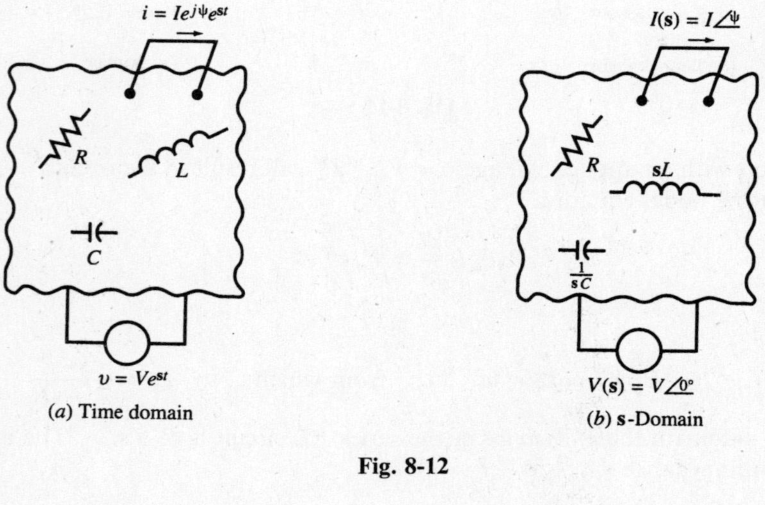

Fig. 8-12

s-domain where only magnitudes and phase angles are shown. In the **s**-domain, inductances are expressed by **s**L and capacitances by $1/(\mathbf{s}C)$. The impedance in the **s**-domain is $\mathbf{Z(s)} = \mathbf{V(s)/I(s)}$.

A *network function* **H(s)** is defined as the ratio of the complex amplitude of an exponential output **Y(s)** to the complex amplitude of an exponential input **X(s)** If, for example, **X(s)** is a driving voltage and **Y(s)** is the output voltage across a pair of terminals, then the ratio **Y(s)/X(s)** is nondimensional.

The network function **H(s)** can be derived from the input-output differential equation

$$a_n \frac{d^n y}{dt^n} + a_{n-1} \frac{d^{n-1} y}{dt^{n-1}} + \cdots + a_1 \frac{dy}{dt} + a_0 y = b_m \frac{d^m x}{dt^m} + b_{m-1} \frac{d^{m-1} x}{dt^{m-1}} + \cdots + b_1 \frac{dx}{dt} + b_0 x$$

When $x(t) = Xe^{\mathbf{s}t}$ and $y(t) = Ye^{\mathbf{s}t}$,

$$(a_n \mathbf{s}^n + a_{n-1}\mathbf{s}^{n-1} + \cdots + a_1 \mathbf{s} + a_0)e^{\mathbf{s}t} = (b_m \mathbf{s}^m + b_{m-1}\mathbf{s}^{m-1} + \cdots + b_1 \mathbf{s} + b_0)e^{\mathbf{s}t}$$

Then,

$$\mathbf{H(s)} = \frac{\mathbf{Y(s)}}{\mathbf{X(s)}} = \frac{a_n \mathbf{s}^n + a_{n-1}\mathbf{s}^{n-1} + \cdots + a_1 \mathbf{s} + a_0}{b_m \mathbf{s}^m + b_{m-1}\mathbf{s}^{m-1} + \cdots + b_1 \mathbf{s} + b_0}$$

In linear circuits made up of lumped elements, the network function **H(s)** is a *rational function* of **s** and can be written in the following general form

$$\mathbf{H(s)} = k \frac{(\mathbf{s} - \mathbf{z}_1)(\mathbf{s} - \mathbf{z}_2) \cdots (\mathbf{s} - \mathbf{z}_\mu)}{(\mathbf{s} - \mathbf{p}_1)(\mathbf{s} - \mathbf{p}_2) \cdots (\mathbf{s} - \mathbf{p}_\nu)}$$

where k is some real number. The complex constants $\mathbf{z}_m$ $(m = 1, 2, \ldots, \mu)$, the *zeros* of **H(s)**, and the $\mathbf{p}_n$ $(n = 1, 2, \ldots, \nu)$ the *poles* of **H(s)**, assume particular importance when **H(s)** is interpreted as the ratio of the *response* (in one part of the **s**-domain network) to the *excitation* (in another part of the network). Thus, when $\mathbf{s} = \mathbf{z}_m$, the response will be zero, no matter how great the excitation; whereas, when $\mathbf{s} = \mathbf{p}_n$, the response will be infinite, no matter how small the excitation.

EXAMPLE 8.8 A passive network in the **s**-domain is shown in Fig. 8-13. Obtain the network function for the current **I(s)** due to an input voltage **V(s)**.

$$\mathbf{H(s)} = \frac{\mathbf{I(s)}}{\mathbf{V(s)}} = \frac{1}{\mathbf{Z(s)}}$$

Since

$$\mathbf{Z(s)} = 2.5 + \frac{\left(\frac{5\mathbf{s}}{3}\right)\left(\frac{20}{\mathbf{s}}\right)}{\frac{5\mathbf{s}}{3} + \frac{20}{\mathbf{s}}} = (2.5)\frac{\mathbf{s}^2 + 8\mathbf{s} + 12}{\mathbf{s}^2 + 12}$$

we have

$$\mathbf{H(s)} = (0.4)\frac{\mathbf{s}^2 + 12}{(\mathbf{s}+2)(\mathbf{s}+6)}$$

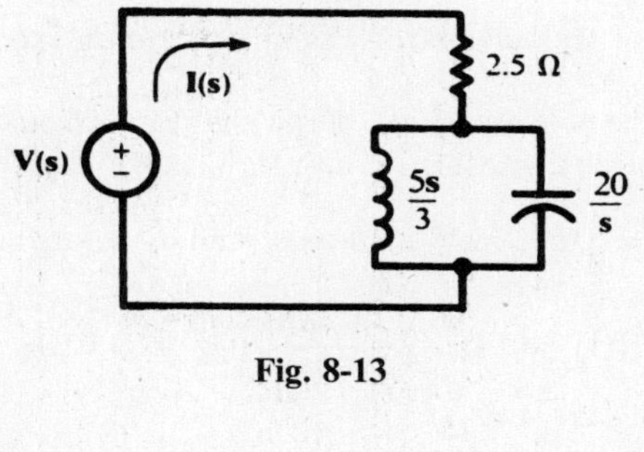

Fig. 8-13

The numerator of $\mathbf{H(s)}$ in Example 8.8 is zero when $\mathbf{s} = \pm j\sqrt{12}$. Consequently, a voltage function at this frequency results in a current of zero. In Chapter 12 where series and parallel resonance are discussed, it will be found that the parallel LC circuit is resonant at $\omega = 1/\sqrt{LC}$. With $L = \frac{5}{3}$ H and $C = \frac{1}{20}$ F, $\omega = \sqrt{12}$ rad/s.

The zeros and poles of a network function $\mathbf{H(s)}$ can be plotted in a complex $\mathbf{s}$-plane. Figure 8-14 shows the poles and zeros of Example 8.8, with zeros marked ⊙ and poles marked ×. The zeros occur in complex conjugate pairs, $\mathbf{s} = \pm j\sqrt{12}$, and the poles are $\mathbf{s} = -2$ and $\mathbf{s} = -6$.

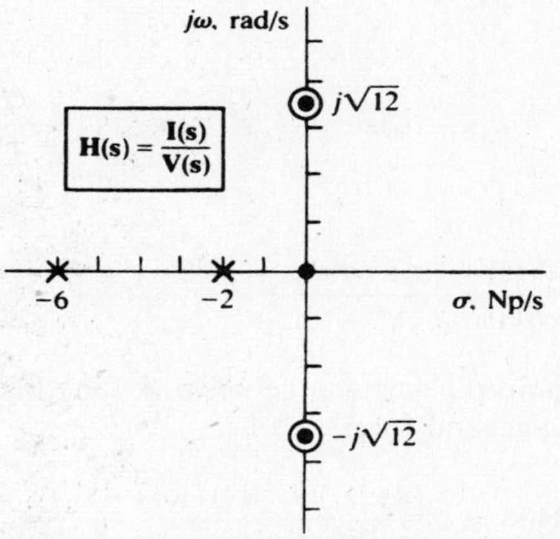

Fig. 8-14

8.8 THE FORCED RESPONSE

The network function can be expressed in polar form and the response obtained graphically. Before starting the development, it is helpful to recall that $\mathbf{H(s)}$ is merely a ratio such as $\mathbf{V}_0(\mathbf{s})/\mathbf{V}_i(\mathbf{s})$, $\mathbf{I}_2(\mathbf{s})/\mathbf{V}_1(\mathbf{s})$, or $\mathbf{I}_2(\mathbf{s})/\mathbf{I}_1(\mathbf{s})$. With the polynomials factored,

$$\mathbf{H(s)} = k\,\frac{(\mathbf{s}-\mathbf{z}_1)(\mathbf{s}-\mathbf{z}_2)\cdots(\mathbf{s}-\mathbf{z}_\mu)}{(\mathbf{s}-\mathbf{p}_1)(\mathbf{s}-\mathbf{z}_2)\cdots(\mathbf{s}-\mathbf{p}_\nu)}$$

Now setting $(\mathbf{s}-\mathbf{z}_m) = N_m\,\underline{/\alpha_m}(m = 1, 2, \ldots, \mu)$ and $(\mathbf{s}-\mathbf{p}_n) = D_n\,\underline{/\beta_n}(n = 1, 2, \ldots, \nu)$, we have

$$\mathbf{H(s)} = k\,\frac{(N_1\,\underline{/\alpha_1})(N_2\,\underline{/\alpha_2})\cdots(N\,\underline{/\alpha_\mu})}{(D_1\,\underline{/\beta_1})(D_2\,\underline{/\beta_2})\cdots(D\,\underline{/\beta_\nu})} = k\,\frac{N_1N_2\cdots N_\mu}{D_1D_2\cdots D_\nu}\,\underline{/(\alpha_1+\cdots+\alpha_\mu)-(\beta_1+\cdots+\beta_\nu)}$$

It follows that the response of the network to an excitation for which $\mathbf{s} = \sigma + j\omega$ is determined by measuring the lengths of the vectors from the zeros and poles to $\mathbf{s}$ as well as the angles these vectors make with the positive σ axis in the pole-zero plot.

EXAMPLE 8.9 Test the response of the network of Example 8.8 to an exponential voltage excitation $v = 1e^{st}$, where $\mathbf{s} = 1$ Np/s.

Locate the test point $1 + j0$ on the pole-zero plot. Draw the vectors from the poles and zeros to the test point and compute the lengths and angles (see Fig. 8-15). Thus,

$$N_1 = N_2 = \sqrt{13},\ D_1 = 3,\ D_2 = 7,\ \beta_1 = \beta_2 = 0,\ \text{and}\ \alpha_1 = -\alpha_2 = \tan^{-1}\sqrt{12} = 73.9^\circ$$

Hence,

$$\mathbf{H}(1) = (0.4)\,\frac{(\sqrt{13})(\sqrt{13})}{(3)(7)}\,\underline{/0^\circ - 0^\circ} = 0.248$$

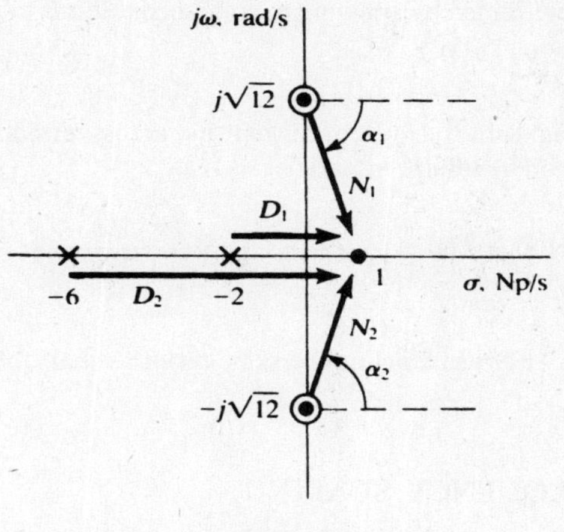

Fig. 8-15

The result implies that, in the time domain, $i(t) = 0.248v(t)$, so that both voltage and current become infinite according to the function e^{1t}. For most practical cases, σ must be either negative or zero.

The above geometrical method does not seem to require knowledge of the analytic expression for **H(s)** as a rational function. It is clear, however, that the expression can be written, to within the constant factor k, from the known poles and zeros of **H(s)** in the pole-zero plot. See Problem 8.37.

8.9 THE NATURAL RESPONSE

This chapter has focused on the forced or steady-state response, and it is in obtaining that response that the complex-frequency method is most helpful. However, the natural frequencies, which characterize the transient response, are easily obtained. They are the poles of the network function.

EXAMPLE 8.10 The same network as in Example 8.8 is shown in Fig. 8-16. Obtain the natural response when a source **V(s)** is inserted at xx'.

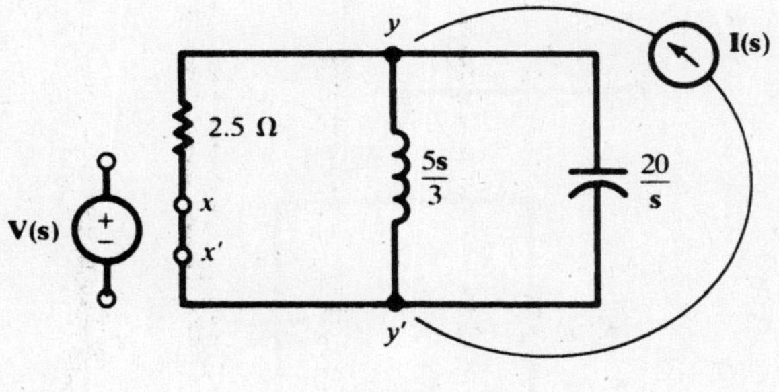

Fig. 8-16

The network function is the same as in Example 8.8:

$$\mathbf{H(s)} = (0.4)\,\frac{\mathbf{s}^2 + 12}{(\mathbf{s}+2)(\mathbf{s}+6)}$$

The natural frequencies are then -2 Np/s and -6 Np/s. Hence, in the time domain, the natural or transient current is of the form

$$i_n = A_1 e^{-2t} + A_2 e^{-6t}$$

where the constants A_1 and A_2 are determined by applying the initial conditions to the complete response, $i = i_n + i_f$, where i_f indicates the forced response.

EXAMPLE 8.11 The network of Fig. 8-16 is driven by current **I**(s) across terminals yy'. The network function is $\mathbf{H}(\mathbf{s}) = \mathbf{V}(\mathbf{s})/\mathbf{I}(\mathbf{s}) = \mathbf{Z}(\mathbf{s})$. The three branches are in parallel so that

$$\mathbf{H}(\mathbf{s}) = \mathbf{Z}(\mathbf{s}) = \frac{1}{\dfrac{1}{2.5} + \dfrac{3}{5\mathbf{s}} + \dfrac{\mathbf{s}}{20}} = \frac{20\mathbf{s}}{(\mathbf{s}+2)(\mathbf{s}+6)}$$

Again the poles are at -2 Np/s and -6 Np/s, which is the same result as that obtained in Example 8.10.

8.10 MAGNITUDE AND FREQUENCY SCALING

Magnitude Scaling

Let a network have input impedance function $\mathbf{Z}_{\text{in}}(\mathbf{s})$, and let K_m be a positive real number. Then, if each resistance R in the network is replaced by $K_m R$, each inductance L by $K_m L$, and each capacitance C by C/K_m, the new input impedance function will be $K_m \mathbf{Z}_{\text{in}}(\mathbf{s})$. We say that the network has been *magnitude-scaled by a factor* K_m.

Frequency Scaling

If, instead of the above changes, we preserve each resistance R, replace each inductance L by L/K_f $(K_f > 0)$, and replace each capacitance C by C/K_f, then the new input impedance function will be $\mathbf{Z}_{\text{in}}(\mathbf{s}/K_f)$. That is, the new network has the same impedance at complex frequency $K_f\mathbf{s}$ as the old had at **s**. We say that the network has been *frequency-scaled by a factor* K_f.

EXAMPLE 8.12 Express **Z**(s) for the circuit shown in Fig. 8-17 and observe the resulting magnitude scaling.

$$\mathbf{Z}(\mathbf{s}) = K_m L\mathbf{s} + \frac{(K_m R)\dfrac{K_m}{C\mathbf{s}}}{K_m R + \dfrac{K_m}{C\mathbf{s}}} = K_m\left[L\mathbf{s} + \frac{R(1/C\mathbf{s})}{R + (1/C\mathbf{s})}\right]$$

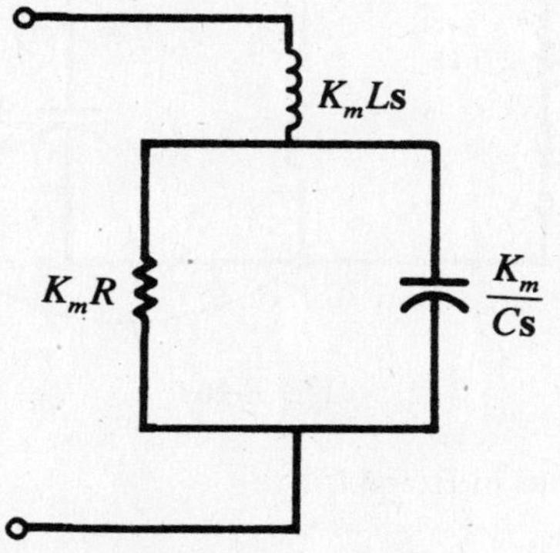

Fig. 8-17

There are practical applications suggested by this brief exposure to magnitude scaling. For example, if the input current to a network were greater than it should be, a factor $K_m = 10$ would reduce the current to 1/10 of the former value.

8.11 HIGHER-ORDER ACTIVE CIRCUITS

Application of circuit laws to circuits which contain op amps and several storage elements produces, in general, several first-order differential equations which may be solved simultaneously or be reduced to a higher-order input-output equation. A convenient tool for developing the equations is the complex frequency s (and generalized impedance in the s-domain) as used throughout Sections 8.5 to 8.10. Again, we assume ideal op amps (see Section 7.16). The method is illustrated in the following examples.

EXAMPLE 8.13 Find $H(s) = V_2/V_1$ in the circuit of Fig. 8-41 and show that the circuit becomes a noninverting integrator if and only if $R_1C_1 = R_2C_2$.

Apply voltage division, in the phasor domain, to the input and feedback paths to find the voltages at the terminals of the op amp.

$$\text{At terminal A:} \qquad V_A = \frac{1}{1 + R_1C_1s} V_1$$

$$\text{At terminal B:} \qquad V_B = \frac{R_2C_2s}{1 + R_2C_2s} V_2$$

But $V_A = V_B$. Therefore,

$$\frac{V_2}{V_1} = \frac{1 + R_2C_2s}{(1 + R_1C_1s)R_2C_2s}$$

Only if $R_1C_1 = R_2C_2 = RC$ do we get an integrator with a gain of $1/RC$

$$\frac{V_2}{V_1} = \frac{1}{RCs}, \qquad v_2 = \frac{1}{RC}\int_{-\infty}^{t} v_1\, dt$$

EXAMPLE 8.14 The circuit of Fig. 8-42 is called an equal-component Sallen-Key circuit. Find $H(s) = V_2/V_1$ and convert it to a differential equation.

Write KCL at nodes A and B.

$$\text{At node A:} \qquad \frac{V_A - V_1}{R} + \frac{V_A - V_B}{R} + (V_A - V_2)Cs = 0$$

$$\text{At node B:} \qquad \frac{V_B - V_A}{R} + V_BCs = 0$$

Let $1 + R_2/R_1 = k$, then $V_2 = kV_B$. Eliminating V_A and V_B between the above equations we get

$$\frac{V_2}{V_1} = \frac{k}{R^2C^2s^2 + (3 - k)RCs + 1}$$

$$R^2C^2\frac{d^2v_2}{dt^2} + (3 - k)RC\frac{dv_2}{dt} + v_2 = kv_1$$

EXAMPLE 8.15 In the circuit of Fig. 8-42 assume $R = 2\,k\Omega$, $C = 10\,nF$, and $R_2 = R_1$. Find v_2 if $v_1 = u(t)$.

By substituting the element values in $H(s)$ found in Example 8.14 we obtain

$$\frac{V_2}{V_1} = \frac{2}{4 \times 10^{-10}s^2 + 2 \times 10^{-5}s + 1}$$

$$\frac{d^2v_2}{dt^2} + 5 \times 10^4\frac{dv_2}{dt} + 25 \times 10^8 v_2 = 5 \times 10^9 v_1$$

The response of the preceding equation for $t > 0$ to $v_1 = u(t)$ is

$$v_2 = 2 + e^{-\alpha t}(2\cos\omega t - 2.31\sin\omega t) = 2 + 3.055e^{-\alpha t}\cos(\omega t + 130.9^\circ)$$

where $\alpha = 25\,000$ and $\omega = 21\,651$ rad/s.

EXAMPLE 8.16 Find conditions in the circuit of Fig. 8-42 for sustained oscillations in $v_2(t)$ (with zero input) and find the frequency of oscillations.

In Example 8.14 we obtained

$$\frac{V_2}{V_1} = \frac{k}{R^2C^2s^2 + (3 - k)RCs + 1}$$

For sustained oscillations the roots of the characteristic equation in Example 8.14 should be imaginary numbers. This happens when $k = 3$ or $R_2 = 2R_1$, in which case $\omega = 1/RC$.

Solved Problems

8.1 A series RLC circuit, with $R = 3\,\mathrm{k}\Omega$, $L = 10\,\mathrm{H}$, and $C = 200\,\mu\mathrm{F}$, has a constant-voltage source, $V = 50\,\mathrm{V}$, applied at $t = 0$. (*a*) Obtain the current transient, if the capacitor has no initial charge. (*b*) Sketch the current and find the time at which it is a maximum.

(*a*)
$$\alpha = \frac{R}{2L} = 150\,\mathrm{s}^{-1} \qquad \omega_0^2 = \frac{1}{LC} = 500\,\mathrm{s}^{-2} \qquad \beta = \sqrt{\alpha^2 - \omega_0^2} = 148.3\,\mathrm{s}^{-1}$$

The circuit is overdamped ($\alpha > \omega_0$).

$$s_1 = -\alpha + \beta = -1.70\,\mathrm{s}^{-1} \qquad s_2 = -\alpha - \beta = -298.3\,\mathrm{s}^{-1}$$

and
$$i = A_1 e^{-1.70t} + A_2 e^{-298.3t}$$

Since the circuit contains an inductance, $i(0^+) = i(0^-) = 0$; also, $Q(0^+) = Q(0^-) = 0$. Thus, at $t = 0^+$, KVL gives

$$0 + 0 + L\left.\frac{di}{dt}\right|_{0^+} = V \qquad \text{or} \qquad \left.\frac{di}{dt}\right|_{0^+} = \frac{V}{L} = 5\,\mathrm{A/s}$$

Applying these initial conditions to the expression for i,

$$0 = A_1(1) + A_2(1)$$
$$5 = -1.70A_1(1) - 298.3A_2(1)$$

from which $A_1 = -A_2 = 16.9\,\mathrm{mA}$.

$$i = 16.9(e^{-1.70t} - e^{-298.3t}) \quad (\mathrm{mA})$$

(*b*) For the time of maximum current,

$$\frac{di}{dt} = 0 = -28.73e^{-1.70t} + 5041.3e^{-298.3t}$$

Solving by logarithms, $t = 17.4\,\mathrm{ms}$. See Fig. 8-18.

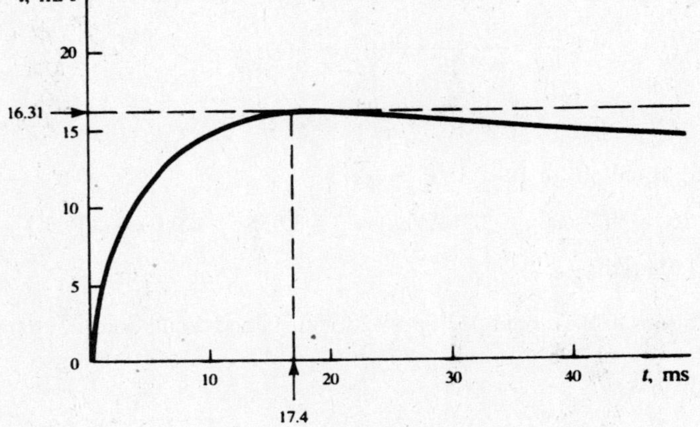

Fig. 8-18

8.2 A series *RLC* circuit, with $R = 50\,\Omega$, $L = 0.1\,\text{H}$, and $C = 50\,\mu\text{F}$, has a constant voltage $V = 100\,\text{V}$ applied at $t = 0$. Obtain the current transient, assuming zero initial charge on the capacitor.

$$\alpha = \frac{R}{2L} = 250\,\text{s}^{-1} \qquad \omega_0^2 = \frac{1}{LC} = 2.0 \times 10^5\,\text{s}^{-2} \qquad \beta = \sqrt{\alpha^2 - \omega_0^2} = j370.8\,\text{rad/s}$$

This is an oscillatory case ($\alpha < \omega_0$), and the general current expression is

$$i = e^{-250t}(A_1 \cos 370.8\,t + A_2 \sin 370.8t)$$

The initial conditions, obtained as in Problem 8.1, are

$$i(0^+) = 0 \qquad \left.\frac{di}{dt}\right|_{0^+} = 1000\,\text{A/s}$$

and these determine the values: $A_1 = 0$, $A_2 = 2.70\,\text{A}$. Then

$$i = e^{-250t}(2.70 \sin 370.8t) \quad (\text{A})$$

8.3 Rework Problem 8.2, if the capacitor has an initial charge $Q_0 = 2500\,\mu\text{C}$.

Everything remains the same as in Problem 8.2 except the second initial condition, which is now

$$0 + L\left.\frac{di}{dt}\right|_{0^+} + \frac{Q_0}{C} = V \qquad \text{or} \qquad \left.\frac{di}{dt}\right|_{0^+} = \frac{100 - (2500/50)}{0.1} = 500\,\text{A/s}$$

The initial values are half those in Problem 8.2, and so, by linearity,

$$i = e^{-250t}(1.35 \sin 370.8t) \quad (\text{A})$$

8.4 A parallel *RLC* network, with $R = 50.0\,\Omega$, $C = 200\,\mu\text{F}$, and $L = 55.6\,\text{mH}$, has an initial charge $Q_0 = 5.0\,\text{mC}$ on the capacitor. Obtain the expression for the voltage across the network.

$$\alpha = \frac{1}{2RC} = 50\,\text{s}^{-1} \qquad \omega_0^2 = \frac{1}{LC} = 8.99 \times 10^4\,\text{s}^{-2}$$

Since $\omega_0^2 > \alpha^2$, the voltage function is oscillatory and so $\omega_d = \sqrt{\omega_0^2 - \alpha^2} = 296\,\text{rad/s}$. The general voltage expression is

$$v = e^{-50t}(A_1 \cos 296t + A_2 \sin 296t)$$

With $Q_0 = 5.0 \times 10^{-3}\,\text{C}$, $V_0 = 25.0\,\text{V}$. At $t = 0$, $v = 25.0\,\text{V}$. Then, $A_1 = 25.0$.

$$\frac{dv}{dt} = -50e^{-50t}(A_1 \cos 296t + A_2 \sin 296t) + 296e^{-50t}(-A_1 \sin 296t + A_2 \cos 296t)$$

At $t = 0$, $dv/dt = -V_0/RC = \omega_d A_2 - \alpha A_1$, from which $A_2 = -4.22$. Thus,

$$v = e^{-50t}(25.0 \cos 296t - 4.22 \sin 296t) \quad (\text{V})$$

8.5 In Fig. 8-19, the switch is closed at $t = 0$. Obtain the current i and capacitor voltage v_C, for $t > 0$.

As far as the natural response of the circuit is concerned, the two resistors are in parallel; hence,

$$\tau = R_{\text{eq}}C = (5\,\Omega)(2\,\mu\text{F}) = 10\,\mu\text{s}$$

By continuity, $v_C(0^+) = v_C(0^-) = 0$. Furthermore, as $t \to \infty$, the capacitor becomes an open circuit, leaving $20\,\Omega$ in series with the 50 V. That is,

$$i(\infty) = \frac{50}{20} = 2.5\,\text{A} \qquad v_C(\infty) = (2.5\,\text{A})(10\,\Omega) = 25\,\text{V}$$

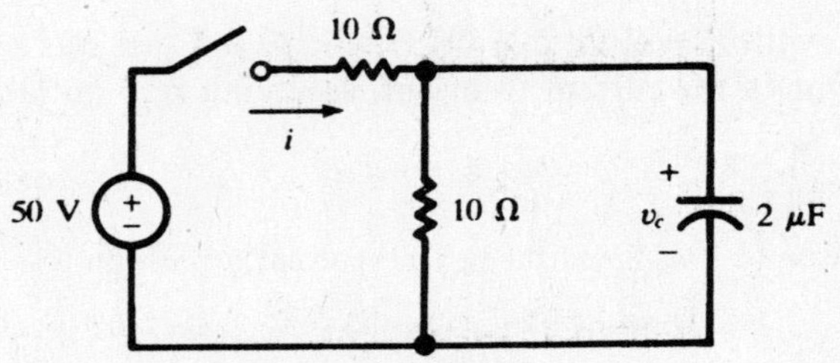

Fig. 8-19

Knowing the end conditions on v_C, we can write

$$v_C = [v_C(0^+) - v_C(\infty)]e^{-t/\tau} + v_C(\infty) = 25(1 - e^{-t/10}) \quad \text{(V)}$$

wherein t is measured in μs.

The current in the capacitor is given by

$$i_C = C\frac{dv_C}{dt} = 5e^{-t/10} \quad \text{(A)}$$

and the current in the parallel 10-Ω resistor is

$$i_{10\Omega} = \frac{v_C}{10\ \Omega} = 2.5(1 - e^{-t/10}) \quad \text{(A)}$$

Hence,
$$i = i_C + i_{10\Omega} = 2.5(1 + e^{-t/10}) \quad \text{(A)}$$

The problem might also have been solved by assigning mesh currents and solving simultaneous differential equations.

8.6 For the time functions listed in the first column of Table 8-2, write the corresponding amplitude and phase angle (cosine-based) and the complex frequency **s**.

See columns 2 and 3 of the table.

Table 8-2

Time Function	$A\angle\phi°$	**s**
$i(t) = 86.6$ A	$86.6\angle 0°$ A	0
$i(t) = 15.0e^{-2\times 10^3 t}$ (A)	$15.0\angle 0°$ A	-2×10^3 Np/s
$v(t) = 25.0\cos(250t - 45°)$ (V)	$25.0\angle -45°$ V	$\pm j250$ rad/s
$v(t) = 0.50\sin(250t + 30°)$ (V)	$0.50\angle -60°$ V	$\pm j250$ rad/s
$i(t) = 5.0e^{-100t}\sin(50t + 90°)$ (A)	$5.0\angle 0°$ A	$-100 \pm j50\ \text{s}^{-1}$
$i(t) = 3\cos 50t + 4\sin 50t$ (A)	$5\angle -53.13°$ A	$\pm j50$ rad/s

8.7 For each amplitude and phase angle in the first column and complex frequency **s** in the second column in Table 8-3, write the corresponding time function.

See column 3 of the table.

Table 8-3

$A\angle\phi°$	**s**	Time Function
$10\angle 0°$	$+j120\pi$	$10\cos 120\pi t$
$2\angle 45°$	$-j120\pi$	$2\cos(120\pi t + 45°)$
$5\angle -90°$	$-2 \pm j50$	$5e^{-2t}\cos(50t - 90°)$
$15\angle 0°$	$-5000 \pm j1000$	$15e^{-5000t}\cos 1000t$
$100\angle 30°$	0	86.6

8.8 An amplitude and phase angle of $10\sqrt{2}\,\underline{/45^\circ}$ V has an associated complex frequency $\mathbf{s} = -50 + j\,100\ \text{s}^{-1}$. Find the voltage at $t = 10$ ms.

$$v(t) = 10\sqrt{2}e^{-50t}\cos(100t + 45^\circ) \quad \text{(V)}$$

At $t = 10^{-2}$ s, $100t = 1$ rad $= 57.3^\circ$, and so

$$v = 10\sqrt{2}e^{-0.5}\cos 102.3^\circ = -1.83\ \text{V}$$

8.9 A passive network contains resistors, a 70-mH inductor, and a 25-μF capacitor. Obtain the respective s-domain impedances for a driving voltage (*a*) $v = 100\sin(300t + 45^\circ)$ (V), (*b*) $v = 100e^{-100t}\cos 300t$ (V).

(*a*) Resistance is independent of frequency. At $\mathbf{s} = j300$ rad/s, the impedance of the inductor is

$$\mathbf{s}L = (j300)(70 \times 10^{-3}) = j21$$

and that of the capacitor is

$$\frac{1}{\mathbf{s}C} = -j133.3$$

(*b*) At $\mathbf{s} = -100 + j300\ \text{s}^{-1}$,

$$\mathbf{s}L = (-100 + j300)(70 \times 10^{-3}) = -7 + j21$$

$$\frac{1}{\mathbf{s}C} = \frac{1}{(-100 + j300)(25 \times 10^{-6})} = -40 - j120$$

8.10 For the circuit shown in Fig. 8-20, obtain v at $t = 0.1$ s for source current (*a*) $i = 10\cos 2t$ (A), (*b*) $i = 10e^{-t}\cos 2t$ (A).

$$\mathbf{Z}_{\text{in}}(\mathbf{s}) = 2 + \frac{2(\mathbf{s}+2)}{\mathbf{s}+4} = (4)\,\frac{\mathbf{s}+3}{\mathbf{s}+4}$$

(*a*) At $\mathbf{s} = j2$ rad/s, $\mathbf{Z}_{\text{in}}(j2) = 3.22\,\underline{/7.13^\circ}\ \Omega$. Then,

$$\mathbf{V} = \mathbf{I}\mathbf{Z}_{\text{in}} = (10\,\underline{/0^\circ})(3.22\,\underline{/7.13^\circ}) = 32.2\,\underline{/7.13^\circ}\ \text{V} \qquad \text{or} \qquad v = 32.2\cos(2t + 7.13^\circ)\ \text{(V)}$$

and $v(0.1) = 32.2\cos(18.59^\circ) = 30.5$ V.

(*b*) At $\mathbf{s} = -1 + j2\ \text{s}^{-1}$, $\mathbf{Z}_{\text{in}}(-1 + j2) = 3.14\,\underline{/11.31^\circ}\ \Omega$. Then

$$\mathbf{V} = \mathbf{I}\mathbf{Z}_{\text{in}} = 31.4\,\underline{/11.31^\circ}\ \text{V} \qquad \text{or} \qquad v = 31.4e^{-}\cos(2t + 11.31^\circ)\ \text{(V)}$$

and $v(0.1) = 31.4e^{-0.1}\cos 22.77^\circ = 26.2$ V.

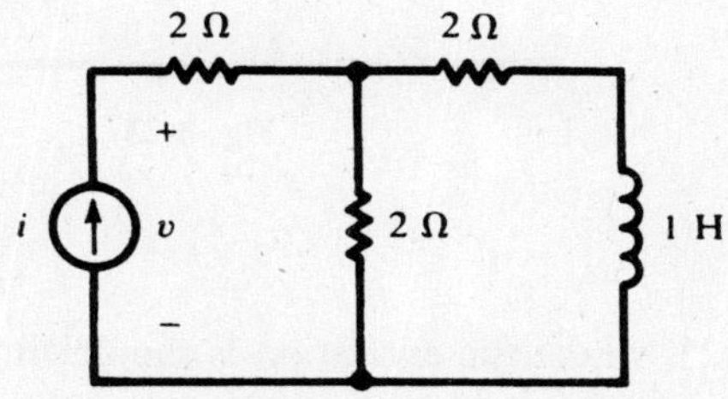

Fig. 8-20

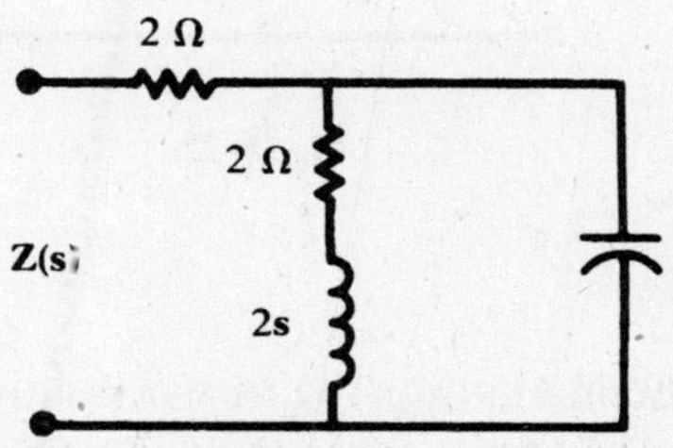

Fig. 8-21

8.11 Obtain the impedance $\mathbf{Z}_{\text{in}}(\mathbf{s})$ for the circuit shown in Fig. 8-21 at (*a*) $\mathbf{s} = 0$, (*b*) $\mathbf{s} = j4$ rad/s, (*c*) $|\mathbf{s}| = \infty$.

$$\mathbf{Z}_{\text{in}}(\mathbf{s}) = 2 + \frac{2(\mathbf{s}+1)\left(\dfrac{4}{\mathbf{s}}\right)}{2(\mathbf{s}+1) + \dfrac{4}{\mathbf{s}}} = (2)\,\frac{\mathbf{s}^2 + 3\mathbf{s} + 4}{\mathbf{s}^2 + \mathbf{s} + 2}$$

(*a*) $\mathbf{Z}_{\text{in}}(0) = 4\,\Omega$, the impedance offered to a constant (dc) source in the steady state.

(*b*)
$$\mathbf{Z}_{\text{in}}(j4) = 2\,\frac{(j4)^2 + 3(j4) + 4}{(j4)^2 + j4 + 2} = 2.33\,\underline{/-29.05^\circ}\;\Omega$$

This is the impedance offered to a source $\sin 4t$ or $\cos 4t$.

(*c*) $\mathbf{Z}_{\text{in}}(\infty) = 2\,\Omega$. At very high frequencies the capacitance acts like a short circuit across the *RL* branch.

8.12 Express the impedance $\mathbf{Z}(\mathbf{s})$ of the parallel combination of $L = 4\,\text{H}$ and $C = 1\,\text{F}$. At what frequencies $\mathbf{s}$ is this impedance zero or infinite?

$$\mathbf{Z}(\mathbf{s}) = \frac{(4\mathbf{s})(1/\mathbf{s})}{4\mathbf{s} + (1/\mathbf{s})} = \frac{\mathbf{s}}{\mathbf{s}^2 + 0.25}$$

By inspection, $\mathbf{Z}(0) = 0$ and $\mathbf{Z}(\infty) = 0$, which agrees with our earlier understanding of parallel *LC* circuits at frequencies of zero (dc) and infinity. For $|\mathbf{Z}(\mathbf{s})| = \infty$,

$$\mathbf{s}^2 + 0.25 = 0 \qquad \text{or} \qquad \mathbf{s} = \pm j0.5\,\text{rad/s}$$

A sinusoidal driving source, of frequency 0.5 rad/s, results in parallel resonance and an infinite impedance.

8.13 The circuit shown in Fig. 8-22 has a voltage source connected at terminals *ab*. The response to the excitation is the input current. Obtain the appropriate network function $\mathbf{H}(\mathbf{s})$.

$$\mathbf{H}(\mathbf{s}) = \frac{\text{response}}{\text{excitation}} = \frac{\mathbf{I}(\mathbf{s})}{\mathbf{V}(\mathbf{s})} \equiv \frac{1}{\mathbf{Z}(\mathbf{s})}$$

$$\mathbf{Z}(\mathbf{s}) = 2 + \frac{(2 + 1/\mathbf{s})(1)}{2 + 1/\mathbf{s} + 1} = \frac{8\mathbf{s} + 3}{3\mathbf{s} + 1} \qquad \text{from which} \qquad \mathbf{H}(\mathbf{s}) = \frac{1}{\mathbf{Z}(\mathbf{s})} = \frac{3\mathbf{s} + 1}{8\mathbf{s} + 3}$$

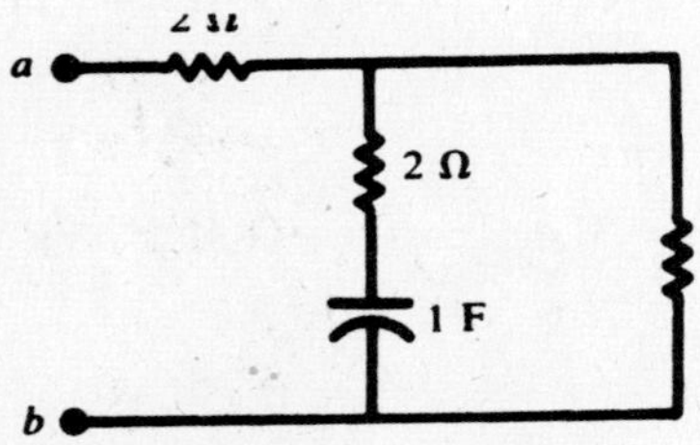

Fig. 8-22

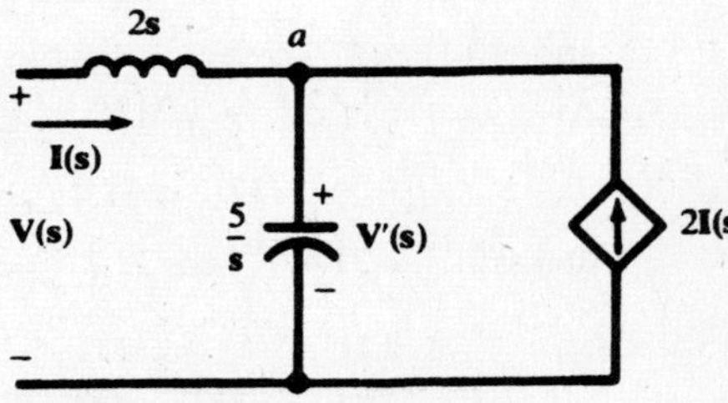

Fig. 8-23

8.14 Obtain $\mathbf{H}(\mathbf{s})$ for the network shown in Fig. 8-23, where the excitation is the driving current $\mathbf{I}(\mathbf{s})$ and the response is the voltage at the input terminals.

Applying KCL at junction *a*,

$$\mathbf{I}(\mathbf{s}) + 2\mathbf{I}(\mathbf{s}) = \frac{\mathbf{s}}{5}\,\mathbf{V}'(\mathbf{s}) \qquad \text{or} \qquad \mathbf{V}'(\mathbf{s}) = \frac{15}{\mathbf{s}}\,\mathbf{I}(\mathbf{s})$$

At the input terminals, KVL gives

$$\mathbf{V}(\mathbf{s}) = 2s\mathbf{I}(\mathbf{s}) + \mathbf{V}'(\mathbf{s}) = \left(2\mathbf{s} + \frac{15}{\mathbf{s}}\right)\mathbf{I}(\mathbf{s})$$

Then $$\mathbf{H}(\mathbf{s}) = \frac{\mathbf{V}(\mathbf{s})}{\mathbf{I}(\mathbf{s})} = \frac{2\mathbf{s}^2 + 15}{\mathbf{s}}$$

8.15 For the two-port network shown in Fig. 8-24 find the values of R_1, R_2, and C given that the voltage transfer function is

$$\mathbf{H}_v(\mathbf{s}) \equiv \frac{\mathbf{V}_o(\mathbf{s})}{\mathbf{V}_i(\mathbf{s})} = \frac{0.2}{\mathbf{s}^2 + 3\mathbf{s} + 2}$$

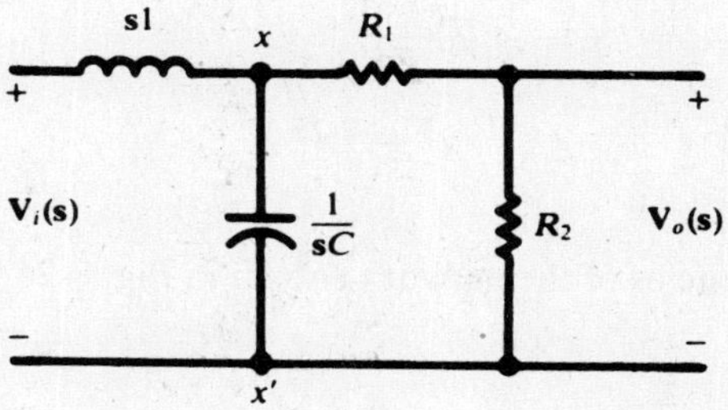

Fig. 8-24

The impedance looking into xx' is

$$\mathbf{Z}' = \frac{(1/\mathbf{s}C)(R_1 + R_2)}{(1/\mathbf{s}C) + R_1 + R_2} = \frac{R_1 + R_2}{1 + (R_1 + R_2)C\mathbf{s}}$$

Then, by repeated voltage division,

$$\frac{\mathbf{V}_o}{\mathbf{V}_i} = \left(\frac{\mathbf{V}_o}{\mathbf{V}_{xx'}}\right)\left(\frac{\mathbf{V}_{xx'}}{\mathbf{V}_i}\right) = \left(\frac{R_2}{R_1 + R_2}\right)\left(\frac{\mathbf{Z}'}{\mathbf{Z}' + \mathbf{s}1}\right) = \frac{R_2/(R_1 + R_2)C}{\mathbf{s}^2 + \dfrac{1}{(R_1 + R_2)C}\mathbf{s} + \dfrac{1}{C}}$$

Equating the coefficients in this expression to those in the given expression for $\mathbf{H}_v(\mathbf{s})$, we find:

$$C = \frac{1}{2}\ \mathrm{F} \qquad R_1 = \frac{3}{5}\ \Omega \qquad R_2 = \frac{1}{15}\ \Omega$$

8.16 Construct the pole-zero plot for the transfer admittance function

$$\mathbf{H}(\mathbf{s}) = \frac{\mathbf{I}_o(\mathbf{s})}{\mathbf{V}_i(\mathbf{s})} = \frac{\mathbf{s}^2 + 2\mathbf{s} + 17}{\mathbf{s}^2 + 3\mathbf{s} + 2}$$

In factored form,

$$\mathbf{H}(\mathbf{s}) = \frac{(\mathbf{s} + 1 + j4)(\mathbf{s} + 1 - j4)}{(\mathbf{s} + 1)(\mathbf{s} + 2)}$$

Poles exist at -1 and -2; zeros at $-1 \pm j4$. See Fig. 8-25.

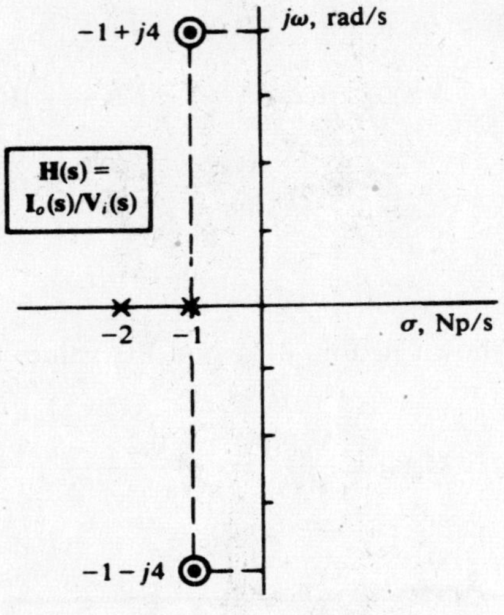

Fig. 8-25

8.17 Obtain the natural frequencies of the network shown in Fig. 8-26 by driving it with a conveniently located current source.

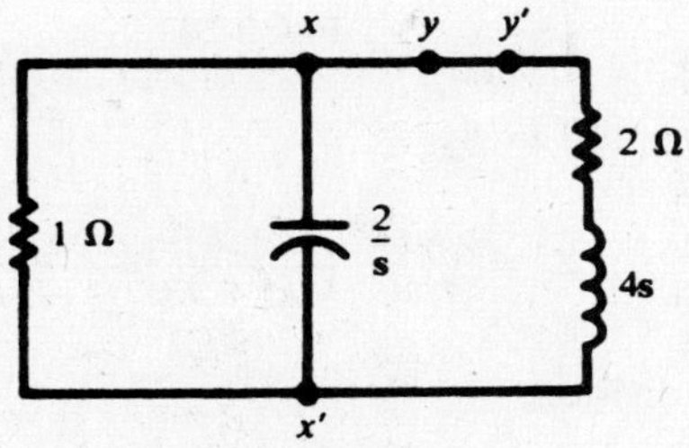

Fig. 8-26

The response to a current source connected at xx' is a voltage across these same terminals; hence the network function $\mathbf{H(s)} = \mathbf{V(s)}/\mathbf{I(s)} = \mathbf{Z(s)}$. Then,

$$\frac{1}{\mathbf{Z(s)}} = \frac{1}{1} + \frac{1}{2/\mathbf{s}} + \frac{1}{2+4\mathbf{s}} = \left(\frac{1}{2}\right)\frac{\mathbf{s}^2 + 2.5\mathbf{s} + 1.5}{\mathbf{s} + 0.5}$$

Thus,

$$\mathbf{Z(s)} = (2)\,\frac{\mathbf{s}+0.5}{\mathbf{s}^2 + 2.5\mathbf{s} + 1.5} = (2)\,\frac{\mathbf{s}+0.5}{(\mathbf{s}+1)(\mathbf{s}+1.5)}$$

The natural frequencies are the poles of the network function, $\mathbf{s} = -1.0$ Np/s = 2 and $\mathbf{s} = -1.5$ Np/s.

8.18 Repeat Problem 8.17, now driving the network with a conveniently located voltage source.

The conductor at yy' in Fig. 8-26 can be opened and a voltage source inserted. Then, $\mathbf{H}(s) = \mathbf{I(s)}/\mathbf{V(s)} = 1/\mathbf{Z(s)}$.

The impedance of the netework at terminals yy' is

$$\mathbf{Z(s)} = 2 + 4\mathbf{s} + \frac{1(2/\mathbf{s})}{1 + 2/\mathbf{s}} = (4)\frac{\mathbf{s}^2 + 2.5\mathbf{s} + 1.5}{\mathbf{s}+2}$$

Then,

$$\mathbf{H(s)} = \frac{1}{\mathbf{Z(s)}} = \left(\frac{1}{4}\right)\frac{\mathbf{s}+2}{\mathbf{s}^2 + 2.5\mathbf{s} + 1.5}$$

The denominator is the same as that in Problem 8.17, with the same roots and corresponding natural frequencies.

8.19 A 5000-rad/s sinusoidal source, $\mathbf{V} = 100\,\underline{/0^\circ}\,\text{V}$ in phasor form, is applied to the circuit of Fig. 8-27. Obtain the magnitude-scaling factor K_m and the element values which will limit the current to 89 mA (maximum value).

At $\omega = 5000$ rad/s,

$$\mathbf{Z}_{\text{in}} = j\omega L_1 + \frac{(j\omega L_2)\left(R + \dfrac{1}{j\omega C}\right)}{j\omega L_2 + R + \dfrac{1}{j\omega C}}$$

$$= j0.250 + \frac{(j0.500)(0.40 - j0.80)}{0.40 - j0.30} = 1.124\,\underline{/69.15^\circ}\,\Omega$$

For $|\mathbf{V}| = 100$ V, $|\mathbf{I}| = 100/1.124 = 89.0$ A. Thus, to limit the current to 89×10^{-3} A, the impedance must be increased by the factor $K_m = 10^3$.

The scaled element values are as follows: $R = 10^3(0.4\,\Omega) = 400\,\Omega$, $L_1 = 10^3(50\,\mu\text{H}) = 50$ mH, $L_2 = 10^3(100\,\mu\text{H}) = 100$ mH, and $C = (250\,\mu\text{F})/10^3 = 0.250\,\mu\text{F}$.

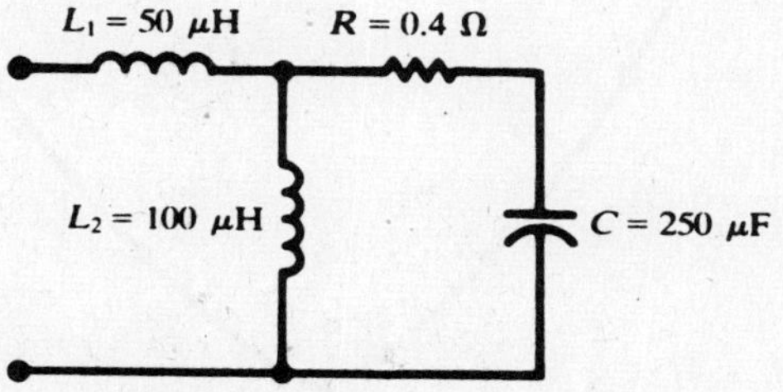

Fig. 8-27

8.20 Refer to Fig. 8-28. Obtain $\mathbf{H(s)} = \mathbf{V}_o/\mathbf{V}_i$ for $\mathbf{s} = j4 \times 10^6$ rad/s. Scale the network with $K_m = 10^{-3}$ and compare $\mathbf{H(s)}$ for the two networks.

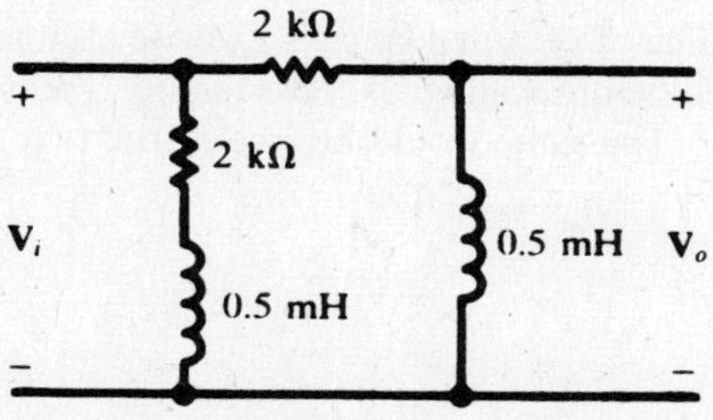

Fig. 8-28

At $\omega = 4 \times 10^6$ rad/s, $X_L = (4 \times 10^6)(0.5 \times 10^{-3}) = 2000\,\Omega$. Then,

$$\mathbf{H(s)} = \frac{\mathbf{V}_o}{\mathbf{V}_i} = \frac{j2000}{2000 + j2000} = \frac{1}{\sqrt{2}}\,\underline{/45^\circ}$$

After magnitude scaling, the inductive reactance is $10^{-3}(2000\,\Omega) = 2\,\Omega$ and the resistance is $10^{-3}(2\,\text{k}\Omega) = 2\,\Omega$. Thus

$$\mathbf{H(s)} = \frac{j2}{2 + j2} = \frac{1}{\sqrt{2}}\,\underline{/45^\circ}$$

The voltage transfer function remains unchanged by magnitude scaling. In general, any dimensionless transfer function is unaffected by magnitude scaling; a transfer function having units Ω is multiplied by K_m; and a function having units S is multiplied by $1/K_m$.

8.21 A three-element series circuit contains $R = 5\,\Omega$, $L = 4\,\text{H}$, and $C = 3.91\,\text{mF}$. Obtain the series resonant frequency, in rad/s, and then frequency-scale the circuit with $K_f = 1000$. Plot $|\mathbf{Z}(\omega)|$ for both circuits.

Before scaling,

$$\omega_0 = \frac{1}{\sqrt{LC}} = 8\,\text{rad/s} \qquad \text{and} \qquad \mathbf{Z}(\omega_0) = R = 5\,\Omega$$

After scaling,

$$R = 5\,\Omega \qquad L = \frac{4\,\text{H}}{1000} = 4\,\text{mH} \qquad C = \frac{3.91\,\text{mF}}{1000} = 3.91\,\mu\text{F}$$

$$\omega_0 = 1000(8\,\text{rad/s}) = 8000\,\text{rad/s} \qquad \mathbf{Z}(\omega_0) = R = 5\,\Omega$$

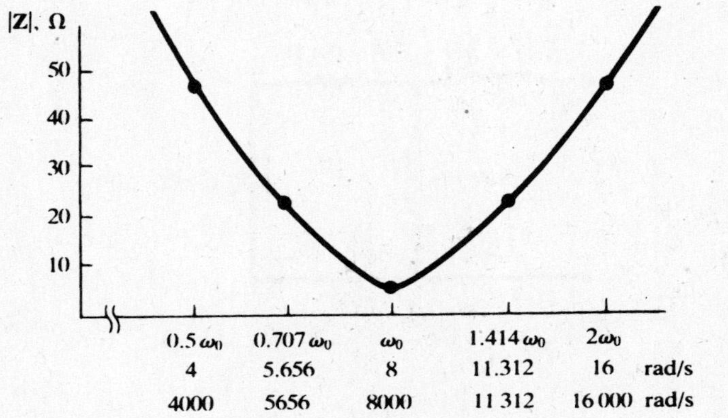

Fig. 8-29

Thus, frequency scaling by a factor of 1000 results in the 5-Ω impedance value being attained at 8000 rad/s instead of 8 rad/s. Any other value of the impedance is likewise attained, after scaling, at a frequency 1000 times that at which it was attained before scaling. Consequently, the two graphs of $|\mathbf{Z}(\omega)|$ differ only in the horizontal scale—see Fig. 8-29. (The same would be true of the two graphs of $\theta_{\mathbf{Z}(\omega)}$.)

Supplementary Problems

8.22 In the *RLC* circuit of Fig. 8-30, the capacitor is initially charged to $V_0 = 200\,\text{V}$. Find the current transient after the switch is closed at $t = 0$. *Ans.* $-2e^{-1000t}\sin 1000t$ (A)

8.23 A series *RLC* circuit, with $R = 200\,\Omega$, $L = 0.1\,\text{H}$, and $C = 100\,\mu\text{F}$, has a voltage source of 200 V applied at $t = 0$. Find the current transient, assuming zero initial charge on the capacitor.
Ans. $1.055(e^{-52t} - e^{-1948t})$ (A)

8.24 What value of capacitance, in place of the 100 μF in Problem 8.23, results in the critically damped case?
Ans. 10 μF

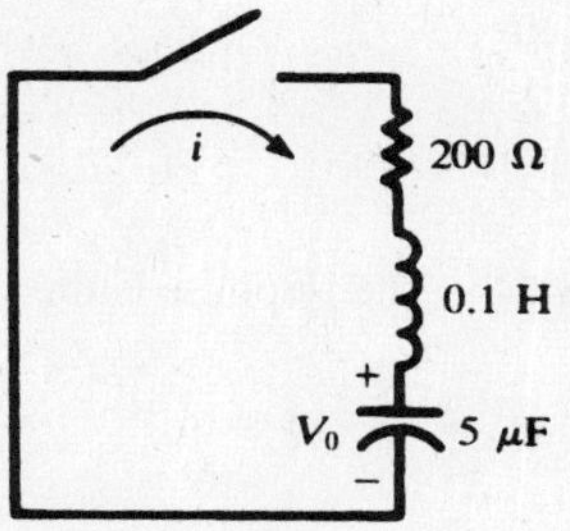

Fig. 8-30

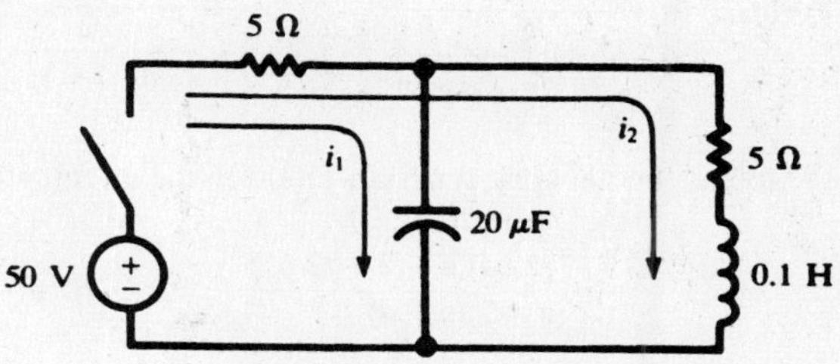

Fig. 8-31

8.25 Find the natural resonant frequency, $|\beta|$, of a series *RLC* circuit with $R = 200\,\Omega$, $L = 0.1\,\text{H}$, $C = 5\,\mu\text{F}$. *Ans.* 1000 rad/s

8.26 A voltage of 10 V is applied at $t = 0$ to a series *RLC* circuit with $R = 5\,\Omega$, $L = 0.1\,\text{H}$, $C = 500\,\mu\text{F}$. Find the transient voltage across the resistance. *Ans.* $3.60e^{-25t}\sin 139t$ (V)

8.27 In the two-mesh circuit shown in Fig. 8-31, the switch is closed at $t = 0$. Find i_1 and i_2, for $t > 0$. *Ans.* $i_1 = 0.101e^{-100t} + 9.899e^{-9950t}$ (A), $i_2 = -5.05e^{-100t} + 5.00 + 0.05e^{-9950t}$ (A)

8.28 A voltage has the **s**-domain representation $100\,\underline{/30^\circ}$ V. Express the time function for (*a*) $\mathbf{s} = -2$ Np/s, (*b*) $\mathbf{s} = -1 + j5\,\text{s}^{-1}$. *Ans.* (*a*) $86.6\,e^{-2t}$ (V); (*b*) $100\,e^{-t}\cos(5t + 30^\circ)$ (V)

8.29 Give the complex frequencies associated with the current $i(t) = 5.0 + 10e^{-3t}\cos(50t + 90^\circ)$ (A). *Ans.* $0, -3 \pm j50\,\text{s}^{-1}$

8.30 A phasor current $25\,\underline{/40^\circ}$ A has complex frequency $\mathbf{s} = -2 + j3\,\text{s}^{-1}$. What is the magnitude of $i(t)$ at $t = 0.2$ s? *Ans.* 4.51 A

8.31 Calculate the impedance **Z(s)** for the circuit shown in Fig. 8-32, at (*a*) $\mathbf{s} = 0$, (*b*) $\mathbf{s} = j1$ rad/s, (*c*) $\mathbf{s} = j2$ rad/s, (*d*) $|\mathbf{s}| = \infty$. *Ans.* (*a*) $1\,\Omega$; (*b*) $1.58\,\underline{/18.43^\circ}\,\Omega$; (*c*) $1.84\,\underline{/12.53^\circ}\,\Omega$; (*d*) $2\,\Omega$

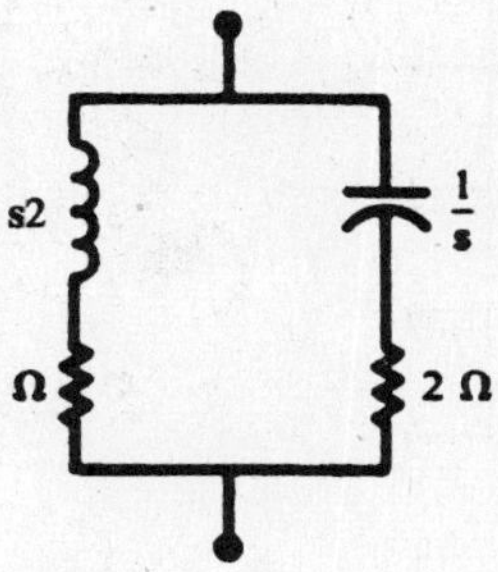

Fig. 8-32

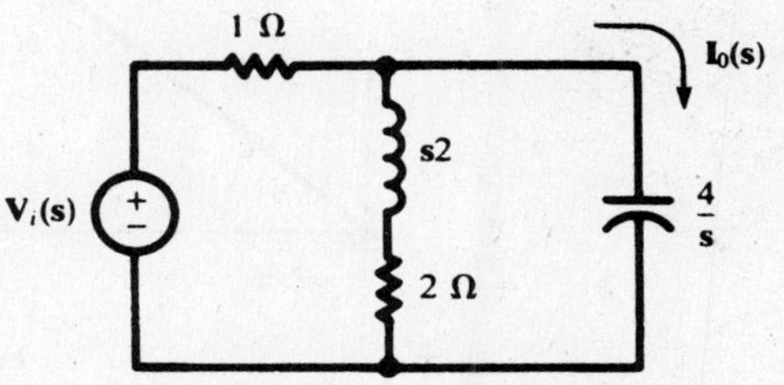

Fig. 8-33

8.32 The voltage source in the **s**-domain circuit shown in Fig. 8-33 has the time-domain expression

$$v_i(t) = 10e^{-t}\cos 2t \quad \text{(V)}$$

Obtain $i_o(t)$. *Ans.* $7.07e^{-t}\cos(2t + 98.13^\circ)$ (A)

8.33 In the time domain, a series circuit of *R*, *L*, and *C* has an applied voltage v_i and element voltages v_R, v_L, and v_C. Obtain the voltage transfer functions (*a*) $\mathbf{V}_R(\mathbf{s})/\mathbf{V}_i(\mathbf{s})$, (*b*) $\mathbf{V}_C(\mathbf{s})/\mathbf{V}_i(\mathbf{s})$.

Ans. (*a*) $\dfrac{R\mathbf{s}/L}{\mathbf{s}^2 + \dfrac{R}{L}\mathbf{s} + \dfrac{1}{LC}}$; (*b*) $\dfrac{1/LC}{\mathbf{s}^2 + \dfrac{R}{L}\mathbf{s} + \dfrac{1}{LC}}$

8.34 Obtain the network function **H(s)** for the circuit shown in Fig. 8-34. The response is the voltage $\mathbf{V}_i(\mathbf{s})$.

Ans. $\dfrac{(\mathbf{s}+7-j2.65)(\mathbf{s}+7+j2.65)}{(\mathbf{s}+2)(\mathbf{s}+4)}$

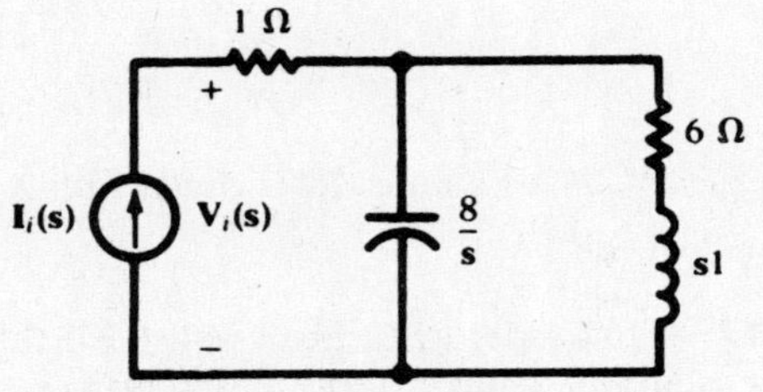

Fig. 8-34

8.35 Construct the **s**-plane plot for the transfer function of Problem 8.34. Evaluate $\mathbf{H}(j3)$ from the plot.
Ans. See Fig. 8-35.

$$\frac{(7.02)(9.0)\ \underline{/2.86 + 38.91^\circ}}{(3.61)(5.0)\ \underline{/56.31^\circ + 36.87^\circ}} = 3.50\ \underline{/-51.41^\circ}\ \Omega$$

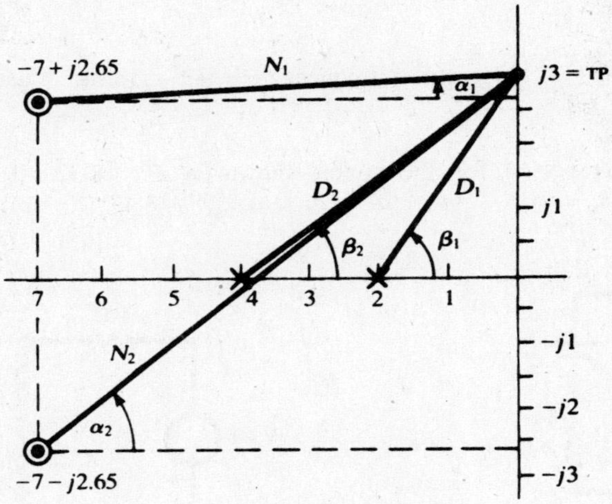

Fig. 8-35

8.36 Obtain $\mathbf{H(s)} = \mathbf{V}_i(\mathbf{s})/\mathbf{I}_i(\mathbf{s})$ for the circuit shown in Fig. 8-36 and construct the pole-zero plot.

Ans. $\mathbf{H(s)} = \dfrac{\mathbf{s}(\mathbf{s}^2+1.5)}{\mathbf{s}^2+1}$. See Fig. 8-37.

8.37 Write the transfer function **H(s)** whose pole-zero plot is given in Fig. 8-38.

Ans. $\mathbf{H(s)} = k\,\dfrac{\mathbf{s}^2 + 50\mathbf{s} + 400}{\mathbf{s}^2 + 40\mathbf{s} + 2000}$

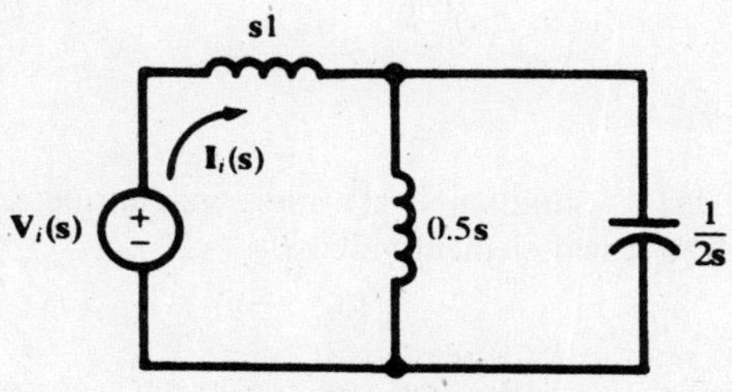

Fig. 8-36

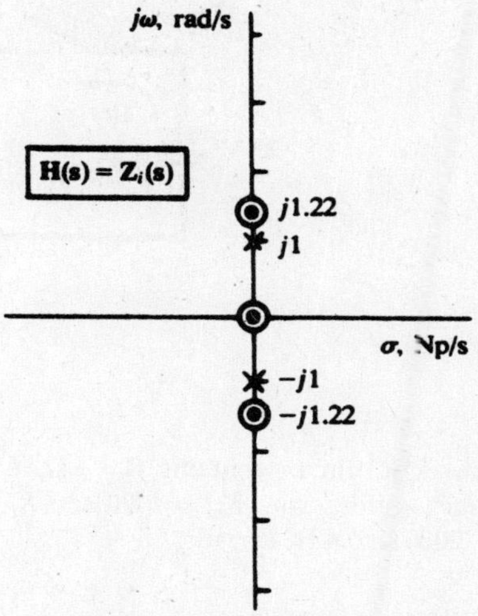

Fig. 8-37

8.38 The pole-zero plot in Fig. 8-39 shows a pole at $\mathbf{s} = 0$ and zeros at $\mathbf{s} = -50 \pm j50$ Use the geometrical method to evaluate the transfer function at the test point $j100$.
Ans. $\mathbf{H}(j100) = 223.6 \underline{/26.57^\circ}$

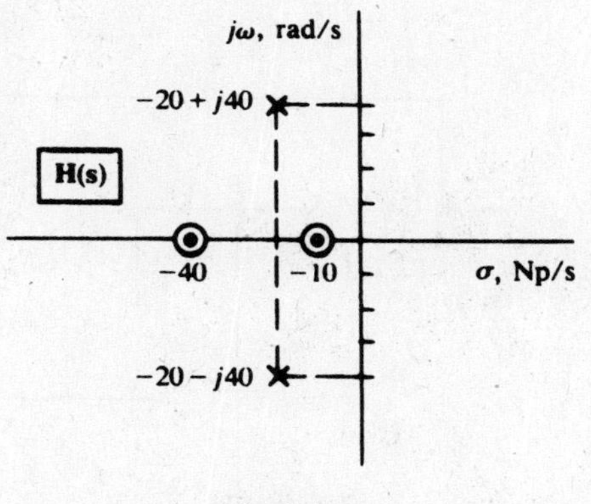

Fig. 8-38

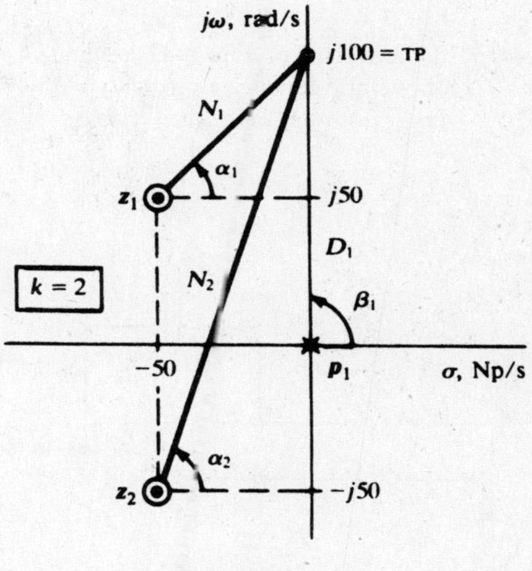

Fig. 8-39

8.39 A two-branch parallel circuit has a resistance of 20 Ω in one branch and the series combination of $R = 10\,\Omega$ and $L = 0.1\,\text{H}$ in the other. First, apply an excitation, $\mathbf{I}_i(\mathbf{s})$, and obtain the natural frequency from the denominator of the network function. Try different locations for applying the current source. Second, insert a voltage source, $\mathbf{V}_i(\mathbf{s})$, and obtain the natural frequency. *Ans.* − 300 Np/s in all cases

8.40 In the network shown in Fig. 8-40, the switch is closed at $t = 0$. At $t = 0^+$, $i = 0$ and

$$\frac{di}{dt} = 25\,\text{A/s}$$

Obtain the natural frequencies and the complete current, $i = i_n + i_f$.
Ans. − 8.5 Np/s, −23.5 Np/s; $i = -2.25e^{-8.5t} - 0.25e^{-23.5t} + 2.5$ (A)

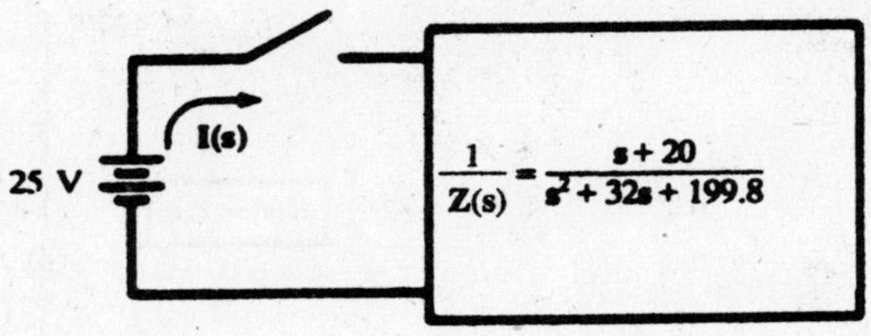

Fig. 8-40

8.41 A series RLC circuit contains $R = 1\,\Omega$, $L = 2\,\text{H}$, and $C = 0.25\,\text{F}$. Simultaneously apply magnitude and frequency scaling, with $K_m = 2000$ and $K_f = 10^4$. What are the scaled element values?
Ans. $2000\,\Omega$, 0.4 H, 12.5 μF

8.42 At a certain frequency ω_1, a voltage $\mathbf{V}_1 = 25\,\angle 0^\circ$ V applied to a passive network results in a current $\mathbf{I}_1 = 3.85\,\angle -30^\circ$ (A). The network elements are magnitude-scaled with $K_m = 10$. Obtain the current which results from a second voltage source, $\mathbf{V}_2 = 10\,\angle 45^\circ$ V, replacing the first, if the second source frequency is $\omega_2 = 10^3\omega_1$. *Ans.* $0.154\,\angle 15^\circ$ A

8.43 In the circuit of Fig. 8-41 let $R_1C_1 = R_2C_2 = 10^{-3}$. Find v_2 for $t > 0$ if: (*a*) $v_1 = \cos(1000t)u(t)$, (*b*) $v_1 = \sin(1000t)u(t)$. *Ans.* (*a*) $v_2 = \sin(1000t)$; (*b*) $v_2 = 1 - \cos(1000t)$

8.44 In the circuit of Fig. 8-42 assume $R = 2\,k\Omega$, $C = 10\,nF$, and $R_2 = R_1$ and $v_1 = \cos\omega t$. Find v_2 for the following frequencies: (*a*) $\omega_0 = 5 \times 10^4$ rad/s, (*b*) $\omega_1 = 10^5$ rad/s.
Ans. (*a*) $v_2 = 2\sin\omega_0 t$; (*b*) $v_2 = 0.555\cos(\omega_1 t - 146.3^\circ)$

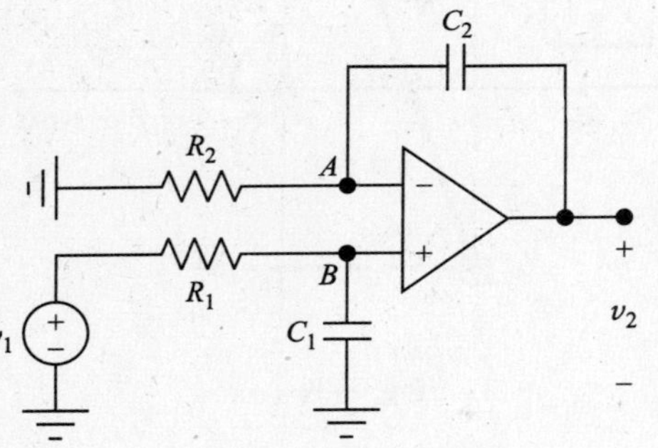

Fig. 8-41

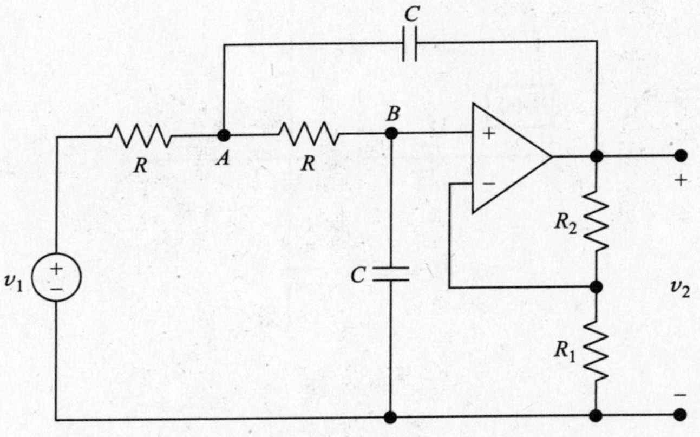

Fig. 8-42

8.45 **Noninverting integrators.** In the circuits of Fig. 8-43(*a*) and 8-43(*b*) find the relationship between v_2 and v_1.
Ans. (*a*) $v_1 = (RC/2)dv_2/dt$; (*b*) $v_1 = 2RCdv_2/dt$

8.46 In the circuit of Fig. 8-44 find the relationship between v_2 and v_1. Show that for $R_1C_1 = R_2C_2$ we obtain $v_2 = R_2v_1/(R_1 + R_2)$.

Ans. $R_1R_2(C_1 + C_2)\dfrac{dv_2}{dt} + (R_1 + R_2)v_2 = R_1R_2C_1\dfrac{dv_1}{dt} + R_2v_1$

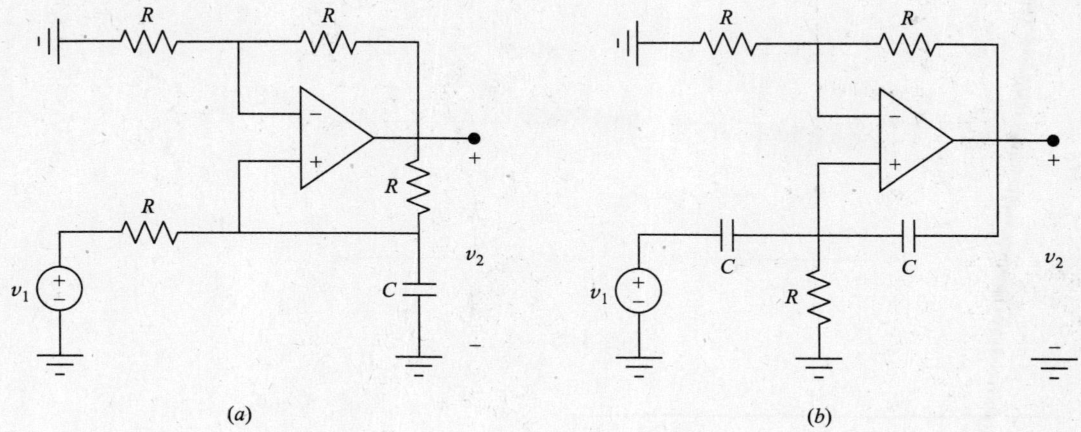

Fig. 8-43

8.47 In the circuit of Fig. 8-44 let $R_1 = 9\,k\Omega = 9R_2$, $C_2 = 100\,pF = 9C_1$, and $v_1 = 10^4 t$ V. Find i at 1 ms after the switch is closed. *Ans.* $i = 1.0001$ mA

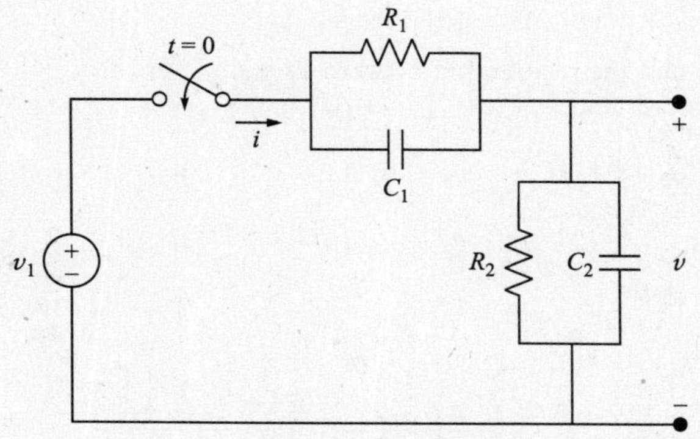

Fig. 8-44

8.48 **Lead network.** The circuit of Fig. 8-45(*a*) is called a *lead network.* (*a*) Find the differential equation relating v_2 to v_1. (*b*) Find the unit-step response of the network with $R_1 = 10\,k\Omega$, $R_2 = 1\,k\Omega$, and $C = 1\,\mu F$. (*c*) Let $v_1 = \cos\omega t$ be the input and $v_2 = A\cos(\omega t + \theta)$ be the output of the network of Part (*b*). Find A and θ for ω at 1, 100, 331.6, 1100, and 10^5, all in rad/s. At what ω is the phase at a maximum?

Ans. (*a*) $\dfrac{dv_2}{dt} + \left(\dfrac{R_1 + R_2}{R_1 R_2 C}\right) v_2 = \dfrac{dv_1}{dt} + \dfrac{1}{R_1 C} v_1$, (*b*) $v_2 = \dfrac{1}{11}(1 + 10e^{-1100t})u(t)$

(*c*)

ω	1	100	331.6	1100	10^5
A	0.091	0.128	0.3015	0.71	1
θ	0.5°	39.8°	56.4°	39.8°	0.5°

Phase is maximum at $\omega = 100\sqrt{11} = 331.6$ rad/s

8.49 **Lag network.** The circuit of Fig. 8-45(*b*) is called a lag network. (*a*) Find the differential equation relating v_2 to v_1. (*b*) Find the unit-step response of the network with $R_1 = 10\,k\Omega$, $R_2 = 1\,k\Omega$, and $C = 1\,\mu F$. (*c*) Let $v_1 = \cos\omega t$ be the input and $v_2 = A\cos(\omega t - \theta)$ be the output of the network of Part (*b*). Find A and θ for ω at 1, 90.9, 301.5, 1000, and 10^5, all in rad/s. At what ω is the phase at a minimum?

Ans. (*a*) $v_2 + (R_1 + R_2)C\dfrac{dv_2}{dt} = v_1 + R_2 C\dfrac{dv_1}{dt}$, (*b*) $v_2 = \left(1 - \dfrac{10}{11}e^{-90.91t}\right)u(t)$

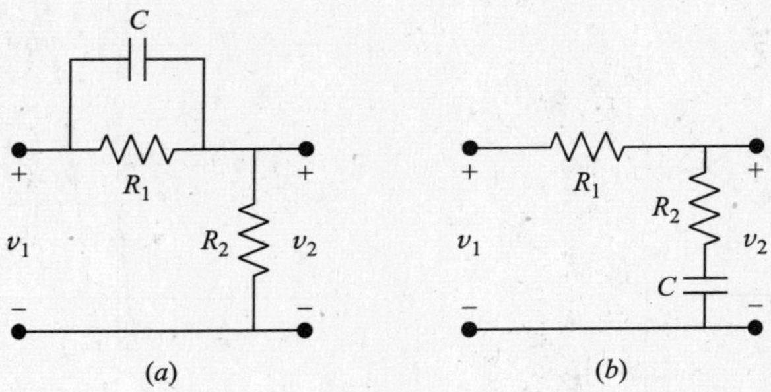

Fig. 8-45

(c)

ω	1	90.9	301.5	1000	10^5
A	1	0.71	0.3015	0.128	0.091
θ	0.5°	39.8°	56.4°	39.8°	0.5°

Phase is minimum at $\omega = 1000/\sqrt{11} = 301.5$ rad/s

8.50 In the circuit of Fig. 8-46 find the relationship between v_2 and v_1 for (a) $k = 10^3$, (b) $k = 10^5$. In each case find its unit-step response; that is, v_2 for $v_1 = u(t)$.

Ans. (a) $\dfrac{dv_2}{dt} + 4 \times 10^6 v_2 = -4 \times 10^7 v_1, \quad v_2 = -10\left(1 - e^{-4\times10^6 t}\right)u(t)$

(b) $\dfrac{dv_2}{dt} + 4 \times 10^8 v_2 = -4 \times 10^9 v_1, \quad v_2 = -10\left(1 - e^{-4\times10^9 t}\right)u(t)$

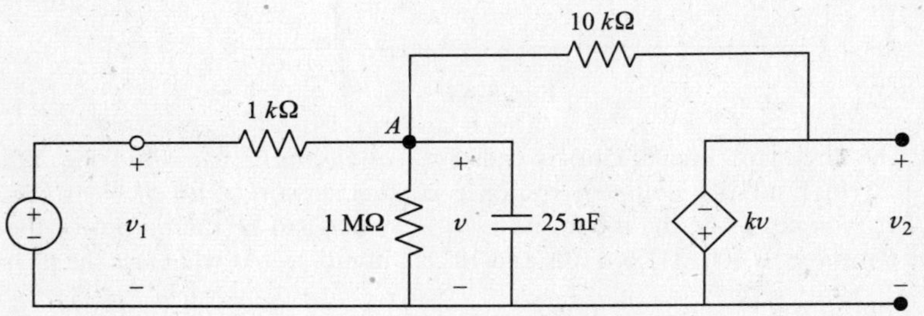

Fig. 8-46

CHAPTER 9

Sinusoidal Steady-State Circuit Analysis

9.1 INTRODUCTION

This chapter will concentrate on the steady-state response of circuits driven by sinusoidal sources. The response will also be sinusoidal. For a linear circuit, the assumption of a *sinusoidal* source represents no real restriction, since a source that can be described by a periodic function can be replaced by an equivalent combination (Fourier series) of sinusoids. This matter will be treated in Chapter 17.

9.2 ELEMENT RESPONSES

The voltage-current relationships for the single elements R, L, and C were examined in Chapter 2 and summarized in Table 2-1. In this chapter, the functions of v and i will be sines or cosines with the argument ωt. ω is the angular frequency and has the unit rad/s. Also, $\omega = 2\pi f$, where f is the frequency with unit cycle/s, or more commonly hertz (Hz).

Consider an inductance L with $i = I\cos(\omega t + 45^\circ)$ A [see Fig. 9-1(a)]. The voltage is

$$v_L = L\frac{di}{dt} = \omega LI[-\sin(\omega t + 45^\circ)] = \omega LI\cos(\omega t + 135^\circ) \quad \text{(V)}$$

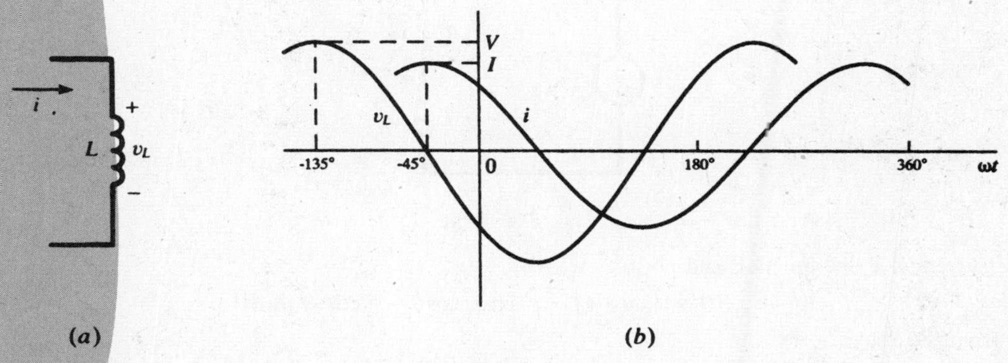

Fig. 9-1

A comparison of v_L and i shows that the current *lags* the voltage by 90° or $\pi/2$ rad. The functions are sketched in Fig. 9-1(b). Note that the current function i is to the right of v, and since the horizontal scale is ωt, events displaced to the right occur later in time. This illustrates that *i lags v*. The horizontal scale is in radians, but note that it is also marked in degrees (−135°, 180°, etc.). This is a case of mixed units just as with $\omega t + 45°$. It is not mathematically correct but is the accepted practice in circuit analysis. The vertical scale indicates two different quantities, that is, v and i, so there should be two scales rather than one.

While examining this sketch, it is a good time to point out that a sinusoid is completely defined when its magnitude (V or I), frequency (ω or f), and phase (45° or 135°) are specified.

In Table 9-1 the responses of the three basic circuit elements are shown for applied current $i = I\cos\omega t$ and voltage $v = V\cos\omega t$. If sketches are made of these responses, they will show that for a resistance R, *v and i are in phase*. For an inductance L, *i lags v* by 90° or $\pi/2$ rad. And for a capacitance C, *i leads v* by 90° or $\pi/2$ rad.

Table 9-1

	$i = I\cos\omega t$	$v = V\cos\omega t$
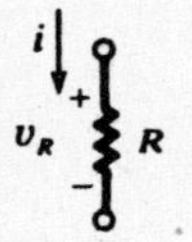	$v_r = RI\cos\omega t$	$i_R = \dfrac{V}{R}\cos\omega t$
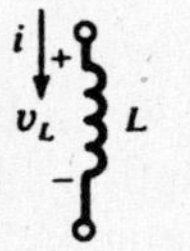	$v_L = \omega LI\cos(\omega t + 90°)$	$i_L = \dfrac{V}{\omega L}\cos(\omega t - 90°)$
	$v_C = \dfrac{I}{\omega C}\cos(\omega t - 90°)$	$i_C = \omega CV\cos(\omega t + 90°)$

EXAMPLE 9.1 The RL series circuit shown in Fig. 9-2 has a current $i = I\sin\omega t$. Obtain the voltage v across the two circuit elements and sketch v and i.

$$v_R = RI\sin\omega t \qquad v_L = L\frac{di}{dt} = \omega LI\sin(\omega t + 90°)$$

$$v = v_R + v_L = RI\sin\omega t + \omega LI\sin(\omega t + 90°)$$

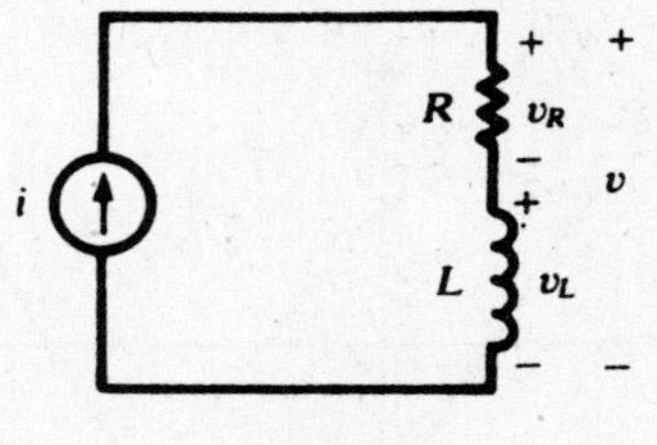

Fig. 9-2

Since the current is a sine function and

$$v = V\sin(\omega t + \theta) = V\sin\omega t\cos\theta + V\cos\omega t\sin\theta \tag{1}$$

we have from the above

$$v = RI\sin\omega t + \omega LI\sin\omega t\cos 90° + \omega LI\cos\omega t\sin 90° \tag{2}$$

Equating coefficients of like terms in (*1*) and (*2*),

$$V \sin\theta = \omega L I \qquad \text{and} \qquad V\cos\theta = RI$$

Then
$$v = I\sqrt{R^2 + (\omega L)^2}\,\sin\,[\omega t + \arctan\,(\omega L/R)]$$

and
$$V = I\sqrt{R^2 + (\omega L)^2} \qquad \text{and} \qquad \theta = \tan^{-1}\frac{\omega L}{R}$$

The functions i and v are sketched in Fig. 9-3. The phase angle θ, the angle by which i lags v, lies within the range $0° \le \theta \le 90°$, with the limiting values attained for $\omega L \ll R$ and $\omega L \gg R$, respectively. If the circuit had an applied voltage $v = V\sin\omega t$, the resulting current would be

$$i = \frac{V}{\sqrt{R^2 + (\omega L)^2}}\sin\,(\omega t - \theta)$$

where, as before, $\theta = \tan^{-1}\,(\omega L/R)$.

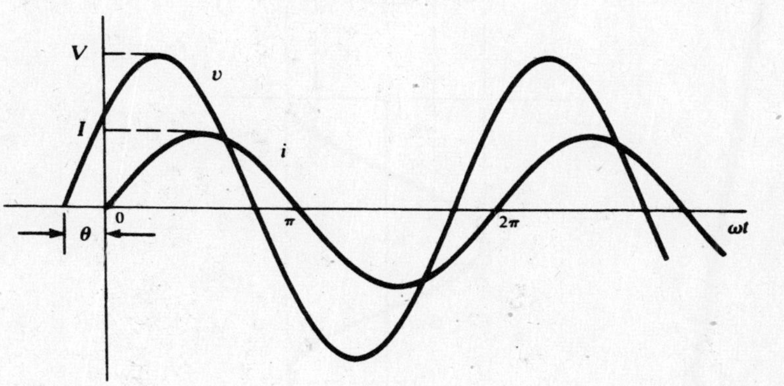

Fig. 9-3

EXAMPLE 9.2 If the current driving a series RC circuit is given by $i = I\sin\omega t$, obtain the total voltage across the two elements.

$$v_R = RI\sin\omega t \qquad v_C = (1/\omega C)\sin\,(\omega t - 90°)$$
$$v = v_R + v_C = V\sin\,(\omega t - \theta)$$

where
$$V = I\sqrt{R^2 + (1/\omega C)^2} \qquad \text{and} \qquad \theta = \tan^{-1}\,(1/\omega CR)$$

The negative phase angle shifts v to the right of the current i. Consequently i *leads* v for a series RC circuit. The phase angle is constrained to the range $0° \le \theta \le 90°$. For $(1/\omega C) \ll R$, the angle $\theta = 0°$, and for $(1/\omega C) \gg R$, the angle $\theta = 90°$. See Fig. 9-4.

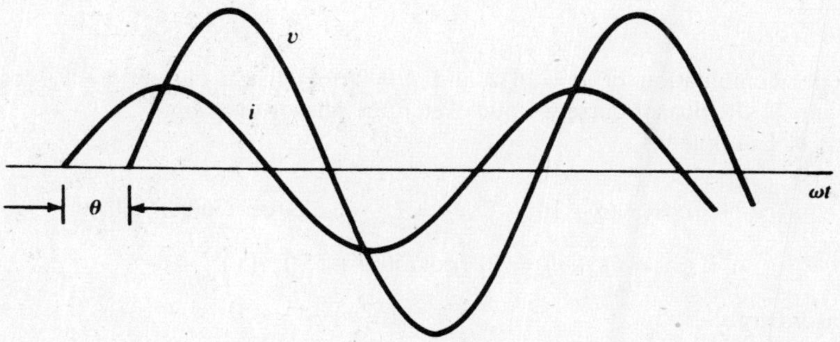

Fig. 9-4

9.3 PHASORS

A brief look at the voltage and current sinusoids in the preceding examples shows that the amplitudes and phase differences are the two principal concerns. A directed line segment, or *phasor*, such as that shown rotating in a counterclockwise direction at a constant angular velocity ω (rad/s) in Fig. 9-5, has a projection on the horizontal which is a cosine function. The length of the phasor or its magnitude is the amplitude or maximum value of the cosine function. The angle between two positions of the phasor is the phase difference between the corresponding points on the cosine function.

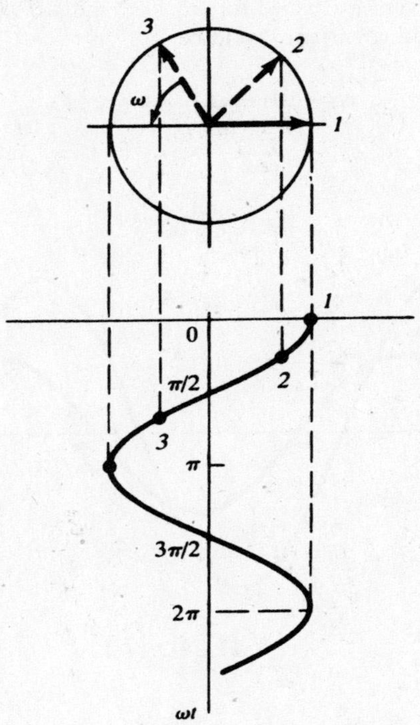

Fig. 9-5

Throughout this book *phasors will be defined from the cosine function.* If a voltage or current is expressed as a sine, it will be changed to a cosine by subtracting 90° from the phase.

Consider the examples shown in Table 9-2. Observe that the phasors, which are directed line segments and vectorial in nature, are indicated by boldface capitals, for example, **V**, **I**. The phase angle of the cosine function is the angle on the phasor. The phasor diagrams here and all that follow may be considered as a snapshot of the counterclockwise-rotating directed line segment taken at $t = 0$. The frequency f (Hz) and ω (rad/s) generally do not appear but they should be kept in mind, since they are implicit in any sinusoidal steady-state problem.

EXAMPLE 9.3 A series combination of $R = 10\,\Omega$ and $L = 20\,\text{mH}$ has a current $i = 5.0\cos(500t + 10^\circ)$ (A). Obtain the voltages v and **V**, the phasor current **I** and sketch the phasor diagram.

Using the methods of Example 9.1,

$$v_R = 50.0\cos(500t + 10^\circ) \qquad v_L = L\frac{di}{dt} = 50.0\cos(500t + 100^\circ)$$

$$v = v_R + v_L = 70.7\cos(500t + 55^\circ) \quad \text{(V)}$$

The corresponding phasors are

$$\mathbf{I} = 5.0\ \underline{/10^\circ}\ \text{A} \qquad \text{and} \qquad \mathbf{V} = 70.7\ \underline{/55^\circ}\ \text{V}$$

Table 9-2

Function	Phasor Representation
$v = 150\cos(500t + 45°)$ (V)	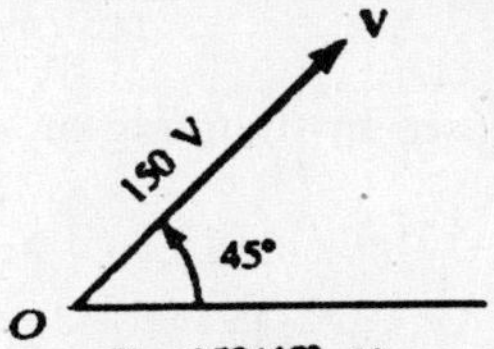 $\mathbf{V} = 150\,\underline{/45°}$ V
$i = 3.0\sin(2000t + 30°)$ (mA) $= 3.0\cos(2000t - 60°)$ (mA)	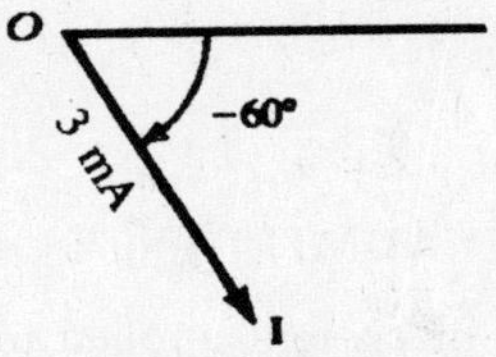 $\mathbf{I} = 3.0\,\underline{/-60°}$ mA

The phase angle of 45° can be seen in the time-domain graphs of i and v shown in Fig. 9-6(a), and the phasor diagram with **I** and **V** shown in Fig. 9-6(b).

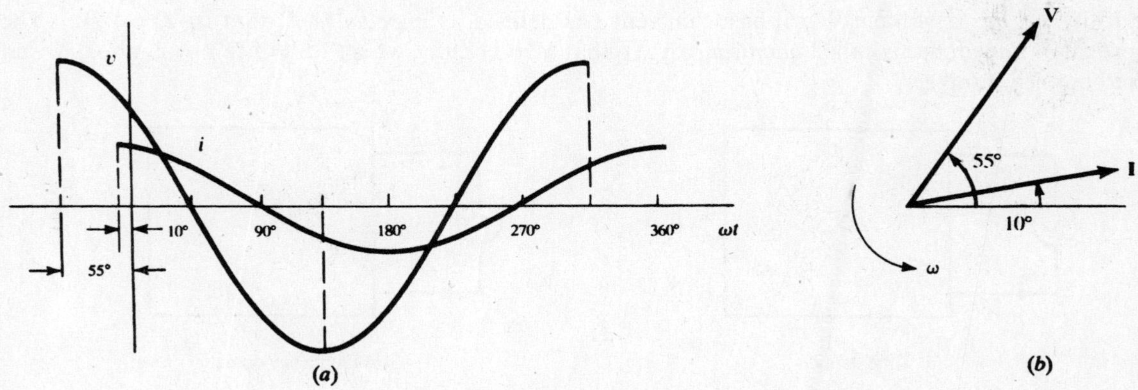

Fig. 9-6

Phasors can be treated as complex numbers. When the horizontal axis is identified as the real axis of a complex plane, the phasors become complex numbers and the usual rules apply. In view of *Euler's identity*, there are three equivalent notations for a phasor.

$$\text{polar form} \qquad \mathbf{V} = V\,\underline{/\theta}$$

$$\text{rectangular form} \qquad \mathbf{V} = V(\cos\theta + j\sin\theta)$$

$$\text{exponential form} \qquad \mathbf{V} = Ve^{j\theta}$$

The cosine expression may also be written as

$$v = V\cos(\omega t + \theta) = \text{Re}\,[Ve^{j(\omega t + \theta)}] = \text{Re}\,[\mathbf{V}e^{j\omega t}]$$

The exponential form suggests how to treat the product and quotient of phasors. Since $(V_1 e^{j\theta_1})(V_2 e^{j\theta_2}) + V_1 V_2 e^{j(\theta_1+\theta_2)}$,

$$(V_1\,\underline{/\theta_1})(V_2\,\underline{/\theta_2}) = V_1 V_2\,\underline{/\theta_1 + \theta_2}$$

and, since $(V_1 e^{j\theta_1})/(V_2 e^{j\theta_2}) = (V_1/V_2)e^{j(\theta_1-\theta_2)}$,

$$\frac{V_1 \underline{/\theta_1}}{V_2 \underline{/\theta}} = V_1/V_2 \underline{/\theta_1 - \theta_2}$$

The rectangular form is used in summing or subtracting phasors.

EXAMPLE 9.4 Given $\mathbf{V}_1 = 25.0 \underline{/143.13^\circ}$ and $\mathbf{V}_2 = 11.2 \underline{/26.57^\circ}$, find the ratio $\mathbf{V}_1/\mathbf{V}_2$ and the sum $\mathbf{V}_1 + \mathbf{V}_2$.

$$\mathbf{V}_1/\mathbf{V}_2 = \frac{25.0 \underline{/143.13^\circ}}{11.2 \underline{/26.57^\circ}} = 2.23 \underline{/116.56^\circ} = -1.00 + j1.99$$

$$\mathbf{V}_1 + \mathbf{V}_2 = (-20.0 + j15.0) + (10.0 + j5.0) = -10.0 + j20.0 = 23.36 \underline{/116.57^\circ}$$

9.4 IMPEDANCE AND ADMITTANCE

A sinusoidal voltage or current applied to a passive *RLC* circuit produces a sinusoidal response. With time functions, such as $v(t)$ and $i(t)$, the circuit is said to be in the *time domain*, Fig. 9-7(*a*); and when the circuit is analyzed using phasors, it is said to be in the *frequency domain*, Fig. 9-7(*b*). The voltage and current may be written, respectively,

$$v(t) = V\cos(\omega t + \theta) = \text{Re}\,[\mathbf{V}e^{j\omega t}] \quad \text{and} \quad \mathbf{V} = V \underline{/\theta}$$

$$i(t) = I\cos(\omega t + \phi) = \text{Re}\,[\mathbf{I}e^{j\omega t}] \quad \text{and} \quad \mathbf{I} = I \underline{/\phi}$$

The ratio of phasor voltage $\mathbf{V}$ to phasor current $\mathbf{I}$ is defined as *impedance* $\mathbf{Z}$, that is, $\mathbf{Z} = \mathbf{V}/\mathbf{I}$. The reciprocal of impedance is called *admittance* $\mathbf{Y}$, so that $\mathbf{Y} = 1/\mathbf{Z}$ (S), where $1\,\text{S} = 1\,\Omega^{-1} = 1\,\text{mho}$. $\mathbf{Y}$ and $\mathbf{Z}$ are complex numbers.

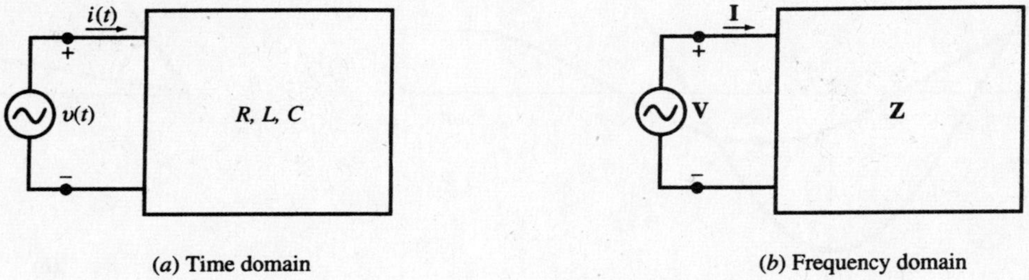

Fig. 9-7

When impedance is written in Cartesian form the real part is the resistance R and the imaginary part is the *reactance* X. The sign on the imaginary part may be positive or negative: When positive, X is called the *inductive reactance*, and when negative, X is called the *capacitive reactance*. When the admittance is written in Cartesian form, the real part is *admittance* G and the imaginary part is *susceptance* B. A positive sign on the susceptance indicates a *capacitive susceptance*, and a negative sign indicates an *inductive susceptance*. Thus,

$$\mathbf{Z} = R + jX_L \quad \text{and} \quad \mathbf{Z} = R - jX_C$$

$$\mathbf{Y} = G - jB_L \quad \text{and} \quad \mathbf{Y} = G + jB_C$$

The relationships between these terms follow from $\mathbf{Z} = 1/\mathbf{Y}$. Then,

$$R = \frac{G}{G^2 + B^2} \quad \text{and} \quad X = \frac{-B}{G^2 + B^2}$$

$$G = \frac{R}{R^2 + X^2} \quad \text{and} \quad B = \frac{-X}{R^2 + X^2}$$

These expressions are not of much use in a problem where calculations can be carried out with the numerical values as in the following example.

EXAMPLE 9.5 The phasor voltage across the terminals of a network such as that shown in Fig. 9-7(*b*) is $100.0\ \angle 45^\circ$ V and the resulting current is $5.0\ \angle 15^\circ$ A. Find the equivalent impedance and admittance.

$$\mathbf{Z} = \frac{\mathbf{V}}{\mathbf{I}} \frac{100.0\ \angle 45^\circ}{5.0\ \angle 15^\circ} = 20.0\ \angle 30^\circ = 17.32 + j10.0\ \Omega$$

$$\mathbf{Y} = \frac{\mathbf{I}}{\mathbf{V}} = \frac{1}{\mathbf{Z}} = 0.05\ \angle -30 = (4.33 - j2.50) \times 10^{-2}\ \mathrm{S}$$

Thus, $R = 17.32\ \Omega$, $X_L = 10.0\ \Omega$, $G = 4.33 \times 10^{-2}$ S, and $B_L = 2.50 \times 10^{-2}$ S.

Combinations of Impedances

The relation $\mathbf{V} = \mathbf{IZ}$ (in the frequency domain) is formally identical to Ohm's law, $v = iR$, for a resistive network (in the time domain). Therefore, impedances combine exactly like resistances:

$$\text{impedances in series} \qquad \mathbf{Z}_{\mathrm{eq}} = \mathbf{Z}_1 + \mathbf{Z}_2 + \cdots$$

$$\text{impedances in parallel} \qquad \frac{1}{\mathbf{Z}_{\mathrm{eq}}} = \frac{1}{\mathbf{Z}_1} + \frac{1}{\mathbf{Z}_2} + \cdots$$

In particular, for two parallel impedances, $\mathbf{Z}_{\mathrm{eq}} = \mathbf{Z}_1\mathbf{Z}_2/(\mathbf{Z}_1 + \mathbf{Z}_2)$.

Impedance Diagram

In an *impedance diagram*, an impedance $\mathbf{Z}$ is represented by a point in the right half of the complex plane. Figure 9-8 shows two impedances; $\mathbf{Z}_1$, in the first quadrant, exhibits inductive reactance, while $\mathbf{Z}_2$, in the fourth quadrant, exhibits capacitive reactance. Their series equivalent, $\mathbf{Z}_1 + \mathbf{Z}_2$, is obtained by vector addition, as shown. Note that the "vectors" are shown without arrowheads, in order to distinguish these complex numbers from phasors.

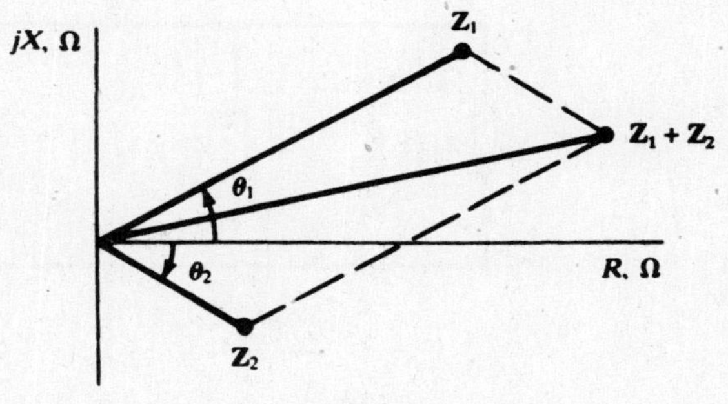

Fig. 9-8

Combinations of Admittances

Replacing $\mathbf{Z}$ by $1/\mathbf{Y}$ in the formulas above gives

$$\text{admittances in series} \qquad \frac{1}{\mathbf{Y}_{\mathrm{eq}}} = \frac{1}{\mathbf{Y}_1} + \frac{1}{\mathbf{Y}_2} + \cdots$$

$$\text{admittances in parallel} \qquad \mathbf{Y}_{\mathrm{eq}} = \mathbf{Y}_1 + \mathbf{Y}_2 + \cdots$$

Thus, series circuits are easiest treated in terms of impedance; parallel circuits, in terms of admittance.

Admittance Diagram

Figure 9-9, an *admittance diagram*, is analogous to Fig. 9-8 for impedance. Shown are an admittance $\mathbf{Y}_1$ having capacitive susceptance and an admittance $\mathbf{Y}_2$ having inductive susceptance, together with their vector sum, $\mathbf{Y}_1 + \mathbf{Y}_2$, which is the admittance of a parallel combination of $\mathbf{Y}_1$ and $\mathbf{Y}_2$.

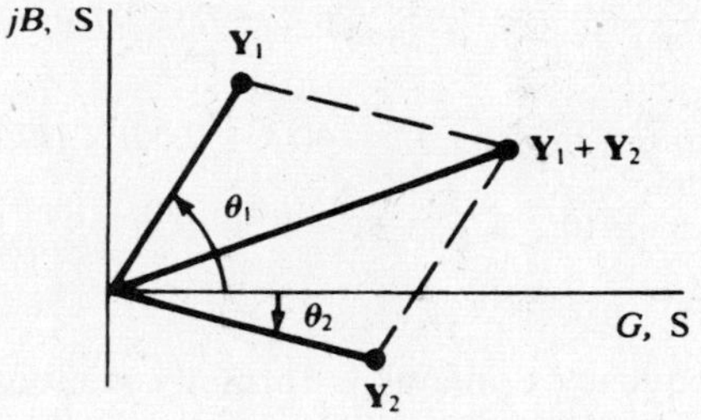

Fig. 9-9

9.5 VOLTAGE AND CURRENT DIVISION IN THE FREQUENCY DOMAIN

In view of the analogy between impedance in the frequency domain and resistance in the time domain, Sections 3.6 and 3.7 imply the following results.

(1) Impedances in series divide the total voltage in the ratio of the impedances:

$$\frac{\mathbf{V}_r}{\mathbf{V}_s} = \frac{\mathbf{Z}_r}{\mathbf{Z}_s} \qquad \text{or} \qquad \mathbf{V}_r = \frac{\mathbf{Z}_r}{\mathbf{Z}_{\text{eq}}} \mathbf{V}_T$$

See Fig. 9-10.

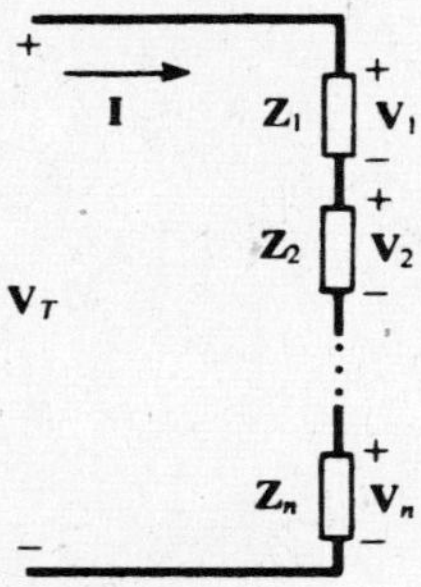

Fig. 9-10

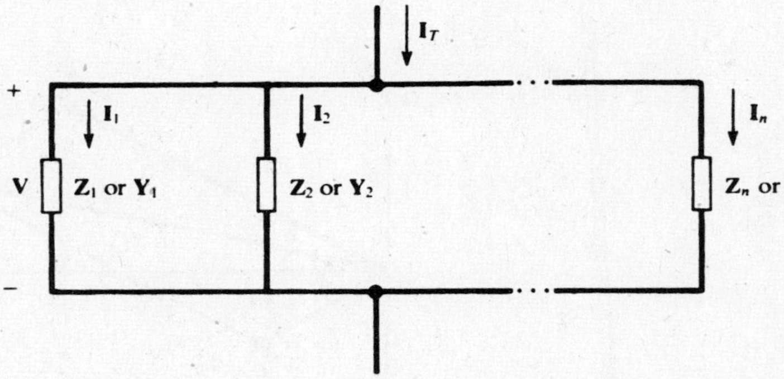

Fig. 9-11

(2) Impedances in parallel (admittances in series) divide the total current in the inverse ratio of the impedances (direct ratio of the admittances):

$$\frac{\mathbf{I}_r}{\mathbf{I}_s} = \frac{\mathbf{Z}_s}{\mathbf{Z}_r} = \frac{\mathbf{Y}_r}{\mathbf{Y}_s} \qquad \text{or} \qquad \mathbf{I}_r = \frac{\mathbf{Z}_{\text{eq}}}{\mathbf{Z}_r} \mathbf{I}_T = \frac{\mathbf{Y}_r}{\mathbf{Y}_{\text{eq}}} \mathbf{I}_T$$

See Fig. 9-11.

9.6 THE MESH CURRENT METHOD

Consider the frequency-domain network of Fig. 9-12. Applying KVL, as in Section 4.3, or simply by inspection, we find the matrix equation

$$\begin{bmatrix} \mathbf{Z}_{11} & \mathbf{Z}_{12} & \mathbf{Z}_{13} \\ \mathbf{Z}_{21} & \mathbf{Z}_{22} & \mathbf{Z}_{23} \\ \mathbf{Z}_{31} & \mathbf{Z}_{32} & \mathbf{Z}_{33} \end{bmatrix} \begin{bmatrix} \mathbf{I}_1 \\ \mathbf{I}_2 \\ \mathbf{I}_3 \end{bmatrix} = \begin{bmatrix} \mathbf{V}_1 \\ \mathbf{V}_2 \\ \mathbf{V}_3 \end{bmatrix}$$

for the unknown mesh currents $\mathbf{I}_1, \mathbf{I}_2, \mathbf{I}_3$. Here, $\mathbf{Z}_{11} \equiv \mathbf{Z}_A + \mathbf{Z}_B$, the *self-impedance* of mesh 1, is the sum of all impedances through which $\mathbf{I}_1$ passes. Similarly, $\mathbf{Z}_{22} \equiv \mathbf{Z}_B + \mathbf{Z}_C + \mathbf{Z}_D$ and $\mathbf{Z}_{33} \equiv \mathbf{Z}_D + \mathbf{Z}_E$ are the self-impedances of meshes 2 and 3.

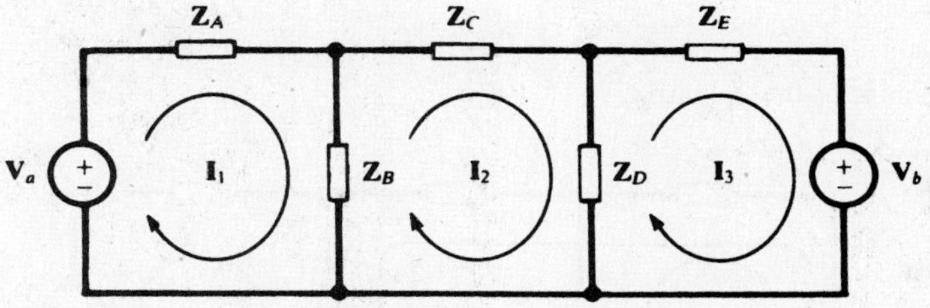

Fig. 9-12

The 1,2-element of the Z-matrix is defined as:

$$\mathbf{Z}_{12} \equiv \sum \pm \text{ (impedance common to } \mathbf{I}_1 \text{ and } \mathbf{I}_2)$$

where a summand takes the plus sign if the two currents pass through the impedance in the same direction, and takes the minus sign in the opposite case. It follows that, invariably, $\mathbf{Z}_{12} = \mathbf{Z}_{21}$. In Fig. 9-12, $\mathbf{I}_1$ and $\mathbf{I}_2$ thread $\mathbf{Z}_B$ in opposite directions, whence

$$\mathbf{Z}_{12} = \mathbf{Z}_{21} = -\mathbf{Z}_B$$

Similarly,

$$\mathbf{Z}_{13} = \mathbf{Z}_{31} \equiv \sum \pm \text{ (impedance common to } \mathbf{I}_1 \text{ and } \mathbf{I}_3) = 0$$

$$\mathbf{Z}_{23} = \mathbf{Z}_{23} \equiv \sum \pm \text{ (impedance common to } \mathbf{I}_2 \text{ and } \mathbf{I}_3 = -\mathbf{Z}_D$$

The **Z**-matrix is symmetric.

In the **V**-column on the right-hand side of the equation, the entries $\mathbf{V}_k$ $(k = 1, 2, 3)$ are defined exactly as in Section 4.3:

$$\mathbf{V}_k \equiv \sum \pm \text{ (driving voltage in mesh } k)$$

where a summand takes the plus sign if the voltage drives in the direction of $\mathbf{I}_k$, and takes the minus sign in the opposite case. For the network of Fig. 9-12,

$$\mathbf{V}_1 = +\mathbf{V}_a \qquad \mathbf{V}_2 = 0 \qquad \mathbf{V}_3 = -\mathbf{V}_b$$

Instead of using the meshes, or "windows" of the (planar) network, it is sometimes expedient to choose an appropriate set of *loops*, each containing one or more meshes in its interior. It is easy to see that two loop currents might have the same direction in one impedance and opposite directions in another. Nevertheless, the preceding rules for writing the **Z**-matrix and the **V**-column have been formulated in such a way as to apply either to meshes or to loops. These rules are, of course, identical to those used in Section 4.3 to write the **R**-matrix and **V**-column.

EXAMPLE 9.6 Suppose that the phasor voltage across $\mathbf{Z}_B$, with polarity as indicated in Fig. 9-13 is sought. Choosing meshes as in Fig. 9-12 would entail solving for both $\mathbf{I}_1$ and $\mathbf{I}_2$, then obtaining the voltage as $\mathbf{V}_B = (\mathbf{I}_2 - \mathbf{I}_1)\mathbf{Z}_B$. In Fig. 9-13 three loops (two of which are meshes) are chosen so as to make $\mathbf{I}_1$ the only current in $\mathbf{Z}_B$. Furthermore, the direction of $\mathbf{I}_1$ is chosen such that $\mathbf{V}_B = \mathbf{I}_1\mathbf{Z}_B$. Setting up the matrix equation:

$$\begin{bmatrix} \mathbf{Z}_A+\mathbf{Z}_B & -\mathbf{Z}_A & 0 \\ -\mathbf{Z}_A & \mathbf{Z}_A+\mathbf{Z}_C+\mathbf{Z}_D & \mathbf{Z}_D \\ 0 & \mathbf{Z}_D & \mathbf{Z}_D+\mathbf{Z}_E \end{bmatrix} \begin{bmatrix} \mathbf{I}_1 \\ \mathbf{I}_2 \\ \mathbf{I}_3 \end{bmatrix} = \begin{bmatrix} -\mathbf{V}_a \\ \mathbf{V}_a \\ \mathbf{V}_b \end{bmatrix}$$

from which

$$\mathbf{V}_B = \mathbf{Z}_B\mathbf{I}_1 = \frac{\mathbf{Z}_B}{\Delta_z} \begin{vmatrix} -\mathbf{V}_a & -\mathbf{Z}_A & 0 \\ \mathbf{V}_a & \mathbf{Z}_A+\mathbf{Z}_B+\mathbf{Z}_C & \mathbf{Z}_D \\ \mathbf{V}_b & \mathbf{Z}_D & \mathbf{Z}_D+\mathbf{Z}_E \end{vmatrix}$$

where Δ_z is the determinant of the **Z**-matrix.

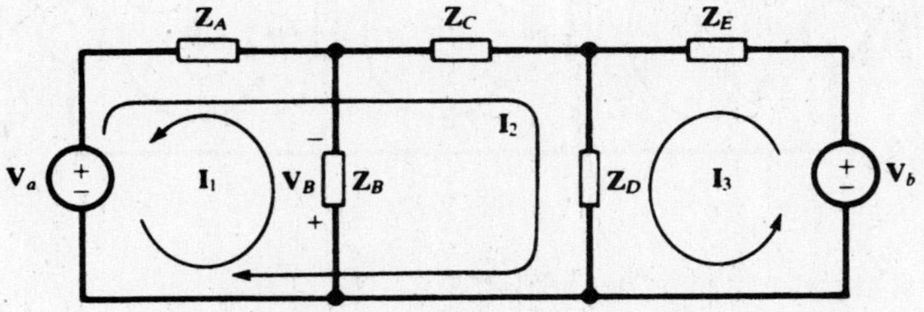

Fig. 9-13

Input and Transfer Impedances

The notions of input resistance (Section 4.5) and transfer resistance (Section 4.6) have their exact counterparts in the frequency domain. Thus, for the single-source network of Fig. 9-14, the *input impedance* is

$$\mathbf{Z}_{\text{input},r} \equiv \frac{\mathbf{V}_r}{\mathbf{I}_r} = \frac{\Delta_z}{\Delta_{rr}}$$

where ${}_{rr}$ is the cofactor of $\mathbf{Z}_{rr}$ in Δ_z; and the *transfer impedance* between mesh (or loop) r and mesh (loop) s is

$$\mathbf{Z}_{\text{transfer},rs} \equiv \frac{\mathbf{V}_r}{\mathbf{I}_s} = \frac{\Delta_z}{\Delta_{rs}}$$

where Δ_{rs} is the cofactor of $\mathbf{Z}_{rs}$ in Δ_z.

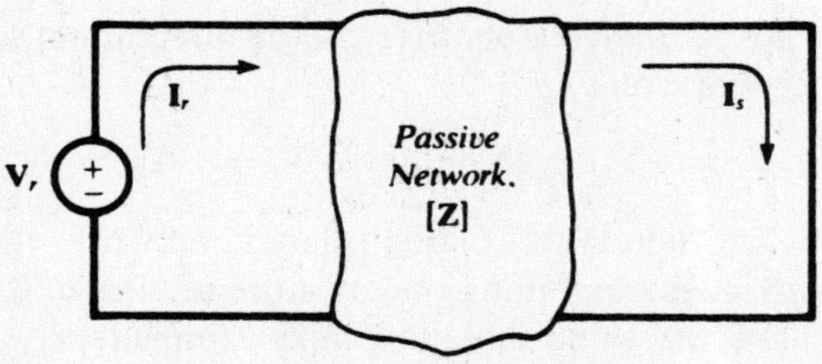

Fig. 9-14

As before, the superposition principle for an arbitrary n-mesh or n-loop network may be expressed as

$$\mathbf{I}_k = \frac{\mathbf{V}_1}{\mathbf{Z}_{\text{transfer},1k}} + \cdots + \frac{\mathbf{V}_{k-1}}{\mathbf{Z}_{\text{transfer},(k-1)k}} + \frac{\mathbf{V}_k}{\mathbf{Z}_{\text{input},k}} + \frac{\mathbf{V}_{k+1}}{\mathbf{Z}_{\text{transfer},(k+1)k}} + \cdots + \frac{\mathbf{V}_n}{\mathbf{Z}_{\text{transfer},nk}}$$

9.7 THE NODE VOLTAGE METHOD

The procedure is exactly as in Section 4.4, with admittances replacing reciprocal resistances. A frequency-domain network with n principal nodes, one of them designated as the reference node, requires $n-1$ node voltage equations. Thus, for $n=4$, the matrix equation would be

$$\begin{bmatrix} \mathbf{Y}_{11} & \mathbf{Y}_{12} & \mathbf{Y}_{13} \\ \mathbf{Y}_{21} & \mathbf{Y}_{22} & \mathbf{Y}_{23} \\ \mathbf{Y}_{31} & \mathbf{Y}_{32} & \mathbf{Y}_{33} \end{bmatrix} \begin{bmatrix} \mathbf{V}_1 \\ \mathbf{V}_2 \\ \mathbf{V}_3 \end{bmatrix} = \begin{bmatrix} \mathbf{I}_1 \\ \mathbf{I}_2 \\ \mathbf{I}_3 \end{bmatrix}$$

in which the unknowns, $\mathbf{V}_1$, $\mathbf{V}_2$, and $\mathbf{V}_3$, are the voltages of principal nodes 1, 2, and 3 with respect to principal node 4, the reference node.

$\mathbf{Y}_{11}$ is the *self-admittance* of node 1, given by the sum of all admittances connected to node 1. Similarly, $\mathbf{Y}_{22}$ and $\mathbf{Y}_{33}$ are the self-admittances of nodes 2 and 3.

$\mathbf{Y}_{12}$, the *coupling admittance* between nodes 1 and 2, is given by *minus* the sum of all admittances connecting nodes 1 and 2. It follows that $\mathbf{Y}_{12} = \mathbf{Y}_{21}$. Similarly, for the other coupling admittances: $\mathbf{Y}_{13} = \mathbf{Y}_{31}$, $\mathbf{Y}_{23} = \mathbf{Y}_{32}$. The $\mathbf{Y}$-matrix is therefore symmetric.

On the right-hand side of the equation, the $\mathbf{I}$-column is formed just as in Section 4.4; i.e.,

$$\mathbf{I}_k = \sum \text{(current driving into node } k) \qquad (k = 1, 2, 3)$$

in which a current driving *out of* node k is counted as negative.

Input and Transfer Admittances

The matrix equation of the node voltage method,

$$[\mathbf{Y}][\mathbf{V}] = [\mathbf{I}]$$

is identical in form to the matrix equation of the mesh current method,

$$[\mathbf{Z}][\mathbf{I}] = [\mathbf{V}]$$

Therefore, in theory at least, *input* and *transfer admittances* can be defined by analogy with input and transfer impedances:

$$\mathbf{Y}_{\text{input},r} \equiv \frac{\mathbf{I}_r}{\mathbf{V}_r} = \frac{\Delta_\mathbf{Y}}{\Delta_{rr}}$$

$$\mathbf{Y}_{\text{transfer},rs} \equiv \frac{\mathbf{I}_r}{\mathbf{V}_s} = \frac{\Delta_\mathbf{Y}}{\Delta_{rs}}$$

where now Δ_{rr} and Δ_{rs} are the cofactors of $\mathbf{Y}_{rr}$ and $\mathbf{Y}_{rs}$ in $\Delta_\mathbf{Y}$. In practice, these definitions are often of limited use. However, they are valuable in providing an expression of the superposition principle (for voltages);

$$\mathbf{V}_k = \frac{\mathbf{I}_1}{\mathbf{Y}_{\text{transfer},1k}} + \cdots + \frac{\mathbf{I}_{k-1}}{\mathbf{Y}_{\text{transfer},(k-1)k}} + \frac{\mathbf{I}_k}{Y_{\text{input},k}} + \frac{\mathbf{I}_{k+1}}{\mathbf{Y}_{\text{transfer},(k+1)k}} + \cdots + \frac{\mathbf{I}_{n-1}}{Y_{\text{transfer},(n-\text{If})k}}$$

for $k = 1, 2, \ldots, n-1$. In words: the voltage at any principal node (relative to the reference node) is obtained by adding the voltages produced at that node by the various driving currents, these currents acting one at a time.

9.8 THÉVENIN'S AND NORTON'S THEOREMS

These theorems are exactly as given in Section 4.9, with the open-circuit voltage V', short-circuit current I', and representative resistance R' replaced by the open-circuit phasor voltage $\mathbf{V}'$, short-circuit phasor current $\mathbf{I}'$, and representative impedance $\mathbf{Z}'$. See Fig. 9-15.

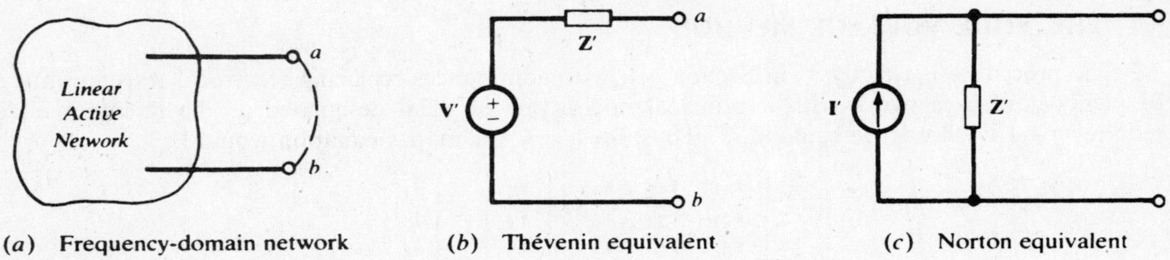

Fig. 9-15

9.9 SUPERPOSITION OF AC SOURCES

How do we apply superposition to circuits with more than one sinusoidal source? If all sources have the same frequency, superposition is applied in the phasor domain. Otherwise, the circuit is solved for each source, and time-domain responses are added.

EXAMPLE 9.7 A practical coil is connected in series between two voltage sources $v_1 = 5\cos\omega_1 t$ and $v_2 = 10\cos(\omega_2 t + 60°)$ such that the sources share the same reference node. See Fig. 9-54. The voltage difference across the terminals of the coil is therefore $v_1 - v_2$. The coil is modeled by a 5-mH inductor in series with a 10-Ω resistor. Find the current $i(t)$ in the coil for (*a*) $\omega_1 = \omega_2 = 2000$ rad/s and (*b*) $\omega_1 = 2000$ rad/s, $\omega_2 = 2\omega_1$.

(*a*) The impedance of the coil is $R + jL\omega = 10 + j10 = 10\sqrt{2}\ \angle 45°\ \Omega$. The phasor voltage between its terminals is $\mathbf{V} = \mathbf{V}_1 - \mathbf{V}_2 = 5 - 10\ \angle 60° = -j5\sqrt{3}$ V. The current is

$$\mathbf{I} = \frac{\mathbf{V}}{Z} = \frac{-j5\sqrt{3}}{10\sqrt{2}\ \angle 45°} \approx \frac{-j8.66}{14.14\ \angle 45°} = 0.61\ \angle -135°\ \text{A}$$
$$i = 0.61\cos(2000t - 135°)$$

(*b*) Because the coil has different impedances at $\omega_1 = 2000$ and $\omega_2 = 4000$ rad/s, the current may be represented in the time domain only. By applying superposition, we get $i = i_1 - i_2$, where i_1 and i_2 are currents due to v_1 and v_2, respectively.

$$\mathbf{I}_1 = \frac{\mathbf{V}_1}{Z_1} = \frac{5}{10 + j10} = 0.35\ \angle -45°\ \text{A}, \qquad i_1(t) = 0.35\cos(2000t - 45°)$$
$$\mathbf{I}_2 = \frac{\mathbf{V}_2}{Z_2} = \frac{10\ \angle 60°}{10 + j20} = 0.45\ \angle -3.4°\ \text{A}, \qquad i_2(t) = 0.45\cos(4000t - 3.4°)$$
$$i = i_1 - i_2 = 0.35\cos(2000t - 45°) - 0.45\cos(4000t - 3.4°)$$

Solved Problems

9.1 A 10-mH inductor has current $i = 5.0\cos 2000t$ (A). Obtain the voltage v_L.

From Table 9-1, $v_L = \omega LI\cos(\omega t + 90°) = 100\cos(2000t + 90°)$ (V). As a sine function,

$$v_L = 100\sin(2000t + 180°) = -100\sin 2000t \quad \text{(V)}$$

9.2 A series circuit, with $R = 10\ \Omega$ and $L = 20$ mH, has current $i = 2.0\sin 500t$ (A). Obtain total voltage v and the angle by which i lags v.

By the methods of Example 9.1,

$$\theta = \arctan \frac{500(20 \times 10^{-3})}{10} = 45^\circ$$

$$v = I\sqrt{R^2 + (\omega L)^2} \sin(\omega t + \theta) = 28.3 \sin(500t + 45^\circ) \quad \text{(V)}$$

It is seen that i lags v by 45°.

9.3 Find the two elements in a series circuit, given that the current and total voltage are

$$i = 10 \cos(5000t - 23.13^\circ) \quad \text{(A)} \qquad v = 50 \cos(5000t + 30^\circ) \quad \text{(V)}$$

Since i lags v (by 53.13°), the elements are R and L. The ratio of V_{max} to I_{max} is 50/10. Hence,

$$\frac{50}{10} = \sqrt{R^2 + (5000L)^2} \qquad \text{and} \qquad \tan 53.13^\circ = 1.33 = \frac{5000L}{R}$$

Solving, $R = 3.0\ \Omega$, $L = 0.8$ mH.

9.4 A series circuit, with $R = 2.0\ \Omega$ and $C = 200$ pF, has a sinusoidal applied voltage with a frequency of 99.47MHz. If the maximum voltage across the capacitance is 24 V, what is the maximum voltage across the series combination?

$$\omega = 2\pi f = 6.25 \times 10^8 \text{ rad/s}$$

From Table 9-1, $I_{max} = \omega C V_{C,max} = 3.0$ A. Then, by the methods of Example 9.2,

$$V_{max} = I_{max}\sqrt{R^2 + (1/\omega C)^2} = \sqrt{(6)^2 + (24)^2} = 24.74 \text{ V}$$

9.5 The current in a series circuit of $R = 5\ \Omega$ and $L = 30$ mH lags the applied voltage by 80°. Determine the source frequency and the impedance **Z**.

From the impedance diagram, Fig. 9-16,

$$5 + jX_L = Z\ \underline{/80^\circ} \qquad X_L = 5 \tan 80^\circ = 28.4\ \Omega$$

Then $28.4 = \omega(30 \times 10^{-3})$, whence $\omega = 945.2$ rad/s and $f = 150.4$ Hz.

$$\mathbf{Z} = 5 + j28.4\ \Omega$$

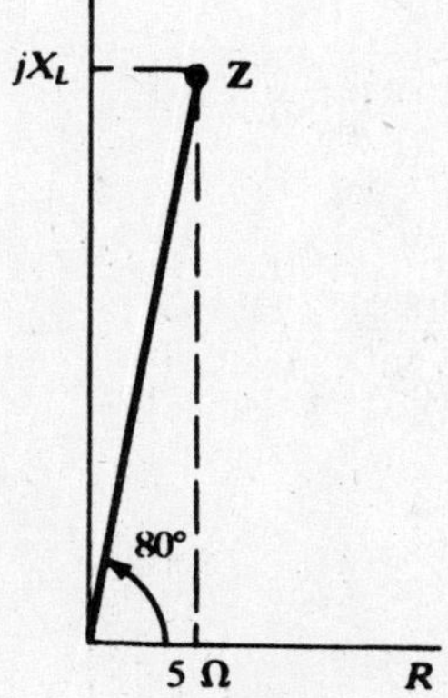

Fig. 9-16

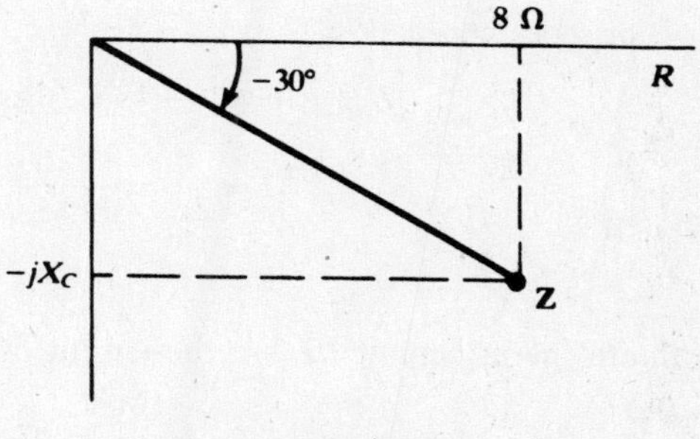

Fig. 9-17

9.6 At what frequency will the current lead the voltage by 30° in a series circuit with $R = 8\ \Omega$ and $C = 30\ \mu\text{F}$?

From the impedance diagram, Fig. 9-17,

$$8 - jX_C = Z\ \angle{-30^\circ} \qquad -X_C = 8\tan(-30^\circ) = -4.62\ \Omega$$

Then
$$4.62 = \frac{1}{2\pi f(30 \times 10^{-6})} \qquad \text{or} \qquad f = 1149\ \text{Hz}$$

9.7 A series *RC* circuit, with $R = 10\ \Omega$, has an impedance with an angle of -45° at $f_1 = 500\ \text{Hz}$. Find the frequency for which the magnitude of the impedance is (*a*) twice that at f_1, (*b*) one-half that at f_1.

From $10 - jX_C = Z_1\ \angle{45^\circ}$, $X_C = 10\ \Omega$ and $Z_1 = 14.14\ \Omega$.

(*a*) For twice the magnitude,

$$10 - jX_C = 28.28\ \angle{\theta^\circ} \qquad \text{or} \qquad X_C = \sqrt{(28.28)^2 - (10)^2} = 26.45\ \Omega$$

Then, since X_C is inversely proportional to f,

$$\frac{10}{26.45} = \frac{f_2}{500} \qquad \text{or} \qquad f_2 = 189\ \text{Hz}$$

(*b*) A magnitude $Z_3 = 7.07\ \Omega$ is impossible; the smallest magnitude possible is $Z = R = 10\ \Omega$.

9.8 A two-element series circuit has voltage $\mathbf{V} = 240\ \angle{0^\circ}\ \text{V}$ and current $\mathbf{I} = 50\ \angle{-60^\circ}\ \text{A}$. Determine the current which results when the resistance is reduced to (*a*) 30 percent, (*b*) 60 percent, of its former value.

$$\mathbf{Z} = \frac{\mathbf{V}}{\mathbf{I}} = \frac{240\ \angle{0^\circ}}{50\ \angle{-60^\circ}} = 4.8\ \angle{60^\circ} = 2.40 + j4.16 \quad \Omega$$

(*a*)
$$30\% \times 2.40 = 0.72 \qquad \mathbf{Z}_1 = 0.72 + j4.16 = 4.22\ \angle{80.2^\circ} \quad \Omega$$

$$\mathbf{I}_1 = \frac{240\ \angle{0^\circ}}{4.22\ \angle{80.2^\circ}} = 56.8\ \angle{-80.2^\circ} \quad \text{A}$$

(*b*)
$$60\% \times 2.40 = 1.44 \qquad \mathbf{Z}_2 = 1.44 + j4.16 = 4.40\ \angle{70.9^\circ} \quad \Omega$$

$$\mathbf{I}_2 = \frac{240\ \angle{0^\circ}}{4.40\ \angle{70.9^\circ}} = 54.5\ \angle{-70.9^\circ} \quad \text{A}$$

9.9 For the circuit shown in Fig. 9-18, obtain $\mathbf{Z}_{eq}$ and compute $\mathbf{I}$.

For series impedances,

$$\mathbf{Z}_{eq} = 10\ \angle{0^\circ} + 4.47\ \angle{63.4^\circ} = 12.0 + j4.0 = 12.65\ \angle{18.43^\circ} \quad \Omega$$

Then
$$\mathbf{I} = \frac{\mathbf{V}}{\mathbf{Z}_{eq}} = \frac{100\ \angle{0^\circ}}{12.65\ \angle{18.43^\circ}} = 7.91\ \angle{-18.43^\circ} \quad \text{A}$$

9.10 Evaluate the impedance $\mathbf{Z}_1$ in the circuit of Fig. 9-19.

$$\mathbf{Z} = \frac{\mathbf{V}}{\mathbf{I}} = 20\ \angle{60^\circ} = 10.0 + j17.3 \quad \Omega$$

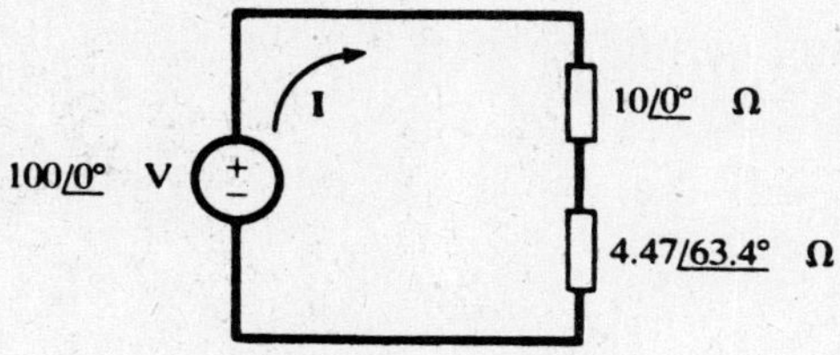

Fig. 9-18

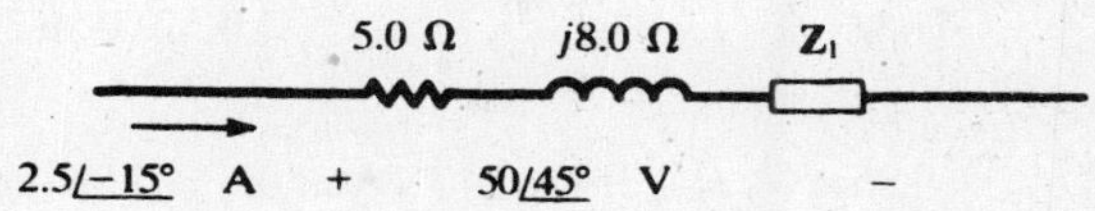

Fig. 9-19

Then, since impedances in series add,

$$5.0 + j8.0 + \mathbf{Z}_1 = 10.0 + j17.3 \qquad \text{or} \qquad \mathbf{Z}_1 = 5.0 + j9.3 \ \Omega$$

9.11 Compute the equivalent impedance $\mathbf{Z}_{eq}$ and admittance $\mathbf{Y}_{eq}$ for the four-branch circuit shown in Fig. 9-20.

Using admittances,

$$\mathbf{Y}_1 = \frac{1}{j5} = -j0.20 \text{ S} \qquad \mathbf{Y}_3 = \frac{1}{15} = 0.067 \text{ S}$$

$$\mathbf{Y}_2 = \frac{1}{5 + j8.66} = 0.05 - j0.087 \text{ S} \qquad \mathbf{Y}_4 = \frac{1}{-j10} = j0.10 \text{ S}$$

Then $$\mathbf{Y}_{eq} = \mathbf{Y}_1 + \mathbf{Y}_2 + \mathbf{Y}_3 + \mathbf{Y}_4 = 0.117 - j0.187 = 0.221 \ \underline{/-58.0^\circ} \text{ S}$$

and $$\mathbf{Z}_{eq} = \frac{1}{\mathbf{Y}_{eq}} = 4.53 \ \underline{/58.0^\circ} \ \Omega$$

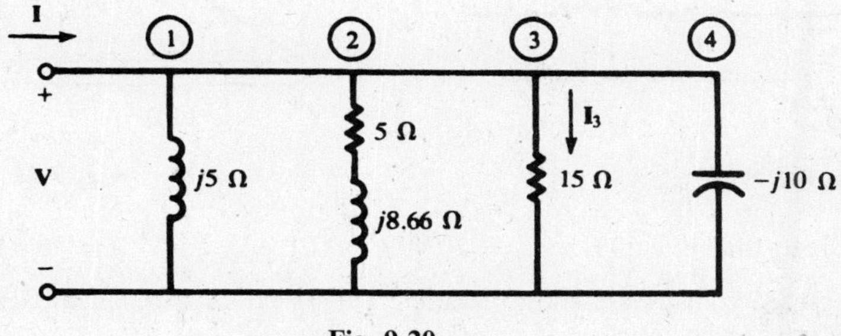

Fig. 9-20

9.12 The total current **I** entering the circuit shown in Fig. 9-20 is $33.0 \ \underline{/-13.0^\circ}$ A. Obtain the branch current $\mathbf{I}_3$ and the voltage **V**.

$$\mathbf{V} = \mathbf{I}\mathbf{Z}_{eq} = (33.0 \ \underline{/-13.0^\circ})(4.53 \ \underline{/58.0^\circ}) = 149.5 \ \underline{/45.0^\circ} \ \text{V}$$

$$\mathbf{I}_3 = \mathbf{V}\mathbf{Y}_3 = (149.5 \ \underline{/45.0^\circ})\left(\frac{1}{15} \ \underline{/0^\circ}\right) = 9.97 \ \underline{/45.0^\circ} \ \text{A}$$

9.13 Find $\mathbf{Z}_1$ in the three-branch network of Fig. 9-21, if $\mathbf{I} = 31.5 \ \underline{/24.0^\circ}$ A for an applied voltage $\mathbf{V} = 50.0 \ \underline{/60.0^\circ}$ V.

$$\mathbf{Y} = \frac{\mathbf{I}}{\mathbf{V}} = 0.630 \angle -36.0^\circ = 0.510 - j0.370 \quad \text{S}$$

Then
$$0.510 - j0.370 = \mathbf{Y}_1 + \frac{1}{10} + \frac{1}{4.0 + j3.0}$$

whence $\mathbf{Y}_1 = 0.354 \angle -45^\circ$ S and $\mathbf{Z}_1 = 2.0 + j2.0\ \Omega$.

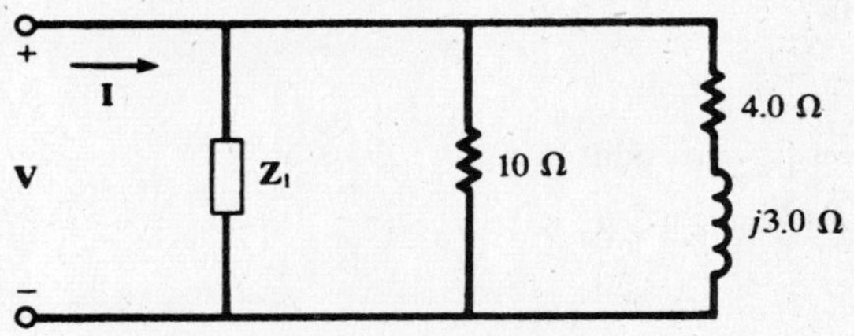

Fig. 9-21

9.14 The constants R and L of a coil can be obtained by connecting the coil in series with a known resistance and measuring the coil voltage V_x, the resistor voltage V_1, and the total voltage V_T (Fig. 9-22). The frequency must also be known, but the phase angles of the voltages are not known. Given that $f = 60$ Hz, $V_1 = 20$ V, $V_x = 22.4$ V, and $V_T = 36.0$ V, find R and L.

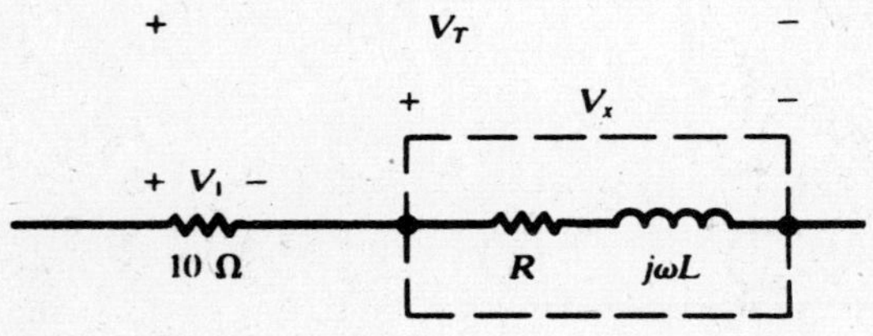

Fig. 9-22

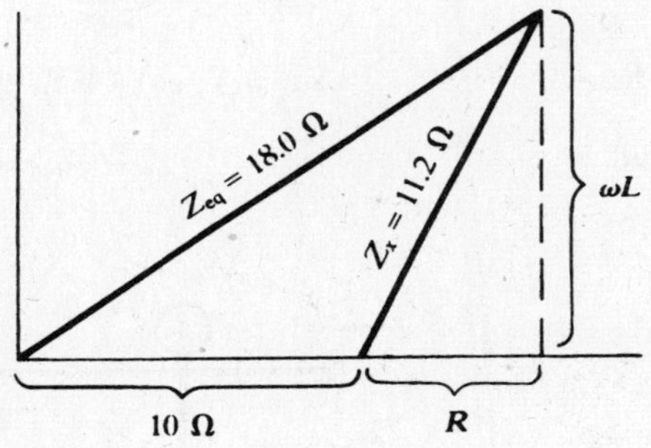

Fig. 9-23

The measured voltages are effective values; but, as far as impedance calculations are concerned, it makes no difference whether effective or peak values are used.

The (effective) current is $I = V_1/10 = 2.0$ A. Then

$$Z_x = \frac{22.4}{2.0} = 11.2\ \Omega \qquad Z_{eq} = \frac{36.0}{2.0} = 18.0\ \Omega$$

From the impedance diagram, Fig. 9-23,

$$(18.0)^2 = (10 + R)^2 + (\omega L)^2$$
$$(11.2)^2 = R^2 + (\omega L)^2$$

where $\omega = 2\pi 60 = 377$ rad/s. Solving simultaneously,

$$R = 4.92\ \Omega \qquad L = 26.7\ \text{mH}$$

9.15 In the parallel circuit shown in Fig. 9-24, the effective values of the currents are: $I_x = 18.0$ A, $I_1 = 15.0$ A, $I_T = 30.0$ A. Determine R and X_L.

The problem can be solved in a manner similar to that used in Problem 9.14 but with the admittance diagram.

The (effective) voltage is $V = I_1(4.0) = 60.0\,\text{V}$. Then

$$Y_x = \frac{I_x}{V} = 0.300\,\text{S} \qquad Y_{eq} = \frac{I_T}{V} = 0.500\,\text{S} \qquad Y_1 = \frac{1}{4.0} = 0.250\,\text{S}$$

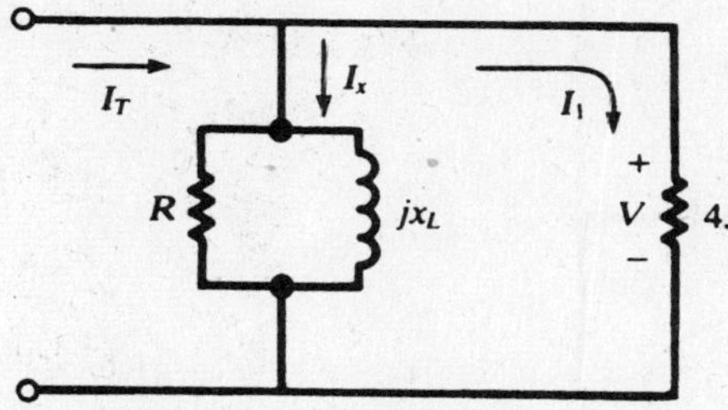

Fig. 9-24

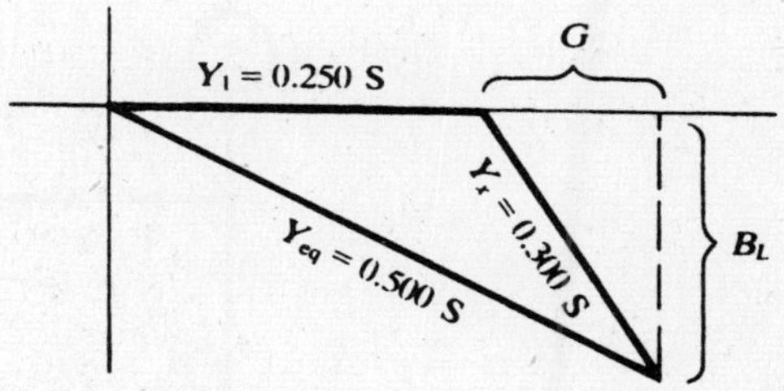

Fig. 9-25

From the admittance diagram, Fig. 9-25,

$$(0.500)^2 = (0.250 + G)^2 + B_L^2$$
$$(0.300)^2 = G^2 + B_L^2$$

which yield $G = 0.195\,\text{S}$, $B_L = 0.228\,\text{S}$. Then

$$R = \frac{1}{G} = 5.13\,\Omega \qquad \text{and} \qquad jX_L = \frac{1}{-jB_L} = j4.39\,\Omega$$

i.e., $X_L = 4.39\,\Omega$.

9.16 Obtain the phasor voltage $\mathbf{V}_{AB}$ in the two-branch parallel circuit of Fig. 9-26.

By current-division methods, $\mathbf{I}_1 = 4.64\ \underline{/120.1^\circ}$ A and $\mathbf{I}_2 = 17.4\ \underline{/30.1^\circ}$ A. Either path AXB or path AYB may be considered. Choosing the former,

$$\mathbf{V}_{AB} = \mathbf{V}_{AX} + \mathbf{V}_{XB} = \mathbf{I}_1(20) - \mathbf{I}_2(j6) = 92.8\ \underline{/120.1^\circ} + 104.4\ \underline{/-59.9^\circ} = 11.6\ \underline{/-59.9^\circ}\ \text{V}$$

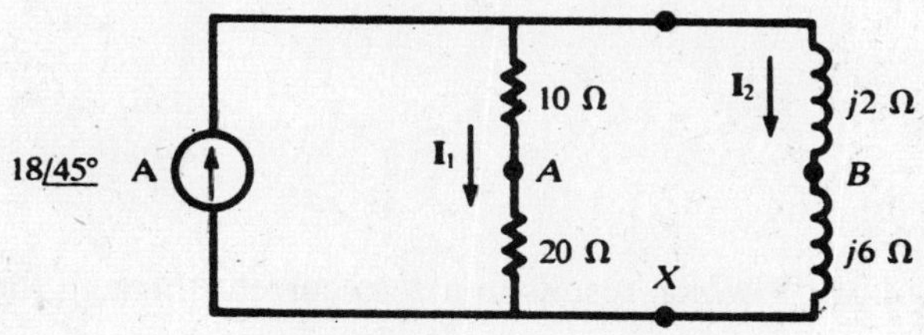

Fig. 9-26

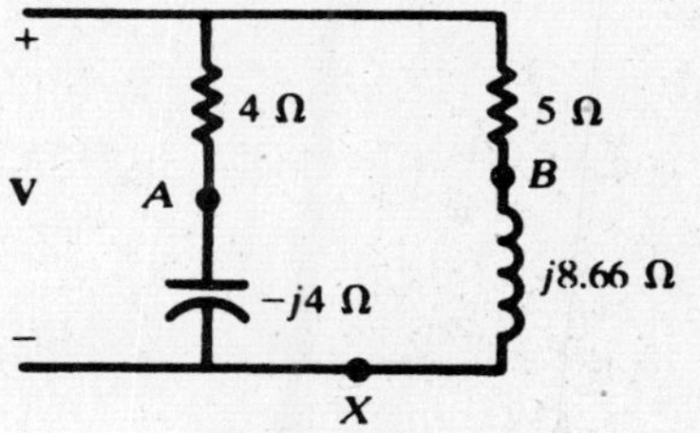

Fig. 9-27

9.17 In the parallel circuit shown in Fig. 9-27, $\mathbf{V}_{AB} = 48.3\ \underline{/30^\circ}$ V. Find the applied voltage $\mathbf{V}$.

By voltage division in the two branches:

$$\mathbf{V}_{AX} = \frac{-j4}{4 - j4}\mathbf{V} = \frac{1}{1+j}\mathbf{V} \qquad \mathbf{V}_{BX} = \frac{j8.66}{5 + j8.66}\mathbf{V}$$

and so
$$\mathbf{V}_{AB} = \mathbf{V}_{AX} - \mathbf{V}_{BX} = \left(\frac{1}{1+j} - \frac{j8.66}{5+j8.66}\right)\mathbf{V} = \frac{1}{-0.268 + j1}\mathbf{V}$$

or
$$\mathbf{V} = (-0.268 + j1)\mathbf{V}_{AB} = (1.035\ \underline{/105^\circ})(48.3\ \underline{/30^\circ}) = 50.0\ \underline{/135^\circ}\ \text{V}$$

9.18 Obtain the voltage $\mathbf{V}_x$ in the network of Fig. 9-28, using the mesh current method.

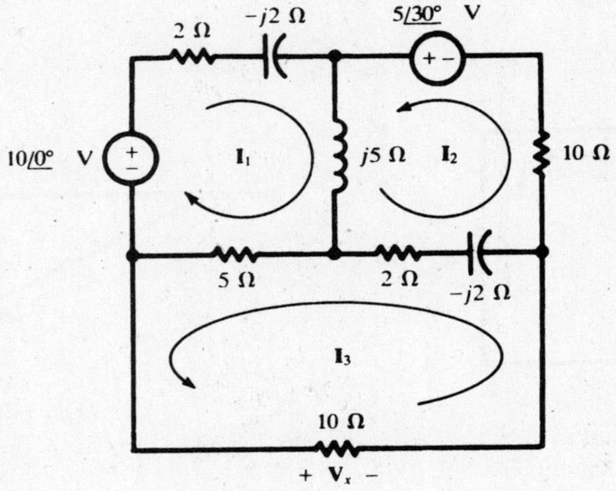

Fig. 9-28

One choice of mesh currents is shown on the circuit diagram, with $\mathbf{I}_3$ passing through the 10-Ω resistor in a direction such that $\mathbf{V}_x = \mathbf{I}_3(10)$ (V). The matrix equation can be written by inspection:

$$\begin{bmatrix} 7+j3 & j5 & 5 \\ j5 & 12+j3 & -(2-j2) \\ 5 & -(2-j2) & 17-j2 \end{bmatrix} \begin{bmatrix} \mathbf{I}_1 \\ \mathbf{I}_2 \\ \mathbf{I}_3 \end{bmatrix} = \begin{bmatrix} 10\,\underline{/0^\circ} \\ 5\,\underline{/30^\circ} \\ 0 \end{bmatrix}$$

Solving by determinants,

$$\mathbf{I}_3 = \frac{\begin{vmatrix} 7+j3 & j5 & 10\,\underline{/0^\circ} \\ j5 & 12+j3 & 5\,\underline{/30^\circ} \\ 5 & -2+j2 & 0 \end{vmatrix}}{\begin{vmatrix} 7+j3 & j5 & 5 \\ j5 & 12+j3 & -2+j2 \\ 5 & -2+j2 & 17-j2 \end{vmatrix}} = \frac{667.96\,\underline{/-169.09^\circ}}{1534.5\,\underline{/25.06^\circ}} = 0.435\,\underline{/-194.15^\circ}\ \text{A}$$

and $\mathbf{V}_x = \mathbf{I}_3(10) = 4.35\,\underline{/-194.15^\circ}$ V.

9.19 In the netwrok of Fig. 9-29, determine the voltage **V** which results in a zero current through the $2 + j3\ \Omega$ impedance.

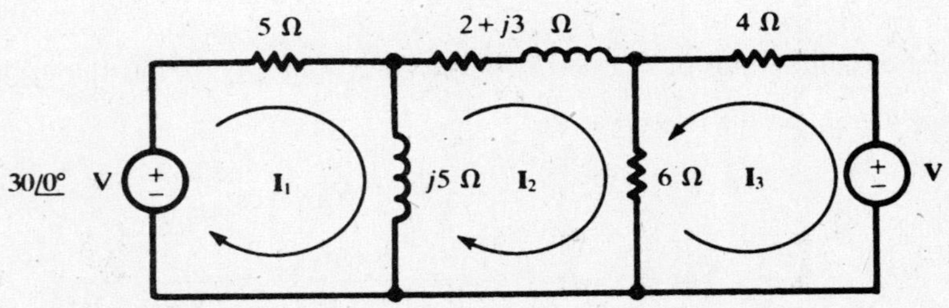

Fig. 9-29

Choosing mesh currents as shown on the circuit diagram,

$$\mathbf{I}_2 = \frac{1}{\Delta z}\begin{vmatrix} 5+j5 & 30\,\underline{/0^\circ} & 0 \\ -j5 & 0 & 6 \\ 0 & \mathbf{V} & 10 \end{vmatrix} = 0$$

Expanding the numerator determinant by cofactors of the second column,

$$-(30\,\underline{/0^\circ})\begin{vmatrix} -j5 & 6 \\ 0 & 10 \end{vmatrix} - \mathbf{V}\begin{vmatrix} 5+j5 & 0 \\ -j5 & 6 \end{vmatrix} = 0 \qquad \text{whence} \qquad \mathbf{V} = 35.4\,\underline{/45.0^\circ}\ \text{V}$$

9.20 Solve Problem 9.19 by the node voltage method.

The network is redrawn in Fig. 9-30 with one end of the $2 + j3$ impedance as the reference node. By the rule of Section 9.7 the matrix equation is

$$\begin{bmatrix} \frac{1}{5}+\frac{1}{j5}+\frac{1}{2+j3} & -\left(\frac{1}{5}+\frac{1}{j5}\right) \\ -\left(\frac{1}{5}+\frac{1}{j5}\right) & \frac{1}{5}+\frac{1}{j5}+\frac{1}{4}+\frac{1}{6} \end{bmatrix}\begin{bmatrix} \mathbf{V}_1 \\ \mathbf{V}_2 \end{bmatrix} = \begin{bmatrix} \frac{30\,\underline{/0^\circ}}{5} \\ \frac{-30\,\underline{/0^\circ}}{5} - \frac{\mathbf{V}}{4} \end{bmatrix}$$

For node voltage $\mathbf{V}_1$ to be zero, it is necessary that the numerator determinant in the solution for $\mathbf{V}_1$ vanish.

$$N_1 = \begin{vmatrix} \frac{30\,\underline{/0^\circ}}{5} & -0.200 + j0.200 \\ \frac{-30\,\underline{/0^\circ}}{5} - \frac{\mathbf{V}}{4} & 0.617 - j0.200 \end{vmatrix} = 0 \qquad \text{from which} \qquad \mathbf{V} = 35.4\,\underline{/45^\circ}\ \text{V}$$

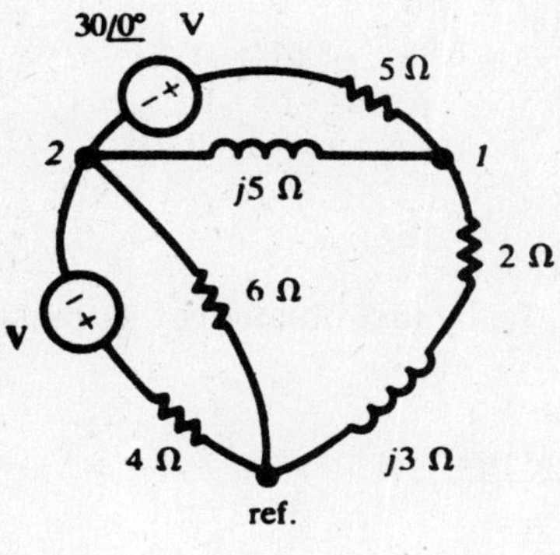

Fig. 9-30

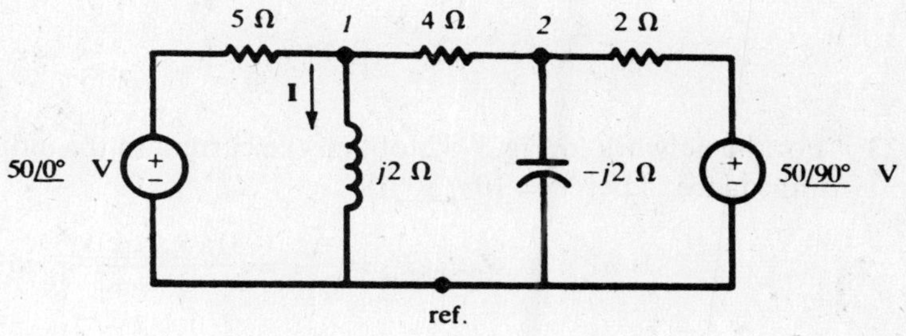

Fig. 9-31

9.21 Use the node voltage method to obtain the current **I** in the network of Fig. 9-31.

There are three principal nodes in the network. The reference and node *1* are selected so that the node *1* voltage is the voltage across the $j2$-Ω reactance.

$$\begin{bmatrix} \frac{1}{5}+\frac{1}{j2}+\frac{1}{4} & -\frac{1}{4} \\ -\frac{1}{4} & \frac{1}{4}+\frac{1}{-j2}+\frac{1}{2} \end{bmatrix}\begin{bmatrix} \mathbf{V}_1 \\ \mathbf{V}_2 \end{bmatrix} = \begin{bmatrix} \frac{50\,\underline{/0^\circ}}{5} \\ \frac{50\,\underline{/90^\circ}}{2} \end{bmatrix}$$

from which

$$\mathbf{V}_1 = \frac{\begin{vmatrix} 10 & -0.250 \\ j25 & 0.750 + j0.500 \end{vmatrix}}{\begin{vmatrix} 0.450 - j0.500 & -0.250 \\ -0.250 & 0.750 + j0.500 \end{vmatrix}} = \frac{13.52 \,\angle 56.31^\circ}{0.546 \,\angle -15.94^\circ} = 24.76 \,\angle 72.25^\circ \text{ V}$$

and
$$\mathbf{I} = \frac{24.76 \,\angle 72.25^\circ}{2 \,\angle 90^\circ} = 12.38 \,\angle -17.75^\circ \text{ A}$$

9.22 Find the input impedance at terminals *ab* for the network of Fig. 9-32.

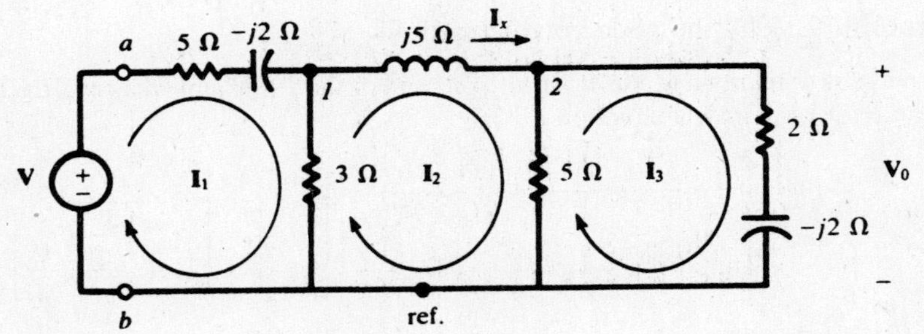

Fig. 9-32

With mesh current $\mathbf{I}_1$ selected as shown on the diagram.

$$\mathbf{Z}_{\text{input},1} = \frac{\Delta_\mathbf{z}}{\Delta_{11}} = \frac{\begin{vmatrix} 8 - j2 & -3 & 0 \\ -3 & 8 + j5 & -5 \\ 0 & -5 & 7 - j2 \end{vmatrix}}{\begin{vmatrix} 8 + j5 & -5 \\ -5 & 7 - j2 \end{vmatrix}} = \frac{315.5 \,\angle 16.19^\circ}{45.2 \,\angle 24.86^\circ} = 6.98 \,\angle -8.67^\circ \;\Omega$$

9.23 For the network in Fig. 9-32, obtain the current in the inductor, $\mathbf{I}_x$, by first obtaining the transfer impedance. Let $\mathbf{V} = 10 \,\angle 30^\circ$ V.

$$\mathbf{Z}_{\text{transfer},12} = \frac{\Delta_\mathbf{z}}{\Delta_{12}} = \frac{315.5 \,\angle 16.19^\circ}{-\begin{vmatrix} -3 & -5 \\ 0 & 7 - j2 \end{vmatrix}} = 14.45 \,\angle 32.14^\circ \;\Omega$$

Then
$$\mathbf{I}_x = \mathbf{I}_2 = \frac{\mathbf{V}}{\mathbf{Z}_{\text{transfer},12}} = \frac{10 \,\angle 30^\circ}{14.45 \,\angle 32.14^\circ} = 0.692 \,\angle -2.14^\circ \text{ A}$$

9.24 For the network in Fig. 9-32, find the value of the source voltage $\mathbf{V}$ which results in $\mathbf{V}_0 = 5.0 \,\angle 0^\circ$ V.

The transfer impedance can be used to compute the current in the $2 - j2\ \Omega$ impedance, from which $\mathbf{V}_0$ is readily obtained.

$$\mathbf{Z}_{\text{transfer},13} = \frac{\Delta_\mathbf{z}}{\Delta_{13}} = \frac{315.5 \,\angle 16.19^\circ}{15 \,\angle 0^\circ} = 21.0 \,\angle 16.19^\circ \;\Omega$$

$$\mathbf{V}_0 = \mathbf{I}_3(2 - j2) = \frac{\mathbf{V}}{\mathbf{Z}_{\text{transfer},13}}(2 - j2) = \mathbf{V}(0.135 \,\angle -61.19^\circ)$$

Thus, if $\mathbf{V}_0 = 5.0\ \angle 0^\circ$ V,

$$\mathbf{V} = \frac{5.0\ \angle 0^\circ}{0.135\ \angle -61.19^\circ} = 37.0\ \angle 61.19^\circ\ \text{V}$$

Alternate Method

The node voltage method may be used. $\mathbf{V}_0$ is the node voltage $\mathbf{V}_2$ for the selection of nodes indicated in Fig. 9-32.

$$\mathbf{V}_0 = \mathbf{V}_2 = \frac{\begin{vmatrix} \dfrac{1}{5-j2}+\dfrac{1}{3}+\dfrac{1}{j5} & \dfrac{\mathbf{V}}{5-j2} \\ -\dfrac{1}{j5} & 0 \end{vmatrix}}{\begin{vmatrix} \dfrac{1}{5-j2}+\dfrac{1}{3}+\dfrac{1}{j5} & -\dfrac{1}{j5} \\ -\dfrac{1}{j5} & \dfrac{1}{j5}+\dfrac{1}{5}+\dfrac{1}{2-j2} \end{vmatrix}} = \mathbf{V}(0.134\ \angle -61.15^\circ)$$

For $\mathbf{V}_0 = 5.0\ \angle 0^\circ$ V, $\mathbf{V} = 37.3\ \angle 61.15^\circ$ V, which agrees with the previous answer to within roundoff errors.

9.25 For the network shown in Fig. 9-33, obtain the input admittance and use it to compute node voltage $\mathbf{V}_1$.

$$\mathbf{Y}_{\text{input},1} = \frac{\Delta_\mathbf{Y}}{\Delta_{11}} = \frac{\begin{vmatrix} \dfrac{1}{10}+\dfrac{1}{j5}+\dfrac{1}{2} & -\dfrac{1}{2} \\ -\dfrac{1}{2} & \dfrac{1}{2}+\dfrac{1}{3+j4}+\dfrac{1}{-j10} \end{vmatrix}}{\dfrac{1}{2}+\dfrac{1}{3+j4}+\dfrac{1}{-j10}} = 0.311\ \angle -49.97^\circ\ \text{S}$$

$$\mathbf{V}_1 = \frac{\mathbf{I}_1}{\mathbf{Y}_{\text{input},1}} = \frac{5.0\ \angle 0^\circ}{0.311\ \angle -49.97^\circ} = 16.1\ \angle 49.97^\circ\ \text{V}$$

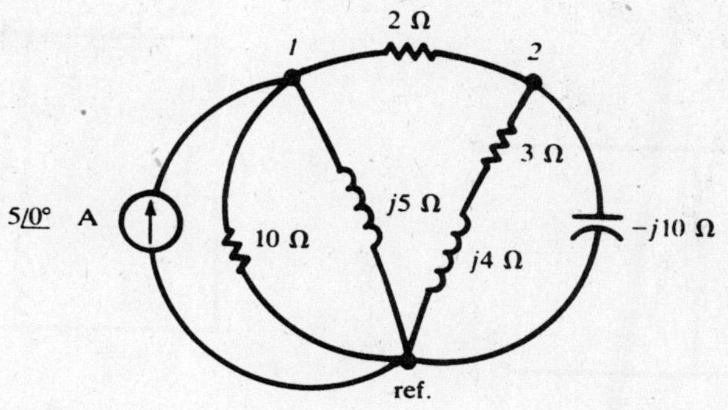

Fig. 9-33

9.26 For the network of Problem 9.25, compute the transfer admittance $\mathbf{Y}_{\text{transfer},12}$ and use it to obtain node voltage $\mathbf{V}_2$.

$$\mathbf{Y}_{\text{transfer},12} = \frac{\Delta_\mathbf{Y}}{\Delta_{12}} = \frac{0.194\ \angle -55.49^\circ}{-(-0.50)} = 0.388\ \angle -55.49^\circ\ \text{S}$$

$$\mathbf{V}_2 = \frac{\mathbf{I}_1}{\mathbf{Y}_{\text{transfer},12}} = 12.9\ \angle 55.49^\circ\ \text{V}$$

9.27 Replace the active network in Fig. 9-34(*a*) at terminals *ab* with a Thévenin equivalent.

$$\mathbf{Z}' = j5 + \frac{5(3+j4)}{5+3+j4} = 2.50 + j6.25 \quad \Omega$$

The open-circuit voltage $\mathbf{V}'$ at terminals *ab* is the voltage across the $3 + j4\ \Omega$ impedance:

$$\mathbf{V}' = \left(\frac{10\ \angle 0^\circ}{8+j4}\right)(3+j4) = 5.59\ \angle 26.56^\circ \quad \text{V}$$

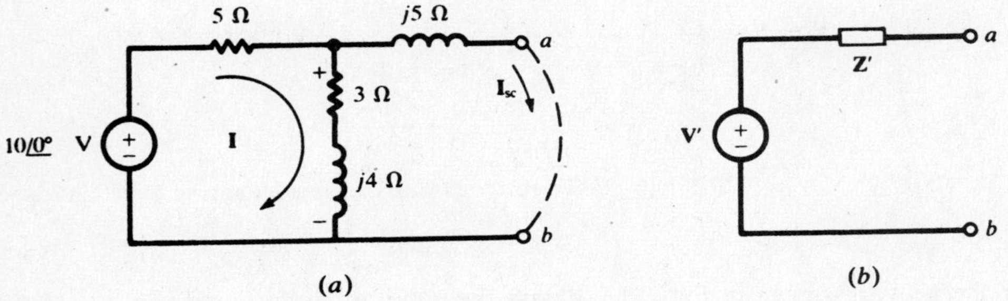

Fig. 9-34

9.28 For the network of Problem 9.27, obtain a Norton equivalent circuit (Fig. 9-35).

At terminals *ab*, $\mathbf{I}_{sc}$ is the Norton current $\mathbf{I}'$. By current division,

$$\mathbf{I}' = \frac{10\ \angle 0^\circ}{5 + \dfrac{j5(3+j4)}{3+j9}}\left(\frac{3+j4}{3+j9}\right) = 0.830\ \angle -41.63^\circ \quad \text{A}$$

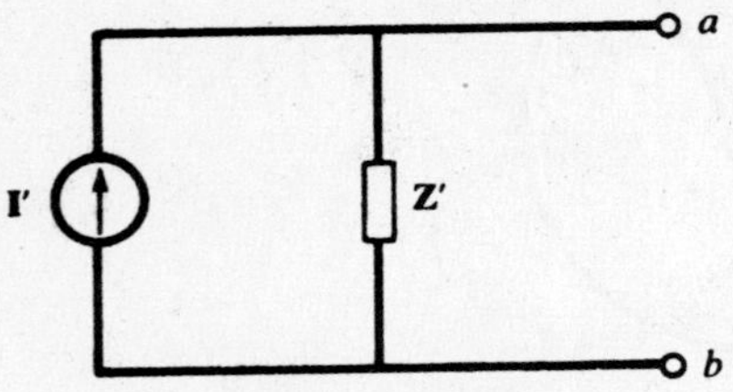

Fig. 9-35

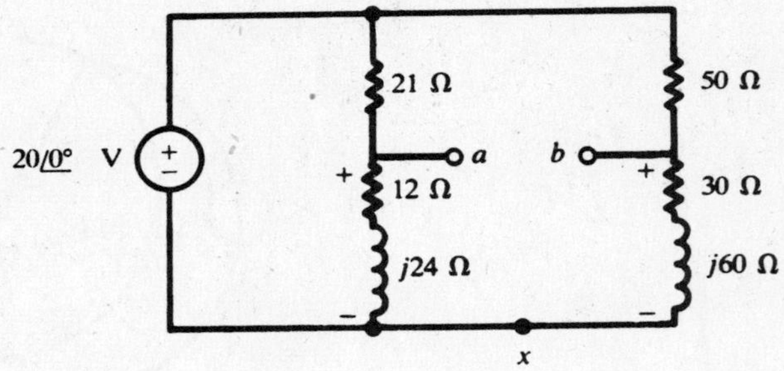

Fig. 9-36

The shunt impedance $\mathbf{Z}'$ is as found in Problem 9.27, $\mathbf{Z}' = 2.50 + j6.25\ \Omega$.

9.29 Obtain the Thévenin equivalent for the bridge circuit of Fig. 9-36. Make $\mathbf{V}'$ the voltage of *a* with respect to *b*.

By voltage division in either branch,

$$\mathbf{V}_{ax} = \frac{12+j24}{33+j24}(20\ \angle 0^\circ) \qquad \mathbf{V}_{bx} = \frac{30+j60}{80+j60}(20\ \angle 0^\circ)$$

Hence, $$\mathbf{V}_{ab} = \mathbf{V}_{ax} - \mathbf{V}_{bx} = (20\ \underline{/0^\circ})\left(\frac{12+j24}{33+j24} - \frac{30+j60}{80+j60}\right) = 0.326\ \underline{/169.4^\circ}\quad V = \mathbf{V}'$$

Viewed from *ab* with the voltage source shorted out, the circuit is two parallel combinations in series, and so

$$\mathbf{Z}' = \frac{21(12+j24)}{33+j24} + \frac{50(30+j60)}{80 - +j60)} = 47.35\ \underline{/26.81^\circ}\quad \Omega$$

9.30 Replace the network of Fig. 9-37 at terminals *ab* with a Norton equivalent and with a Thévenin equivalent.

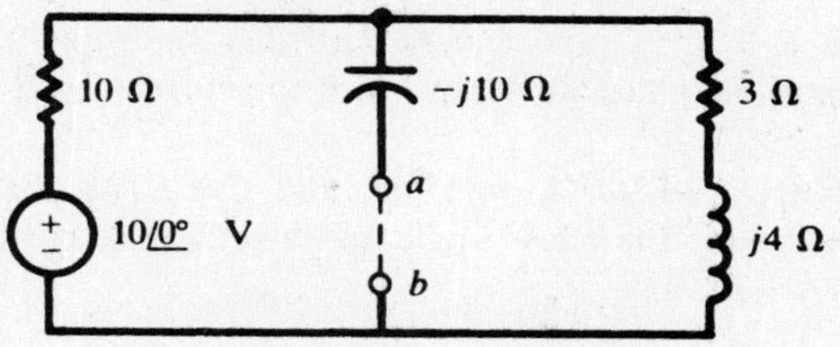

Fig. 9-37

By current division,

$$\mathbf{I}_{sc} = \mathbf{I}' = \left[\frac{10\ \underline{/0^\circ}}{10 + \dfrac{(-j10)(3+j4)}{3-j6}}\right]\left(\frac{3+j4}{3-j6}\right) = 0.439\ \underline{/105.26^\circ}\quad \mathrm{A}$$

and by voltage division in the open circuit,

$$\mathbf{V}_{ab} = \mathbf{V}' = \frac{3+j4}{13+j4}(10\ \underline{/0^\circ}) = 3.68\ \underline{/36.03^\circ}\quad \mathrm{V}$$

Then $$\mathbf{Z}' = \frac{\mathbf{V}'}{\mathbf{I}'} = \frac{3.68\ \underline{/36.03^\circ}}{0.439\ \underline{/105.26^\circ}} = 8.37\ \underline{/-69.23^\circ}\quad \Omega$$

See Fig. 9-38.

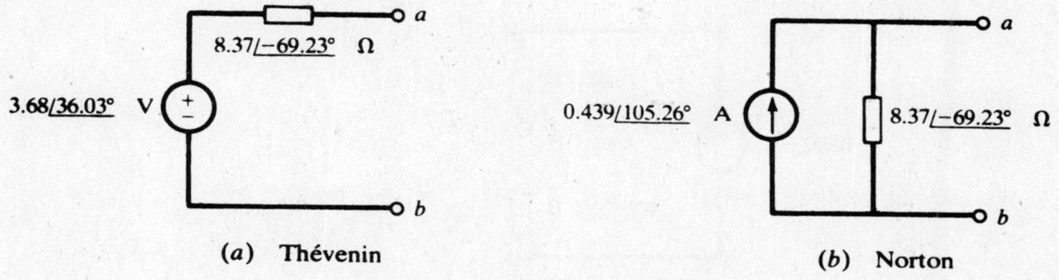

Fig. 9-38

Supplementary Problems

9.31 Two circuit elements in a series connection have current and total voltage

$$i = 13.42 \sin(500t - 53.4^\circ) \quad \text{(A)} \qquad v = 150 \sin(500t + 10^\circ) \quad \text{(V)}$$

Identify the two elements. *Ans.* $R = 5\,\Omega$, $L = 20\,\text{mH}$

9.32 Two circuit elements in a series connection have current and total voltage

$$i = 4.0 \cos(2000t + 13.2^\circ) \quad \text{(A)} \qquad v = 200 \sin(2000t + 50.0^\circ) \quad \text{(V)}$$

Identify the two elements. *Ans.* $R = 30\,\Omega$, $C = 12.5\,\mu\text{F}$

9.33 A series *RC* circuit, with $R = 27.5\,\Omega$ and $C = 66.7\,\mu\text{F}$, has sinusoidal voltages and current, with angular frequency 1500 rad/s. Find the phase angle by which the current leads the voltage. *Ans.* 20°

9.34 A series *RLC* circuit, with $R = 15\,\Omega$, $L = 80\,\text{mH}$, and $C = 30\,\mu\text{F}$, has a sinusoidal current at angular frequency 500 rad/s. Determine the phase angle and whether the current leads or lags the total voltage. *Ans.* 60.6°, leads

9.35 A capacitance $C = 35\,\mu\text{F}$ is in parallel with a certain element. Identify the element, given that the voltage and total current are

$$v = 150 \sin 3000t \quad \text{(V)} \qquad i_T = 16.5 \sin(3000t + 72.4^\circ) \quad \text{(A)}$$

Ans. $R = 30.1\,\Omega$

9.36 A two-element series circuit, with $R = 20\,\Omega$ and $L = 20\,\text{mH}$, has an impedance $40.0\,\underline{/\theta}\,\Omega$. Determine the angle θ and the frequency. *Ans.* 60°, 276 Hz

9.37 Determine the impedance of the series *RL* circuit, with $R = 25\,\Omega$ and $L = 10\,\text{mH}$, at (*a*) 100 Hz, (*b*) 500 Hz, (*c*) 1000 Hz. *Ans.* (*a*) $25.8\,\underline{/14.1^\circ}\,\Omega$; (*b*) $40.1\,\underline{/51.5^\circ}\,\Omega$; (*c*) $67.6\,\underline{/68.3^\circ}\,\Omega$

9.38 Determine the circuit constants of a two-element series circuit if the applied voltage

$$v = 150 \sin(5000t + 45^\circ) \quad \text{(V)}$$

results in a current $i = 3.0 \sin(5000t - 15^\circ)$ (A). *Ans.* $25\,\Omega$, 8.66 mH

9.39 A series circuit of $R = 10\,\Omega$ and $C = 40\,\mu\text{F}$ has an applied voltage $v = 500 \cos(2500t - 20^\circ)$ (V). Find the resulting current i. *Ans.* $25\sqrt{2} \cos(2500t + 25^\circ)$ (A)

9.40 Three impedances are in series: $\mathbf{Z}_1 = 3.0\,\underline{/45^\circ}\,\Omega$, $\mathbf{Z}_2 = 10\sqrt{2}\,\underline{/45^\circ}\,\Omega$, $\mathbf{Z}_3 = 5.0\,\underline{/-90^\circ}\,\Omega$. Find the applied voltage $\mathbf{V}$, if the voltage across $\mathbf{Z}_1$ is $27.0\,\underline{/-10^\circ}$ V. *Ans.* $126.5\,\underline{/-24.6^\circ}$ V

9.41 For the three-element series circuit in Fig. 9-39, (*a*) find the current $\mathbf{I}$; (*b*) find the voltage across each impedance and construct the voltage phasor diagram which shows that $\mathbf{V}_1 + \mathbf{V}_2 + \mathbf{V}_3 = 100\,\underline{/0^\circ}$ V. *Ans.* (*a*) $6.28\,\underline{/-9.17^\circ}$ A; (*b*) see Fig. 9-40.

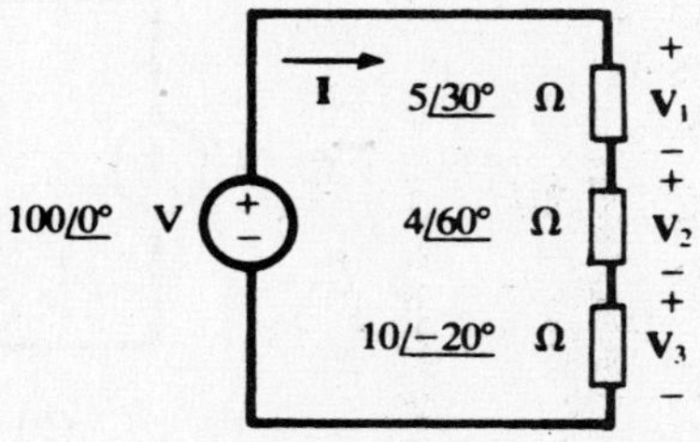

Fig. 9-39

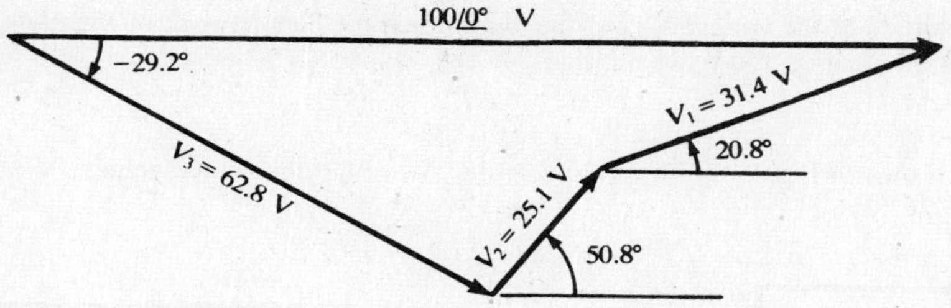

Fig. 9-40

9.42 Find **Z** in the parallel circuit of Fig. 9-41, if $\mathbf{V} = 50.0\ \underline{/30.0^\circ}$ V and $\mathbf{I} = 27.9\ \underline{/57.8^\circ}$ A.
Ans. $5.0\ \underline{/-30^\circ}\ \Omega$

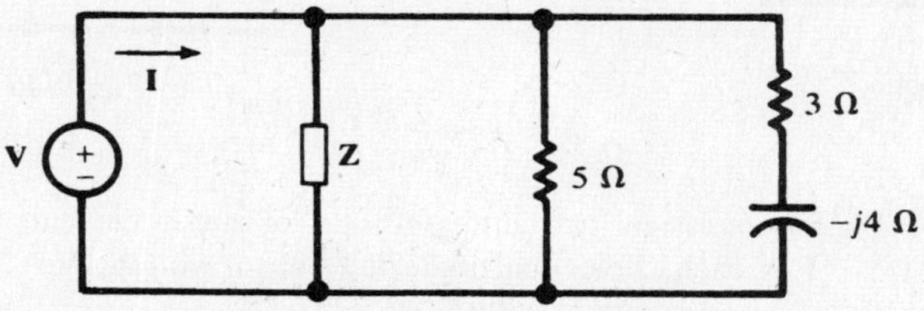

Fig. 9-41

9.43 Obtain the conductance and susceptance corresponding to a voltage $\mathbf{V} = 85.0\ \underline{/205^\circ}$ V and a resulting current $\mathbf{I} = 41.2\ \underline{/-141.0^\circ}$ A. *Ans.* 0.471 S, 0.117 S (capacitive)

9.44 A practical coil contains resistance as well as inductance and can be represented by either a series or parallel circuit, as suggested in Fig. 9-42. Obtain R_p and L_p in terms of R_s and L_s.

Ans. $R_p = R_s + \dfrac{(\omega L_s)^2}{R_s},\ L_p = L_s + \dfrac{R_s^2}{\omega^2 L_s}$

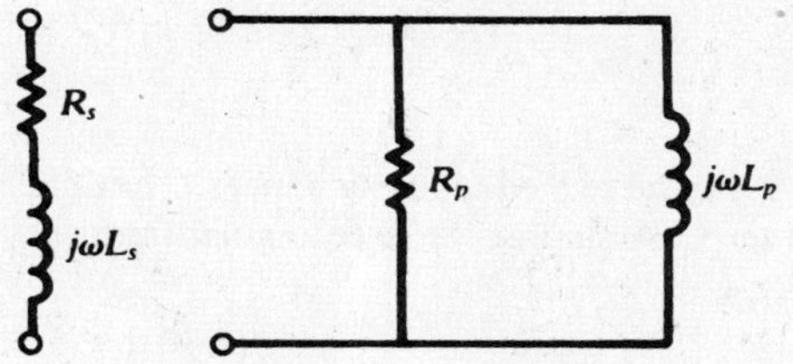

Fig. 9-42

9.45 In the network shown in Fig. 9-43 the 60-Hz current magnitudes are known to be: $I_T = 29.9$ A, $I_1 = 22.3$ A, and $I_2 = 8.0$ A. Obtain the circuit constants R and L. *Ans.* 5.8 Ω, 38.5 mH

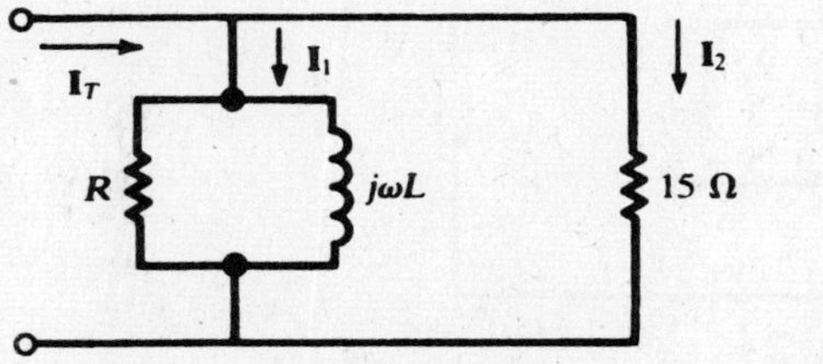

Fig. 9-43

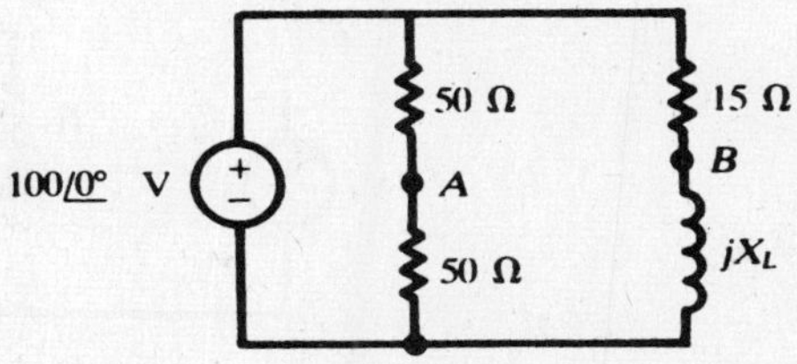

Fig. 9-44

9.46 Obtain the magnitude of the voltage $\mathbf{V}_{AB}$ in the two-branch parallel network of Fig. 9-44, if X_L is (*a*) $5\,\Omega$, (*b*) $15\,\Omega$, (*c*) $0\,\Omega$. *Ans.* 50 V, whatever X_L

9.47 In the network shown in Fig. 9-45, $\mathbf{V}_{AB} = 36.1\,\underline{/3.18^\circ}$ V. Find the source voltage $\mathbf{V}$.
Ans. $75\,\underline{/-90^\circ}$ V

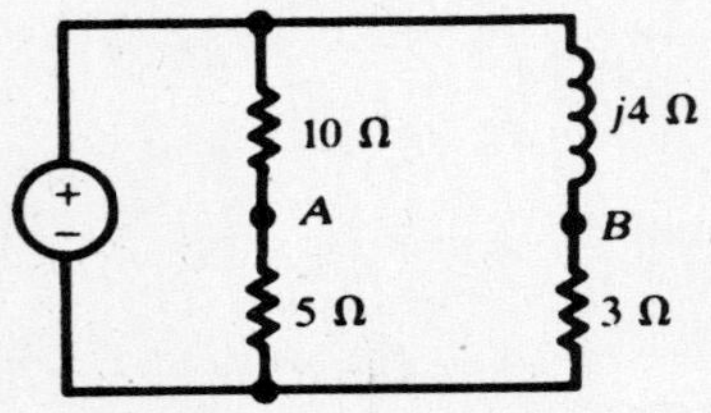

Fig. 9-45

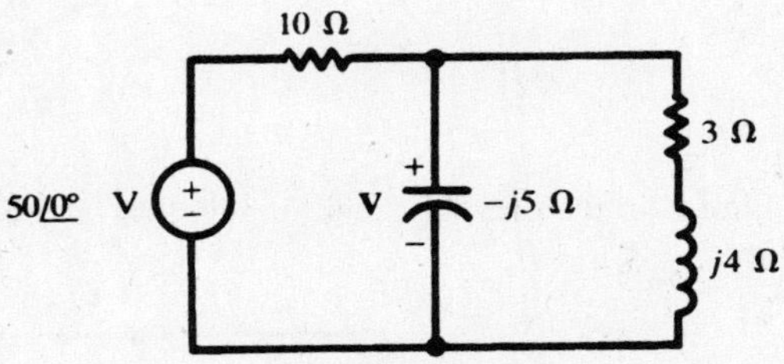

Fig. 9-46

9.48 For the network of Fig. 9-46 assign two different sets of mesh currents and show that for each, $\Delta_z = 55.9\,\underline{/-26.57^\circ}\,\Omega^2$. For each choice, calculate the phasor voltage $\mathbf{V}$. Obtain the phasor voltage across the $3 + j4\,\Omega$ impedance and compare with $\mathbf{V}$. *Ans.* $\mathbf{V} = \mathbf{V}_{3+j4} = 22.36\,\underline{/-10.30^\circ}$ V

9.49 For the network of Fig. 9-47, use the mesh current method to find the current in the $2 + j3\,\Omega$ impedance due to each of the sources $\mathbf{V}_1$ and $\mathbf{V}_2$. *Ans.* $2.41\,\underline{/6.45^\circ}$ A, $1.36\,\underline{/141.45^\circ}$ A

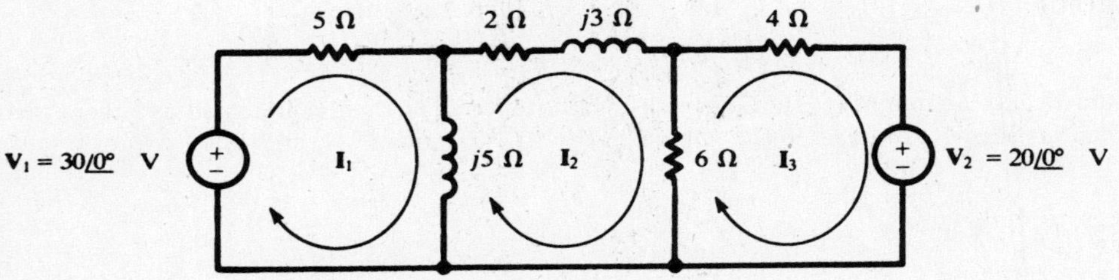

Fig. 9-47

9.50 In the network shown in Fig. 9-48, the two equal capacitances C and the shunting resistance R are adjusted until the detector current $\mathbf{I}_D$ is zero. Assuming a source angular frequency ω, determine the values of R_x and L_x. *Ans.* $R_x = 1/(\omega^2 C^2 R)$, $L_x = 1/(2\omega C)$

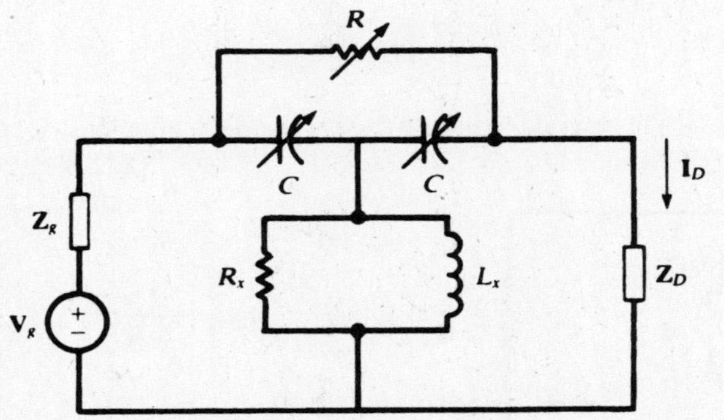

Fig. 9-48

9.51 For the network of Fig. 9-49, obtain the current ratio $\mathbf{I}_1/\mathbf{I}_3$. *Ans.* $3.3\ \underline{/-90^\circ}$

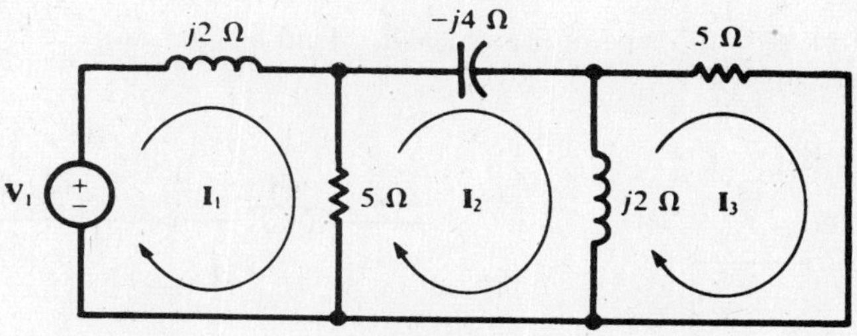

Fig. 9-49

9.52 For the network of Fig. 9-49, obtain $\mathbf{Z}_{\text{input},1}$ and $\mathbf{Z}_{\text{transfer},13}$. Show that $\mathbf{Z}_{\text{transfer},31} = \mathbf{Z}_{\text{transfer},13}$.
Ans. $1.31\ \underline{/21.8^\circ}\ \Omega,\ 4.31\ \underline{/-68.2^\circ}\ \Omega$

9.53 In the network of Fig. 9-50, obtain the voltage ratio $\mathbf{V}_1/\mathbf{V}_2$ by application of the node voltage method.
Ans. $\dfrac{\Delta_{11}}{\Delta_{12}} = 1.61\ \underline{/-29.8^\circ}$

9.54 For the network of Fig. 9-50, obtain the driving-point impedance $\mathbf{Z}_{\text{input},1}$. *Ans.* $5.59\ \underline{/17.35^\circ}\ \Omega$

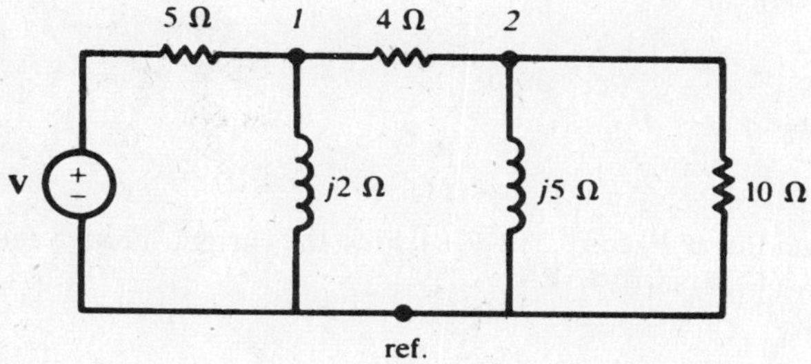

Fig. 9-50

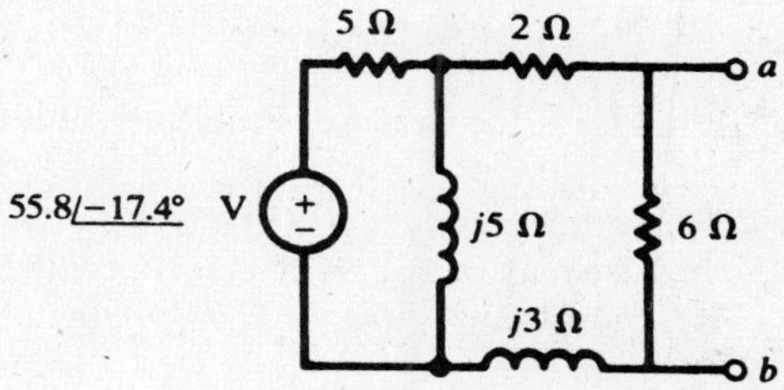

Fig. 9-51

9.55 Obtain the Thévenin and Norton equivalent circuits at terminals *ab* for the network of Fig. 9-51. Choose the polarity such that $\mathbf{V}' = \mathbf{V}_{ab}$. *Ans.* $\mathbf{V}' = 20.0\ \underline{/0^\circ}\ \text{V},\ \mathbf{I}' = 5.56\ \underline{/-23.06^\circ}\ \text{A},\ \mathbf{Z}' = 3.60\ \underline{/23.06^\circ}\ \Omega$

9.56 Obtain the Thévenin and Norton equivalent circuits at terminals *ab* for the network of Fig. 9-52.
Ans. $\mathbf{V}' = 11.5\ \underline{/-95.8^\circ}\ \text{V},\ \mathbf{I}' = 1.39\ \underline{/-80.6^\circ}\ \text{A},\ \mathbf{Z}' = 8.26\ \underline{/-15.2^\circ}\ \Omega$

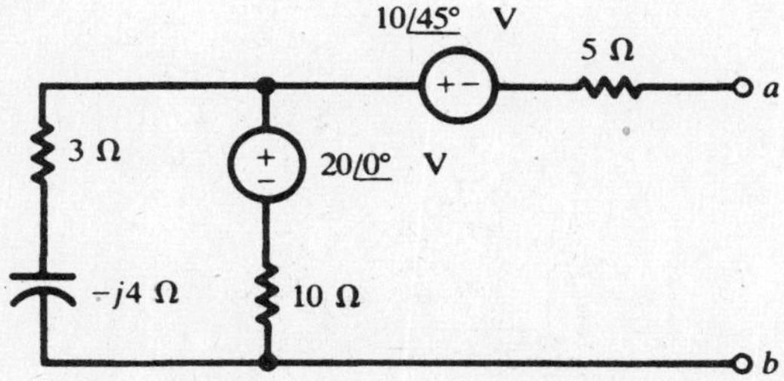

Fig. 9-52

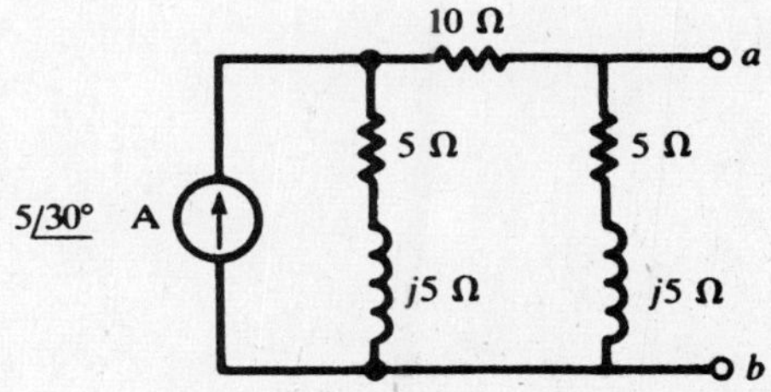

Fig. 9-53

9.57 Obtain the Thévenin and Norton equivalent circuits at terminals *ab* for the network of Fig. 9-53.
Ans. $\mathbf{V}' = 11.18 \angle 93.43^\circ$ V, $\mathbf{I}' = 2.24 \angle 56.56^\circ$ A, $\mathbf{Z}' = 5.0 \angle 36.87^\circ\ \Omega$

9.58 In the circuit of Fig. 9-54, $v_1 = 10$ V and $v_2 = 5 \sin 2000t$. Find i.
Ans. $i = 1 - 0.35 \sin(2000t - 45^\circ)$

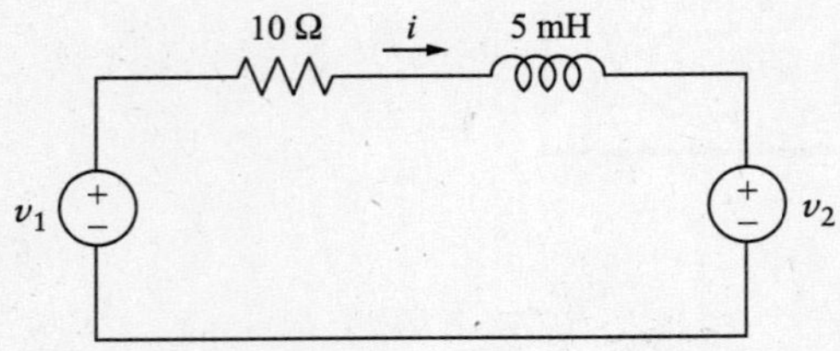

Fig. 9-54

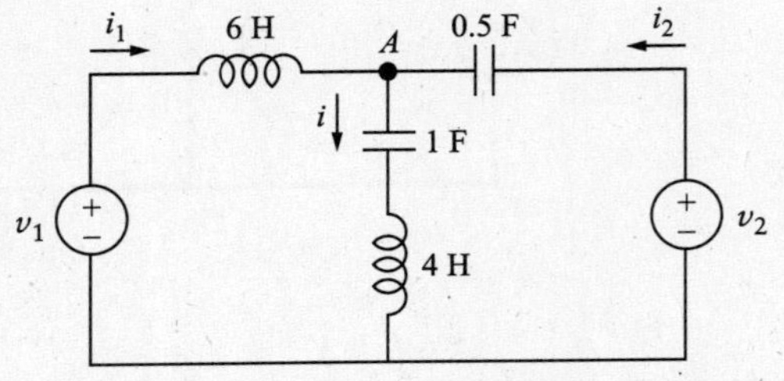

Fig. 9-55

9.59 In the circuit of Fig. 9-55, $v_1 = 6 \cos \omega t$ and $v_2 = \cos(\omega t + 60^\circ)$. Find v_A if $\omega = 2$ rad/sec. *Hint*: Apply KCL at node A in the phasor domain. *Ans.* $v_A = 1.11 \sin 2t$

9.60 In the circuit of Problem 9.59 find phasor currents $\mathbf{I}_1$ and $\mathbf{I}_2$ drawn from the two sources. *Hint*: Apply phasor KVL to the loops on the left and right sides of the circuit.
Ans. $\mathbf{I}_1 = 508 \angle -100.4^\circ$, $\mathbf{I}_2 = 1057 \angle -145^\circ$, both in mA

9.61 Find v_A in the circuit of Problem 9.59 if $\omega = 0.5$ rad/s. *Ans.* $V_a = 0$

9.62 In the circuit of Fig. 9-55, $v_1 = V_1 \cos(0.5t + \theta_1)$ and $v_2 = V_2 \cos(0.5t + \theta_2)$. Find the current through the 4 H inductor. *Ans.* $i = (V_2/4) \sin(0.5t + \theta_2) - (V_1/3) \sin(0.5t + \theta_1)$

9.63 In the circuit of Fig. 9-55, $v_1 = V_1 \cos(t + \theta_1)$ and $v_2 = V_2 \cos(t + \theta_2)$. Find v_A.
Ans. $v_A = \infty$, unless $V_1 = V_2 = 0$, in which case $v_A = 0$

9.64 In the circuit of Fig. 9-55, $v_1 = V_1 \cos(2t)$ and $v_2 = V_2 \cos(0.25t)$. Find v_A.
Ans. $v_A = -0.816 V_1 \cos(2t) - 0.6 V_2 \cos(0.25t)$

CHAPTER 10

AC Power

10.1 POWER IN THE TIME DOMAIN

The *instantaneous power* entering a two-terminal circuit N (Fig. 10-1) is defined by

$$p(t) = v(t)i(t) \tag{1}$$

where $v(t)$ and $i(t)$ are terminal voltage and current, respectively. If p is positive, energy is delivered to the circuit. If p is negative, energy is returned from the circuit to the source.

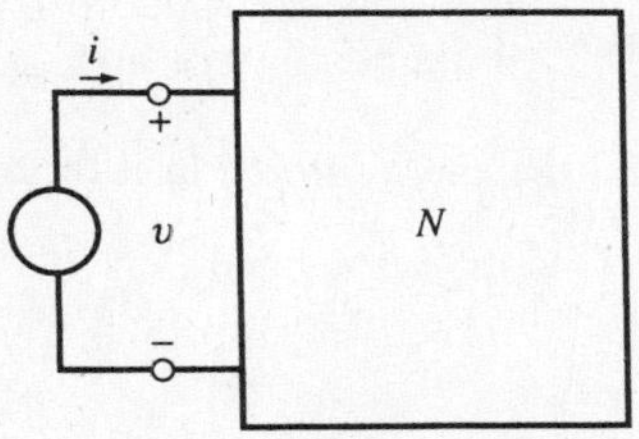

Fig. 10-1

In this chapter, we consider periodic currents and voltages, with emphasis on the sinusoidal steady state in linear *RLC* circuits. Since the storage capacity of an inductor or a capacitor is finite, these passive elements cannot continue receiving energy without returning it. Therefore, in the steady state and during each cycle, all of the energy received by an inductor or capacitor is returned. The energy received by a resistor is, however, dissipated in the form of thermal, mechanical, chemical, and/or electromagnetic energies. The net energy flow to a passive circuit during one cycle is, therefore, positive or zero.

EXAMPLE 10.1 Figure 10-2(*a*) shows the graph of a current in a resistor of 1 kΩ. Find and plot the instantaneous power $p(t)$.

From $v = Ri$, we have $p(t) = vi = Ri^2 = 1000 \times 10^{-6} = 10^{-3}\,\text{W} = 1\,\text{mW}$. See Fig. 10-2(*b*).

EXAMPLE 10.2 The current in Example 10.1 passes through a 0.5-μF capacitor. Find the power $p(t)$ entering the capacitor and the energy $w(t)$ stored in it. Assume $v_C(0) = 0$. Plot $p(t)$ and $w(t)$.

Figure 10-2(*a*) indicates that the current in the capacitor is a periodic function with a period $T = 2\,\text{ms}$. During one period the current is given by

$$i = \begin{cases} 1\,\text{mA} & (0 < t < 1\,\text{ms}) \\ -1\,\text{mA} & (1 < t < 2\,\text{ms}) \end{cases}$$

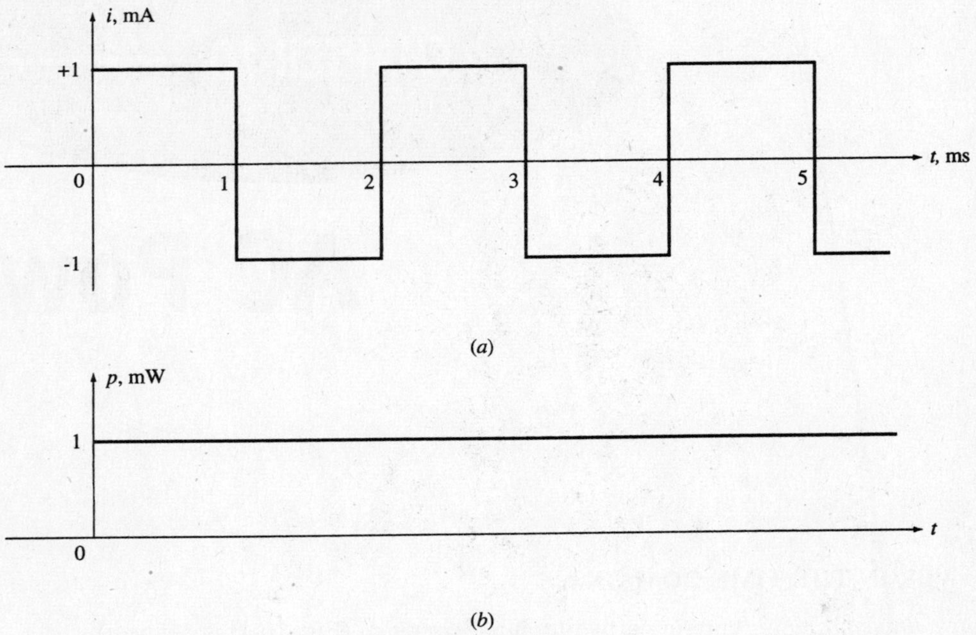

Fig. 10-2

The voltage across the capacitor is also a periodic function with the same period T [Fig. 10-3(*a*)]. During one period the voltage is

$$v(t) = \frac{1}{C}\int_0^t i\,dt = \begin{cases} 2000t \text{ (V)} & (0 < t < 1 \text{ ms}) \\ 4 - 2000t \text{ (V)} & (1 < t < 2 \text{ ms}) \end{cases}$$

Finally, the power entering the capacitor and the energy stored in it (both also periodic with period T) are

$$p(t) = vi = \begin{cases} 2000t \text{ (mW)} & (0 < t < 1 \text{ ms}) \\ 2000t - 4 \text{ (mW)} & (1 < t < 2 \text{ ms}) \end{cases} \qquad \text{[Fig. 10-3(b)]}$$

$$w(t) = \frac{1}{2}Cv^2 = \begin{cases} t^2 \text{ (J)} & (0 < t < 1 \text{ ms}) \\ t^2 + 4 \times 10^{-6} - 4 \times 10^{-3}t \text{ (J)} & (1 < t < 2 \text{ ms}) \end{cases} \qquad \text{[Fig. 10-3(c)]}$$

Alternatively, $w(t)$ may be obtained by integrating $p(t)$. Power entering the capacitor during one period is equally positive and negative [see Fig. 10-3(*b*)]. The energy stored in the capacitor is always positive as shown in Fig. 10-3(*c*). The maximum stored energy is $W_{max} = 10^{-6}$ J $= 1\ \mu$J at $t = 1, 3, 4, \ldots$ ms.

10.2 POWER IN SINUSOIDAL STEADY STATE

A sinusoidal voltage $v = V_m \cos \omega t$, applied across an impedance $Z = |Z| \angle \theta$, establishes a current $i = I_m \cos(\omega t - \theta)$. The power delivered to the impedance at time t is

$$\begin{aligned} p(t) = vi &= V_m I_m \cos\ \omega t\ \cos(\omega t - \theta) = \tfrac{1}{2} V_m I_m [\cos\theta + \cos(2\omega t - \theta)] \\ &= V_{\text{eff}} I_{\text{eff}} [\cos\theta + \cos(2\omega t - \theta)] \\ &= V_{\text{eff}} I_{\text{eff}} \cos\theta + V_{\text{eff}} I_{\text{eff}} \cos(2\omega t - \theta) \end{aligned} \qquad (2)$$

where $V_{\text{eff}} = V_m/\sqrt{2}$, $I_{\text{eff}} = I_m/\sqrt{2}$, and $I_{\text{eff}} = V_{\text{eff}}/|Z|$. The instantaneous power in (*2*) consists of a sinusoidal component $V_{\text{eff}} I_{\text{eff}} \cos(2\omega t - \theta)$ plus a constant value $V_{\text{eff}} I_{\text{eff}} \cos\theta$ which becomes the average power P_{avg}. This is illustrated in Fig. 10-4. During a portion of one cycle, the instantaneous power is positive which indicates that the power flows into the load. During the rest of the cycle, the instan-

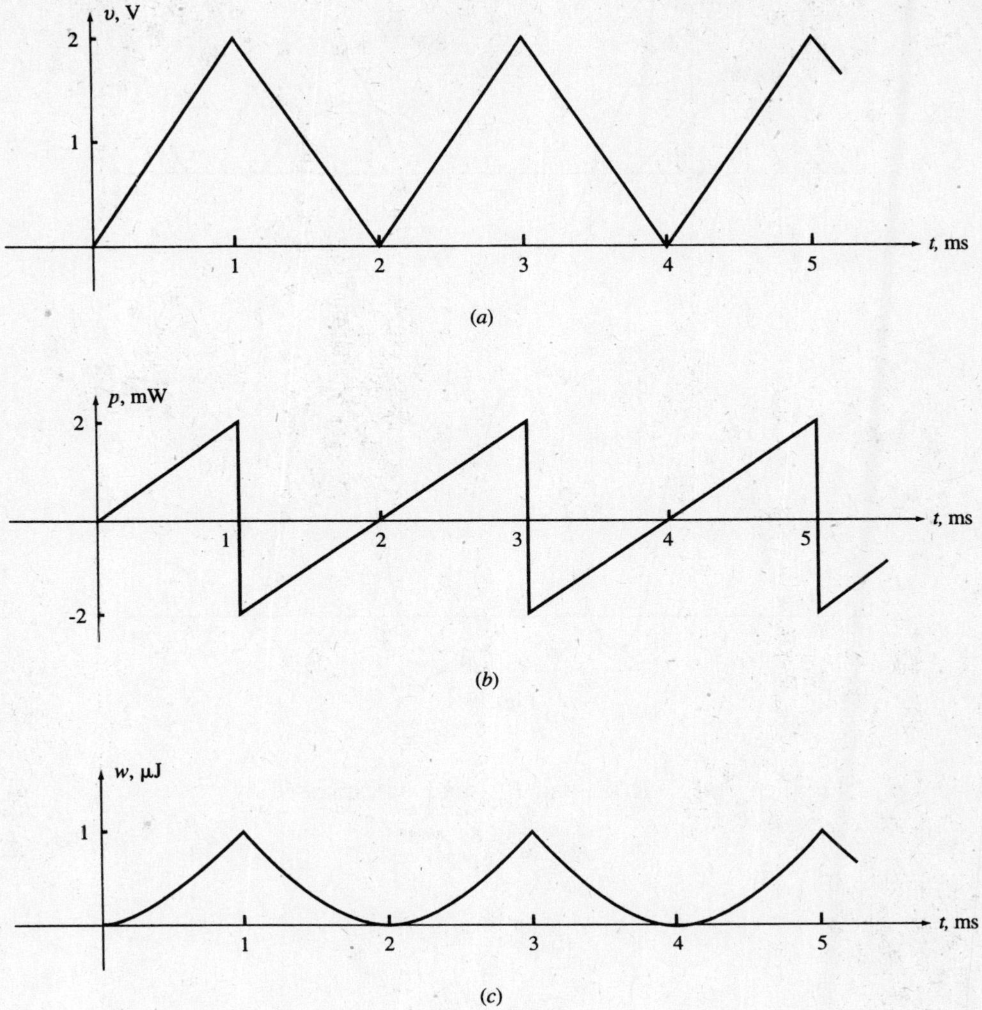

Fig. 10-3

taneous power may be negative, which indicates that the power flows out of the load. The net flow of power during one cycle is, however, nonnegative and is called the *average power*.

EXAMPLE 10.3 A voltage $v = 140 \cos \omega t$ is connected across an impedance $Z = 5 \angle -60^\circ$. Find $p(t)$.
The voltage v results in a current $i = 28 \cos(\omega t + 60^\circ)$. Then,

$$p(t) = vi = 140(28) \cos \omega t \cos(\omega t + 60^\circ) = 980 + 1960 \cos(2\omega t + 60^\circ)$$

The instantaneous power has a constant component of 980 W and a sinusoidal component with twice the frequency of the source. The plot of p vs. t is similar to that in Fig. 10-4 with $\theta = -\pi/3$.

10.3 AVERAGE OR REAL POWER

The net or average power $P_{avg} = \langle p(t) \rangle$ entering a load during one period is called the *real power*. Since the average of $\cos(2\omega t - \theta)$ over one period is zero, from (*2*) we get

$$P_{avg} = V_{eff} I_{eff} \cos \theta \tag{3}$$

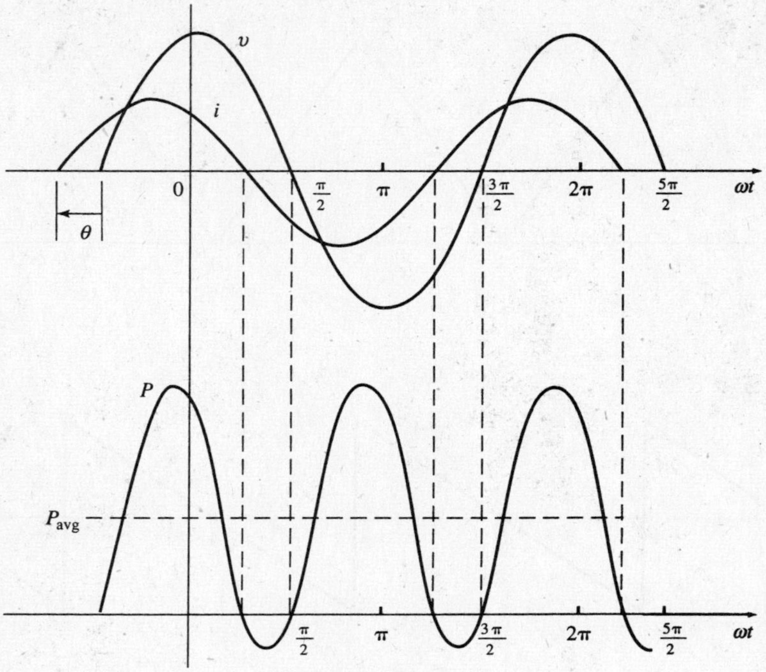

Fig. 10-4

If $Z = R + jX = |Z| \underline{/\theta}$, then $\cos\theta = R/|Z|$ and P_{avg} may be expressed by

$$P_{avg} = V_{eff} I_{eff} \frac{R}{|Z|} \tag{4}$$

or

$$P_{avg} = \frac{V_{eff}^2}{|Z|^2} R \tag{5}$$

or

$$P_{avg} = RI_{eff}^2 \tag{6}$$

The average power is nonnegative. It depends on V, I, and the phase angle between them. When V_{eff} and I_{eff} are given, P is maximum for $\theta = 0$. This occurs when the load is purely resistive. For a purely reactive load, $|\theta| = 90^\circ$ and $P_{avg} = 0$. The ratio of P_{avg} to $V_{eff} I_{eff}$ is called the *power factor* pf. From (*3*), the ratio is equal to $\cos\theta$ and so

$$\text{pf} = \frac{P_{avg}}{V_{eff} I_{eff}} \qquad 0 \leq \text{pf} \leq 1 \tag{7}$$

The subscript avg in the average power P_{avg} is often omitted and so in the remainder of this chapter P will denote average power.

EXAMPLE 10.4 Find P delivered from a sinusoidal voltage source with $V_{eff} = 110\,\text{V}$ to an impedance of $Z = 10 + j8$. Find the power factor.

$$Z = 10 + j8 = 12.81 \underline{/38.7^\circ}$$

$$I_{eff} = \frac{V_{eff}}{Z} = \frac{110}{12.81 \underline{/38.7^\circ}} = 8.59 \underline{/-38.7^\circ} \quad \text{A}$$

$$P = V_{eff} I_{eff} \cos\theta = 110(8.59 \cos 38.7^\circ) = 737.43 \quad \text{W}$$

$$\text{pf} = \cos 38.7^\circ = 0.78$$

Alternative Solution

We have $|Z|^2 = 100 + 64 = 164$. Then,

$$P = V_{\text{eff}}^2 R/|Z|^2 = 110^2(10)/164 = 737.8 \text{ W}$$

The alternative solution gives a more accurate answer.

10.4 REACTIVE POWER

If a passive network contains inductors, capacitors, or both, a portion of energy entering it during one cycle is stored and then returned to the source. During the period of energy return, the power is negative. The power involved in this exchange is called *reactive* or *quadrature power*. Although the net effect of reactive power is zero, it degrades the operation of power systems. Reactive power, indicated by Q, is defined as

$$Q = V_{\text{eff}} I_{\text{eff}} \sin\theta \tag{8}$$

If $Z = R + jX = |Z| \angle\theta$, then $\sin\theta = X/|Z|$ and Q may be expressed by

$$Q = V_{\text{eff}} I_{\text{eff}} \frac{X}{|Z|} \tag{9}$$

or

$$Q = \frac{V_{\text{eff}}^2}{|Z|^2} X \tag{10}$$

or

$$Q = X I_{\text{eff}}^2 \tag{11}$$

The unit of reactive power is the *volt-amperes reactive* (var).

The reactive power Q depends on V, I, and the phase angle between them. It is the product of the voltage and that component of the current which is 90° out of phase with the voltage. Q is zero for $\theta = 0°$. This occurs for a purely resistive load, when **V** and **I** are in phase. When the load is purely reactive, $|\theta| = 90°$ and Q attains its maximum magnitude for given V and I. Note that, while P is always nonnegative, Q can assume positive values (for an inductive load where the current lags the voltage) or negative values (for a capacitive load where the current leads the voltage). It is also customary to specify Q by it magnitude and load type. For example, 100-kvar inductive means $Q = 100$ kvar and 100-kvar capacitive indicates $Q = -100$ kvar.

EXAMPLE 10.5 The voltage and current across a load are given by $\mathbf{V}_{\text{eff}} = 110$ V and $\mathbf{I}_{\text{eff}} = 20 \angle -50°$ A. Find P and Q.

$$P = 110(20 \cos 50°) = 1414 \text{ W} \qquad Q = 110(20 \sin 50°) = 1685 \text{ var}$$

10.5 SUMMARY OF AC POWER IN *R*, *L*, AND *C*

AC power in resistors, inductors, and capacitors, is summarized in Table 10-1. We use the notation $\mathbf{V}_{\text{eff}}$ and $\mathbf{I}_{\text{eff}}$ to include the phase angles. The last column of Table 10-1 is $S = VI$ where S is called *apparent power*. S is discussed in Section 10.7 in more detail.

EXAMPLE 10.6 Find the power delivered from a sinusoidal source to a resistor R. Let the effective values of voltage and current be V and I, respectively.

$$p_R(t) = v i_R = (V\sqrt{2}) \cos\omega t (I\sqrt{2}) \cos\omega t = 2VI \cos^2 \omega t = VI(1 + \cos 2\omega t)$$

$$= RI^2(1 + \cos 2\omega t) = \frac{V^2}{R}(1 + \cos 2\omega t)$$

Thus,

$$P_R = \frac{V^2}{R} = RI^2 \qquad Q = 0$$

Table 10-1

$v = (V\sqrt{2})\cos\omega t \qquad \mathbf{V}_{\text{eff}} = V\angle 0°$
$i = (I\sqrt{2})\cos(\omega t - \theta) \qquad \mathbf{I}_{\text{eff}} = I\angle -\theta°$
$P = VI\cos\theta$, $Q = VI\sin\theta$ and $S = VI$ (apparent power)

	Z	i	$\mathbf{I}_{\text{eff}}$	$p(t)$	P	Q	S
R	R	$\dfrac{V\sqrt{2}}{R}\cos\omega t$	$\dfrac{V}{R}\angle 0°$	$\dfrac{V^2}{R}(1+\cos 2\omega t)$	$\dfrac{V^2}{R}$	0	$\dfrac{V^2}{R}$
L	$jL\omega$	$\dfrac{V\sqrt{2}}{L\omega}\cos(\omega t - 90°)$	$\dfrac{V}{L\omega}\angle -90°$	$\dfrac{V^2}{L\omega}\sin 2\omega t$	0	$\dfrac{V^2}{L\omega}$	$\dfrac{V^2}{L\omega}$
C	$\dfrac{-j}{C\omega}$	$V\sqrt{2}C\omega\cos(\omega t + 90°)$	$VC\omega\angle 90°$	$-V^2C\omega\sin 2\omega t$	0	$-V^2C\omega$	$V^2C\omega$

The instantaneous power entering a resistor varies sinusoidally between zero and $2RI^2$, with twice the frequency of excitation, and with an average value of $P = RI^2$. $v(t)$ and $p_R(t)$ are plotted in Fig. 10-5(a).

EXAMPLE 10.7 Find the ac power entering an inductor L.

$$p_L(t) = vi_L = (V\sqrt{2})\cos\omega t(I\sqrt{2})\cos(\omega t - 90°) = 2VI\cos\omega t\sin\omega t = VI\sin 2\omega t = L\omega I^2\sin 2\omega t$$
$$= \frac{V^2}{L\omega}\sin 2\omega t$$

Thus,
$$P = 0 \qquad Q = VI = \frac{V^2}{L\omega} = L\omega I^2$$

The instantaneous power entering an inductor varies sinusoidally between Q and $-Q$, with twice the frequency of the source, and an average value of zero. See Fig. 10-5(b).

EXAMPLE 10.8 Find the ac power delivered in a capacitor C.

$$p_C(t) = vi_C = (V\sqrt{2})\cos\omega t\,(I\sqrt{2})\cos(\omega t + 90°) = -2VI\cos\omega t\sin\omega t = -VI\sin 2\omega t = -C\omega V^2\sin 2\omega t$$
$$= -\frac{I^2}{C\omega}\sin 2\omega t$$

Thus,
$$P = 0 \qquad Q = -VI = -\frac{I^2}{C\omega} = -C\omega V^2$$

Like an inductor, the instantaneous power entering a capacitor varies sinusoidally between $-Q$ and Q, with twice the frequency of the source, and an average value of zero. See Fig. 10-5(c).

10.6 EXCHANGE OF ENERGY BETWEEN AN INDUCTOR AND A CAPACITOR

If an inductor and a capacitor are fed in parallel by the same ac voltage source or in series by the same current source, the power entering the capacitor is 180° out of phase with the power entering the inductor. This is explicitly reflected in the opposite signs of reactive power Q for the inductor and the capacitor. In such cases, the inductor and the capacitor will exchange some energy with each other, bypassing the ac source. This reduces the reactive power delivered by the source to the LC combination and consequently improves the power factor. See Sections 10.8 and 10.9.

EXAMPLE 10.9 Find the total instantaneous power $p(t)$, the average power P, and the reactive power Q, delivered from $v = (V\sqrt{2})\cos\omega t$ to a parallel RLC combination.

The total instantaneous power is

$$p_T = vi = v(i_R + i_L + i_C) = p_R + p_L + p_C$$

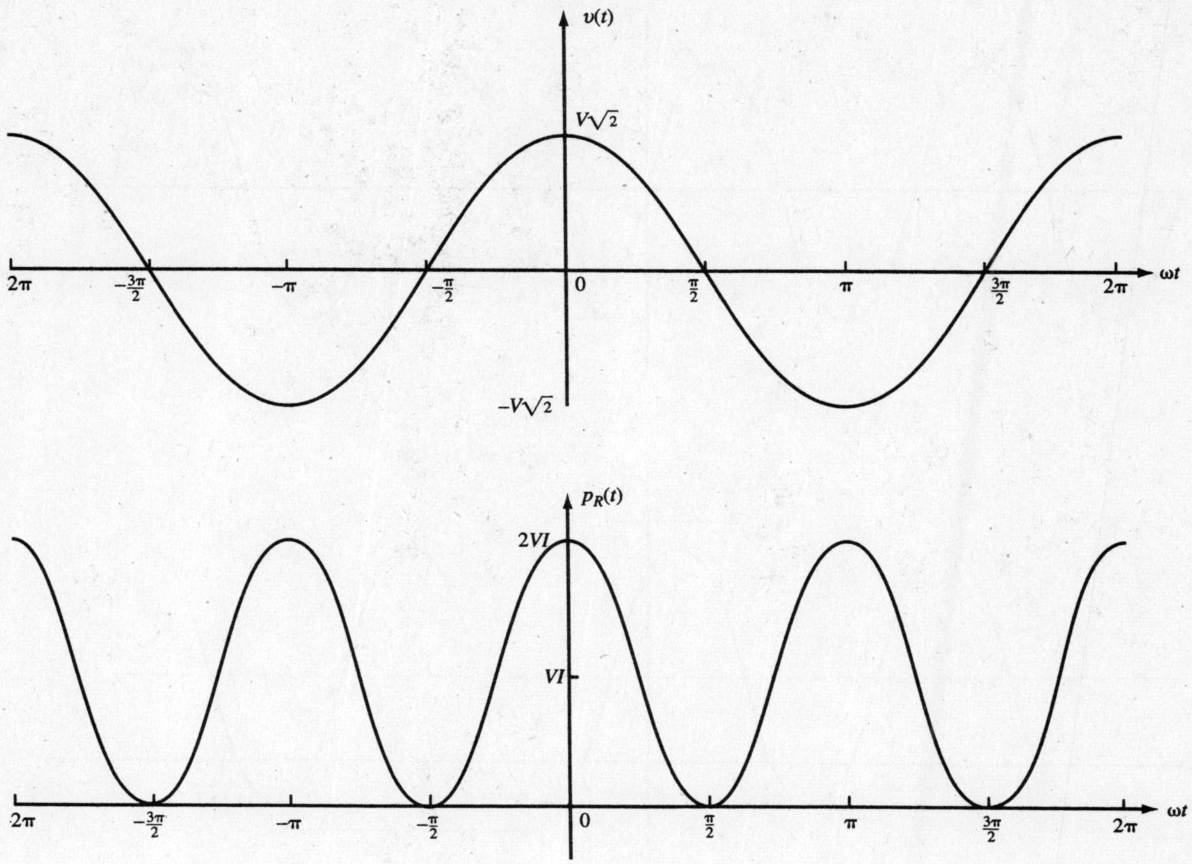

v(t)
V√2
−V√2
2π
−3π/2
−π
−π/2
0
π/2
π
3π/2
2π
ωt
p_R(t)
2VI
VI
2π
−3π/2
−π
−π/2
0
π/2
π
3π/2
2π
ωt

(*a*)

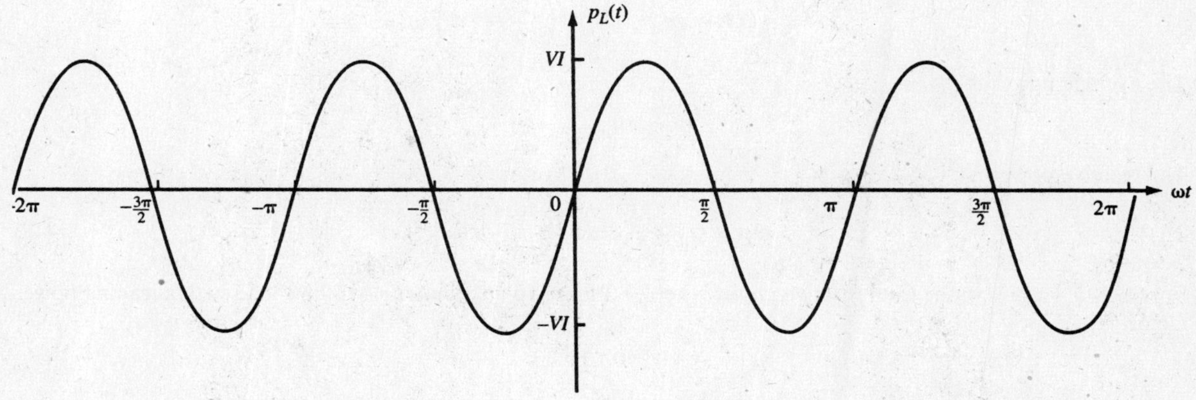

p_L(t)
VI
−VI
·2π
−3π/2
−π
−π/2
0
π/2
π
3π/2
2π
ωt

(*b*)

Fig. 10-5

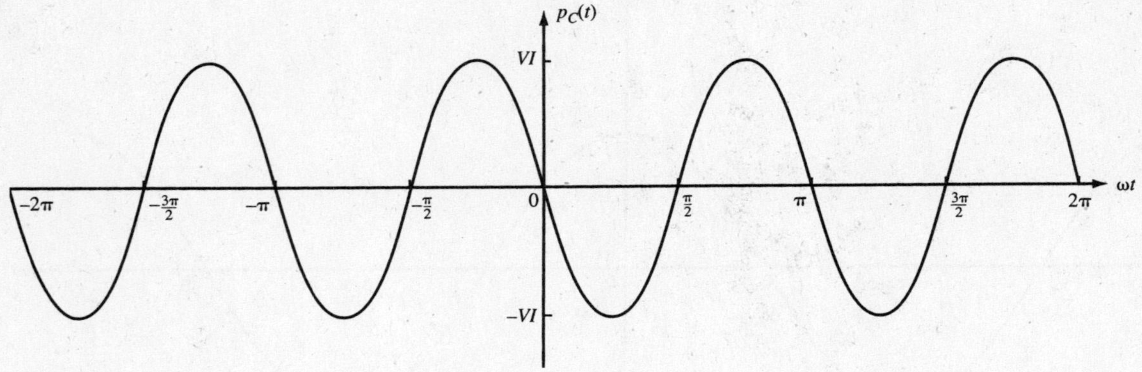

(c)

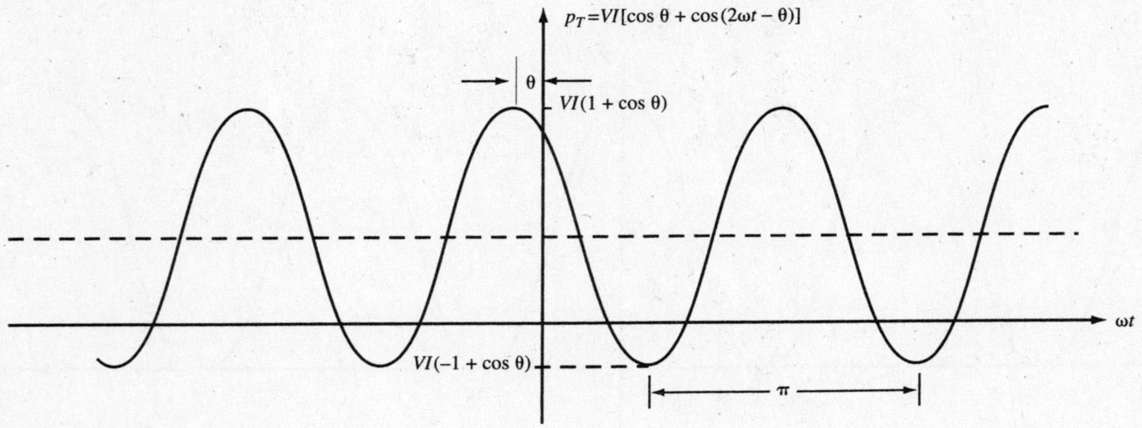

(d)

Fig. 10-5 *(cont.)*

Substituting the values of p_R, p_L, and p_C found in Examples 10.6, 10.7, and 10.8, respectively, we get

$$p_T = \frac{V^2}{R}(1 + \cos 2\omega t) + V^2\left(\frac{1}{L\omega} - C\omega\right)\sin 2\omega t$$

The average power is

$$P_T = P_R = V^2/R$$

The reactive power is

$$Q_T = Q_L + Q_C = V^2\left(\frac{1}{L\omega} - C\omega\right) \tag{12}$$

For $(1/L\omega) - C\omega = 0$, the total reactive power is zero. Figure 10-5(*d*) shows $p_T(t)$ for a load with a leading power factor.

10.7 COMPLEX POWER, APPARENT POWER, AND POWER TRIANGLE

The two components P and Q of power play different roles and may not be added together. However, they may conveniently be brought together in the form of a vector quantity called *complex power* $\mathbf{S}$ and defined by $\mathbf{S} = P + jQ$. The magnitude $|\mathbf{S}| = \sqrt{P^2 + Q^2} = V_{\text{eff}}I_{\text{eff}}$ is called the *apparent power* S and is expressed in units of volt-amperes (VA). The three scalar quantities S, P, and Q may be represented geometrically as the hypotenuse, horizontal and vertical legs, respectively, of a right triangle (called a *power triangle*) as shown in Fig. 10-6(*a*). The power triangle is simply the triangle of the impedance Z scaled by the factor I_{eff}^2 as shown in Fig. 10-6(*b*). Power triangles for an inductive load and a capacitive load are shown in Figs. 10-6(*c*) and (*d*), respectively.

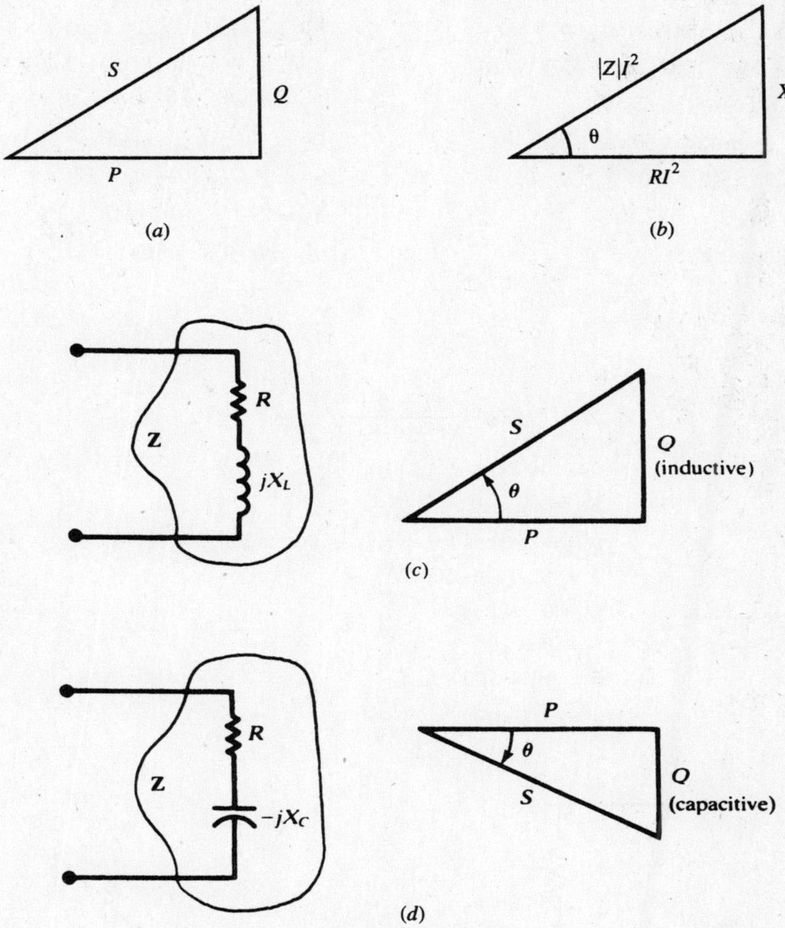

Fig. 10-6

It can be easily proved that $\mathbf{S} = \mathbf{V}_{\text{eff}}\mathbf{I}_{\text{eff}}^*$, where $\mathbf{V}_{\text{eff}}$ is the complex amplitude of effective voltage and $\mathbf{I}_{\text{eff}}^*$ is the complex conjugate of the amplitude of effective current. An equivalent formula is $\mathbf{S} = \mathbf{I}_{\text{eff}}^2\mathbf{Z}$. In summary,

Complex Power: $$\mathbf{S} = \mathbf{V}_{\text{eff}}\mathbf{I}_{\text{eff}}^* = P + jQ = I_{\text{eff}}^2\mathbf{Z} \quad (13)$$

Real Power: $$P = \text{Re}[\mathbf{S}] = V_{\text{eff}}I_{\text{eff}}\cos\theta \quad (14)$$

Reactive Power: $$Q = \text{Im}[\mathbf{S}] = V_{\text{eff}}I_{\text{eff}}\sin\theta \quad (15)$$

Apparent Power: $$S = V_{\text{eff}}I_{\text{eff}} \quad (16)$$

EXAMPLE 10.10 (*a*) A sinusoidal voltage with $V_{\text{eff}} = 10\,\text{V}$ is connected across $Z_1 = 1 + j$ as shown in Fig. 10-7(*a*). Find i_1, $\mathbf{I}_{1,\text{eff}}$, $p_1(t)$, P_1, Q_1, power factor pf_1, and $\mathbf{S}_1$. (*b*) Repeat part (*a*) replacing the load Z_1 in (*a*) by $Z_2 = 1 - j$, as shown in Fig. 10-7(*b*). (*c*) Repeat part (*a*) after connecting in parallel Z_1 in (*a*) and Z_2 in (*b*) as shown in Fig. 10-7(*c*).

Let $v = 10\sqrt{2}\cos\omega t$.

(*a*) See Fig. 10-7(*a*).

$$Z_1 = \sqrt{2}\,\underline{/45^\circ}$$
$$i_1 = 10\cos(\omega t - 45^\circ)$$
$$\mathbf{I}_{1,\text{eff}} = 5\sqrt{2}\,\underline{/-45^\circ}$$
$$p_1(t) = (100\sqrt{2})\cos\omega t\cos(\omega t - 45^\circ)$$
$$= 50 + (50\sqrt{2})\cos(2\omega t - 45^\circ)\text{ W}$$
$$P_1 = V_{\text{eff}} I_{1,\text{eff}} \cos 45^\circ = 50\text{ W}$$
$$Q_1 = V_{\text{eff}} I_{1,\text{eff}} \sin 45^\circ = 50\text{ var}$$
$$\mathbf{S}_1 = P_1 + jQ_1 = 50 + j50$$
$$S_1 = |\mathbf{S}_1| = 50\sqrt{2} = 70.7\text{ VA}$$
$$\text{pf}_1 = 0.707\text{ lagging}$$

(*b*) See Fig. 10-7(*b*)

$$Z_2 = \sqrt{2}\,\underline{/-45^\circ}$$
$$i_2 = 10\cos(\omega t + 45^\circ)$$
$$\mathbf{I}_{2,\text{eff}} = 5\sqrt{2}\,\underline{/45^\circ}$$
$$p_2(t) = (100\sqrt{2})\cos\omega t\cos(\omega t + 45^\circ)$$
$$= 50 + (50\sqrt{2})\cos(2\omega t + 45^\circ)\text{ W}$$
$$P_2 = V_{\text{eff}} I_{2,\text{eff}} \cos 45^\circ = 50\text{ W}$$
$$Q_2 = -V_{\text{eff}} I_{2,\text{eff}} \sin 45^\circ = -50\text{ var}$$
$$\mathbf{S}_2 = P_2 + jQ_2 = 50 - j50$$
$$S_2 = |\mathbf{S}_2| = 50\sqrt{2} = 70.7\text{ VA}$$
$$\text{pf}_2 = 0.707\text{ leading}$$

(*c*) See Fig. 10-7(*c*).

$$Z = Z_1 \| Z_2 = \frac{(1+j)(1-j)}{(1+j)+(1-j)} = 1$$
$$i = 10\sqrt{2}\cos\omega t$$
$$\mathbf{I}_{\text{eff}} = 10$$
$$p(t) = 200\cos^2\omega t = 100 + 100\cos 2\omega t\text{ W}$$
$$P = V_{\text{eff}} I_{\text{eff}} = 100\text{ W}$$
$$Q = 0$$
$$\mathbf{S} = P = 100$$
$$S = |\mathbf{S}| = 100\text{ VA}$$
$$\text{pf} = 1$$

(*a*)

(*b*)

Fig. 10-7

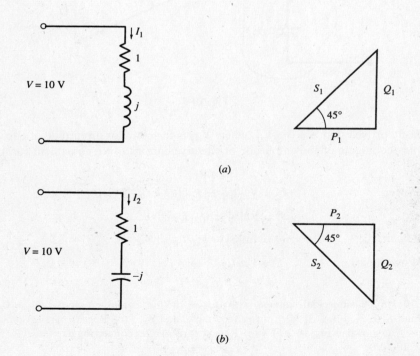

(c)

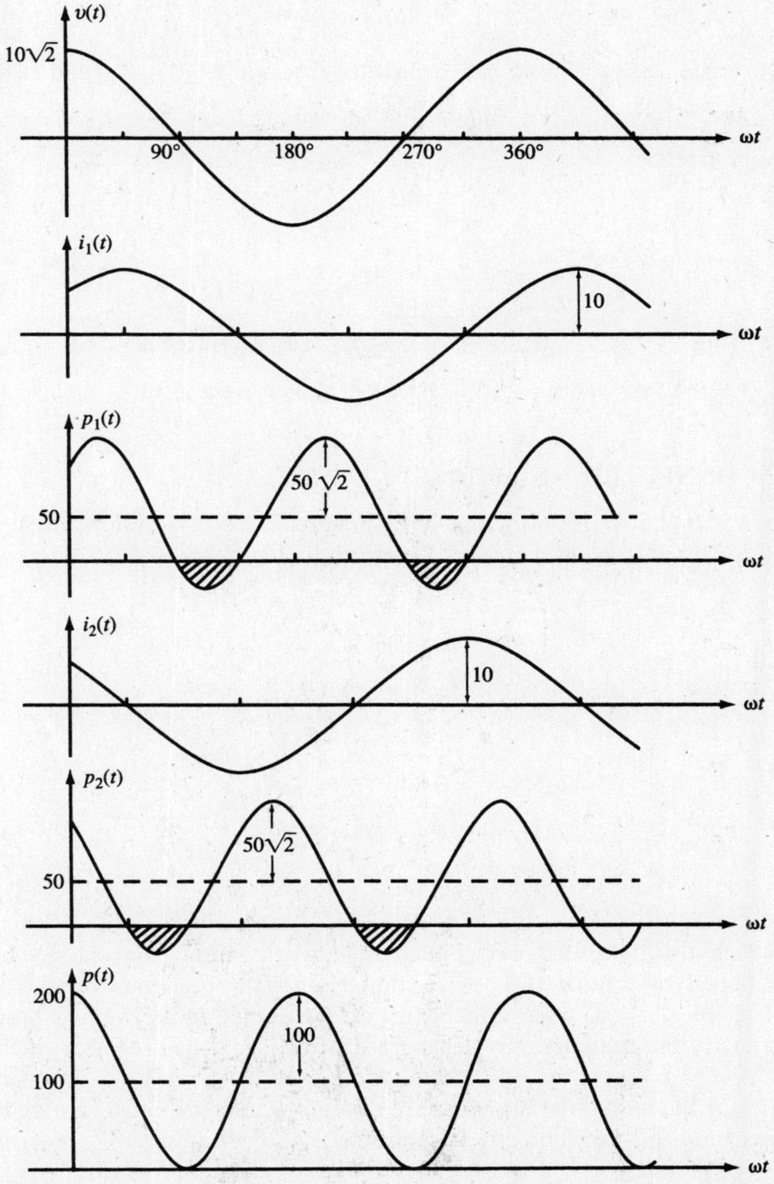

(d)

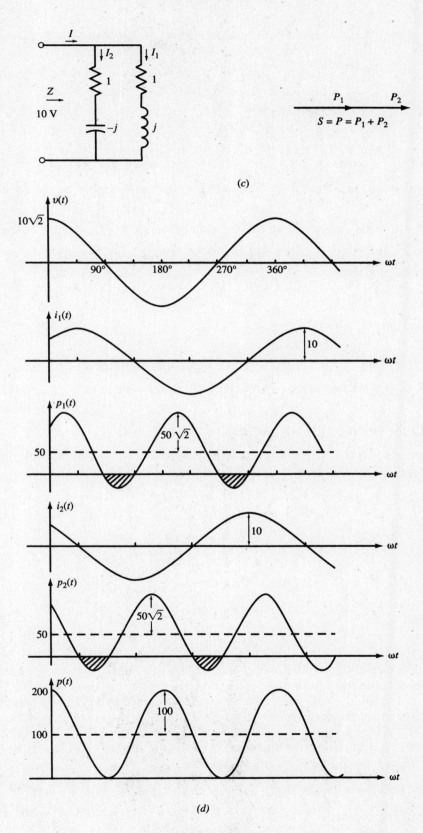

Fig. 10-7 *(Cont.)*

The results of part (c) may also be derived from the observation that for the $Z_1 \| Z_2$ combination, $i = i_1 + i_2$ and, consequently,

$$\begin{aligned} p(t) &= p_1(t) + p_2(t) \\ &= [50 + (50\sqrt{2})\cos(2\omega t - 45^\circ)] + [50 + (50\sqrt{2})\cos(2\omega t + 45^\circ)] \\ &= 100 + 100\cos 2\omega t \text{ W} \end{aligned}$$

$$\begin{aligned} P &= P_1 + P_2 = 50 + 50 = 100 \text{ W} \\ Q &= Q_1 + Q_2 = 50 - 50 = 0 \\ S &= 100 < S_1 + S_2 \end{aligned}$$

The power triangles are shown in Figs. 10-7(a), (b), and (c). Figure 10-7(d) shows the plots of v, i, and p for the three loads.

EXAMPLE 10.11 A certain passive network has equivalent impedance $\mathbf{Z} = 3 + j4\,\Omega$ and an applied voltage

$$v = 42.5\cos(1000t + 30^\circ) \quad \text{(V)}$$

Give complete power information.

$$\mathbf{V}_{\text{eff}} = \frac{42.5}{\sqrt{2}} \angle 30^\circ \quad \text{V}$$

$$\mathbf{I}_{\text{eff}} = \frac{\mathbf{V}_{\text{eff}}}{\mathbf{Z}} = \frac{(42.5/\sqrt{2})\angle 30^\circ}{5\angle 53.13^\circ} = \frac{8.5}{\sqrt{2}} \angle -23.13^\circ \quad \text{A}$$

$$\mathbf{S} = \mathbf{V}_{\text{eff}}\mathbf{I}^*_{\text{eff}} = 180.6\angle 53.13^\circ = 108.4 + j144.5$$

Hence, $P = 108.4$ W, $Q = 144.5$ var (inductive), $S = 180.6$ VA, and pf $= \cos 53.13^\circ = 0.6$ lagging.

10.8 PARALLEL-CONNECTED NETWORKS

The complex power $\mathbf{S}$ is also useful in analyzing practical networks, for example, the collection of households drawing on the same power lines. Referring to Fig. 10-8,

$$\begin{aligned} \mathbf{S}_T &= \mathbf{V}_{\text{eff}}\mathbf{I}^*_{\text{eff}} = \mathbf{V}_{\text{eff}}(\mathbf{I}^*_{1,\text{eff}} + \mathbf{I}^*_{2,\text{eff}} + \cdots + \mathbf{I}^*_{n,\text{eff}}) \\ &= \mathbf{S}_1 + \mathbf{S}_2 + \cdots + \mathbf{S}_n \end{aligned}$$

from which

$$\begin{aligned} P_T &= P_1 + P_2 + \cdots + P_n \\ Q_T &= Q_1 + Q_2 + \cdots + Q_n \\ S_T &= \sqrt{P_T^2 + Q_T^2} \\ \text{pf}_T &= \frac{P_T}{S_T} \end{aligned}$$

These results (which also hold for series-connected networks) imply that the power triangle for the network may be obtained by joining the power triangles for the branches vertex to vertex. In the example shown in Fig. 10-9, $n = 3$, with branches 1 and 3 assumed inductive and branch 2 capacitive. In such diagrams, some of the triangles may degenerate into straight-line segments if the corresponding R or X is zero.

If the power data for the individual branches are not important, the network may be replaced by its equivalent admittance, and this used directly to compute $\mathbf{S}_T$.

EXAMPLE 10.12 Three loads are connected in parallel to a 6-kV$_{\text{eff}}$ ac line, as shown in Fig. 10-8. Given

$$P_1 = 10\text{ kW, pf}_1 = 1; \qquad P_2 = 20\text{ kW, pf}_2 = 0.5 \text{ lagging}; \qquad P_3 = 15\text{ kW, pf}_3 = 0.6 \text{ lagging}$$

Find P_T, Q_T, S_T, pf$_T$, and the current $\mathbf{I}_{\text{eff}}$.

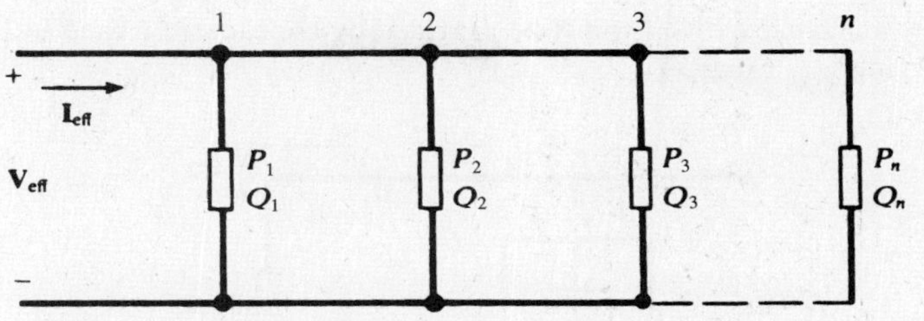

Fig. 10-8

We first find the reactive power for each load:

$$\begin{array}{lll} \text{pf}_1 = \cos\theta_1 = 1 & \tan\theta_1 = 0 & Q_1 = P_1 \tan\theta_1 = 0\,\text{kvar} \\ \text{pf}_2 = \cos\theta_2 = 0.5 & \tan\theta_2 = 1.73 & Q_2 = P_2 \tan\theta_2 = 34.6\,\text{kvar} \\ \text{pf}_3 = \cos\theta_3 = 0.6 & \tan\theta_3 = 1.33 & Q_3 = P_3 \tan\theta_3 = 20\,\text{kvar} \end{array}$$

Then P_T, Q_T, S_T, and pf_T, are

$$P_T = P_1 + P_2 + P_3 = 10 + 20 + 15 = 45\,\text{kW}$$
$$Q_T = Q_1 + Q_2 + Q_3 = 0 + 34.6 + 20 = 54.6\,\text{kvar}$$
$$S_T = \sqrt{P^2 + Q^2} = \sqrt{45^2 + 54.6^2} = 70.75\,\text{kVA}$$
$$\text{pf}_T = P_T/S_T = 0.64 = \cos\theta, \theta = 50.5^\circ \text{ lagging}$$
$$I_{\text{eff}} = S/V_{\text{eff}} = (70.75\,\text{kVA})/(6\,\text{kV}) = 11.8\text{ A}$$
$$\mathbf{I}_{\text{eff}} = 11.8\,\underline{/-50.5^\circ}\text{ A}$$

The current could also be found from $\mathbf{I} = \mathbf{I}_1 + \mathbf{I}_2 + \mathbf{I}_3$. However, this approach is more time-consuming.

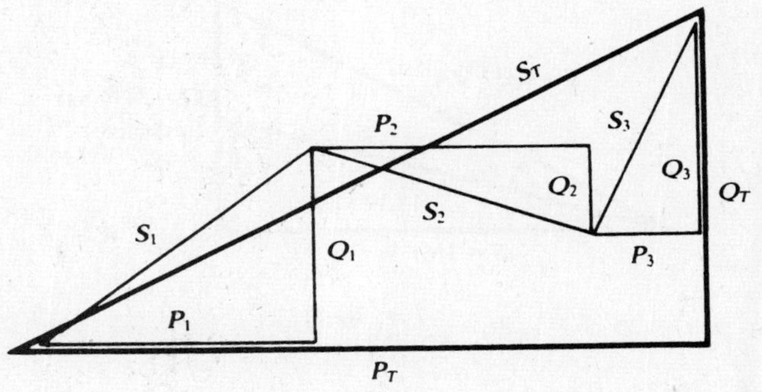

Fig. 10-9

10.9 POWER FACTOR IMPROVEMENT

Electrical service to industrial customers is three-phase, as opposed to the single-phase power supplied to residential and small commercial customers. While metering and billing practices vary among the utilities, the large consumers will always find it advantageous to reduce the quadrature component of their power triangle; this is called "improving the power factor." Industrial systems generally have an overall inductive component because of the large number of motors. Each individual load tends to be either pure resistance, with unity power factor, or resistance and inductive reactance, with a lagging power factor. All of the loads are parallel-connected, and the equivalent impedance results in a lagging current and a corresponding inductive quadrature power Q. To improve the power factor, capacitors, in three-phase banks, are connected to the system either on the primary or secondary side of the main transformer, such that the combination of the plant load and the capacitor banks presents a load to the serving utility which is nearer to unit power factor.

EXAMPLE 10.13 How much capacitive Q must be provided by the capacitor bank in Fig. 10-10 to improve the power factor to 0.95 lagging?

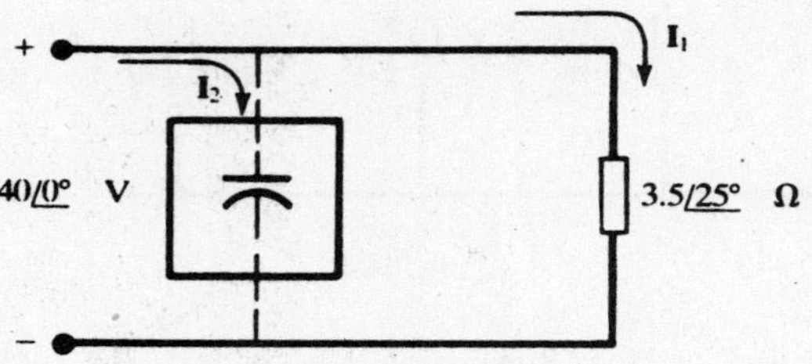

Fig. 10-10

Before addition of the capacitor bank, pf = cos 25°C = 0.906 lagging, and

$$\mathbf{I}_1 = \frac{240\,\underline{/0^\circ}}{3.5\,\underline{/25^\circ}} = 68.6\,\underline{/-25^\circ}\ \text{A}$$

$$\mathbf{S} = \mathbf{V}_{\text{eff}}\mathbf{I}^*_{\text{eff}} = \left(\frac{240}{\sqrt{2}}\,\underline{/0^\circ}\right)\left(\frac{68.6}{\sqrt{2}}\,\underline{/+25^\circ} = 8232\,\underline{/25^\circ}\right) = 7461 + j3479$$

After the improvement, the triangle has the same P, but its angle is $\cos^{-1} 0.95 = 18.19^\circ$. Then (see Fig. 10-11),

$$\frac{3479 - Q_c}{7461} = \tan 18.19^\circ \qquad \text{or} \qquad Q_c = 1027 \text{ var (capacitive)}$$

The new value of apparent power is $S' = 7854$ VA, as compared to the original $S = 8232$ VA. The decrease, 378 VA, amounts to 4.6 percent.

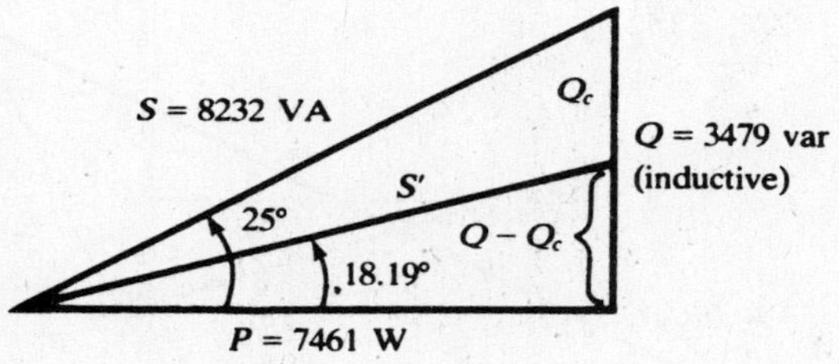

Fig. 10-11

The transformers, the distribution systems, and the utility company alternators are all rated in kVA or MVA. Consequently, an improvement in the power factor, with its corresponding reduction in kVA, releases some of this generation and transmission capability so that it can be used to serve other customers. This is the reason behind the rate structures which, in one way or another, make it more costly for an industrial customer to operate with a lower power factor. An economic study comparing the cost of the capacitor bank to the savings realized is frequently made. The results of such a study will show whether the improvement should be made and also what final power factor should be attained.

EXAMPLE 10.14 A load of $P = 1000$ kW with pf $= 0.5$ lagging is fed by a 5-kV source. A capacitor is added in parallel such that the power factor is improved to 0.8. Find the reduction in current drawn from the generator.

Before improvement:

$$P = 1000\,\text{kW},\ \cos\theta = 0.5,\ S = P/\cos\theta = 2000\,\text{kVA},\ I = 400\ \text{A}$$

After improvement:

$$P = 1000\,\text{kW},\ \cos\theta = 0.8,\ S = P/\cos\theta = 1250\,\text{kVA},\ I = 250\ \text{A}$$

Hence, for the same amount of real power, the current is reduced by (400 − 250)/400 = 0.375 or 37.5 percent.

EXAMPLE 10.15 A fourth load Q_4 is added in parallel to the three parallel loads of Example 10.12 such that the total power factor becomes 0.8 lagging while the total power remains the same. Find Q_4 and the resulting S. discuss the effect on the current.

In Example 10.12 we found total real and reactive powers to be $P = P_1 + P_2 + P_3 = 45\,\text{kW}$ and $Q = Q_1 + Q_2 + Q_3 = 54.6\,\text{kvar}$, respectively. For compensation, we now add load Q_4 (with $P_4 = 0$) such that the new power factor is pf $= \cos\theta = 0.8$ lagging, $\theta = 36.87°$.

Then, $$\tan 36.87° = (Q + Q_4)/P = (54.6 + Q_4)/45 = 0.75 \qquad Q_4 = -20.85\,\text{kvar}$$

The results are summarized in Table 10-2. Addition of the compensating load Q_4 reduces the reactive power from 54.6 kvar to 33.75 kvar and improves the power factor. This reduces the apparent power S from 70.75 kVA to 56.25 kVA. The current is proportionally reduced.

Table 10-2

Load	P, kW	pf	Q, kvar	S, kVA
#1	10	1	0	10
#2	20	0.5 lagging	34.6	40
#3	15	0.6 lagging	20	25
#(1 + 2 + 3)	45	0.64 lagging	54.6	70.75
#4	0	0 leading	−20.85	20.85
Total	45	0.8 lagging	33.75	56.25

10.10 MAXIMUM POWER TRANSFER

The average power delivered to a load $\mathbf{Z}_1$ from a sinusoidal signal generator with open circuit voltage $\mathbf{V}_g$ and internal impedance $\mathbf{Z}_g = R + jX$ is maximum when $\mathbf{Z}_1$ is equal to the complex conjugate of $\mathbf{Z}_g$ so that $\mathbf{Z}_1 = R - jX$. The maximum average power delivered to $\mathbf{Z}_1$ is $P_{max} = V_g^2/4R$.

EXAMPLE 10.16 A generator, with $V_g = 100\,\text{V(rms)}$ and $\mathbf{Z}_g = 1 + j$, feeds a load $\mathbf{Z}_1 = 2$ (Fig. 10-12). (*a*) Find the average power P_{Z1} (absorbed by $\mathbf{Z}_1$), the power P_g (dissipated in $\mathbf{Z}_g$) and P_T (provided by the generator). (*b*) Compute the value of a second load $\mathbf{Z}_2$ such that, when in parallel with $\mathbf{Z}_1$, the equivalent impedance is $\mathbf{Z} = \mathbf{Z}_1 \| \mathbf{Z}_2 = \mathbf{Z}*_g$. (*c*) Connect in parallel $\mathbf{Z}_2$ found in (*b*) with $\mathbf{Z}_1$ and then find the powers P_Z, P_{Z1}, P_{Z2} (absorbed by $\mathbf{Z}$, $\mathbf{Z}_1$, and $\mathbf{Z}_2$, respectively), P_g (dissipated in $\mathbf{Z}_g$) and P_T (provided by the generator).

(*a*) $|\mathbf{Z}_1 + \mathbf{Z}_g| = |2 + 1 + j| = \sqrt{10}$. Thus $\mathbf{I} = \mathbf{V}_g/(\mathbf{Z}_1 + \mathbf{Z}_g) = 100/(2 + 1 + j)$ and $|I| = 10\sqrt{10}$ A. The required powers are

$$P_{Z1} = \text{Re}[\mathbf{Z}_1] \times |I|^2 = 2(10\sqrt{10})^2 = 2000\text{ W}$$
$$P_g = \text{Re}[\mathbf{Z}_g] \times |I|^2 = 1(10\sqrt{10})^2 = 1000\text{ W}$$
$$P_T = P_{Z1} + P_g = 2000 + 1000 = 3000\text{ W}$$

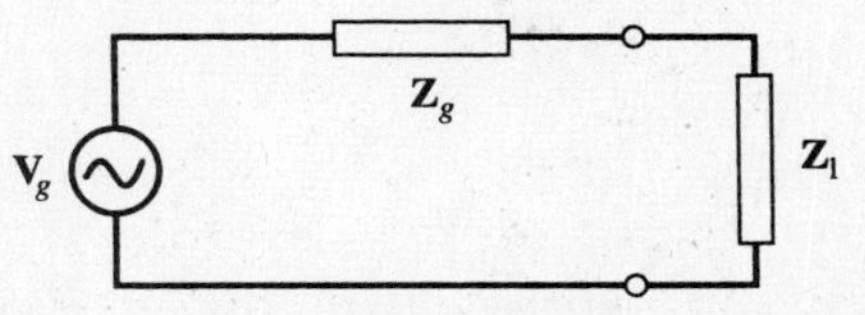

Fig. 10-12

(*b*) Let $\mathbf{Z}_2 = a + jb$. To find a and b, we set $\mathbf{Z}_1 \| \mathbf{Z}_2 = \mathbf{Z}_g^* = 1 - j$. Then,

$$\frac{\mathbf{Z}_1\mathbf{Z}_2}{\mathbf{Z}_1 + \mathbf{Z}_2} = \frac{2(a + jb)}{2 + a + jb} = 1 - j$$

from which $a - b - 2 = 0$ and $a + b + 2 = 0$. Solving these simultaneous equations, $a = 0$ and $b = -2$; substituting into the equation above, $\mathbf{Z}_2 = -j2$.

(*b*) $\mathbf{Z} = \mathbf{Z}_1 \| \mathbf{Z}_2 = 1 - j$ and $\mathbf{Z} + \mathbf{Z}_g = 1 - j + 1 + j = 2$. Then, $I = V_g/(\mathbf{Z} + \mathbf{Z}_g) = 100/(1 - j + 1 + j) = 100/2 = 50$ A, and so

$$P_Z = \text{Re}[\mathbf{Z}] \times \mathbf{I}^2 = 1 \times 50^2 = 2500 \text{ W} \qquad P_g = \text{Re}[\mathbf{Z}_g] \times \mathbf{I}^2 = 1 \times 50^2 = 2500 \text{ W}$$

To find P_{Z1} and P_{Z2}, we first find $\mathbf{V}_Z$ across $\mathbf{Z}$: $\mathbf{V}_Z = \mathbf{IZ} = 50(1 - j)$. Then $\mathbf{I}_{Z1} = \mathbf{V}_Z/\mathbf{Z}_1 = 50(1 - j)/2 = (25\sqrt{2})\,\underline{/-45^\circ}$, and

$$P_{Z1} = \text{Re}[\mathbf{Z}_1] \times |I_{Z1}|^2 = 2(25\sqrt{2})^2 = 2500 \text{ W} \qquad P_{Z2} = 0 \text{ W} \qquad P_T = P_g + P_{Z1} = 5000 \text{ W}$$

Alternatively, we can state that

$$P_{Z2} = 0 \qquad \text{and} \qquad P_{Z1} = P_Z = 2500 \text{ W}$$

10.11 SUPERPOSITION OF AVERAGE POWERS

Consider a network containing two AC sources with two different frequencies, ω_1 and ω_2. If a common period T may be found for the sources (i.e., $\omega_1 = m\omega$, $\omega_2 = n\omega$, where $\omega = 2\pi/T$ and $m \neq n$), then superposition of individual powers applies (i.e., $P = P_1 + P_2$), where P_1 and P_2 are average powers due to application of each source. The preceding result may be generalized to the case of any n number of sinusoidal sources operating simultaneously on a network. If the n sinusoids form harmonics of a fundamental frequency, then superposition of powers applies.

$$P = \sum_{k=1}^{n} P_k$$

EXAMPLE 10.17 A practical coil is placed between two voltage sources $v_1 = 5\cos\omega_1 t$ and $v_2 = 10\cos(\omega_2 t = 60^\circ)$, which share the same common reference terminal (see Fig. 9-54). The coil is modeled by a 5-mH inductor in series with a 10-Ω resistor. Find the average power in the coil for (*a*) $\omega_2 = 2\omega_1 = 4000$, (*b*) $\omega_1 = \omega_2 = 2000$, (*c*) $\omega_1 = 2000$ and $\omega_2 = 1000\sqrt{2}$, all in rad/s.

Let v_1 by itself produce i_1. Likewise, let v_2 produce i_2. Then $i = i_1 - i_2$. The instantaneous power p and the average power P are

$$p = Ri^2 = R(i_1 - i_2)^2 = Ri_1^2 + Ri_2^2 - 2Ri_1 i_2$$
$$P = \langle p \rangle = R\langle i_1^2 \rangle + R\langle i_2^2 \rangle - 2R\langle i_1 i_2 \rangle = P_1 + P_2 - 2R\langle i_1 i_2 \rangle$$

where $\langle p \rangle$ indicates the average of p. Note that in addition to P_1 and P_2, we need to take into account a third term $\langle i_1 i_2 \rangle$ which, depending on ω_1 and ω_2, may or may not be zero.

(*a*) By applying superposition in the phasor domain we find the current in the coil (see Example 9.7).

$$\mathbf{I}_1 = \frac{\mathbf{V}_1}{Z_1} = \frac{5}{10 + j10} = 0.35\,\underline{/-45^\circ} \text{ A}, \; i_1 = 0.35\cos(2000t - 45^\circ)$$

$$P_1 = \frac{RI_1^2}{2} = \frac{10 \times 0.35^2}{2} = 0.625 \text{ W}$$

$$\mathbf{I}_2 = \frac{\mathbf{V}_2}{Z_2} = \frac{10\,\underline{/60^\circ}}{10 + j20} = 0.45\,\underline{/-3.4^\circ} \text{ A}, \; i_2 = 0.45\cos(4000t - 3.4^\circ)$$

$$P_2 = \frac{RI_2^2}{2} = \frac{10 \times 0.45^2}{2} = 1 \text{ W}$$

$$i = i_1 - i_2 = 0.35\cos(2000t - 45^\circ) - 0.45\cos(4000t - 3.4^\circ)$$

In this case $\langle i_1 i_2 \rangle = 0$ because $\langle \cos(2000t - 45^\circ)\cos(4000t - 3.4^\circ)\rangle = 0$. Therefore, superposition of power applies and $P = P_1 + P_2 = 0.625 + 1 = 1.625$ W.

(*b*) The current in the coil is $i = 0.61\cos(2000t - 135^\circ)$ (see Example 9.7). The average power dissipated in the coil is $P = RI^2/2 = 5 \times (0.61)^2 = 1.875$ W. Note that $P > P_1 + P_2$.

(*c*) By applying superposition in the time domain we find

$$i_1 = 0.35\cos(2000t - 45^\circ),\ P_1 = 0.625 \text{ W}$$
$$i_2 = 0.41\cos(1000\sqrt{2}t - 35.3^\circ),\ P_2 = 0.833 \text{ W}$$
$$i = i_1 - i_2,\ P = \langle Ri^2/2\rangle = P_1 + P_2 - 1.44\langle\cos(2000t - 45^\circ)\cos(1000\sqrt{2}t - 35.3^\circ)\rangle$$

The term $\langle\cos(2000t - 45^\circ)\cos(1000\sqrt{2}t - 35.3^\circ)\rangle$ is not determined because a common period can't be found. The average power depends on the duration of averaging.

Solved Problems

10.1 The current plotted in Fig. 10-2(*a*) enters a 0.5-μF capacitor in series with a 1-kΩ resistor. Find and plot (*a*) v across the series RC combination and (*b*) the instantaneous power p entering RC. (*c*) Compare the results with Examples 10.1 and 10.2.

(*a*) Referring to Fig. 10-2(*a*), in one cycle of the current the voltages are

$$v_R = \begin{cases} 1 \text{ V} & (0 < t < 1 \text{ ms}) \\ -1 \text{ V} & (1 < t < 2 \text{ ms}) \end{cases}$$

$$v_C = \frac{1}{C}\int_0^t i\,dt = \begin{cases} 2000t \text{ (V)} & (0 < t < 1 \text{ ms}) \\ 4 - 2000t \text{ (V)} & (1 < t < 2 \text{ ms}) \end{cases}$$

$$v = v_R + v_C = \begin{cases} 1 + 2000t \text{ (V)} & (0 < t < 1 \text{ ms}) \\ 3 - 2000t \text{ (V)} & (1 < t < 2 \text{ ms}) \end{cases} \qquad \text{[See Fig. 10-13(}a\text{)]}$$

(*b*) During one cycle,

$$p_R = Ri^2 = 1 \text{ mW}$$

$$p_C = v_C i = \begin{cases} 2000t \text{ (mW)} & (0 < t < 1 \text{ ms}) \\ 2000t - 4 \text{ (mW)} & (1 < t < 2 \text{ ms}) \end{cases}$$

$$p = vi = p_R + p_C = \begin{cases} 1 + 2000t \text{ (mW)} & (0 < t < 1 \text{ ms}) \\ 2000t - 3 \text{ (mW)} & (1 < t < 2 \text{ ms}) \end{cases} \qquad \text{[(See Fig. 10-13(}b\text{)]}$$

(*c*) The average power entering the circuit during one cycle is equal to the average power absorbed by the resistor. It is the same result obtained in Example 10.1. The power exchanged between the source and the circuit during one cycle also agrees with the result obtained in Example 10.2.

10.2 A 1-V ac voltage feeds (*a*) a 1-Ω resistor, (*b*) a load $\mathbf{Z} = 1 + j$, and (*c*) a load $\mathbf{Z} = 1 - j$. Find P in each of the three cases.

(*a*) $P = V^2/R = 1/1 = 1$ W

(*b*) and (*c*) $|Z| = |1 \pm j| = \sqrt{2}$. $I = V/|Z| = 1/\sqrt{2}$. $P = RI^2 = 0.5$ W

10.3 Obtain the complete power information for a passive circuit with an applied voltage $v = 150\cos(\omega t + 10^\circ)$ V and a resulting current $i = 5.0\cos(\omega t - 50^\circ)$ A.

Using the complex power

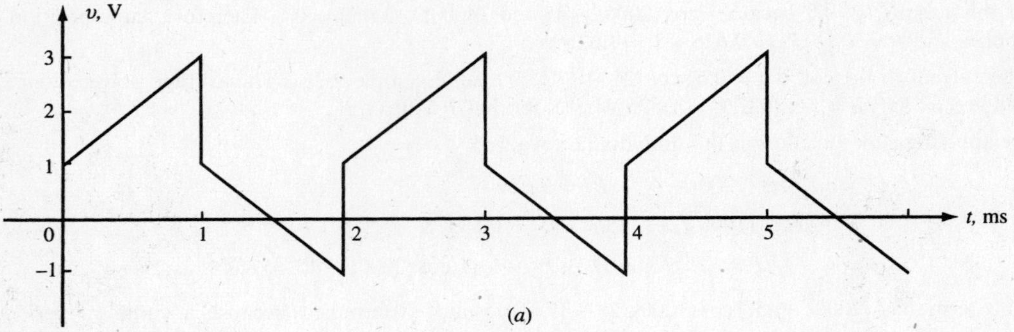

(a)

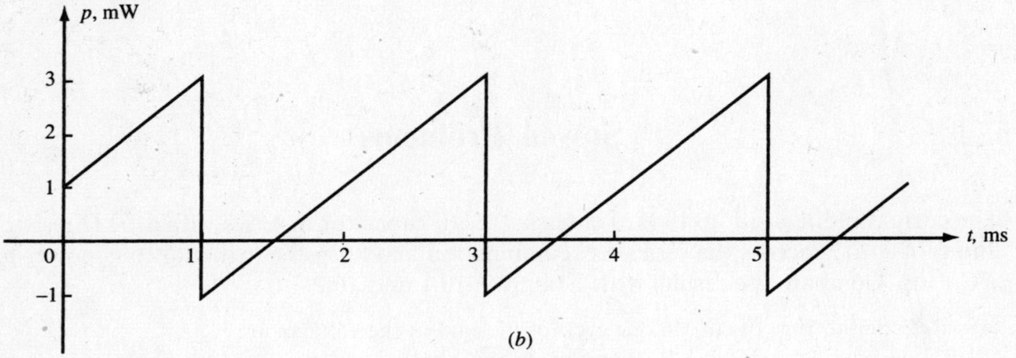

(b)

Fig. 10-13

$$\mathbf{S} = \mathbf{V}_{\text{eff}}\mathbf{I}^*_{\text{eff}} = \left(\frac{150}{\sqrt{2}}\angle 10^\circ\right)\left(\frac{5.0}{\sqrt{2}}\angle 50^\circ\right) = 375\angle 60^\circ = 187.5 + j342.8$$

Thus, $P = 187.5$ W, $Q = 324.8$ var (inductive), $S = 375$ VA, and pf $= \cos 60^\circ = 0.50$ lagging.

10.4 A two-element series circuit has average power 940 W and power factor 0.707 leading. Determine the circuit elements if the applied voltage is $v = 99.0\cos(6000t + 30^\circ)$ V.

The effective applied voltage is $99.0/\sqrt{2} = 70.0$ V. Substituting in $P = V_{\text{eff}}I_{\text{eff}}\cos\theta$,

$$940 = (70.0)I_{\text{eff}}(0.707) \qquad \text{or} \qquad I_{\text{eff}} = 19.0 \text{ A}$$

Then, $(19.0)^2R = 940$, from which $R = 2.60\ \Omega$. For a leading pf, $\theta = \cos^{-1} 0.707 = -45^\circ$, and so

$$\mathbf{Z} = R - jX_C \qquad \text{where} \qquad X_C = R\tan 45^\circ = 2.60\ \Omega$$

Finally, from $2.60 = 1/\omega C$, $C = 64.1\ \mu\text{F}$.

10.5 Find the two elements of a series circuit having current $i = 4.24\cos(5000t + 45^\circ)$ A, power 180 W, and power factor 0.80 lagging.

The effective value of the current is $I_{\text{eff}} = 4.24/\sqrt{2} = 3.0$ A. Then,

$$180 = (3.0)^2R \qquad \text{or} \qquad R = 20.0\ \Omega$$

The impedance angle is $\theta = \cos^{-1} 0.80 = +36.87^\circ$, wherefore the second element must be an inductor. From the power triangle,

$$\frac{Q}{P} = \frac{I^2_{\text{eff}}X_L}{180} = \tan 36.87^\circ \qquad \text{or} \qquad X_L = 15.0\ \Omega$$

Finally, from $15.0 = 5000L$, $L = 3.0$ mH.

10.6 Obtain the power information for each element in Fig. 10-14 and construct the power triangle.

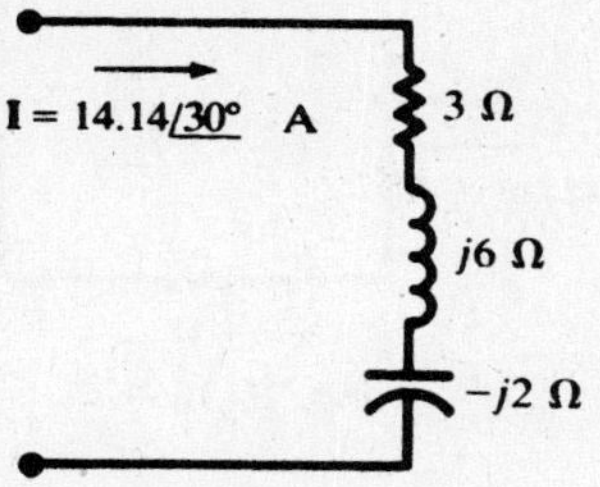

Fig. 10-14

The effective current is $14.14/\sqrt{2} = 10$ A.

$$P = (10)^2 3 = 300 \text{ W} \qquad Q_{j6\Omega} = (10)^2 6 = 600 \text{ var (inductive)} \qquad Q_{-j2\Omega} = (10)^2 2 = 200 \text{ var (capacitive)}$$

$$S = \sqrt{(300)^2 + (600 - 200)^2} = 500 \text{ VA} \qquad \text{pf} = P/S = 0.6 \text{ lagging}$$

The power triangle is shown in Fig. 10-15.

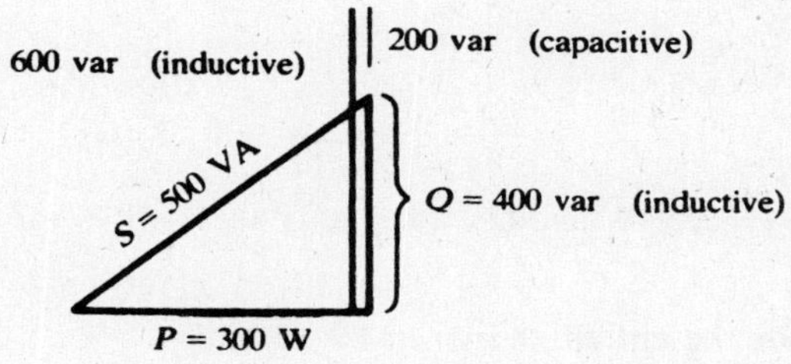

Fig. 10-15

10.7 A series circuit of $R = 10\ \Omega$ and $X_C = 5\ \Omega$ has an effective applied voltage of 120 V. Determine the complete power information.

$$Z = \sqrt{10^2 + 5^2} = 11.18\ \Omega \qquad I_{\text{eff}} = \frac{120}{11.18} = 10.73 \text{ A}$$

Then:

$$P = I_{\text{eff}}^2 R = 1152 \text{ W} \qquad Q = I_{\text{eff}}^2 X_C = 576 \text{ var (capacitive)} \qquad S = \sqrt{(1152)^2 + (576)^2} = 1288 \text{ VA}$$

and pf $= 1152/1288 = 0.894$ leading

10.8 Impedances $Z_i = 5.83\angle{-59.0^\circ}\ \Omega$ and $\mathbf{Z}_2 = 8.94\angle{63.43^\circ}\ \Omega$ are in series and carry an effective current of 5.0 A. Determine the complete power information.

$$\mathbf{Z}_T = \mathbf{Z}_1 + \mathbf{Z}_2 = 7.0 + j3.0\ \Omega$$

Hence, $$P_T = (5.0)^2(7.0) = 175 \text{ W} \qquad Q_T = (5.0)^2(3.0) = 75 \text{ var (inductive)}$$

$$S_T = \sqrt{(175)^2 + (75)^2} = 190.4 \text{ VA} \qquad \text{pf} = \frac{175}{190.4} = 0.919 \text{ lagging}$$

10.9 Obtain the total power information for the parallel circuit shown in Fig. 10-16.

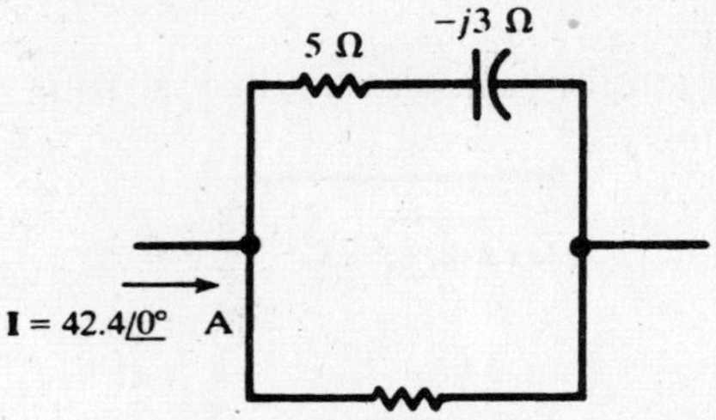

Fig. 10-16

By current division,

$$\mathbf{I}_5 = 17.88\,\underline{/18.43^\circ}\text{ A} \qquad \mathbf{I}_4 = 26.05\,\underline{/-12.53^\circ}\text{ A}$$

Then,

$$P_T = \left(\frac{17.88}{\sqrt{2}}\right)^2(5) + \left(\frac{26.05}{\sqrt{2}}\right)^2(4) = 2156\text{ W}$$

$$Q_T = \left(\frac{17.88}{\sqrt{2}}\right)^2(3) = 480\text{ var (capacitive)}$$

$$S_T = \sqrt{(2156)^2 + (480)^2} = 2209\text{ VA}$$

$$\text{pf} = \frac{2156}{2209} = 0.976\text{ leading}$$

Alternate Method

$$\mathbf{Z}_{\text{eq}} = \frac{4(5-j3)}{9-j3} = 2.40 - j0.53\quad\Omega$$

Then, $P = (42.4/\sqrt{2})^2(2.40) = 2157$ W and $Q = (42.4/\sqrt{2})^2(0.53) = 476$ var (capacitive).

10.10 Find the power factor for the circuit shown in Fig. 10-17.

With no voltage or current specified, P, Q, and S cannot be calculated. However, the power factor is the cosine of the angle on the equivalent impedance.

$$\mathbf{Z}_{\text{eq}} = \frac{(3+j4)(10)}{13+j4} = 3.68\,\underline{/36.03^\circ}\quad\Omega$$

$$\text{pf} = \cos 36.03^\circ = 0.809\text{ lagging}$$

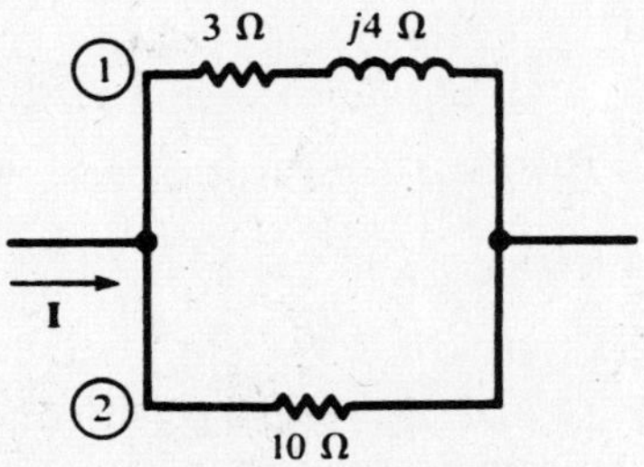

Fig. 10-17

10.11 If the total power in the circuit Fig. 10-17 is 1100 W, what are the powers in the two resistors?

By current division,

$$\frac{I_{1,\text{eff}}}{I_{2,\text{eff}}} = \frac{Z_2}{Z_1} = \frac{10}{\sqrt{3^2+4^2}} = 2$$

and so
$$\frac{P_{3\Omega}}{P_{10\Omega}} = \frac{I_{1,\text{eff}}^2(3)}{I_{2,\text{eff}}^2(10)} = \frac{6}{5}$$

Solving simultaneously with $P_{3\Omega} + P_{10\Omega} = 1100$ W gives $P_{3\Omega} = 600$ W, $P_{10\Omega} = 500$ W.

10.12 Obtain the power factor of a two-branch parallel circuit where the first branch has $\mathbf{Z}_1 = 2 + j4\ \Omega$, and the second $\mathbf{Z}_2 = 6 + j0\ \Omega$. To what value must the 6-Ω resistor be changed to result in the overall power factor 0.90 lagging?

Since the angle on the equivalent admittance is the negative of the angle on the equivalent impedance, its cosine also gives the power factor.

$$\mathbf{Y}_{\text{eq}} = \frac{1}{2+j4} + \frac{1}{6} = 0.334\ \underline{/-36.84^\circ}\quad \text{S}$$
$$\text{pf} = \cos(-36.84^\circ) = 0.80 \text{ lagging}$$

The pf is lagging because the impedance angle is positive.

Now, for a change in power factor to 0.90, the admittance angle must become $\cos^{-1} 0.90 = -25.84^\circ$. Then,

$$\mathbf{Y}'_{\text{eq}} = \frac{1}{2+j4} + \frac{1}{R} = \left(\frac{1}{10} + \frac{1}{R}\right) - j\frac{1}{5}$$

requires
$$\frac{1/5}{\dfrac{1}{10} + \dfrac{1}{R}} = \tan 25.84^\circ \qquad \text{or} \qquad R = 3.20\ \Omega$$

10.13 A voltage, $28.28\ \underline{/60^\circ}$ V, is applied to a two-branch parallel circuit in which $\mathbf{Z}_1 = 4\ \underline{/30^\circ}$ and $\mathbf{Z}_1 = 5\ \underline{/60^\circ}\ \Omega$. Obtain the power triangles for the branches and combine them into the total power triangle.

$$\mathbf{I}_1 = \frac{\mathbf{V}}{\mathbf{Z}_1} = 7.07\ \underline{/30^\circ}\ \text{A} \qquad \mathbf{I}_2 = \frac{\mathbf{V}}{\mathbf{Z}_2} = 5.66\ \underline{/0^\circ}\ \text{A}$$
$$\mathbf{S}_1 = \left(\frac{28.28}{\sqrt{2}}\ \underline{/60^\circ}\right)\left(\frac{7.07}{\sqrt{2}}\ \underline{/-30^\circ}\right) = 100\ \underline{/30^\circ} = 86.6 + j50.0$$
$$\mathbf{S}_2 = \left(\frac{28.28}{\sqrt{2}}\ \underline{/60^\circ}\right)\left(\frac{5.66}{\sqrt{2}}\ \underline{/0^\circ}\right) = 80.0\ \underline{/60^\circ} = 40.0 + j69.3$$
$$\mathbf{S}_T = \mathbf{S}_1 + \mathbf{S}_2 = 126.6 + j119.3 = 174.0\ \underline{/43.3^\circ}\quad \text{VA}$$

The power triangles and their summation are shown in Fig. 10-18.

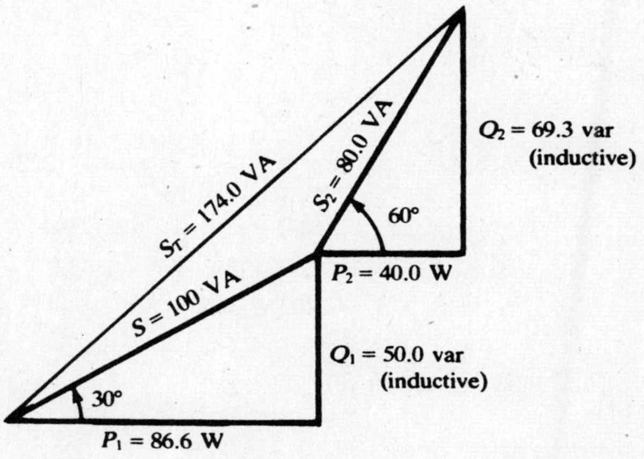

Fig. 10-18

10.14 Determine the total power information for three parallel-connected loads: load #1, 250 VA, pf = 0.50 lagging; load #2, 180 W, pf = 0.80 leading; load #3, 300 VA, 100 var (inductive).

Calculate the average power P and the reactive power Q for each load.

Load #1 Given $s = 250$ VA, $\cos\theta = 0.50$ lagging. Then,

$$P = 250(0.50) = 125\ \text{W} \qquad Q = \sqrt{(250)^2 - (125)^2} = 216.5\ \text{var (inductive)}$$

Load #2 Given $P = 180$ W, $\cos\theta = 0.80$ leading. Then, $\theta = \cos^{-1} 0.80 = -36.87^\circ$ and

$$Q = 180\tan(-36.87^\circ) = 135\ \text{var (capacitive)}$$

Load #3 Given $S = 300$ VA, $Q = 100$ var (inductive). Then,

$$P = \sqrt{(300)^2 - (100)^2} = 282.8\ \text{W}$$

Combining componentwise:

$$P_T = 125 + 180 + 282.8 = 587.8\ \text{W}$$
$$Q_T = 216.5 - 135 + 100 = 181.5\ \text{var (inductive)}$$
$$\mathbf{S}_T = 587.8 + j181.5 = 615.2 \underline{/17.16^\circ}$$

Therefore, $S_T = 615.2$ VA and pf $= \cos 17.16^\circ = 0.955$ lagging.

10.15 Obtain the complete power triangle and the total current for the parallel circuit shown in Fig. 10-19, if for branch 2, $S_2 = 1490$ VA.

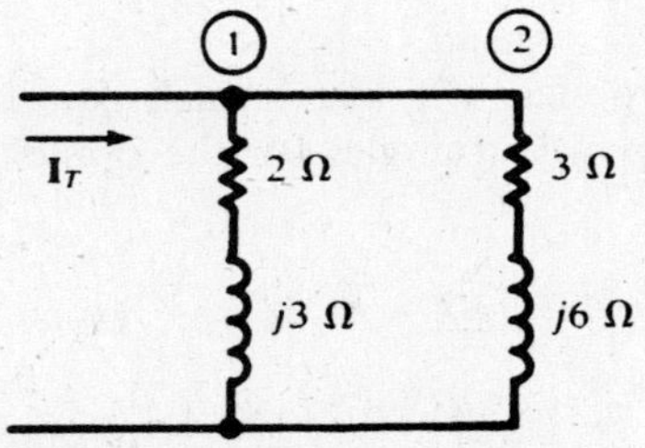

Fig. 10-19

From $S_2 = I_{2,\text{eff}}^2 Z_2$,

$$I_{2,\text{eff}}^2 = \frac{1490}{\sqrt{3^2 + 6^2}} = 222\ \text{A}^2$$

and, by current division,

$$\frac{\mathbf{I}_1}{\mathbf{I}_2} = \frac{3 + j6}{2 + j3} \qquad \text{whence} \qquad I_{1,\text{eff}}^2 = \frac{3^2 + 6^2}{2^2 + 3^2} I_{2,\text{eff}}^2 = \frac{45}{13}(222) = 768\ \text{A}^2$$

Then,

$$\mathbf{S}_1 = I_{1,\text{eff}}^2 \mathbf{Z}_1 = 768(2 + j3) = 1536 + j2304$$
$$\mathbf{S}_2 = I_{2,\text{eff}}^2 \mathbf{Z}_2 = 222(3 + j6) = 666 + j1332$$
$$\mathbf{S}_T = \mathbf{S}_1 + \mathbf{S}_2 = 2202 + j3636$$

that is, $P_T = 2202$ W, $Q_T = 3636$ var (inductive),

$$S_T = \sqrt{(2202)^2 + (3636)^2} = 4251\ \text{VA} \qquad \text{and} \qquad \text{pf} = \frac{2202}{4251} = 0.518\ \text{lagging}$$

Since the phase angle of the voltage is unknown, only the magnitude of $\mathbf{I}_T$ can be given. By current division,

$$\mathbf{I}_2 = \frac{2+j3}{5+j9}\mathbf{I}_T \qquad \text{or} \qquad I^2_{2,\text{eff}} = \frac{2^2+3^2}{5^2+9^2} I^2_{T,\text{eff}} = \frac{13}{106} I^2_{T,\text{eff}}$$

and so

$$I^2_{T,\text{eff}} = \frac{106}{13}(222) = 1811 \text{ A}^2 \qquad \text{or} \qquad I_{T,\text{eff}} = 42.6 \text{ A}$$

10.16 Obtain the complete power triangle for the circuit shown in Fig. 10-20, if the total reactive power is 2500 var (inductive). Find the branch powers P_1 and P_2.

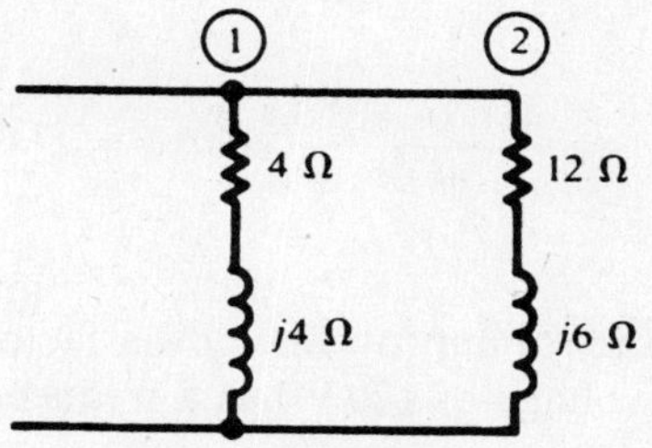

Fig. 10-20

The equivalent admittance allows the calculation of the total power triangle.

$$\mathbf{Y}_{eq} = \mathbf{Y}_1 + \mathbf{Y}_2 = 0.2488 \,\underline{/-39.57^\circ} \text{ S}$$

Then,

$$P_T = 2500 \cot 39.57^\circ = 3025 \text{ W}$$

$$\mathbf{S}_T = 3025 + j2500 = 3924 \,\underline{/39.57^\circ} \text{ VA}$$

and pf $= P_T/S_T = 0.771$ lagging.

The current ratio is $I_1/I_2 = Y_1/Y_2 = 0.177/0.0745$.

$$\frac{P_1}{P_2} = \frac{I_1^2(4)}{I_2^2(12)} = 1.88 \qquad \text{and} \qquad P_1 + P_2 = 3025 \text{ W}$$

from which $P_1 = 1975$ W and $P_2 = 1050$ W.

10.17 A load of 300 kW, with power factor 0.65 lagging, has the power factor improved to 0.90 lagging by parallel capacitors. How many kvar must these capacitors furnish, and what is the resulting percent reduction in apparent power?

The angles corresponding to the power factors are first obtained:

$$\cos^{-1} 0.65 = 49.46^\circ \qquad \cos^{-1} 0.90 = 25.84^\circ$$

Then (see Fig. 10-21),

$$Q = 300 \tan 49.46^\circ = 350.7 \text{ kvar (inductive)}$$

$$Q - Q_c = 300 \tan 25.84^\circ = 145.3 \text{ kvar (inductive)}$$

whence $Q_c = 205.4$ kvar (capacitive). Since

$$S = \frac{300}{0.65} = 461.5 \text{ kVA} \qquad S' = \frac{300}{0.90} = 333.3 \text{ kVA}$$

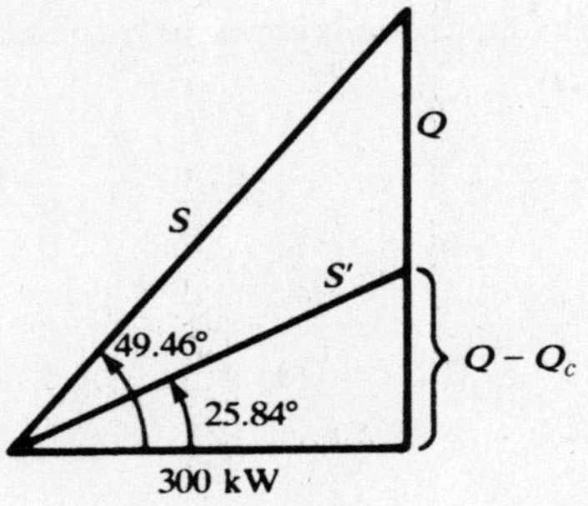

Fig. 10-21

the reduction is

$$\frac{461.5 - 333.3}{461.5}(100\%) = 27.8\%$$

10.18 Find the capacitance C necessary to improve the power factor to 0.95 lagging in the circuit shown in Fig. 10-22, if the effective voltage of 120 V has a frequency of 60 Hz.

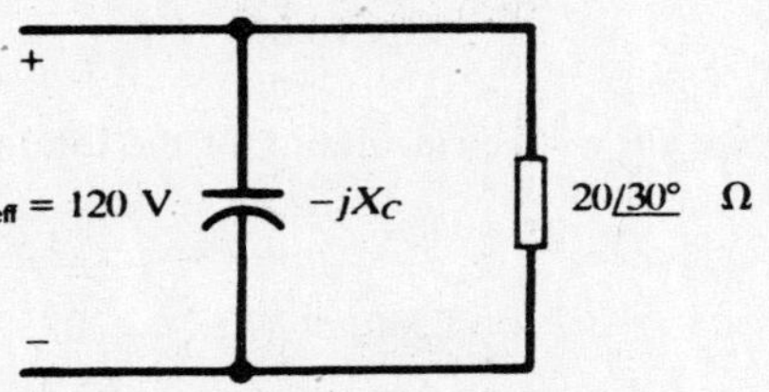

Fig. 10-22

Admittance provides a good approach.

$$Y_{eq} = j\omega C + \frac{1}{20\,\underline{/30^\circ}} = 0.0433 - j(0.0250 - \omega C) \quad \text{(S)}$$

The admittance diagram, Fig. 10-23, illustrates the next step.

$$\begin{aligned}\theta &= \cos^{-1} 0.95 = 18.19^\circ \\ 0.0250 - \omega C &= (0.0433)(\tan 18.19^\circ) \\ \omega C &= 0.0108 \\ C &= 28.6\,\mu\text{F}\end{aligned}$$

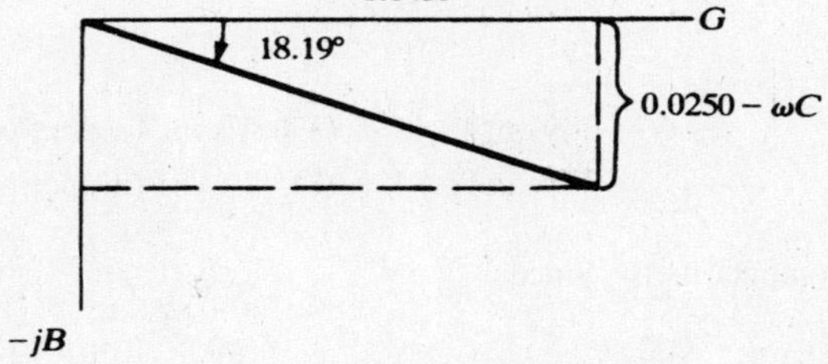

Fig. 10-23

10.19 A circuit with impedance $\mathbf{Z} = 10.0\,\angle 60^\circ\ \Omega$ has the power factor improved by a parallel capacitive reactance of $20\ \Omega$. What percent reduction in current results?

Since $\mathbf{I} = \mathbf{VY}$, the current reduction can be obtained from the ratio of the admittances after and before addition of the capacitors.

$$\mathbf{Y}_{\text{before}} = 0.100\,\angle -60^\circ\ \text{S} \qquad \text{and} \qquad \mathbf{Y}_{\text{after}} = 0.050\,\angle 90^\circ + 0.100\,\angle -60^\circ = 0.062\,\angle -36.20^\circ\ \text{S}$$

$$\frac{I_{\text{after}}}{I_{\text{before}}} = \frac{0.062}{0.100} = 0.620$$

so the reduction is 38 percent.

10.20 A transformer rated at a maximum of 25 kVA supplies a 12-kW load at power factor 0.60 lagging. What percent of the transformer rating does this load represent? How many kW in additional load may be added at unity power factor before the transformer exceeds its rated kVA?

For the 12-kW load, $S = 12/060 = 20\,\text{kVA}$. The transformer is at $(20/25)(100\%) = 80\%$ of full rating.

The additional load at unity power factor does not change the reactive power,

$$Q = \sqrt{(20)^2 - (12)^2} = 16 \text{ kvar (inductive)}$$

Then, at full capacity,

$$\theta' = \sin^{-1}(16/25) = 39.79^\circ$$
$$P' = 25\cos 39.79^\circ = 19.2\,\text{kW}$$
$$P_{\text{add}} = 19.2 - 12.0 = 7.2\,\text{kW}$$

Note that the full-rated kVA is shown by an arc in Fig. 10-24, of radius 25.

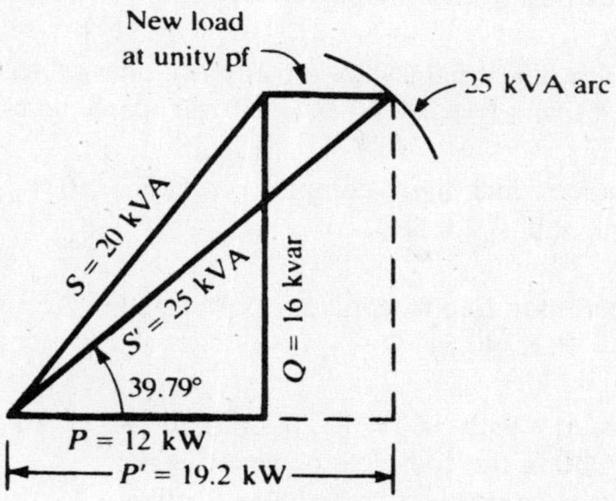

Fig. 10-24

10.21 Referring to Problem 10.20, if the additional load has power factor 0.866 leading, how many kVA may be added without exceeding the transformer rating?

The original load is $\mathbf{S} = 12 + j16$ kVA and the added load is

$$\mathbf{S}_2 = S_2\,\angle -30^\circ = S_2(0.866) - jS_2(0.500) \quad (\text{kVA})$$

The total is $\mathbf{S}_t = (12 + 0.866S_2) + j(16 - 0.500S_2)$ (kVA). Then,

$$S_T^2 = (12 + 0.866S_2)^2 + (16 - 0.500S_2)^2 = (25)^2$$

gives $S_2 = 12.8$ kVA.

10.22 An induction motor with a shaft power output of 1.56 kW has an efficiency of 85 percent. At this load, the power factor is 0.80 lagging. Give complete input power information.

$$\frac{P_{\text{out}}}{P_{\text{in}}} = 0.85 \qquad \text{or} \qquad P_{\text{in}} = \frac{1.5}{0.85} = 1.765 \text{ kW}$$

Then, from the power triangle,

$$S_{\text{in}} = \frac{1.765}{0.80} = 2.206 \text{ kVA} \qquad Q_{\text{in}} = \sqrt{(2.206)^2 - (1.765)^2} = 1.324 \text{ kvar (inductive)}$$

The equivalent circuit of an induction motor contains a variable resistance which is a function of the shaft load. The power factor is therefore variable, ranging from values near 0.30 at starting to 0.85 at full load.

Supplementary Problems

10.23 Given a circuit with an applied voltage $v = 14.14 \cos \omega t$ (V) and a resulting current $i = 17.1 \cos(\omega t - 14.05°)$ (mA), determine the complete power triangle.
Ans. $P = 117$ mW, $Q = 29.3$ mvar (inductive), pf $= 0.970$ lagging

10.24 Given a circuit with an applied voltage $v = 340 \sin(\omega t - 60°)$ (V) and a resulting current $i = 13.3 \sin(\omega t - 48.7°)$ (A), determine the complete power triangle.
Ans. $P = 2217$ W, $Q = 443$ var (capacitive), pf $= 0.981$ leading

10.25 A two-element series circuit with $R = 5.0\,\Omega$ and $X_L = 15.0\,\Omega$, has an effective voltage 31.6 V across the resistance. Find the complex power and the power factor. *Ans.* $200 + j600$ Va, 0.316 lagging

10.26 A circuit with impedance $\mathbf{Z} = 8.0 - j6.0\,\Omega$ has an applied phasor voltage $70.7\,\angle{-90.0°}$ V. Obtain the complete power triangle. *Ans.* $P = 200$ W, $Q = 150$ var (capacitive), pf $= 0.80$ leading

10.27 Determine the circuit impedance which has a complex power $\mathbf{S} = 5031\,\angle{-26.57°}$ VA for an applied phasor voltage $212.1\,\angle{0°}$ V. *Ans.* $4.0 - j2.0\,\Omega$

10.28 Determine the impedance corresponding to apparent power 3500 VA, power factor 0.76 lagging, and effective current 18.0 A. *Ans.* $10.8\,\angle{40.54°}\,\Omega$

10.29 A two-branch parallel circuit, with $\mathbf{Z}_1 = 10\,\angle{0°}\,\Omega$ and $\mathbf{Z}_2 = 8.0\,\angle{-30.0°}\,\Omega$, has a total current $i = 7.07 \cos(\omega t - 90°)$ (A). Obtain the complete power triangle.
Ans. $P = 110$ W, $Q = 32.9$ var (capacitive), pf $= 0.958$ leading

10.30 A two-branch parallel circuit has branch impedances $\mathbf{Z}_1 = 2.0 - j5.0\,\Omega$ and $\mathbf{Z}_2 = 1.0 + j1.0\,\Omega$. Obtain the complete power triangle for the circuit if the 2.0-Ω resistor consumes 20 W.
Ans. $P = 165$ W, $Q = 95$ var (inductive), pf $= 0.867$ lagging

10.31 A two-branch parallel circuit, with impedances $\mathbf{Z}_1 = 4.0\,\angle{-30°}\,\Omega$ and $\mathbf{Z}_2 = 5.0\,\angle{60°}\,\Omega$, has an applied effective voltage of 20 V. Obtain the power triangles for the branches and combine them to obtain the total power triangle. *Ans.* $S_T = 128.1$ VA, pf $= 0.989$ lagging

10.32 Obtain the complex power for the complete circuit of Fig. 10-25 if branch 1 takes 8.0 kvar.
Ans. $\mathbf{S} = 8 + j12$ kVA, pf $= 0.555$ lagging

10.33 In the circuit of Fig. 10-26, find $\mathbf{Z}$ if $S_T = 3373$ Va, pf $= 0.938$ leading, and the 3-Ω resistor has an average power of 666 W. *Ans.* $2 - j2\,\Omega$

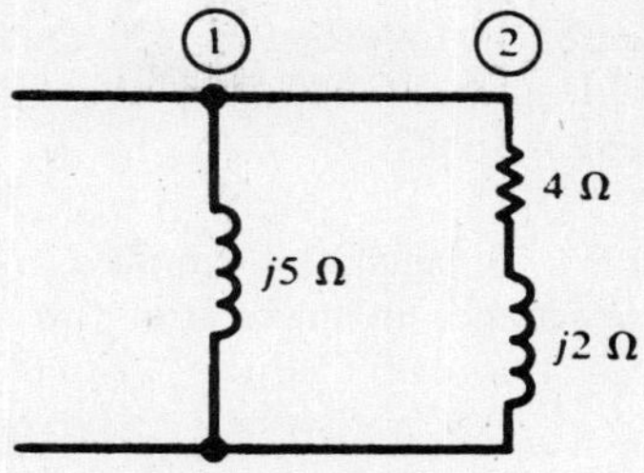

Fig. 10-25

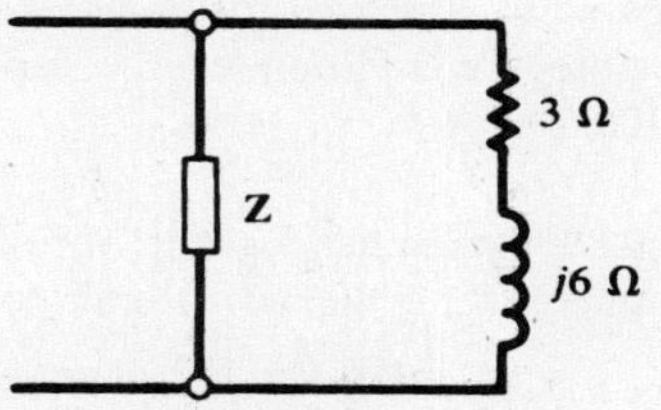

Fig. 10-26

10.34 The parallel circuit in Fig. 10-27 has a total average power of 1500 W. Obtain the total power-triangle information. *Ans.* $\mathbf{S} = 1500 + j2471$ VA, pf = 0.519 lagging

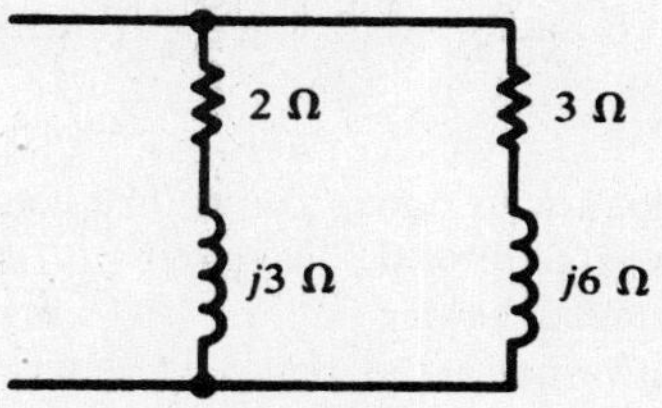

Fig. 10-27

10.35 Determine the average power in the 15-Ω and 8-Ω resistances in Fig. 10-28 if the total average power in the circuit is 2000 W. *Ans.* 723 W, 1277 W

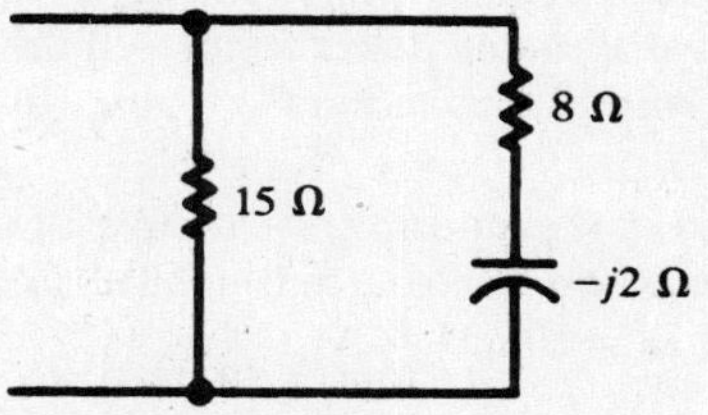

Fig. 10-28

10.36 A three-branch parallel circuit, with $\mathbf{Z}_1 = 25\,\angle 15^\circ\ \Omega$, $\mathbf{Z}_2 = 15\,\angle 60^\circ$, and $\mathbf{Z}_3 = 15\,\angle 90^\circ\ \Omega$, has an applied voltage $\mathbf{V} = 339.4\,\angle -30^\circ$ V. Obtain the total apparent power and the overall power factor.
Ans. 4291 VA, 0.966 lagging

10.37 Obtain the complete power triangle for the following parallel-connected loads: load #1, 5 kW, pf = 0.80 lagging; load #2, 4 kVA, 2 kvar (capacitive); load #3, 6 kVA, pf = 0.90 lagging.
Ans. 14.535 kVA, pf = 0.954 lagging

10.38 Obtain the complete power triangle for the following parallel-connected loads: load #1, 200 VA, pf = 0.70 lagging; load #2, 350 VA, pf = 0.50 lagging; load #3, 275 VA, pf = 1.00.
Ans. $\mathbf{S} = 590 + j444$ VA, pf = 0.799 lagging

10.39 A 4500-VA load at power factor 0.75 lagging is supplied by a 60-Hz source at effective voltage 240 V. Determine the parallel capacitance in microfarads necessary to improve the power factor to (*a*) 0.90 lagging, (*b*) 0.90 leading. *Ans.* (*a*) 61.8 μF; (*b*) 212 μF

10.40 In Problem 10.39, what percent reduction in line current and total voltamperes was achieved in part (*a*)? What further reduction was achieved in part (*b*)? *Ans.* 16.1 percent, none

10.41 The addition of a 20-kvar capacitor bank improved the power factor of a certain load to 0.90 lagging. Determine the complex power before the addition of the capacitors, if the final apparent power is 185 kVA. *Ans.* S = 166.5 + j100.6 kVA

10.42 A 25-kVA load with power factor 0.80 lagging has a group of resistive heating units added at unity power factor. How many kW do these units take, if the new overall power factor is 0.85 lagging?
Ans. 4.2 kW

10.43 A 500-kVA transformer is at full load and 0.60 lagging power faactor. A capacitor bank is added, improving the power factor to 0.90 lagging. After improvement, what percent of rated kVA is the transformer carrying? *Ans.* 66.7 percent

10.44 A 100-kVA transformer is at 80 percent of rated load at power factor 0.85 lagging. How many kVA in additional load at 0.60 lagging power factor will bring the transformer to full rated load?
Ans. 21.2 kVA

10.45 A 250-kVA transformer is at full load with power factor 0.80 lagging. (*a*) How many kvar of capacitors must be added to improve this power factor to 0.90 lagging? (*b*) After improvement of the power factor, a new load is to be added at 0.50 lagging power factor. How many kVA of this new load will bring the transformer back to rated kVA, and what is the final power factor?
Ans. (*a*) 53.1 kvar (capacitive); (*b*) 33.35 kVA, 0.867 lagging

10.46 A 65-kVA load with a lagging power factor is combined with a 25-kVA synchronous motor load which operates at pf = 0.60 leading. Find the power factor of the 65-kVA load, if the overall power factor is 0.85 lagging. *Ans.* 0.585 lagging

10.47 An induction motor load of 2000 kVA has power factor 0.80 lagging. Synchronous motors totaling 500 kVA are added and operated at a leading power factor. If the overall power factor is then 0.90 lagging, what is the power factor of the synchronous motors? *Ans.* 0.92 leading

10.48 Find maximum energy (E) stored in the inductor of Example 10.17(*a*) and show that it is greater than the sum of maximum stored energies when each source is applied alone (E_1 and E_2).
Ans. $E = 1.6$ mJ, $E_1 = 306$ μJ, $E_2 = 506$ μJ

10.49 The terminal voltage and current of a two-terminal circuit are $\overline{V}_{rms} = 120$ V and $\overline{I}_{rms} = 30\angle -60^\circ$ A at $f = 60$ Hz. Compute the complex power. Find the impedance of the circuit and its equivalent circuit made of two series elements.
Ans. $\overline{S} = 1800 + j3117.7$ VA, $Z = 2 + j3.464 = R + jL\omega$, $R = 2\ \Omega$, $L = 9.2$ mH

10.50 In the circuit of Fig. 10-29 the voltage source has effective value 10 V at $\omega = 1$ rad/s and the current source is zero. (*a*) Find the average and reactive powers delivered by the voltage source. (*b*) Find the effective value of the current in the resistor and the average power absorbed by it and the reactive powers in L and C. Show the balance sheet for the average and reactive powers between the source and R, L, and C.
Ans. (*a*) $P = 80$ W, $Q = -60$ var, (*b*) $I_R = 5\sqrt{2}$ A, $P_R = 80$ W, $Q_C = -160$ var, $Q_L = 100$ var, $P_R = P$ and $Q_L + Q_C = Q$

10.51 In the circuit of Fig. 10-29, $v_a = 10\sqrt{2}\cos t$ and $i_b = 10\sqrt{2}\cos 2t$. (*a*) Find the average power delivered by each source. (*b*) Find the current in the resistor and the average power absorbed by it.
Ans. (*a*) $P_a = P_b = 80$ W, (*b*) $i_R = 2\sqrt{10}\cos(t - 26.5^\circ) + 2\sqrt{10}\cos(2t - 63.4^\circ)$, $P_R = 160$ W

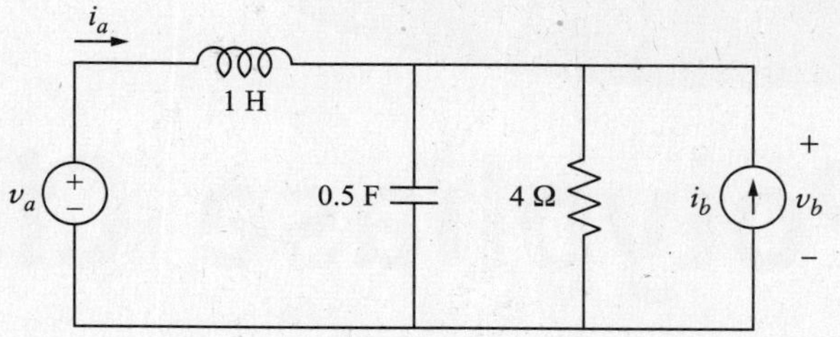

Fig. 10-29

10.52 A single-phase AC source having effective value 6 kV delivers 100 kW at a power factor 0.8 lagging to two parallel loads. The individual power factors of the loads are $pf_1 = 0.7$ lagging and $pf_2 = 0.1$ leading. (*a*) Find powers P_1 and P_2 delivered to each load. (*b*) Find the impedance of each load and their combination.
Ans. (*a*) $P_1 = 97.54$ kW, $P_2 = 2.46$ kW, (*b*) $Z_1 = 0.244\,\underline{/-84.26}$ Ω, $Z_2 = 0.043\,\underline{/45.57}$ Ω, $Z = 0.048\,\underline{/36.87}$ Ω

10.53 A practical voltage source is modeled by an ideal voltage source V_g with an open-circuited effective value of 320 V in series with an output impedance $Z_g = 50 + j100$ Ω. The source feeds a load $Z_\ell = 200 + j100$ Ω. See Fig. 10-30. (*a*) Find the average power and reactive power delivered by V_g. (*b*) Find the average power and reactive power absorbed by the load. (*c*) A reactive element jX is added in parallel to Z_ℓ. Find the X such that power delivered to Z_ℓ is maximized.
Ans. (*a*) $P_g = 250$ W and $Q_g = 200$ var, (*b*) $P_\ell = 200$ W and $Q_\ell = 100$ var, (*c*) $X = -100$ Ω

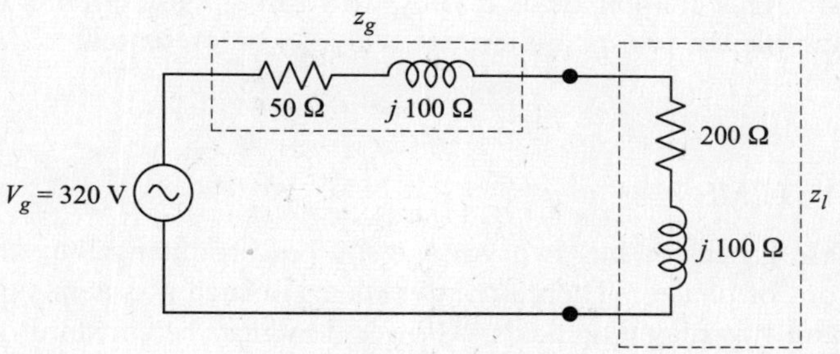

Fig. 10-30

CHAPTER 11

Polyphase Circuits

11.1 INTRODUCTION

The instantaneous power delivered from a sinusoidal source to an impedance is

$$p(t) = v(t)i(t) = V_pI_p \cos\theta + V_pI_p \cos(2\omega t - \theta) \tag{1}$$

where V_p and I_p are the rms values of v and i, respectively, and θ is the angle between them. The power oscillates between $V_pI_p(1+\cos\theta)$ and $V_pI_p(-1+\cos\theta)$. In power systems, especially at higher levels, it is desirable to have a steady flow of power from source to load. For this reason, polyphase systems are used. Another advantage is having more than one voltage value on the lines. In polyphase systems, V_p and I_p indicate voltage and current, respectively, in a phase which may be different from voltages and currents in other phases. This chapter deals mainly with three-phase circuits which are the industry standard. However, examples of two-phase circuits will also be presented.

11.2 TWO-PHASE SYSTEMS

A balanced two-phase generator has two voltage sources producing the same amplitude and frequency but 90° or 180° out of phase. There are advantages in such a system since it gives the user the option of two voltages and two magnetic fields. Power flow may be constant or pulsating.

EXAMPLE 11.1 An ac generator contains two voltage sources with voltages of the same amplitude and frequency, but 90° out of phase. The references of the sources are connected together to form the generator's reference terminal n. The system feeds two identical loads [Fig. 11-1(*a*)]. Find currents, voltages, the instantaneous and average powers delivered.

Terminal voltages and currents at generator's terminal are

$$\begin{aligned} v_a(t) &= V_p\sqrt{2}\cos\omega t & v_b(t) &= V_p\sqrt{2}\cos(\omega t - 90^\circ) \\ i_a(t) &= I_p\sqrt{2}\cos(\omega t - \theta) & i_b(t) &= I_p\sqrt{2}\cos(\omega t - 90^\circ - \theta) \end{aligned} \tag{2}$$

In the phasor domain, let $\mathbf{Z} = |Z|\underline{/\theta}$ and $I_p = V_p/|Z|$. Then,

$$\begin{aligned} &\mathbf{V}_{\text{AN}} = V_p \underline{/0} \quad && \mathbf{V}_{BN} = V_p \underline{/-90^\circ} \quad && \mathbf{V}_{AB} = \mathbf{V}_{AN} - \mathbf{V}_{BN} = \sqrt{2}V_p \underline{/45^\circ} \\ &\mathbf{I}_A = I_p \underline{/-\theta} && \mathbf{I}_B = I_p \underline{/-90^\circ - \theta} && \mathbf{I}_N = \mathbf{I}_A + \mathbf{I}_B = I_p\sqrt{2}\underline{/-45^\circ - \theta} \end{aligned} \tag{3}$$

The voltage and current phasors are shown in Fig. 11-1(*b*).

Instantaneous powers $p_A(t)$ and $p_B(t)$ delivered by the two sources are

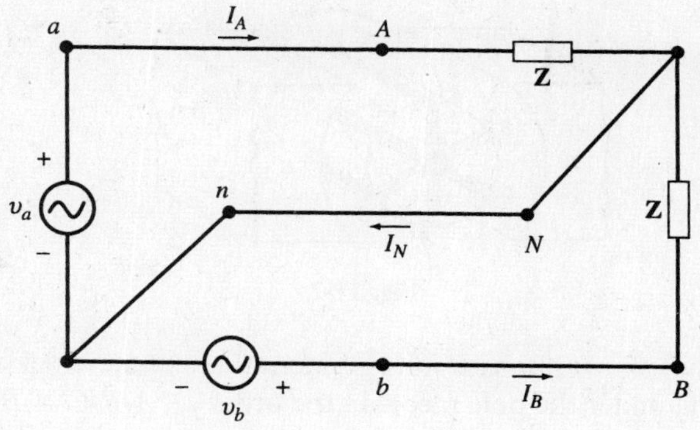

(a)

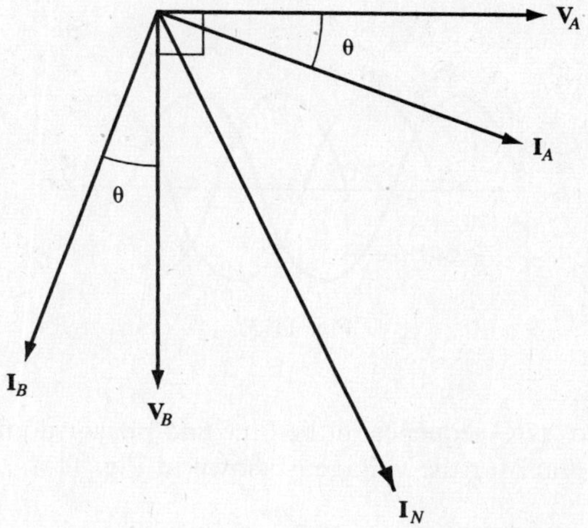

(b)

Fig. 11-1

$$p_a(t) = v_a(t)i_a(t) = V_pI_p \cos\theta + V_pI_p \cos(2\omega t - \theta)$$
$$p_b(t) = v_b(t)i_b(t) = V_pI_p \cos\theta - V_pI_p \cos(2\omega t - \theta)$$

The total instantaneous power $p_T(t)$ delivered by the generator is

$$p_T(t) = p_a(t) + p_b(t) = V_pI_p \cos\theta + V_pI_p \cos(2\omega t - \theta) + V_pI_p \cos\theta - V_pI_p \cos(2\omega t - \theta) = 2V_pI_p \cos\theta$$

Thus,
$$p_T(t) = P_{avg} = 2V_pI_p \cos\theta \tag{4}$$

In the system of Fig. 11-1(*a*), two voltage values V_p and $\sqrt{2}V_p$ are available to the load and the power flow is constant. In addition, the 90°-phase shift between the two voltages may be used to produce a special rotating magnetic field needed in some applications.

11.3 THREE-PHASE SYSTEMS

Three-phase generators contain three sinusoidal voltage sources with voltages of the same frequency but a 120°-phase shift with respect to each other. This is realized by positioning three coils at 120° electrical angle separations on the same rotor. Normally, the amplitudes of the three phases are also equal. The generator is then balanced. In Fig. 11-2, three coils are equally distributed about the circumference of the rotor; that is, the coils are displaced from one another by 120 mechanical degrees.

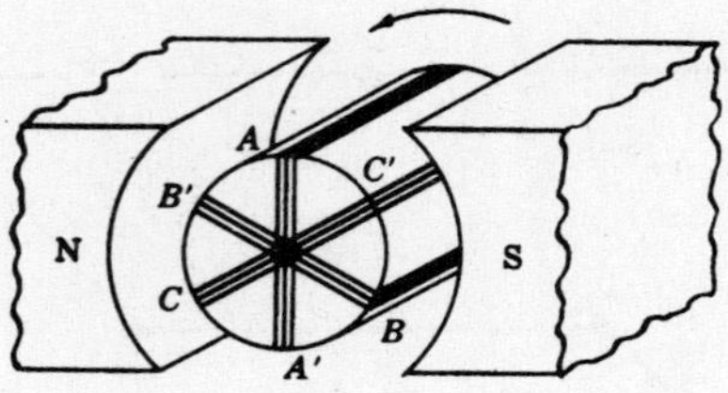

Fig. 11-2

Coil ends and slip rings are not shown; however, it is evident that counterclockwise rotation results in the coil sides *A*, *B*, and *C* passing under the pole pieces in the order ... *A-B-C-A-B-C* ... Voltage polarities reverse for each change of pole. Assuming that the pole shape and corresponding magnetic flux density are such that the induced voltages are sinusoidal, the result for the three coils is as shown in Fig. 11-3. Voltage *B* is 120 electrical degrees later than *A*, and *C* is 240° later. This is referred to as the *ABC sequence*. Changing the direction of rotation would result in ... *A-C-B-A-C-B* ..., which is called the *CBA sequence*.

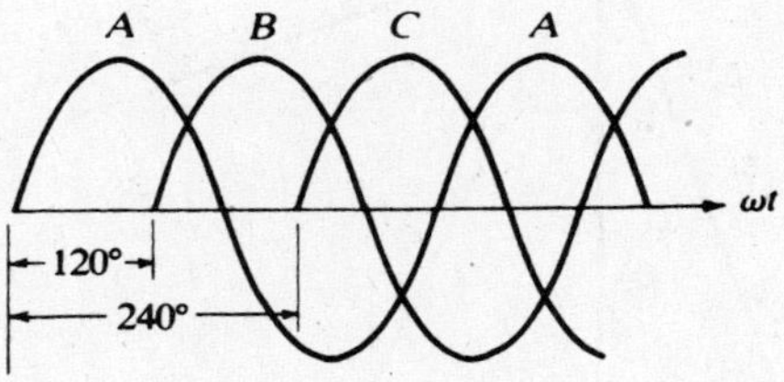

Fig. 11-3

The voltages of a balanced *ABC* sequence in the time and phasor domains are given in (5) and (6), respectively. The phasor diagram for the voltage is shown in Fig. 11-4.

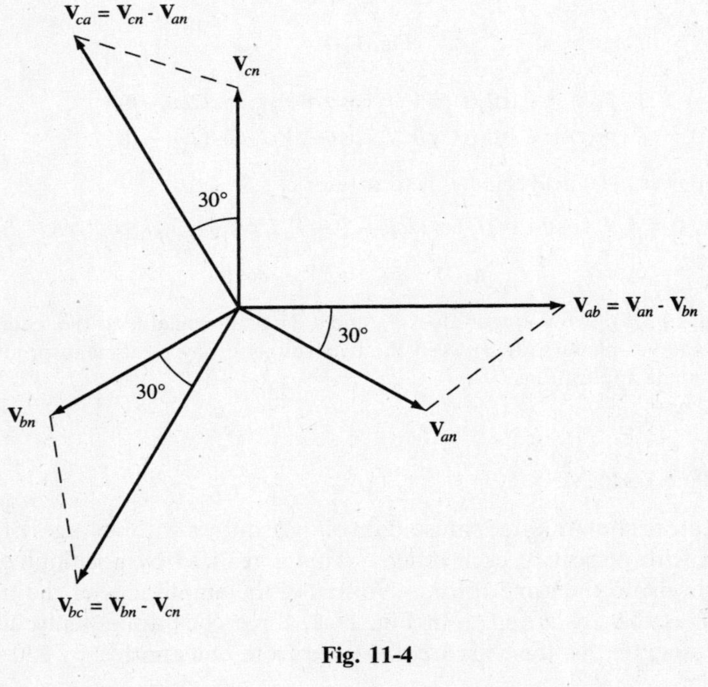

Fig. 11-4

$$v_{an}(t) = (V_p\sqrt{2})\cos\omega t \qquad v_{bn}(t) = (V_p\sqrt{2})\cos(\omega t - 120°) \qquad v_{cn}(t) = (V_p\sqrt{2})\cos(\omega t - 240°) \quad (5)$$

$$\mathbf{V}_{an} = V_p \angle 0 \qquad \mathbf{V}_{bn} = V_p \angle -120° \qquad \mathbf{V}_{cn} = V_p \angle -240° \quad (6)$$

11.4 WYE AND DELTA SYSTEMS

The ends of the coils can be connected in wye (also designated Y; see Section 11.8), with ends A', B', and C' joined at a common point designated the *neutral*, N; and with ends A, B, and C brought out to become the *lines* A, B, and C of the three-phase system. If the neutral point is carried along with the lines, it is a *three-phase, four-wire* system. In Fig. 11-5, the lines are designated by lowercase a, b, and c at the supply, which could either be a transformer bank or a three-phase alternator, and by uppercase A, B, and C at the load. If line impedances must be considered, then the current direction through, for example, line aA would be $\mathbf{I}_{aA}$, and the phasor line voltage drop $\mathbf{V}_{aA}$.

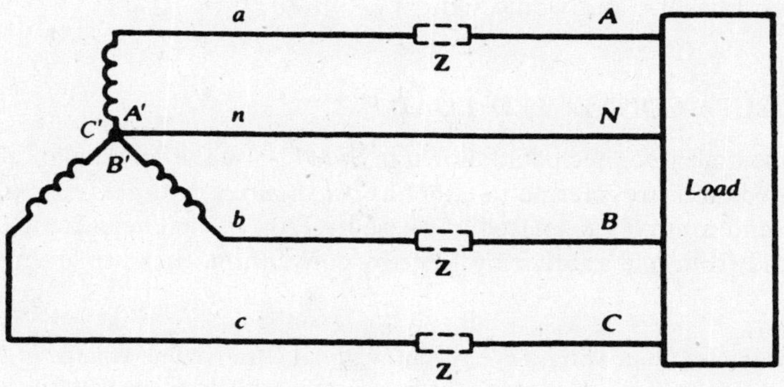

Fig. 11-5

The generator coil ends can be connected as shown in Fig. 11-6, making a delta-connected (or Δ-connected), three-phase system with lines a, b, and c. A delta-connected set of coils has no neutral point to produce a four-wire system, except through the use of Δ-Y transformers.

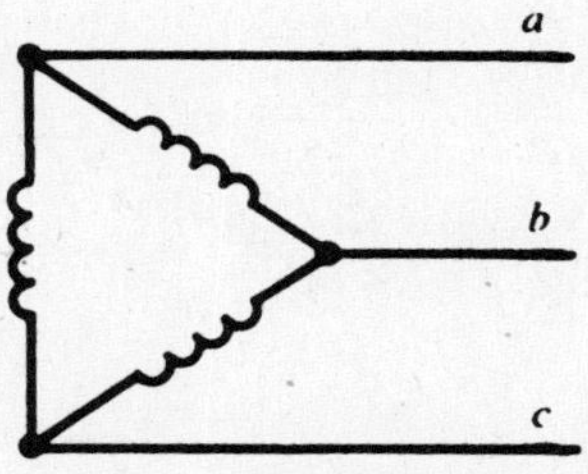

Fig. 11-6

11.5 PHASOR VOLTAGES

The selection of a phase angle for one voltage in a three-phase system fixes the angles of all other voltages. This is tantamount to fixing the $t = 0$ point on the horizontal axis of Fig. 11-3, which can be done quite arbitrarily. In this chapter, an angle of zero will always be associated with the phasor voltage of line B with respect to line C: $\mathbf{V}_{BC} \equiv V_L \angle 0°$.

It is shown in Problem 11.4 that the line-to-line voltage V_L is $\sqrt{3}$ times the line-to-neutral voltage. All ABC-sequence voltages are shown in Fig. 11-7(a) and CBA voltages in Fig. 11-7(b). These phasor

$$\mathbf{V}_{AB} = V_L\,\underline{/120°}$$
$$\mathbf{V}_{BC} = V_L\,\underline{/0°}$$
$$\mathbf{V}_{CA} = V_L\,\underline{/240°}$$
$$\mathbf{V}_{AN} = (V_L/\sqrt{3})\underline{/90°}$$
$$\mathbf{V}_{BN} = (V_L/\sqrt{3})\underline{/-30°}$$
$$\mathbf{V}_{CN} = (V_L/\sqrt{3})\underline{/-150°}$$

$$\mathbf{V}_{AB} = V_L\,\underline{/240°}$$
$$\mathbf{V}_{BC} = V_L\,\underline{/0°}$$
$$\mathbf{V}_{CA} = V_L\,\underline{/120°}$$
$$\mathbf{V}_{AN} = (V_L/\sqrt{3})\underline{/-90°}$$
$$\mathbf{V}_{BN} = (V_L/\sqrt{3})\underline{/30°}$$
$$\mathbf{V}_{CN} = (V_L/\sqrt{3})\underline{/150°}$$

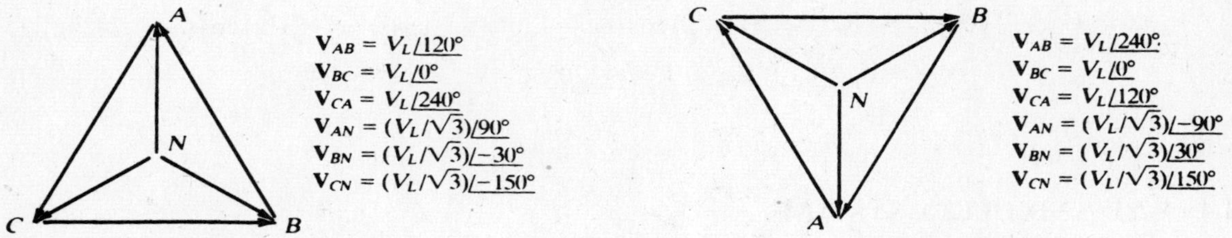

(a) Sequence ABC (b) Sequence CBA

Fig. 11-7

voltages, in keeping with the previous chapters, reflect maximum values. In the three-phase, four-wire, 480-volt system, widely used for industrial loads, and the 208-volt system, common in commercial buildings, effective values are specified. In this chapter, a line-to-line voltage in the former system would be $\mathbf{V}_{BC} = 678.8\,\underline{/0°}$ V, making $V_{BC\,\text{eff}} = 678.8/\sqrt{2} = 480$ V. People who regularly work in this field use effective-valued phasors, and would write $\mathbf{V}_{BC} = 480\,\underline{/0°}$ V.

11.6 BALANCED DELTA-CONNECTED LOAD

Three identical impedances connected as shown in Fig. 11-8 make up a balanced Δ-connected load. The currents in the impedances are referred to either as *phase currents* or *load currents*, and the three will be equal in magnitude and mutually displaced in phase by 120°. The line currents will also be equal in magnitude and displaced from one another by 120°; by convention, they are given a direction from the source to the load.

EXAMPLE 11.2 A three-phase, three-wire, ABC system, with an effective line voltage of 120 V, has three impedances of $5.0\,\underline{/45°}\,\Omega$ in a Δ-connection. Determine the line currents and draw the voltage-current phasor diagram.

The maximum line voltage is $120\sqrt{2} = 169.7$ V. Referring to Fig. 11-7(a), the voltages are:

$$\mathbf{V}_{AB} = 169.7\,\underline{/120°}\ \text{V} \qquad \mathbf{V}_{BC} = 169.7\,\underline{/0°}\ \text{V} \qquad \mathbf{V}_{CA} = 169.7\,\underline{/240°}\ \text{V}$$

Double subscripts give the phase-current directions; for example, $\mathbf{I}_{AB}$ passes through the impedance from line A to line B. All current directions are shown in Fig. 11-8. Then the phase currents are

$$\mathbf{I}_{AB} = \frac{\mathbf{V}_{AB}}{\mathbf{Z}} = \frac{169.7\,\underline{/120°}}{5\,\underline{/45°}} = 33.9\,\underline{/75°}\ \text{A}$$

$$\mathbf{I}_{BC} = \frac{\mathbf{V}_{BC}}{\mathbf{Z}} = \frac{169.7\,\underline{/0°}}{5\,\underline{/45°}} = 33.9\,\underline{/-45°}\ \text{A}$$

$$\mathbf{I}_{CA} = \frac{\mathbf{V}_{CA}}{\mathbf{Z}} = \frac{169.7\,\underline{/240°}}{5\,\underline{/45°}} = 33.9\,\underline{/195°}\ \text{A}$$

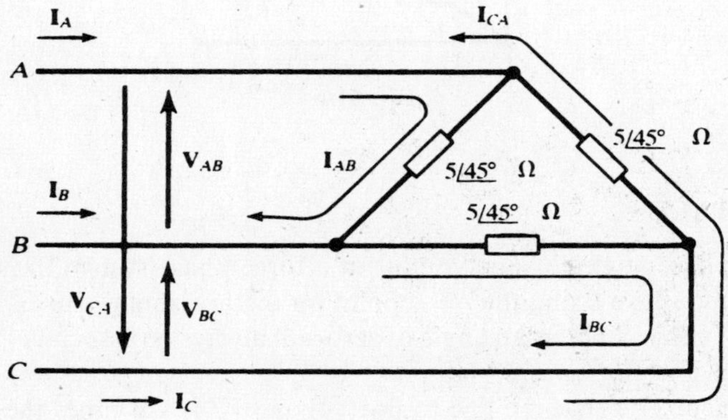

Fig. 11-8

By KCL, line current $\mathbf{I}_A$ is given by

$$\mathbf{I}_A = \mathbf{I}_{AB} + \mathbf{I}_{AC} = 33.9\,\angle 75^\circ - 33.9\,\angle 195^\circ = 58.7\,\angle 45^\circ \quad \text{A}$$

Similarly, $\mathbf{I}_B = 58.7\,\angle -75^\circ$ A and $\mathbf{I}_C = 58.7\,\angle 165^\circ$ A.

The line-to-line voltages and all currents are shown on the phasor diagram, Fig. 11-9. Note particularly the balanced currents. After one phase current has been computed, all other currents may be obtained through the symmetry of the phasor diagram. Note also that $33.9 \times \sqrt{3} = 58.7$; that is, $I_L = \sqrt{3} I_{Ph}$ for a balanced delta load.

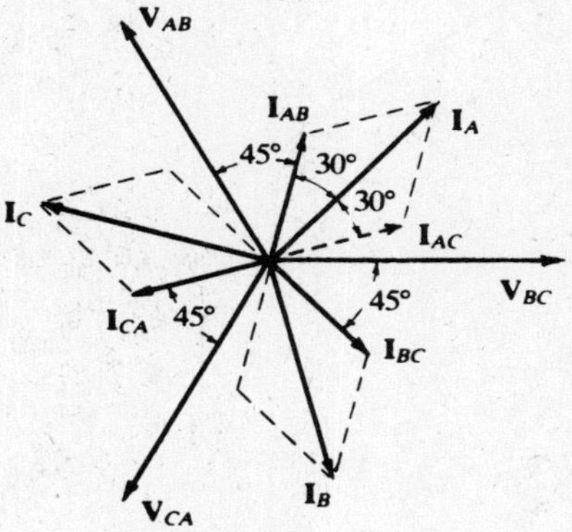

Fig. 11-9

11.7 BALANCED FOUR-WIRE, WYE-CONNECTED LOAD

Three identical impedances connected as shown in Fig. 11-10 make up a balanced Y-connected load. the currents in the impedances are also the line currents; so the directions are chosen from the source to the load, as before.

EXAMPLE 11.3 A three-phase, four-wire, *CBA* system, with an effective line voltage of 120 V, has three impedances of $20\,\angle -30^\circ\ \Omega$ in a Y-connection (Fig. 11-10). Determine the line currents and draw the voltage-current phasor diagram.

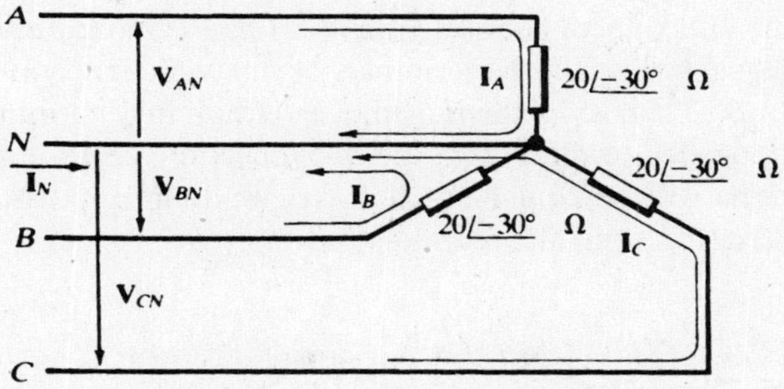

Fig. 11-10

The maximum line voltage is 169.7 V, and the line-to-neutral magnitude, $169.7/\sqrt{3} = 98.0$ V. From Fig. 11-7(*b*),

$$\mathbf{V}_{AN} = 98.0\,\angle -90^\circ \text{ V} \qquad \mathbf{V}_{BN} = 98.0\,\angle 30^\circ \text{ V} \qquad \mathbf{V}_{CN} = 98.0\,\angle 150^\circ \text{ V}$$

Then
$$\mathbf{I}_A = \frac{\mathbf{V}_{AN}}{\mathbf{Z}} = \frac{98.01\,\angle -90^\circ}{20\,\angle -30^\circ} = 4.90\,\angle -60^\circ \text{ A}$$

and, similarly, $\mathbf{I}_B = 4.90\,\angle 60^\circ$ A, $\mathbf{I}_C = 4.90\,\angle 180^\circ$ A.

The voltage-current phasor diagram is shown in Fig. 11-11. Note that with one line current calculated, the other two can be obtained through the symmetry of the phasor diagram. All three line currents return through the neutral. Therefore, the neutral current is the negative sum of the line currents:

$$\mathbf{I}_N = -(\mathbf{I}_A + \mathbf{I}_B + \mathbf{I}_C) = 0$$

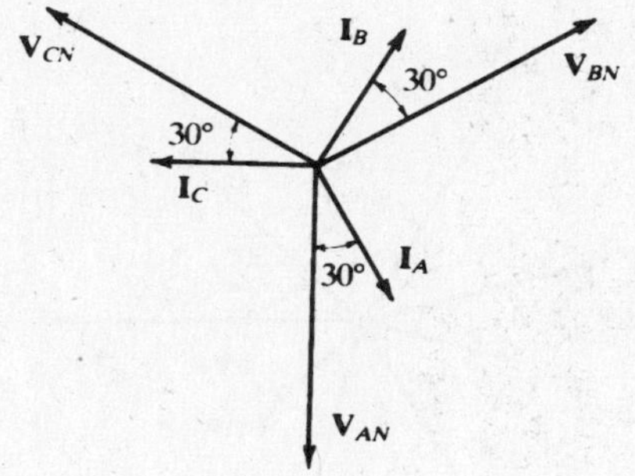

Fig. 11-11

Since the neutral current of a balanced, Y-connected, three-phase load is always zero, the neutral conductor may, for computation purposes, be removed, with no change in the results. In actual power circuits, it must not be physically removed, since it carries the (small) unbalance of the currents, carries short-circuit or fault currents for operation of protective devices, and prevents overvoltages on the phases of the load. Since the computation in Example 11.3 proceeded without difficulty, the neutral will be included when calculating line currents in balanced loads, even when the system is actually three-wire.

11.8 EQUIVALENT Y- AND Δ-CONNECTIONS

Figure 11-12 shows three impedances connected in a Δ (*delta*) configuration, and three impedances connected in a Y (*wye*) configuration. Let the terminals of the two connections be identified in pairs as indicated by the labels α, β, γ. Then $\mathbf{Z}_1$ is the impedance "adjoining" terminal α in the Y-connection, and $\mathbf{Z}_C$ is the impedance "opposite" terminal α in the Δ-connection, and so on. Looking into any two terminals, the two connections will be equivalent if corresponding input, output, and transfer impedances are equal. The criteria for equivalence are as follows:

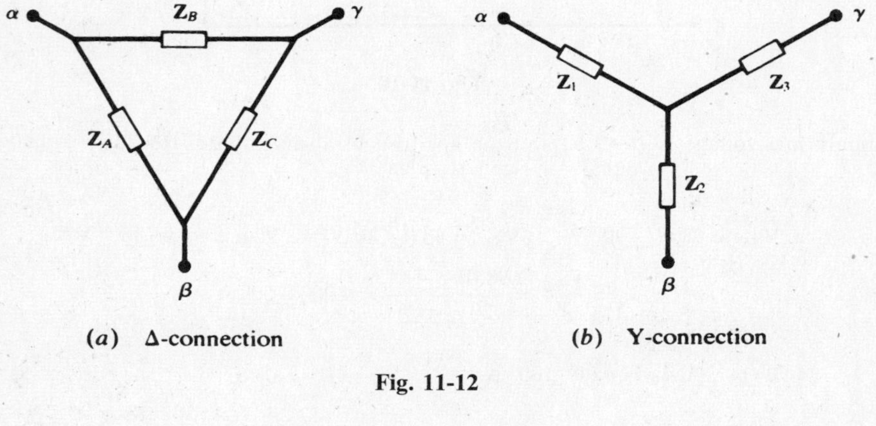

Fig. 11-12

Y-to-Δ Transformation	Δ-to-Y Transformation
$\mathbf{Z}_A = \dfrac{\mathbf{Z}_1\mathbf{Z}_2 + \mathbf{Z}_1\mathbf{Z}_3 + \mathbf{Z}_2\mathbf{Z}_3}{\mathbf{Z}_3}$	$\mathbf{Z}_1 = \dfrac{\mathbf{Z}_A\mathbf{Z}_B}{\mathbf{Z}_A + \mathbf{Z}_B + \mathbf{Z}_C}$
$\mathbf{Z}_B = \dfrac{\mathbf{Z}_1\mathbf{Z}_2 + \mathbf{Z}_1\mathbf{Z}_3 + \mathbf{Z}_2\mathbf{Z}_3}{\mathbf{Z}_2}$	$\mathbf{Z}_2 = \dfrac{\mathbf{Z}_A\mathbf{Z}_C}{\mathbf{Z}_A + \mathbf{Z}_B + \mathbf{Z}_C}$
$\mathbf{Z}_C = \dfrac{\mathbf{Z}_1\mathbf{Z}_2 + \mathbf{Z}_1\mathbf{Z}_3 + \mathbf{Z}_2\mathbf{Z}_3}{\mathbf{Z}_1}$	$\mathbf{Z}_3 = \dfrac{\mathbf{Z}_B\mathbf{Z}_C}{\mathbf{Z}_A + \mathbf{Z}_B + \mathbf{Z}_C}$

It should be noted that if the three impedances of one connection are equal, so are those of the equivalent connection, with $\mathbf{Z}_\Delta/\mathbf{Z}_Y = 3$.

11.9 SINGLE-LINE EQUIVALENT CIRCUIT FOR BALANCED THREE-PHASE LOADS

Figure 11-13(*a*) shows a balanced Y-connected load. In many cases, for instance, in power calculations, only the common magnitude, I_L, of the three line currents is needed. This may be obtained from the *single-line equivalent*, Fig. 11-13(*b*), which represents one phase of the original system, with the line-to-neutral voltage arbitrarily given a zero phase angle. This makes $\mathbf{I}_L = I_L \angle -\theta$, where θ is the impedance angle. If the actual line currents $\mathbf{I}_A$, $\mathbf{I}_B$, and $\mathbf{I}_C$ are desired, their phase angles may be found by adding $-\theta$ to the phase angles of $\mathbf{V}_{AN}$, $\mathbf{V}_{BN}$, and $\mathbf{V}_{CN}$ as given in Fig. 11-7. Observe that the angle on $\mathbf{I}_L$ gives the power factor for each phase, pf $= \cos\theta$.

The method may be applied to a balanced Δ-connected load if the load is replaced by its Y-equivalent, where $\mathbf{Z}_Y = \frac{1}{3}\mathbf{Z}_\Delta$ (Section 11.8).

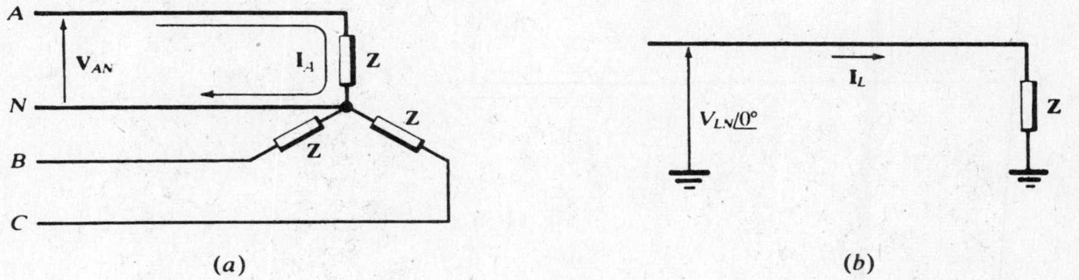

Fig. 11-13

EXAMPLE 11.4 Rework Example 11.3 by the single-line equivalent method.

Referring to Fig. 11-14 (in which the symbol Y indicates the type of connection of the original load),

$$\mathbf{I}_L = \frac{\mathbf{V}_{LN}}{\mathbf{Z}} = \frac{98.0\angle 0°}{20\angle -30°} = 4.90\angle 30° \text{ A}$$

From Fig. 11-7(*b*), the phase angles of $\mathbf{V}_{AN}$, $\mathbf{V}_{BN}$, and $\mathbf{V}_{CN}$ are −90°, 30°, and 150°. Hence,

$$\mathbf{I}_A = 4.90\angle -60° \text{ A} \qquad \mathbf{I}_B = 4.90\angle 60° \text{ A} \qquad \mathbf{I}_C = 4.90\angle 180° \text{ A}$$

11.10 UNBALANCED DELTA-CONNECTED LOAD

The solution of the unbalanced delta-connected load consists in computing the phase currents and then applying KCL to obtain the line currents. The currents will be unequal and will not have the symmetry of the balanced case.

EXAMPLE 11.5 A three-phase, 339.4-V, *ABC* system [Fig. 11-15(*a*)] has a Δ-connected load, with

$$\mathbf{Z}_{AB} = 10\angle 0° \ \Omega \qquad \mathbf{Z}_{BC} = 10\angle 30° \ \Omega \qquad \mathbf{Z}_{CA} = 15\angle -30° \ \Omega$$

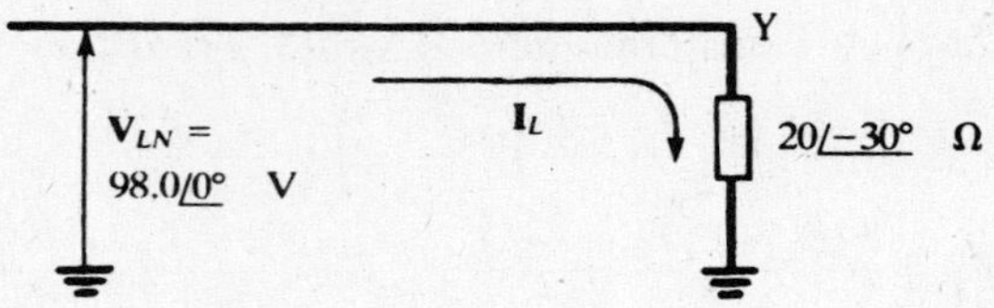

Fig. 11-14

Obtain phase and line currents and draw the phasor diagram.

$$\mathbf{I}_{AB} = \frac{\mathbf{V}_{AB}}{\mathbf{Z}_{AB}} = \frac{339.4 \angle 120^\circ}{10 \angle 0^\circ} = 33.94 \angle 120^\circ \quad \text{A}$$

Similarly, $\mathbf{I}_{BC} = 33.94 \angle -30^\circ$ A and $\mathbf{I}_{CA} = 22.63 \angle 270^\circ$ A. Then,

$$\mathbf{I}_A = \mathbf{I}_{AB} + \mathbf{I}_{AC} = 33.94 \angle 120^\circ - 22.63 \angle 270^\circ = 54.72 \angle 108.1^\circ \quad \text{A}$$

Also, $\mathbf{I}_B = 65.56 \angle -45^\circ$ A and $\mathbf{I}_C = 29.93 \angle -169.1^\circ$ A.

The voltage-current phasor diagram is shown in Fig. 11-15(*b*), with magnitudes and angles to scale.

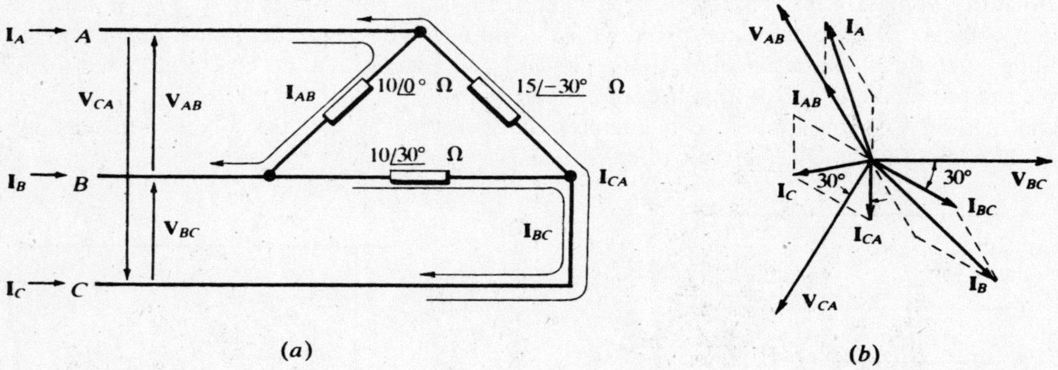

Fig. 11-15

11.11 UNBALANCED WYE-CONNECTED LOAD

Four-Wire

The neutral conductor carries the unbalanced current of a wye-connected load and maintains the line-to-neutral voltage magnitude across each phase of the load. The line currents are unequal and the currents on the phasor diagram have no symmetry.

EXAMPLE 11.6 A three-phase, four-wire, 150-V, *CBA* system has a Y-connected load, with

$$\mathbf{Z}_A = 6 \angle 0^\circ \ \Omega \qquad \mathbf{Z}_B = 6 \angle 30^\circ \ \Omega \qquad \mathbf{Z}_C = 5 \angle 45^\circ \ \Omega$$

Obtain all line currents and draw the phasor diagram. See Figure 11-16(*a*).

$$\mathbf{I}_A = \frac{\mathbf{V}_{AN}}{\mathbf{Z}_A} = \frac{86.6 \angle -90^\circ}{6 \angle 0^\circ} = 14.43 \angle -90^\circ \quad \text{A}$$

$$\mathbf{I}_B = \frac{\mathbf{V}_{BN}}{\mathbf{Z}_B} = \frac{86.6 \angle 30^\circ}{6 \angle 30^\circ} = 14.43 \angle 0^\circ \quad \text{A}$$

$$\mathbf{I}_C = \frac{\mathbf{V}_{CN}}{\mathbf{Z}_C} = \frac{86.6 \angle 150^\circ}{5 \angle 45^\circ} = 17.32 \angle 105^\circ \quad \text{A}$$

$$\mathbf{I}_N = -(14.43 \angle -90^\circ + 14.43 \angle 0^\circ + 17.32 \angle 105^\circ) = 10.21 \angle -167.0^\circ \quad \text{A}$$

Figure 11-16(*b*) gives the phasor diagram.

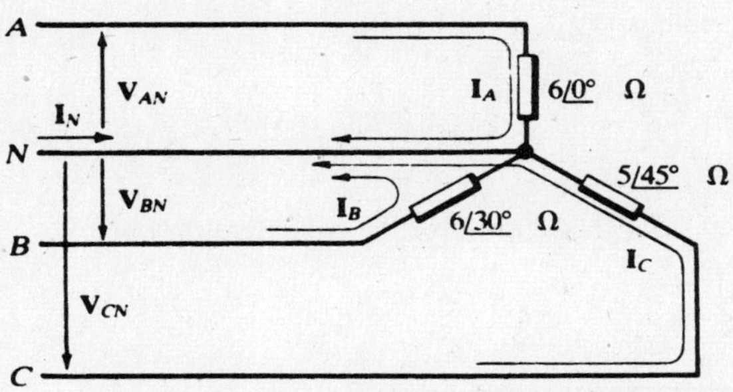

(a)

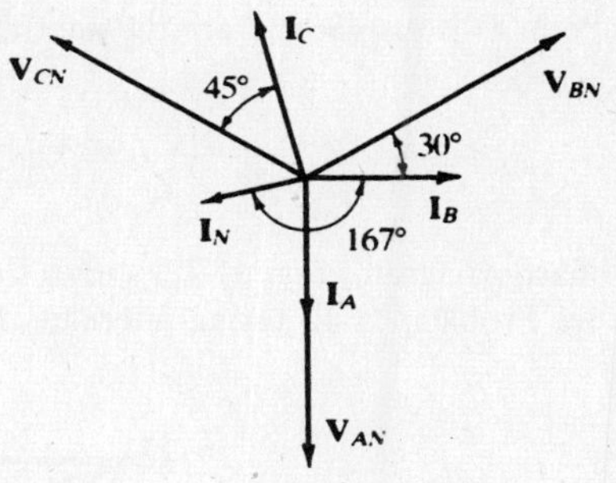

(b)

Fig. 11-16

Three-Wire

Without the neutral conductor, the Y-connected impedances will have voltages which vary considerably from the line-to-neutral magnitude.

EXAMPLE 11.7 Figure 11-17(*a*) shows the same system as treated in Example 11.6 except that the neutral wire is no longer present. Obtain the line currents and find the *displacement neutral voltage*, $\mathbf{V}_{ON}$.

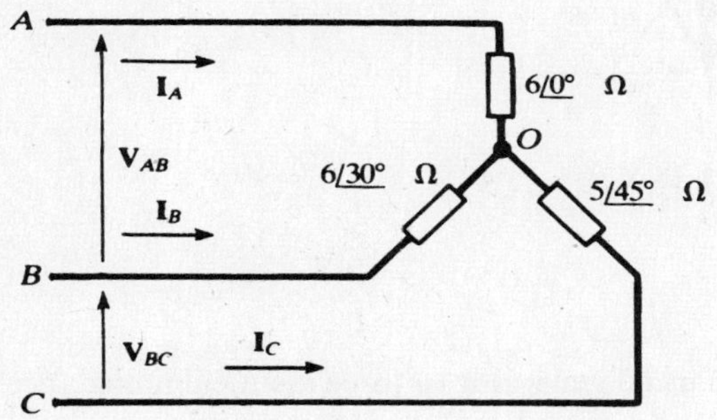

(a)

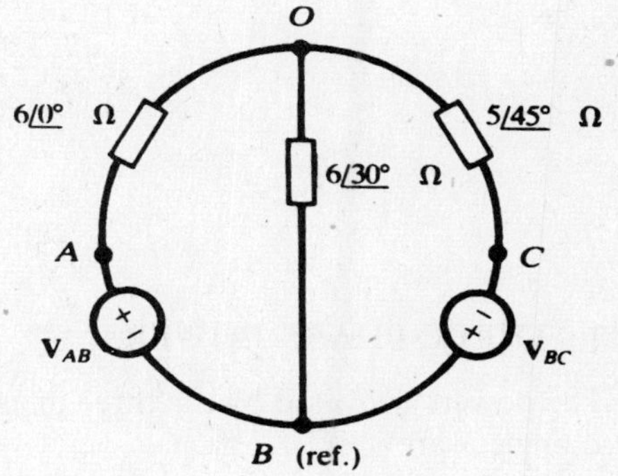

(b)

Fig. 11-17

The circuit is redrawn in Fig. 11-17(*b*) so as to suggest a single node-voltage equation with $\mathbf{V}_{OB}$ as the unknown.

$$\frac{\mathbf{V}_{OB}-\mathbf{V}_{AB}}{\mathbf{Z}_A}+\frac{\mathbf{V}_{OB}}{\mathbf{Z}_B}+\frac{\mathbf{V}_{OB}+\mathbf{V}_{BC}}{\mathbf{Z}_C}=0$$

$$\mathbf{V}_{OB}\left(\frac{1}{6\,\underline{/0^\circ}}+\frac{1}{6\,\underline{/30^\circ}}+\frac{1}{5\,\underline{/45^\circ}}\right)=\frac{150\,\underline{/240^\circ}}{6\,\underline{/0^\circ}}-\frac{150\,\underline{/0^\circ}}{5\,\underline{/45^\circ}}$$

from which $\mathbf{V}_{OB}=66.76\,\underline{/-152.85^\circ}$ V. Then,

$$\mathbf{I}_B=-\frac{\mathbf{V}_{OB}}{\mathbf{Z}_B}=11.13\,\underline{/-2.85^\circ}\quad\text{A}$$

From $\mathbf{V}_{OA}+\mathbf{V}_{AB}=\mathbf{V}_{OB}$, $\mathbf{V}_{OA}=100.7\,\underline{/81.08^\circ}$ V, and

$$\mathbf{I}_A=-\frac{\mathbf{V}_{OA}}{\mathbf{Z}_A}=16.78\,\underline{/-98.92^\circ}\quad\text{A}$$

Similarly, $\mathbf{V}_{OC}=\mathbf{V}_{OB}-\mathbf{V}_{CB}=95.58\,\underline{/-18.58^\circ}$ V, and

$$\mathbf{I}_C=19.12\,\underline{/116.4^\circ}\quad\text{A}$$

Point O is displaced from the neutral N by a phasor voltage $\mathbf{V}_{ON}$, given by

$$\mathbf{V}_{ON} = \mathbf{V}_{OA} + \mathbf{V}_{AN} = 100.7\,\underline{/81.08^\circ} + \frac{150}{\sqrt{3}}\,\underline{/-90^\circ} = 20.24\,\underline{/39.53^\circ}\quad \text{V}$$

The phasor diagram, Fig. 11-18, shows the shift of point O from the centroid of the equilateral triangle. See Problem 11-13 for an alternate method.

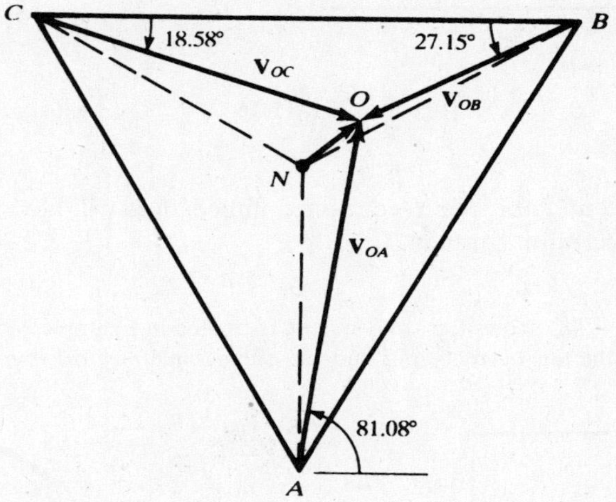

Fig. 11-18

11.12 THREE-PHASE POWER

The powers delivered by the three phases of a balanced generator to three identical impedances with phase angle θ are

$$p_a(t) = V_p I_p \cos\theta + V_p I_p \cos(2\omega t - \theta)$$
$$p_b(t) = V_p I_p \cos\theta + V_p I_p \cos(2\omega t - 240^\circ - \theta)$$
$$p_c(t) = V_p I_p \cos\theta + V_p I_p \cos(2\omega t - 480^\circ - \theta)$$

$$\begin{aligned} p_T(t) &= p_a(t) + p_b(t) + p_c(t) \\ &= 3V_p I_p \cos\theta + V_p I_p[\cos(2\omega t - \theta) + \cos(2\omega t - 240^\circ - \theta) + \cos(2\omega t - 480^\circ - \theta)] \end{aligned}$$

But $\cos(2\omega t - \theta) + \cos(2\omega t - 240^\circ - \theta) + \cos(2\omega t - 480^\circ - \theta) = 0$ for all t. Therefore,

$$p_T(t) = 3V_p I_p \cos\theta = P$$

The total instantaneous power is the same as the total average power. It may be written in terms of line voltage V_L and line current I_L. Thus,

In the delta system, $V_L = V_p$ and $I_L = \sqrt{3}I_p$. Therefore, $P = \sqrt{3}V_L I_L \cos\theta$.

In the wye system, $V_L = \sqrt{3}V_p$ and $I_L = I_p$. Therefore, $P = \sqrt{3}V_L I_L \cos\theta$.

The expression $\sqrt{3}V_L I_L \cos\theta$ gives the power in a three-phase balanced system, regardless of the connection configuration. The power factor of the three-phase system is $\cos\theta$. The line voltage V_L in industrial systems is always known. If the load is balanced, the total power can then be computed from the line current and power factor.

In summary, power, reactive power, apparent power, and power factor in a three-phase system are

$$P = \sqrt{3}V_L I_L \cos\theta \qquad Q = \sqrt{3}V_L I_L \sin\theta \qquad S = \sqrt{3}V_L I_L \qquad \text{pf} = \frac{P}{S}$$

Of course, all voltage and currents are effective values.

11.13 POWER MEASUREMENT AND THE TWO-WATTMETER METHOD

An ac wattmeter has a potential coil and a current coil and responds to the product of the effective voltage, the effective current, and the cosine of the phase angle between them. Thus, in Fig. 11-19, the wattmeter will indicate the average power supplied to the passive network,

$$P = V_{\text{eff}} I_{\text{eff}} \cos\theta = \text{Re}\,(\mathbf{V}_{\text{eff}}\mathbf{I}^*_{\text{eff}})$$

(see Section 10.7).

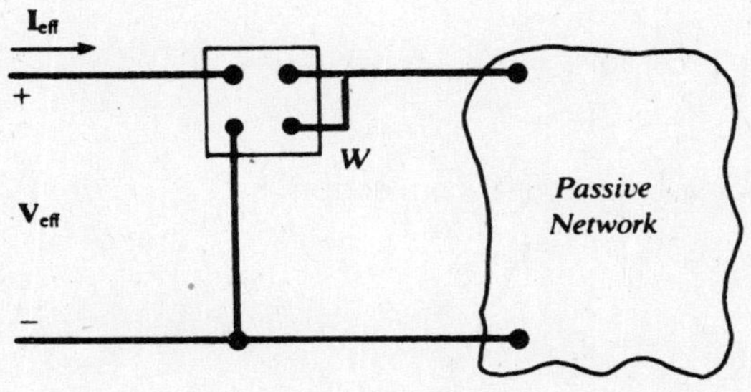

Fig. 11-19

Two wattmeters connected in any two lines of a three-phase, three-wire system will correctly indicate the total three-phase power by the sum of the two meter readings. A meter will attempt to go downscale if the phase angle between the voltage and current exceeds 90°. In this event, the current-coil connections can be reversed and the upscale meter reading treated as negative in the sum. In Fig. 11-20 the meters are inserted in lines A and C, with the potential-coil reference connections in line B. Their readings will be

$$W_A = \text{Re}\,(\mathbf{V}_{AB\,\text{eff}}\mathbf{I}^*_{A\,\text{eff}}) = \text{Re}\,(\mathbf{V}_{AB\,\text{eff}}\mathbf{I}^*_{AB\,\text{eff}}) + \text{Re}\,(\mathbf{V}_{AB\,\text{eff}}\mathbf{I}^*_{AC\,\text{eff}})$$
$$W_C = \text{Re}\,(\mathbf{V}_{CB\,\text{eff}}\mathbf{I}^*_{C\,\text{eff}}) = \text{Re}\,(\mathbf{V}_{CB\,\text{eff}}\mathbf{I}^*_{CA\,\text{eff}}) + \text{Re}\,(\mathbf{V}_{CB\,\text{eff}}\mathbf{I}^*_{CB\,\text{eff}})$$

in which the KCL expressions $\mathbf{I}_A = \mathbf{I}_{AB} + \mathbf{I}_{AC}$ and $\mathbf{I}_C = \mathbf{I}_{CA} + \mathbf{I}_{CB}$ have been used to replace line currents by phase currents. The first term in W_A is recognized as P_{AB}, the average power in phase AB of the delta

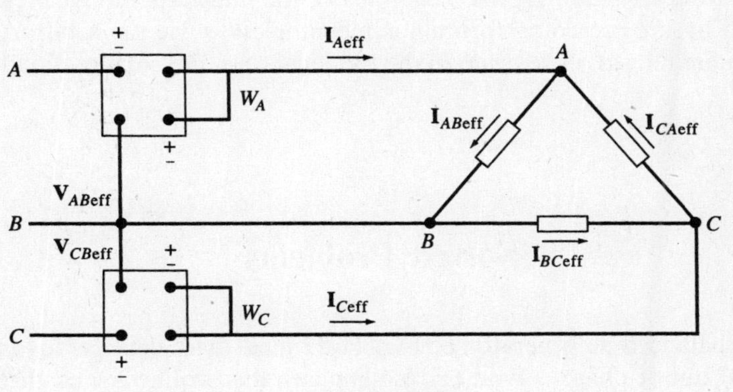

Fig. 11-20

load; likewise, the second term in W_C is P_{CB}. Adding the two equations and recombining the middle terms then yields

$$W_A + W_C = P_{AB} + \text{Re}\,[(\mathbf{V}_{AB\text{eff}} - \mathbf{V}_{CB\text{eff}})\mathbf{I}^*_{AC\text{eff}}] + P_{CB} = P_{AB} + P_{AC} + P_{CB}$$

since, by KVL, $\mathbf{V}_{AB} - \mathbf{V}_{CB} = \mathbf{V}_{AC}$.

The same reasoning establishes the analogous result for a Y-connected load.

Balanced Loads

When three equal impedances $Z\,\underline{/\theta}$ are connected in delta, the phase currents make 30° angles with their resultant line currents. Figure 11-21 corresponds to Fig. 11-20 under the assumption of *ABC* sequencing. It is seen that $\mathbf{V}_{AB}$ leads $\mathbf{I}_A$ by $\theta + 30°$, while $\mathbf{V}_{CB}$ leads $\mathbf{I}_C$ by $\theta - 30°$. Consequently, the two wattmeters will read

$$W_A = V_{AB\text{eff}} I_{A\text{eff}} \cos(\theta + 30°) \qquad W_C = V_{CB\text{eff}} I_{C\text{eff}} \cos(\theta - 30°)$$

or, since in general we do not know the relative order in the voltage sequence of the two lines chosen for the wattmeters,

$$W_1 = V_{L\text{eff}} I_{L\text{eff}} \cos(\theta + 30°)$$
$$W_2 = V_{L\text{eff}} I_{L\text{eff}} \cos(\theta - 30°)$$

These expressions also hold for a balanced Y-connection.

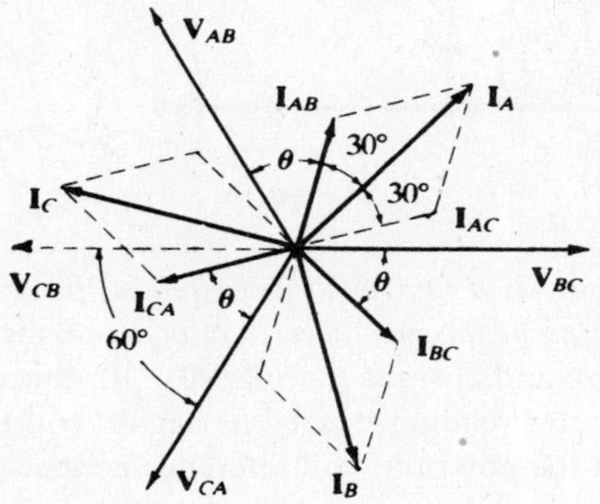

Fig. 11-21

Elimination of $V_{L\text{eff}} I_{L\text{eff}}$ between the two readings leads to

$$\tan\theta = \sqrt{3}\left(\frac{W_2 - W_1}{W_2 + W_1}\right)$$

Thus, from the two wattmeter readings, the *magnitude* of the impedance angle θ can be inferred. The sign of $\tan\theta$ suggested by the preceding formula is meaningless, since the arbitrary subscripts 1 and 2 might just as well be interchanged. However, in the practical case, the balanced load is usually known to be inductive ($\theta > 0$).

Solved Problems

11.1 The two-phase balanced ac generator of Fig. 11-22 feeds two identical loads. The two voltage sources are 180° out of phase. Find (*a*) the line currents, voltages, and their phase angles, and (*b*) the instantaneous and average powers delivered by the generator.

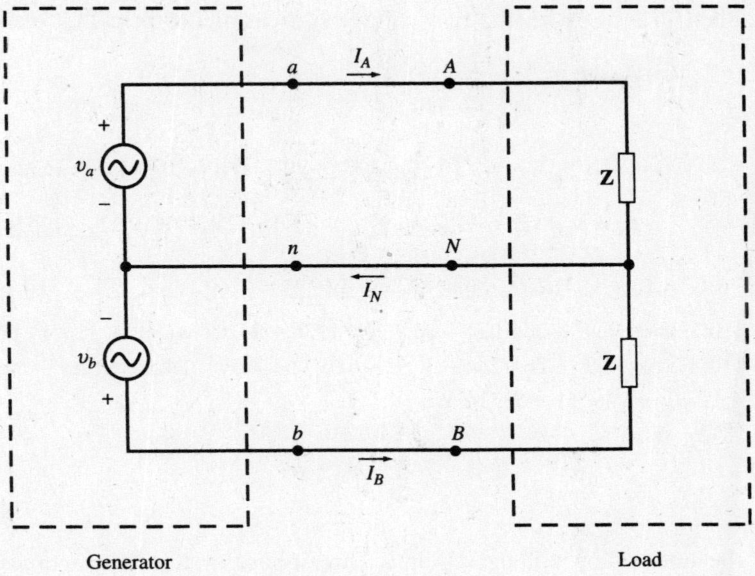

Fig. 11-22

Let $\mathbf{Z} = |Z| \underline{/\theta}$ and $I_p = V_p/|Z|$.

(*a*) The voltages and currents in phasor domain are

$$\mathbf{V}_{AN} = V_p \underline{/0} \qquad V_{BN} = V_p \underline{/-180^\circ} = -V_p \underline{/0} \qquad \mathbf{V}_{AB} = \mathbf{V}_{AN} - \mathbf{V}_{BN} = 2V_p \underline{/0}$$

Now, from I_p and $\mathbf{Z}$ given above, we have

$$\mathbf{I}_A = I_p \underline{/-\theta} \qquad \mathbf{I}_B = I_p \underline{/-180^\circ - \theta} = -I_p \underline{/-\theta} \qquad \mathbf{I}_N = \mathbf{I}_A + \mathbf{I}_B = 0$$

(*b*) The instantaneous powers delivered are

$$p_a(t) = v_a(t)i_a(t) = V_pI_p \cos\theta + V_pI_p \cos(2\omega t - \theta)$$
$$p_b(t) = v_b(t)i_b(t) = V_pI_p \cos\theta + V_pI_p \cos(2\omega t - \theta)$$

The total instantaneous power $p_T(t)$ is

$$p_T(t) = p_a(t) + p_b(t) = 2V_pI_p \cos\theta + 2V_pI_p \cos(2\omega t - \theta)$$

The average power is $P_{\text{avg}} = 2V_PI_p \cos\theta$.

11.2 Solve Problem 11.1 given $V_p = 110$ Vrms and $\mathbf{Z} = 4 + j3\ \Omega$.

(*a*) In phasor form, $\mathbf{Z} = 4 + j3 = 5 \underline{/36.9^\circ}\ \Omega$. Then,

$$\mathbf{V}_{AN} = 110 \underline{/0} \text{ V} \qquad \mathbf{V}_{BN} = 110 \underline{/-180^\circ} \text{ V}$$
$$\mathbf{V}_{AB} = \mathbf{V}_{AN} - \mathbf{V}_{BN} = 110 \underline{/0} - 110 \underline{/-180^\circ} = 220 \underline{/0} \text{ V}$$

and $\quad \mathbf{I}_A = \mathbf{V}_{AN}/Z = 22 \underline{/-36.9^\circ} \text{ A} \qquad \mathbf{I}_B = \mathbf{V}_{BN}/Z = 22 \underline{/-216.9^\circ} = -22 \underline{/-36.9^\circ} \text{ A}$

$$\mathbf{I}_N = \mathbf{I}_A + \mathbf{I}_B = 0$$

(*b*) $p_a(t) = 110(22)[\cos 36.9^\circ + \cos(2\omega t - 36.9^\circ)] = 1936 + 2420\cos(2\omega t - 36.9^\circ) \quad \text{(W)}$
$p_b(t) = 110(22)[\cos 36.9^\circ + \cos(2\omega t - 36.9^\circ - 360^\circ)] = 1936 + 2420\cos(2\omega t - 36.9^\circ) \quad \text{(W)}$
$p(t) = p_a(t) + p_b(t) = 3872 + 4840\cos(2\omega t - 36.9^\circ) \quad \text{(W)}$
$P_{\text{avg}} = 3872 \text{ W}$

11.3 Repeat Problem 11.2 but with the two voltage sources of Problem 11.1 90° out of phase.

(*a*) Again, $\mathbf{Z} = 5\,\underline{/36.9^\circ}$. Then,

$$\mathbf{V}_{AN} = 110\,\underline{/0}\ \ \text{V} \qquad \mathbf{V}_{BN} = 110\,\underline{/-90^\circ}\ \ \text{V}$$

$$\mathbf{V}_{AB} = \mathbf{V}_{AN} - \mathbf{V}_{BN} = 110\,\underline{/0} - 110\,\underline{/-90^\circ} = 110(\sqrt{2}\,\underline{/-45^\circ} = 155.6\,\underline{/-45^\circ}\ \ \text{V}$$

and $$\mathbf{I}_A = \mathbf{V}_{AN}/\mathbf{Z} = 22\,\underline{/-36.9^\circ}\ \ \text{A} \qquad \mathbf{I}_B = \mathbf{V}_{BN}/\mathbf{Z} = 22\,\underline{/-126.9^\circ}\ \ \text{A}$$

$$\mathbf{I}_N = \mathbf{I}_A + \mathbf{I}_B = 22\,\underline{/-36.9^\circ} + 22\,\underline{/-126.9^\circ} = 22(\sqrt{2}\,\underline{/-81.9^\circ}) = 31.1\,\underline{/-81.9^\circ}\ \ \text{A}$$

(*b*) $p_a(t) = 110(22)[\cos 36.9^\circ + \cos(2\omega t - 36.9^\circ)] = 1936 + 2420\cos(2\omega t - 36.9^\circ)$ (W)
$p_b(t) = 110(22)[\cos 36.9^\circ + \cos(2\omega t - 36.9^\circ - 180^\circ)] = 1936 - 2420\cos(2\omega t - 36.9^\circ)$ (W)
$p(t) = P_a + P_b = 2(1936) = 3872$ W
$P_{\text{avg}} = 3872$ W

11.4 Show that the line-to-line voltage V_L in a three-phase system is $\sqrt{3}$ times the line-to-neutral voltage V_{Ph}.

See the voltage phasor diagram (for the *ABC* sequence), Fig. 11-23.

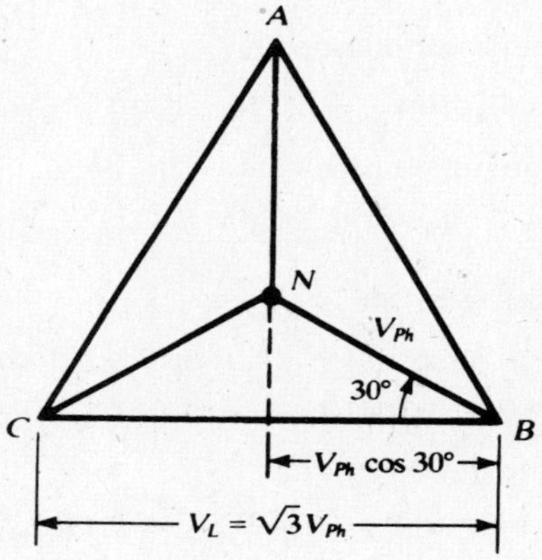

Fig. 11-23

11.5 A three-phase, *ABC* system, with an effective voltage 70.7 V, has a balanced Δ-connected load with impedances $20\,\underline{/45^\circ}\ \Omega$. Obtain the line currents and draw the voltage-current phasor diagram.

The circuit is shown in Fig. 11-24. The phase voltages have magnitudes $V_{\max} = \sqrt{2}V_{\text{eff}} = 100$ V. Phase angles are obtained from Fig. 11-7(*a*). Then,

$$\mathbf{I}_{AB} = \frac{\mathbf{V}_{AB}}{\mathbf{Z}} = \frac{100\,\underline{/120^\circ}}{20\,\underline{/45^\circ}} = 5.0\,\underline{/75^\circ}\ \ \text{A}$$

Similarly, $\mathbf{I}_{BC} = 5.0\,\underline{/-45^\circ}$ A and $\mathbf{I}_{CA} = 5.0\,\underline{/195^\circ}$ A. The line currents are:

$$\mathbf{I}_A = \mathbf{I}_{AB} + \mathbf{I}_{AC} = 5\,\underline{/75^\circ} - 5\,\underline{/195^\circ} = 8.65\,\underline{/45^\circ}\ \ \text{A}$$

Similarly, $\mathbf{I}_B = 8.65\,\underline{/-75^\circ}$ A, $\mathbf{I}_C = 8.65\,\underline{/165^\circ}$ A.

The voltage-current phasor diagram is shown in Fig. 11-25.

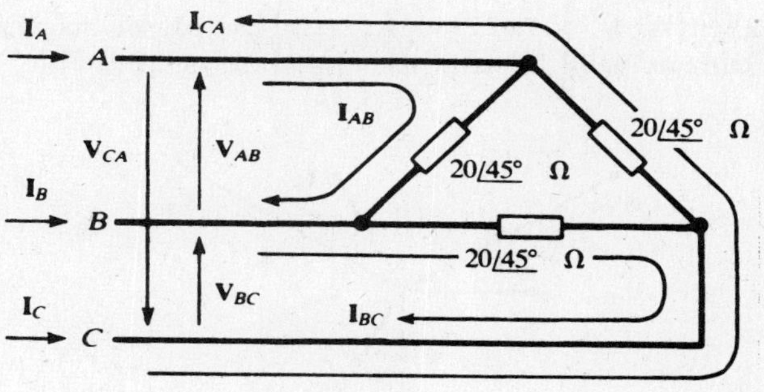

Fig. 11-24

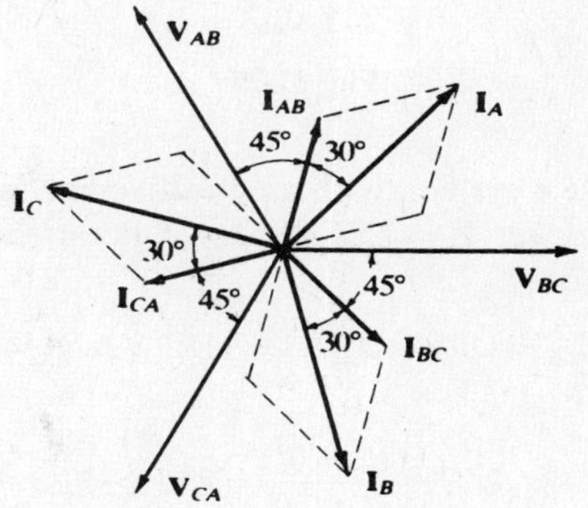

Fig. 11-25

11.6 A three-phase, three-wire *CBA* system, with an effective line voltage 106.1 V, has a balanced Y-connected load with impedances $5\,\underline{/-30^\circ}\ \Omega$ (Fig. 11-26). Obtain the currents and draw the voltage-current phasor diagram.

With balanced Y-loads, the neutral conductor carries no current. Even though this system is three-wire, the neutral may be added to simplify computation of the line currents. The magnitude of the line voltage is $V_L = \sqrt{2}(106.1) = 150$ V. Then the line-to-neutral magnitude is $V_{LN} = 150/\sqrt{3} = 86.6$ V.

$$\mathbf{I}_A = \frac{\mathbf{V}_{AN}}{\mathbf{Z}} = \frac{86.6\,\underline{/-90^\circ}}{5\,\underline{/-30^\circ}} = 17.32\,\underline{/-60^\circ}\ \ \text{A}$$

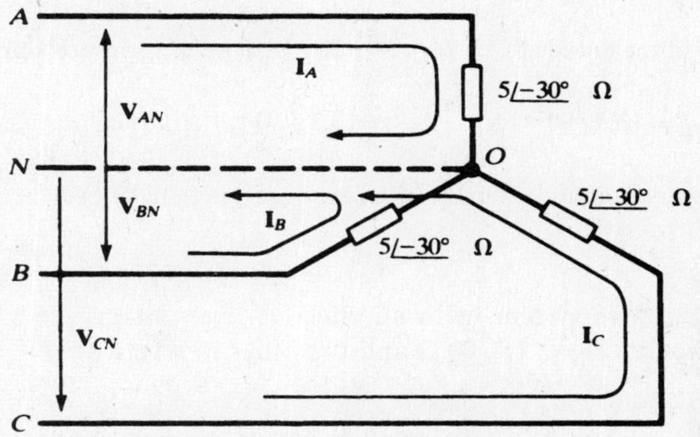

Fig. 11-26

Similarly, $\mathbf{I}_B = 17.32\,\underline{/60^\circ}$ A, $\mathbf{I}_C = 17.32\,\underline{/180^\circ}$ A. See the phasor diagram, Fig. 11-27, in which the balanced set of line currents leads the set of line-to-neutral voltages by 30°, the negative of the angle of the impedances.

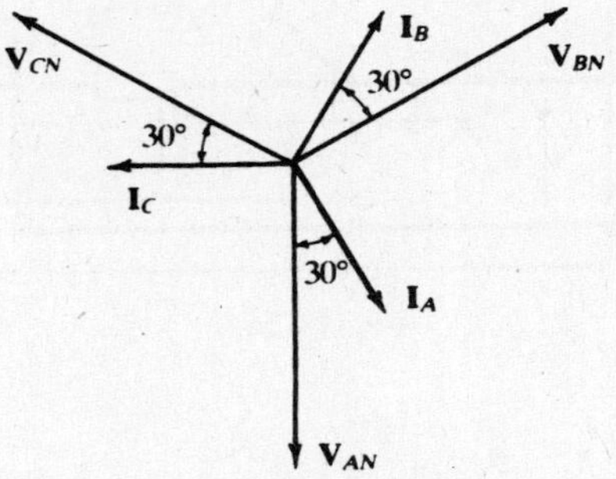

Fig. 11-27

11.7 A three-phase, three-wire *CBA* system, with an effective line voltage 106.1 V, has a balanced Δ-connected load with impedances $\mathbf{Z} = 15\,\underline{/30^\circ}\ \Omega$. Obtain the line and phase currents by the single-line equivalent method.

Referring to Fig. 11-28, $V_{LN} = (141.4\sqrt{2})/\sqrt{3} = 115.5$ V, and so

$$\mathbf{I}_L = \frac{115.5\,\underline{/0^\circ}}{(15/3)\,\underline{/30^\circ}} = 23.1\,\underline{/-30^\circ}\quad \text{A}$$

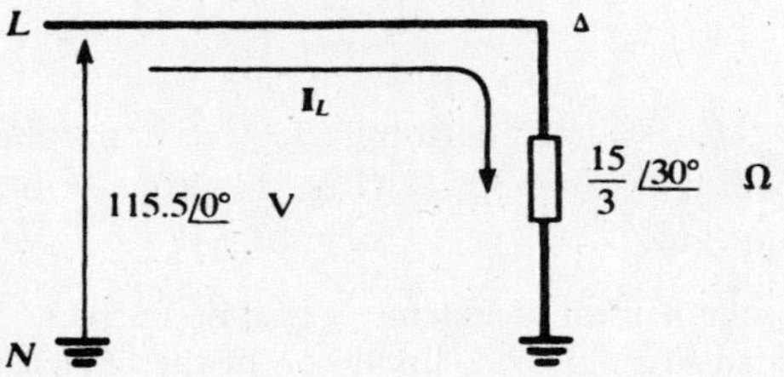

Fig. 11-28

The line currents lag the *ABC*-sequence, line-to-neutral voltages by 30°:

$$\mathbf{I}_A = 23.1\,\underline{/60^\circ}\ \text{A} \qquad \mathbf{I}_B = 23.1\,\underline{/-60^\circ}\ \text{A} \qquad \mathbf{I}_C = 23.1\,\underline{/180^\circ}\ \text{A}$$

The phase currents, of magnitude $I_{Ph} = I_L/\sqrt{3} = 13.3$ A, lag the corresponding line-to-line voltages by 30°:

$$\mathbf{I}_{AB} = 13.3\,\underline{/90^\circ}\ \text{A} \qquad \mathbf{I}_{BC} = 13.3\,\underline{/-30^\circ}\ \text{A} \qquad \mathbf{I}_{CA} = 13.3\,\underline{/210^\circ}\ \text{A}$$

A sketch of the phasor diagram will make all of the foregoing angles evident.

11.8 A three-phase, three-wire system, with an effective line voltage 176.8 V, supplies two balanced loads, one in delta with $\mathbf{Z}_\Delta = 15\,\underline{/0^\circ}\ \Omega$ and the other in wye with $\mathbf{Z}_Y = 10\,\underline{/30^\circ}\ \Omega$. Obtain the total power.

First convert the Δ-load to Y, and then use the single-line equivalent circuit, Fig. 11-29, to obtain the line current.

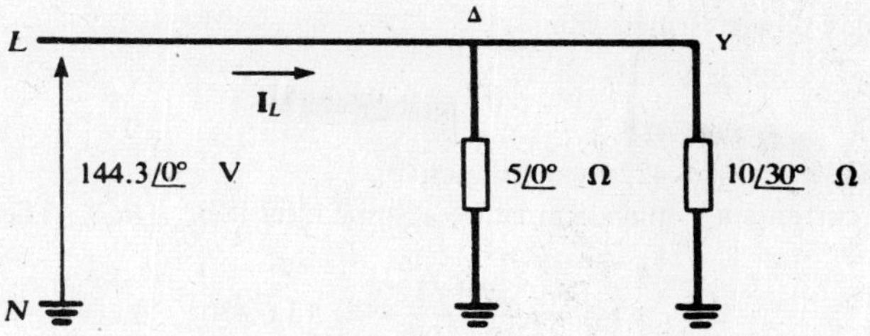

Fig. 11-29

$$\mathbf{I}_L = \frac{144.3\,/0^\circ}{5\,/0^\circ} + \frac{144.3\,/0^\circ}{10\,/30^\circ} = 42.0\,/-9.9^\circ \quad \text{A}$$

Then $$P = \sqrt{3}V_{L\text{eff}}I_{L\text{eff}}\cos\theta = \sqrt{3}(176.8)(29.7)\cos 9.9^\circ = 8959 \quad \text{W}$$

11.9 Obtain the readings when the two-wattmeter method is applied to the circuit of Problem 11.8.

The angle on $\mathbf{I}_L$, -9.9°, is the negative of the angle on the equivalent impedance of the parallel combination of $5\,/0^\circ\ \Omega$ and $10\,/30^\circ\ \Omega$. Therefore, $\theta = 9.9^\circ$ in the formulas of Section 11.13.

$$W_1 = V_{L\text{eff}}I_{L\text{eff}}\cos(\theta + 30^\circ) = (176.8)(29.7)\cos 39.9^\circ = 4028 \quad \text{W}$$
$$W_2 = V_{L\text{eff}}I_{L\text{eff}}\cos(\theta - 30^\circ) = (176.8)(29.7)\cos(-20.1^\circ) = 4931 \quad \text{W}$$

As a check, $W_1 + W_2 = 8959$ W, in agreement with Problem 11.8.

11.10 A three-phase supply, with an effective line voltage 240 V, has an unbalanced Δ-connected load shown in Fig. 11-30. Obtain the line currents and the total power.

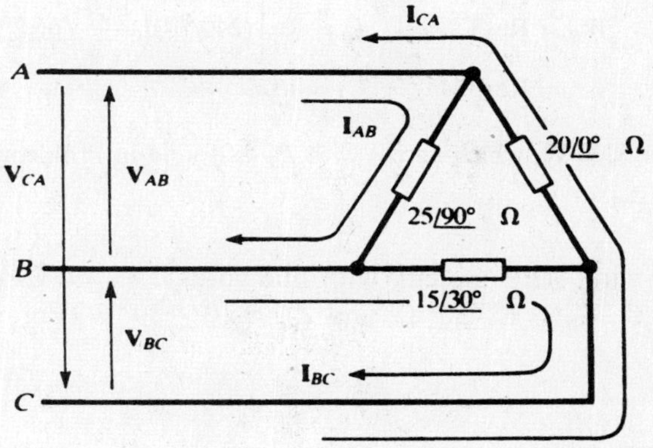

Fig. 11-30

The power calculations can be performed without knowledge of the sequence of the system. The effective values of the phase currents are

$$I_{AB\text{eff}} = \frac{240}{25} = 9.6\ \text{A} \qquad I_{BC\text{eff}} = \frac{240}{15} = 16\ \text{A} \qquad I_{CA\text{eff}} = \frac{240}{20} = 12\ \text{A}$$

Hence, the complex powers in the three phases are

$$\mathbf{S}_{AB} = (9.6)^2(25\,/90^\circ) = 2304\,/90^\circ = 0 + j2304$$
$$\mathbf{S}_{BC} = (16)^2(15\,/30^\circ) = 3840\,/30^\circ = 3325 + j1920$$
$$\mathbf{S}_{CA} = (12)^2(20\,/0^\circ) = 2880\,/0^\circ = 2880 + j0$$

and the total complex power is their sum,

$$\mathbf{S}_T = 6205 + j4224$$

that is, $P_T = 6205$ W and $Q_T = 4224$ var (inductive).

To obtain the currents a sequence must be assumed; let it be ABC. Then, using Fig. 11-7(a),

$$\mathbf{I}_{AB} = \frac{339.4\,\underline{/120^\circ}}{25\,\underline{/90^\circ}} = 13.6\,\underline{/30^\circ} \quad \text{A}$$

$$\mathbf{I}_{BC} = \frac{339.4\,\underline{/0^\circ}}{15\,\underline{/30^\circ}} = 22.6\,\underline{/-30^\circ} \quad \text{A}$$

$$\mathbf{I}_{CA} = \frac{339.4\,\underline{/240^\circ}}{20\,\underline{/0^\circ}} = 17.0\,\underline{/240^\circ} \quad \text{A}$$

The line currents are obtained by applying KCL at the junctions.

$$\mathbf{I}_A = \mathbf{I}_{AB} + \mathbf{I}_{AC} = 13.6\,\underline{/30^\circ} - 17.0\,\underline{/240^\circ} = 29.6\,\underline{/46.7^\circ} \quad \text{A}$$

$$\mathbf{I}_B = \mathbf{I}_{BC} + \mathbf{I}_{BA} = 22.6\,\underline{/-30^\circ} - 13.6\,\underline{/30^\circ} = 19.7\,\underline{/-66.7^\circ} \quad \text{A}$$

$$\mathbf{I}_C = \mathbf{I}_{CA} + \mathbf{I}_{CB} = 17.0\,\underline{/240^\circ} - 22.6\,\underline{/-30^\circ} = 28.3\,\underline{/-173.1^\circ} \quad \text{A}$$

11.11 Obtain the readings of wattmeters placed in lines A and B of the circuit of Problem 11.10 (Line C is the potential reference for both meters.)

$$W_A = \text{Re}\,(\mathbf{V}_{AC\,\text{eff}}\mathbf{I}^*_{A\,\text{eff}}) = \text{Re}\left[(240\underline{/60^\circ})\left(\frac{29.6}{\sqrt{2}}\,\underline{/-46.7^\circ}\right)\right]$$

$$= \text{Re}\,(5023\,\underline{/13.3^\circ}) = 4888 \quad \text{W}$$

$$W_B = \text{Re}\,(\mathbf{V}_{BC\,\text{eff}}\mathbf{I}^*_{B\,\text{eff}}) = \text{Re}\left[(240\underline{/0^\circ})\left(\frac{19.7}{\sqrt{2}}\,\underline{/66.7^\circ}\right)\right]$$

$$= \text{Re}\,(3343\,\underline{/66.7^\circ}) = 1322 \quad \text{W}$$

Note that $W_A + W_B = 6210$ W, which agrees with P_T as found in Problem 11.10.

11.12 A three-phase, four-wire, ABC system, with line voltage $\mathbf{V}_{BC} = 294.2\,\underline{/0^\circ}$ V, has a Y-connected load of $\mathbf{Z}_A = 10\,\underline{/0^\circ}\ \Omega$, $\mathbf{Z}_B = 15\,\underline{/30^\circ}\ \Omega$, and $\mathbf{Z}_C = 10\underline{/-30^\circ}\ \Omega$ (Fig. 11-31). Obtain the line and neutral currents.

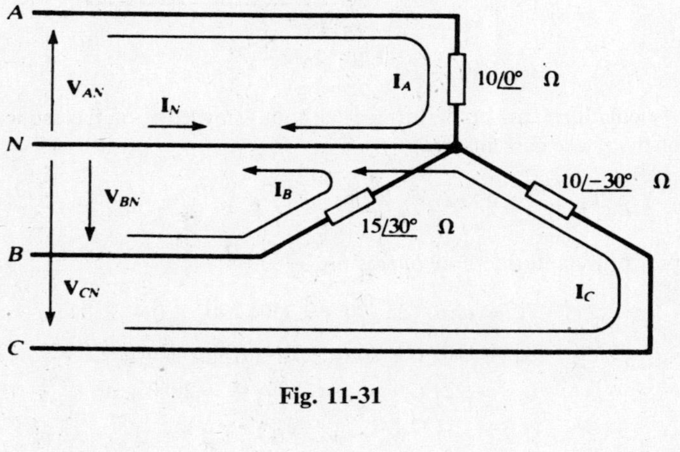

Fig. 11-31

$$\mathbf{I}_A = \frac{169.9\angle 90^\circ}{10\angle 0^\circ} = 16.99\angle 90^\circ \quad \text{A}$$

$$\mathbf{I}_B = \frac{169.9\angle -30^\circ}{15\angle 30^\circ} = 11.33\angle -60^\circ \quad \text{A}$$

$$\mathbf{I}_C = \frac{169.9\angle -150^\circ}{10\angle -30^\circ} = 16.99\angle -120^\circ \quad \text{A}$$

$$\mathbf{I}_N = -(\mathbf{I}_A + \mathbf{I}_B + \mathbf{I}_C) = 8.04\angle 69.5^\circ \quad \text{A}$$

11.13 The Y-connected load impedances $\mathbf{Z}_A = 10\angle 0^\circ\ \Omega$, $\mathbf{Z} = 15\angle 30^\circ\ \Omega$, and $\mathbf{Z}_C = 10\angle -30^\circ\ \Omega$, in Fig. 11-32, are supplied by a three-phase, three-wire, *ABC* system in which $\mathbf{V}_{BC} = 208\angle 0^\circ$ V. Obtain the voltages across the impedances and the displacement neutral voltage $\mathbf{V}_{ON}$.

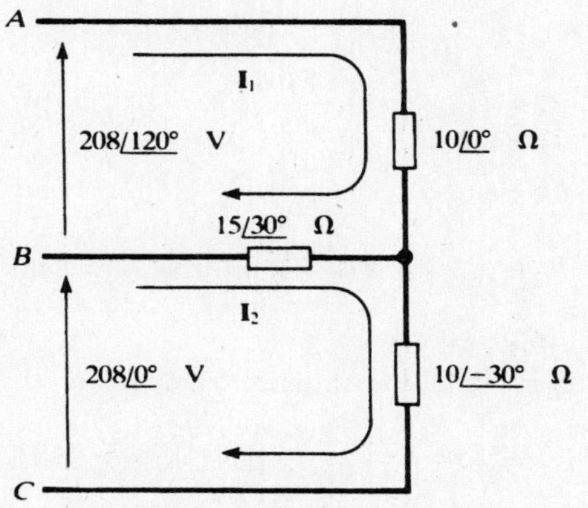

Fig. 11-32

The method of Example 11.7 could be applied here and one node-voltage equation solved. However, the mesh currents $\mathbf{I}_1$ and $\mathbf{I}_2$ suggested in Fig. 11-32 provide another approach.

$$\begin{bmatrix} 10\angle 0^\circ + 15\angle 30^\circ & -15\angle 30^\circ \\ -15\angle 30^\circ & 15\angle 30^\circ + 10\angle -30^\circ \end{bmatrix} \begin{bmatrix} \mathbf{I}_1 \\ \mathbf{I}_2 \end{bmatrix} = \begin{bmatrix} 208\angle 120^\circ \\ 208\angle 0^\circ \end{bmatrix}$$

Solving, $\mathbf{I}_1 = 14.16\angle 86.09^\circ$ A and $\mathbf{I}_2 = 10.21\angle 52.41^\circ$ A. The line currents are then

$$\mathbf{I}_A = \mathbf{I}_1 = 14.16\angle 86.09^\circ \quad \text{A} \qquad \mathbf{I}_B = \mathbf{I}_2 - \mathbf{I}_1 = 8.01\angle -48.93^\circ \quad \text{A} \qquad \mathbf{I}_C = -\mathbf{I}_2 = 10.21\angle -127.59^\circ \quad \text{A}$$

Now the phasor voltages at the load may be computed.

$$\mathbf{V}_{AO} = \mathbf{I}_A\mathbf{Z}_A = 141.6\angle 86.09^\circ \quad \text{V}$$

$$\mathbf{V}_{BO} = \mathbf{I}_B\mathbf{Z}_B = 120.2\angle -18.93^\circ \quad \text{V}$$

$$\mathbf{V}_{CO} = \mathbf{I}_C\mathbf{Z}_C = 102.1\angle -157.59^\circ \quad \text{V}$$

$$\mathbf{V}_{ON} = \mathbf{V}_{OA} + \mathbf{V}_{AN} = 141.6\angle -93.91^\circ + 120.1\angle 90^\circ = 23.3\angle -114.53^\circ \quad \text{V}$$

The phasor diagram is given in Fig. 11-33.

11.14 Obtain the total average power for the unbalanced, Y-connected load in Problem 11.13, and compare with the readings of wattmeters in lines *B* and *C*.

The phase powers are

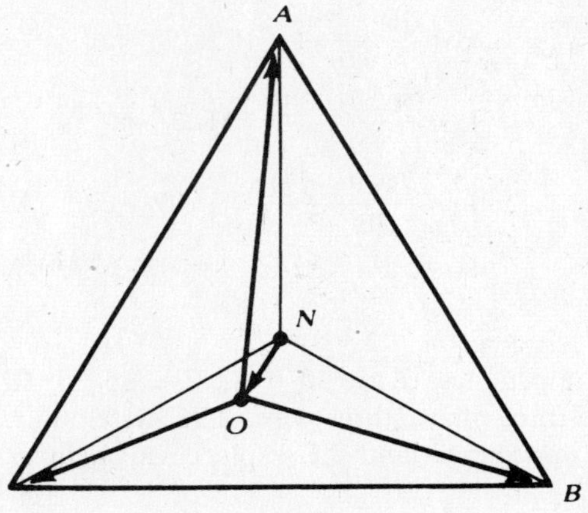

Fig. 11-33

$$P_A = I^2_{A\,\text{eff}} R_A = \left(\frac{14.16}{\sqrt{2}}\right)(10) = 1002.5 \quad \text{W}$$

$$P_B = I^2_{B\,\text{eff}} R_B = \left(\frac{8.01}{\sqrt{2}}\right)(15\cos 30^\circ) = 417.0 \quad \text{W}$$

$$P_C = I^2_{C\,\text{eff}} R_C = \left(\frac{10.21}{\sqrt{2}}\right)^2 (10\cos 30^\circ) = 451.4 \quad \text{W}$$

and so the total average power is 1870.9 W.

From the results of Problem 11.13, the wattmeter readings are:

$$W_B = \text{Re}\,(\mathbf{V}_{BA\,\text{eff}}\mathbf{I}^*_{B\,\text{eff}}) = \text{Re}\left[\left(\frac{208}{\sqrt{2}}\angle{-60^\circ}\right)\left(\frac{8.01}{\sqrt{2}}\angle 48.93^\circ\right)\right] = 817.1 \quad \text{W}$$

$$W_C = \text{Re}\,(\mathbf{V}_{CA\,\text{eff}}\mathbf{I}^*_{C\,\text{eff}} = \text{Re}\left[\left(\frac{208}{\sqrt{2}}\angle 2400^\circ\right)\left(\frac{10.21}{\sqrt{2}}\angle 127.59^\circ\right)\right] = 1052.8 \quad \text{W}$$

The total power read by the two wattmeters is 1869.9 W.

11.15 A three-phase, three-wire, balanced, Δ-connected load yields wattmeter readings of 1154 W and 557 W. Obtain the load impedance, if the line voltage is 141.4 V.

$$\pm\tan\theta = \sqrt{3}\left(\frac{W_2 - W_1}{W_2 + W_1}\right) = \sqrt{3}\left(\frac{577}{1731}\right) = 0.577 \qquad \theta = \pm 30.0^\circ$$

and, using $P_T = \sqrt{3}V_{L\,\text{eff}} I_{L\,\text{eff}} \cos\theta$,

$$Z_\Delta = \frac{V_{L\text{eff}}}{I_{Ph\,\text{eff}}} = \frac{\sqrt{3}V_{L\,\text{eff}}}{I_{L\,\text{eff}}} = \frac{3V^2_{L\,\text{eff}}\cos\theta}{P_T} = \frac{3(100)^2\cos 30.0^\circ}{1154 + 577}\ \Omega = 15.0\ \Omega$$

Thus, $\mathbf{Z}_\Delta = 15.0\angle{\pm 30.0^\circ}\ \Omega$.

11.16 A balanced Δ-connected load, with $\mathbf{Z}_\Delta = 30\angle 30^\circ\ \Omega$, is connected to a three-phase, three-wire, 250-V system by conductors having impedances $\mathbf{Z}_c = 0.4 + j0.3\ \Omega$. Obtain the line-to-line voltage at the load.

The single-line equivalent circuit is shown in Fig. 11-34. By voltage division, the voltage across the substitute Y-load is

$$\mathbf{V}_{AN} = \left(\frac{10\angle 30^\circ}{0.4 + j0.3 + 10\angle 30^\circ}\right)\left(\frac{250}{\sqrt{3}}\angle 0^\circ\right) = 137.4\angle{-0.33^\circ} \quad \text{V}$$

whence $V_L = (137.4)(\sqrt{3}) = 238.0$ V.

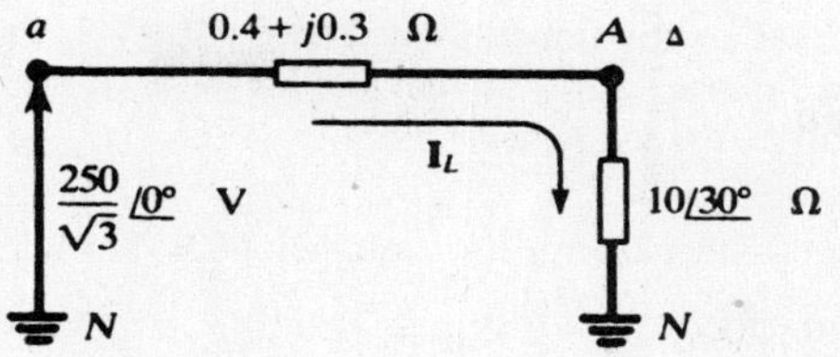

Fig. 11-34

Considering the magnitudes only, the line voltage at the load, 238.0 V, represents a drop of 12.0 V. The wire size and total length control the resistance in $\mathbf{Z}_c$, while the enclosing conduit material (e.g., steel, aluminum, or fiber), as well as the length, affects the inductive reactance.

Supplementary Problems

In the following, the voltage-current phasor diagram will not be included in the answer, even though the problem may ask specifically for one. As a general rule, a phasor diagram should be constructed for every polyphase problem.

11.17 Three impedances of $10.0\angle 53.13^\circ$ Ω are connected in delta to a three-phase, *CBA* system with an affective line voltage 240 V. Obtain the line currents.
Ans. $\mathbf{I}_A = 58.8\angle -143.13^\circ$ A, $\mathbf{I}_B = 58.8\angle -23.13^\circ$ A, $\mathbf{I}_C = 58.8\angle 96.87^\circ$ A

11.18 Three impedances of $4.20\angle -35^\circ$ Ω are connected in delta to a three-phase, *ABC* system having $\mathbf{V}_{BC} = 495.0\angle 0^\circ$ V. Obtain the line currents.
Ans. $\mathbf{I}_A = 20.41\angle 125^\circ$ A, $\mathbf{I}_B = 20.41\angle 5^\circ$ A, $\mathbf{I}_C = 20.41\angle -115^\circ$ A

11.19 A three-phase, three-wire system, with an effective line voltage 100 V, has currents

$$\mathbf{I}_A = 15.41\angle -160^\circ \text{ A} \qquad \mathbf{I}_B = 15.41\angle -40^\circ \text{ A} \qquad \mathbf{I}_C = 15.41\angle 80^\circ \text{ A}$$

What is the sequence of the system and what are the impedances, if the connection is delta?
Ans. *CBA*, $15.9\angle 70^\circ$ Ω

11.20 A balanced Y-connected load, with impedances $6.0\angle 45^\circ$ Ω, is connected to a three-phase, four-wire *CBA* system having effective line voltage 208 V. Obtain the four line currents.
Ans. $\mathbf{I}_A = 28.31\angle -135^\circ$ A, $\mathbf{I}_B = 28.31\angle -15^\circ$ A, $\mathbf{I}_C = 28.31\angle 105^\circ$ A, $\mathbf{I}_N = 0$

11.21 A balanced Y-connected load, with impedances $65.0\angle -20^\circ$ Ω, is connected to a three-phase, three-wire, *CBA* system, where $\mathbf{V}_{AB} = 678.8\angle -120^\circ$ V. Obtain the three line currents.
Ans. $\mathbf{I}_A = 6.03\angle -70^\circ$ A, $\mathbf{I}_B = 6.03\angle 50^\circ$ A, $\mathbf{I}_C = 6.03\angle 170^\circ$ A

11.22 A balanced Δ-connected load, with $\mathbf{Z}_\Delta = 9.0\angle -30^\circ$, and a balanced Y-connected load, with $\mathbf{Z}_Y = 5.0\angle 45^\circ$ Ω, are supplied by the same three-phase, *ABC* system, with effective line voltage 480 V. Obtain the line currents, using the single-line equivalent method.
Ans. $\mathbf{I}_A = 168.9\angle 93.36^\circ$ A, $\mathbf{I}_B = 168.9\angle -26.64^\circ$ A, $\mathbf{I}_C = 168.9\angle -146.64^\circ$ A

11.23 A balanced Δ-connected load having impedances $27.0\angle -25^\circ$ Ω and a balanced Y-connected load having impedances $10.0\angle -30^\circ$ Ω are supplied by the same three-phase, *ABC* system, with $\mathbf{V}_{CN} = 169.8\angle -150^\circ$ V. Obtain the line currents.
Ans. $\mathbf{I}_A = 35.8\angle 117.36^\circ$ A, $\mathbf{I}_B = 35.8\angle -2.64^\circ$ A, $\mathbf{I}_C = 35.8\angle -122.64^\circ$ A

11.24 A balanced Δ-connected load, with impedances $10.0\angle{-36.9^\circ}\ \Omega$, and a balanced Y-connected load are supplied by the same three-phase, *ABC* system having $\mathbf{V}_{CA} = 141.4\angle{240^\circ}$ V. If $\mathbf{I}_B = 40.44\angle{13.41^\circ}$ A, what are the impedances of the Y-connected load? *Ans.* $5.0\angle{-53.3^\circ}$

11.25 A three-phase, *ABC* system, with effective line voltage 500 V, has a Δ-connected load for which

$$\mathbf{Z}_{AB} = 10.0\angle{30^\circ}\ \Omega \qquad \mathbf{Z}_{BC} = 25.0\angle{0^\circ}\ \Omega \qquad \mathbf{Z}_{CA} = 20.0\angle{-30^\circ}\ \Omega$$

Obtain the line currents.
Ans. $\mathbf{I}_A = 106.1\angle{90.0^\circ}$ A, $\mathbf{I}_B = 76.15\angle{-68.20^\circ}$ A, $\mathbf{I}_C = 45.28\angle{-128.65^\circ}$ A

11.26 A three-phase, *ABC* system, with $\mathbf{V}_{BC} = 294.2\angle{0^\circ}$ V, has the Δ-connected load

$$\mathbf{Z}_{AB} = 5.0\angle{0^\circ}\ \Omega \qquad \mathbf{Z}_{BC} = 4.0\angle{30^\circ}\ \Omega \qquad \mathbf{Z}_{CA} = 6.0\angle{-15^\circ}$$

Obtain the line currents.
Ans. $\mathbf{I}_A = 99.7\angle{99.7^\circ}$ A, $\mathbf{I}_B = 127.9\angle{-43.3}$ A, $\mathbf{I}_C = 77.1\angle{-172.1^\circ}$ A

11.27 A three-phase, four-wire, *CBA* system, with effective line voltage 100 V, has Y-connected impedances

$$\mathbf{Z}_A = 3.0\angle{0^\circ}\ \Omega \qquad \mathbf{Z}_B = 3.61\angle{56.31^\circ}\ \Omega \qquad \mathbf{Z}_C = 2.24\angle{-26.57^\circ}\ \Omega$$

Obtain the currents $\mathbf{I}_A, \mathbf{I}_B, \mathbf{I}_C$, and $\mathbf{I}_N$.
Ans. $27.2\angle{-90^\circ}$ A, $22.6\angle{-26.3^\circ}$ A, $36.4\angle{176.6^\circ}$ A, $38.6\angle{65.3^\circ}$ A

11.28 A three-phase, four-wire, *ABC* system, with $\mathbf{V}_{BC} = 294.2\angle{0^\circ}$ V, has Y-connected impedances

$$\mathbf{Z}_A = 12.0\angle{45^\circ}\ \Omega \qquad \mathbf{Z}_B = 10.0\angle{30^\circ}\ \Omega \qquad \mathbf{Z}_C = 8.0\angle{0^\circ}\ \Omega$$

Obtain the currents $\mathbf{I}_A, \mathbf{I}_B, \mathbf{I}_C$, and $\mathbf{I}_N$.
Ans. $14.16\angle{45^\circ}$ A, $16.99\angle{-60^\circ}$ A, $21.24\angle{-150^\circ}$ A, $15.32\angle{90.4^\circ}$ A

11.29 A Y-connected load, with $\mathbf{Z}_A = 10\angle{0^\circ}\ \Omega$, $\mathbf{Z}_B = 10\angle{60^\circ}$, and $\mathbf{Z}_C = 10\angle{-60^\circ}\ \Omega$, is connected to a three-phase, three-wire, *ABC* system having effective line voltage 141.4 V. Find the load voltages $\mathbf{V}_{AO}, \mathbf{V}_{BO}, \mathbf{V}_{CO}$ and the displacement neutral voltage $\mathbf{V}_{ON}$. Construct a phasor diagram similar to Fig. 11-18.
Ans. $173.2\angle{90^\circ}$ V, $100\angle{0^\circ}$ V, $100\angle{180^\circ}$ V, $57.73\angle{-90^\circ}$ V

11.30 A Y-connected load, with $\mathbf{Z}_A = 10\angle{-60^\circ}\ \Omega$, $\mathbf{Z}_B = 10\angle{0^\circ}\ \Omega$, and $\mathbf{Z}_C = 10\angle{60^\circ}\ \Omega$, is connected to a three-phase, three-wire, *CBA* system having effective line voltage 147.1 V. Obtain the line currents $\mathbf{I}_A$, $\mathbf{I}_B$, and $\mathbf{I}_C$.
Ans. $20.8\angle{-60^\circ}$ A, 0, $20.8\angle{120^\circ}$ A

11.31 A three-phase, three-wire, *ABC* system with a balanced load has effective line voltage 200 V and (maximum) line current $\mathbf{I}_A = 13.61\angle{60^\circ}$ A. Obtain the total power. *Ans.* 2887 W

11.32 Two balanced Δ-connected loads, with impedances $20\angle{-60^\circ}\ \Omega$ and $18\angle{45^\circ}$, respectively, are connected to a three-phase system for which a line voltage is $\mathbf{V}_{BC} = 212.1\angle{0^\circ}$ V. Obtain the phase power of each load. After using the single-line equivalent method to obtain the total line current, compute the total power, and compare with the sum of the phase powers.
Ans. 562.3 W, 883.6 W, 4337.5 W = 3(562.3 W) + 3(883.6 W)

11.33 In Problem 11.5, a balanced Δ-connected load with $\mathbf{Z} = 20\angle{45^\circ}\ \Omega$ resulted in line currents 8.65 A for line voltages 100 V, both maximum values. Find the readings of two wattmeters used to measure the total average power. *Ans.* 111.9 W, 417.7 W

11.34 Obtain the readings of two wattmeters in a three-phase, three-wire system having effective line voltage 240 V and balanced, Δ-connected load impedances $20\angle{80^\circ}\ \Omega$. *Ans.* − 1706 W, 3206 W

11.35 A three-phase, three-wire, *ABC* system, with line voltage $\mathbf{V}_{BC} = 311.1\angle{0^\circ}$ V, has line currents

$$\mathbf{I}_A = 61.5\angle{116.6^\circ}\ \text{A} \qquad \mathbf{I}_B = 61.2\angle{-48.0^\circ}\ \text{A} \qquad \mathbf{I}_C = 16.1\angle{218^\circ}\ \text{A}$$

Find the readings of wattmeters in lines (*a*) A and B, (*b*) B and C, and (*c*) A and C.
Ans. (*a*) 5266 W, 6370 W; (*b*) 9312 W, 2322 W; (*c*) 9549 W, 1973 W

11.36 A three-phase, three-wire, ABC system has an effective line voltage 440 V. The line currents are

$$\mathbf{I}_A = 27.9\angle 90^\circ \text{ A} \qquad \mathbf{I}_B = 81.0\angle -9.9^\circ \text{ A} \qquad \mathbf{I}_C = 81.0\angle 189.9^\circ \text{ A}$$

Obtain the readings of wattmeters in lines (*a*) A and B, (*b*) B and C.
Ans. (*a*) 7.52 kW, 24.8 kW; (*b*) 16.16 kW, 16.16 kW

11.37 Two wattmeters in a three-phase, three-wire system with effective line voltage 120 V read 1500 W and 500 W. What is the impedance of the balanced Δ-connected load? *Ans.* $16.3\angle \pm 40.9^\circ\ \Omega$

11.38 A three-phase, three-wire, ABC system has effective line voltage 173.2 V. Wattmeters in lines A and B read −301 W and 1327 W, respectively. Find the impedance of the balanced Y-connected load. (Since the sequence is specified, the sign of the impedance angle can be determined.)
Ans. $10\angle -70^\circ\ \Omega$

11.39 A three-phase, three-wire system, with a line voltage $\mathbf{V}_{BC} = 339.4\angle 0^\circ$ V, has a balanced Y-connected load of $\mathbf{Z}_Y = 15\angle 60^\circ\ \Omega$. The lines between the system and the load have impedances $2.24\angle 26.57^\circ\ \Omega$. Find the line-voltage magnitude at the load. *Ans.* 301.1 V

11.40 Repeat Problem 11.39 with the load impedance $\mathbf{Z}_Y = 15\angle -60^\circ\ \Omega$. By drawing the voltage phasor diagrams for the two cases, illustrate the effect of load impedance angle on the voltage drop for a given line impedance. *Ans.* 332.9 V

11.41 A three-phase generator with an effective line voltage of 6000 V supplies the following four balanced loads in parallel: 16 kW at pf = 0.8 lagging, 24 kW at pf = 0.6 lagging, 4 kW at pf = 1, and 1 kW at pf = 0.1 leading. (*a*) Find the total average power (P) supplied by the generator, reactive power (Q), apparent power (S), power factor, and effective value of line current. (*b*) Find the amount of reactive load Q_c to be added in parallel to produce an overall power factor of 0.9 lagging, then find apparent power and effective value of line current.
Ans. (*a*) $P = 45$ kW, $Q = 34.05$ kvar, $S = 56.43$ kVA, pf = 0.8 lagging, $I_L = 5.43$ A, (*b*) $Q_C = -12.25$ kvar, $S = 50$ kVA, $I_L = 5.35$ A

11.42 A balanced Δ-connected load with impedances $Z_\Delta = 6 + j9\ \Omega$ is connected to a three-phase generator with an effective line voltage of 400 V. The lines between the load and the generator have resistances of 1 Ω each. Find the effective line current, power delivered by the generator, and power absorbed by the load.
Ans. $I_L = 54.43$ A, $P_g = 26666$ W, $P_\ell = 17777$ W

11.43 In Problem 11.42, find the effective line voltage at the load.
Ans. $V_L = 340$ V

11.44 A three-phase generator feeds two balanced loads (9 kW at pf = 0.8 and 12 kW at pf = 0.6, both lagging) through three cables (0.1 Ω each). The generator is regulated such that the effective line voltage at the load is 220 V. Find the effective line voltage at the generator. *Ans.* 230 V

11.45 A balanced Δ-connected load has impedances $45 + j60\ \Omega$. Find the average power delivered to it at an effective line voltage of: (*a*) 400 V, (*b*) 390 V.
Ans. (*a*) 3.84 kW, (*b*) 3.65 kW

11.46 Obtain the change in average power delivered to a three-phase balanced load if the line voltage is multiplied by a factor α. *Ans.* Power is multiplied by the factor α^2

11.47 A three-phase, three-wire source supplies a balanced load rated for 15 kW with pf = 0.8 at an effective line voltage of 220 V. Find the power absorbed by the load if the three wires connecting the source to the load have resistances of 0.05 Ω each and the effective line voltage at the source is 220 V. Use both a simplified

approximation and also an exact method.
Ans. 14.67 kW (by an approximate method), 14.54 kW (by an exact method)

11.48 In Problem 11.47 determine the effective value of line voltage such that the load operates at its rated values.
Ans. 222.46 V (by an approximate method), 221.98 V (by an exact method)

11.49 What happens to the quantity of power supplied by a three-phase, three-wire system to a balanced load if one phase is disconnected? *Ans.* Power is halved.

11.50 A three-phase, three-wire generator with effective line voltage 6000 V is connected to a balanced load by three lines with resistances of 1 Ω each, delivering a total of 200 kW. Find the efficiency (the ratio of power absorbed by the load to power delivered by the system) if the power factor of the generator is (*a*) 0.6, (*b*) 0.9 *Ans.* (*a*) 98.5 percent (*b*) 99.3 percent.

11.51 A 60-Hz three-phase, three-wire system with terminals labeled 1, 2, 3 has an effective line voltage of 220 V. To determine if the system is ABC or CBA, the circuit of Fig. 11-35 is tested. Find the effective voltage between node 4 and line 2 if the system is (*a*) ABC, (*b*) CBA.
Ans. (*a*) 80.5 V; (*b*) 300.5 V

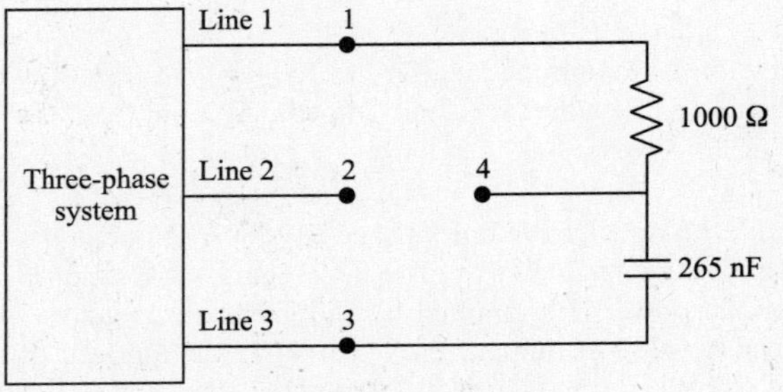

Fig. 11-35

CHAPTER 12

Frequency Response, Filters, and Resonance

12.1 FREQUENCY RESPONSE

The response of linear circuits to a sinusoidal input is also a sinusoid, with the same frequency but possibly a different amplitude and phase angle. This response is a function of the frequency. We have already seen that a sinusoid can be represented by a phasor which shows its magnitude and phase. The *frequency response* is defined as the ratio of the output phasor to the input phasor. It is a real function of $j\omega$ and is given by

$$\mathbf{H}(j\omega) = \text{Re}\,[\mathbf{H}] + j\,\text{Im}\,[\mathbf{H}] = |\mathbf{H}|e^{j\theta} \tag{1a}$$

where Re [**H**] and Im [**H**] are the real and imaginary parts of $\mathbf{H}(j\omega)$ and $|\mathbf{H}|$ and θ are its magnitude and phase angle. Re [**H**], Im [**H**], $|\mathbf{H}|$, and θ are, in general, functions of ω. They are related by

$$|\mathbf{H}|^2 = |\mathbf{H}(j\omega)|^2 = \text{Re}^2\,[\mathbf{H}] + \text{Im}^2\,[\mathbf{H}] \tag{1b}$$

$$\theta = \underline{/\mathbf{H}(j\omega)} = \tan^{-1}\frac{\text{Im}\,[\mathbf{H}]}{\text{Re}\,[\mathbf{H}]} \tag{1c}$$

The frequency response, therefore, depends on the choice of input and output variables. For example, if a current source is connected across the network of Fig. 12-1(a), the terminal current is the input and the terminal voltage may be taken as the output. In this case, the input impedance $\mathbf{Z} = \mathbf{V}_1/\mathbf{I}_1$ constitutes the frequency response. Conversely, if a voltage source is applied to the input and

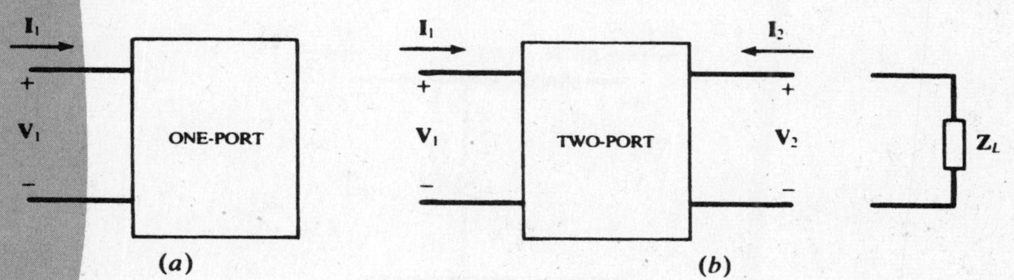

Fig. 12-1

the terminal current is measured, the input admittance $\mathbf{Y} = \mathbf{I}_1/\mathbf{V}_1 = 1/\mathbf{Z}$ represents the frequency response.

For the two-port network of Fig. 12-1(b), the following frequency responses are defined:

Input impedance $\mathbf{Z}_{in}(j\omega) = \mathbf{V}_1/\mathbf{I}_1$
Input admittance $\mathbf{Y}_{in}(j\omega) = 1/\mathbf{Z}_{in}(j\omega) = \mathbf{I}_1/\mathbf{V}_1$
Voltage transfer ratio $\mathbf{H}_v(j\omega) = \mathbf{V}_2/\mathbf{V}_1$
Current transfer ratio $\mathbf{H}_i(j\omega) = \mathbf{I}_2/\mathbf{I}_1$
Transfer impedances $\mathbf{V}_2/\mathbf{I}_1$ and $\mathbf{V}_1/\mathbf{I}_2$

EXAMPLE 12.1 Find the frequency response $\mathbf{V}_2/\mathbf{V}_1$ for the two-port circuit shown in Fig. 12-2.

Let $\mathbf{Y}_{RC}$ be the admittance of the parallel RC combination. Then, $\mathbf{Y}_{RC} = 10^{-6}j\omega + 1/1250$. $\mathbf{V}_2/\mathbf{V}_1$ is obtained by dividing $\mathbf{V}_1$ between $\mathbf{Z}_{RC}$ and the 5-kΩ resistor.

$$\mathbf{H}(j\omega) = \frac{\mathbf{V}_2}{\mathbf{V}_1} = \frac{\mathbf{Z}_{RC}}{\mathbf{Z}_{RC} + 5000} = \frac{1}{1 + 5000\mathbf{Y}_{RC}} = \frac{1}{5(1 + 10^{-3}j\omega)} \tag{2a}$$

$$|\mathbf{H}| = \frac{1}{5\sqrt{1 + 10^{-6}\omega^2}} \qquad \theta = -\tan^{-1}(10^{-3}\omega) \tag{2b}$$

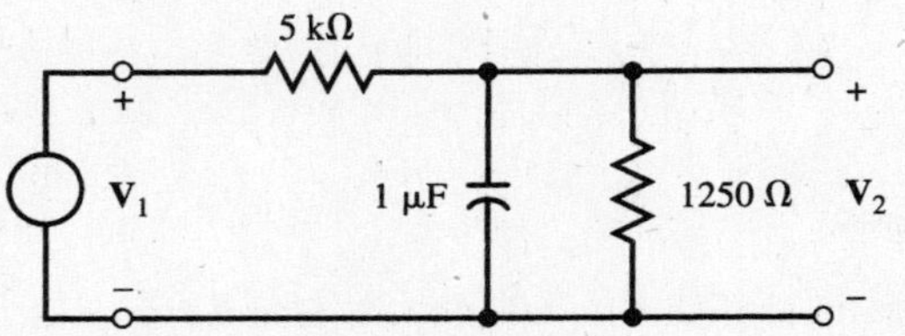

Fig. 12-2

Alternative solution: First we find the Thévenin equivalent of the resistive part of the circuit, $V_{Th} = V_1/5$ and $R_{Th} = 1\ \text{k}\Omega$, and then divide V_{Th} between R_{Th} and the 1-μF capacitor to obtain ($2a$).

12.2 HIGH-PASS AND LOW-PASS NETWORKS

A resistive voltage divider under a no-load condition is shown in Fig. 12-3, with the standard two-port voltages and currents. The voltage transfer function and input impedance are

$$\mathbf{H}_{v\infty}(\omega) = \frac{R_2}{R_1 + R_2} \qquad \mathbf{H}_{z\infty}(\omega) = R_1 + R_2$$

The ∞ in subscripts indicates no-load conditions. Both $\mathbf{H}_{v\infty}$ and $\mathbf{H}_{z\infty}$ are real constants, independent of frequency, since no reactive elements are present. If the network contains either an inductance or a capacitance, then $\mathbf{H}_{v\infty}$ and $\mathbf{H}_{z\infty}$ will be complex and will vary with frequency. If $|\mathbf{H}_{v\infty}|$ decreases as

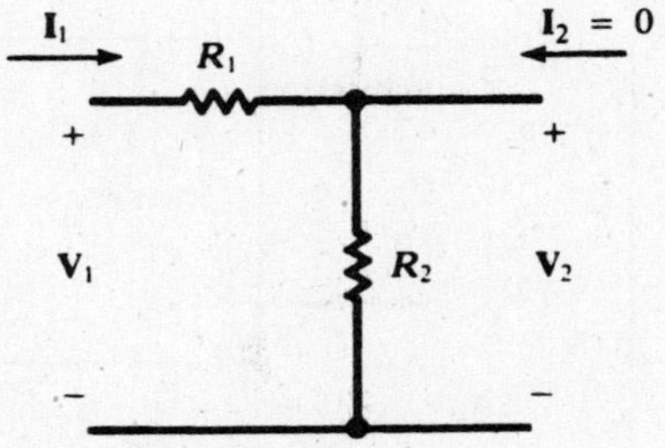

Fig. 12-3

frequency increases, the performance is called *high-frequency roll-off* and the circuit is a *low-pass network*. On the contrary, a *high-pass network* will have low-frequency roll-off, with $|\mathbf{H}_{v\infty}|$ decreasing as the frequency decreases. Four two-element circuits are shown in Fig. 12-4, two high-pass and two low-pass.

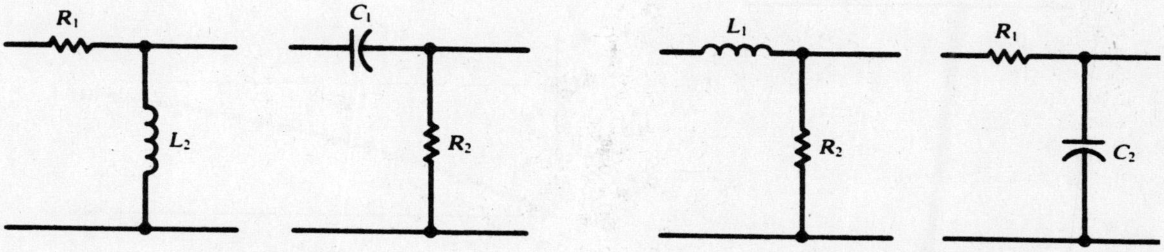

Fig. 12-4

The RL high-pass circuit shown in Fig. 12-5 is open-circuited or under no-load. The input impedance frequency response is determined by plotting the magnitude and phase angle of

$$\mathbf{H}_{z\infty}(\omega) = R_1 + j\omega L_2 \equiv |\mathbf{H}_z| \underline{/\theta_{\mathbf{H}}}$$

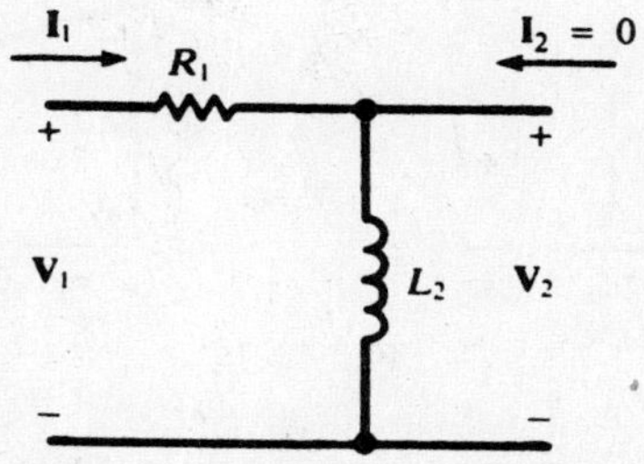

Fig. 12-5

or, normalizing and writing $\omega_x \equiv R_1/L_2$,

$$\frac{\mathbf{H}_{z\infty}(\omega)}{R_1} = 1 + j(\omega/\omega_x) = \sqrt{1 + (\omega/\omega_x)^2}\,\underline{/\tan^{-1}(\omega/\omega_x)}$$

Five values of ω provide sufficient data to plot $|\mathbf{H}_z|/R_1$ and $\theta_{\mathbf{H}}$, as shown in Fig. 12-6. The magnitude approaches infinity with increasing frequency, and so, at very high frequencies, the network current $\mathbf{I}_1$ will be zero.

In a similar manner, the frequency response of the output-to-input voltage ratio can be obtained. Voltage division under no-load gives

$$\mathbf{H}_{v\infty}(\omega) = \frac{j\omega L_2}{R_1 + j\omega L_2} = \frac{1}{1 - j(\omega_x/\omega)}$$

so that
$$|\mathbf{H}_v| = \frac{1}{\sqrt{1 + (\omega_x/\omega)^2}} \qquad \text{and} \qquad \theta_{\mathbf{H}} = \tan^{-1}(\omega_x/\omega)$$

ω	$\lvert\mathbf{H}_z\rvert/R_1$	θ_H
0	1	0°
$0.5\omega_x$	$0.5\sqrt{5}$	26.6°
ω_x	$\sqrt{2}$	45°
$2\omega_x$	$\sqrt{5}$	63.4°
∞	∞	90°

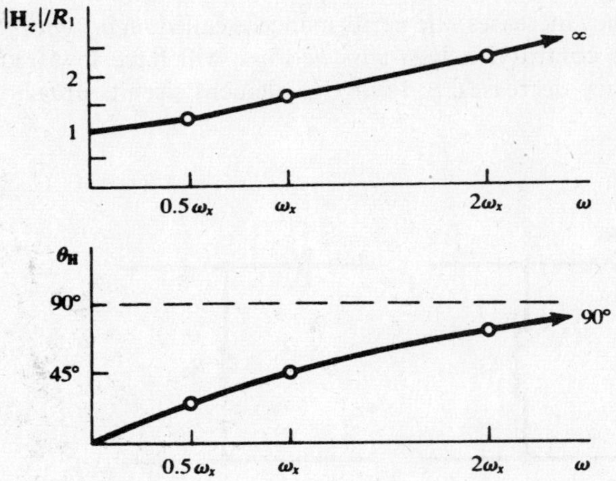

Fig. 12-6

The magnitude and angle are plotted in Fig. 12-7. This transfer function approaches unity at high frequency, where the output voltage is the same as the input. Hence the description "low-frequency roll-off" and the name "high-pass."

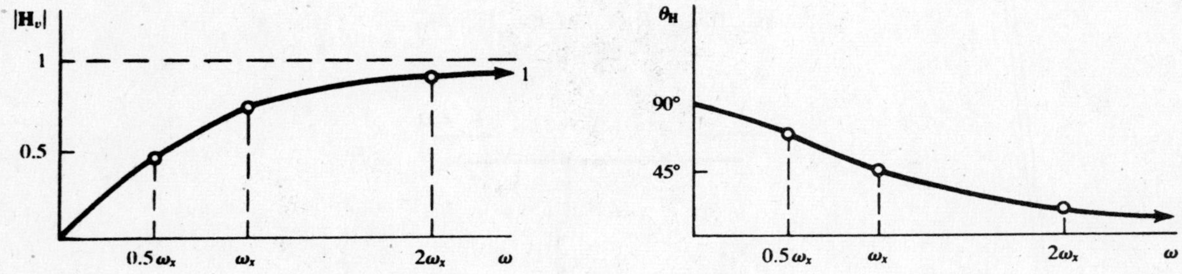

Fig. 12-7

A transfer impedance of the *RL* high-pass circuit under no-load is

$$\mathbf{H}_\infty(\omega) = \frac{\mathbf{V}_2}{\mathbf{I}_1} = j\omega L_2 \qquad \text{or} \qquad \frac{\mathbf{H}_\infty(\omega)}{R_1} = j\,\frac{\omega}{\omega_x}$$

The angle is constant at 90°; the graph of magnitude versus ω is a straight line, similar to a reactance plot of ωL versus ω. See Fig. 12-8.

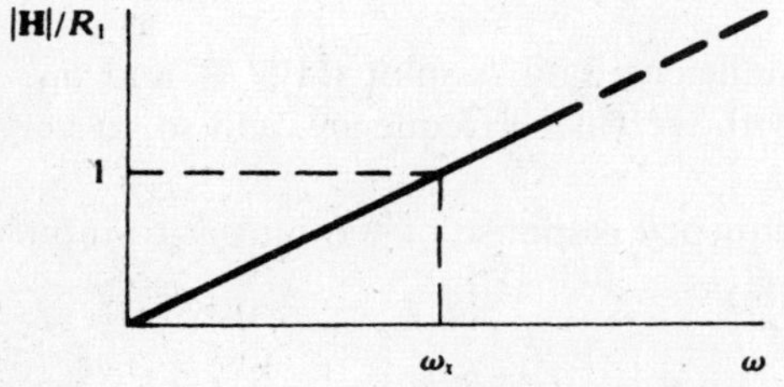

Fig. 12-8

Interchanging the positions of *R* and *L* results in a low-pass network with high-frequency roll-off (Fig. 12-9). For the open-circuit condition

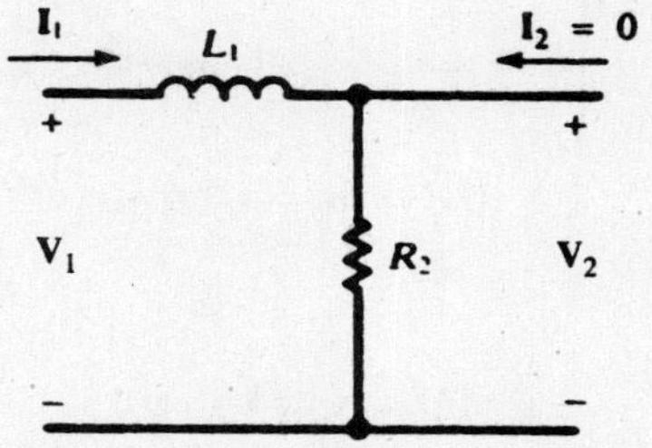

Fig. 12-9

$$\mathbf{H}_{v\infty}(\omega) = \frac{R_2}{R_2 + j\omega L_1} = \frac{1}{1 + j(\omega/\omega_x)}$$

with $\omega_x \equiv R_2/L_1$; that is,

$$|\mathbf{H}_v| = \frac{1}{\sqrt{1 + (\omega/\omega_x)^2}} \qquad \text{and} \qquad \theta_{\mathbf{H}} = \tan^{-1}(-\omega/\omega_x)$$

The magnitude and angle plots are shown in Fig. 12-10. The voltage transfer function $\mathbf{H}_{v\infty}$ approaches zero at high frequencies and unity at $\omega = 0$. Hence the name "low-pass."

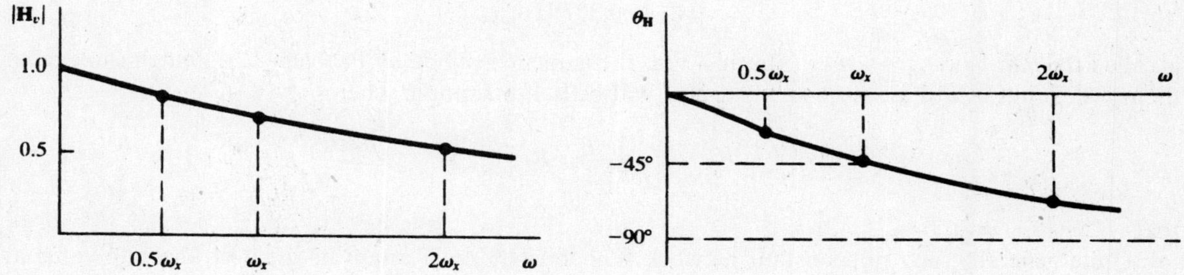

Fig. 12-10

The other network functions of this low-pass network are obtained in the Solved Problems.

EXAMPLE 12.2 Obtain the voltage transfer function $\mathbf{H}_{v\infty}$ for the open circuit shown in Fig. 12-11. At what frequency, in hertz, does $|\mathbf{H}_v| = 1/\sqrt{2}$ if (*a*) $C_2 = 10$ nF, (*b*) $C_2 = 1$ nF?

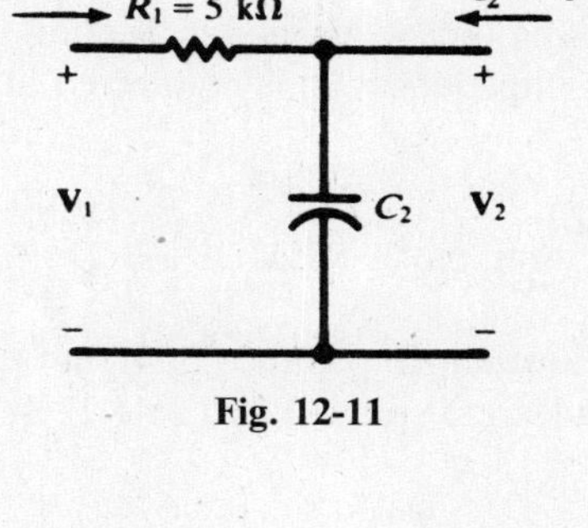

Fig. 12-11

$$\mathbf{H}_{v\infty}(\omega) = \frac{1/j\omega C_2}{R_1 + (1/j\omega C_2)} = \frac{1}{1 + j(\omega/\omega_x)} \qquad \text{where} \qquad \omega_x \equiv \frac{1}{R_1 C_2} = \frac{2 \times 10^{-4}}{C_2} \quad \text{(rad/s)}$$

(*a*)

$$|\mathbf{H}_v| = \frac{1}{\sqrt{1 + (\omega/\omega_x)^2}}$$

and so $|\mathbf{H}_v| = 1/\sqrt{2}$ when

$$\omega = \omega_x = \frac{2 \times 10^{-4}}{10 \times 10^{-9}} = 2 \times 10^4 \text{ rad/s}$$

or when $f = (2 \times 10^4)/2\pi = 3.18$ kHz.

(*b*)

$$f = \frac{10}{1}(3.18) = 31.8 \text{ kHz}$$

Comparing (*a*) and (*b*), it is seen that the greater the value of C_2, the lower is the frequency at which $|\mathbf{H}_\mathbf{v}|$ drops to 0.707 of its peak value, 1; in other words, the more is the graph of $|\mathbf{H}_\mathbf{v}|$, shown in Fig. 12-10, shifted to the left. Consequently, any stray shunting capacitance, in parallel with C_2, serves to reduce the response of the circuit.

12.3 HALF-POWER FREQUENCIES

The frequency ω_x calculated in Example 12.2, the frequency at which

$$|\mathbf{H}_v| = 0.707|\mathbf{H}_v|_{\max}$$

is called the *half-power frequency*. In this case, the name is justified by Problem 12.5, which shows that the power input into the circuit of Fig. 12-11 will be half-maximum when

$$\left|\frac{1}{j\omega C_2}\right| = R_1$$

that is, when $\omega = \omega_x$.

Quite generally, any nonconstant network function $\mathbf{H}(\omega)$ will attain its greatest absolute value at some unique frequency ω_x. We shall call a frequency at which

$$|\mathbf{H}(\omega)| = 0.707|\mathbf{H}(\omega_x)|$$

a *half-power frequency* (or *half-power point*), whether or not this frequency actually corresponds to 50 percent power. In most cases, $0 < \omega_x < \infty$, so that there are two half-power frequencies, one above and one below the peak frequency. These are called the *upper* and *lower* half-power frequencies (points), and their separation, the *bandwidth*, serves as a measure of the sharpness of the peak.

12.4 GENERALIZED TWO-PORT, TWO-ELEMENT NETWORKS

The basic *RL* or *RC* network of the type examined in Section 12.2 can be generalized with $\mathbf{Z}_1$ and $\mathbf{Z}_2$, as shown in Fig. 12-12; the load impedance $\mathbf{Z}_L$ is connected at the output port.

By voltage division,

$$\mathbf{V}_2 = \frac{\mathbf{Z}'}{\mathbf{Z}_1 + \mathbf{Z}'}\mathbf{V}_1 \qquad \text{or} \qquad \mathbf{H}_v = \frac{\mathbf{V}_2}{\mathbf{V}_1} = \frac{\mathbf{Z}'}{\mathbf{Z}_1 + \mathbf{Z}'}$$

where $\mathbf{Z}' = \mathbf{Z}_2\mathbf{Z}_L/(\mathbf{Z}_2 + \mathbf{Z}_L)$, the equivalent impedance of $\mathbf{Z}_2$ and $\mathbf{Z}_L$ in parallel. The other transfer functions are calculated similarly, and are displayed in Table 12-1.

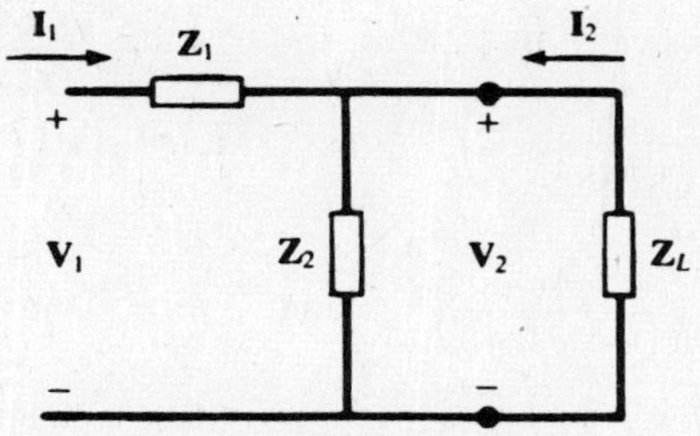

Fig. 12-12

Table 12-1

Output Condition \ Network Function	$H_z = \frac{V_1}{I_1}$ (Ω)	$H_v = \frac{V_2}{V_1}$	$H_i = \frac{I_2}{I_1}$	$H_vH_z = \frac{V_2}{I_1}$ (Ω)	$\frac{H_i}{H_z} = \frac{I_2}{V_1}$ (S)
Short-circuit, $Z_L = 0$	Z_1	0	−1	0	$-\frac{1}{Z_1}$
Open-circuit $Z_L = \infty$	$Z_1 + Z_2$	$\frac{Z_2}{Z_1 + Z_2}$	0	Z_2	0
Load, Z_L	$Z_1 + Z'$	$\frac{Z'}{Z_1 + Z'}$	$\frac{-Z_2}{Z_2 + Z_L}$	Z'	$\frac{-Z'}{Z_L(Z_1 + Z')}$

12.5 THE FREQUENCY RESPONSE AND NETWORK FUNCTIONS

The frequency response of a network may be found by substituting $j\omega$ for s in its network function. This useful method is illustrated in the following example.

EXAMPLE 12.3 Find (*a*) the network function $\mathbf{H(s)} = \mathbf{V}_2/\mathbf{V}_1$ in the circuit shown in Fig. 12-13, (*b*) $\mathbf{H}(j\omega)$ for $LC = 2/\omega_0^2$ and $L/C = R^2$, and (*c*) the magnitude and phase angle of $\mathbf{H}(j\omega)$ in (*b*) for $\omega_0 = 1$ rad/s.

(*a*) Assume $\mathbf{V}_2$ is known. Use generalized impedances $L\mathbf{s}$ and $1/C\mathbf{s}$ and solve for $\mathbf{V}_1$.
From $\mathbf{I}_R = \mathbf{V}_2/R$,

$$\mathbf{V}_A = (R + L\mathbf{s})\mathbf{I}_R = \frac{R + L\mathbf{s}}{R}\mathbf{V}_2 \tag{3}$$

$$\mathbf{I}_C = C\mathbf{s}\mathbf{V}_A = \frac{C\mathbf{s}(R + L\mathbf{s})}{R}\mathbf{V}_2 \quad \text{and} \quad \mathbf{I}_1 = \mathbf{I}_R + \mathbf{I}_C = \frac{\mathbf{V}_2}{R} + \frac{C\mathbf{s}(R + L\mathbf{s})}{R}\mathbf{V}_2 = \frac{1 + C\mathbf{s}(R + L\mathbf{s})}{R}\mathbf{V}_2$$

Then,

$$\mathbf{V}_1 = \mathbf{V}_A + R\mathbf{I}_1 = \frac{R + L\mathbf{s}}{R}\mathbf{V}_2 + [1 + C\mathbf{s}(R + L\mathbf{s})]\mathbf{V}_2$$

and

$$\mathbf{H(s)} = \frac{\mathbf{V}_2}{\mathbf{V}_1} = \frac{1}{2 + (L/R + CR)\mathbf{s} + LC\mathbf{s}^2} \tag{4a}$$

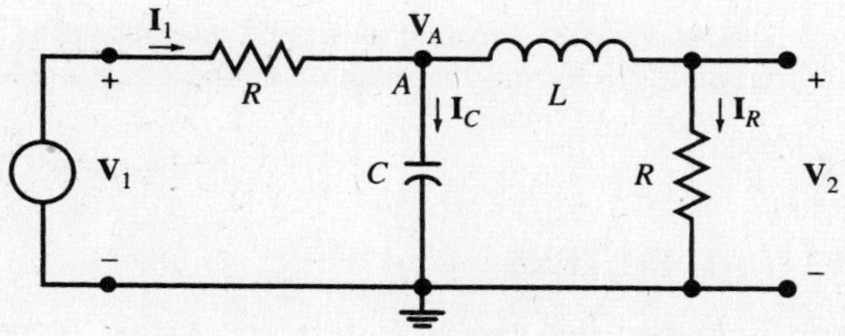

Fig. 12-13

(*b*) From $LC = 2/\omega_0^2$ and $L/C = R^2$ we get $L = \sqrt{2}R/\omega_0$ and $C = \sqrt{2}/R\omega_0$. Substituting L and C into (*4a*) gives

$$\mathbf{H}(\mathbf{s}) = \frac{1}{2}\left(\frac{1}{1+\sqrt{2}(\mathbf{s}/\omega_0)+(\mathbf{s}/\omega_0)^2}\right) \quad \text{or} \quad \mathbf{H}(j\omega) = \frac{1}{2}\left(\frac{1}{1+j\sqrt{2}(\omega/\omega_0)-(\omega/\omega_0)^2}\right) \tag{4b}$$

$$|\mathbf{H}|^2 = \frac{1}{4}\left(\frac{1}{1+(\omega/\omega_0)^4}\right) \quad \text{and} \quad \theta = -\tan^{-1}\left(\frac{\sqrt{2}\omega_0\omega}{\omega_0^2-\omega^2}\right)$$

Note that $\mathbf{H}(j\omega)$ is independent of R. The network passes low-frequency sinusoids and rejects, or attenuates, the high-frequency sinusoids. It is a low-pass filter with a half-power frequency of $\omega = \omega_0$ and, in this case, the magnitude of the frequency response is $|\mathbf{H}(j\omega_0)| = |\mathbf{H}(0)|/\sqrt{2} = \sqrt{2}/4$ and its phase angle is $\underline{/\mathbf{H}(j\omega_0)} = -\pi/2$.

(*c*) For $\omega_0 = 1$,

$$\mathbf{H}(\mathbf{s}) = \frac{1}{2}\left(\frac{1}{1+\sqrt{2}\mathbf{s}+\mathbf{s}^2}\right) \quad \text{or} \quad \mathbf{H}(j\omega) = \frac{1}{2}\left(\frac{1}{1+j\sqrt{2}\omega-\omega^2}\right) \tag{4c}$$

$$|\mathbf{H}|^2 = \frac{1}{4}\frac{1}{1+\omega^4} \quad \text{and} \quad \theta = -\tan^{-1}\left(\frac{\sqrt{2}\omega}{1-\omega^2}\right)$$

The RC network of Fig 12-4(*b*) was defined as a first-order low-pass filter with half-power frequency at $\omega_0 = 1/R_1C_2$. The circuit of Fig. 12-13 is called a *second-order Butterworth filter*. It has a sharper cutoff.

12.6 FREQUENCY RESPONSE FROM POLE-ZERO LOCATION

The frequency response of a network is the value of the network function $\mathbf{H}(\mathbf{s})$ at $\mathbf{s} = j\omega$. This observation can be used to evaluate $\mathbf{H}(j\omega)$ graphically. The graphical method can produce a quick sketch of $\mathbf{H}(j\omega)$ and bring to our attention its behavior near a pole or a zero without the need for a complete solution.

EXAMPLE 12.4 Find poles and zeros of $\mathbf{H}(\mathbf{s}) = 10\mathbf{s}/(\mathbf{s}^2 + 2\mathbf{s} + 26)$. Place them in the **s**-domain and use the pole-zero plot to sketch $\mathbf{H}(j\omega)$.

$\mathbf{H}(\mathbf{s})$ has a zero at $z_1 = 0$. Its poles $\mathbf{p}_1$ and $\mathbf{p}_2$ are found from $\mathbf{s}^2 + 2\mathbf{s} + 26 = 0$ so that $\mathbf{p}_1 = -1 + j5$ and $\mathbf{p}_2 = -1 - j5$. The pole-zero plot is shown in Fig. 12-14(*a*). The network function can then be written as

$$\mathbf{H}(\mathbf{s}) = (10)\frac{\mathbf{s}-\mathbf{z}_1}{(\mathbf{s}-\mathbf{p}_1)(\mathbf{s}-\mathbf{p}_2)}$$

For each value of **s**, the term $(\mathbf{s} - \mathbf{z}_1)$ is a vector originating from the zero $\mathbf{z}_1$ and ending at point **s** in the **s**-domain. Similarly, $\mathbf{s} - \mathbf{p}_1$ and $\mathbf{s} - \mathbf{p}_2$ are vectors drawn from poles $\mathbf{p}_1$ and $\mathbf{p}_2$, respectively, to the point **s**. Therefore, for any value of **s**, the network function may be expressed in terms of three vectors **A**, **B**, and **C** as follows:

$$\mathbf{H}(\mathbf{s}) = (10)\frac{\mathbf{A}}{\mathbf{B}\times\mathbf{C}} \quad \text{where } \mathbf{A} = (\mathbf{s}-\mathbf{z}_1), \mathbf{B} = (\mathbf{s}-\mathbf{p}_1), \text{ and } \mathbf{C} = (\mathbf{s}-\mathbf{p}_2)$$

The magnitude and phase angle of $\mathbf{H}(\mathbf{s})$ at any point on the **s**-plane may be found from:

$$|\mathbf{H}(\mathbf{s})| = (10)\frac{|\mathbf{A}|}{|\mathbf{B}|\times|\mathbf{C}|} \tag{5a}$$

$$\underline{/\mathbf{H}(\mathbf{s})} = \underline{/\mathbf{A}} - \underline{/\mathbf{B}} - \underline{/\mathbf{C}} \tag{5b}$$

By placing s on the $j\omega$ axis [Fig. 12-14(*a*)], varying ω from 0 to ∞, and measuring the magnitudes and phase angles of vectors **A**, **B**, and **C**, we can use (*5a*) and (*5b*) to find the magnitude and phase angle plots. Figure 12-14(*b*) shows the magnitude plot.

12.7 IDEAL AND PRACTICAL FILTERS

In general, networks are frequency selective. Filters are a class of networks designed to possess specific frequency selectivity characteristics. They pass certain frequencies unaffected (the pass-band)

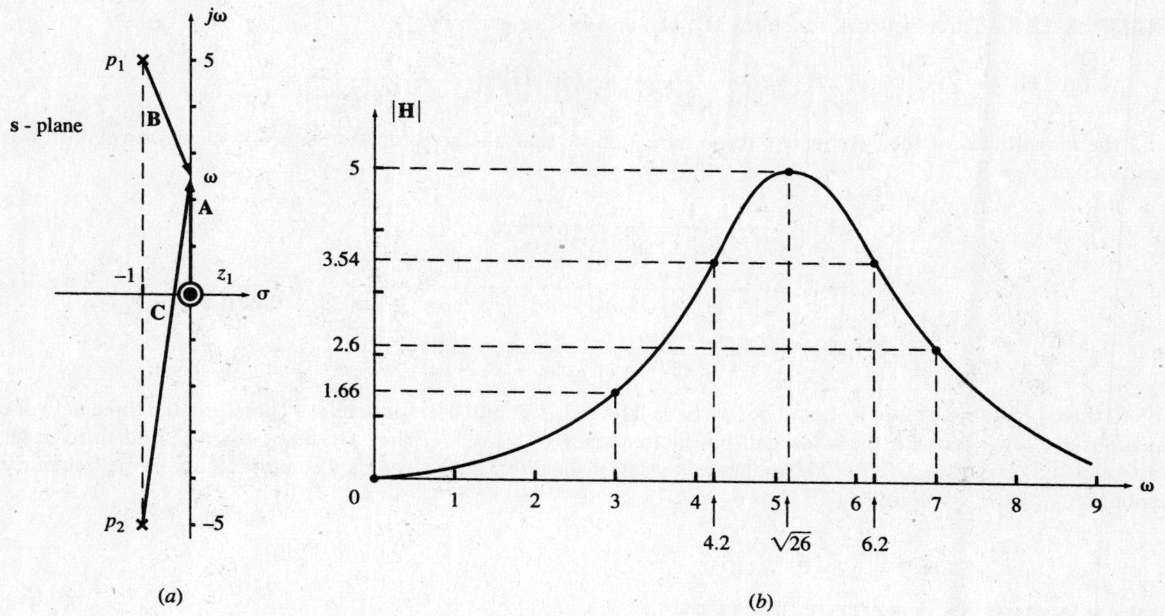

Fig. 12-14

and stop others (the stop-band). Ideally, in the pass-band, $\mathbf{H}(j\omega) = 1$ and in the stop-band, $\mathbf{H}(j\omega) = 0$. We therefore recognize the following classes of filters: low-pass [Fig. 12-15(*a*)], high-pass [Fig. 12-15(*b*)], bandpass [Fig. 12-15(*c*)], and bandstop [Fig. 12-15(*d*)]. Ideal filters are not physically realizable, but we can design and build practical filters as close to the ideal one as desired. The closer to the ideal characteristic, the more complex the circuit of a practical filter will be.

The *RC* or *RL* circuits of Section 12.2 are first-order filters. They are far from ideal filters. As illustrated in the following example, the frequency response can approach that of the ideal filters if we increase the order of the filter.

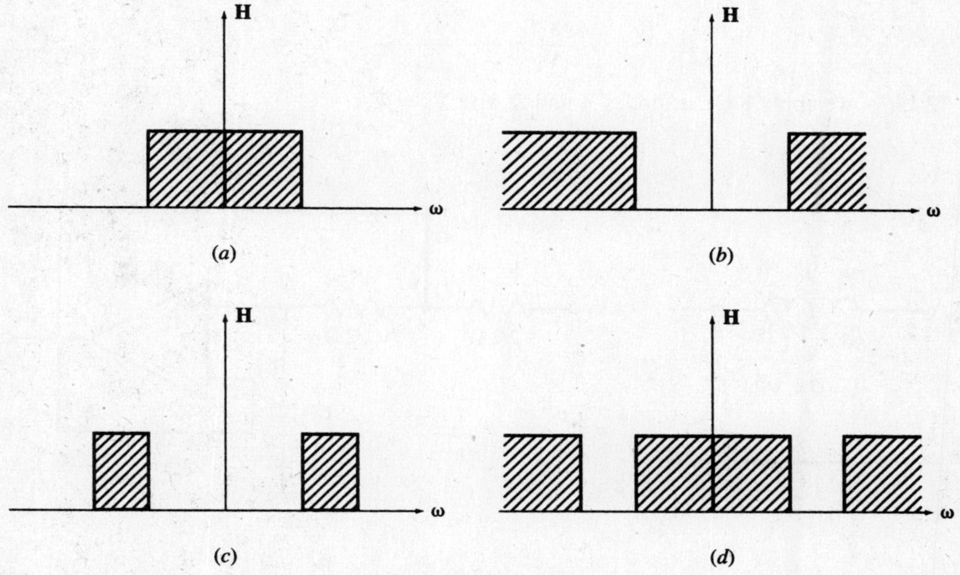

Fig. 12-15

EXAMPLE 12.5 Three network functions $\mathbf{H}_1$, $\mathbf{H}_2$, and $\mathbf{H}_3$ are given by

$$(a)\quad \mathbf{H}_1 = \frac{1}{\mathbf{s}+1} \qquad (b)\quad \mathbf{H}_2 = \frac{1}{\mathbf{s}^2+\sqrt{2}\mathbf{s}+1} \qquad (c)\quad \mathbf{H}_3 = \frac{1}{\mathbf{s}^3+2\mathbf{s}^2+2\mathbf{s}+1} = \frac{1}{(\mathbf{s}+1)(\mathbf{s}^2+\mathbf{s}+1)}$$

Find the magnitudes of their frequency responses. Show that all three functions are low-pass with half-power frequency at $\omega_0 = 1$.

(a)
$$|\mathbf{H}_1|^2 = \frac{1}{(1+j\omega)(1-j\omega)} = \frac{1}{1+\omega^2}$$

(b)
$$|\mathbf{H}_2|^2 = \frac{1}{(1-\omega^2+j\sqrt{2}\omega)(1-\omega^2-j\sqrt{2}\omega)} = \frac{1}{1+\omega^4}$$

(c)
$$|\mathbf{H}_3|^2 = \frac{1}{(1+\omega^2)(1-\omega^2+j\omega)(1-\omega^2-j\omega)} = \frac{1}{1+\omega^6}$$

For all three functions, at $\omega = 0$, 1, and ∞, we have $|\mathbf{H}|^2 = 1$, 1/2, and 0, respectively. Therefore, the three network functions are low-pass with the same half-power frequency of $\omega_0 = 1$. They are first-, second-, and third-order Butterworth filters, respectively. The higher the order of the filter, the sharper is the cutoff region in the frequency response.

12.8 PASSIVE AND ACTIVE FILTERS

Filters which contain only resistors, inductors, and capacitors are called *passive*. Those containing additional dependent sources are called *active*. Passive filters do not require external energy sources and they can last longer. Active filters are generally made of RC circuits and amplifiers. The circuit in Fig. 12-16(*a*) shows a second-order low-pass passive filter. The circuit in Fig. 12-16(*b*) shows an active filter with a frequency response $\mathbf{V}_2/\mathbf{V}_1$ equivalent to that of the circuit in Fig. 12-16(*a*).

EXAMPLE 12.6 Find the network function $\mathbf{V}_2/\mathbf{V}_1$ in the circuits shown in (*a*) Fig. 12-16(*a*) and (*b*) Fig. 12-16(*b*).

(*a*) In Fig. 12-16(*a*), we find $\mathbf{V}_2$ from $\mathbf{V}_1$ by voltage division.

$$\mathbf{V}_2 = \frac{1}{C\mathbf{s}}\frac{\mathbf{V}_1}{R+L\mathbf{s}+1/C\mathbf{s}} = \frac{\mathbf{V}_1}{LC\mathbf{s}^2+RC\mathbf{s}+1} = \frac{1}{LC}\frac{\mathbf{V}_1}{\mathbf{s}^2+(R/L)\mathbf{s}+(1/LC)}$$

Substituting for $R = 1$, $L = 1/\sqrt{2}$, and $C = \sqrt{2}$, and dividing by $\mathbf{V}_1$, we get

$$\frac{\mathbf{V}_2}{\mathbf{V}_1} = \frac{1}{\mathbf{s}^2+\sqrt{2}\mathbf{s}+1}$$

(*b*) In Fig. 12-16(*b*), we apply KCL at nodes A and B with $\mathbf{V}_B = \mathbf{V}_2$.

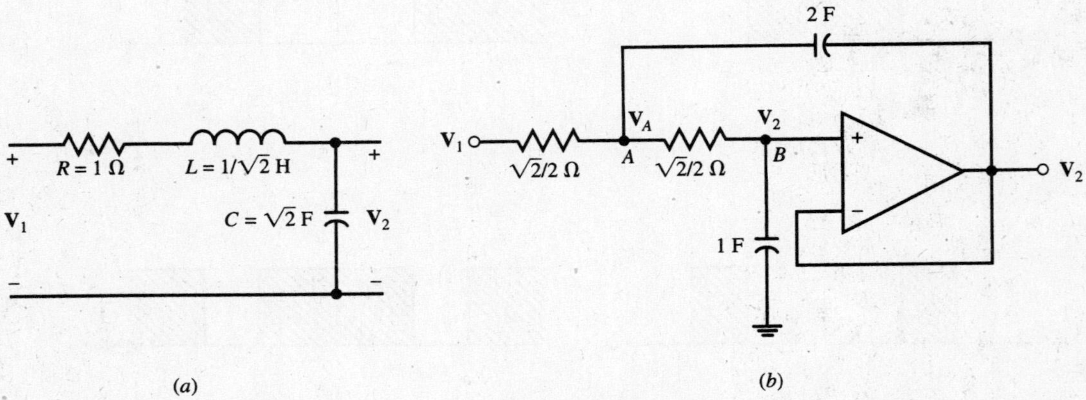

Fig. 12-16

Node A: $$(\mathbf{V}_A - \mathbf{V}_1)\sqrt{2} + (\mathbf{V}_A - \mathbf{V}_2)\sqrt{2} + (\mathbf{V}_A - \mathbf{V}_2)2\mathbf{s} = 0 \quad (6a)$$

Node B: $$\mathbf{V}_2\mathbf{s} + (\mathbf{V}_2 - \mathbf{V}_A)\sqrt{2} = 0 \quad (6b)$$

By eliminating $\mathbf{V}_A$ in (*6a*) and (*6b*), the network function $\mathbf{H(s)} = \mathbf{V}_2/\mathbf{V}_1$ is obtained. Thus,

$$\frac{\mathbf{V}_2}{\mathbf{V}_1} = \frac{1}{\mathbf{s}^2 + \sqrt{2}\mathbf{s} + 1}$$

Note that the circuits of Figs. 12-16(*a*) and (*b*) have identical network functions. They are second-order Butterworth low-pass filters with half-power frequencies at $\omega = 1$ rad/s.

12.9 BANDPASS FILTERS AND RESONANCE

The following network function is called a *bandpass function*.

$$\mathbf{H}(s) = \frac{k\mathbf{s}}{\mathbf{s}^2 + a\mathbf{s} + b} \qquad \text{where } a > 0, b > 0, k > 0 \quad (7)$$

The name is especially appropriate when the poles are complex, close to the $j\omega$ axis, and away from the origin in the s-domain. The frequency response of the bandpass function is

$$\mathbf{H}(j\omega) = \frac{kj\omega}{b - \omega^2 + aj\omega} \qquad |\mathbf{H}|^2 = \frac{k^2\omega^2}{(b-\omega^2)^2 + a^2\omega^2} = \frac{k^2}{a^2 + (b-\omega^2)^2/\omega^2} \quad (8)$$

The maximum of $|\mathbf{H}|$ occurs when $b - \omega^2 = 0$ or $\omega = \sqrt{b}$, which is called the *center frequency* ω_0. At the center frequency, we have $|\mathbf{H}|_{\max} = |\mathbf{H}(\omega_0)| = k/a$. The half-power frequencies are at ω_l and ω_h, where

$$|\mathbf{H}(\omega_l)|^2 = |\mathbf{H}(\omega_h)|^2 = \tfrac{1}{2}|\mathbf{H}(\omega_0)|^2 \quad (9a)$$

By applying (*8*) to (*9a*), ω_l and ω_h are found to be roots of the following equation:

$$\frac{(b-\omega^2)^2}{\omega^2} = a^2 \quad (9b)$$

Solving, $$\omega_l = \sqrt{a^2/4 + b} - a/2 \quad (9c)$$

$$\omega_h = \sqrt{a^2/4 + b} + a/2 \quad (9d)$$

From (*9c*) and (*9d*) we have

$$\omega_h - \omega_l = a \qquad \text{and} \qquad \omega_h\omega_l = b = \omega_0^2 \quad (10a)$$

The *bandwidth* β is defined by

$$\beta = \omega_h - \omega_l = a \quad (10b)$$

The *quality factor* Q is defined by

$$Q = \omega_0/\beta = \sqrt{b}/a \quad (10c)$$

The quality factor measures the sharpness of the frequency response around the center frequency. This behavior is also called *resonance* (see Sections 12.11 to 12.15). When the quality factor is high, ω_l and ω_h may be approximated by $\omega_0 - \beta/2$ and $\omega_0 + \beta/2$, respectively.

EXAMPLE 12.7 Consider the network function $\mathbf{H(s)} = 10\mathbf{s}/(\mathbf{s}^2 + 300\mathbf{s} + 10^6)$. Find the center frequency, lower and upper half-power frequencies, the bandwidth, and the quality factor.

Since $\omega_0^2 = 10^6$, the center frequency $\omega_0 = 1000$ rad/s.

The lower and upper half-power frequencies are, respectively,

$$\omega_l = \sqrt{a^2/4 + b} - a/2 = \sqrt{300^2/4 + 10^6} - 300/2 = 861.2 \text{ rad/s}$$
$$\omega_h = \sqrt{a^2/4 + b} + a/2 = \sqrt{300^2/4 + 10^6} + 300/2 = 1161.2 \text{ rad/s}$$

The bandwidth $\beta = \omega_h - \omega_l = 1161.2 - 861.2 = 300$ rad/s.
The quality factor $Q = 1000/300 = 3.3$.

EXAMPLE 12.8 Repeat Example 12.7 for $\mathbf{H(s)} = 10\mathbf{s}/(\mathbf{s}^2 + 30\mathbf{s} + 10^6)$. Again, from $\omega_0^2 = 10^6$, $\omega_0 = 1000$ rad/s. Then,

$$\omega_l = \sqrt{30^2/4 + 10^6} - 30/2 = 985.1 \text{ rad/s}$$
$$\omega_h = \sqrt{30^2/4 + 10^6} + 30/2 = 1015.1 \text{ rad/s}$$
$$\beta = a = 30 \text{ rad/s} \qquad \text{and} \qquad Q = 1000/30 = 33.3$$

Note that ω_l and ω_h can also be approximated with good accuracy by

$$\omega_l = \omega_0 - \beta/2 = 1000 - 30/2 = 985 \text{ rad/s} \qquad \text{and} \qquad \omega_h = \omega_0 + \beta/2 = 1000 + 30/2 = 1015 \text{ rad/s}$$

12.10 NATURAL FREQUENCY AND DAMPING RATIO

The denominator of the bandpass function given in (*7*) may be written as

$$\mathbf{s}^2 + a\,\mathbf{s} + b = \mathbf{s}^2 + 2\xi\omega_0\mathbf{s} + \omega_0^2$$

where $\omega_0 = \sqrt{b}$ is called the *natural frequency* and $\xi = a/(2\sqrt{b})$ is called the *damping ratio*. For $\xi > 1$, the circuit has two distinct poles on the negative real axis and is called *overdamped*. For $\xi = 1$, the circuit has a real pole of order two at $-\omega_0$ and is *critically damped*. For $\xi < 1$, the circuit has a pair of conjugate poles at $-\xi\omega_0 + j\omega_0\sqrt{1-\xi^2}$ and $-\xi\omega_0 - j\omega_0\sqrt{1-\xi^2}$. The poles are positioned on a semicircle in the left half plane with radius ω_0. The placement angle of the poles is $\phi = \sin^{-1}\xi$ (see Fig. 12-17). The circuit is *underdamped* and can contain damped oscillations. Note that the damping ratio is equal to half of the inverse of the quality factor.

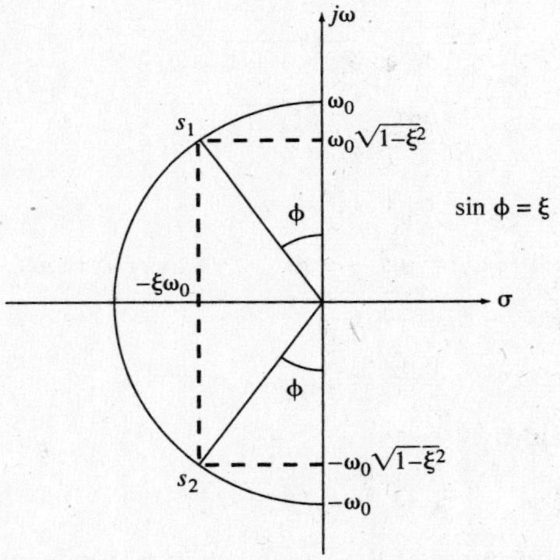

Fig. 12-17

12.11 *RLC* SERIES CIRCUIT; SERIES RESONANCE

The *RLC* circuit shown in Fig. 12-18 has, under open-circuit condition, an input or driving-point impedance

$$\mathbf{Z}_{\text{in}}(\omega) = R + j\left(\omega L - \frac{1}{\omega C}\right)$$

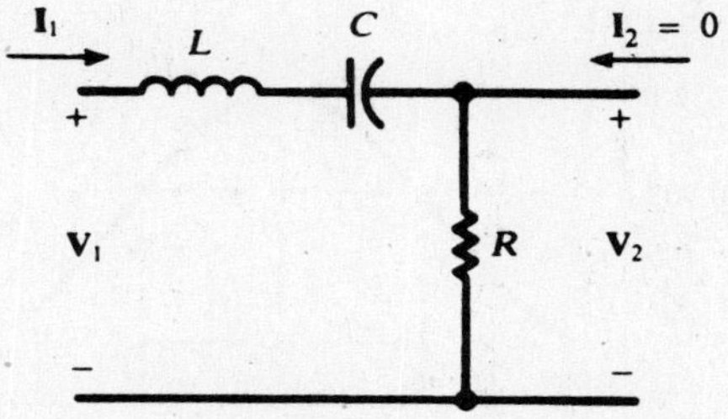

Fig. 12-18

The circuit is said to be in *series resonance* (or *low-impedance resonance*) when $\mathbf{Z}_{\text{in}}(\omega)$ is real (and so $|\mathbf{Z}_{\text{in}}(\omega)|$ is a minimum); that is, when

$$\omega L - \frac{1}{\omega C} = 0 \qquad \text{or} \qquad \omega = \omega_0 \equiv \frac{1}{\sqrt{LC}}$$

Figure 12-19 shows the frequency response. The capacitive reactance, inversely proportional to ω, is higher at low frequencies, while the inductive reactance, directly proportional to ω, is greater at the higher frequencies. Consequently, the net reactance at frequencies below ω_0 is capacitive, and the angle on $\mathbf{Z}_{\text{in}}$ is negative. At frequencies above ω_0, the circuit appears inductive, and the angle on $\mathbf{Z}_{\text{in}}$ is positive.

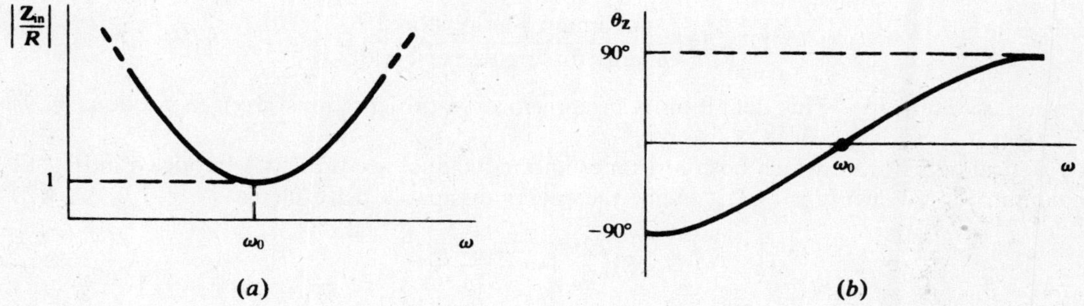

Fig. 12-19

By voltage division, the voltage transfer function for Fig. 12-18 is

$$\mathbf{H}_{v\infty}(\omega) = \frac{R}{\mathbf{Z}_{\text{in}}(\omega)} = R\mathbf{Y}_{\text{in}}(\omega)$$

The frequency response (magnitude only) is plotted in Fig. 12-20; the curve is just the reciprocal of that in Fig. 12-19(a). Note that roll-off occurs both below and above the series resonant frequency ω_0. The points where the response is 0.707, the half-power points (Section 12.3), are at frequencies ω_l and ω_h. The bandwidth is the width between these two frequencies: $\beta = \omega_h - \omega_l$.

A *quality factor*, $Q_0 = \omega_0 L/R$, may be defined for the series RLC circuit at resonance. (See Section 12.12 for the general development of Q.) The half-power frequencies can be expressed in terms of the circuit elements, or in terms of ω_0 and Q_0, as follows:

$$\omega_h = \frac{R}{2L} + \sqrt{\left(\frac{R}{2L}\right)^2 + \frac{1}{LC}} = \omega_0\left(\sqrt{1 + \frac{1}{4Q_0^2}} + \frac{1}{2Q_0}\right)$$

$$\omega_l = -\frac{R}{2L} + \sqrt{\left(\frac{R}{2L}\right)^2 + \frac{1}{LC}} = \omega_0\left(\sqrt{1 + \frac{1}{4Q_0^2}} - \frac{1}{2Q_0}\right)$$

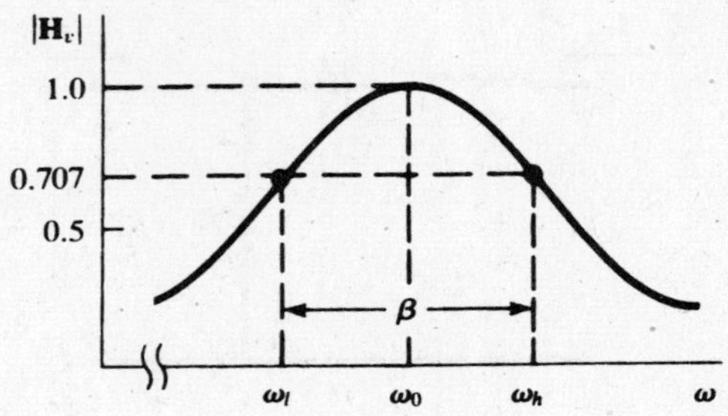

Fig. 12-20

See Problem 12.5. Subtraction of the expressions gives

$$\beta = \frac{R}{L} = \frac{\omega_0}{Q_0}$$

which suggests that the higher the "quality," the narrower the bandwidth.

12.12 QUALITY FACTOR

A *quality factor* or *figure of merit* can be assigned to a component or to a complete circuit. It is defined as

$$Q \equiv 2\pi\left(\frac{\text{maximum energy stored}}{\text{energy dissipated per cycle}}\right)$$

a dimensionless number. This definition is in agreement with definitions given in Sections 12.9 and 12.11.

A practical inductor, in which both resistance and inductance are present, is modeled in Fig. 12-21. The maximum stored energy is $\frac{1}{2}LI_{\max}^2$, while the energy dissipated per cycle is

$$(I_{\text{eff}}^2 R)\left(\frac{2\pi}{\omega}\right) = \frac{I_{\max}^2 R\pi}{\omega}$$

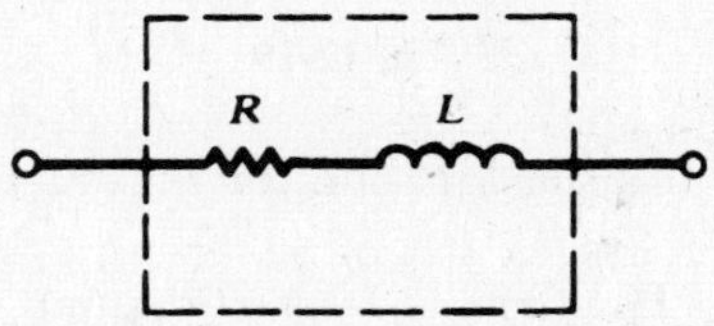

Fig. 12-21

Hence,

$$Q_{\text{ind}} = \frac{\omega L}{R}$$

A practical capacitor can be modeled by a parallel combination of R and C, as shown in Fig. 12.22. The maximum stored energy is $\frac{1}{2}CV_{\max}^2$ and the energy dissipated per cycle is $V_{\max}^2\pi/R\omega$. Thus, $Q_{\text{cap}} = \omega CR$.

The Q of the series RLC circuit is derived in Problem 12.6(a). It is usually applied at resonance, in which case it has the equivalent forms

$$Q_0 = \frac{\omega_0 L}{R} = \frac{1}{\omega_0 CR} = \frac{1}{R}\sqrt{\frac{L}{C}}$$

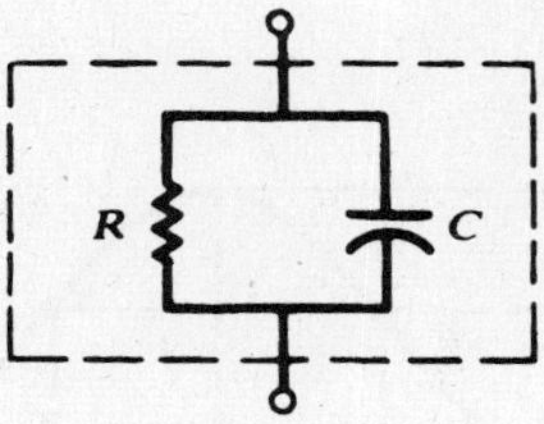

Fig. 12-22

12.13 *RLC* PARALLEL CIRCUIT; PARALLEL RESONANCE

A parallel *RLC* network is shown in Fig. 12-23. Observe that $\mathbf{V}_2 = \mathbf{V}_1$. Under the open-circuit condition, the input admittance is

$$\mathbf{Y}_{\text{in}}(\omega) = \frac{1}{R} + \frac{1}{j\omega L} + j\omega C = \frac{1}{\mathbf{Z}_{\text{in}}(\omega)}$$

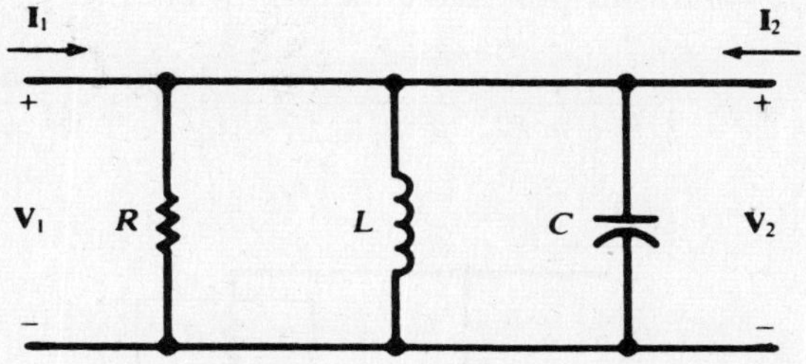

Fig. 12-23

The network will be in *parallel resonance* (or *high-impedance resonance*) when $\mathbf{Y}_{\text{in}}(\omega)$, and thus $\mathbf{Z}_{\text{in}}(\omega)$, is real (and so $|\mathbf{Y}_{\text{in}}(\omega)|$ is a minimum and $|\mathbf{Z}_{\text{in}}(\omega)|$ is a maximum); that is, when

$$-\frac{1}{\omega L} + \omega C = 0 \qquad \text{or} \qquad \omega = \omega_a \equiv \frac{1}{\sqrt{LC}}$$

The symbol ω_a is now used to denote the quantity $1/\sqrt{LC}$ in order to distinguish the resonance from a low-impedance resonance. Complex series-parallel networks may have several high-impedance resonant frequencies ω_a and several low-impedance resonant frequencies ω_0.

The normalized input impedance

$$\frac{\mathbf{Z}_{\text{in}}(\omega)}{R} = \frac{1}{1 + jR\left(\omega C - \dfrac{1}{\omega L}\right)}$$

is plotted (magnitude only) in Fig. 12-24. Half-power frequencies ω_l and ω_h are indicated on the plot. Analogous to series resonance, the bandwidth is given by

$$\beta = \frac{\omega_a}{Q_a}$$

where Q_a, the quality factor of the parallel circuit at $\omega = \omega_a$, has the equivalent expressions

$$Q_a = \frac{R}{\omega_a L} = \omega_a RC = R\sqrt{\frac{C}{L}}$$

See Problem 12.6(*b*).

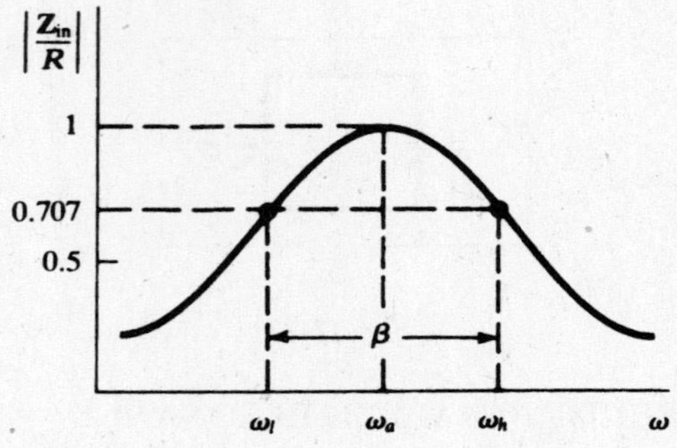

Fig. 12-24

12.14 PRACTICAL *LC* PARALLEL CIRCUIT

A parallel *LC* "tank" circuit has frequency applications in electronics as a tuning or frequency selection device. While the capacitor may often be treated as "pure C," the losses in the inductor should be included. A reasonable model for the practical tank is shown in Fig. 12-25. The input admittance is

$$\mathbf{Y}_{\text{in}}(\omega) = j\omega C + \frac{1}{R + j\omega L} = \frac{R}{R^2 + (\omega L)^2} + j\left[\omega C - \frac{\omega L}{R^2 + (\omega L)^2}\right]$$

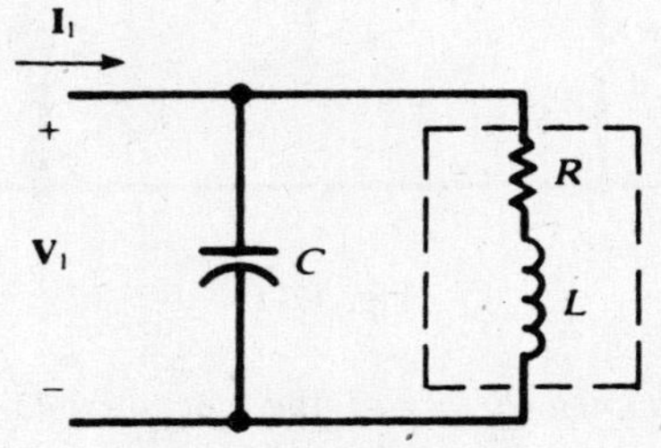

Fig. 12-25

For resonance,

$$\omega_a C = \frac{\omega_a L}{R^2 + (\omega_a L)^2} \qquad \text{or} \qquad \omega_a = \frac{1}{\sqrt{LC}}\sqrt{1 - \frac{R^2 C}{L}}$$

At the resonant frequency, $\mathbf{Y}_{\text{in}}(\omega_a) = RC/L$ and, from Section 12.11, the Q of the inductance at ω_a is

$$Q_{\text{ind}} = \frac{\omega_a L}{R} = \sqrt{\frac{L}{CR^2} - 1}$$

If $Q_{\text{ind}} \geq 10$, then $\omega_a \approx 1/\sqrt{LC}$ and

$$\left|\frac{\mathbf{Z}_{\text{in}}(\omega_a)}{R}\right| \approx Q_{\text{ind}}^2$$

The frequency response is similar to that of the parallel *RLC* circuit, except that the high-impedance resonance occurs at a lower frequency for low Q_{ind}. This becomes evident when the expression for ω_a above is rewritten as

$$\omega_a = \left(\frac{1}{\sqrt{LC}}\right)\frac{1}{\sqrt{1 + (1/Q_{\text{ind}}^2)}}$$

12.15 SERIES-PARALLEL CONVERSIONS

It is often convenient in the analysis of circuits to convert the series RL to the parallel form (see Fig. 12-26). Given R_s, L_s, *and the operating frequency* ω, the elements R_p, L_p of the equivalent parallel circuit are determined by equating the admittances

$$\mathbf{Y}_s = \frac{R_s - j\omega L_s}{R_s^2 + (\omega L_s)^2} \qquad \text{and} \qquad \mathbf{Y}_p = \frac{1}{R_p} + \frac{1}{j\omega L_p}$$

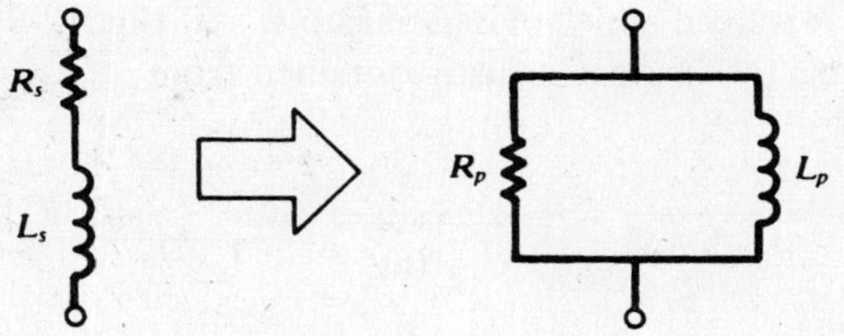

Fig. 12-26

The results are

$$R_p = R_s\left[1 + \left(\frac{\omega L_s}{R_s}\right)^2\right] = R_s(1 + Q_s^2)$$

$$L_p = L_s\left[1 + \left(\frac{R_s}{\omega L_s}\right)^2\right] = L_s\left(1 + \frac{1}{Q_s^2}\right)$$

If $Q_s \geq 10$, $R_p \approx R_s Q_s^2$ and $L_p \approx L_s$.

There are times when the RC circuit in either form should be converted to the other form (see Fig. 12-27). Equating either the impedances or the admittances, one finds

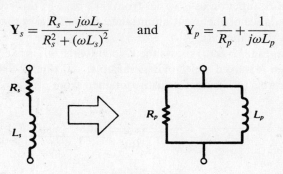

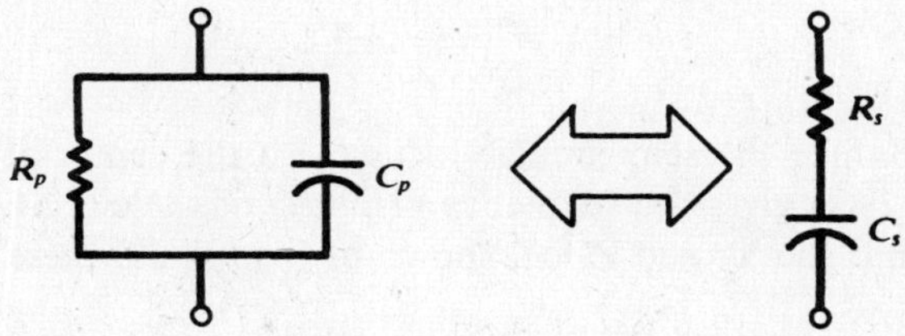

Fig. 12-27

$$R_s = \frac{R_p}{1 + (\omega C_p R_p)^2} = \frac{R_p}{1 + Q_p^2}$$

$$C_s = C_p\left[1 + \frac{1}{(\omega C_p R_p)^2}\right] = C_p\left(1 + \frac{1}{Q_p^2}\right)$$

as the parallel-to-series transformation, and

$$R_p = R_s\left[1 + \frac{1}{(\omega C_s R_s)^2}\right] = R_s(1 + Q_s^2)$$

$$C_p = \frac{C_s}{1 + (\omega C_s R_s)^2} = \frac{C_s}{1 + (1/Q_s)^2}$$

as the series-to-parallel transformation. Again, the equivalence depends on the operating frequency.

12.16 LOCUS DIAGRAMS

Heretofore, the frequency response of a network has been exhibited by plotting separately the magnitude and the angle of a suitable network function against frequency ω. This same information can be presented in a single plot: one finds the curve (*locus diagram*) in the complex plane traced by the point representing the network function as ω varies from 0 to ∞. In this section we shall discuss locus diagrams for the input impedance or the input admittance; in some cases the variable will not be ω, but another parameter (such as resistance R).

For the series RL circuit, Fig. 12-28(*a*) shows the **Z**-locus when ωL is fixed and R is variable; Fig. 12-28(*b*) shows the **Z**-locus when R is fixed and L or ω is variable; and Fig. 12-28(*c*) shows the **Y**-locus when R is fixed and L or ω is variable. This last locus is obtained from

$$\mathbf{Y} = \frac{1}{R + j\omega L} = \frac{1}{\sqrt{R^2 + (\omega L)^2}} \underline{/\tan^{-1}(-\omega L/R)}$$

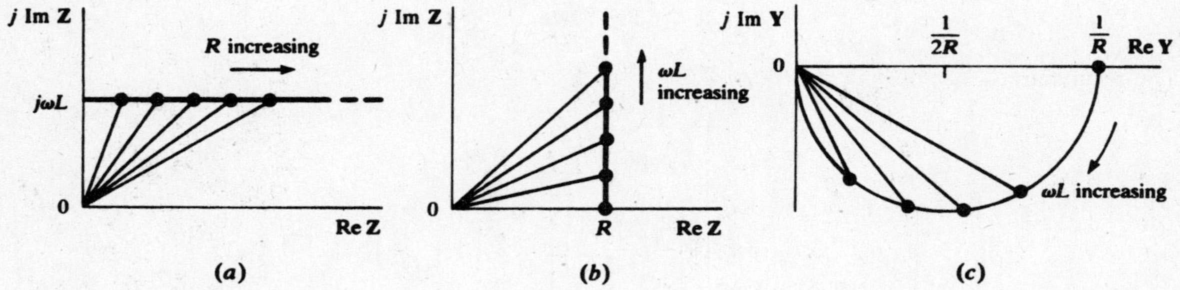

Fig. 12-28

Note that for $\omega L = 0$, $\mathbf{Y} = (1/R)\underline{/0^\circ}$; and for $\omega L \to \infty$, $\mathbf{Y} \to 0\underline{/-90^\circ}$. When $\omega L = R$,

$$\mathbf{Y} = \frac{1}{R\sqrt{2}} \underline{/-45^\circ}$$

A few other points will confirm the semicircular locus, with the center at $1/2R$ and the radius $1/2R$. Either Fig. 12-28(*b*) or 12-28(*c*) gives the frequency response of the circuit.

A parallel RC circuit has the **Y**- and **Z**-loci shown in Fig. 12-29; these are derived from

$$\mathbf{Y} = \frac{1}{R} + j\omega C \qquad \text{and} \qquad \mathbf{Z} = \frac{R}{\sqrt{1 + (\omega CR)^2}} \underline{/\tan^{-1}(-\omega CR)}$$

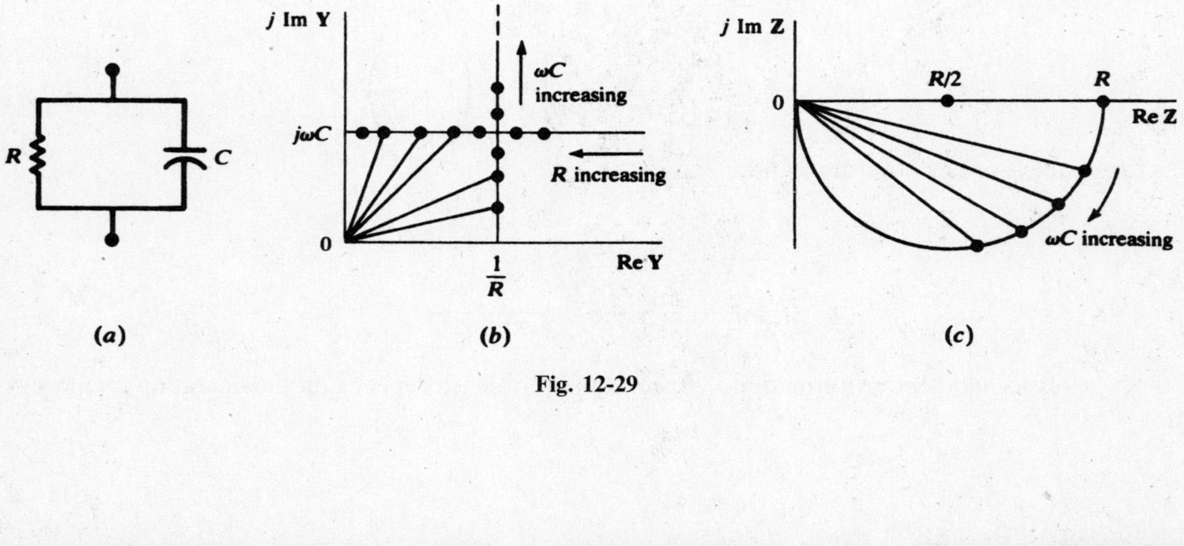

Fig. 12-29

For the RLC series circuit, the **Y**-locus, with ω as the variable, may be determined by writing

$$\mathbf{Y} = G + jB = \frac{1}{R + jX} = \frac{R - jX}{R^2 + X^2}$$

whence $$G = \frac{R}{R^2 + X^2} \qquad B = -\frac{X}{R^2 + X^2}$$

Both G and B depend on ω via X. Eliminating X between the two expressions yields the equation of the locus in the form

$$G^2 + B^2 = \frac{G}{R} \qquad \text{or} \qquad \left(G - \frac{1}{2R}\right)^2 + B^2 = \left(\frac{1}{2R}\right)^2$$

which is the circle shown in Fig. 12-30. Note the points on the locus corresponding to $\omega = \omega_l$, $\omega = \omega_0$, and $\omega = \omega_h$.

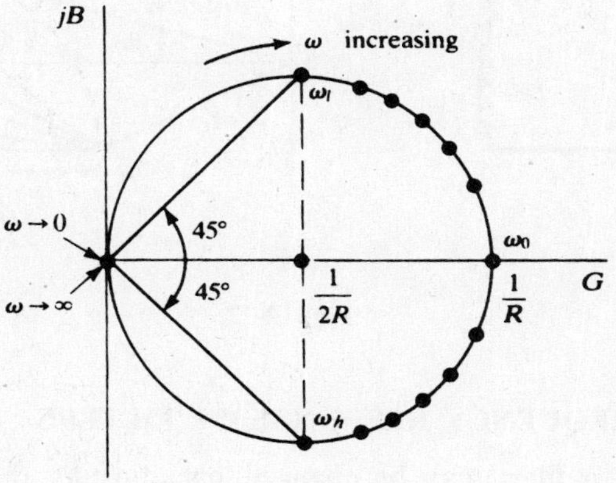

Fig. 12-30

For the practical "tank" circuit examined in Section 12.14, the **Y**-locus may be constructed by combining the C-branch locus and the RL-branch locus. To illustrate the addition, the points corresponding to frequencies $\omega_1 < \omega_2 < \omega_3$ are marked on the individual loci and on the sum, shown in Fig. 12-31(c). It is seen that $|\mathbf{Y}|_{\min}$ occurs at a frequency greater than ω_a; that is, the resonance is high-impedance but not maximum-impedance. This comes about because G varies with ω (see Section 12.14), and varies in such a way that forcing $B = 0$ does not automatically minimize $G^2 + B^2$. The separation of

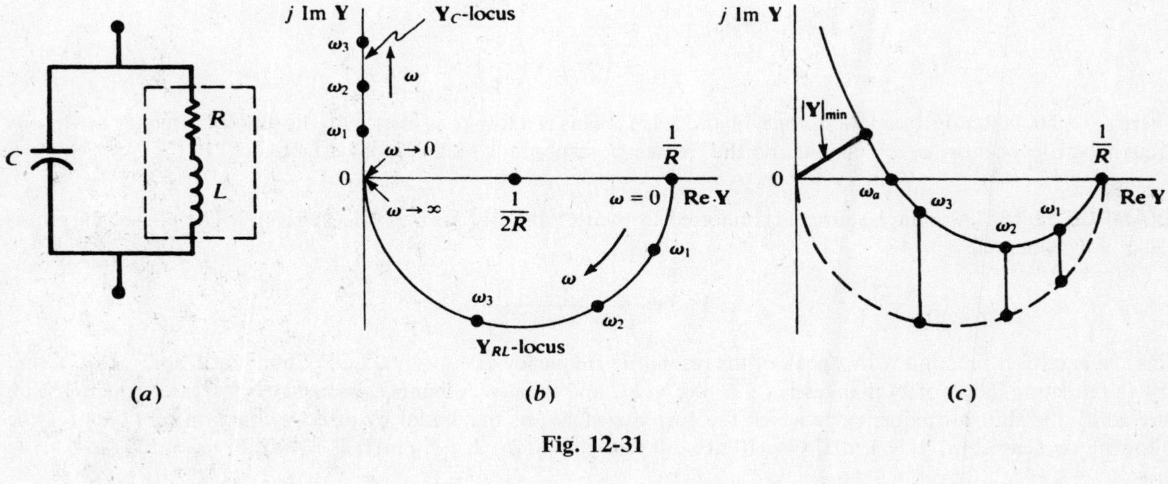

Fig. 12-31

the resonance and minimum-admittance frequencies is governed by the Q of the coil. Higher Q_{ind} corresponds to lower values of R. It is seen from Fig. 12-31(b) that low R results in a larger semicircle, which when combined with the $\mathbf{Y}_C$-locus, gives a higher ω_a and a lower minimum-admittance frequency. When $Q_{\text{ind}} \geq 10$, the two frequencies may be taken as coincident.

The case of the two-branch RC and RL circuit shown in Fig. 12-32(a) can be examined by adding the admittance loci of the two branches. For fixed $\mathbf{V} = V\angle 0°$, this amounts to adding the loci of the two branch currents. Consider C variable without limit, and R_1, R_2, L, and ω constant. Then current $\mathbf{I}_L$ is fixed as shown in Fig. 12-32(b). The semicircular locus of $\mathbf{I}_C$ is added to $\mathbf{I}_L$ to result in the locus of $\mathbf{I}_T$.

Resonance of the circuit corresponds to $\theta_T = 0$. This may occur for two values of the real, positive parameter C [the case illustrated in Fig. 12.32(b)], for one value, or for no value—depending on the number of real positive roots of the equation Im $\mathbf{Y}_T(C) = 0$.

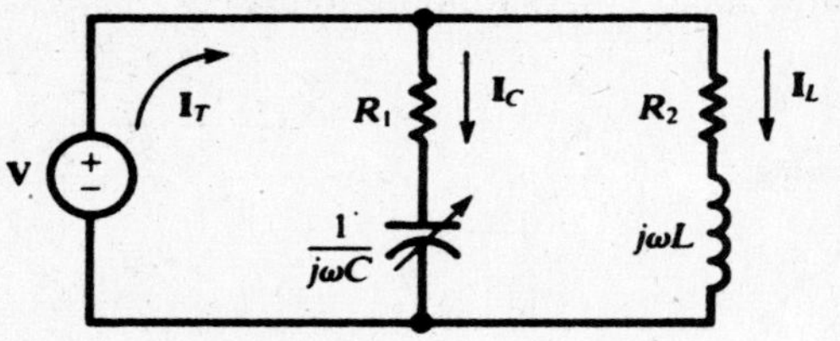

(a)

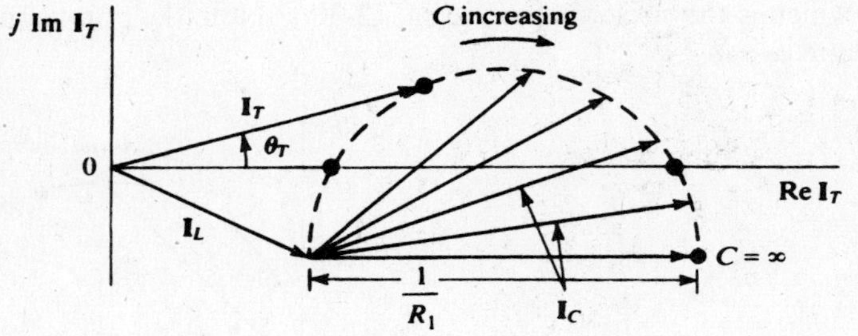

(b)

Fig. 12-32

12.17 SCALING THE FREQUENCY RESPONSE OF FILTERS

The frequency scale of a filter may be changed by adjusting the values of its inductors and capacitors. Here we summarize the method (see also Section 8.10). Inductors and capacitors affect the frequency behavior of circuits through $L\omega$ and $C\omega$; that is, always as a product of element values and the frequency. Dividing inductor and capacitor values in a circuit by a factor k will scale-up the ω-axis of the frequency response by a factor k. For example, a 1-mH inductor operating at 1 kHz has the same impedance as a 1-μH inductor operating at 1 MHz. Similarly, a 1-μF capacitor at 1 MHz behaves similar to a 1-nF capacitor at 1 GHz. This is called *frequency scaling* and is a useful property of linear circuits. The following two examples illustrate its application in filter design.

EXAMPLE 12.17 The network function of the circuit of Fig. 8-42 with $R = 2\ k\Omega$, $C = 10\ nF$, and $R_2 = R_1$ is

$$H(s) = \frac{V_2}{V_1} = \frac{2}{\left(\frac{s}{\omega_0}\right)^2 + \left(\frac{s}{\omega_0}\right) + 1}$$

where $\omega_0 = 50,000$ rad/s (see Examples 8.14 and 8.15). This is a low-pass filter with the cutoff frequency at ω_0. By using a 1-nF capacitor, $\omega_0 = 500{,}000$ and the frequency response is scaled up by a factor of 10.

EXAMPLE 12.18 A voltage source is connected to the terminals of a series RLC circuit. The phasor current is $I = Y \times V$, where

$$Y(s) = \frac{Cs}{LCs^2 + RCs + 1}$$

This is a bandpass function with a peak of the resonance frequency of $\omega_0 = 1/\sqrt{LC}$. Changing L and C to L/k and C/k (a reduction factor of k) changes $1/\sqrt{LC}$ to $k/\sqrt{LC}$ and the new resonance frequency is increased to $k\omega_0$. You may verify the shift in frequency at which the current reaches its maximum by direct evaluation of $Y(j\omega)$ for the following two cases: (a) L = 1 mH, C = 10 nF, $\omega_0 = 10^6$ rad/s; (b) L = 10 mH, C = 100 nF, $\omega_0 = 10^5$ rad/s.

Solved Problems

12.1 In the two-port network shown in Fig. 12-33, $R_1 = 7\ \text{k}\Omega$ and $R_2 = 3\ \text{k}\Omega$. Obtain the voltage ratio $\mathbf{V}_2/\mathbf{V}_1$ (*a*) at no-load, (*b*) for $R_L = 20\ \text{k}\Omega$.

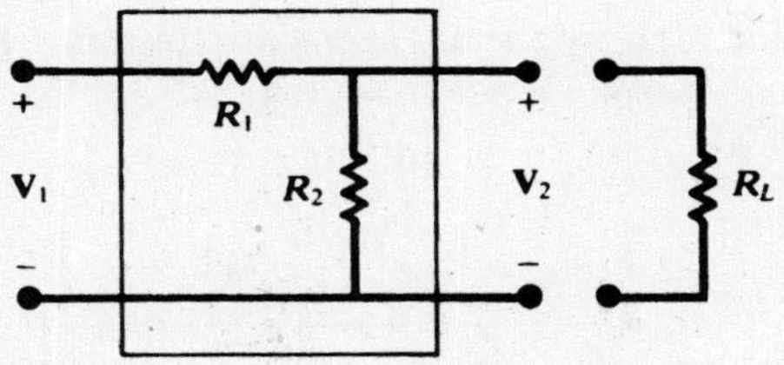

Fig. 12-33

(*a*) At no-load, voltage division gives

$$\frac{\mathbf{V}_2}{\mathbf{V}_1} = \frac{R_2}{R_1 + R_2} = \frac{3}{7+3} = 0.30$$

(*b*) With $R_L = 20\ \text{k}\Omega$,

$$R_p = \frac{R_2 R_L}{R_2 + R_L} = \frac{60}{23}\ \text{k}\Omega$$

and
$$\frac{\mathbf{V}_2}{\mathbf{V}_1} = \frac{R_p}{R_1 + R_p} = \frac{60}{221} = 0.27$$

The voltage ratio is independent of frequency. The load resistance, $20\ \text{k}\Omega$, reduced the ratio from 0.30 to 0.27.

12.2 (*a*) Find L_2 in the high-pass circuit shown in Fig. 12-34, if $|\mathbf{H}_v(\omega)| = 0.50$ at a frequency of 50 MHz. (*b*) At what frequency is $|\mathbf{H}_v| = 0.90$?

(*a*) From Section 12.2, with $\omega_x \equiv R_1/L_2$,

$$|\mathbf{H}_v(\omega)| = \frac{1}{\sqrt{1 + (\omega_x/\omega)^2}}$$

Then,
$$0.50 = \frac{1}{\sqrt{1 + (f_x/50)^2}} \qquad \text{or} \qquad f_x = 50\sqrt{3}\ \text{MHz}$$

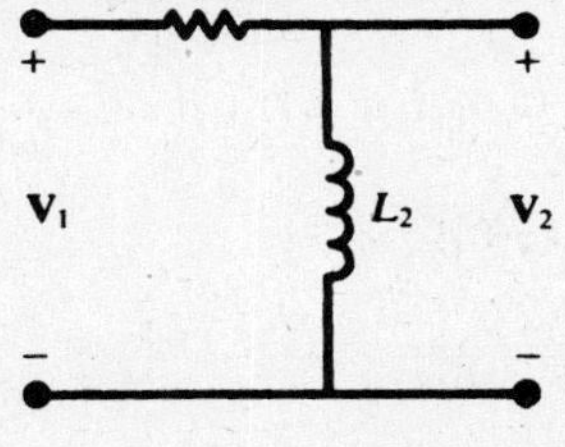

Fig. 12-34

and $$L_2 = \frac{R_1}{2\pi f_x} = \frac{50 \times 10^3}{2\pi(50\sqrt{3} \times 10^6)} = 91.9\ \mu\text{H}$$

(*b*) $$0.90 = \frac{1}{1 + (50\sqrt{3}/f)^2} \qquad \text{or} \qquad f = 179\ \text{MHz}$$

12.3 A voltage divider, useful for high-frequency applications, can be made with two capacitors C_1 and C_2 in the generalized two-port network Fig. 12-2. Under open-circuit, find C_2 if $C_1 = 0.01\ \mu\text{F}$ and $|\mathbf{H}_v| = 0.20$.

From Table 12-1,

$$\mathbf{H}_v = \frac{\mathbf{Z}_2}{\mathbf{Z}_1 + \mathbf{Z}_2} = \frac{1/j\omega C_2}{\dfrac{1}{j\omega C_1} + \dfrac{1}{j\omega C_2}} = \frac{C_1}{C_1 + C_2}$$

Hence, $$0.20 = \frac{0.01}{0.01 + C_2} \qquad \text{or} \qquad C_2 = 0.04\ \mu\text{F}$$

The voltage ratio is seen to be frequency-independent under open-circuit.

12.4 Find the frequency at which $|\mathbf{H}_v| = 0.50$ for the low-pass RC network shown in Fig. 12-35.

$$\mathbf{H}_v(\omega) = \frac{1}{1 + j(\omega/\omega_x)} \qquad \text{where} \qquad \omega_x \equiv \frac{1}{R_1 C_2}$$

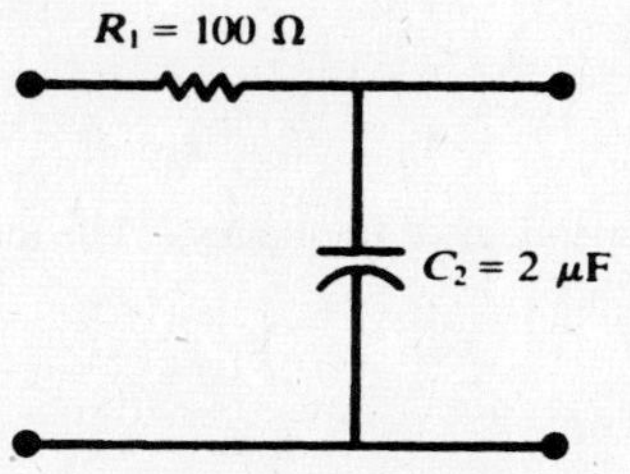

Fig. 12-35

Then, $$(0.50)^2 = \frac{1}{1 + (\omega/\omega_x)^2} \qquad \text{from which} \qquad \frac{\omega}{\omega_x} = \sqrt{3}$$

and $$\omega = \sqrt{3}\left(\frac{1}{R_1 C_2}\right) = 8660\ \text{rad/s} \qquad \text{or} \qquad f = 1378\ \text{Hz}$$

12.5 For the series RLC circuit shown in Fig. 12-36, find the resonant frequency $\omega_0 = 2\pi f_0$. Also obtain the half-power frequencies and the bandwidth β.

$$\mathbf{Z}_{\text{in}}(\omega) = R + j\left(\omega L - \frac{1}{\omega C}\right)$$

At resonance, $\mathbf{Z}_{\text{in}}(\omega) = R$ and $\omega_0 = 1/\sqrt{LC}$.

$$\omega_0 = \frac{1}{\sqrt{0.5(0.4 \times 10^{-6})}} = 2236.1\ \text{rad/s} \qquad f_0 = \frac{\omega_0}{2\pi} = 355.9\ \text{Hz}$$

The power formula

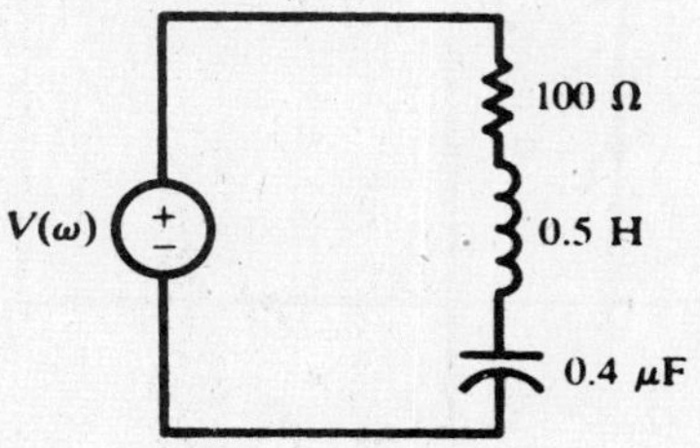

Fig. 12-36

$$P = I_{\text{eff}}^2 R = \frac{V_{\text{eff}}^2 R}{|\mathbf{Z}_{\text{in}}|^2}$$

shows that $P_{\max} = V_{\text{eff}}^2/R$, achieved at $\omega = \omega_0$, and that $P = \frac{1}{2}P_{\max}$ when $|\mathbf{Z}_{\text{in}}|^2 = 2R^2$; that is, when

$$\omega L - \frac{1}{\omega C} = \pm R \qquad \text{or} \qquad \omega^2 \mp \frac{R}{L}\omega - \frac{1}{LC} = 0$$

Corresponding to the upper sign, there is a single real positive root:

$$\omega_h = \frac{R}{2L} + \sqrt{\left(\frac{R}{2L}\right)^2 + \frac{1}{LC}} = 2338.3 \text{ rad/s} \qquad \text{or} \qquad f_h = 372.1 \text{ Hz}$$

and corresponding to the lower sign, the single real positive root

$$\omega_l = -\frac{R}{2L} + \sqrt{\left(\frac{R}{2L}\right)^2 + \frac{1}{LC}} = 2138.3 \text{ rad/s} \qquad \text{or} \qquad f_l = 340.3 \text{ Hz}$$

12.6 Derive the Q of (*a*) the series RLC circuit, (*b*) the parallel RLC circuit.

(*a*) In the time domain, the instantaneous stored energy in the circuit is given by

$$W_s = \frac{1}{2}Li^2 + \frac{q^2}{2C}$$

For a maximum,

$$\frac{dW_s}{dt} = Li\frac{di}{dt} + \frac{q}{C}\frac{dq}{dt} = i\left(L\frac{di}{dt} + \frac{q}{C}\right) = i(v_L + v_C) = 0$$

Thus, the maximum stored energy is W_s at $i = 0$ or W_s at $v_L + v_C = 0$, whichever is the larger. Now the capacitor voltage, and therefore the charge, lags the current by 90°; hence, $i = 0$ implies $q = \pm Q_{\max}$ and

$$W_s|_{i=0} = \frac{Q_{\max}^2}{2C} = \frac{1}{2}CV_{C\max}^2 = \frac{1}{2}C\left(\frac{I_{\max}}{\omega C}\right)^2 = \frac{I_{\max}^2}{2C\omega^2}$$

On the other hand, $v_L + v_C = 0$ implies $v_L = v_C = 0$ and $i = \pm I_{\max}$ (see the phasor diagram, Fig. 12-37), so that

$$W_s|_{v_L+v_C=0} = \tfrac{1}{2}LI_{\max}^2$$

It follows that

$$W_{s\max} = \begin{cases} I_{\max}^2/2C\omega^2 & (\omega \le \omega_0) \\ LI_{\max}^2/2 & (\omega \ge \omega_0) \end{cases}$$

The energy dissipated per cycle (in the resistor) is $W_d = I_{\max}^2 R\pi/\omega$. Consequently,

$$Q = 2\pi\frac{W_{s\max}}{W_d} = \begin{cases} 1/\omega CR & (\omega \le \omega_0) \\ \omega L/R & (\omega \ge \omega_0) \end{cases}$$

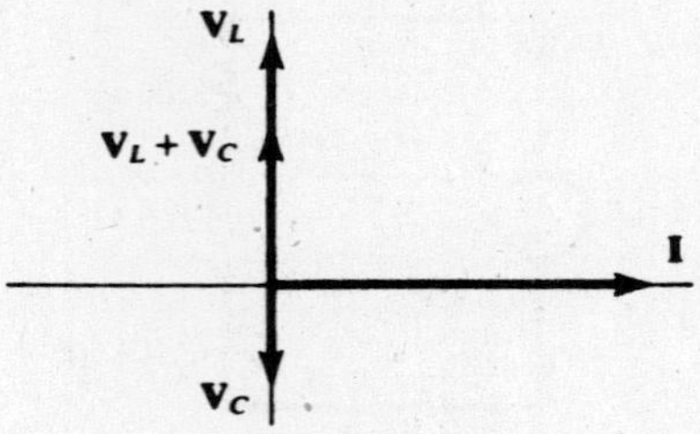

Fig. 12-37

(*b*) For the parallel combination with applied voltage $v(t)$,

$$W_s = \frac{1}{2} L I_L^2 + \frac{1}{2C} q_C^2$$

and $$\frac{dW_s}{dt} = L i_L \frac{di_L}{dt} + \frac{q_C}{C} i_C = v(i_L + i_C) = 0$$

If $v = 0$, then $q_C = 0$ and

$$i_L = \pm I_{L\max} = \pm \frac{V_{\max}}{\omega L}$$

giving $$W_{s|v=0} = \frac{V_{\max}^2}{2L\omega^2}$$

If $i_L + i_C = 0$, then (see Fig. 12-38) $i_L = i_C = 0$ and $q_C = \pm C V_{\max}$, giving

$$W_s|_{i_L + i_C = 0} = \tfrac{1}{2} C V_{\max}^2$$

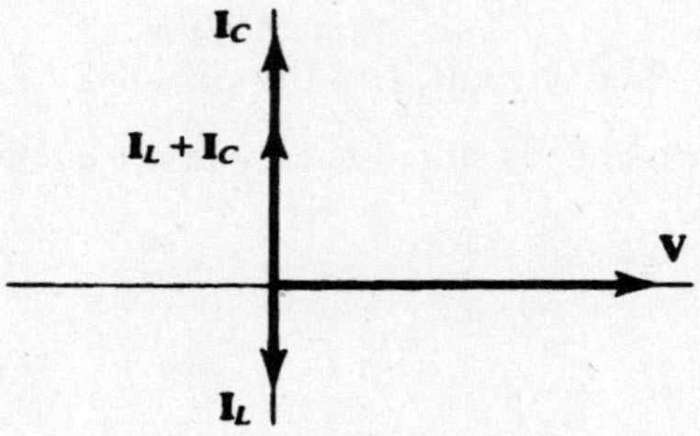

Fig. 12-38

Therefore $$W_{s\max} = \begin{cases} V_{\max}^2/2L\omega^2 & (\omega \le \omega_a) \\ C V_{\max}^2/2 & (\omega \ge \omega_a) \end{cases}$$

The energy dissipated per cycle in R is $W_d = V_{\max}^2 \pi / R\omega$. Consequently,

$$Q = 2\pi \frac{W_{s\max}}{W_d} = \begin{Bmatrix} R/L\omega & (\omega \le \omega_a) \\ \omega CR & (\omega \ge \omega_a) \end{Bmatrix}$$

12.7 A three-element series circuit contains $R = 10\ \Omega$, $L = 5$ mH, and $C = 12.5\ \mu$F. Plot the magnitude and angle of $\mathbf{Z}$ as functions of ω for values of ω from $0.8\ \omega_0$ through $1.2\ \omega_0$.

$\omega_0 = 1/\sqrt{LC} = 4000$ rad/s. At ω_0,

$$X_L = (4000)(5 \times 10^{-3}) = 20\ \Omega \qquad X_C = \frac{1}{(4000)(12.5 \times 10^{-6})} = 20\ \Omega$$

$$\mathbf{Z} = 10 + j(X_L - X_C) = 10 + j0\ \Omega$$

The values of the reactances at other frequencies are readily obtained. A tabulation of reactances and impedances appear in Fig. 12-39(a), and Fig. 12-39(b) shows the required plots.

ω	X_L	X_C	**Z**	
3200	16	25	$10 - j9$	$13.4\angle -42^\circ$
3600	18	22.2	$10 - j4.2$	$10.8\angle -22.8^\circ$
4000	20	20	10	$10\angle 0^\circ$
4400	22	18.2	$10 + j3.8$	$10.7\angle 20.8^\circ$
4800	24	16.7	$10 + j7.3$	$12.4\angle 36.2^\circ$

(a)

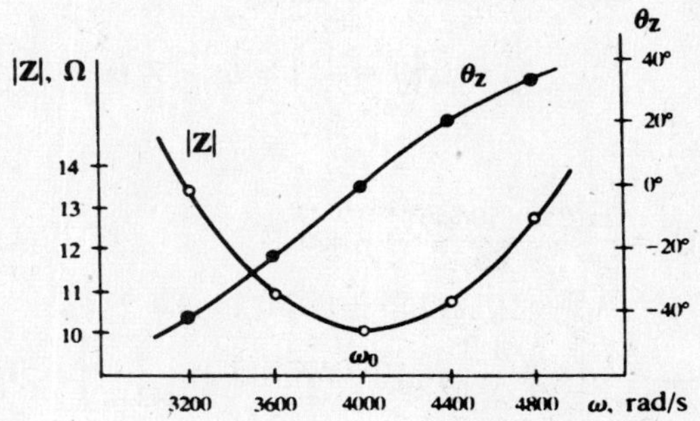

(b)

Fig. 12-39

12.8 Show that $\omega_0 = \sqrt{\omega_l \omega_h}$ for the series RLC circuit.

By the results of Problem 12.5,

$$\omega_l \omega_h = \left(\sqrt{\left(\frac{R}{2L}\right)^2 + \frac{1}{LC}} - \frac{R}{2L}\right)\left(\sqrt{\left(\frac{R}{2L}\right)^2 + \frac{1}{LC}} + \frac{R}{2L}\right) = \frac{1}{LC} = \omega_0^2$$

12.9 Compute the quality factor of an RLC series circuit, with $R = 20\ \Omega$, $L = 50$ mH, and $C = 1\ \mu$F, using (a) $Q = \omega_0 L/R$, (b) $Q = 1/\omega_0 CR$, and (c) $Q = \omega_0/\beta$.

$$\omega_0 = \frac{1}{\sqrt{0.05 \times 10^{-6}}} = 4472 \text{ rad/s}$$

$$\omega_l = -\frac{R}{2L} + \sqrt{\left(\frac{R}{2L}\right)^2 + \frac{1}{LC}} = 4276.6 \text{ rad/s} \qquad \omega_h = \frac{R}{2L} + \sqrt{\left(\frac{R}{2L}\right)^2 + \frac{1}{LC}} = 4676.6 \text{ rad/s}$$

and $\beta = \omega_h - \omega_l = 400$ rad/s.

(a) $$Q = \frac{\omega_0 L}{R} = \frac{4472(0.050)}{20} = 11.2$$

(b) $$Q = \frac{1}{\omega_0 CR} = \frac{1}{4472(10^{-6})20} = 11.2$$

(c) $$Q = \frac{\omega_0}{\beta} = \frac{4472}{400} = 11.2$$

12.10 A coil is represented by a series combination of $L = 50$ mH and $R = 15\ \Omega$. Calculate the quality factor at (*a*) 10 kHz, (*b*) 50 kHz.

(*a*) $$Q_{\text{coil}} = \frac{\omega L}{R} = \frac{2\pi(10\times 10^3)(50\times 10^{-3})}{15} = 209$$

(*c*) $$Q_{\text{coil}} = 209\left(\frac{50}{10}\right) = 1047$$

12.11 Convert the circuit constants of Problem 12.10 to the parallel form (*a*) at 10 kHz, (*b*) at 250 Hz.

(*a*) $$R_p = R_s\left[1 + \left(\frac{\omega L_s}{R_s}\right)^2\right] = R_s[1 + Q_s^2] = 15[1 + (209)^2] = 655\ \text{k}\Omega$$

or, since $Q_s \gg 10$, $R_p \approx R_s Q_s^2 = 15(209)^2 = 655\ \text{k}\Omega$.

$$L_p = L_s\left(1 + \frac{1}{Q_s^2}\right) \approx L_s = 50\ \text{mH}$$

(*b*) At 250 Hz,

$$Q_s = \frac{2\pi(250)(50\times 10^{-3})}{15} = 5.24$$
$$R_p = R_s[1 + Q_s^2] = 15[1 + (5.24)^2] = 426.9\ \Omega$$
$$L_p = L_s\left[1 + \frac{1}{Q_s^2}\right] = (50\times 10^{-3})\left[1 + \frac{1}{(5.24)^2}\right] = 51.8\ \text{mH}$$

Conversion of circuit elements from series to parallel can be carried out at a specific frequency, the equivalence holding only at that frequency. Note that in (*b*), where $Q_s < 10$, L_p differs significantly from L_s.

12.12 For the circuit shown in Fig. 12-40, (*a*) obtain the voltage transfer function $\mathbf{H}_v(\omega)$, and (*b*) find the frequency at which the function is real.

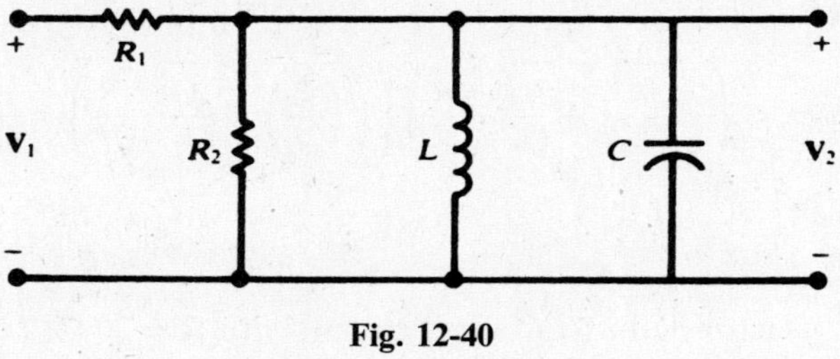

Fig. 12-40

(*a*) Let $\mathbf{Z}_2$ and $\mathbf{Y}_2$ represent the impedance and admittance of the R_2LC parallel tank.

$$\mathbf{H}_v(\omega) = \frac{\mathbf{Z}_2}{R_1 + \mathbf{Z}_2} = \frac{1}{1 + R_1\mathbf{Y}_2} = \frac{1}{1 + R_1\left(\dfrac{1}{R_2} + \dfrac{1}{j\omega L} + j\omega C\right)}$$
$$= \frac{1}{1 + \dfrac{R_1}{R_2} + jR_1\left(\omega C - \dfrac{1}{\omega L}\right)}$$

(*b*) The transfer function is real when $\mathbf{Y}_2$ is real; that is, when

$$\omega = \omega_a \equiv \frac{1}{\sqrt{LC}}$$

At $\omega = \omega_a$, not only are $|\mathbf{Z}_2|$ and $|\mathbf{H}_v|$ maximized, but $|\mathbf{Z}_{\text{in}}| = |R_1 + \mathbf{Z}_2|$ also is maximized (because R_1 is real and positive—see the locus diagram, Fig. 12-41).

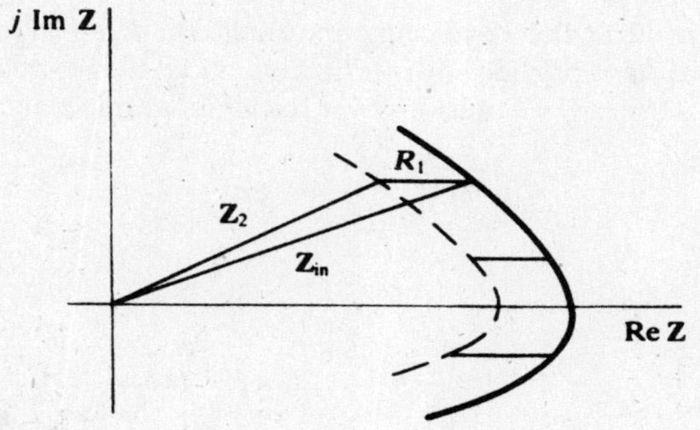

Fig. 12-41

12.13 Obtain the bandwidth β for the circuit of Fig. 12-40 and plot β against the parameter

$$R_x \equiv \frac{R_1 R_2}{R_1 + R_2}$$

Here, the half-power frequencies are determined by the condition $|\mathbf{H}_v(\omega)| = 0.707|\mathbf{H}_v|_{max}$, or, from Problem 12.12(*a*),

$$R_1\left(\omega C - \frac{1}{\omega L}\right) = \pm\left(1 + \frac{R_1}{R_2}\right) \qquad \text{or} \qquad R_x\left(\omega C - \frac{1}{\omega L}\right) = \pm 1$$

But (see Section 12.13) this is just the equation for the half-power frequencies of an R_xLC parallel circuit. Hence,

$$\beta = \frac{\omega_a}{Q_a} = \frac{1}{CR_x}$$

The hyperbolic graph is shown in Fig. 12-42.

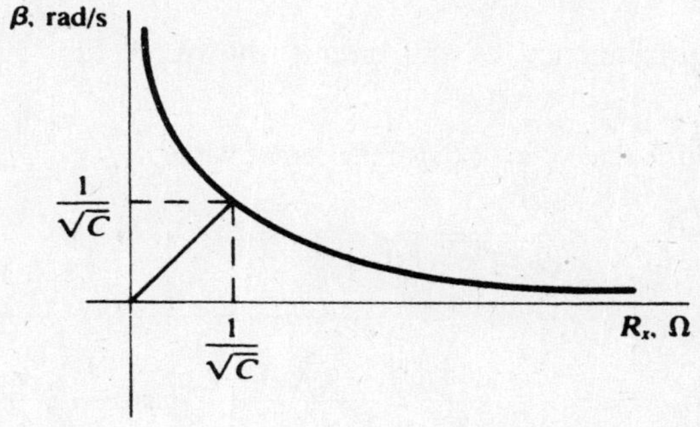

Fig. 12-42

12.14 In the circuit of Fig. 12-40, let $R_1 = R_2 = 2\ \text{k}\Omega$, $L = 10\ \text{mH}$, and $C = 40\ \text{nF}$. Find the resonant frequency and bandwidth, and compare with the results for $R_1 = 0$ (i.e., a pure parallel circuit).

$$\omega_a = \frac{1}{\sqrt{(10 \times 10^{-3})(40 \times 10^{-9})}} = 5 \times 10^4\ \text{rad/s}$$

or $f_a = 7958\ \text{Hz}$. With $R_x = 2^2/4 = 1\ \text{k}\Omega$, Problem 12.13 gives

$$\beta = \frac{1}{(40 \times 10^{-9})(1 \times 10^3)} = 2.5 \times 10^4\ \text{rad/s}$$

The results of Problem 12.12 and 12.13 cannot be applied as $R_1 \to 0$, for, in the limit, the voltage ratio is identically unity and so cannot provide any information about the residual R_2LC parallel circuit. (Note that $\beta \to \infty$ as $R_x \to 0$.) Instead, we must go over to the input impedance function, as in Section 12.13, whereby

$$\omega_a = \frac{1}{\sqrt{LC}} = 5 \times 10^4 \text{ rad/s}$$

as previously, and

$$\beta = \frac{1}{CR_2} = 1.25 \times 10^4 \text{ rad/s}$$

12.15 For the circuit of Fig. 12-40, $R_1 = 5\ \text{k}\Omega$ and $C = 10$ nF. If $\mathbf{V}_2/\mathbf{V}_1 = 0.8\underline{/0^\circ}$ at 15 kHz, calculate R_2, L, and the bandwidth.

An angle of zero on the voltage ratio $\mathbf{H}_v$ indicates that the circuit as a whole, and the parallel rank by itself, is at resonance (see Problem 12.14). Then,

$$\omega_a = \frac{1}{\sqrt{LC}} \qquad L = \frac{1}{\omega_a^2 C} = \frac{1}{[2\pi(15 \times 10^3)]^2(10 \times 10^{-9})} = 11.26 \text{ mH}$$

From Problem 12.12,

$$\mathbf{H}_v(\omega_a) = 0.8\underline{/0^\circ} = \frac{1}{1 + (R_1/R_2)} \qquad \text{whence} \qquad R_2 = \frac{R_1}{0.25} = 20\ \text{k}\Omega$$

Then, $R_x = (5)(20)/25 = 4\ \text{k}\Omega$, and Problem 12.3 gives

$$\beta = \frac{1}{(10 \times 10^{-9})(4 \times 10^3)} = 2.5 \times 10^4 \text{ rad/s}$$

12.16 Compare the resonant frequency of the circuit shown in Fig. 12-43 for $R = 0$ to that for $R = 50\ \Omega$.

For $R = 0$, the circuit is that of an LC parallel tank, with

$$\omega_a = \frac{1}{\sqrt{LC}} = \frac{1}{\sqrt{(0.2)(30 \times 10^{-6})}} = 408.2 \text{ rad/s} \qquad \text{or} \qquad f_a = 65 \text{ Hz}$$

For $R = 50\ \Omega$,

$$\mathbf{Y}_{\text{in}} = j\omega C + \frac{1}{R + j\omega L} = \frac{R}{R^2 + (\omega L)^2} + j\left[\omega C - \frac{\omega L}{R^2 + (\omega L)^2}\right]$$

For resonance, Im $\mathbf{Y}_{\text{in}}$ is zero, so that

$$\omega_a = \frac{1}{\sqrt{LC}}\sqrt{1 - \frac{R^2C}{L}}$$

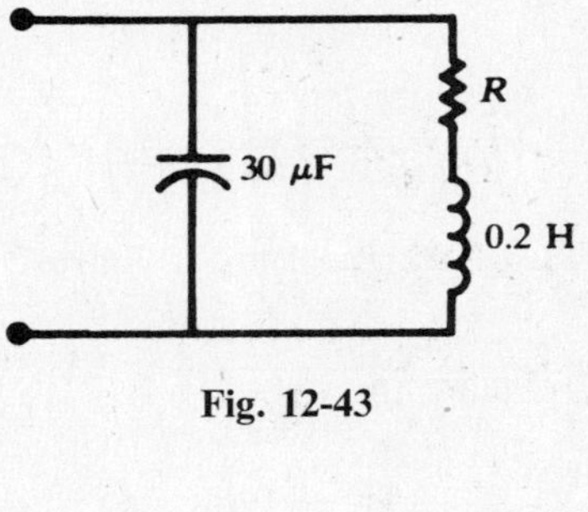

Fig. 12-43

Clearly, as $R \to 0$, this expression reduces to that given for the pure LC tank. Substituting the numerical values produces a value of 0.791 for the radical; hence,

$$\omega_a = 408.2(0.791) = 322.9 \text{ rad/s} \qquad \text{or} \qquad f_a = 51.4 \text{ Hz}$$

12.17 Measurements on a practical inductor at 10 MHz give $L = 8.0\ \mu\text{H}$ and $Q_{\text{ind}} = 40$. (*a*) Find the ideal capacitance C for parallel resonance at 10 MHz and calculate the corresponding bandwidth β. (*b*) Repeat if a practical capacitor, with a *dissipation factor* $D = Q_{\text{cap}}^{-1} = 0.005$ at 10 MHz, is used instead of an ideal capacitance.

(*a*) From Section 12.14,

$$\omega_a = \frac{1}{\sqrt{LC}} \frac{1}{\sqrt{1 + Q_{\text{ind}}^{-2}}}$$

or

$$C = \frac{1}{\omega_a^2 L(1 + Q_{\text{ind}}^{-2})} = \frac{1}{[2\pi(10 \times 10^6)]^2 (8.0 \times 10^{-6})\left(1 + \dfrac{1}{1600}\right)} = 31.6 \text{ pF}$$

Using Section 12.15 to convert the series RL branch of Fig. 12-25 to parallel at the resonant frequency,

$$R_p = R(1 + Q_{\text{ind}}^2) = \frac{\omega_a L}{Q_{\text{ind}}}(1 + Q_{\text{ind}}^2)$$

Then, from Section 12.13,

$$\beta = \frac{\omega_a}{Q_a} = \frac{\omega_a^2 L}{R_p} = \frac{\omega_a Q_{\text{ind}}}{1 + Q_{\text{ind}}^2} = \frac{2\pi(10 \times 10^6)(40)}{1 + 1600} \text{ rad/s}$$

or 0.25 MHz.

(*b*) The circuit is shown in Fig. 12-44; part (*a*) gives the resistance of the practical inductor as

$$R = \frac{\omega_a L}{Q_{\text{ind}}} = 4\pi\ \Omega$$

Also, from the given dissipation factor, it is known that

$$\frac{1}{\omega_a C R_C} = 0.005$$

The input admittance is

$$\mathbf{Y}_{\text{in}} = \frac{1}{R_C} + j\omega C + \frac{1}{R + j\omega L} = \left[\frac{1}{R_C} + \frac{R}{R^2 + (\omega L)^2}\right] + j\left[\omega C - \frac{\omega L}{R^2 + (\omega L)^2}\right]$$

which differs from the input admittance for part (*a*) only in the real part. Since the imaginary part involves the same L and the same R, and must vanish at the same frequency, C must be the same as in part (*a*); namely, $C = 31.6$ pF.

For fixed C, bandwidth is inversely proportional to resistance. With the practical capacitor, the net parallel resistance is

$$R' = \frac{R_p R_C}{R_p + R_C}$$

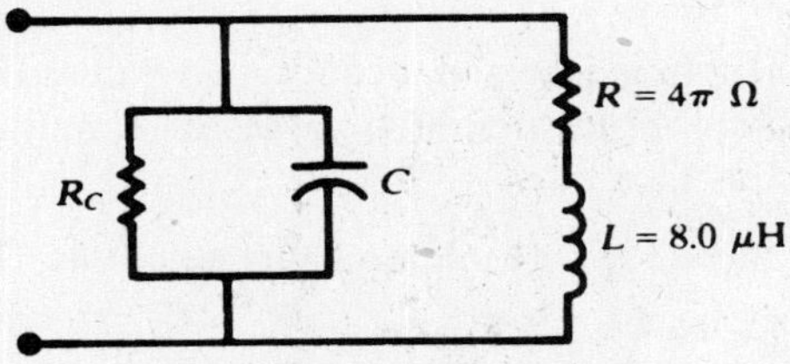

Fig. 12-44

where R_p is as calculated in part (*a*). Therefore,

$$\frac{\beta}{0.25\ \text{MHz}} = \frac{R_p}{R'} = 1 + \frac{R_p}{R_C} = 1 + \frac{(\omega_a L/Q_{\text{ind}})(1+Q_{\text{ind}}^2)}{1/\omega_a C(0.005)}$$

$$= 1 + \frac{(1+Q_{\text{ind}}^2)(0.005)}{Q_{\text{ind}}(1+Q_{\text{ind}}^{-2})}$$

$$= 1 + \frac{(1+1600)(0.005)}{40\left(1+\dfrac{1}{1600}\right)} = 1.2$$

and so $\beta = 0.30$ MHz.

A lossy capacitor has the same effect as any loading resistor placed across the tank; the Q_a is reduced and the bandwidth increased, while f_a is unchanged.

12.18 A lossy capacitor, in the series-circuit model, consists of $R = 25\ \Omega$ and $C = 20$ pF. Obtain the equivalent parallel model at 50 kHz.

From Section 12.15, or by letting $L \to 0$ in Problem 12.6(*a*),

$$Q_s = \frac{1}{\omega C_s R_s} = \frac{1}{2\pi(50\times 10^3)(20\times 10^{-12})(25)} = 6370$$

For this large Q_s-value,

$$R_p \approx R_s Q_s^2 = 1010\ \text{M}\Omega \qquad C_p \approx C_s = 20\ \text{pF}$$

12.19 A variable-frequency source of $\mathbf{V} = 100\angle 0^\circ$ V is applied to a series *RL* circuit having $R = 20\ \Omega$ and $L = 10$ mH. Compute $\mathbf{I}$ for $\omega = 0$, 500, 1000, 2000, 5000 rad/s. Plot all currents on the same phasor diagram and note the locus of the currents.

$$Z = R + jX_L = R + j\omega L$$

Table 12-2 exhibits the required computations. With the phasor voltage at the angle zero, the locus of $\mathbf{I}$ as ω varies is the semicircle shown in Fig. 12-45. Since $\mathbf{I} = \mathbf{VY}$, with constant $\mathbf{V}$, Fig. 12-45 is essentially the same as Fig. 12-28(*c*), the admittance locus diagram for the series *RL* circuit.

Table 12-2

ω, rad/s	X_L, Ω	R, Ω	**Z**, Ω	**I**, A
0	0	20	$20\angle 0^\circ$	$5\angle 0^\circ$
500	5	20	$20.6\angle 14.04^\circ$	$4.85\angle -14.04^\circ$
1000	10	20	$22.4\angle 26.57^\circ$	$4.46\angle -26.57^\circ$
2000	20	20	$28.3\angle 45^\circ$	$3.54\angle -45^\circ$
5000	50	20	$53.9\angle 68.20^\circ$	$1.86\angle -68.20^\circ$

12.20 The circuit shown in Fig. 12-46 is in resonance for two values of C when the frequency of the driving voltage is 5000 rad/s. Find these two values of C and construct the admittance locus diagram which illustrates this fact.

At the given frequency, $X_L = 3\ \Omega$. Then the admittance of this fixed branch is

$$\mathbf{Y}_1 = \frac{1}{5+j3} = 0.147 - j0.088 \quad \text{S}$$

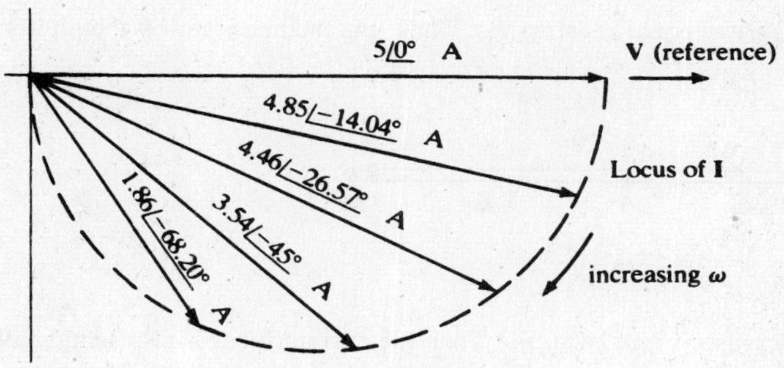

Fig. 12-45

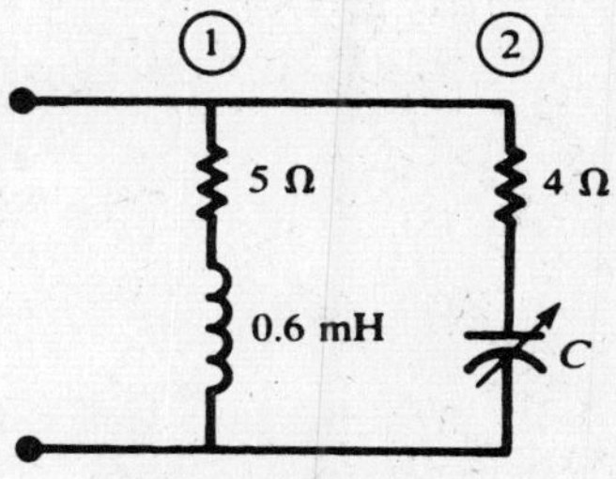

Fig. 12-46

The semicircular admittance locus of branch 2 has the radius $r = 1/2R = 0.125$ S. The total admittance is the sum of the fixed admittance $\mathbf{Y}_1$ and the variable admittance $\mathbf{Y}_2$. In Fig. 12-47, the semicircular locus is added to the fixed complex number $\mathbf{Y}_1$. The circuit resonance occurs at points a and b, where $\mathbf{Y}_T$ is real.

$$\mathbf{Y}_T = 0.417 - j0.088 + \frac{1}{4 - jX_C}$$

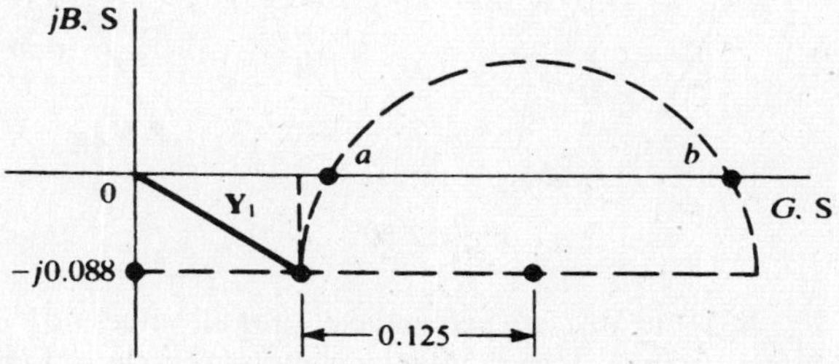

Fig. 12-47

which is real if

$$X_C^2 - 11.36X_C + 16 = 0$$

or $X_{C_1} = 9.71\ \Omega$, $X_{C_2} = 1.65\ \Omega$. With $\omega = 5000$ rad/s,

$$C_1 = 20.6\ \mu\text{F} \qquad C_2 = 121\ \mu\text{F}$$

12.21 Show by locus diagrams that the magnitude of the voltage between points A and B in Fig. 12-48 is always one-half the magnitude of the applied voltage $\mathbf{V}$ as L is varied.

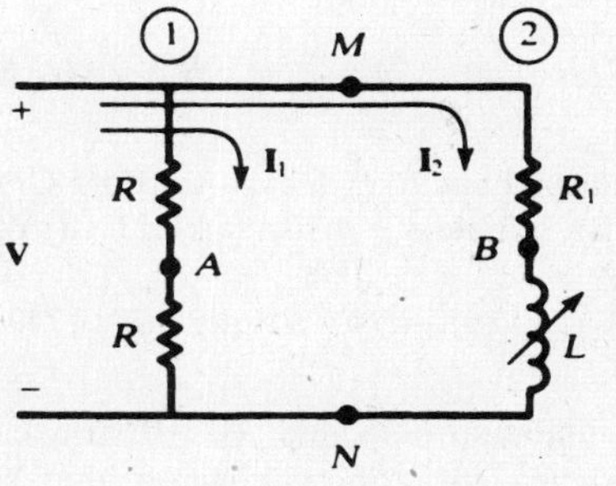

Fig. 12-48

Branch-1 current $\mathbf{I}_1$ passes through two equal resistors R. Thus A is the midpoint on the phasor $\mathbf{V}$, as shown in Fig. 12-49.

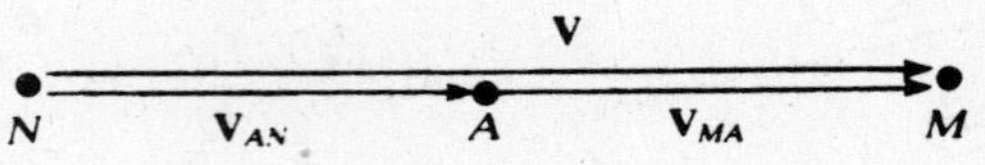

Fig. 12-49

Branch 2 has a semicircular **Y**-locus [see Fig. 12-28(*c*)]. Then the current locus is also semicircular, as shown in Fig. 12.50(*a*). The voltage phasor diagram, Fig. 12-50(*b*), consists of the voltage across the inductance, $\mathbf{V}_{BN}$, and the voltage across R_1, $\mathbf{V}_{MB}$. The two voltages add vectorially,

$$\mathbf{V} = \mathbf{V}_{MN} = \mathbf{V}_{BN} + \mathbf{V}_{MB}$$

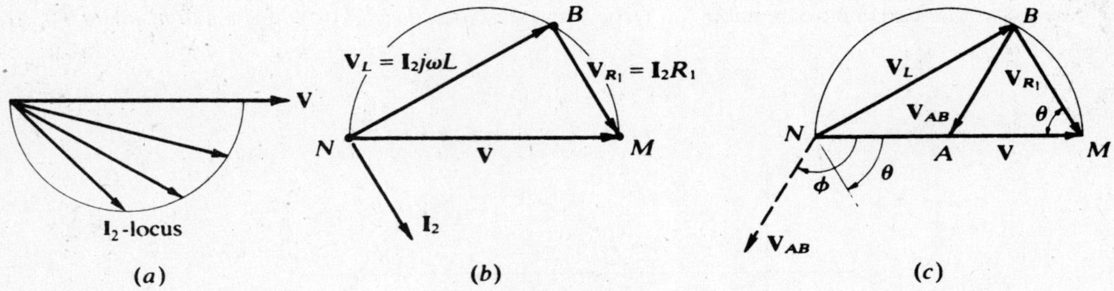

Fig. 12-50

Because $\mathbf{I}_2$ lags $\mathbf{V}_{BN}$ by 90°, $\mathbf{V}_{BN}$ and $\mathbf{V}_{MB}$ are perpendicular for all values of L in Fig. 12-50(*b*). As L varies from 0 to ∞, point B moves from N toward M along the semicircle. Figures 12-49 and 12-50(*b*) are superimposed in Fig. 12-50(*c*). It is clear that $\mathbf{V}_{AB}$ is a radius of the semicircle and therefore,

$$|\mathbf{V}_{AB}| = \tfrac{1}{2}|\mathbf{V}|$$

Further, the angle ϕ by which $\mathbf{V}_{AB}$ lags $\mathbf{V}$ is equal to 2θ, where $\theta = \tan^{-1} \omega L/R_1$.

Supplementary Problems

12.22 A high-pass RL circuit has $R_1 = 50\ \text{k}\Omega$ and $L_2 = 0.2$ mH. (*a*) Find ω if the magnitude of the voltage transfer function is $|\mathbf{H}_{v\infty}| = 0.90$. (*b*) With a load $R = 1\ \text{M}\Omega$ across L_2, find $|\mathbf{H}_v|$ at $\omega = 7.5 \times 10^8$ rad/s.
Ans. (*a*) 5.16×10^8 rad/s; (*b*) 0.908

12.23 Obtain $\mathbf{H}_{v\infty}$ for a high-pass RL circuit at $\omega = 2.5\omega_x$, $R = 2\ \text{k}\Omega$, $L = 0.05$ H. *Ans.* $0.928\underline{/21.80^\circ}$

12.24 A low-pass RC circuit under no-load has $R_1 = 5\ \text{k}\Omega$. (*a*) Find C_2 if $|\mathbf{H}_v| = 0.5$ at 10 kHz. (*b*) Obtain $\mathbf{H}_v$ at 5 kHz. (*c*) What value of C_2 results in $|\mathbf{H}_v| = 0.90$ at 8 kHz? (*d*) With C_2 as in (*a*), find a new value for R_1 to result in $|\mathbf{H}_v| = 0.90$ at 8 kHz.
Ans. (*a*) 5.51 μF; (*b*) $0.756\underline{/-40.89^\circ}$; (*c*) 1.93 μF; (*d*) 1749 Ω

12.25 A simple voltage divider would consist of R_1 and R_2. If stray capacitance C_s is present, then the divider would generally be frequency-dependent. Show, however, that $\mathbf{V}_2/\mathbf{V}_1$ is independent of frequency for the circuit of Fig. 12-51 if the compensating capacitance C_1 has a certain value. *Ans.* $C_1 = (R_2/R_1)C_s$

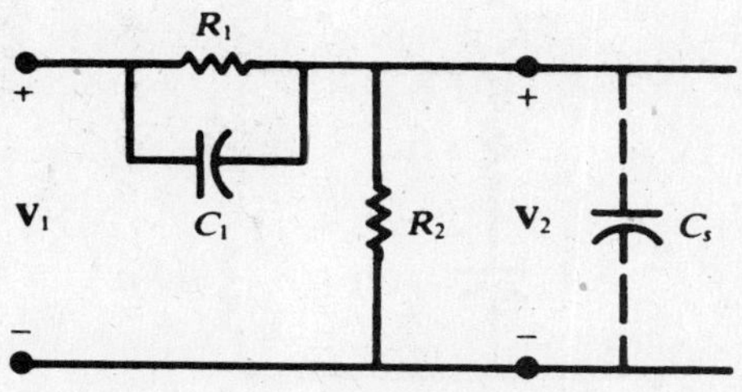

Fig. 12-51

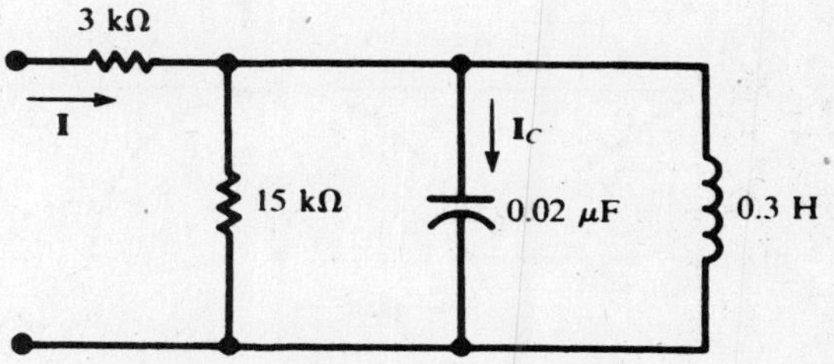

Fig. 12-52

12.26 Assume that a sinusoidal voltage source with a variable frequency and $V_{max} = 50$ V is applied to the circuit shown in Fig. 12-52. (*a*) At what frequency f is $|\mathbf{I}|$ a minimum? (*b*) Calculate this minimum current. (*c*) What is $|\mathbf{I}_C|$ at this frequency? *Ans.* (*a*) 2.05 kHz; (*b*) 2.78 mA; (*c*) 10.8 mA

12.27 A 20-μF capacitor is in parallel with a practical inductor represented by $L = 1$ mHz in series with $R = 7\ \Omega$. Find the resonant frequency, in rad/s and in Hz, of the parallel circuit. *Ans.* 1000 rad/s, 159.2 Hz

12.28 What must be the relationship between the values of R_L and R_C if the network shown in Fig. 12-53 is to be resonant at all frequencies? *Ans.* $R_L = R_C = 5\ \Omega$

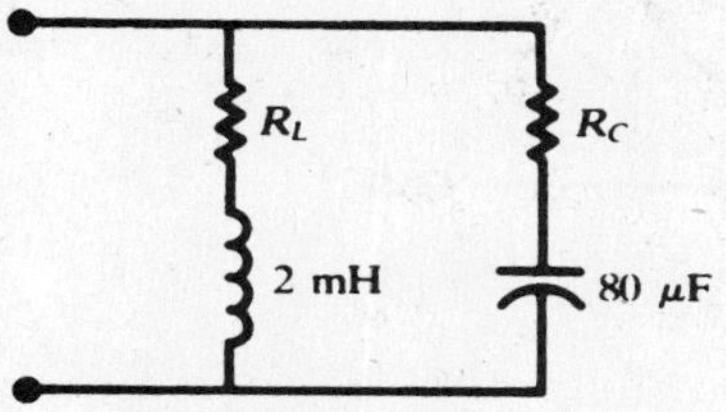

Fig. 12-53

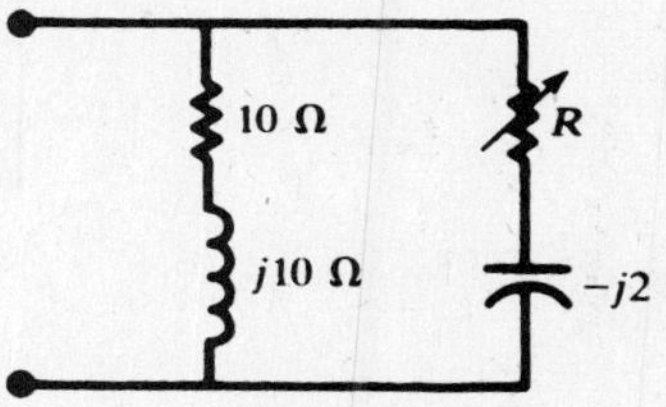

Fig. 12-54

12.29 For the parallel network shown in Fig. 12-54, (*a*) find the value of R for resonance; (*b*) convert the RC branch to a parallel equivalent. *Ans.* (*a*) 6.0 Ω; (*b*) $R_p = 6.67\ \Omega$, $X_{C_p} = 20\ \Omega$

12.30 For the network of Fig. 12-55(*a*), find R for resonance. Obtain the values of R', X_L, and X_C in the parallel equivalent of Fig. 12-55(*b*). *Ans.* $R = 12.25\ \Omega$, $R' = 7.75\ \Omega$, $X_L = 25\ \Omega$, $X_C = 25\ \Omega$

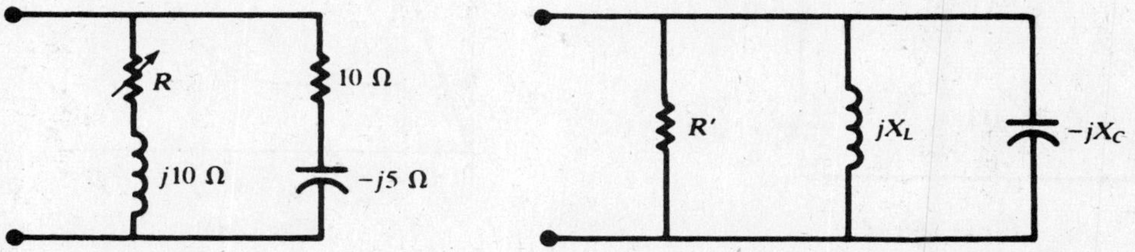

(*a*) (*b*)

Fig. 12-55

12.31 Branch 1 of a two-branch parallel circuit has an impedance $\mathbf{Z}_1 = 8 + j6\ \Omega$ at $\omega = 5000$ rad/s. Branch 2 contains $R = 8.34\ \Omega$ in series with a variable capacitance C. (*a*) Find C for resonance. (*b*) Sketch the admittance locus diagram. *Ans.* (*a*) 24 μF (*b*) See Fig. 12-56

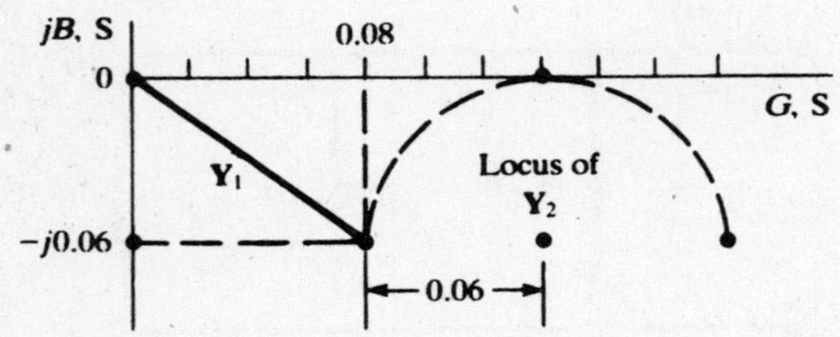

Fig. 12-56

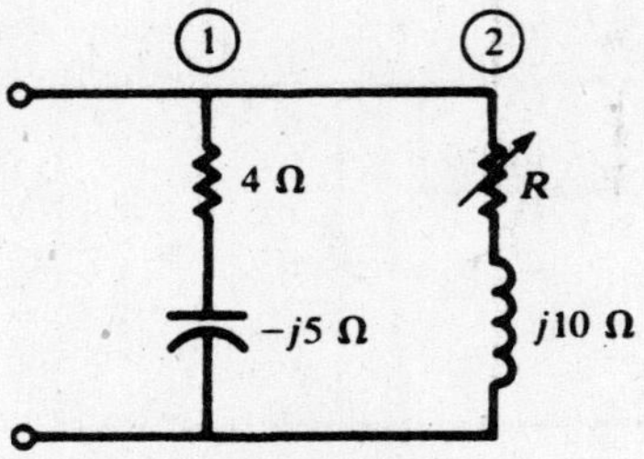

Fig. 12-57

12.32 Find R for resonance of the network shown in Fig. 12-57. Sketch the admittance locus diagram.
Ans. Resonance cannot be achieved by varying R. See Fig. 12-58.

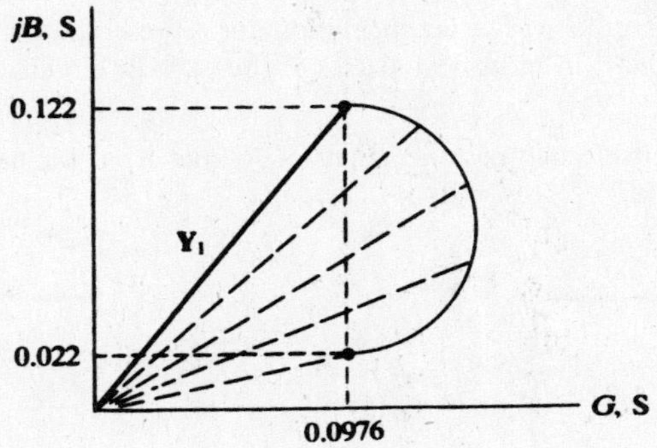

Fig. 12-58

12.33 In Problem 12.32, for what values of the inductive reactance will it be possible to obtain resonance at some value of the variable resistance R? *Ans.* $X_L \leq 8.2\ \Omega$

12.34 (*a*) Construct the admittance locus diagram for the circuit shown in Fig. 12-59. (*b*) For what value of resistance in the RL branch is resonance possible for only one value of X_L?
Ans. (*a*) See Fig. 12-60. (*b*) 6.25 Ω.

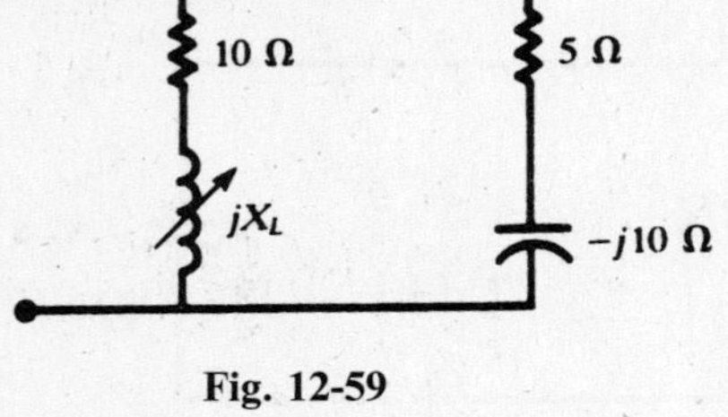

Fig. 12-59

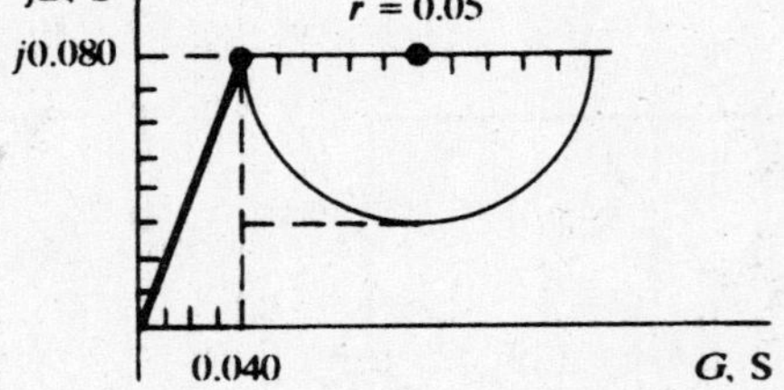

Fig. 12-60

12.35 Determine the value(s) of L for which the circuit shown in Fig. 12-61 is resonant at 5000 rad/s.
Ans. 2.43 mH, 66.0 μH

12.36 A three-branch parallel circuit has fixed elements in two branches; in the third branch, one element is variable. The voltage-current phasor diagram is shown in Fig. 12-62. Identify all the elements if $\omega = 5000$ rad/s.

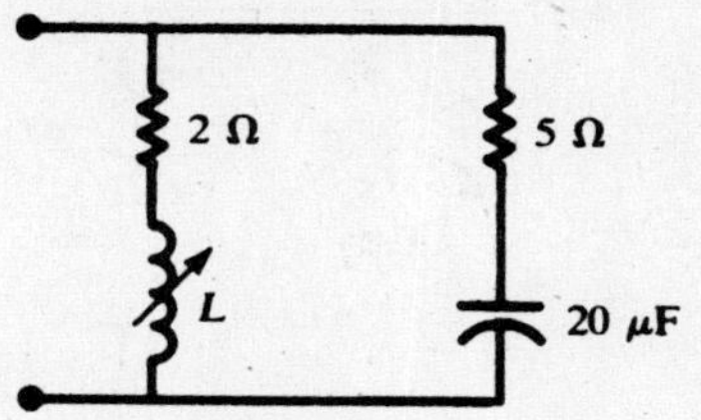

Fig. 12-61

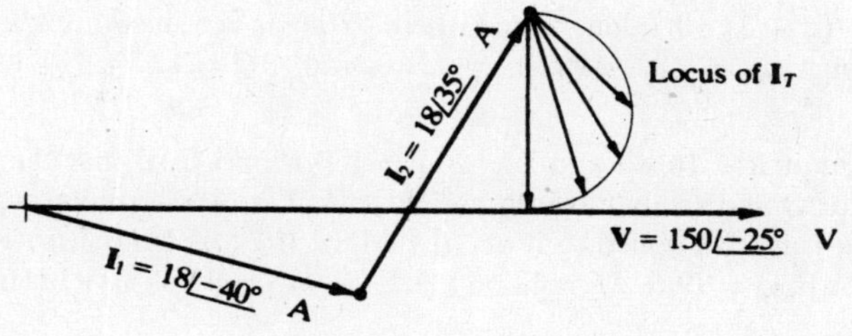

Fig. 12-62

Ans. Branch 1: $R = 8.05\ \Omega$, $L = 0.431$ mH
Branch 2: $R = 4.16\ \Omega$, $C = 27.7\ \mu$F
Branch 3: $L = 2.74$ mH, variable R

12.37 Describe the circuit which corresponds to each locus in Fig. 12-63 if there is only one variable element in each circuit.

Ans. (*a*) A two-branch parallel circuit. Branch 1: fixed R and X_C; branch 2: fixed R and variable X_C.
(*b*) A three-branch parallel circuit. Branch 1: fixed R and X_C; branch 2: fixed X_C; branch 3: fixed R and variable X_L.
(*c*) A two-branch parallel circuit. Branch 1: fixed R and X_C; branch 2: fixed X_L and variable R.

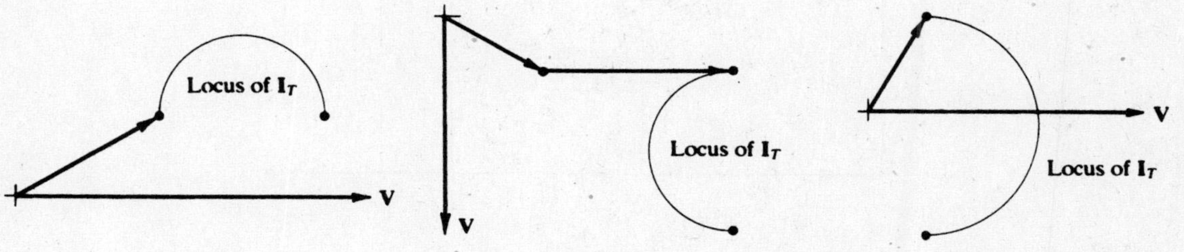

Fig. 12-63

12.38 In the circuit of Fig. 12-64, L = 1 mH. Determine R_1, R_2, and C such that the impedance between the two terminals of the circuit is 100 Ω at all frequencies. *Ans.* $C = 100$ nF, $R_1 = R_2 = 100\ \Omega$

12.39 Given $V_2/V_1 = 10s/(s^2 + 2s + 81)$ and $v_1(t) = \cos(\omega t)$, determine ω such that the amplitude of $v_2(t)$ attains a maximum. Find that maximum. *Ans.* $\omega = 9$ rad/s, $V_2 = 5$ V

12.40 Given $H(s) = s/(s^2 + as + b)$ determine a and b such that the magnitude of the frequency response $|H(\omega)|$ has a maximum at 100 Hz with a half-power bandwidth of 5 Hz. Then find the quality factor Q.
Ans. a = 31.416, b = 394784, Q = 20

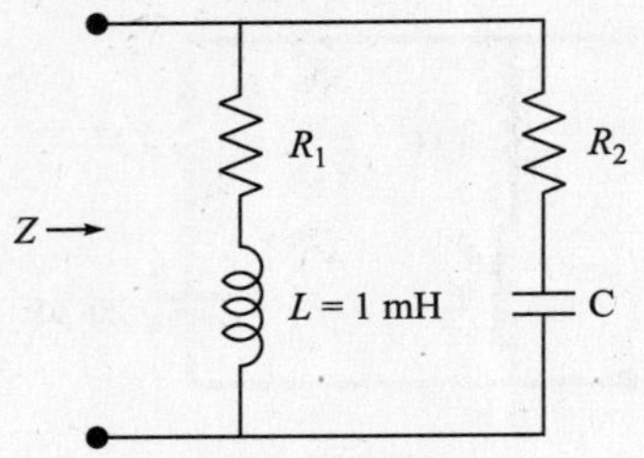

Fig. 12-64

12.41 Given $H(s) = (s+1)/(s^2 + 2s + 82)$, determine where $|H(\omega)|$ is at a maximum, its half-power bandwidth and quality factor. *Ans.* $\omega_0 = \sqrt{82} \approx 9$ rad/s, $\Delta\omega = 2$ rad/s, $Q = 4.53$

12.42 In a parallel RLC circuit $R = 10\ k\Omega$ and $L = 20\ \mu H$. (*a*) Find C so that the circuit resonates at 1 MHz. Find the quality factor Q and the bandwidth in kHz. (*b*) Find the terminal voltage of the circuit if an AC current source of $I = 1$ mA is applied to it at: (*i*) 1 MHz, (*ii*) 1.01 MHz, (*iii*) 1.006 MHz
Ans. (*a*) C = 1.267 nF, Q = 79.6, $\Delta f = 12.56$ kHz; (*b*) $V_2 = 10$ V at 1 MHz, 5.34 V at 1.01 MHz, and 7.24 V at 1.006 MHz

12.43 A coil is modeled as a 50-μH inductor in series with a 5-Ω resistor. Specify the value of a capacitor to be placed in series with the coil so that the circuit would resonate at 600 kHz. Find the quality factor Q and bandwidth Δf in kHz. *Ans.* $C = 1.4\ nF, Q = 37.7, \Delta f = 15.9$ kHz

12.44 The coil of Problem 12.43 placed in parallel with a capacitor C resonates at 600 kHz. Find C, quality factor Q, and bandwidth Δf in kHz. *Hint*: Find the equivalent parallel RLC circuit.
Ans. $C = 1.4\ nF, Q = 37.7, \Delta f = 15.9$ kHz

12.45 The circuit in Fig. 12-65(*a*) is a third-order Butterworth low-pass filter. Find the network function, the magnitude of the frequency response, and its half-power cutoff frequency ω_0.
Ans. $H(s) = 1/(s^3 + 2s^2 + 2s + 1)$, $|H(\omega)|^2 = 1/(1+\omega^6)$, $\omega_0 = 1$ rad/s

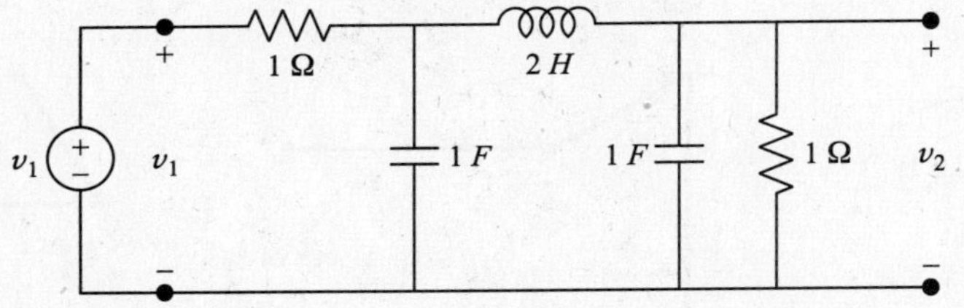

(*a*)

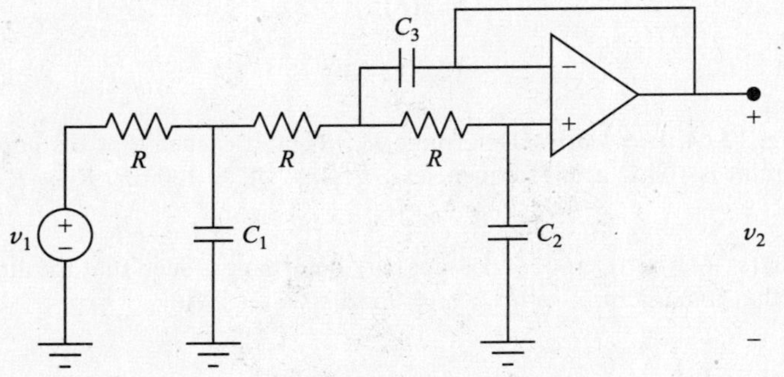

(*b*)

Fig. 12-65

12.46 In the circuit of Fig. 12-65(*b*), let $R = 1\ \Omega, C_1 = 1.394\ F, C_2 = 0.202\ F$, and $C_3 = 3.551\ F$. Find $H(s) = V_2/V_1$ and show that it approximates the passive third-order Butterworth low-pass filter of Fig. 12-65(*a*). *Ans.* $H(s) = 1/(0.99992s^3 + 1.99778s^2 + 2s + 1)$

12.47 Show that the half-power cutoff frequency in the circuit of Fig. 8-42 is $\omega_0 = 1/(RC)$ and, therefore, frequency scaling may be done by changing the value of C or R.

Ans. $$\frac{V_2}{V_1} = \frac{2}{R^2C^2s^2 + RCs + 1} = \frac{2}{\left(\dfrac{s}{\omega_0}\right)^2 + \left(\dfrac{s}{\omega_0}\right) + 1}, \quad \omega_0 = \frac{1}{RC}$$

12.48 Find RLC values in the low-pass filter of Fig. 12-65(*a*) to move its half-power cutoff frequency to 5 kHz. *Ans.* $R = 1\ \Omega, C = 31.83\ \mu F, L = 63.66\ mH$

CHAPTER 13

Two-Port Networks

13.1 TERMINALS AND PORTS

In a two-terminal network, the terminal voltage is related to the terminal current by the impedance $Z = V/I$. In a four-terminal network, if each terminal pair (or port) is connected separately to another circuit as in Fig. 13-1, the four variables i_1, i_2, v_1, and v_2 are related by two equations called the *terminal characteristics*. These two equations, plus the terminal characteristics of the connected circuits, provide the necessary and sufficient number of equations to solve for the four variables.

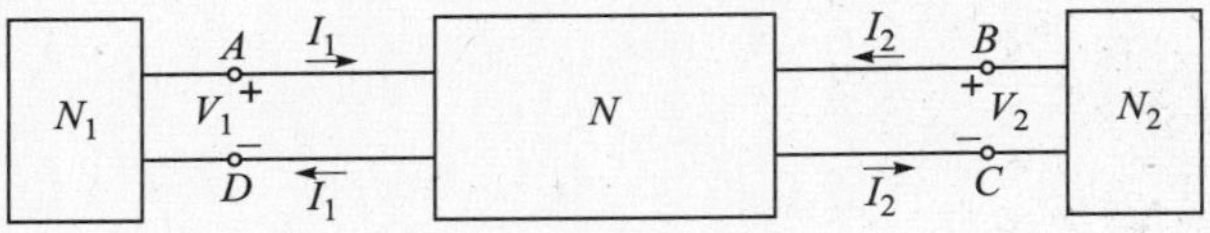

Fig. 13-1

13.2 *Z*-PARAMETERS

The terminal characteristics of a two-port network, having linear elements and dependent sources, may be written in the **s**-domain as

$$\begin{aligned} \mathbf{V}_1 &= \mathbf{Z}_{11}\mathbf{I}_1 + \mathbf{Z}_{12}\mathbf{I}_2 \\ \mathbf{V}_2 &= \mathbf{Z}_{21}\mathbf{I}_1 + \mathbf{Z}_{22}\mathbf{I}_2 \end{aligned} \tag{1}$$

The coefficients $\mathbf{Z}_{ij}$ have the dimension of impedance and are called the *Z-parameters* of the network. The **Z**-parameters are also called *open-circuit impedance parameters* since they may be measured at one terminal while the other terminal is open. They are

$$\begin{aligned} \mathbf{Z}_{11} &= \left.\frac{\mathbf{V}_1}{\mathbf{I}_1}\right|_{\mathbf{I}_2=0} \\ \mathbf{Z}_{12} &= \left.\frac{\mathbf{V}_1}{\mathbf{I}_2}\right|_{\mathbf{I}_1=0} \\ \mathbf{Z}_{21} &= \left.\frac{\mathbf{V}_2}{\mathbf{I}_1}\right|_{\mathbf{I}_2=0} \\ \mathbf{Z}_{22} &= \left.\frac{\mathbf{V}_2}{\mathbf{I}_2}\right|_{\mathbf{I}_1=0} \end{aligned} \tag{2}$$

EXAMPLE 13.1 Find the **Z**-parameters of the two-port circuit in Fig. 13-2.

Apply KVL around the two loops in Fig. 13-2 with loop currents $\mathbf{I}_1$ and $\mathbf{I}_2$ to obtain

$$\begin{aligned}\mathbf{V}_1 &= 2\mathbf{I}_1 + \mathbf{s}(\mathbf{I}_1 + \mathbf{I}_2) = (2 + \mathbf{s})\mathbf{I}_1 + \mathbf{s}\mathbf{I}_2 \\ \mathbf{V}_2 &= 3\mathbf{I}_2 + \mathbf{s}(\mathbf{I}_1 + \mathbf{I}_2) = \mathbf{s}\mathbf{I}_1 + (3 + \mathbf{s})\mathbf{I}_2\end{aligned} \tag{3}$$

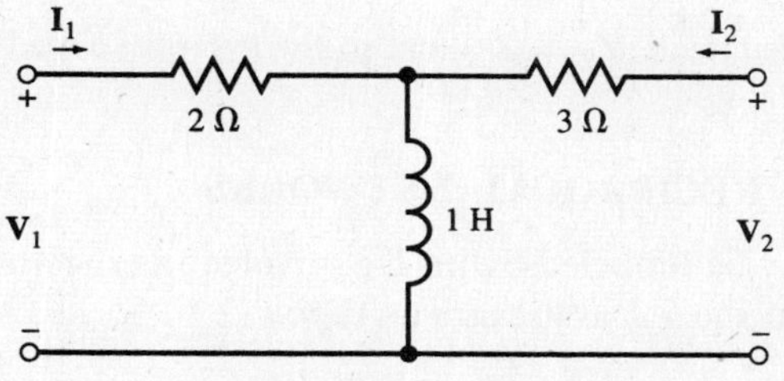

Fig. 13-2

By comparing (*1*) and (*3*), the **Z**-parameters of the circuit are found to be

$$\begin{aligned}\mathbf{Z}_{11} &= \mathbf{s} + 2 \\ \mathbf{Z}_{12} &= \mathbf{Z}_{21} = \mathbf{s} \\ \mathbf{Z}_{22} &= \mathbf{s} + 3\end{aligned} \tag{4}$$

Note that in this example $\mathbf{Z}_{12} = \mathbf{Z}_{21}$.

Reciprocal and Nonreciprocal Networks

A two-port network is called *reciprocal* if the open-circuit transfer impedances are equal; $\mathbf{Z}_{12} = \mathbf{Z}_{21}$. Consequently, in a reciprocal two-port network with current **I** feeding one port, the open-circuit voltage measured at the other port is the same, irrespective of the ports. The voltage is equal to $\mathbf{V} = \mathbf{Z}_{12}\mathbf{I} = \mathbf{Z}_{21}\mathbf{I}$. Networks containing resistors, inductors, and capacitors are generally reciprocal. Networks that additionally have dependent sources are generally nonreciprocal (see Example 13.2).

EXAMPLE 13.2 The two-port circuit shown in Fig. 13-3 contains a current-dependent voltage source. Find its **Z**-parameters.

As in Example 13.1, we apply KVL around the two loops:

$$\begin{aligned}\mathbf{V}_1 &= 2\mathbf{I}_1 - \mathbf{I}_2 + \mathbf{s}(\mathbf{I}_1 + \mathbf{I}_2) = (2 + \mathbf{s})\mathbf{I}_1 + (\mathbf{s} - 1)\mathbf{I}_2 \\ \mathbf{V}_2 &= 3\mathbf{I}_2 + \mathbf{s}(\mathbf{I}_1 + \mathbf{I}_2) = \mathbf{s}\mathbf{I}_1 + (3 + \mathbf{s})\mathbf{I}_2\end{aligned}$$

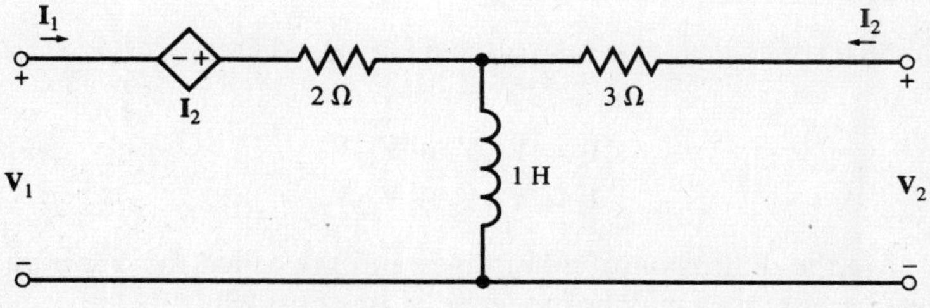

Fig. 13-3

The **Z**-parameters are

$$\begin{aligned} \mathbf{Z}_{11} &= \mathbf{s}+2 \\ \mathbf{Z}_{12} &= \mathbf{s}-1 \\ \mathbf{Z}_{21} &= \mathbf{s} \\ \mathbf{Z}_{22} &= \mathbf{s}+3 \end{aligned} \tag{5}$$

With the dependent source in the circuit, $\mathbf{Z}_{12} \neq \mathbf{Z}_{21}$ and so the two-port circuit is nonreciprocal.

13.3 T-EQUIVALENT OF RECIPROCAL NETWORKS

A reciprocal network may be modeled by its T-equivalent as shown in the circuit of Fig. 13-4. $\mathbf{Z}_a$, $\mathbf{Z}_b$, and $\mathbf{Z}_c$ are obtained from the **Z**-parameters as follows.

$$\begin{aligned} \mathbf{Z}_a &= \mathbf{Z}_{11} - \mathbf{Z}_{12} \\ \mathbf{Z}_b &= \mathbf{Z}_{22} - \mathbf{Z}_{21} \\ \mathbf{Z}_c &= \mathbf{Z}_{12} = \mathbf{Z}_{21} \end{aligned} \tag{6}$$

The T-equivalent network is not necessarily realizable.

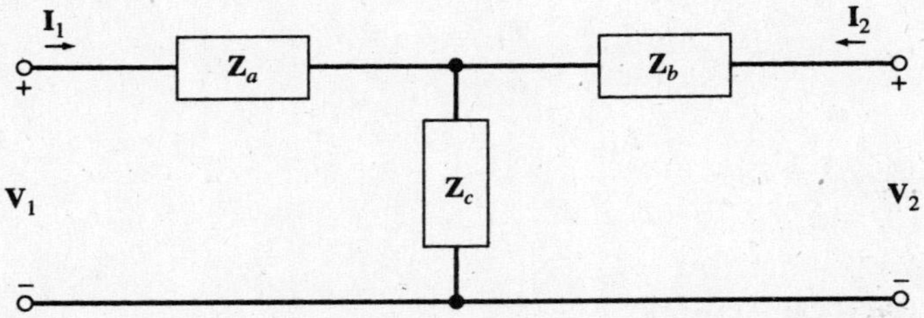

Fig. 13-4

EXAMPLE 13.3 Find the **Z**-parameters of Fig. 13-4.

Again we apply KVL to obtain

$$\begin{aligned} \mathbf{V}_1 &= \mathbf{Z}_a\mathbf{I}_1 + \mathbf{Z}_c(\mathbf{I}_1+\mathbf{I}_2) = (\mathbf{Z}_a+\mathbf{Z}_c)\mathbf{I}_1 + \mathbf{Z}_c\mathbf{I}_2 \\ \mathbf{V}_2 &= \mathbf{Z}_b\mathbf{I}_2 + \mathbf{Z}_c(\mathbf{I}_1+\mathbf{I}_2) = \mathbf{Z}_c\mathbf{I}_1 + (\mathbf{Z}_b+\mathbf{Z}_c)\mathbf{I}_2 \end{aligned} \tag{7}$$

By comparing *(1)* and *(7)*, the **Z**-parameters are found to be

$$\begin{aligned} \mathbf{Z}_{11} &= \mathbf{Z}_a + \mathbf{Z}_c \\ \mathbf{Z}_{12} &= \mathbf{Z}_{21} = \mathbf{Z}_c \\ \mathbf{Z}_{22} &= \mathbf{Z}_b + \mathbf{Z}_c \end{aligned} \tag{8}$$

13.4 Y-PARAMETERS

The terminal characteristics may also be written as in *(9)*, where $\mathbf{I}_1$ and $\mathbf{I}_2$ are expressed in terms of $\mathbf{V}_1$ and $\mathbf{V}_2$.

$$\begin{aligned} \mathbf{I}_1 &= \mathbf{Y}_{11}\mathbf{V}_1 + \mathbf{Y}_{12}\mathbf{V}_2 \\ \mathbf{I}_2 &= \mathbf{Y}_{21}\mathbf{V}_1 + \mathbf{Y}_{22}\mathbf{V}_2 \end{aligned} \tag{9}$$

The coefficients $\mathbf{Y}_{ij}$ have the dimension of admittance and are called the ***Y**-parameters* or *short-circuit admittance parameters* because they may be measured at one port while the other port is short-circuited. The **Y**-parameters are

$$\mathbf{Y}_{11} = \left.\frac{\mathbf{I}_1}{\mathbf{V}_1}\right|_{\mathbf{V}_2=0}$$
$$\mathbf{Y}_{12} = \left.\frac{\mathbf{I}_1}{\mathbf{V}_2}\right|_{\mathbf{V}_1=0}$$
$$\mathbf{Y}_{21} = \left.\frac{\mathbf{I}_2}{\mathbf{V}_1}\right|_{\mathbf{V}_2=0} \qquad (10)$$
$$\mathbf{Y}_{22} = \left.\frac{\mathbf{I}_2}{\mathbf{V}_2}\right|_{\mathbf{V}_1=0}$$

EXAMPLE 13.4 Find the **Y**-parameters of the circuit in Fig. 13-5.

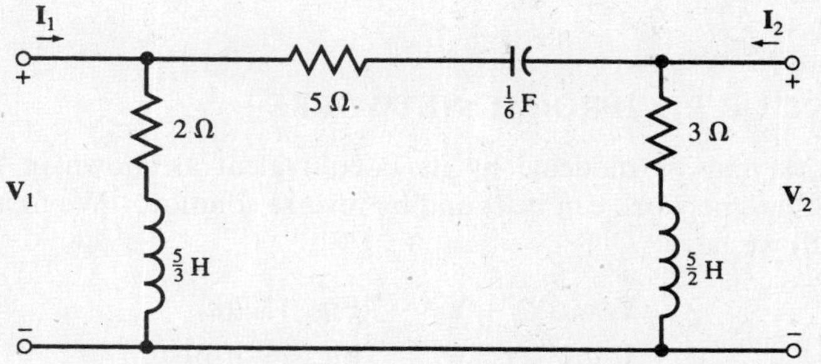

Fig. 13-5

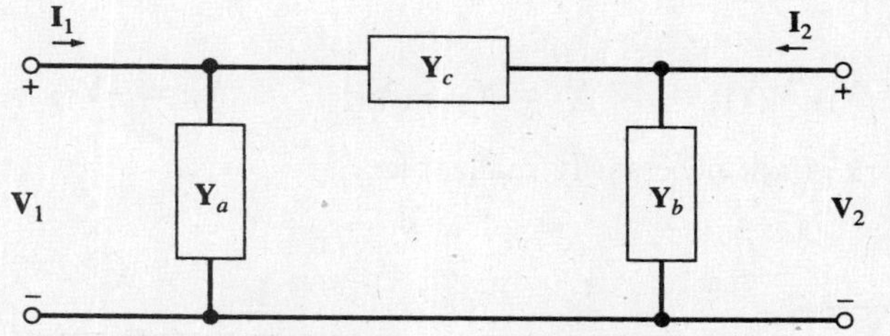

Fig. 13-6

We apply KCL to the input and output nodes (for convenience, we designate the admittances of the three branches of the circuit by $\mathbf{Y}_a$, $\mathbf{Y}_b$, and $\mathbf{Y}_c$ as shown in Fig. 13-6). Thus,

$$\mathbf{Y}_a = \frac{1}{2 + 5\mathbf{s}/3} = \frac{3}{5\mathbf{s} + 6}$$
$$\mathbf{Y}_b = \frac{1}{3 + 5\mathbf{s}/2} = \frac{2}{5\mathbf{s} + 6} \qquad (11)$$
$$\mathbf{Y}_c = \frac{1}{5 + 6/\mathbf{s}} = \frac{\mathbf{s}}{5\mathbf{s} + 6}$$

The node equations are

$$\mathbf{I}_1 = \mathbf{V}_1\mathbf{Y}_a + (\mathbf{V}_1 - \mathbf{V}_2)\mathbf{Y}_c = (\mathbf{Y}_a + \mathbf{Y}_c)\mathbf{V}_1 - \mathbf{Y}_c\mathbf{V}_2$$
$$\mathbf{I}_2 = \mathbf{V}_2\mathbf{Y}_b + (\mathbf{V}_2 - \mathbf{V}_1)\mathbf{Y}_c = -\mathbf{Y}_c\mathbf{V}_1 + (\mathbf{Y}_b + \mathbf{Y}_c)\mathbf{V}_2 \qquad (12)$$

By comparing (*9*) with (*12*), we get

$$\begin{aligned}\mathbf{Y}_{11} &= \mathbf{Y}_a + \mathbf{Y}_c \\ \mathbf{Y}_{12} &= \mathbf{Y}_{21} = -\mathbf{Y}_c \\ \mathbf{Y}_{22} &= \mathbf{Y}_b + \mathbf{Y}_c\end{aligned} \tag{13}$$

Substituting $\mathbf{Y}_a$, $\mathbf{Y}_b$, and $\mathbf{Y}_c$ in *(11)* into *(13)*, we find

$$\begin{aligned}\mathbf{Y}_{11} &= \frac{\mathbf{s}+3}{5\mathbf{s}+6} \\ \mathbf{Y}_{12} &= \mathbf{Y}_{21} = \frac{-\mathbf{s}}{5\mathbf{s}+6} \\ \mathbf{Y}_{22} &= \frac{\mathbf{s}+2}{5\mathbf{s}+6}\end{aligned} \tag{14}$$

Since $\mathbf{Y}_{12} = \mathbf{Y}_{21}$, the two-port circuit is reciprocal.

13.5 PI-EQUIVALENT OF RECIPROCAL NETWORKS

A reciprocal network may be modeled by its Pi-equivalent as shown in Fig. 13-6. The three elements of the Pi-equivalent network can be found by reverse solution. We first find the **Y**-parameters of Fig. 13-6. From *(10)* we have

$$\begin{aligned}\mathbf{Y}_{11} &= \mathbf{Y}_a + \mathbf{Y}_c && \text{[Fig. 13.7}(a)] \\ \mathbf{Y}_{12} &= -\mathbf{Y}_c && \text{[Fig. 13-7}(b)] \\ \mathbf{Y}_{21} &= -\mathbf{Y}_c && \text{[Fig. 13-7}(a)] \\ \mathbf{Y}_{22} &= \mathbf{Y}_b + \mathbf{Y}_c && \text{[Fig. 13-7}(b)]\end{aligned} \tag{15}$$

from which

$$\mathbf{Y}_a = \mathbf{Y}_{11} + \mathbf{Y}_{12} \qquad \mathbf{Y}_b = \mathbf{Y}_{22} + \mathbf{Y}_{12} \qquad \mathbf{Y}_c = -\mathbf{Y}_{12} = -\mathbf{Y}_{21} \tag{16}$$

The Pi-equivalent network is not necessarily realizable.

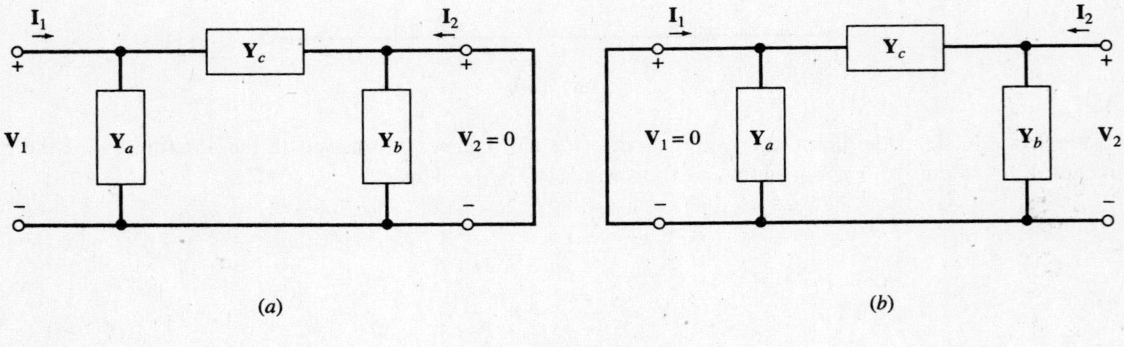

Fig. 13-7

13.6 APPLICATION OF TERMINAL CHARACTERISTICS

The four terminal variables $\mathbf{I}_1$, $\mathbf{I}_2$, $\mathbf{V}_1$, and $\mathbf{V}_2$ in a two-port network are related by the two equations *(1)* or *(9)*. By connecting the two-port circuit to the outside as shown in Fig. 13-1, two additional equations are obtained. The four equations then can determine $\mathbf{I}_1$, $\mathbf{I}_2$, $\mathbf{V}_1$, and $\mathbf{V}_2$ without any knowledge of the inside structure of the circuit.

EXAMPLE 13.5 The **Z**-parameters of a two-port network are given by

$$\mathbf{Z}_{11} = 2\mathbf{s} + 1/\mathbf{s} \qquad \mathbf{Z}_{12} = \mathbf{Z}_{21} = 2\mathbf{s} \qquad \mathbf{Z}_{22} = 2\mathbf{s} + 4$$

The network is connected to a source and a load as shown in Fig. 13-8. Find $\mathbf{I}_1$, $\mathbf{I}_2$, $\mathbf{V}_1$, and $\mathbf{V}_2$.

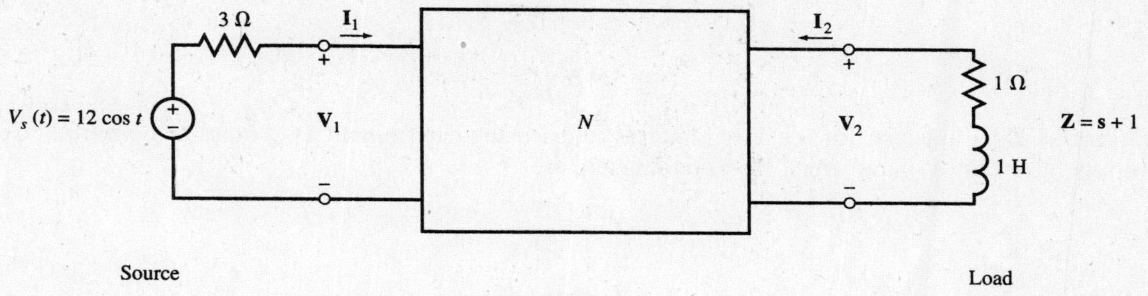

Fig. 13-8

The terminal characteristics are given by

$$\begin{aligned} \mathbf{V}_1 &= (2\mathbf{s} + 1/\mathbf{s})\mathbf{I}_1 + 2\mathbf{s}\mathbf{I}_2 \\ \mathbf{V}_2 &= 2\mathbf{s}\mathbf{I}_1 + (2\mathbf{s} + 4)\mathbf{I}_2 \end{aligned} \tag{17}$$

The phasor representation of voltage $v_s(t)$ is $\mathbf{V}_s = 12$ V with $\mathbf{s} = j$. From KVL around the input and output loops we obtain the two additional equations (*18*)

$$\begin{aligned} \mathbf{V}_s &= 3\mathbf{I}_1 + \mathbf{V}_1 \\ 0 &= (1 + \mathbf{s})\mathbf{I}_2 + \mathbf{V}_2 \end{aligned} \tag{18}$$

Substituting $\mathbf{s} = j$ and $\mathbf{V}_s = 12$ in (*17*) and in (*18*) we get

$$\begin{aligned} \mathbf{V}_1 &= j\mathbf{I}_1 + 2j\mathbf{I}_2 \\ \mathbf{V}_2 &= 2j\mathbf{I}_1 + (4 + 2j)\mathbf{I}_2 \\ 12 &= 3\mathbf{I}_1 + \mathbf{V}_1 \\ 0 &= (1 + j)\mathbf{I}_2 + \mathbf{V}_2 \end{aligned}$$

from which

$$\mathbf{I}_1 = 3.29\,\underline{/-10.2^\circ} \qquad \mathbf{I}_2 = 1.13\,\underline{/-131.2^\circ}$$
$$\mathbf{V}_1 = 2.88\,\underline{/37.5^\circ} \qquad \mathbf{V}_2 = 1.6\,\underline{/93.8^\circ}$$

13.7 CONVERSION BETWEEN Z- AND Y-PARAMETERS

The **Y**-parameters may be obtained from the **Z**-parameters by solving (*1*) for $\mathbf{I}_1$ and $\mathbf{I}_2$. Applying Cramer's rule to (*1*), we get

$$\begin{aligned} \mathbf{I}_1 &= \frac{\mathbf{Z}_{22}}{\mathbf{D}_{\mathbf{ZZ}}}\mathbf{V}_1 - \frac{\mathbf{Z}_{12}}{\mathbf{D}_{\mathbf{ZZ}}}\mathbf{V}_2 \\ \mathbf{I}_2 &= \frac{-\mathbf{Z}_{21}}{\mathbf{D}_{\mathbf{ZZ}}}\mathbf{V}_1 + \frac{\mathbf{Z}_{11}}{\mathbf{D}_{\mathbf{ZZ}}}\mathbf{V}_2 \end{aligned} \tag{19}$$

where $\mathbf{D}_{\mathbf{ZZ}} = \mathbf{Z}_{11}\mathbf{Z}_{22} - \mathbf{Z}_{12}\mathbf{Z}_{21}$ is the determinant of the coefficients in (*1*). By comparing (*19*) with (*9*) we have

$$\mathbf{Y}_{11} = \frac{\mathbf{Z}_{22}}{\mathbf{D}_{\mathbf{ZZ}}}$$
$$\mathbf{Y}_{12} = \frac{-\mathbf{Z}_{12}}{\mathbf{D}_{\mathbf{ZZ}}}$$
$$\mathbf{Y}_{21} = \frac{-\mathbf{Z}_{21}}{\mathbf{D}_{\mathbf{ZZ}}} \tag{20}$$
$$\mathbf{Y}_{22} = \frac{\mathbf{Z}_{11}}{\mathbf{D}_{\mathbf{ZZ}}}$$

Given the **Z**-parameters, for the **Y**-parameters to exist, the determinant $\mathbf{D}_{\mathbf{ZZ}}$ must be nonzero. Conversely, given the **Y**-parameters, the **Z**-parameters are

$$\mathbf{Z}_{11} = \frac{\mathbf{Y}_{22}}{\mathbf{D}_{\mathbf{YY}}}$$
$$\mathbf{Z}_{12} = \frac{-\mathbf{Y}_{12}}{\mathbf{D}_{\mathbf{YY}}}$$
$$\mathbf{Z}_{21} = \frac{-\mathbf{Y}_{21}}{\mathbf{D}_{\mathbf{YY}}} \tag{21}$$
$$\mathbf{Z}_{22} = \frac{\mathbf{Y}_{11}}{\mathbf{D}_{\mathbf{YY}}}$$

where $\mathbf{D}_{\mathbf{YY}} = \mathbf{Y}_{11}\mathbf{Y}_{22} - \mathbf{Y}_{12}\mathbf{Y}_{21}$ is the determinant of the coefficients in (*9*). For the **Z**-parameters of a two-port circuit to be derived from its **Y**-parameters, $\mathbf{D}_{\mathbf{YY}}$ should be nonzero.

EXAMPLE 13.6 Referring to Example 13.4, find the **Z**-parameters of the circuit of Fig. 13-5 from its **Y**-parameters.

The **Y**-parameters of the circuit were found to be [see (*14*)]

$$\mathbf{Y}_{11} = \frac{\mathbf{s}+3}{5\mathbf{s}+6} \qquad \mathbf{Y}_{12} = \mathbf{Y}_{21} = \frac{-\mathbf{s}}{5\mathbf{s}+6} \qquad \mathbf{Y}_{22} = \frac{\mathbf{s}+2}{5\mathbf{s}+6}$$

Substituting into (*21*), where $\mathbf{D}_{\mathbf{YY}} = 1/(5\mathbf{s}+6)$, we obtain

$$\mathbf{Z}_{11} = \mathbf{s}+2$$
$$\mathbf{Z}_{12} = \mathbf{Z}_{21} = \mathbf{s} \tag{22}$$
$$\mathbf{Z}_{22} = \mathbf{s}+3$$

The **Z**-parameters in (*22*) are identical to the **Z**-parameters of the circuit of Fig. 13-2. The two circuits are equivalent as far as the terminals are concerned. This was by design. Figure 13-2 is the T-equivalent of Fig. 13-5. The equivalence between Fig. 13-2 and Fig. 13-5 may be verified directly by applying (*6*) to the **Z**-parameters given in (*22*) to obtain its T-equivalent network.

13.8 h-PARAMETERS

Some two-port circuits or electronic devices are best characterized by the following terminal equations:

$$\mathbf{V}_1 = \mathbf{h}_{11}\mathbf{I}_1 + \mathbf{h}_{12}\mathbf{V}_2$$
$$\mathbf{I}_2 = \mathbf{h}_{21}\mathbf{I}_1 + \mathbf{h}_{22}\mathbf{V}_2 \tag{23}$$

where the $\mathbf{h}_{ij}$ coefficients are called the *hybrid parameters*, or **h**-*parameters*.

EXAMPLE 13.7 Find the **h**-parameters of Fig. 13-9.

This is the simple model of a bipolar junction transistor in its linear region of operation. By inspection, the terminal characteristics of Fig. 13-9 are

$$\mathbf{V}_1 = 50\mathbf{I}_1 \qquad \text{and} \qquad \mathbf{I}_2 = 300\mathbf{I}_1 \tag{24}$$

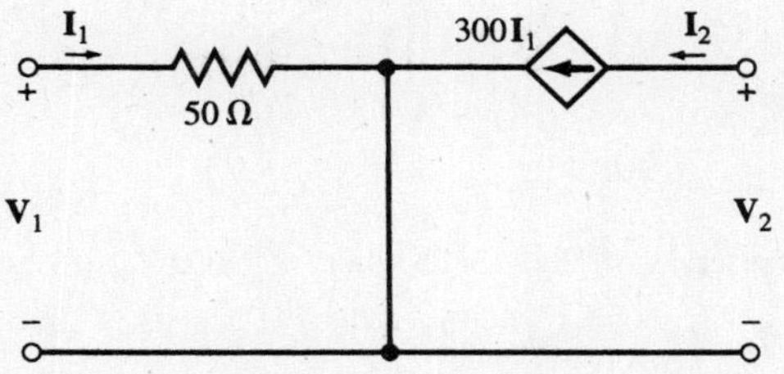

Fig. 13-9

By comparing (*24*) and (*23*) we get

$$\mathbf{h}_{11} = 50 \qquad \mathbf{h}_{12} = 0 \qquad \mathbf{h}_{21} = 300 \qquad \mathbf{h}_{22} = 0 \tag{25}$$

13.9 g-PARAMETERS

The terminal characteristics of a two-port circuit may also be described by still another set of hybrid parameters given in (*26*).

$$\begin{aligned} \mathbf{I}_1 &= \mathbf{g}_{11}\mathbf{V}_1 + \mathbf{g}_{12}\mathbf{I}_2 \\ \mathbf{V}_2 &= \mathbf{g}_{21}\mathbf{V}_1 + \mathbf{g}_{22}\mathbf{I}_2 \end{aligned} \tag{26}$$

where the coefficients $\mathbf{g}_{ij}$ are called *inverse hybrid* or **g**-*parameters*.

EXAMPLE 13.8 Find the **g**-parameters in the circuit shown in Fig. 13-10.

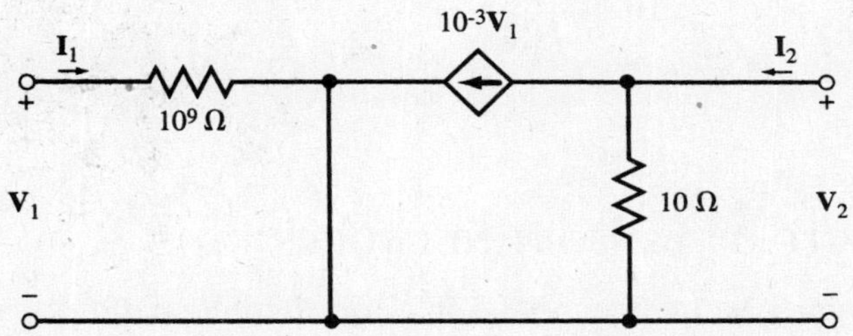

Fig. 13-10

This is the simple model of a field effect transistor in its linear region of operation. To find the **g**-parameters, we first derive the terminal equations by applying Kirchhoff's laws at the terminals:

At the input terminal: $\mathbf{V}_1 = 10^9\mathbf{I}_1$

At the output terminal: $\mathbf{V}_2 = 10(\mathbf{I}_2 - 10^{-3}\mathbf{V}_1)$

or $$\mathbf{I}_1 = 10^{-9}\mathbf{V}_1 \quad \text{and} \quad \mathbf{V}_2 = 10\mathbf{I}_2 - 10^{-2}\mathbf{V}_1 \tag{28}$$

By comparing (*27*) and (*26*) we get

$$\mathbf{g}_{11} = 10^{-9} \qquad \mathbf{g}_{12} = 0 \qquad \mathbf{g}_{21} = -10^{-2} \qquad \mathbf{g}_{22} = 10 \tag{28}$$

13.10 TRANSMISSION PARAMETERS

The transmission parameters **A**, **B**, **C**, and **D** express the required source variables $\mathbf{V}_1$ and $\mathbf{I}_1$ in terms of the existing destination variables $\mathbf{V}_2$ and $\mathbf{I}_2$. They are called **ABCD** or T-parameters and are defined by

$$\begin{aligned} \mathbf{V}_1 &= \mathbf{AV}_2 - \mathbf{BI}_2 \\ \mathbf{I}_1 &= \mathbf{CV}_2 - \mathbf{DI}_2 \end{aligned} \tag{29}$$

EXAMPLE 13.9 Find the T-parameters of Fig. 13-11 where $\mathbf{Z}_a$ and $\mathbf{Z}_b$ are nonzero.

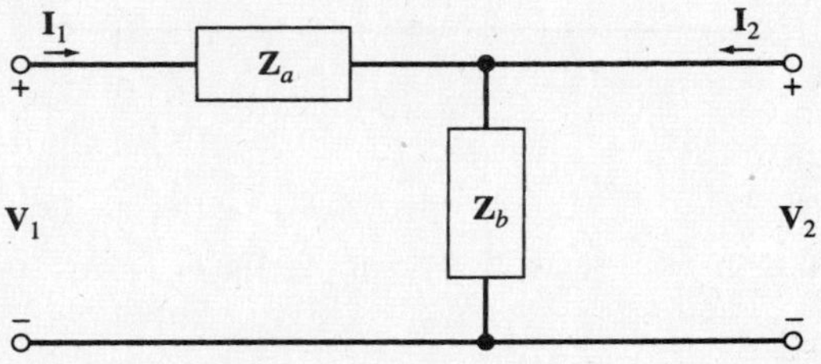

Fig. 13-11

This is the simple lumped model of an incremental segment of a transmission line. From (*29*) we have

$$\begin{aligned} \mathbf{A} &= \left.\frac{\mathbf{V}_1}{\mathbf{V}_2}\right|_{\mathbf{I}_2=0} = \frac{\mathbf{Z}_a + \mathbf{Z}_b}{\mathbf{Z}_b} = 1 + \mathbf{Z}_a\mathbf{Y}_b \\ \mathbf{B} &= -\left.\frac{\mathbf{V}_1}{\mathbf{I}_2}\right|_{\mathbf{V}_2=0} = \mathbf{Z}_a \\ \mathbf{C} &= \left.\frac{\mathbf{I}_1}{\mathbf{V}_2}\right|_{\mathbf{I}_2=0} = \mathbf{Y}_b \\ \mathbf{D} &= -\left.\frac{\mathbf{I}_1}{\mathbf{I}_2}\right|_{\mathbf{V}_2=0} = 1 \end{aligned} \tag{30}$$

13.11 INTERCONNECTING TWO-PORT NETWORKS

Two-port networks may be interconnected in various configurations, such as series, parallel, or cascade connection, resulting in new two-port networks. For each configuration, certain set of parameters may be more useful than others to describe the network.

Series Connection

Figure 13-12 shows a series connection of two two-port networks **a** and **b** with open-circuit impedance parameters $\mathbf{Z}_a$ and $\mathbf{Z}_b$, respectively. In this configuration, we use the **Z**-parameters since they are combined as a series connection of two impedances. The **Z**-parameters of the series connection are (see Problem 13.10):

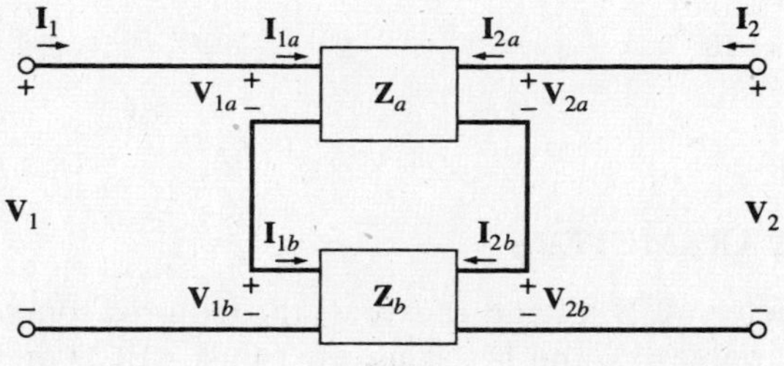

Fig. 13-12

$$\begin{aligned}\mathbf{Z}_{11} &= \mathbf{Z}_{11,a} + \mathbf{Z}_{11,b}\\ \mathbf{Z}_{12} &= \mathbf{Z}_{12,a} + \mathbf{Z}_{12,b}\\ \mathbf{Z}_{21} &= \mathbf{Z}_{21,a} + \mathbf{Z}_{21,b}\\ \mathbf{Z}_{22} &= \mathbf{Z}_{22,a} + \mathbf{Z}_{22,b}\end{aligned} \tag{31a}$$

or, in the matrix form,

$$[\mathbf{Z}] = [\mathbf{Z}_a] + [\mathbf{Z}_b] \tag{31b}$$

Parallel Connection

Figure 13-13 shows a parallel connection of two-port networks **a** and **b** with short-circuit admittance parameters $\mathbf{Y}_a$ and $\mathbf{Y}_b$. In this case, the **Y**-parameters are convenient to work with. The **Y**-parameters of the parallel connection are (see Problem 13.11):

$$\begin{aligned}\mathbf{Y}_{11} &= \mathbf{Y}_{11,a} + \mathbf{Y}_{11,b}\\ \mathbf{Y}_{12} &= \mathbf{Y}_{12,a} + \mathbf{Y}_{12,b}\\ \mathbf{Y}_{21} &= \mathbf{Y}_{21,a} + \mathbf{Y}_{21,b}\\ \mathbf{Y}_{22} &= \mathbf{Y}_{22,a} + \mathbf{Y}_{22,b}\end{aligned} \tag{32a}$$

or, in the matrix form

$$[\mathbf{Y}] = [\mathbf{Y}_a] + [\mathbf{Y}_b] \tag{32b}$$

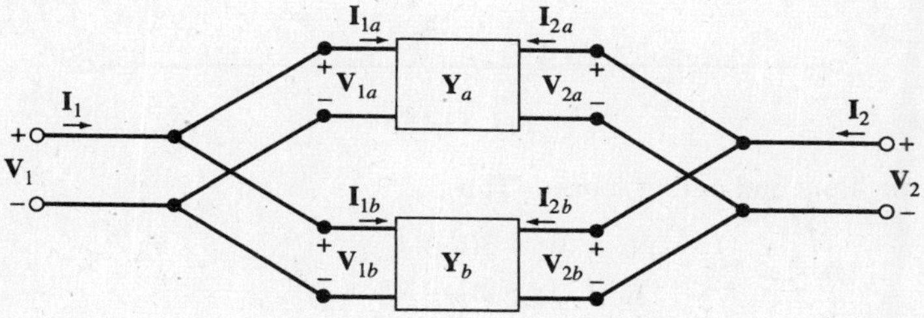

Fig. 13-13

Cascade Connection

The cascade connection of two-port networks **a** and **b** is shown in Fig. 13-14. In this case the T-parameters are particularly convenient. The T-parameters of the cascade combination are

$$\begin{aligned}\mathbf{A} &= \mathbf{A}_a\mathbf{A}_b + \mathbf{B}_a\mathbf{C}_b\\ \mathbf{B} &= \mathbf{A}_a\mathbf{B}_b + \mathbf{B}_a\mathbf{D}_b\\ \mathbf{C} &= \mathbf{C}_a\mathbf{A}_b + \mathbf{D}_a\mathbf{C}_b\\ \mathbf{D} &= \mathbf{C}_a\mathbf{B}_b + \mathbf{D}_a\mathbf{D}_b\end{aligned} \tag{33a}$$

or, in the matrix form,

$$[\mathbf{T}] = [\mathbf{T}_a][\mathbf{T}_b] \tag{33b}$$

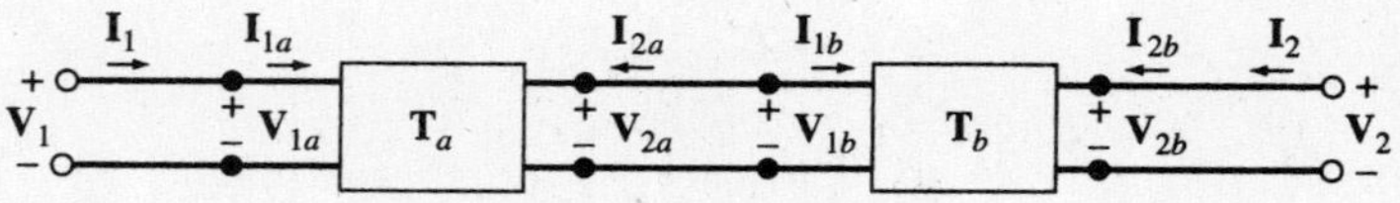

Fig. 13-14

13.12 CHOICE OF PARAMETER TYPE

What types of parameters are appropriate to and can best describe a given two-port network or device? Several factors influence the choice of parameters. (*1*) It is possible that some types of parameters do not exist as they may not be defined at all (see Example 13.10). (*2*) Some parameters are more convenient to work with when the network is connected to other networks, as shown in Section 13.11. In this regard, by converting the two-port network to its T- and Pi-equivalent and then applying the familiar analysis techniques, such as element reduction and current division, we can greatly reduce and simplify the overall circuit. (*3*) For some networks or devices, a certain type of parameter produces better computational accuracy and better sensitivity when used within the interconnected circuit.

EXAMPLE 13.10 Find the **Z**- and **Y**-parameters of Fig. 13-15.

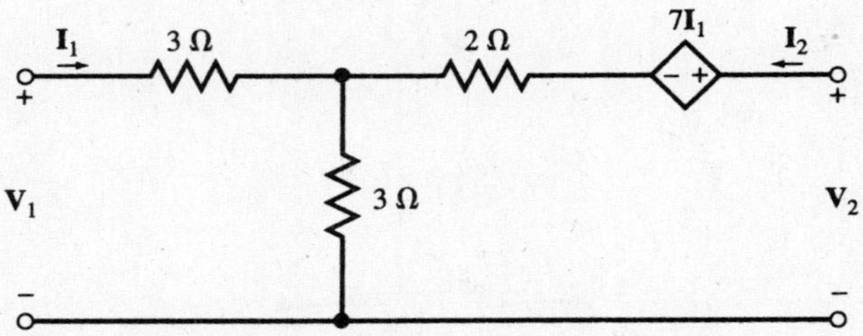

Fig. 13-15

We apply KVL to the input and output loops. Thus,

Input loop: $\mathbf{V}_1 = 3\mathbf{I}_1 + 3(\mathbf{I}_1 + \mathbf{I}_2)$

Output loop: $\mathbf{V}_2 = 7\mathbf{I}_1 + 2\mathbf{I}_2 + 3(\mathbf{I}_1 + \mathbf{I}_2)$

or
$$\mathbf{V}_1 = 6\mathbf{I}_1 + 3\mathbf{I}_2 \quad \text{and} \quad \mathbf{V}_2 = 10\mathbf{I}_1 + 5\mathbf{I}_2 \tag{34}$$

By comparing (*34*) and (*2*) we get

$$\mathbf{Z}_{11} = 6 \qquad \mathbf{Z}_{12} = 3 \qquad \mathbf{Z}_{21} = 10 \qquad \mathbf{Z}_{22} = 5$$

The **Y**-parameters are, however, not defined, since the application of the direct method of (*10*) or the conversion from **Z**-parameters (*19*) produces $\mathbf{D}_{ZZ} = 6(5) - 3(10) = 0$.

13.13 SUMMARY OF TERMINAL PARAMETERS AND CONVERSION

Terminal parameters are defined by the following equations

Z-parameters
$\mathbf{V}_1 = \mathbf{Z}_{11}\mathbf{I}_1 + \mathbf{Z}_{12}\mathbf{I}_2$
$\mathbf{V}_2 = \mathbf{Z}_{21}\mathbf{I}_1 + \mathbf{Z}_{22}\mathbf{I}_2$
$[\mathbf{V}] = [\mathbf{Z}][\mathbf{I}]$

h-parameters
$\mathbf{V}_1 = \mathbf{h}_{11}\mathbf{I}_1 + \mathbf{h}_{12}\mathbf{V}_2$
$\mathbf{I}_2 = \mathbf{h}_{21}\mathbf{I}_1 + \mathbf{h}_{22}\mathbf{V}_2$

T-parameters
$\mathbf{V}_1 = \mathbf{A}\mathbf{V}_2 - \mathbf{B}\mathbf{I}_2$
$\mathbf{I}_1 = \mathbf{C}\mathbf{V}_2 - \mathbf{D}\mathbf{I}_2$

Y-parameters
$\mathbf{I}_1 = \mathbf{Y}_{11}\mathbf{V}_1 + \mathbf{Y}_{12}\mathbf{V}_2$
$\mathbf{I}_2 = \mathbf{Y}_{21}\mathbf{V}_1 + \mathbf{Y}_{22}\mathbf{V}_2$
$[\mathbf{I}] = [\mathbf{Y}][\mathbf{V}]$

g-parameters
$\mathbf{I}_1 = \mathbf{g}_{11}\mathbf{V}_1 + \mathbf{g}_{12}\mathbf{I}_2$
$\mathbf{V}_2 = \mathbf{g}_{21}\mathbf{V}_1 + \mathbf{g}_{22}\mathbf{I}_2$

Table 13-1 summarizes the conversion between the **Z**-, **Y**-, **h**-, **g**-, and T-parameters. For the conversion to be possible, the determinant of the source parameters must be nonzero.

Table 13-1

	Z		Y		h		g		T	
Z	Z_{11}	Z_{12}	$\frac{Y_{22}}{D_{YY}}$	$\frac{-Y_{12}}{D_{YY}}$	$\frac{D_{hh}}{h_{22}}$	$\frac{h_{12}}{h_{22}}$	$\frac{1}{g_{11}}$	$\frac{-g_{12}}{g_{11}}$	$\frac{A}{C}$	$\frac{D_{TT}}{C}$
	Z_{21}	Z_{22}	$\frac{-Y_{21}}{D_{YY}}$	$\frac{Y_{11}}{D_{YY}}$	$\frac{-h_{21}}{h_{22}}$	$\frac{1}{h_{22}}$	$\frac{g_{21}}{g_{11}}$	$\frac{D_{gg}}{g_{11}}$	$\frac{1}{C}$	$\frac{D}{C}$
Y	$\frac{Z_{22}}{D_{zz}}$	$\frac{-Z_{12}}{D_{zz}}$	Y_{11}	Y_{12}	$\frac{1}{h_{11}}$	$\frac{-h_{12}}{h_{11}}$	$\frac{D_{gg}}{g_{22}}$	$\frac{g_{12}}{g_{22}}$	$\frac{D}{B}$	$\frac{-D_{TT}}{B}$
	$\frac{-Z_{21}}{D_{zz}}$	$\frac{Z_{11}}{D_{zz}}$	Y_{21}	Y_{22}	$\frac{h_{21}}{h_{11}}$	$\frac{-D_{nn}}{h_{11}}$	$\frac{-g_{21}}{g_{22}}$	$\frac{1}{g_{22}}$	$\frac{-1}{B}$	$\frac{A}{B}$
h	$\frac{D_{zz}}{Z_{22}}$	$\frac{Z_{12}}{Z_{22}}$	$\frac{1}{Y_{11}}$	$\frac{-Y_{12}}{Y_{11}}$	h_{11}	h_{12}	$\frac{g_{22}}{D_{gg}}$	$\frac{g_{12}}{D_{gg}}$	$\frac{B}{D}$	$\frac{D_{TT}}{D}$
	$\frac{-Z_{21}}{Z_{22}}$	$\frac{1}{Z_{22}}$	$\frac{Y_{21}}{Y_{11}}$	$\frac{D_{yy}}{Y_{11}}$	h_{21}	h_{22}	$\frac{g_{21}}{D_{gg}}$	$\frac{g_{11}}{D_{gg}}$	$\frac{-1}{D}$	$\frac{C}{D}$
g	$\frac{1}{Z_{11}}$	$\frac{-Z_{12}}{Z_{11}}$	$\frac{D_{YY}}{Y_{22}}$	$\frac{Y_{12}}{Y_{22}}$	$\frac{h_{22}}{D_{hh}}$	$\frac{-h_{12}}{D_{hh}}$	g_{11}	g_{12}	$\frac{C}{A}$	$\frac{-D_{TT}}{A}$
	$\frac{Z_{21}}{Z_{11}}$	$\frac{D_{ZZ}}{Z_{11}}$	$\frac{-Y_{21}}{Y_{22}}$	$\frac{1}{Y_{22}}$	$\frac{-h_{21}}{D_{hh}}$	$\frac{h_{11}}{D_{hh}}$	g_{21}	g_{22}	$\frac{1}{A}$	$\frac{B}{A}$
T	$\frac{Z_{11}}{Z_{21}}$	$\frac{D_{ZZ}}{Z_{21}}$	$\frac{-Y_{22}}{Y_{21}}$	$\frac{-1}{Y_{21}}$	$\frac{-D_{hh}}{h_{21}}$	$\frac{-h_{11}}{h_{21}}$	$\frac{1}{g_{21}}$	$\frac{g_{22}}{g_{21}}$	A	B
	$\frac{1}{Z_{21}}$	$\frac{Z_{22}}{Z_{21}}$	$\frac{-D_{YY}}{Y_{21}}$	$\frac{-Y_{11}}{Y_{21}}$	$\frac{-h_{22}}{h_{21}}$	$\frac{-1}{h_{21}}$	$\frac{g_{11}}{g_{21}}$	$\frac{D_{gg}}{g_{21}}$	C	D

$D_{PP} = P_{11}P_{22} - P_{12}P_{21}$ is the determinant of **Z**−, **Y**−, **h**−, **g**−, or T-parameters.

Solved Problems

13.1 Find the **Z**-parameters of the circuit in Fig. 13-16(*a*).

Z_{11} and Z_{21} are obtained by connecting a source to port #1 and leaving port #2 open [Fig. 13-16(*b*)]. The parallel and series combination of resistors produces

$$Z_{11} = \frac{V_1}{I_1}\bigg|_{I_2=0} = 8 \quad \text{and} \quad Z_{21} = \frac{V_2}{I_1}\bigg|_{I_2=0} = \frac{1}{3}$$

Similarly, Z_{22} and Z_{12} are obtained by connecting a source to port #2 and leaving port #1 open [Fig. 13-16(*c*)].

$$Z_{22} = \frac{V_2}{I_2}\bigg|_{I_1=0} = \frac{8}{9} \qquad Z_{12} = \frac{V_1}{I_2}\bigg|_{I_1=0} = \frac{1}{3}$$

The circuit is reciprocal, since $Z_{12} = Z_{21}$.

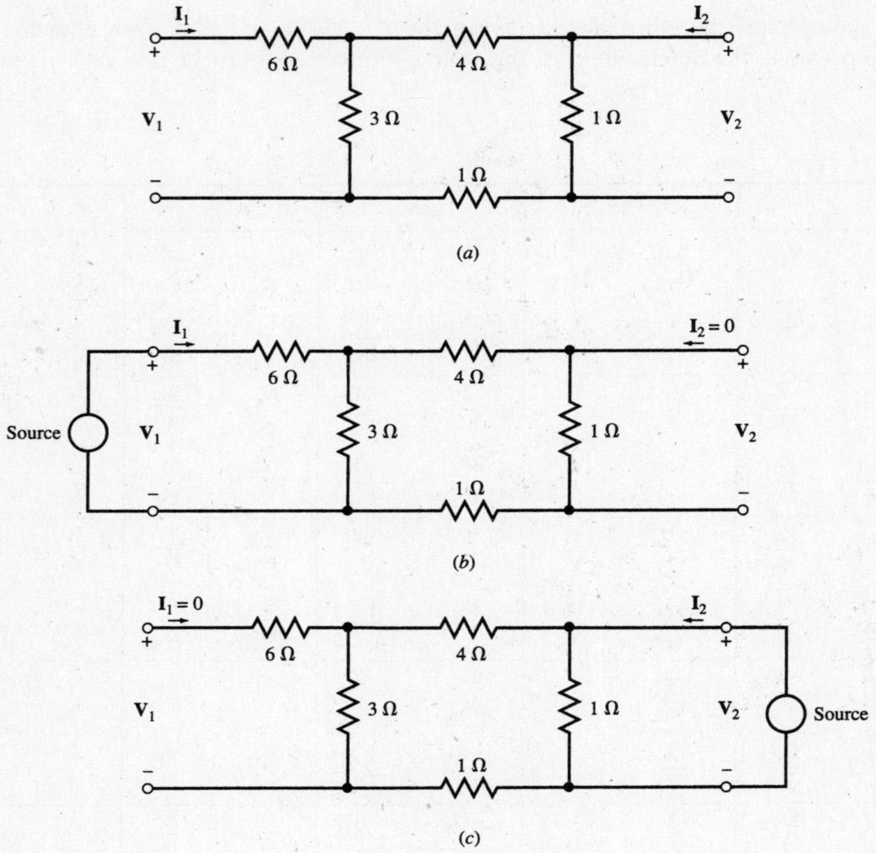

Fig. 13-16

13.2 The **Z**-parameters of a two-port network N are given by

$$\mathbf{Z}_{11} = 2\mathbf{s} + 1/\mathbf{s} \qquad \mathbf{Z}_{12} = \mathbf{Z}_{21} = 2\mathbf{s} \qquad \mathbf{Z}_{22} = 2\mathbf{s} + 4$$

(*a*) Find the T-equivalent of N. (*b*) The network N is connected to a source and a load as shown in the circuit of Fig. 13-8. Replace N by its T-equivalent and then solve for i_1, i_2, v_1, and v_2.

(*a*) The three branches of the T-equivalent network (Fig. 13-4) are

$$\mathbf{Z}_a = \mathbf{Z}_{11} - \mathbf{Z}_{12} = 2\mathbf{s} + \frac{1}{\mathbf{s}} - 2\mathbf{s} = \frac{1}{\mathbf{s}}$$
$$\mathbf{Z}_b = \mathbf{Z}_{22} - \mathbf{Z}_{12} = 2\mathbf{s} + 4 - 2\mathbf{s} = 4$$
$$\mathbf{Z}_c = \mathbf{Z}_{12} = \mathbf{Z}_{21} = 2\mathbf{s}$$

(*b*) The T-equivalent of N, along with its input and output connections, is shown in phasor domain in Fig. 13-17.

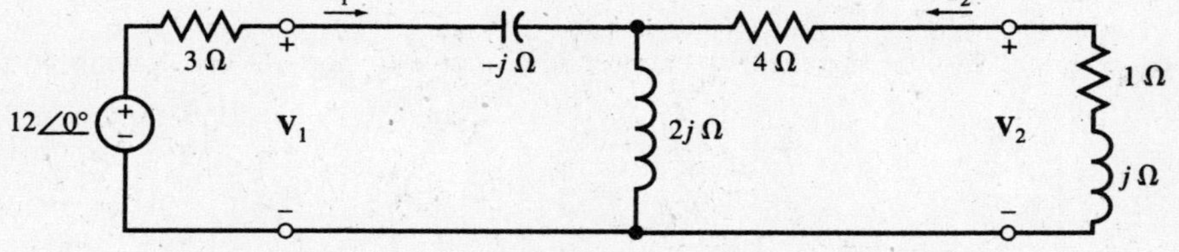

Fig. 13-17

By applying the familiar analysis techniques, including element reduction and current division, to Fig. 13-17, we find i_1, i_2, v_1, and v_2.

In phasor domain	In the time domain:
$\mathbf{I}_1 = 3.29\,\underline{/-10.2^\circ}$	$i_1 = 3.29\cos(t - 10.2^\circ)$
$\mathbf{I}_2 = 1.13\,\underline{/-131.2^\circ}$	$i_2 = 1.13\cos(t - 131.2^\circ)$
$\mathbf{V}_1 = 2.88\,\underline{/37.5^\circ}$	$v_1 = 2.88\cos(t + 37.5^\circ)$
$\mathbf{V}_2 = 1.6\,\underline{/93.8^\circ}$	$v_2 = 1.6\cos(t + 93.8^\circ)$

13.3 Find the **Z**-parameters of the two-port network in Fig. 13-18.

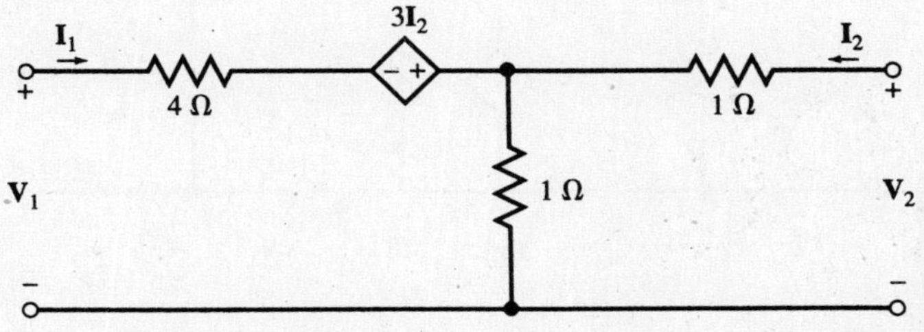

Fig. 13-18

KVL applied to the input and output ports obtains the following:

Input port: $\mathbf{V}_1 = 4\mathbf{I}_1 - 3\mathbf{I}_2 + (\mathbf{I}_1 + \mathbf{I}_2) = 5\mathbf{I}_1 - 2\mathbf{I}_2$

Output port: $\mathbf{V}_2 = \mathbf{I}_2 + (\mathbf{I}_1 + \mathbf{I}_2) = \mathbf{I}_1 + 2\mathbf{I}_2$

By applying (*2*) to the above, $\mathbf{Z}_{11} = 5$, $\mathbf{Z}_{12} = -2$, $\mathbf{Z}_{21} = 1$, and $\mathbf{Z}_{22} = 2$.

13.4 Find the **Z**-parameters of the two-port network in Fig. 13-19 and compare the results with those of Problem 13.3.

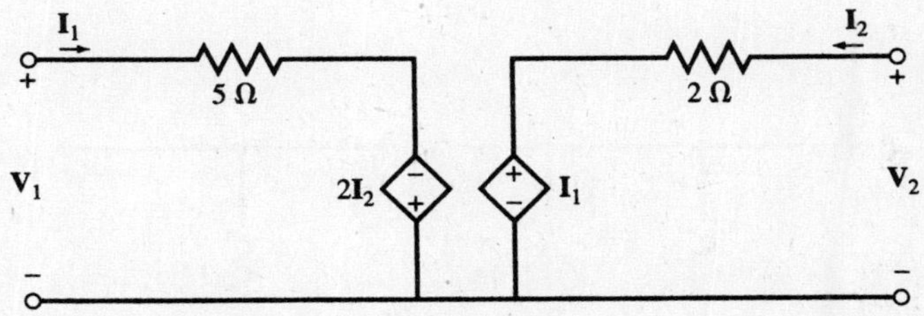

Fig. 13-19

KVL gives

$$\mathbf{V}_1 = 5\mathbf{I}_1 - 2\mathbf{I}_2 \quad \text{and} \quad \mathbf{V}_2 = \mathbf{I}_1 + 2\mathbf{I}_2$$

The above equations are identical with the terminal characteristics obtained for the network of Fig. 13-18. Thus, the two networks are equivalent.

13.5 Find the **Y**-parameters of Fig. 13-19 using its **Z**-parameters.

From Problem 13.4,

$$\mathbf{Z}_{11} = 5, \ \mathbf{Z}_{12} = -2, \ \mathbf{Z}_{21} = 1, \ \mathbf{Z}_{22} = 2$$

Since $\mathbf{D_{ZZ}} = \mathbf{Z}_{11}\mathbf{Z}_{22} - \mathbf{Z}_{12}\mathbf{Z}_{21} = (5)(2) - (-2)(1) = 12$,

$$\mathbf{Y}_{11} = \frac{\mathbf{Z}_{22}}{\mathbf{D_{ZZ}}} = \frac{2}{12} = \frac{1}{6} \qquad \mathbf{Y}_{12} = \frac{-\mathbf{Z}_{12}}{\mathbf{D_{ZZ}}} = \frac{2}{12} = \frac{1}{6} \qquad \mathbf{Y}_{21} = \frac{-\mathbf{Z}_{21}}{\mathbf{D_{ZZ}}} = \frac{-1}{12} \qquad \mathbf{Y}_{22} = \frac{\mathbf{Z}_{11}}{\mathbf{D_{ZZ}}} = \frac{5}{12}$$

13.6 Find the **Y**-parameters of the two-port network in Fig. 13-20 and thus show that the networks of Figs. 13-19 and 13-20 are equivalent.

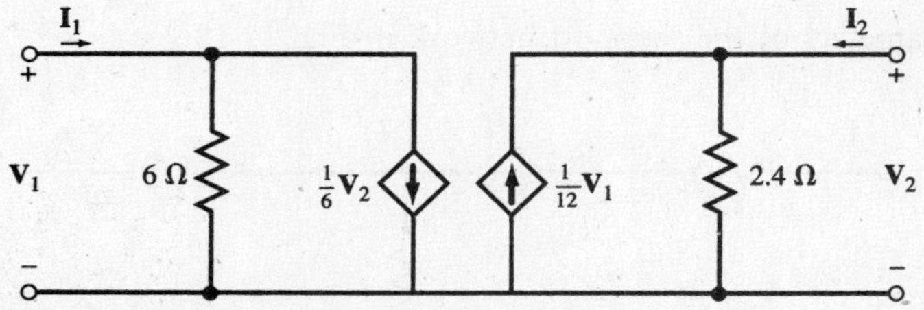

Fig. 13-20

Apply KCL at the ports to obtain the terminal characteristics and **Y**-parameters. Thus,

Input port: $$\mathbf{I}_1 = \frac{\mathbf{V}_1}{6} + \frac{\mathbf{V}_2}{6}$$

Output port: $$\mathbf{I}_2 = \frac{\mathbf{V}_2}{2.4} - \frac{\mathbf{V}_1}{12}$$

and $$\mathbf{Y}_{11} = \frac{1}{6} \qquad \mathbf{Y}_{12} = \frac{1}{6} \qquad \mathbf{Y}_{21} = \frac{-1}{12} \qquad \mathbf{Y}_{22} = \frac{1}{2.4} = \frac{5}{12}$$

which are identical with the **Y**-parameters obtained in Problem 3.5 for Fig. 13-19. Thus, the two networks are equivalent.

13.7 Apply the short-circuit equations (*10*) to find the **Y**-parameters of the two-port network in Fig. 13-21.

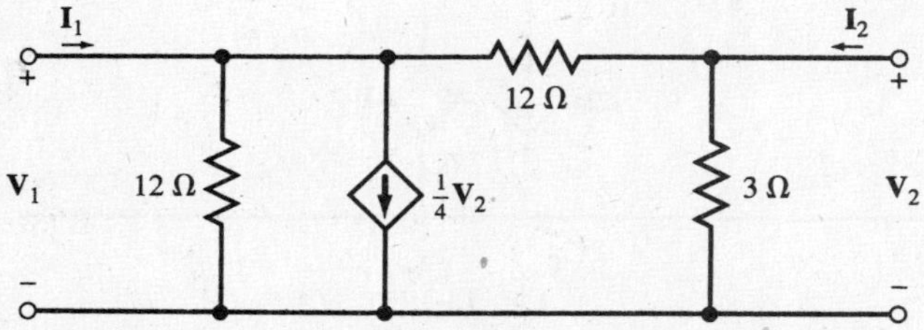

Fig. 13-21

$$\mathbf{I}_1 = \mathbf{Y}_{11}\mathbf{V}_1|_{\mathbf{V}_2=0} = \left(\frac{1}{12} + \frac{1}{12}\right)\mathbf{V}_1 \quad \text{or} \quad \mathbf{Y}_{11} = \frac{1}{6}$$

$$\mathbf{I}_1 = \mathbf{Y}_{12}\mathbf{V}_2|_{\mathbf{V}_1=0} = \frac{\mathbf{V}_2}{4} - \frac{\mathbf{V}_2}{12} = \left(\frac{1}{4} - \frac{1}{12}\right)\mathbf{V}_2 \quad \text{or} \quad \mathbf{Y}_{12} = \frac{1}{6}$$

$$\mathbf{I}_2 = \mathbf{Y}_{21}\mathbf{V}_1|_{\mathbf{V}_2=0} = -\frac{\mathbf{V}_1}{12} \quad \text{or} \quad \mathbf{Y}_{21} = -\frac{1}{12}$$

$$\mathbf{I}_2 = \mathbf{Y}_{22}\mathbf{V}_2|_{\mathbf{V}_1=0} = \frac{\mathbf{V}_2}{3} + \frac{\mathbf{V}_2}{12} = \left(\frac{1}{3} + \frac{1}{12}\right)\mathbf{V}_2 \quad \text{or} \quad \mathbf{Y}_{22} = \frac{5}{12}$$

13.8 Apply KCL at the nodes of the network in Fig. 13-21 to obtain its terminal characteristics and **Y**-parameters. Show that two-port networks of Figs. 13-18 to 13-21 are all equivalent.

Input node: $$\mathbf{I}_1 = \frac{\mathbf{V}_1}{12} + \frac{\mathbf{V}_1 - \mathbf{V}_2}{12} + \frac{\mathbf{V}_2}{4}$$

Output node: $$\mathbf{I}_2 = \frac{\mathbf{V}_2}{3} + \frac{\mathbf{V}_2 - \mathbf{V}_1}{12}$$

$$\mathbf{I}_1 = \frac{1}{6}\mathbf{V}_1 + \frac{1}{6}\mathbf{V}_2 \qquad \mathbf{I}_2 = -\frac{1}{12}\mathbf{V}_1 + \frac{5}{12}\mathbf{V}_2$$

The **Y**-parameters observed from the above characteristic equations are identical with the **Y**-parameters of the circuits in Figs. 13-18, 13-19, and 13-20. Therefore, the four circuits are equivalent.

13.9 **Z**-parameters of the two-port network N in Fig. 13-22(a) are $\mathbf{Z}_{11} = 4\mathbf{s}$, $\mathbf{Z}_{12} = \mathbf{Z}_{21} = 3\mathbf{s}$, and $\mathbf{Z}_{22} = 9\mathbf{s}$. ($a$) Replace N by its T-equivalent. (b) Use part (a) to find input current i_1 for $v_s = \cos 1000t$ (V).

(a) The network is reciprocal. Therefore, its T-equivalent exists. Its elements are found from (6) and shown in the circuit of Fig. 13-22(b).

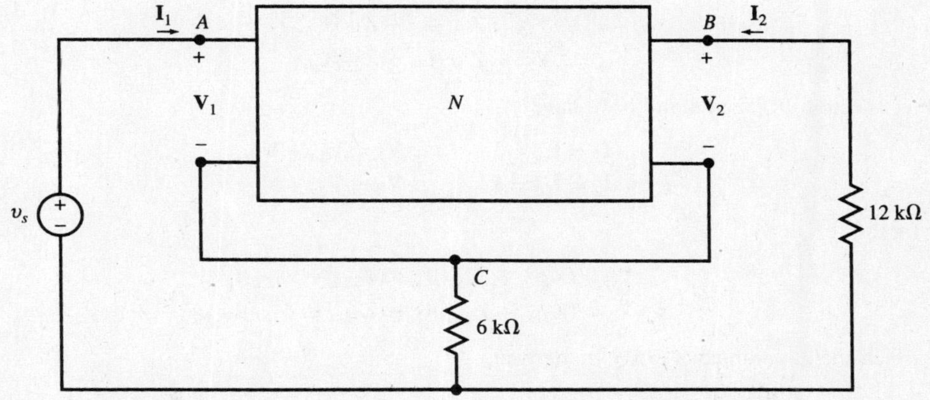

(a)

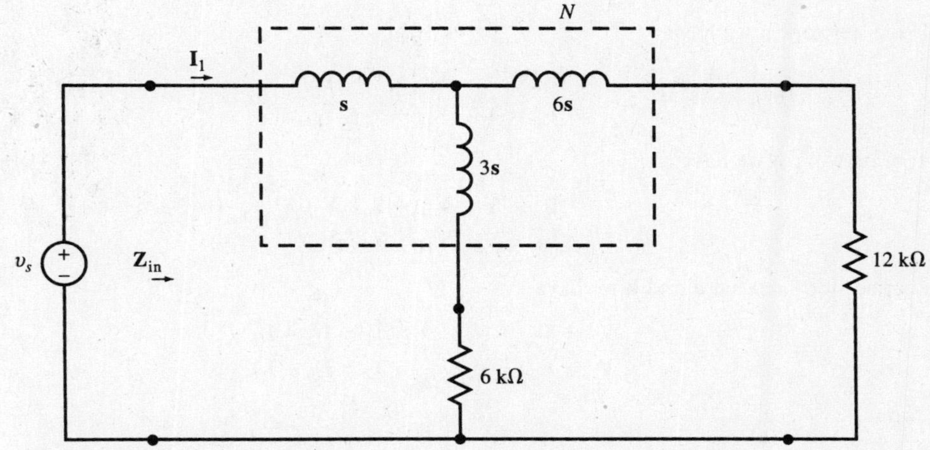

(b)

Fig. 13-22

$$\mathbf{Z}_a = \mathbf{Z}_{11} - \mathbf{Z}_{12} = 4\mathbf{s} - 3\mathbf{s} = \mathbf{s}$$
$$\mathbf{Z}_b = \mathbf{Z}_{22} - \mathbf{Z}_{21} = 9\mathbf{s} - 3\mathbf{s} = 6\mathbf{s}$$
$$\mathbf{Z}_c = \mathbf{Z}_{12} = \mathbf{Z}_{21} = 3\mathbf{s}$$

(*b*) We repeatedly combine the series and parallel elements of Fig. 13-22(*b*), with resistors being in kΩ and **s** in krad/s, to find $\mathbf{Z}_{\text{in}}$ in kΩ as shown in the following.

$$\mathbf{Z}_{\text{in}}(\mathbf{s}) = \mathbf{V}_s/\mathbf{I}_1 = \mathbf{s} + \frac{(3\mathbf{s}+6)(6\mathbf{s}+12)}{9\mathbf{s}+18} = 3\mathbf{s}+4 \qquad \text{or} \qquad \mathbf{Z}_{\text{in}}(j) = 3j + 4 = 5\,\underline{/36.9^\circ} \quad \text{k}\Omega$$

and $i_1 = 0.2\cos(1000t - 36.9^\circ)$ (mA).

13.10 Two two-port networks **a** and **b**, with open-circuit impedances $\mathbf{Z}_a$ and $\mathbf{Z}_b$, are connected in series (see Fig. 13-12). Derive the **Z**-parameters equations (*31a*).

From network **a** we have

$$\mathbf{V}_{1a} = \mathbf{Z}_{11,a}\mathbf{I}_{1a} + \mathbf{Z}_{12,a}\mathbf{I}_{2a}$$
$$\mathbf{V}_{2a} = \mathbf{Z}_{21,a}\mathbf{I}_{1a} + \mathbf{Z}_{22,a}\mathbf{I}_{2a}$$

From network **b** we have

$$\mathbf{V}_{1b} = \mathbf{Z}_{11,b}\mathbf{I}_{1b} + \mathbf{Z}_{12,b}\mathbf{I}_{2b}$$
$$\mathbf{V}_{2b} = \mathbf{Z}_{21,b}\mathbf{I}_{1b} + \mathbf{Z}_{22,b}\mathbf{I}_{2b}$$

From connection between **a** and **b** we have

$$\mathbf{I}_1 = \mathbf{I}_{1a} = \mathbf{I}_{1b} \qquad \mathbf{V}_1 = \mathbf{V}_{1a} + \mathbf{V}_{1b}$$
$$\mathbf{I}_2 = \mathbf{I}_{2a} = \mathbf{I}_{2b} \qquad \mathbf{V}_2 = \mathbf{V}_{2a} + \mathbf{V}_{2b}$$

Therefore,

$$\mathbf{V}_1 = (\mathbf{Z}_{11,a} + \mathbf{Z}_{11,b})\mathbf{I}_1 + (\mathbf{Z}_{12,a} + \mathbf{Z}_{12,b})\mathbf{I}_2$$
$$\mathbf{V}_2 = (\mathbf{Z}_{21,a} + \mathbf{Z}_{21,b})\mathbf{I}_1 + (\mathbf{Z}_{22,a} + \mathbf{Z}_{22,b})\mathbf{I}_2$$

from which the **Z**-parameters (*31a*) are derived.

13.11 Two two-port networks **a** and **b**, with short-circuit admittances $\mathbf{Y}_a$ and $\mathbf{Y}_b$, are connected in parallel (see Fig. 13-13). Derive the **Y**-parameters equations (*32a*).

From network **a** we have

$$\mathbf{I}_{1a} = \mathbf{Y}_{11,a}\mathbf{V}_{1a} + \mathbf{Y}_{12,a}\mathbf{V}_{2a}$$
$$\mathbf{I}_{2a} = \mathbf{Y}_{21,a}\mathbf{V}_{1a} + \mathbf{Y}_{22,a}\mathbf{V}_{2a}$$

and from network **b** we have

$$\mathbf{I}_{1b} = \mathbf{Y}_{11,b}\mathbf{V}_{1b} + \mathbf{Y}_{12,b}\mathbf{V}_{2b}$$
$$\mathbf{I}_{2b} = \mathbf{Y}_{21,b}\mathbf{V}_{1b} + \mathbf{Y}_{22,b}\mathbf{V}_{2b}$$

From connection between **a** and **b** we have

$$\mathbf{V}_1 = \mathbf{V}_{1a} = \mathbf{V}_{1b} \qquad \mathbf{I}_1 = \mathbf{I}_{1a} + \mathbf{I}_{1b}$$
$$\mathbf{V}_2 = \mathbf{V}_{2a} = \mathbf{V}_{2b} \qquad \mathbf{I}_2 = \mathbf{I}_{2a} + \mathbf{I}_{2b}$$

Therefore,

$$\mathbf{I}_1 = (\mathbf{Y}_{11,a} + \mathbf{Y}_{11,b})\mathbf{V}_1 + (\mathbf{Y}_{12,a} + \mathbf{Y}_{12,b})\mathbf{V}_2$$
$$\mathbf{I}_2 = (\mathbf{Y}_{21,a} + \mathbf{Y}_{21,b})\mathbf{V}_1 + (\mathbf{Y}_{22,a} + \mathbf{Y}_{22,b})\mathbf{V}_2$$

from which the **Y**-parameters of (*32a*) result.

13.12 Find (*a*) the **Z**-parameters of the circuit of Fig. 13-23(*a*) and (*b*) an equivalent model which uses three positive-valued resistors and one dependent voltage source.

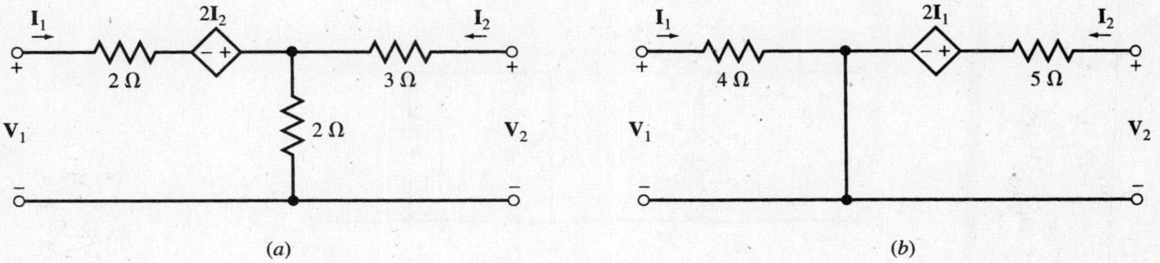

Fig. 13-23

(*a*) From application of KVL around the input and output loops we find, respectively,

$$\mathbf{V}_1 = 2\mathbf{I}_1 - 2\mathbf{I}_2 + 2(\mathbf{I}_1 + \mathbf{I}_2) = 4\mathbf{I}_1$$
$$\mathbf{V}_2 = 3\mathbf{I}_2 + 2(\mathbf{I}_1 + \mathbf{I}_2) = 2\mathbf{I}_1 + 5\mathbf{I}_2$$

The **Z**-parameters are $\mathbf{Z}_{11} = 4$, $\mathbf{Z}_{12} = 0$, $\mathbf{Z}_{21} = 2$, and $\mathbf{Z}_{22} = 5$.

(*b*) The circuit of Fig. 13-23(*b*), with two resistors and a voltage source, has the same **Z**-parameters as the circuit of Fig. 13-23(*a*). This can be verified by applying KVL to its input and output loops.

13.13 (*a*) Obtain the **Y**-parameters of the circuit in Fig. 13-23(*a*) from its **Z**-parameters. (*b*) Find an equivalent model which uses two positive-valued resistors and one dependent current source.

(*a*) From Problem 13.12, $\mathbf{Z}_{11} = 4$, $\mathbf{Z}_{12} = 0$, $\mathbf{Z}_{21} = 2$, $\mathbf{Z}_{22} = 5$, and so $\mathbf{D}_{\mathbf{ZZ}} = \mathbf{Z}_{11}\mathbf{Z}_{22} - \mathbf{Z}_{12}\mathbf{Z}_{21} = 20$. Hence,

$$\mathbf{Y}_{11} = \frac{\mathbf{Z}_{22}}{\mathbf{D}_{\mathbf{ZZ}}} = \frac{5}{20} = \frac{1}{4} \qquad \mathbf{Y}_{12} = \frac{-\mathbf{Z}_{12}}{\mathbf{D}_{\mathbf{ZZ}}} = 0 \qquad \mathbf{Y}_{21} = \frac{-\mathbf{Z}_{21}}{\mathbf{D}_{\mathbf{ZZ}}} = \frac{-2}{20} = -\frac{1}{10} \qquad \mathbf{Y}_{22} = \frac{\mathbf{Z}_{11}}{\mathbf{D}_{\mathbf{ZZ}}} = \frac{4}{20} = \frac{1}{5}$$

(*b*) Figure 13-24, with two resistors and a current source, has the same **Y**-parameters as the circuit in Fig. 13-23(*a*). This can be verified by applying KCL to the input and output nodes.

13.14 Referring to the network of Fig. 13-23(*b*), convert the voltage source and its series resistor to its Norton equivalent and show that the resulting network is identical with that in Fig. 13-24.

The Norton equivalent current source is $\mathbf{I}_N = 2I_1/5 = 0.4I_1$. But $I_1 = V_1/4$. Therefore, $I_N = 0.4I_1 = 0.1V_1$. The 5-Ω resistor is then placed in parallel with I_N. The circuit is shown in Fig. 13-25 which is the same as the circuit in Fig. 13-24.

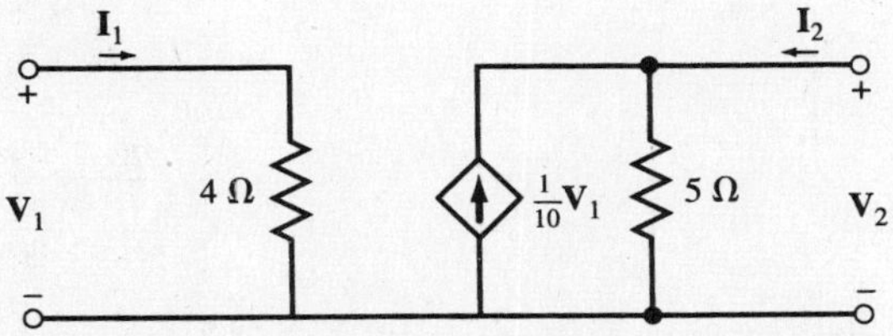

Fig. 13-24

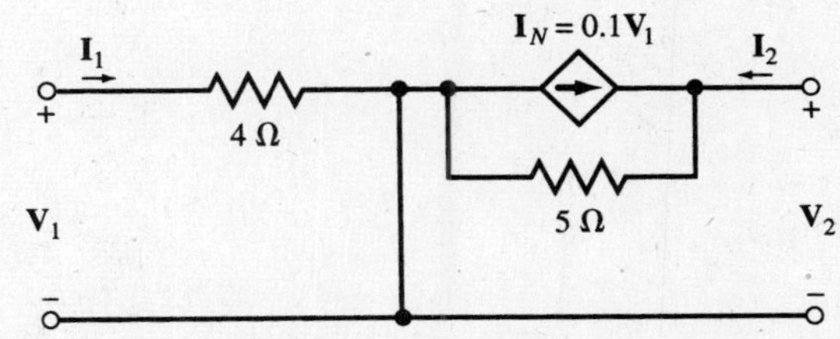

Fig. 13-25

13.15 The **h**-parameters of a two-port network are given. Show that the network may be modeled by the network in Fig. 13-26 where $\mathbf{h}_{11}$ is an impedance, $\mathbf{h}_{12}$ is a voltage gain, $\mathbf{h}_{21}$ is a current gain, and $\mathbf{h}_{22}$ is an admittance.

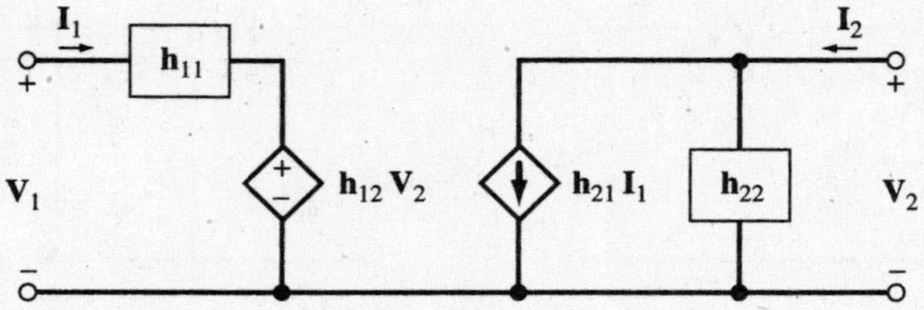

Fig. 13-26

Apply KVL around the input loop to get

$$\mathbf{V}_1 = \mathbf{h}_{11}\mathbf{I}_1 + \mathbf{h}_{12}\mathbf{V}_2$$

Apply KCL at the output node to get

$$\mathbf{I}_2 = \mathbf{h}_{21}\mathbf{I}_1 + \mathbf{h}_{22}\mathbf{V}_2$$

These results agree with the definition of **h**-parameters given in *(23)*.

13.16 Find the **h**-parameters of the circuit in Fig. 13-25.

By comparing the circuit in Fig. 13-25 with that in Fig. 13-26, we find

$$\mathbf{h}_{11} = 4\ \Omega, \qquad \mathbf{h}_{12} = 0, \qquad \mathbf{h}_{21} = -0.4, \qquad \mathbf{h}_{22} = 1/5 = 0.2\ \ \Omega^{-1}$$

13.17 Find the **h**-parameters of the circuit in Fig. 13-25 from its **Z**-parameters and compare with results of Problem 13.16.

Refer to Problem 13.13 for the values of the **Z**-parameters and $\mathbf{D}_{ZZ}$. Use Table 13-1 for the conversion of the **Z**-parameters to the **h**-parameters of the circuit. Thus,

$$\mathbf{h}_{11} = \frac{\mathbf{D}_{ZZ}}{\mathbf{Z}_{22}} = \frac{20}{5} = 4 \qquad \mathbf{h}_{12} = \frac{\mathbf{Z}_{12}}{\mathbf{Z}_{22}} = 0 \qquad \mathbf{h}_{21} = \frac{-\mathbf{Z}_{21}}{\mathbf{Z}_{22}} = \frac{-2}{5} = -0.4 \qquad \mathbf{h}_{22} = \frac{1}{\mathbf{Z}_{22}} = \frac{1}{5} = 0.2$$

The above results agree with the results of Problem 13.16.

13.18 The simplified model of a bipolar junction transistor for small signals is shown in the circuit of Fig. 13-27. Find its **h**-parameters.

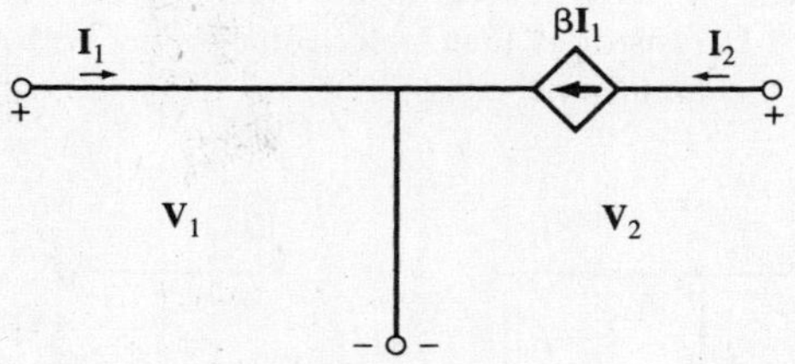

Fig. 13-27

The terminal equations are $\mathbf{V}_1 = 0$ and $\mathbf{I}_2 = \beta\mathbf{I}_1$. By comparing these equations with *(23)*, we conclude that $\mathbf{h}_{11} = \mathbf{h}_{12} = \mathbf{h}_{22} = 0$ and $\mathbf{h}_{21} = \beta$.

13.19 **h**-parameters of a two-port device H are given by

$$\mathbf{h}_{11} = 500\ \Omega \qquad \mathbf{h}_{12} = 10^{-4} \qquad \mathbf{h}_{21} = 100 \qquad \mathbf{h}_{22} = 2(10^{-6})\ \Omega^{-1}$$

Draw a circuit model of the device made of two resistors and two dependent sources including the values of each element.

From comparison with Fig. 13-26, we draw the model of Fig. 13-28.

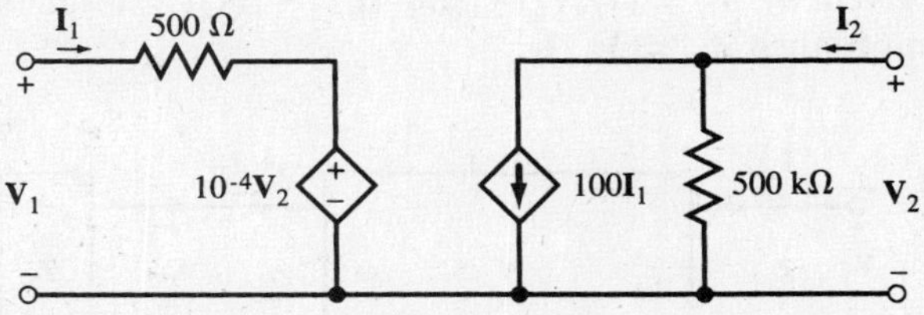

Fig. 13-28

13.20 The device H of Problem 13-19 is placed in the circuit of Fig. 13-29(*a*). Replace H by its model of Fig. 13-28 and find $\mathbf{V}_2/\mathbf{V}_s$.

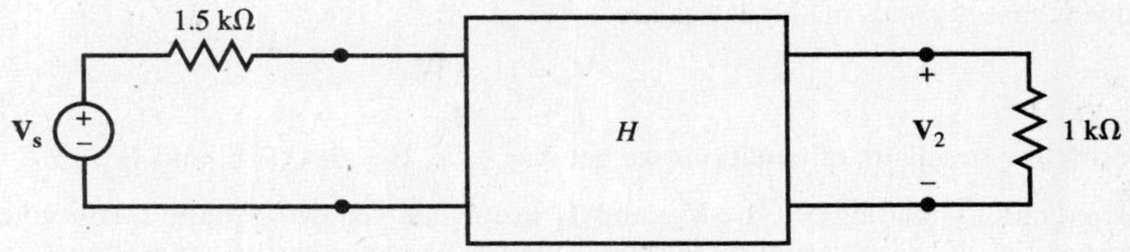

(*a*)

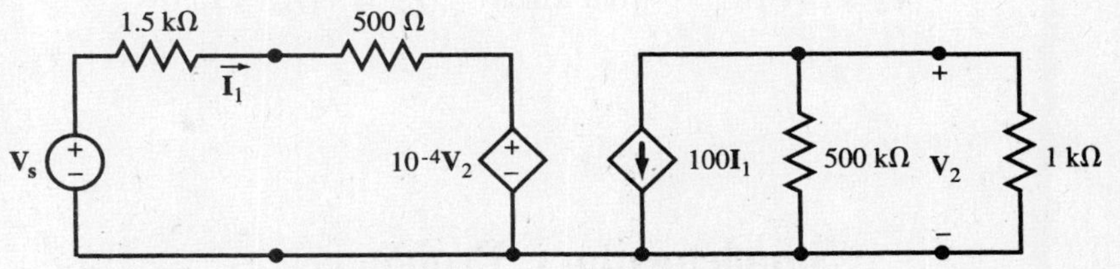

(*b*)

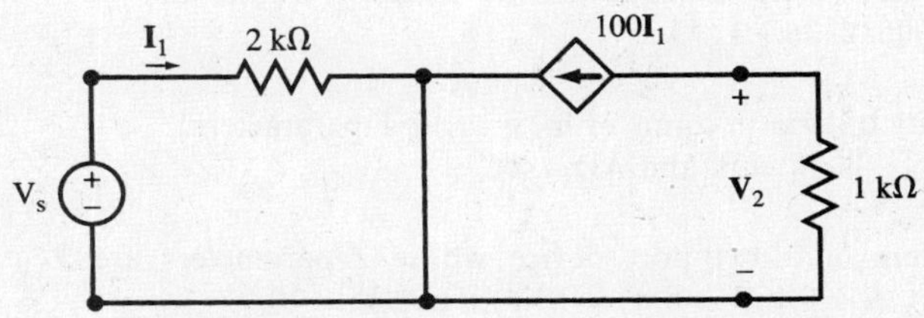

(*c*)

Fig. 13-29

The circuit of Fig. 13-29(*b*) contains the model. With good approximation, we can reduce it to Fig. 13-29(*c*) from which

$$\mathbf{I}_1 = \mathbf{V}_s/2000 \qquad \mathbf{V}_2 = -1000(100\mathbf{I}_1) = -1000(100\mathbf{V}_s/2000) = -50\mathbf{V}_s$$

Thus, $\mathbf{V}_2/\mathbf{V}_s = -50$.

13.21 A load $\mathbf{Z}_L$ is connected to the output of a two-port device N (Fig. 13-30) whose terminal characteristics are given by $\mathbf{V}_1 = (1/N)\mathbf{V}_2$ and $\mathbf{I}_1 = -N\mathbf{I}_2$. Find (*a*) the T-parameters of N and (*b*) the input impedance $\mathbf{Z}_{in} = \mathbf{V}_1/\mathbf{I}_1$.

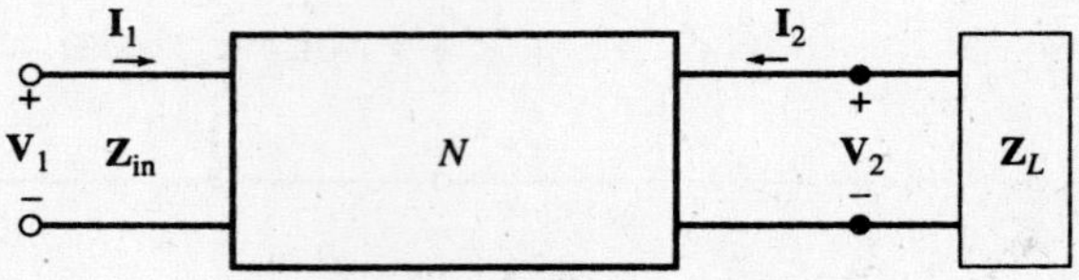

Fig. 13-30

(*a*) The T-parameters are defined by [see (*29*)]

$$\mathbf{V}_1 = \mathbf{A}\mathbf{V}_2 - \mathbf{B}\mathbf{I}_2$$
$$\mathbf{I}_1 = \mathbf{C}\mathbf{V}_2 - \mathbf{D}\mathbf{I}_2$$

The terminal characteristics of the device are

$$\mathbf{V}_1 = (1/N)\mathbf{V}_2$$
$$\mathbf{I}_1 = -N\mathbf{I}_2$$

By comparing the two pairs of equations we get $\mathbf{A} = 1/N$, $\mathbf{B} = 0$, $\mathbf{C} = 0$, and $\mathbf{D} = N$.

(*b*) Three equations relating $\mathbf{V}_1$, $\mathbf{I}_1$, $\mathbf{V}_2$, and $\mathbf{I}_2$ are available: two equations are given by the terminal characteristics of the device and the third equation comes from the connection to the load,

$$\mathbf{V}_2 = -\mathbf{Z}_L\mathbf{I}_2$$

By eliminating $\mathbf{V}_2$ and $\mathbf{I}_2$ in these three equations, we get

$$\mathbf{V}_1 = \mathbf{Z}_L\mathbf{I}_1/N^2 \qquad \text{from which} \qquad \mathbf{Z}_{in} = \mathbf{V}_1/\mathbf{I}_1 = \mathbf{Z}_L/N^2$$

Supplementary Problems

13.22 The **Z**-parameters of the two-port network N in Fig. 13-22(*a*) are $\mathbf{Z}_{11} = 4\mathbf{s}$, $\mathbf{Z}_{12} = \mathbf{Z}_{21} = 3\mathbf{s}$, and $\mathbf{Z}_{22} = 9\mathbf{s}$. Find the input current i_1 for $v_s = \cos 1000t$ (V) by using the open circuit impedance terminal characteristic equations of N, together with KCL equations at nodes A, B, and C.
Ans. $i_1 = 0.2\cos(1000t - 36.9°)$ (A)

13.23 Express the reciprocity criteria in terms of **h**-, **g**-, and T-parameters.
Ans. $\mathbf{h}_{12} + \mathbf{h}_{21} = 0$, $\mathbf{g}_{12} + \mathbf{g}_{21} = 0$, and $\mathbf{AD} - \mathbf{BC} = 1$

13.24 Find the T-parameters of a two-port device whose **Z**-parameters are $\mathbf{Z}_{11} = \mathbf{s}$, $\mathbf{Z}_{12} = \mathbf{Z}_{21} = 10\mathbf{s}$, and $\mathbf{Z}_{22} = 100\mathbf{s}$. *Ans.* $\mathbf{A} = 0.1$, $\mathbf{B} = 0$, $\mathbf{C} = 10^{-1}/\mathbf{s}$, and $\mathbf{D} = 10$

13.25 Find the T-parameters of a two-port device whose **Z**-parameters are $\mathbf{Z}_{11} = 10^6\mathbf{s}$, $\mathbf{Z}_{12} = \mathbf{Z}_{21} = 10^7\mathbf{s}$, and $\mathbf{Z}_{22} = 10^8\mathbf{s}$. Compare with the results of Problem 13.21.

Ans. $\mathbf{A} = 0.1, \mathbf{B} = 0, \mathbf{C} = 10^{-7}/\mathbf{s}$ and $\mathbf{D} = 10$. For high frequencies, the device is similar to the device of Problem 13.21, with $N = 10$.

13.26 The **Z**-parameters of a two-port device N are $\mathbf{Z}_{11} = k\mathbf{s}$, $\mathbf{Z}_{12} = \mathbf{Z}_{21} = 10k\mathbf{s}$, and $\mathbf{Z}_{22} = 100k\mathbf{s}$. A 1-Ω resistor is connected across the output port (Fig. 13-30). (*a*) Find the input impedance $\mathbf{Z}_{\text{in}} = \mathbf{V}_1/\mathbf{I}_1$ and construct its equivalent circuit. (*b*) Give the values of the elements for $k = 1$ and 10^6.

Ans. (*a*) $\mathbf{Z}_{\text{in}} = \dfrac{k\mathbf{s}}{1 + 100k\mathbf{s}} = \dfrac{1}{100 + 1/k\mathbf{s}}$

The equivalent circuit is a parallel RL circuit with $R = 10^{-2}$ Ω and $L = 1$ kH.

(*b*) For $k = 1$, $R = \dfrac{1}{100}$ Ω and $L = 1$ H. For $k = 10^6$, $R = \dfrac{1}{100}$ Ω and $L = 10^6$ H

13.27 The device N in Fig. 13-30 is specified by its following **Z**-parameters: $\mathbf{Z}_{22} = N^2\mathbf{Z}_{11}$ and $\mathbf{Z}_{12} = \mathbf{Z}_{21} = \sqrt{\mathbf{Z}_{11}\mathbf{Z}_{22}} = N\mathbf{Z}_{11}$. Find $\mathbf{Z}_{\text{in}} = \mathbf{V}_1/\mathbf{I}_1$ when a load $\mathbf{Z}_L$ is connected to the output terminal. Show that if $\mathbf{Z}_{11} \gg \mathbf{Z}_L/N^2$ we have impedance scaling such that $\mathbf{Z}_{\text{in}} = \mathbf{Z}_L/N^2$.

Ans. $\mathbf{Z}_{\text{in}} = \dfrac{\mathbf{Z}_L}{N^2 + \mathbf{Z}_L/\mathbf{Z}_{11}}$. For $\mathbf{Z}_L \ll N^2\mathbf{Z}_{11}$, $\mathbf{Z}_{\text{in}} = \mathbf{Z}_L/N^2$

13.28 Find the **Z**-parameters in the circuit of Fig. 13-31. *Hint*: Use the series connection rule.
Ans. $\mathbf{Z}_{11} = \mathbf{Z}_{22} = \mathbf{s} + 3 + 1/\mathbf{s}$, $\mathbf{Z}_{12} = \mathbf{Z}_{21} = \mathbf{s} + 1$

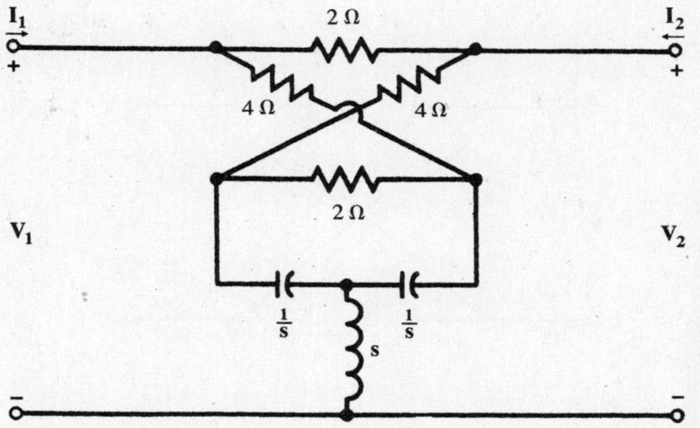

Fig. 13-31

13.29 Find the **Y**-parameters in the circuit of Fig. 13-32. *Hint*: Use the parallel connection rule.
Ans. $\mathbf{Y}_{11} = \mathbf{Y}_{22} = 9(\mathbf{s} + 2)/16$, $\mathbf{Y}_{12} = \mathbf{Y}_{21} = -3(\mathbf{s} + 2)/16$

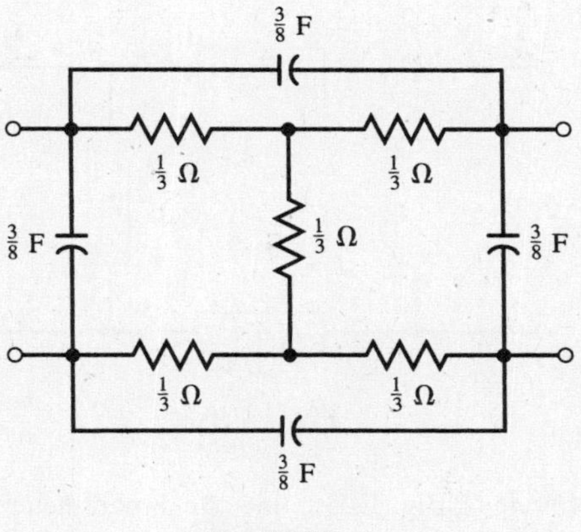

Fig. 13-32

13.30 Two two-port networks **a** and **b** with transmission parameters T_a and T_b are connected in cascade (Fig. 13-14). Given $\mathbf{I}_{2a} = -\mathbf{I}_{1b}$ and $\mathbf{V}_{2a} = \mathbf{V}_{1b}$, find the T-parameters of the resulting two-port network.
Ans. $\mathbf{A} = \mathbf{A}_a\mathbf{A}_b + \mathbf{B}_a\mathbf{C}_b$, $\mathbf{B} = \mathbf{A}_a\mathbf{B}_b + \mathbf{B}_a\mathbf{D}_b$, $\mathbf{C} = \mathbf{C}_a\mathbf{A}_b + \mathbf{D}_a\mathbf{C}_b$, $\mathbf{D} = \mathbf{C}_a\mathbf{B}_b + \mathbf{D}_a\mathbf{D}_b$

13.31 Find the T- and **Z**-parameters of the network in Fig. 13-33. The impedances of capacitors are given. *Hint*: Use the cascade connection rule.
Ans. $\mathbf{A} = 5j - 4$, $\mathbf{B} = 4j + 2$, $\mathbf{C} = 2j - 4$, and $\mathbf{D} = j3$, $\mathbf{Z}_{11} = 1.3 - 0.6j$, $\mathbf{Z}_{22} = 0.3 - 0.6j$, $\mathbf{Z}_{12} = \mathbf{Z}_{21} = -0.2 - 0.1j$

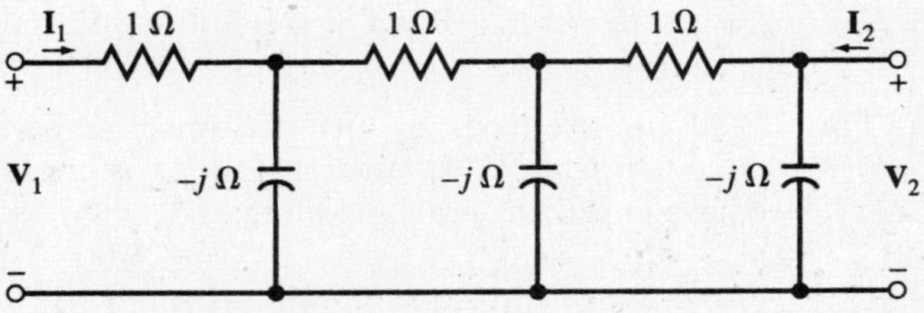

Fig. 13-33

13.32 Find the **Z**-parameters of the two-port circuit of Fig. 13-34.
Ans. $\mathbf{Z}_{11} = \mathbf{Z}_{22} = \frac{1}{2}(\mathbf{Z}_b + \mathbf{Z}_a)$, $\mathbf{Z}_{12} = \mathbf{Z}_{21} = \frac{1}{2}(\mathbf{Z}_b - \mathbf{Z}_a)$

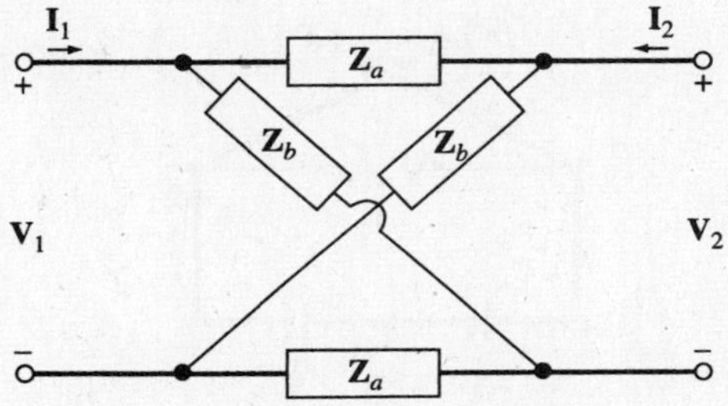

Fig. 13-34

13.33 Find the **Z**-parameters of the two-port circuit of Fig. 13-35.

Ans. $\mathbf{Z}_{11} = \mathbf{Z}_{22} = \frac{1}{2}\frac{\mathbf{Z}_b(2\mathbf{Z}_a + \mathbf{Z}_b)}{\mathbf{Z}_a + \mathbf{Z}_b}$, $\mathbf{Z}_{12} = \mathbf{Z}_{21} = \frac{1}{2}\frac{\mathbf{Z}_b^2}{\mathbf{Z}_a + \mathbf{Z}_b}$

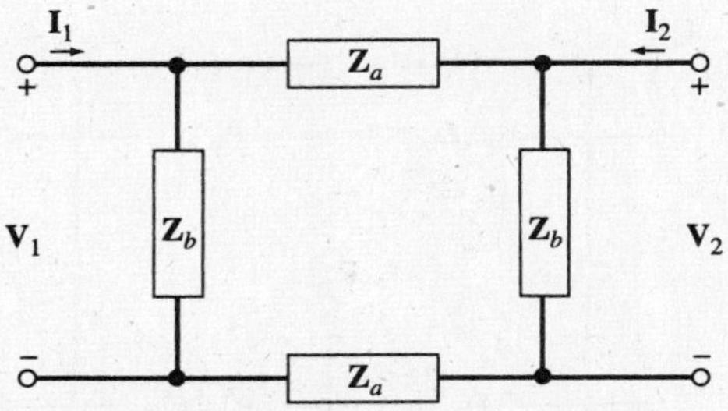

Fig. 13-35

13.34 Referring to the two-port circuit of Fig. 13-36, find the T-parameters as a function of ω and specify their values at $\omega = 1$, 10^3, and 10^6 rad/s.

Ans. $\mathbf{A} = 1 - 10^{-9}\omega^2 + j10^{-9}\omega$, $\mathbf{B} = 10^{-3}(1 + j\omega)$, $\mathbf{C} = 10^{-6}j\omega$, and $\mathbf{D} = 1$. At $\omega = 1$ rad/s, $\mathbf{A} = 1$, $\mathbf{B} = 10^{-3}(1 + j)$, $\mathbf{C} = 10^{-6}j$, and $\mathbf{D} = 1$. At $\omega = 10^3$ rad/s, $\mathbf{A} \approx 1$, $\mathbf{B} \approx j$, $\mathbf{C} = 10^{-3}j$, and $D = 1$. At $\omega = 10^6$ rad/s, $\mathbf{A} \approx -10^3$, $\mathbf{B} \approx 10^3 j$, $\mathbf{C} = j$, and $\mathbf{D} = 1$

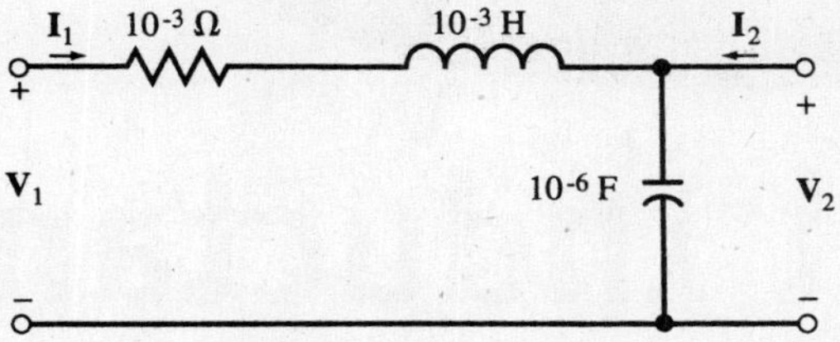

Fig. 13-36

13.35 A two-port network contains resistors, capacitors, and inductors only. With port #2 open [Fig. 13-37(*a*)], a unit step voltage $v_1 = u(t)$ produces $i_1 = e^{-t}u(t)$ (μA) and $v_2 = (1 - e^{-t})u(t)$ (V). With port #2 short-circuited [Fig. 13-37(*b*)], a unit step voltage $v_1 = u(t)$ delivers a current $i_1 = 0.5(1 + e^{-2t})u(t)$ (μA). Find i_2 and the T-equivalent network. *Ans.* $i_2 = 0.5(-1 + e^{-2t})u(t)$ [see Fig. 13-37(*c*)]

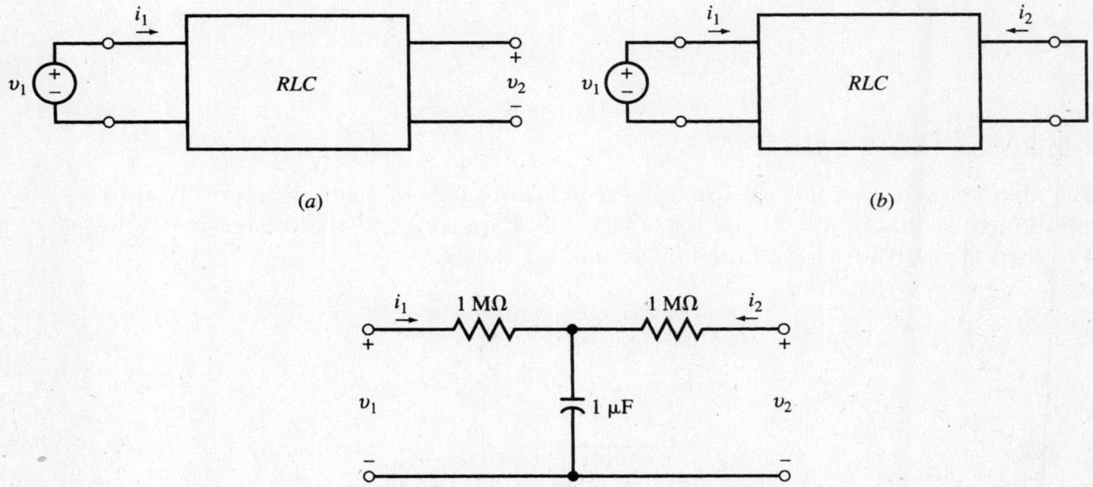

Fig. 13-37

13.36 The two-port network N in Fig. 13-38 is specified by $Z_{11} = 2$, $Z_{12} = Z_{21} = 1$, and $Z_{22} = 4$. Find I_1, I_2, and I_3. *Ans.* $I_1 = 24$ A, $I_2 = 1.5$ A, and $I_3 = 6.5$ A

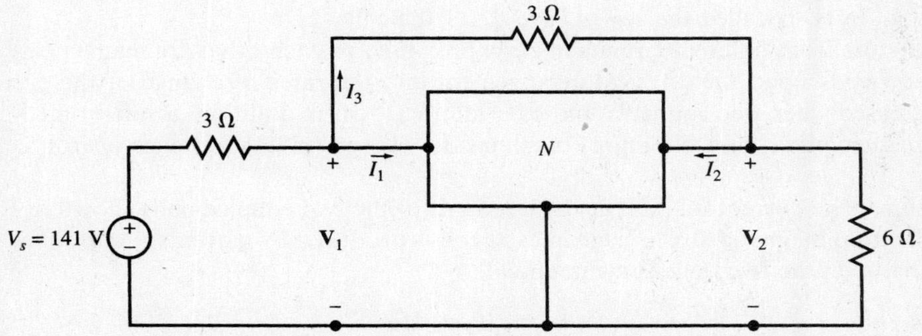

Fig. 13-38

CHAPTER 14

Mutual Inductance and Transformers

14.1 MUTUAL INDUCTANCE

The *total magnetic flux linkage* λ in a linear inductor made of a coil is proportional to the current passing through it; that is, $\lambda = Li$ (see Fig. 14-1). By Faraday's law, the voltage across the inductor is equal to the time derivative of the total influx linkage; that is,

$$v = \frac{d\lambda}{dt} = L\frac{di}{dt}$$

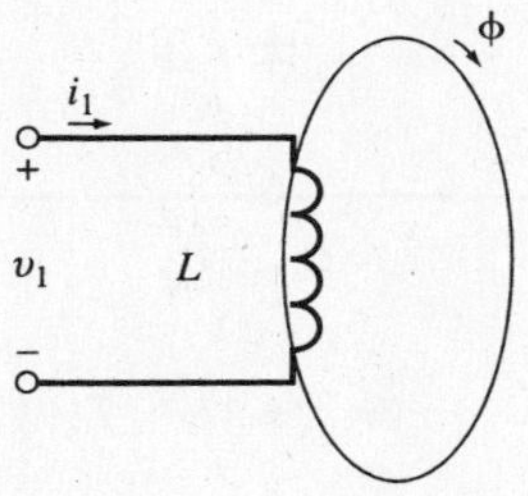

Fig. 14-1

The coefficient L, in H, is called the *self-inductance* of the coil.

Two conductors from different circuits in close proximity to each other are magnetically coupled to a degree that depends upon the physical arrangement and the rates of change of the currents. This coupling is increased when one coil is wound over another. If, in addition, a soft-iron core provides a path for the magnetic flux, the coupling is maximized. (However, the presence of iron can introduce nonlinearity.)

To find the voltage-current relation at the terminals of the two coupled coils shown in Fig. 14-2, we observe that the total magnetic flux linkage in each coil is produced by currents i_1 and i_2 and the mutual linkage effect between the two coils is symmetrical.

$$\begin{aligned} \lambda_1 &= L_1 i_1 + M i_2 \\ \lambda_2 &= M i_1 + L_2 i_2 \end{aligned} \tag{1}$$

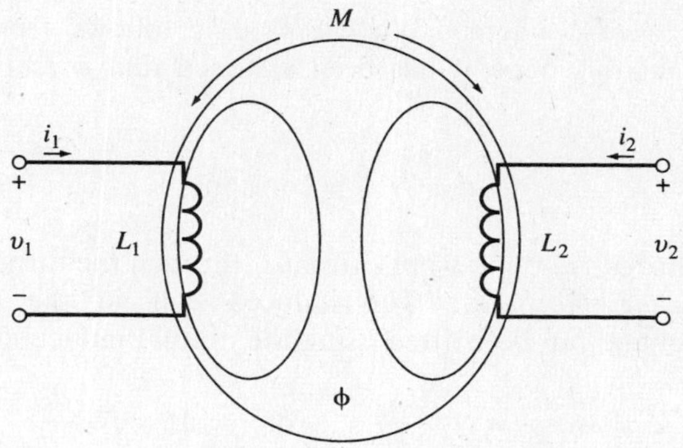

Fig. 14-2

where M is the *mutual inductance* (in H).

The terminal voltages are time derivatives of the flux linkages.

$$\begin{aligned} v_1(t) &= \frac{d\lambda_1}{dt} = L_1\frac{di_1}{dt} + M\frac{di_2}{dt} \\ v_2(t) &= \frac{d\lambda_2}{dt} = M\frac{di_1}{dt} + L_2\frac{di_2}{dt} \end{aligned} \tag{2}$$

The coupled coils constitute a special case of a two-port network discussed in Chapter 13. The terminal characteristics (*2*) may also be expressed in the frequency domain or in the **s**-domain as follows.

Frequency Domain		**s**-*Domain*	
$\mathbf{V}_1 = j\omega L_1\mathbf{I}_1 + j\omega M\mathbf{I}_2$ $\mathbf{V}_2 = j\omega M\mathbf{I}_1 + j\omega L_2\mathbf{I}_2$	(*3*)	$\mathbf{V}_1 = L_1\mathbf{s}\mathbf{I}_1 + M\mathbf{s}\mathbf{I}_2$ $\mathbf{V}_2 = M\mathbf{s}\mathbf{I}_1 + L_2\mathbf{s}\mathbf{I}_2$	(*4*)

The coupling coefficient M is discussed in Section 14.2. The frequency domain equations (*3*) deal with the sinusoidal steady state. The **s**-domain equations (*4*) assume exponential sources with complex frequency **s**.

EXAMPLE 14.1 Given $L_1 = 0.1$ H, $L_2 = 0.5$ H, and $i_1(t) = i_2(t) = \sin\omega t$ in the coupled coils of Fig. 14-2. Find $v_1(t)$ and $v_2(t)$ for (*a*) $M = 0.01$ H, (*b*) $M = 0.2$ H, and (*c*) $M = -0.2$ H.

(*a*)
$$v_1(t) = 0.1\,\omega\cos\omega t + 0.01\,\omega\cos\omega t = 0.11\,\omega\cos\omega t \quad \text{(V)}$$
$$v_2(t) = 0.01\,\omega\cos\omega t + 0.5\,\omega\cos\omega t = 0.51\,\omega\cos\omega t \quad \text{(V)}$$

(*b*)
$$v_1(t) = 0.1\,\omega\cos\omega t + 0.2\,\omega\cos\omega t = 0.3\,\omega\cos\omega t \quad \text{(V)}$$
$$v_2(t) = 0.2\,\omega\cos\omega t + 0.5\,\omega\cos\omega t = 0.7\,\omega\cos\omega t \quad \text{(V)}$$

(*c*)
$$v_1(t) = 0.1\,\omega\cos\omega t - 0.2\,\omega\cos\omega t = -0.1\,\omega\cos\omega t \quad \text{(V)}$$
$$v_2(t) = -0.2\,\omega\cos\omega t + 0.5\,\omega\cos\omega t = 0.3\,\omega\cos\omega t \quad \text{(V)}$$

14.2 COUPLING COEFFICIENT

A coil containing N turns with magnetic flux ϕ linking each turn has total magnetic flux linkage $\lambda = N\phi$. By Faraday's law, the induced *emf* (voltage) in the coil is $e = d\lambda/dt = N(d\phi/dt)$. A negative sign is frequently included in this equation to signal that the voltage polarity is established according to Lenz's law. By definition of self-inductance this voltage is also given by $L(di/dt)$; hence,

$$L\frac{di}{dt} = N\frac{d\phi}{dt} \qquad \text{or} \qquad L = N\frac{d\phi}{di} \tag{5a}$$

The unit of ϕ being the *weber*, where $1\text{ Wb} = 1\text{ V}\cdot\text{s}$, it follows from the above relation that $1\text{ H} = 1\text{ Wb/A}$. Throughout this book it has been assumed that ϕ and i are proportional to each other, making

$$L = N\frac{\phi}{i} = \text{constant} \tag{5b}$$

In Fig. 14-3, the total flux ϕ_1 resulting from current i_1 through the turns N_1 consists of *leakage flux*, ϕ_{11}, and *coupling* or *linking flux*, ϕ_{12}. The induced emf in the coupled coil is given by $N_2(d\phi_{12}/dt)$. This same voltage can be written using the mutual inductance M:

$$e = M\frac{di_1}{dt} = N_2\frac{d\phi_{12}}{dt} \qquad \text{or} \qquad M = N_2\frac{d\phi_{12}}{di_1} \tag{6}$$

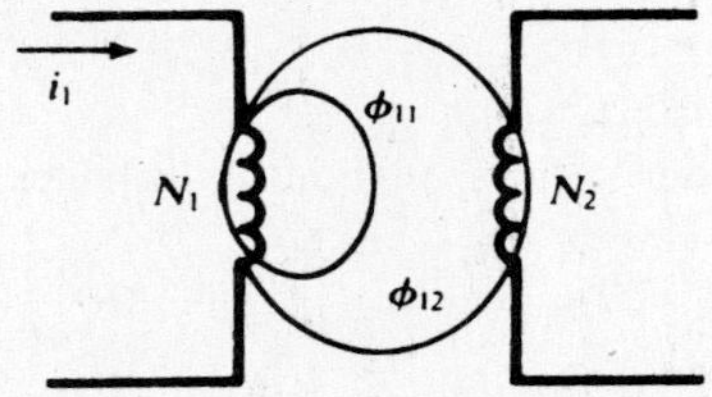

Fig. 14-3

Also, as the coupling is bilateral,

$$M = N_1\frac{d\phi_{21}}{di_2} \tag{7}$$

The *coupling coefficient*, k, is defined as the ratio of linking flux to total flux:

$$k \equiv \frac{\phi_{12}}{\phi_1} = \frac{\phi_{21}}{\phi_2}$$

where $0 \le k \le 1$. Taking the product of (*6*) and (*7*) and assuming that k depends only on the geometry of the system,

$$M^2 = \left(N_2\frac{d\phi_{12}}{di_1}\right)\left(N_1\frac{d\phi_{21}}{di_2}\right) = \left(N_2\frac{d(k\phi_1)}{di_1}\right)\left(N_1\frac{d(k\phi_2)}{di_2}\right) = k^2\left(N_1\frac{d\phi_1}{di_1}\right)\left(N_2\frac{d\phi_2}{di_2}\right) = k^2L_1L_2$$

from which

$$M = k\sqrt{L_1L_2} \qquad \text{or} \qquad X_M = k\sqrt{X_1X_2} \tag{8}$$

Note that (*8*) implies that $M \le \sqrt{L_1L_2}$, a bound that may be independently derived by an energy argument.

If all of the flux links the coils without any leakage flux, then $k = 1$. On the other extreme, the coil axes may be oriented such that no flux from one can induce a voltage in the other, which results in $k = 0$. The term *close coupling* is used to describe the case where most of the flux links the coils, either by way of a magnetic core to contain the flux or by interleaving the turns of the coils directly over one another. Coils placed side-by-side without a core are loosely coupled and have correspondingly low values of k.

14.3 ANALYSIS OF COUPLED COILS

Polarities in Close Coupling

In Fig. 14-4, two coils are shown on a common core which channels the magnetic flux ϕ. This arrangement results in *close coupling*, which was mentioned in Section 14.2. To determine the proper signs on the voltages of mutual inductance, apply the right-hand rule to each coil: If the fingers wrap

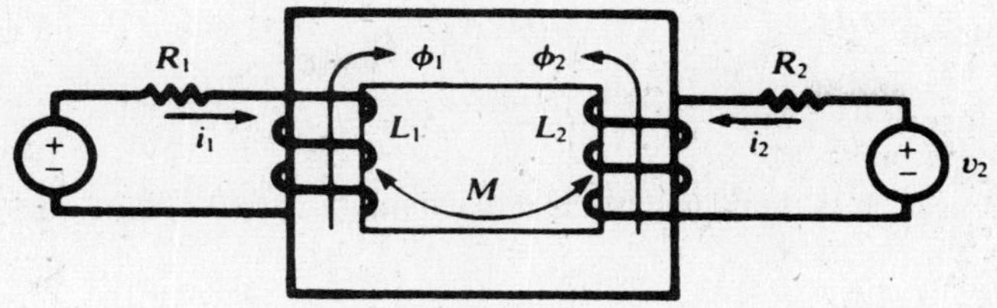

Fig. 14-4

around in the direction of the assumed current, the thumb points in the direction of the flux. Resulting positive directions for ϕ_1 and ϕ_2 are shown on the figure. If fluxes ϕ_1 and ϕ_2 aid one another, then the signs on the voltages of mutual inductance are the same as the signs on the voltages of self-inductance. Thus, the plus sign would be written in all four equations (*2*) and (*3*). In Fig. 14-4, ϕ_1 and ϕ_2 oppose each other; consequently, the equations (*2*) and (*3*) would be written with the minus sign.

Natural Current

Further understanding of coupled coils is achieved from consideration of a passive second loop as shown in Fig. 14-5. Source v_1 drives a current i_1, with a corresponding flux ϕ_1 as shown. Now *Lenz's law* implies that the polarity of the induced voltage in the second circuit is such that if the circuit is completed, a current will pass through the second coil in such a direction as to create a flux opposing the main flux established by i_1. That is, when the switch is closed in Fig. 14-5, flux ϕ_2 will have the direction shown. The right-hand rule, with the thumb pointing in the direction of ϕ_2, provides the direction of the *natural current* i_2. The induced voltage is the driving voltage for the second circuit, as suggested in Fig. 14-6; this voltage is present whether or not the circuit is closed. When the switch is closed, current i_2 is established, with a positive direction as shown.

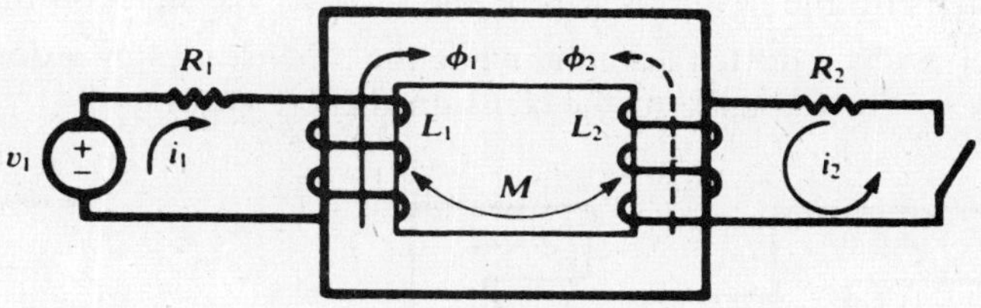

Fig. 14-5

EXAMPLE 14.2 Suppose the switch in the passive loop to be closed at an instant ($t = 0$) when $i_1 = 0$. For $t > 0$, the sequence of the passive loop is (see Fig. 14-6).

$$R_2 i_2 + L_2 \frac{di_2}{dt} - M \frac{di_1}{dt} = 0$$

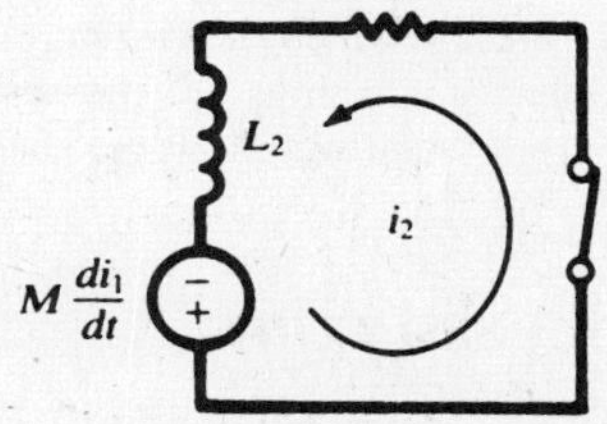

Fig. 14-6

while that of the active loop is

$$R_1 i_1 + L_1 \frac{di_1}{dt} - M \frac{di_2}{dt} = v_1$$

Writing the above equations in the s-domain with the initial conditions $i_1(0^+) = i_2(0^+) = 0$ and eliminating $\mathbf{I}_1(\mathbf{s})$, we find

$$\mathbf{H}(\mathbf{s}) \equiv \frac{\text{response}}{\text{excitation}} = \frac{\mathbf{I}_2(\mathbf{s})}{\mathbf{V}_1(\mathbf{s})} = \frac{M\mathbf{s}}{(L_1L_2 - M^2)\mathbf{s}^2 + (R_1L_2 + R_2L_1)\mathbf{s} + R_1R_2}$$

and from the poles of $\mathbf{H}(\mathbf{s})$ we have the natural frequencies of i_2.

14.4 DOT RULE

The sign on a voltage of mutual inductance can be determined if the winding sense is shown on the circuit diagram, as in Figs. 14-4 and 14-5. To simplify the problem of obtaining the correct sign, the coils are marked with dots at the terminals which are instantaneously of the same polarity.

To assign the dots to a pair of coupled coils, select a current direction in one coil and place a dot at the terminal where this current *enters* the winding. Determine the corresponding flux by application of the right-hand rule [see Fig. 14-7(*a*)]. The flux of the other winding, according to Lenz's law, opposes the first flux. Use the right-hand rule to find the natural current direction corresponding to this second flux [see Fig. 14-7(*b*)]. Now place a dot at the terminal of the second winding where the natural current *leaves* the winding. This terminal is positive simultaneously with the terminal of the first coil where the initial current entered. With the instantaneous polarity of the coupled coils given by the dots, the pictorial representation of the core with its winding sense is no longer needed, and the coupled coils may be illustrated as in Fig. 14-7(*c*). The following *dot rule* may now be used:

(1) when the assumed currents both enter or both leave a pair of coupled coils by the dotted terminals, the signs on the M-terms will be the same as the signs on the L-terms; but

(2) if one current enters by a dotted terminal while the other leaves by a dotted terminal, the signs on the M-terms will be opposite to the signs on the L-terms.

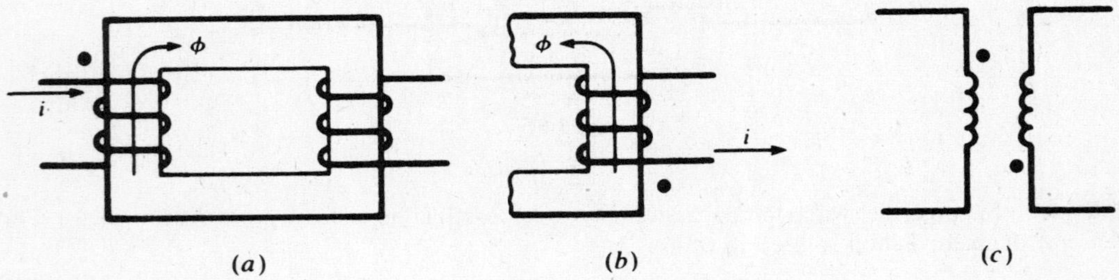

Fig. 14-7

EXAMPLE 14.3 The current directions chosen in Fig. 14-8(*a*) are such that the signs on the M-terms are opposite to the signs on the L-terms and the dots indicate the terminals with the same instantaneous polarity. Compare this to the conductively coupled circuit of Fig. 14-8(*b*), in which the two mesh currents pass through the common element in opposite directions, and in which the polarity markings are the same as the dots in the magnetically coupled circuit. The similarity becomes more apparent when we allow the shading to suggest two black boxes.

14.5 ENERGY IN A PAIR OF COUPLED COILS

The energy stored in a single inductor L carrying current i is $0.5Li^2$ J. The energy stored in a pair of coupled coils is given by

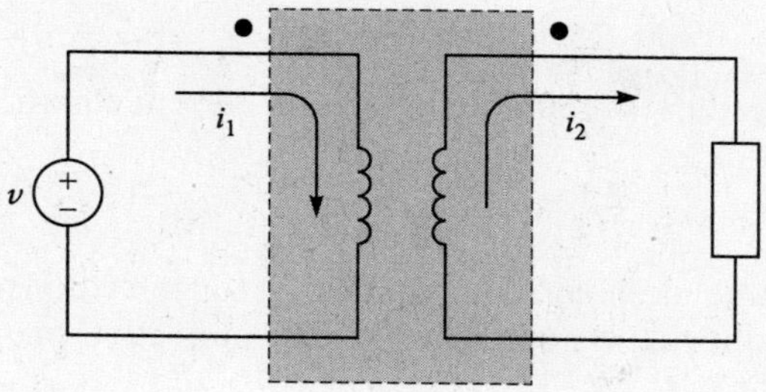

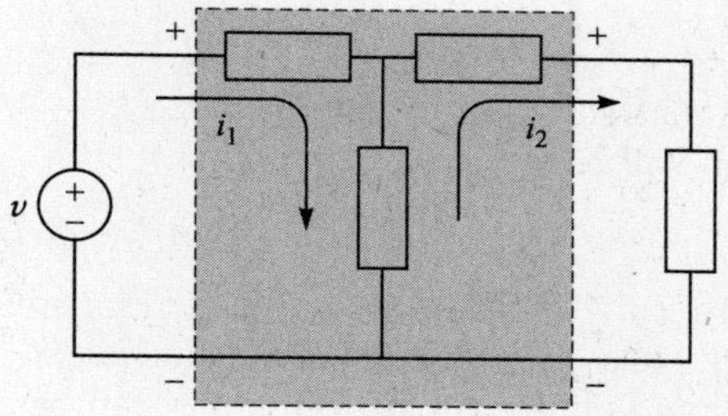

Fig. 14-8

$$W = \frac{1}{2}L_1 i_1^2 + \frac{1}{2}L_2 i_2^2 + M i_1 i_2 \quad \text{(J)} \tag{9}$$

where L_1 and L_2 are the inductances of the two coils and M is their mutual inductance. The term $Mi_1 i_2$ in *(9)* represents the energy due to the effect of the mutual inductance. The sign of this term is *(a)* positive if both currents i_1 and i_2 enter either at the dotted or undotted terminals, or *(b)* negative if one of the currents enters at the dotted terminal and the other enters the undotted end.

EXAMPLE 14.4 In a pair of coils, with $L_1 = 0.1$ H and $L_2 = 0.2$ H, at a certain moment, $i_1 = 4$ A and $i_2 = 10$ A. Find the total energy in the coils if the coupling coefficient M is *(a)* 0.1 H, *(b)* $\sqrt{2}/10$ H, *(c)* −0.1 H, and *(d)* $-\sqrt{2}/10$ H.

From (9),

(a) $W = (0.5)(0.1)4^2 + (0.5)(0.2)10^2 + (0.1)(10)(4) = 14.8$ J

(b) $W = 16.46$ J

(c) $W = 6.8$ J

(d) $W = 5.14$ J

The maximum and minimum energies occur in conjunction with perfect positive coupling ($M = \sqrt{2}/10$) and perfect negative coupling ($M = -\sqrt{2}/10$).

14.6 CONDUCTIVELY COUPLED EQUIVALENT CIRCUITS

From the mesh current equations written for magnetically coupled coils, a conductively coupled equivalent circuit can be constructed. Consider the sinusoidal steady-state circuit of Fig. 14-9*(a)*, with the mesh currents as shown. The corresponding equations in matrix form are

$$\begin{bmatrix} R_1 + j\omega L_1 & -j\omega M \\ -j\omega M & R_2 + j\omega L_2 \end{bmatrix} \begin{bmatrix} \mathbf{I}_1 \\ \mathbf{I}_2 \end{bmatrix} = \begin{bmatrix} \mathbf{V}_1 \\ 0 \end{bmatrix}$$

In Fig. 14-9*(b)*, an inductive reactance, $X_M = \omega M$, carries the two mesh currents in opposite directions, whence

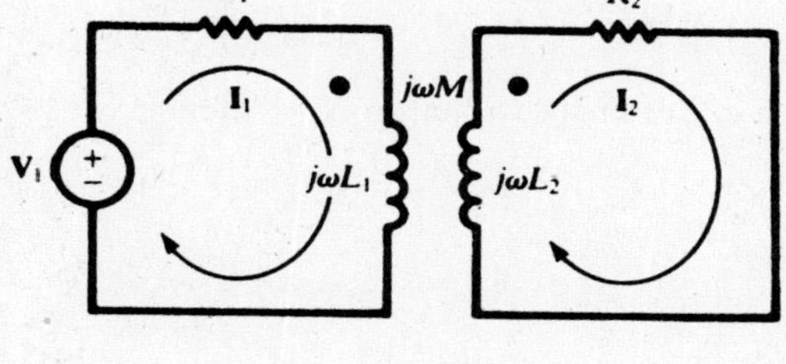

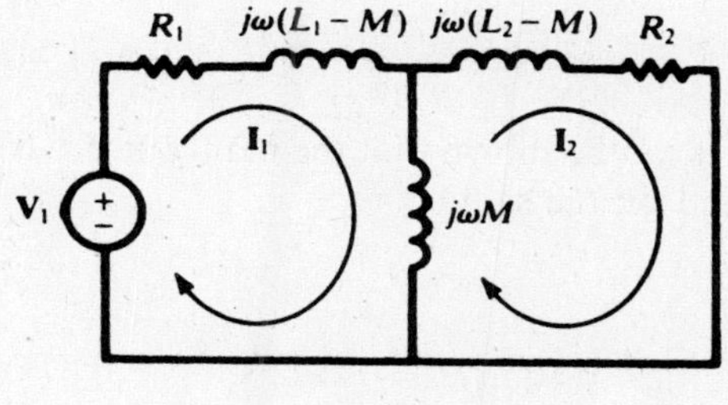

Fig. 14-9

$$\mathbf{Z}_{12} = \mathbf{Z}_{21} = -j\omega M$$

in the **Z**-matrix. If now an inductance $L_1 - M$ is placed in the first loop, the mesh current equation for this loop will be

$$(R_1 + j\omega L_1)\mathbf{I}_1 - j\omega M\mathbf{I}_2 = \mathbf{V}_1$$

Similarly, $L_2 - M$ in the second loop results in the same mesh current equation as for the coupled-coil circuit. Thus, the two circuits are equivalent. The dot rule is not needed in the conductively coupled circuit, and familiar circuit techniques can be applied.

14.7 LINEAR TRANSFORMER

A *transformer* is a device for introducing mutual coupling between two or more electric circuits. The term *iron-core transformer* identifies the coupled coils which are wound on a magnetic core of laminated specialty steel to confine the flux and maximize the coupling. *Air-core* transformers are found in electronic and communications applications. A third group consists of coils wound over one another on a nonmetallic form, with a movable slug of magnetic material within the center for varying the coupling.

Attention here is directed to iron-core transformers where the permeability μ of the iron is assumed to be constant over the operating range of voltage and current. The development is restricted to two-winding transformers, although three and more windings on the same core are not uncommon.

In Fig. 14-10, the *primary winding*, of N_1 turns, is connected to the source voltage $\mathbf{V}_1$, and the *secondary winding*, of N_2 turns, is connected to the load impedance $\mathbf{Z}_L$. The coil resistances are shown by lumped parameters R_1 and R_2. Natural current $\mathbf{I}_2$ produces flux $\phi_2 = \phi_{21} + \phi_{22}$, while $\mathbf{I}_1$ produces $\phi_1 = \phi_{12} + \phi_{11}$. In terms of the coupling coefficient k,

$$\phi_{11} = (1-k)\phi_1 \qquad \phi_{22} = (1-k)\phi_2$$

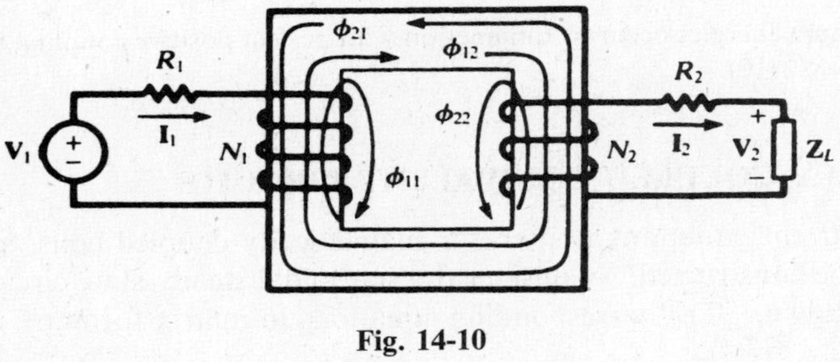

Fig. 14-10

From these flux relationships, *leakage inductances* can be related to the self-inductances:

$$L_{11} \equiv (1-k)L_1 \qquad L_{22} \equiv (1-k)L_2$$

The corresponding leakage reactances are:

$$X_{11} \equiv (1-k)X_1 \qquad X_{22} \equiv (1-k)X_2$$

It can be shown that the inductance L of an N-turn coil is proportional to N^2. Hence, for two coils wound on the same core,

$$\frac{L_1}{L_2} = \left(\frac{N_1}{N_2}\right)^2 \tag{10}$$

The flux common to both windings in Fig. 14-10 is the *mutual flux*, $\phi_m = \phi_{12} - \phi_{21}$. This flux induces the coil emfs by Faraday's law,

$$e_1 = N_1 \frac{d\phi_m}{dt} \qquad e_2 = N_2 \frac{d\phi_m}{dt}$$

Defining the *turns ratio*, $a \equiv N_1/N_2$, we obtain from these the basic equation of the linear transformer:

$$\frac{e_1}{e_2} = a \tag{11}$$

In the frequency domain, $\mathbf{E}_1/\mathbf{E}_2 = a$.

The relationship between the mutual flux and the mutual inductance can be developed by analysis of the secondary induced emf, as follows:

$$e_2 = N_2 \frac{d\phi_m}{dt} = N_2 \frac{d\phi_{12}}{dt} - N_2 \frac{d\phi_{21}}{dt} = N_2 \frac{d\phi_{12}}{dt} - N_2 \frac{d(k\phi_2)}{dt}$$

By use of (*6*) and (*5a*), this may be rewritten as

$$e_2 = M \frac{di_1}{dt} - kL_2 \frac{di_2}{dt} = M \frac{di_1}{dt} - \frac{M}{a} \frac{di_2}{dt}$$

where the last step involved (*8*) and (*10*):

$$M = k\sqrt{(a^2L_2)(L_2)} = kaL_2$$

Now, defining the *magnetizing current* i_ϕ by the equation

$$i_1 = \frac{i_2}{a} + i_\phi \qquad \text{or} \qquad \mathbf{I}_1 = \frac{\mathbf{I}_2}{a} + \mathbf{I}_\phi \tag{12}$$

we have

$$e_2 = M \frac{di_\phi}{dt} \qquad \text{or} \qquad \mathbf{E}_2 = jX_M \mathbf{I}_\phi \tag{13}$$

According to (*13*), the magnetizing current may be considered to set up the mutual flux ϕ_m in the core.

In terms of coil emfs and leakage reactances, an equivalent circuit for the linear transformer may be drawn, in which the primary and secondary are effectively decoupled. This is shown in Fig. 14-11(*a*); for comparison, the dotted equivalent circuit is shown in Fig. 14-11(*b*).

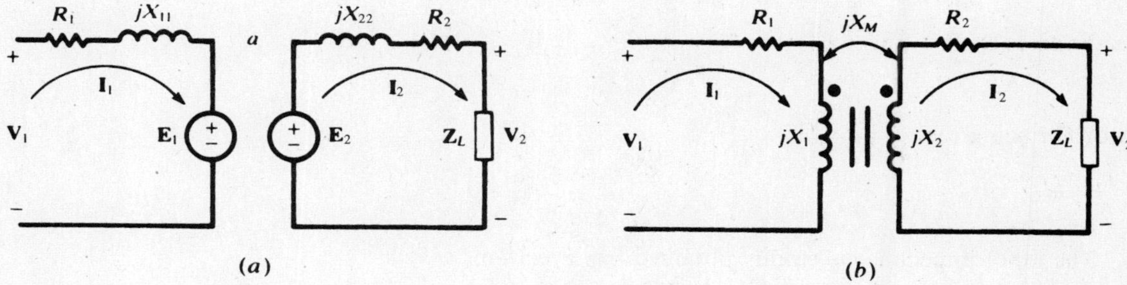

Fig. 14-11

EXAMPLE 14.5 Draw the voltage-current phasor diagram corresponding to Fig. 14-11(*a*), and from it derive the input impedance of the transformer.

The diagram is given in Fig. 14-12, in which θ_L denotes the phase angle of $\mathbf{Z}_L$. Note that, in accordance with (*13*), the induced emfs $\mathbf{E}_1$ and $\mathbf{E}_2$ lead the magnetizing current $\mathbf{I}_\phi$ by 90°. The diagram yields the three phasor equations

$$\mathbf{V}_1 = ajX_M\mathbf{I}_\phi + (R_1 + jX_{11})\mathbf{I}_1$$
$$jX_M\mathbf{I}_\phi = (\mathbf{Z}_L + R_2 + jX_{22})\mathbf{I}_2$$
$$\mathbf{I}_1 = \frac{1}{a}\mathbf{I}_2 + \mathbf{I}_\phi$$

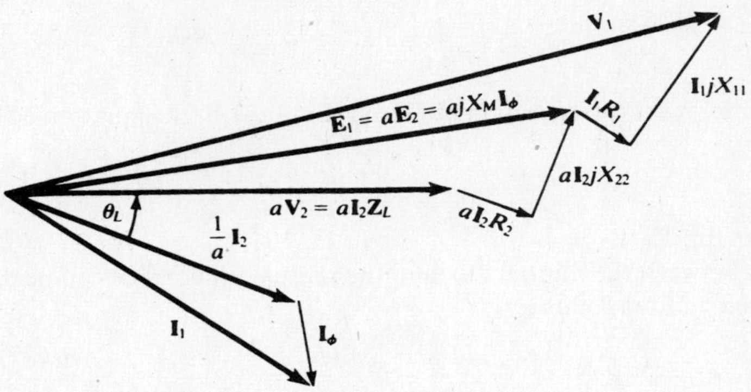

Fig. 14-12

Elimination of $\mathbf{I}_2$ and $\mathbf{I}_\phi$ among these equations results in

$$\frac{\mathbf{V}_1}{\mathbf{I}_1} \equiv \mathbf{Z}_{\text{in}} = (R_1 + jX_{11}) + a^2 \frac{(jX_M/a)(R_2 + jX_{22} + \mathbf{Z}_L)}{(jX_M/a) + (R_2 + jX_{22} + \mathbf{Z}_L)} \tag{14a}$$

If, instead, the mesh current equations for Fig. 14-11(*b*) are used to derive $\mathbf{Z}_{\text{in}}$, the result is

$$\mathbf{Z}_{\text{in}} = R_1 + jX_1 + \frac{X_M^2}{R_2 + jX_2 + \mathbf{Z}_L} \tag{14b}$$

The reader may verify the equivalence of (*14a*) and (*14b*)—see Problem 14.36.

14.8 IDEAL TRANSFORMER

An *ideal transformer* is a hypothetical transformer in which there are no losses and the core has infinite permeability, resulting in perfect coupling with no leakage flux. In large power transformers the losses are so small relative to the power transferred that the relationships obtained from the ideal transformer can be very useful in engineering applications.

Referring to Fig. 14-13, the lossless condition is expressed by

$$\tfrac{1}{2}\mathbf{V}_1\mathbf{I}_1^* = \tfrac{1}{2}\mathbf{V}_2\mathbf{I}_2^*$$

(see Section 10.7). But

$$\mathbf{V}_1 = \mathbf{E}_1 = a\mathbf{E}_2 = a\mathbf{V}_2$$

and so, a being real,

$$\frac{\mathbf{V}_1}{\mathbf{V}_2} = \frac{\mathbf{I}_2}{\mathbf{I}_1} = a \tag{15}$$

The input impedance is readily obtained from relations (*15*):

$$\mathbf{Z}_{\text{in}} = \frac{\mathbf{V}_1}{\mathbf{I}_1} = \frac{a\mathbf{V}_2}{\mathbf{I}_2/a} = a^2 \frac{\mathbf{V}_2}{\mathbf{I}_2} = a^2\mathbf{Z}_L \tag{16}$$

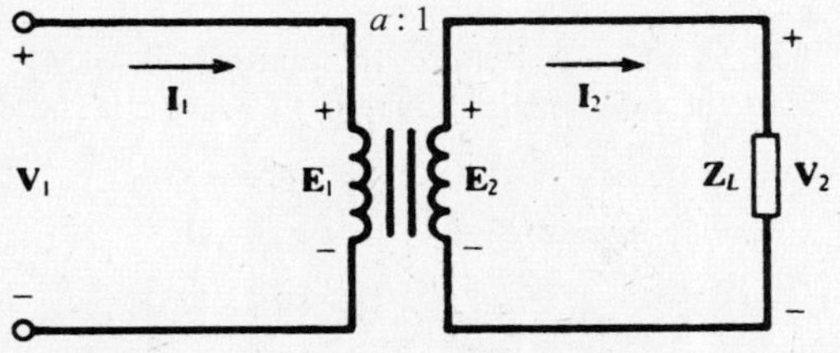

Fig. 14-13

EXAMPLE 14.6 The ideal transformer may be considered as the limiting case of the linear transformer of Section 14.7. Thus, in (*14a*) set

$$R_1 = R_2 = X_{11} = X_{22} = 0$$

(no losses) and then let $X_M \to \infty$ (infinite core permeability), to obtain

$$\mathbf{Z}_{\text{in}} = \lim_{X_M \to \infty} \left[a^2 \frac{(jX_M/a)(\mathbf{Z}_L)}{(jX_M/a) + \mathbf{Z}_L} \right] = a^2 \mathbf{Z}_L$$

in agreement with (*16*)

Ampere-Turn Dot Rule

Since $a = N_1/N_2$ in (*15*),

$$N_1 \mathbf{I}_1 = N_2 \mathbf{I}_2$$

that is, the *ampere turns* of the primary equal the *ampere turns* of the secondary. A rule can be formulated which extends this result to transformers having more than two windings. A positive sign is applied to an ampere-turn product if the current enters the winding by the dotted terminal; a negative sign is applied if the current leaves by the dotted terminal. The *ampere-turn dot rule* then states that the algebraic sum of the ampere-turns for a transformer is zero.

EXAMPLE 14.7 The three-winding transformer shown in Fig. 14-14 has turns $N_1 = 20$, $N_2 = N_3 = 10$. Find $\mathbf{I}_1$ given that $\mathbf{I}_2 = 10.0\angle{-53.13^\circ}$ A, $\mathbf{I}_3 = 10.0\angle{-45^\circ}$ A.

With the dots and current directions as shown on the diagram,

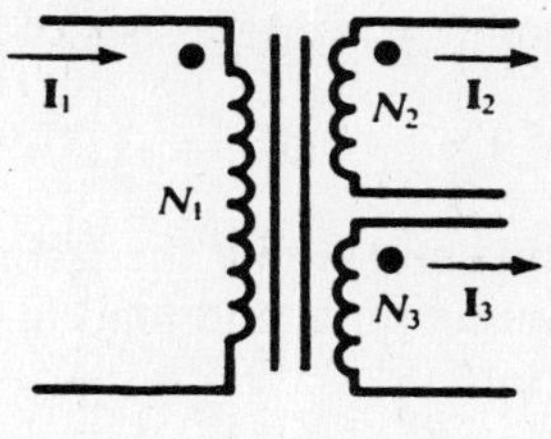

Fig. 14-14

$$N_1\mathbf{I}_1 - N_2\mathbf{I}_2 - N_3\mathbf{I}_3 = 0$$

from which

$$20\mathbf{I}_1 = 10(10.0\angle{-53.13^\circ}) + 10(10.0\angle{-45^\circ})$$
$$\mathbf{I}_1 = 6.54 - j7.54 = 9.98\angle{-49.06^\circ} \text{ A}$$

14.9 AUTOTRANSFORMER

An *autotransformer* is an electrically continuous winding, with one or more taps, on a magnetic core. One circuit is connected to the end terminals, while the other is connected to one end terminal and to a tap, part way along the winding.

Referring to Fig. 14-15(*a*), the transformation ratio is

$$\frac{\mathbf{V}_1}{\mathbf{V}_2} = \frac{N_1 + N_2}{N_2} \equiv a + 1$$

which exceeds by unity the transformation ratio of an ideal two-winding transformer having the same turns ratio. Current $\mathbf{I}_1$ through the upper or series part of the winding, of N_1 turns, produces the flux ϕ_1. By Lenz's law the natural current in the lower part of the winding produces an opposing flux

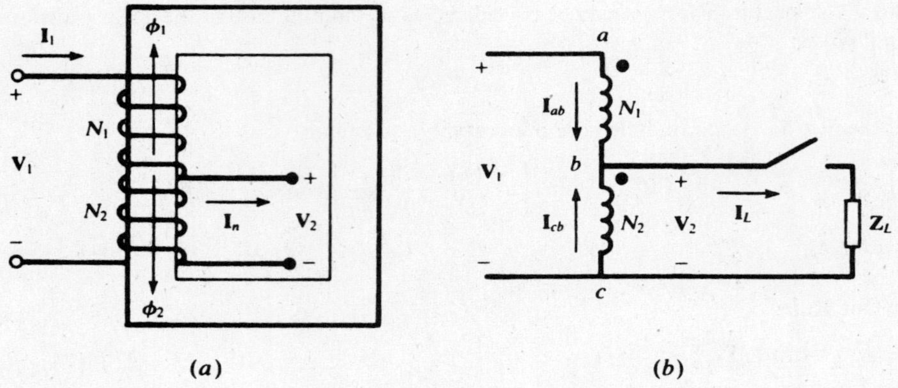

Fig. 14-15

ϕ_2. Therefore, current $\mathbf{I}_n$ leaves the lower winding by the tap. The dots on the winding are as shown in Fig. 14-15(*b*). In an ideal autotransformer, as in an ideal transformer, the input and output complex powers must be equal.

$$\tfrac{1}{2}\mathbf{V}_1\mathbf{I}_1^* = \tfrac{1}{2}\mathbf{V}_1\mathbf{I}_{ab}^* = \tfrac{1}{2}\mathbf{V}_2\mathbf{I}_L^*$$

whence
$$\frac{\mathbf{I}_L}{\mathbf{I}_{ab}} = a + 1$$

That is, the currents also are in the transformation ratio.

Since $\mathbf{I}_L = \mathbf{I}_{ab} + \mathbf{I}_{cb}$, the output complex power consists of two parts:

$$\tfrac{1}{2}\mathbf{V}_2\mathbf{I}_L^* = \tfrac{1}{2}\mathbf{V}_2\mathbf{I}_{ab}^* + \tfrac{1}{2}\mathbf{V}_2\mathbf{I}_{cb}^* = \tfrac{1}{2}\mathbf{V}_2\mathbf{I}_{ab}^* + a(\tfrac{1}{2}\mathbf{V}_2\mathbf{I}_{ab}^*)$$

The first term on the right is attributed to conduction; the second to induction. Thus, there exist both conductive and magnetic coupling between source and load in an autotransformer.

14.10 REFLECTED IMPEDANCE

A load $\mathbf{Z}_2$ connected to the secondary port of a transformer, as shown in Fig. 14-16, contributes to its input impedance. This contribution is called *reflected impedance*. Using the terminal characteristics of the coupled coils and applying KVL around the secondary loop, we find

$$\mathbf{V}_1 = L_1\mathbf{s}\mathbf{I}_1 + M\mathbf{s}\mathbf{I}_2$$
$$0 = M\mathbf{s}\mathbf{I}_1 + L_2\mathbf{s}\mathbf{I}_2 + \mathbf{Z}_2\mathbf{I}_2$$

By eliminating $\mathbf{I}_2$, we get

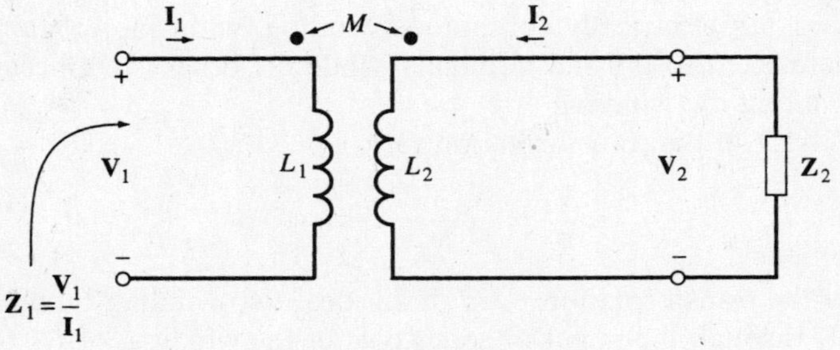

Fig. 14-16

$$\mathbf{Z}_1 = \frac{\mathbf{V}_1}{\mathbf{I}_1} = L_1\mathbf{s} - \frac{M^2\mathbf{s}^2}{\mathbf{Z}_2 + L_2\mathbf{s}} \tag{17}$$

For the ac steady state where $\mathbf{s} = j\omega$, we have

$$\mathbf{Z}_1 = j\omega L_1 + \frac{M^2\omega^2}{\mathbf{Z}_2 + j\omega L_2} \tag{18}$$

The reflected impedance is

$$\mathbf{Z}_{\text{reflected}} = \frac{M^2\omega^2}{\mathbf{Z}_2 + j\omega L_2} \tag{19}$$

The load $\mathbf{Z}_2$ is seen by the source as $M^2\omega^2/(\mathbf{Z}_2 + j\omega L_2)$. The technique is often used to change an impedance to a certain value; for example, to match a load to a source.

EXAMPLE 14.8 Given $L_1 = 0.2$ H, $L_2 = 0.1$ H, $M = 0.1$ H, and $R = 10\ \Omega$ in the circuit of Fig. 14-17. Find i_1 for $v_1 = 142.3 \sin 100t$.

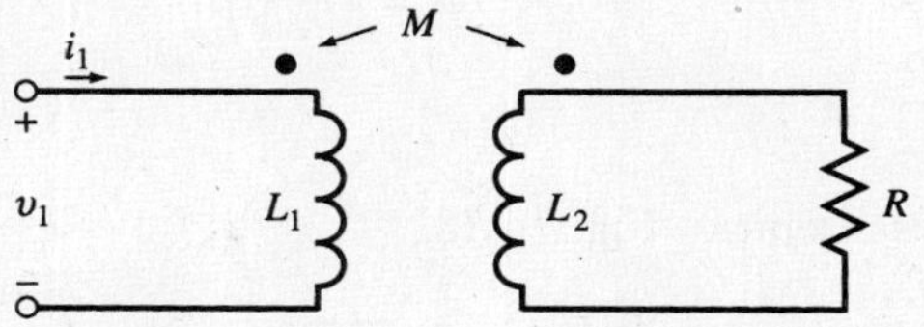

Fig. 14-17

The input impedance $\mathbf{Z}_1$ at $\omega = 100$ is [see (*18*)]

$$\mathbf{Z}_1 = \frac{\mathbf{V}_1}{\mathbf{I}_1} = j\omega L_1 + \frac{M^2\omega^2}{\mathbf{Z}_2 + j\omega L_2} = j20 + \frac{0.01(10\,000)}{10 + j10} = 5 + j15 = 5\sqrt{10}\underline{/71.6^\circ}$$

Then,
$$\mathbf{I}_1 = \mathbf{V}_1/\mathbf{Z}_1 = 9\underline{/-71.6^\circ}\ \text{A}$$

or
$$i_1 = 9\sin(100t - 71.6^\circ)\ \text{(A)}$$

EXAMPLE 14.9 Referring to Example 14.8, let $v_1 = u(t)$. Find $i_{1,\text{f}}$, the forced response.

The input impedance is [see (*17*)]

$$\mathbf{Z}_1(\mathbf{s}) = L_1\mathbf{s} - \frac{M^2\mathbf{s}^2}{R + L_2\mathbf{s}}$$

Substituting the given values for the elements, we get

$$\mathbf{Z}_1(\mathbf{s}) = \frac{\mathbf{s}(\mathbf{s}+200)}{10(\mathbf{s}+100)} \qquad \text{or} \qquad \mathbf{Y}_1(\mathbf{s}) = \frac{10(\mathbf{s}+100)}{\mathbf{s}(\mathbf{s}+200)}$$

For $t > 0$, the input $v_1 = 1$ V is an exponential $e^{\mathbf{s}t}$ whose exponent $\mathbf{s} = 0$ is a pole of $\mathbf{Y}_1(\mathbf{s})$. Therefore, $i_{1,\text{f}} = kt$ with $k = 1/L_1 = 5$. This result may also be obtained directly by dc analysis of the circuit in Fig. 14-17.

Solved Problems

14.1 When one coil of a magnetically coupled pair has a current 5.0 A the resulting fluxes ϕ_{11} and ϕ_{12} are 0.2 mWb and 0.4 mWb, respectively. If the turns are $N_1 = 500$ and $N_2 = 1500$, find L_1, L_2, M, and the coefficient of coupling k.

$$\phi_1 = \phi_{11} + \phi_{12} = 0.6 \text{ mWb} \qquad L_1 = \frac{N_1\phi_1}{I_1} = \frac{500(0.6)}{5.0} = 60 \text{ mH}$$

$$M = \frac{N_2\phi_{12}}{I_1} = \frac{1500(0.4)}{5.0} = 120 \text{ mH} \qquad k = \frac{\phi_{12}}{\phi_1} = 0.667$$

Then, from $M = k\sqrt{L_1 L_2}$, $L_2 = 540$ mH.

14.2 Two coupled coils have self-inductances $L_1 = 50$ mH and $L_2 = 200$ mH, and a coefficient of coupling $k = 0.50$. If coil 2 has 1000 turns, and $i_1 = 5.0 \sin 400t$ (A), find the voltage at coil 2 and the flux ϕ_1.

$$M = k\sqrt{L_1 L_2} = 0.50\sqrt{(50)(200)} = 50 \text{ mH}$$

$$v_2 = M\frac{di_1}{dt} = 0.05\frac{d}{dt}(5.0 \sin 400t) = 100 \cos 400t \quad \text{(V)}$$

Assuming, as always, a linear magnetic circuit,

$$M = \frac{N_2\phi_{12}}{i_1} = \frac{N_2(k\phi_1)}{i_1} \qquad \text{or} \qquad \phi_1 = \left(\frac{M}{N_2 k}\right) i_1 = 5.0 \times 10^{-4} \sin 400t \quad \text{(Wb)}$$

14.3 Apply KVL to the series circuit of Fig. 14-18.

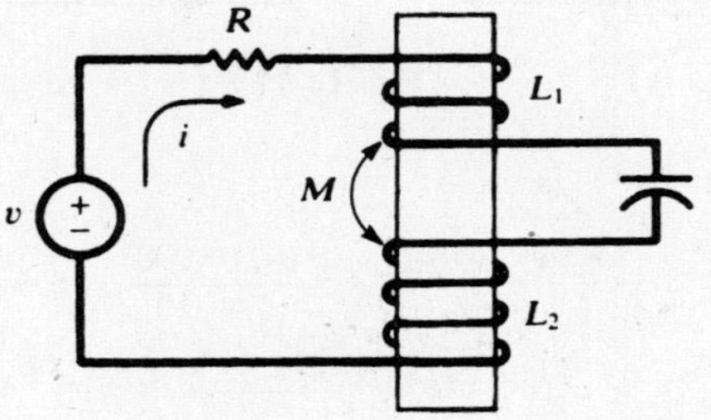

Fig. 14-18

Examination of the winding sense shows that the signs of the M-terms are opposite to the signs on the L-terms.

$$Ri + L_1\frac{di}{dt} - M\frac{di}{dt} + \frac{1}{C}\int i\,dt + L_2\frac{di}{dt} - M\frac{di}{dt} = v$$

or

$$Ri + L'\frac{di}{dt} + \frac{1}{C}\int i\,dt = v$$

where $L' \equiv L_1 + L_2 - 2M$. Because

$$M \le \sqrt{L_1 L_2} \le \frac{L_1 + L_2}{2}$$

L' is nonnegative.

14.4 In a series aiding connection, two coupled coils have an equivalent inductance L_A; in a series opposing connection, L_B. Obtain an expression for M in terms of L_A and L_B.

As in Problem 14.3,

$$L_1 + L_2 + 2M = L_A \qquad L_1 + L_2 - 2M = L_B$$

which give

$$M = \frac{1}{4}(L_A - L_B)$$

This problem suggests a method by which M can be determined experimentally.

14.5 (*a*) Write the mesh current equations for the coupled coils with currents i_1 and i_2 shown in Fig. 14-19. (*b*) Repeat for i_2 as indicated by the dashed arrow.

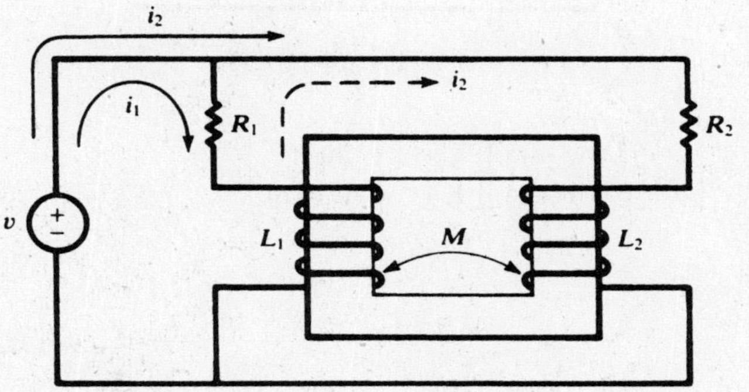

Fig. 14-19

(*a*) The winding sense and selected directions result in signs on the M-terms as follows:

$$R_1 i_1 + L_1 \frac{di_1}{dt} + M \frac{di_2}{dt} = v$$
$$R_2 i_2 + L_2 \frac{di_2}{dt} + M \frac{di_1}{dt} = v$$

(*b*)
$$R_1(i_1 - i_2) + L_1 \frac{d}{dt}(i_1 - i_2) + M \frac{di_2}{dt} = v$$
$$R_1(i_2 - i_1) + R_2 i_2 + L_2 \frac{di_2}{dt} - M \frac{d}{dt}(i_2 - i_1) + L_1 \frac{d}{dt}(i_2 - i_1) - M \frac{di_2}{dt} = 0$$

14.6 Obtain the dotted equivalent circuit for the coupled circuit shown in Fig. 14-20, and use it to find the voltage **V** across the 10-Ω capacitive reactance.

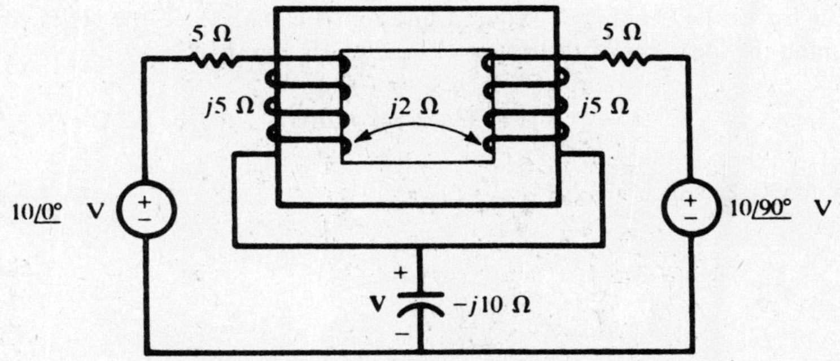

Fig. 14-20

To place the dots on the circuit, consider only the coils and their winding sense. Drive a current into the top of the left coil and place a dot at this terminal. The corresponding flux is upward. By Lenz's law, the flux at the right coil must be upward-directed to oppose the first flux. Then the natural current leaves this winding by the upper terminal, which is marked with a dot. See Fig. 14-21 for the complete dotted equivalent circuit, with currents $\mathbf{I}_1$ and $\mathbf{I}_2$ chosen for calculation of **V**.

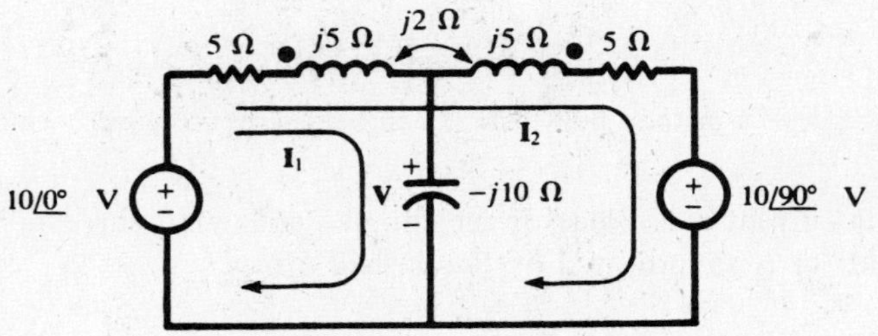

Fig. 14-21

$$\begin{bmatrix} 5-j5 & 5+j3 \\ 5+j3 & 10+j6 \end{bmatrix} \begin{bmatrix} \mathbf{I}_1 \\ \mathbf{I}_2 \end{bmatrix} = \begin{bmatrix} 10\underline{/0^\circ} \\ 10-j10 \end{bmatrix}$$

$$\mathbf{I}_1 = \frac{\begin{vmatrix} 10 & 5+j3 \\ 10-j10 & 10+j6 \end{vmatrix}}{\Delta_\mathbf{Z}} = 1.015\underline{/113.96^\circ} \text{ A}$$

and $\mathbf{V} = \mathbf{I}_1(-j10) = 10.15\underline{/23.96^\circ}$ V.

14.7 Obtain the dotted equivalent for the circuit shown in Fig. 14-22 and use the equivalent to find the equivalent inductive reactance.

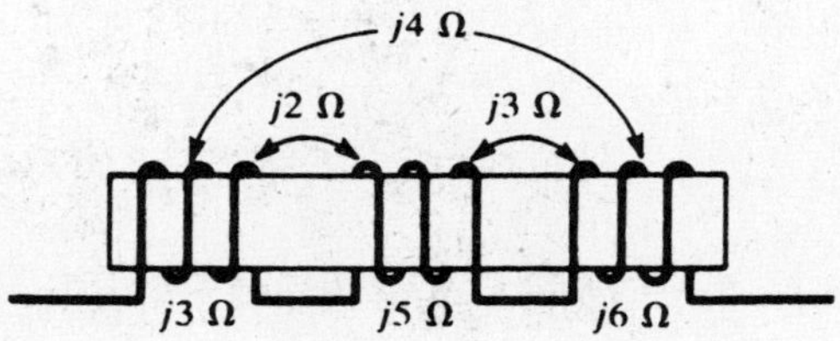

Fig. 14-22

Drive a current into the first coil and place a dot where this current enters. The natural current in both of the other windings establishes an opposing flux to that set up by the driven current. Place dots where the natural current leaves the windings. (Some confusion is eliminated if the series connections are ignored while determining the locations of the dots.) The result is Fig. 14-23.

$$\mathbf{Z} = j3 + j5 + j6 - 2(j2) + 2(j4) - 2(j3) = j12\ \Omega$$

that is, an inductive reactance of 12 Ω.

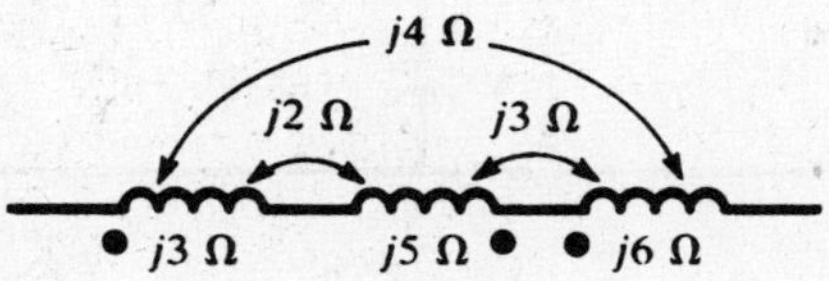

Fig. 14-23

14.8 (*a*) Compute the voltage **V** for the coupled circuit shown in Fig. 14-24. (*b*) Repeat with the polarity of one coil reversed.

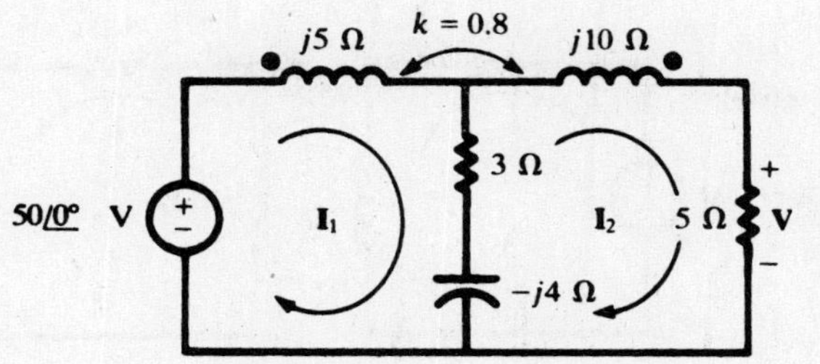

Fig. 14-24

(*a*) $X_M = (0.8)\sqrt{5(10)} = 5.66\ \Omega$, and so the **Z**-matrix is

$$[\mathbf{Z}] = \begin{bmatrix} 3 + j1 & -3 - j1.66 \\ -3 - j1.66 & 8 + j6 \end{bmatrix}$$

Then,
$$\mathbf{I}_2 = \frac{\begin{vmatrix} 3 + j1 & 50 \\ -3 - j1.66 & 0 \end{vmatrix}}{\Delta_{\mathbf{Z}}} = 8.62\underline{/-24.79^\circ}\ \text{A}$$

and $\mathbf{V} = \mathbf{I}_2(5) = 43.1\underline{/-24.79^\circ}$ V.

(*b*)
$$[\mathbf{Z}] = \begin{bmatrix} 3 + j1 & -3 + j9.66 \\ -3 + j9.66 & 8 + j6 \end{bmatrix}$$

$$\mathbf{I}_2 = \frac{\begin{vmatrix} 3 + j1 & 50 \\ -3 + j9.66 & 0 \end{vmatrix}}{\Delta_{\mathbf{Z}}} = 3.82\underline{/-112.12^\circ}\ \text{A}$$

and $\mathbf{V} = \mathbf{I}_2(5) = 19.1\underline{/-112.12^\circ}$ V.

14.9 Obtain the equivalent inductance of the parallel-connected, coupled coils shown in Fig. 14-25.

Currents $\mathbf{I}_1$ and $\mathbf{I}_2$ are selected as shown on the diagram; then $\mathbf{Z}_{\text{in}} = \mathbf{V}_1/\mathbf{I}_1$.

$$[\mathbf{Z}] = \begin{bmatrix} j\omega 0.3 & j\omega 0.043 \\ j\omega 0.043 & j\omega 0.414 \end{bmatrix}$$

and
$$\mathbf{Z}_{\text{in}} = \frac{\Delta_{\mathbf{Z}}}{\Delta_{11}} = \frac{(j\omega 0.3)(j\omega 0.414) - (j\omega 0.043)^2}{j\omega 0.414} = j\omega 0.296$$

or L_{eq} is 0.296 H.

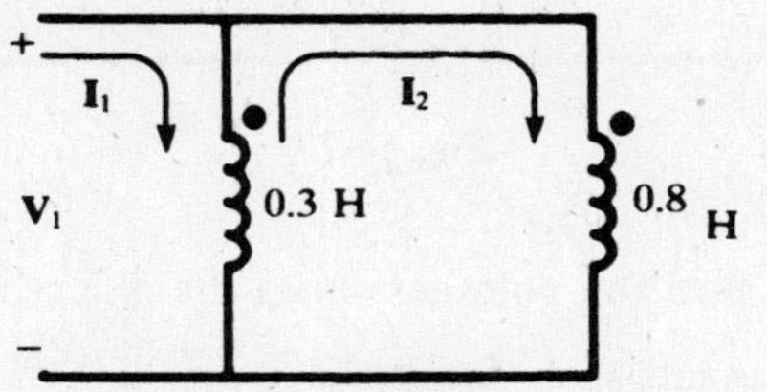

Fig. 14-25

14.10 For the coupled circuit shown in Fig. 14-26, show that dots are not needed so long as the second loop is passive.

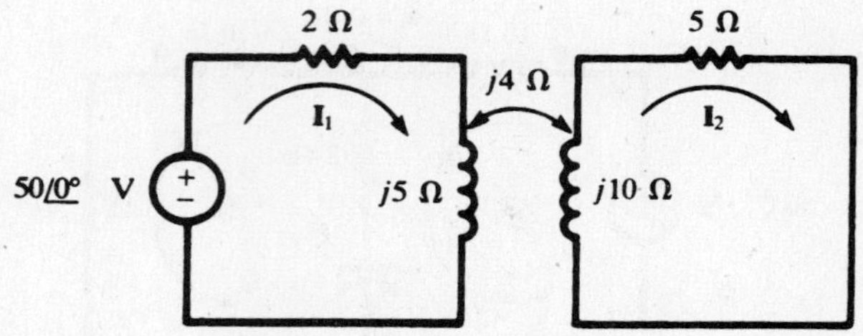

Fig. 14-26

Currents $\mathbf{I}_1$ and $\mathbf{I}_2$ are selected as shown.

$$\mathbf{I}_1 = \frac{\begin{vmatrix} 50 & \pm j4 \\ 0 & 5+j10 \end{vmatrix}}{\begin{vmatrix} 2+j5 & \pm j4 \\ \pm j4 & 5+j10 \end{vmatrix}} = \frac{250+j500}{-24+j45} = 10.96\underline{/-54.64^\circ} \text{ A}$$

$$\mathbf{I}_2 = \frac{\begin{vmatrix} 2+j5 & 50 \\ \pm j4 & 0 \end{vmatrix}}{\Delta_\mathbf{Z}} = 3.92\underline{/-118.07 \mp 90^\circ} \text{ A}$$

The value of $\Delta_\mathbf{Z}$ is unaffected by the sign on M. Since the numerator determinant for $\mathbf{I}_1$ does not involve the coupling impedance, $\mathbf{I}_1$ is also unaffected. The expression for $\mathbf{I}_2$ shows that a change in the coupling polarity results in a 180° phase shift. With no other phasor voltage present in the second loop, this change in phase is of no consequence.

14.11 For the coupled circuit shown in Fig. 14-27, find the ratio $\mathbf{V}_2/\mathbf{V}_1$ which results in zero current $\mathbf{I}_1$.

$$\mathbf{I}_1 = 0 = \frac{\begin{vmatrix} \mathbf{V}_1 & j2 \\ \mathbf{V}_2 & 2+j2 \end{vmatrix}}{\Delta_\mathbf{Z}}$$

Then, $\mathbf{V}_1(2+j2) - \mathbf{V}_2(j2) = 0$, from which $\mathbf{V}_2/\mathbf{V}_1 = 1 - j1$.

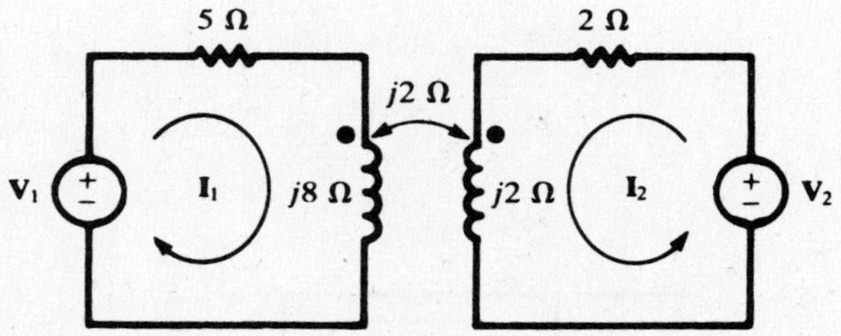

Fig. 14-27

14.12 In the circuit of Fig. 14-28, find the voltage across the 5 Ω reactance with the polarity shown.

For the choice of mesh currents shown on the diagram,

$$\mathbf{I}_1 = \frac{\begin{vmatrix} 50\underline{/45^\circ} & j8 \\ 0 & -j3 \end{vmatrix}}{\begin{vmatrix} 3+j15 & j8 \\ j8 & -j3 \end{vmatrix}} = \frac{150\underline{/-45^\circ}}{109-j9} = 1.37\underline{/-40.28^\circ} \text{ A}$$

Similarly, $\mathbf{I}_2 = 3.66\underline{/-40.28^\circ}$ A.

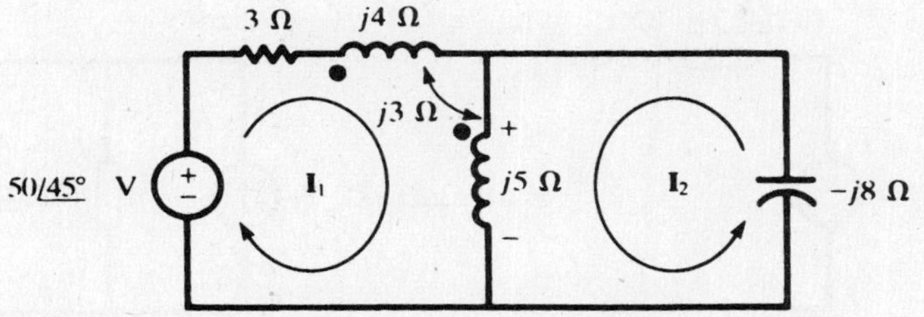

Fig. 14-28

The voltage across the $j5$ is partly conductive, from the currents $\mathbf{I}_1$ and $\mathbf{I}_2$, and partly mutual, from current $\mathbf{I}_1$ in the 4 Ω reactance.

$$\mathbf{V} = (\mathbf{I}_1 + \mathbf{I}_2)(j5) + \mathbf{I}_1(j3) = 29.27\underline{/49.72^\circ} \quad \text{V}$$

Of course, the same voltage must exist across the capacitor:

$$\mathbf{V} = -\mathbf{I}_2(-j8) = 29.27\underline{/49.72^\circ} \quad \text{V}$$

14.13 Obtain Thévenin and Norton equivalent circuits at terminals *ab* for the coupled circuit shown in Fig. 14-29.

In open circuit, a single clockwise loop current **I** is driven by the voltage source.

$$\mathbf{I} = \frac{10\underline{/0^\circ}}{8 + j3} = 1.17\underline{/-20.56^\circ} \quad \text{A}$$

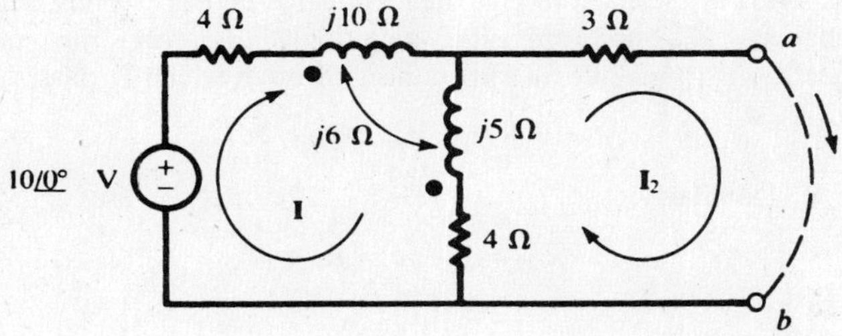

Fig. 14-29

Then $\mathbf{V}' = \mathbf{I}(j5 + 4) - \mathbf{I}(j6) = 4.82\underline{/-34.60^\circ}$ V.

To find the short-circuit current $\mathbf{I}'$, two clockwise mesh currents are assumed, with $\mathbf{I}_2 = \mathbf{I}'$.

$$\mathbf{I}' = \frac{\begin{vmatrix} 8 + j3 & 10 \\ -4 + j1 & 0 \end{vmatrix}}{\begin{vmatrix} 8 + j3 & -4 + j1 \\ -4 + j1 & 7 + j5 \end{vmatrix}} = 0.559\underline{/-83.39^\circ} \quad \text{A}$$

and
$$\mathbf{Z}' = \frac{\mathbf{V}'}{\mathbf{I}'} = \frac{4.82\underline{/-34.60^\circ}}{0.559\underline{/-83.39^\circ}} = 8.62\underline{/48.79^\circ} \quad \Omega$$

The equivalent circuits are pictured in Fig. 14-30.

14.14 Obtain a conductively coupled equivalent circuit for the magnetically coupled circuit shown in Fig. 14-31.

Select mesh currents $\mathbf{I}_1$ and $\mathbf{I}_2$ as shown on the diagram and write the KVL equations in matrix form.

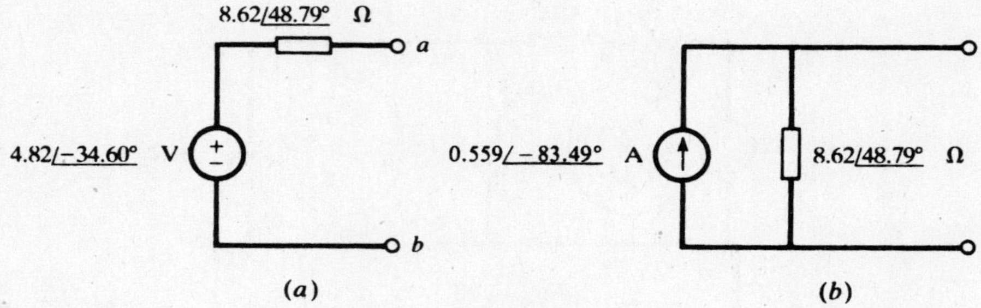

Fig. 14-30

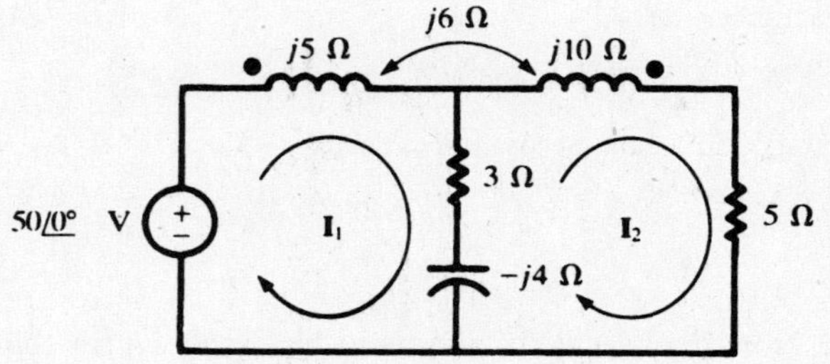

Fig. 14-31

$$\begin{bmatrix} 3+j1 & -3-j2 \\ -3-j2 & 8+j6 \end{bmatrix}\begin{bmatrix} \mathbf{I}_1 \\ \mathbf{I}_2 \end{bmatrix} = \begin{bmatrix} 50\underline{/0^\circ} \\ 0 \end{bmatrix}$$

The impedances in Fig. 14-32 are selected to give the identical **Z**-matrix. Thus, since $\mathbf{I}_1$ and $\mathbf{I}_2$ pass through the common impedance, $\mathbf{Z}_b$, in opposite directions, $\mathbf{Z}_{12}$ in the matrix is $-\mathbf{Z}_b$. Then, $\mathbf{Z}_b = 3 + j2 \quad \Omega$. Since $\mathbf{Z}_{11}$ is to include all impedances through which $\mathbf{I}_1$ passes,

$$3 + j1 = \mathbf{Z}_a + (3 + j2)$$

from which $\mathbf{Z}_a = -j1\ \Omega$. Similarly,

$$\mathbf{Z}_{22} = 8 + j6 = \mathbf{Z}_b + \mathbf{Z}_c$$

and $\mathbf{Z}_c = 5 + j4 \quad \Omega$.

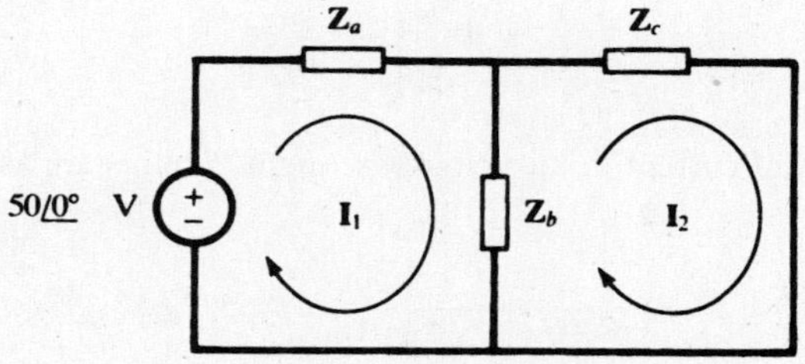

Fig. 14-32

14.15 For the transformer circuit of Fig. 14-11(*b*), $k = 0.96$, $R_1 = 1.2\ \Omega$, $R_2 = 0.3\ \Omega$, $X_1 = 20\ \Omega$, $X_2 = 5\ \Omega$, $\mathbf{Z}_L = 5.0\underline{/36.87^\circ}\ \Omega$, and $\mathbf{V}_2 = 100\underline{/0^\circ}$ V. Obtain the coil emfs $\mathbf{E}_1$ and $\mathbf{E}_2$, and the magnetizing current $\mathbf{I}_\phi$.

$$X_{11} = (1-k)X_1 = (1-0.96)(20) = 0.8\ \Omega \qquad X_{22} = (1-k)X_2 = 0.2\ \Omega$$

$$a = \sqrt{\frac{X_1}{X_2}} = 2 \qquad X_M = k\sqrt{X_1 X_2} = 9.6\ \Omega$$

Now a circuit of the form Fig. 14-11(*a*) can be constructed, starting from the phasor voltage-current relationship at the load, and working back through $\mathbf{E}_2$ to $\mathbf{E}_1$.

$$\mathbf{I}_2 = \frac{\mathbf{V}_2}{\mathbf{Z}_L} = \frac{100\angle 0^\circ}{5.0\angle 36.87^\circ} = 20\angle -36.87^\circ \quad \text{A}$$

$$\mathbf{E}_2 = \mathbf{I}_2(R_2 + jX_{22}) + \mathbf{V}_2 = (20\angle -36.87^\circ)(0.3 + j0.2) + 100\angle 0^\circ = 107.2 - j0.4 \quad \text{V}$$

$$\mathbf{E}_1 = a\mathbf{E}_2 = 214.4 - j0.8 \quad \text{V}$$

$$\mathbf{I}_\phi = \frac{\mathbf{E}_2}{jX_M} = -0.042 - j11.17 \quad \text{A}$$

14.16 For the linear transformer of Problem 14.15, calculate the input impedance at the terminals where $\mathbf{V}_1$ is applied.

Method 1

Completing the construction begun in Problem 14.15,

$$\mathbf{I}_1 = \mathbf{I}_\phi + \frac{1}{a}\mathbf{I}_2 = (-0.042 - j11.17) + 10\angle -36.87^\circ = 18.93\angle -65.13^\circ \quad \text{A}$$

$$\mathbf{V}_1 = \mathbf{I}_1(R_1 + jX_{11}) + \mathbf{E}_1 = (18.93\angle -65.13^\circ)(1.2 + j0.8) + (214.4 - j0.8)$$
$$= 238.2\angle -3.62^\circ \quad \text{V}$$

Therefore,

$$\mathbf{Z}_{\text{in}} = \frac{\mathbf{V}_1}{\mathbf{I}_1} = \frac{238.2\angle -3.62^\circ}{18.93\angle -65.13^\circ} = 12.58\angle 61.51^\circ \quad \Omega$$

Method 2

By (*14a*) of Example 14.5,

$$\mathbf{Z}_{\text{in}} = (1.2 + j0.8) + 2^2\,\frac{(j4.8)(0.3 + j0.2 + 5.0\angle 36.87^\circ)}{0.3 + j5.0 + 5.0\angle 36.87^\circ}$$
$$= \frac{114.3\angle 123.25^\circ}{9.082\angle 61.75^\circ} = 12.58\angle 61.50^\circ \quad \Omega$$

Method 3

By (*14b*) of Example 14.5,

$$\mathbf{Z}_{\text{in}} = (1.2 + j20) + \frac{(9.6)^2}{0.3 + j5 + 5.0\angle 36.87^\circ}$$
$$= (1.2 + j20) + (4.80 - j8.94) = 12.58\angle 61.53^\circ \quad \Omega$$

14.17 In Fig. 14-33, three identical transformers are primary wye-connected and secondary delta-connected. A single load impedance carries current $\mathbf{I}_L = 30\angle 0^\circ$ A. Given

$$\mathbf{I}_{b2} = 20\angle 0^\circ \ \text{A} \qquad \mathbf{I}_{a2} = \mathbf{I}_{c2} = 10\angle 0^\circ \ \text{A}$$

and $N_1 = 10N_2 = 100$, find the primary currents $\mathbf{I}_{a1}$, $\mathbf{I}_{b1}$, $\mathbf{I}_{c1}$.

The ampere-turn dot rule is applied to each transformer.

$$N_1\mathbf{I}_{a1} + N_2\mathbf{I}_{a2} = 0 \quad \text{or} \quad \mathbf{I}_{a1} = -\frac{10}{100}(10\angle 0^\circ) = -1\angle 0^\circ \ \text{A}$$

$$N_1\mathbf{I}_{b1} - N_2\mathbf{I}_{b2} = 0 \quad \text{or} \quad \mathbf{I}_{b1} = \frac{10}{100}(20\angle 0^\circ) = 2\angle 0^\circ \ \text{A}$$

$$N_1\mathbf{I}_{c1} + N_2\mathbf{I}_{c2} = 0 \quad \text{or} \quad \mathbf{I}_{c1} = -\frac{10}{100}(10\angle 0^\circ) = -1\angle 0^\circ \ \text{A}$$

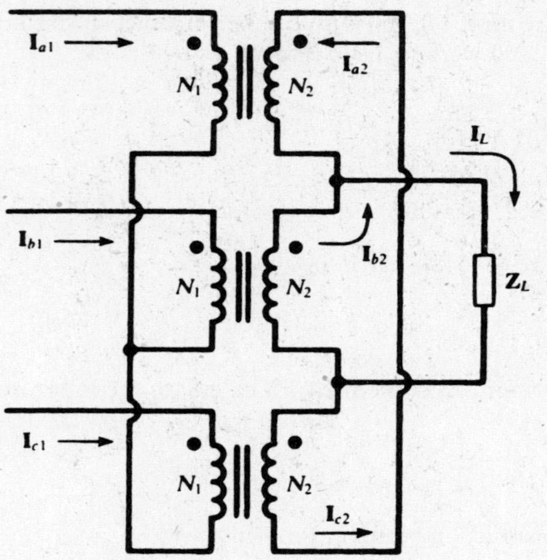

Fig. 14-33

The sum of the primary currents provides a check:

$$\mathbf{I}_{a1} + \mathbf{I}_{b1} + \mathbf{I}_{c1} = 0$$

14.18 For the ideal autotransformer shown in Fig. 14-34, find $\mathbf{V}_2$, $\mathbf{I}_{cb}$, and the input current $\mathbf{I}_1$.

$$a \equiv \frac{N_1}{N_2} = \frac{1}{2}$$

$$\mathbf{V}_2 = \frac{\mathbf{V}_1}{a+1} = 100\underline{/0^\circ} \quad \text{V} \qquad \mathbf{I}_L = \frac{\mathbf{V}_2}{\mathbf{Z}_L} = 10\underline{/-60^\circ} \quad \text{A}$$

$$\mathbf{I}_{cb} = \mathbf{I}_L - \mathbf{I}_{ab} = 3.33\underline{/-60^\circ} \quad \text{A} \qquad \mathbf{I}_{ab} = \frac{\mathbf{I}_L}{a+1} = 6.67\underline{/-60^\circ} \quad \text{A}$$

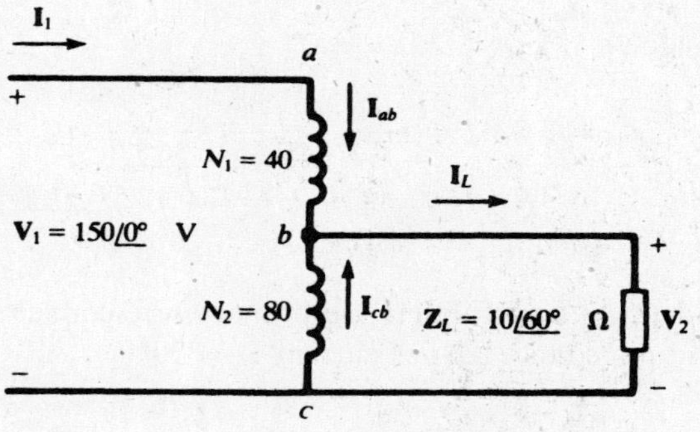

Fig. 14-34

14.19 In Problem 14.18, find the apparent power delivered to the load by transformer action and that supplied by conduction.

$$\mathbf{S}_{\text{cond}} = \tfrac{1}{2}\mathbf{V}_2\mathbf{I}^*_{ab} = \tfrac{1}{2}(100\underline{/0^\circ})(6.67\underline{/60^\circ}) = 333\underline{/60^\circ} \quad \text{VA}$$

$$\mathbf{S}_{\text{trans}} = a\mathbf{S}_{\text{cond}} = 167\underline{/60^\circ} \quad \text{VA}$$

14.20 In the coupled circuit of Fig. 14-35, find the input admittance $\mathbf{Y}_1 = \mathbf{I}_1/\mathbf{V}_1$ and determine the current $i_1(t)$ for $v_1 = 2\sqrt{2}\cos t$.

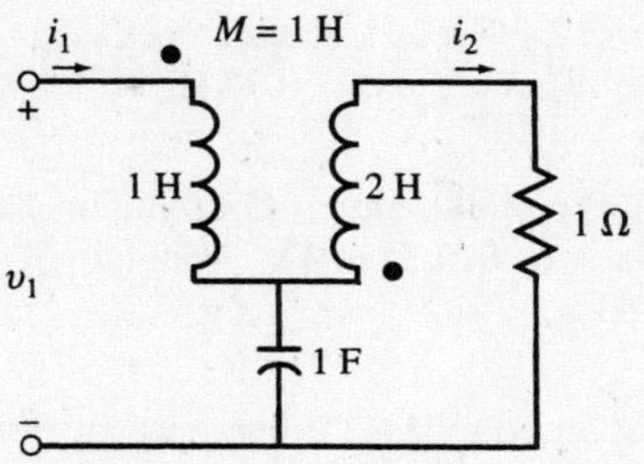

Fig. 14-35

Apply KVL around loops 1 and 2 in the **s**-domain.

$$\mathbf{V}_1 = \mathbf{s}\mathbf{I}_1 + \mathbf{s}\mathbf{I}_2 + \frac{\mathbf{I}_1 - \mathbf{I}_2}{\mathbf{s}}$$

$$0 = \mathbf{s}\mathbf{I}_1 + (2\mathbf{s} + 1)\mathbf{I}_2 + \frac{\mathbf{I}_2 - \mathbf{I}_1}{\mathbf{s}}$$

Eliminating $\mathbf{I}_2$ in these equations results in

$$\mathbf{Y}_1 = \frac{\mathbf{I}_1}{\mathbf{V}_1} = \frac{2\mathbf{s}^2 + \mathbf{s} + 1}{\mathbf{s}^3 + \mathbf{s}^2 + 5\mathbf{s} + 1}$$

For $\mathbf{s} = j$, the input admittance is $\mathbf{Y}_1 = (1 + j)/4 = \sqrt{2}/4\underline{/45^\circ}$. Therefore, $i_1(t) = \cos(t + 45^\circ)$.

14.21 Find the input impedance $\mathbf{Z}_1 = \mathbf{V}_1/\mathbf{I}_1$ in the coupled circuit of Fig. 14-36.

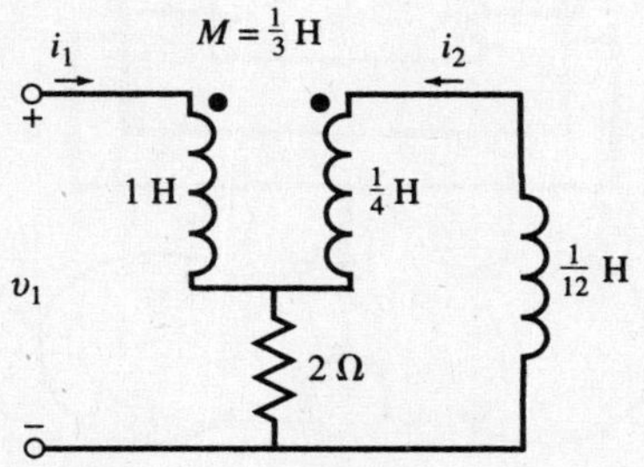

Fig. 14-36

Apply KVL around loops 1 and 2 in the **s**-domain.

$$\begin{cases} \mathbf{V}_1 = \mathbf{s}\mathbf{I}_1 + \frac{1}{3}\mathbf{s}\mathbf{I}_2 + 2(\mathbf{I}_1 + \mathbf{I}_2) \\ 0 = \frac{1}{3}\mathbf{s}\mathbf{I}_1 + \frac{1}{4}\mathbf{s}\mathbf{I}_2 + 2(\mathbf{I}_1 + \mathbf{I}_2) + \frac{1}{12}\mathbf{s}\mathbf{I}_2 \end{cases}$$

or

$$\begin{cases} \mathbf{V}_1 = (2 + \mathbf{s})\mathbf{I}_1 + (2 + \frac{1}{3}\mathbf{s})\mathbf{I}_2 \\ 0 = (2 + \frac{1}{3}\mathbf{s})\mathbf{I}_1 + (2 + \frac{1}{3}\mathbf{s})\mathbf{I}_2 \end{cases}$$

The result is

$$\mathbf{I}_2 = -\mathbf{I}_1 \qquad \text{and} \qquad \mathbf{Z}_1 = \frac{\mathbf{V}_1}{\mathbf{I}_1} = \frac{2}{3}\mathbf{s}$$

The current through the resistor is $\mathbf{I}_1 + \mathbf{I}_2 = 0$ and the resistor has no effect on $\mathbf{Z}_1$. The input impedance is purely inductive.

Supplementary Problems

14.22 Two coupled coils, $L_1 = 0.8$ H and $L_2 = 0.2$ H, have a coefficient of coupling $k = 0.90$. Find the mutual inductance M and the turns ratio N_1/N_2. *Ans.* 0.36 H, 2

14.23 Two coupled coils, $N_1 = 100$ and $N_2 = 800$, have a coupling coefficient $k = 0.85$. With coil 1 open and a current of 5.0 A in coil 2, the flux is $\phi_2 = 0.35$ mWb. Find L_1, L_2, and M.
Ans. 0.875 mH, 56 mH, 5.95 mH

14.24 Two identical coupled coils have an equivalent inductance of 80 mH when connected series aiding, and 35 mH in series opposing. Find L_1, L_2, M, and k. *Ans.* 28.8 mH, 28.8 mH, 11.25 mH, 0.392

14.25 Two coupled coils, with $L_1 = 20$ mH, $L_2 = 10$ mH, and $k = 0.50$, are connected four different ways: series aiding, series opposing, and parallel with both arrangements of winding sense. Obtain the equivalent inductances of the four connections. *Ans.* 44.1 mH, 15.9 mH, 9.47 mH, 3.39 mH

14.26 Write the mesh current equations for the coupled circuit shown in Fig. 14-37. Obtain the dotted equivalent circuit and write the same equations.

Ans. $$(R_1 + R_3)i_1 + L_1 \frac{di_1}{dt} + R_3 i_2 + M \frac{di_2}{dt} = v$$
$$(R_2 + R_3)i_2 + L_2 \frac{di_2}{dt} + R_3 i_1 + M \frac{di_1}{dt} = v$$

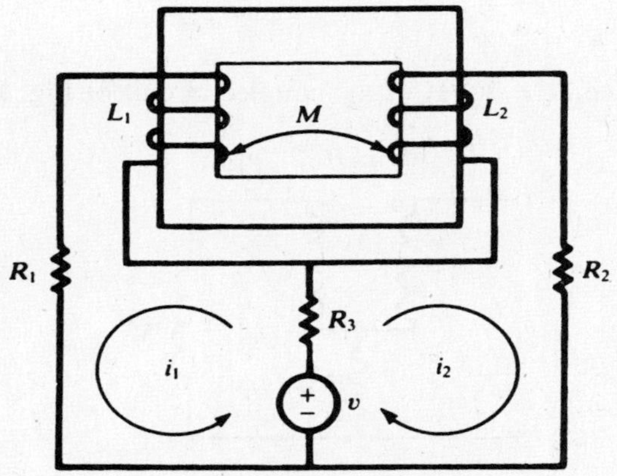

Fig. 14-37

14.27 Write the phasor equation for the single-loop, coupled circuit shown in Fig. 14-38.
Ans. $(j5 + j3 - j5.03 - j8 + 10)\mathbf{I} = 50\underline{/0^\circ}$

14.28 Obtain the dotted equivalent circuit for the coupled circuit of Fig. 14-38. *Ans.* See Fig. 14-39.

14.29 The three coupled coils shown in Fig. 14-40 have coupling coefficients of 0.50. Obtain the equivalent inductance between the terminals AB. *Ans.* 239 mH

14.30 Obtain two forms of the dotted equivalent circuit for the coupled coils shown in Fig. 14-40.
Ans. See Fig. 14-41.

14.31 (*a*) Obtain the equivalent impedance at terminals AB of the coupled circuit shown in Fig. 14-42. (*b*) Reverse the winding sense of one coil and repeat. *Ans.* (*a*) $3.40\underline{/41.66^\circ}$ Ω; (*b*) $2.54\underline{/5.37^\circ}$ Ω

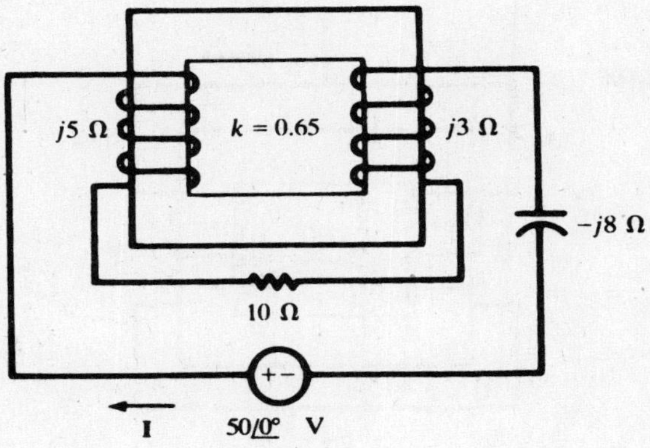

Fig. 14-38

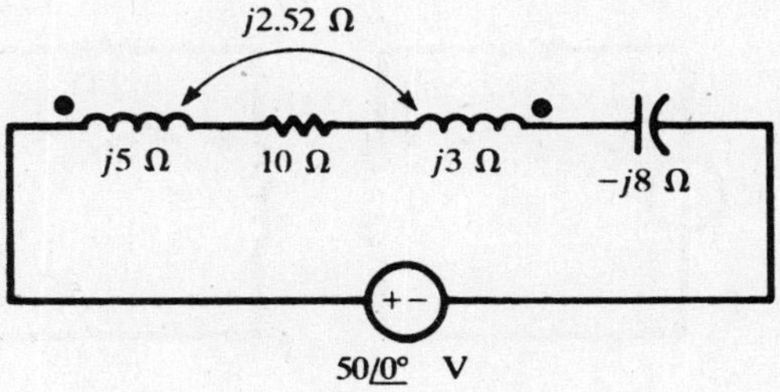

Fig. 14-39

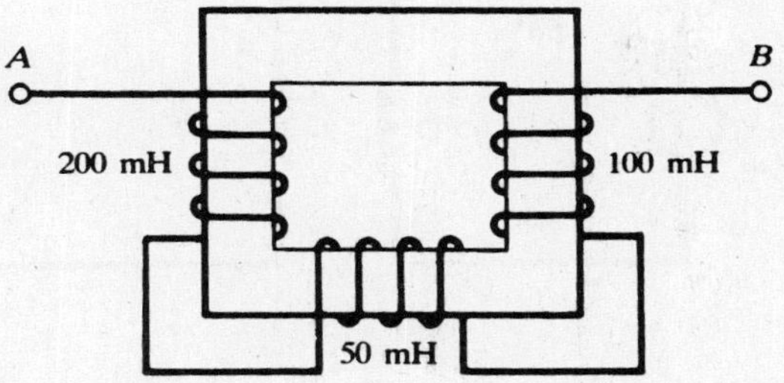

Fig. 14-40

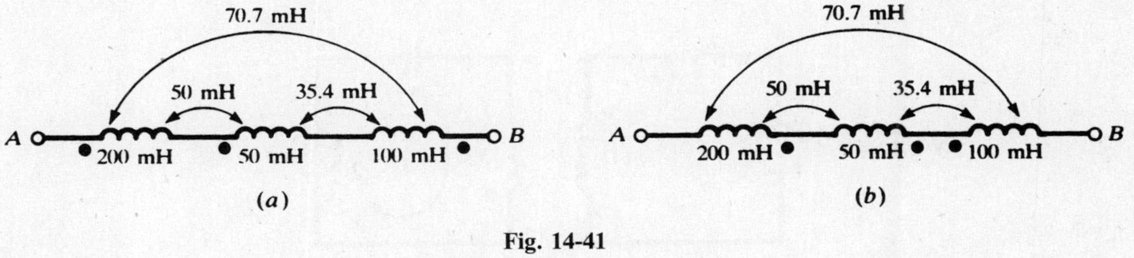

Fig. 14-41

14.32 In the coupled circuit shown in Fig. 14-43, find $\mathbf{V}_2$ for which $\mathbf{I}_1 = 0$. What voltage appears at the 8 Ω inductive reactance under this condition? *Ans.* $141.4\angle{-45^\circ}$ V, $100\angle{0^\circ}$ V (+ at dot)

14.33 Find the mutual reactance X_M for the coupled circuit of Fig. 14-44, if the average power in the 5-Ω resistor is 45.24 W. *Ans.* 4 Ω

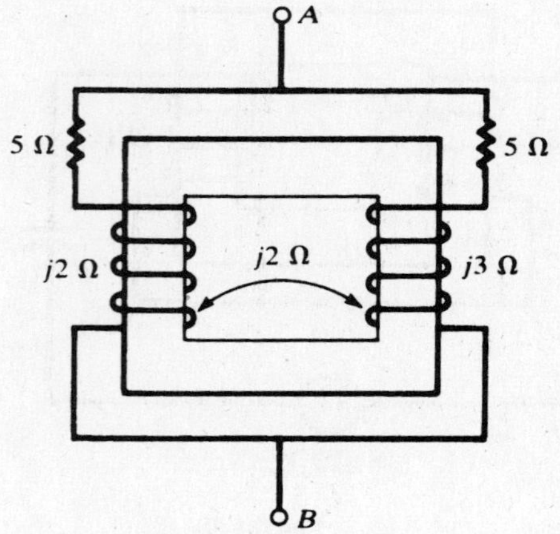

Fig. 14-42

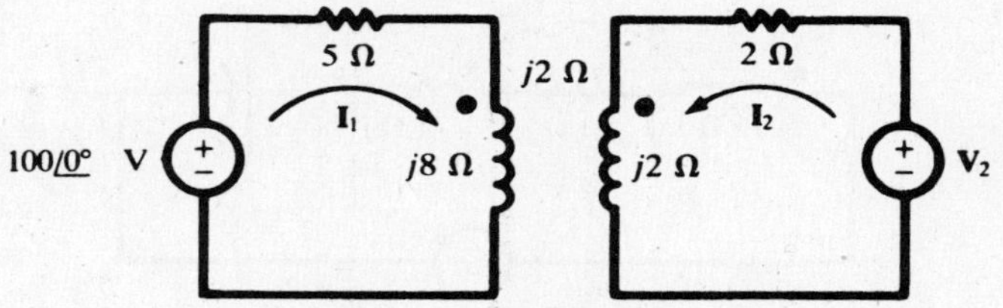

Fig. 14-43

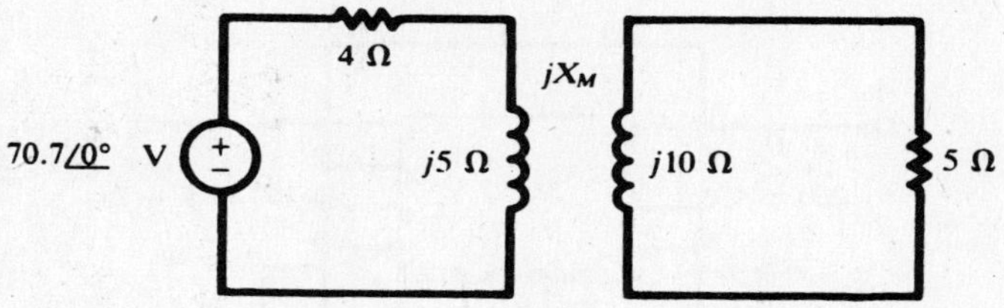

Fig. 14-44

14.34 For the coupled circuit shown in Fig. 14-45, find the components of the current $\mathbf{I}_2$ resulting from each source $\mathbf{V}_1$ and $\mathbf{V}_2$. *Ans.* $0.77\angle 112.6^\circ$ A, $1.72\angle 86.05^\circ$ A

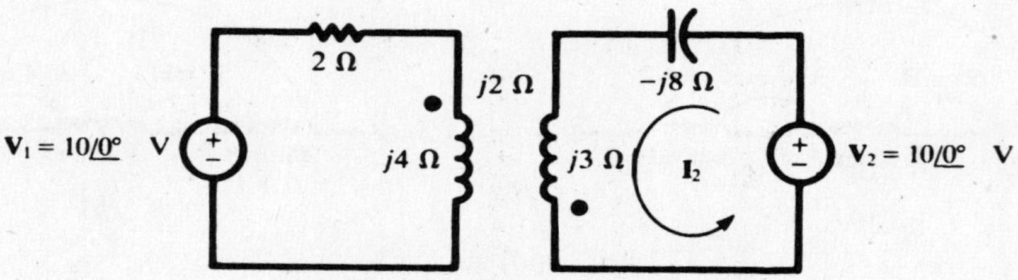

Fig. 14-45

14.35 Determine the coupling coefficient k in the circuit shown in Fig. 14-46, if the power in the 10-Ω resistor is 32 W. *Ans.* 0.791

14.36 In (*14a*), replace a, X_{11}, X_{22}, and X_M by their expressions in terms of X_1, X_2, and k, thereby obtaining (*14b*).

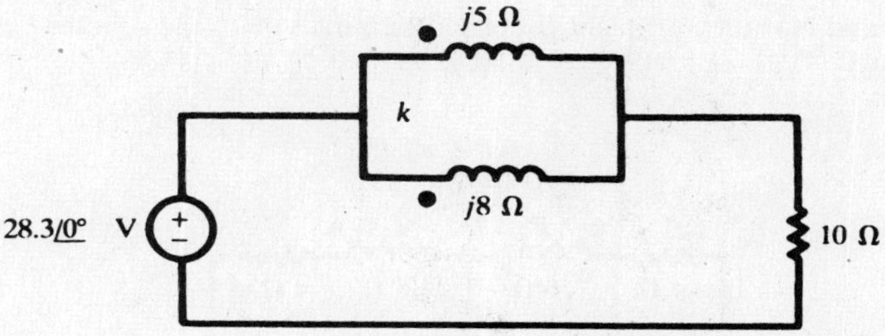

Fig. 14-46

14.37 For the coupled circuit shown in Fig. 14-47, find the input impedance at terminals *ab*.
Ans. $3 + j36.3\ \Omega$

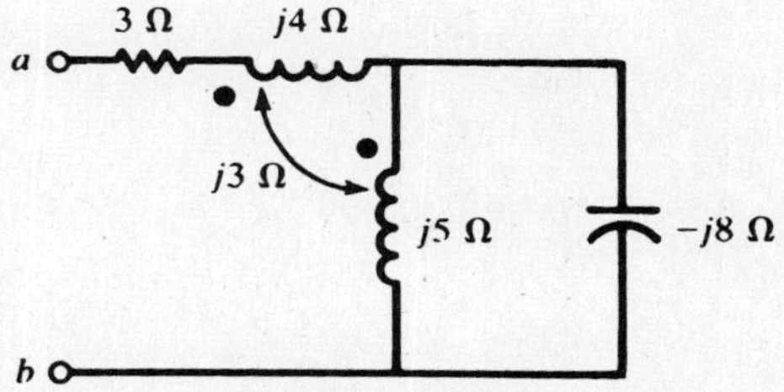

Fig. 14-47

14.38 Find the input impedance at terminals *ab* of the coupled circuit shown in Fig. 14-48.
Ans. $1 + j1.5\ \Omega$

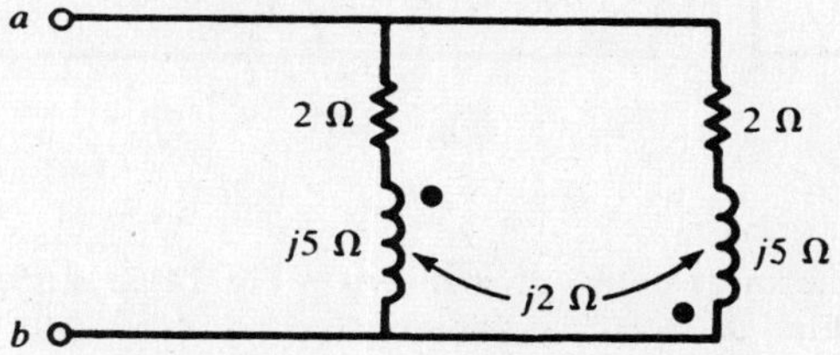

Fig. 14-48

14.39 Find the input impedance at terminals *ab* of the coupled circuit shown in Fig. 14-49.
Ans. $6.22 + j4.65\ \Omega$

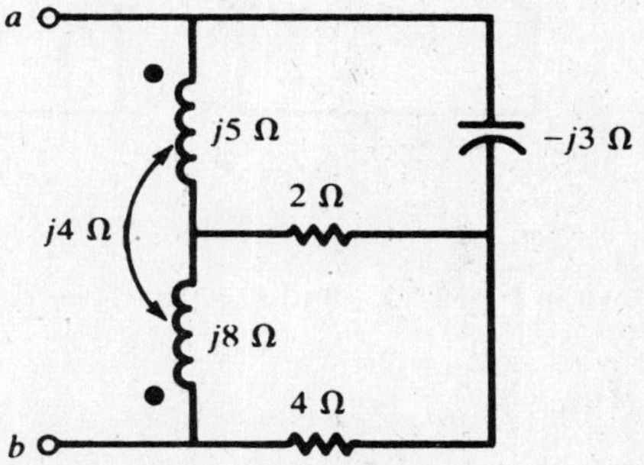

Fig. 14-49

14.40 Obtain Thévenin and Norton equivalent circuits at terminals *ab* of the coupled circuit shown in Fig. 14-50.
Ans. $\mathbf{V}' = 7.07\underline{/45^\circ}$ V, $\mathbf{I}' = 1.04\underline{/-27.9^\circ}$ A, $\mathbf{Z}' = 6.80\underline{/72.9^\circ}$ Ω

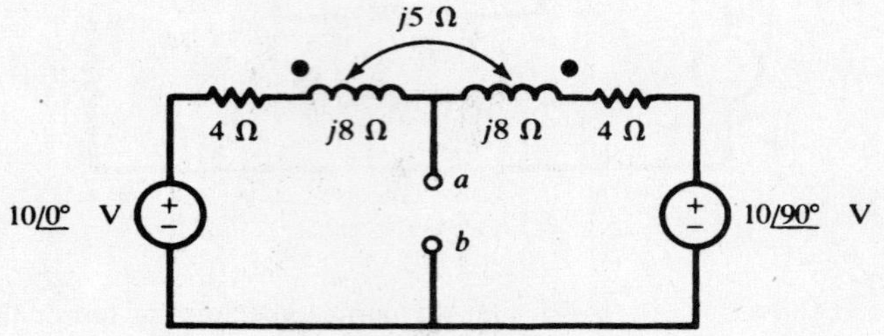

Fig. 14-50

14.41 For the ideal transformer shown in Fig. 14-51, find $\mathbf{I}_1$, given

$$\mathbf{I}_{L1} = 10.0\underline{/0^\circ}\ \text{A} \qquad \mathbf{I}_{L2} = 10.0\underline{/-36.87^\circ}\ \text{A} \qquad \mathbf{I}_{L3} = 4.47\underline{/-26.57^\circ}\ \text{A}$$

Ans. $16.5\underline{/-14.04^\circ}$ A

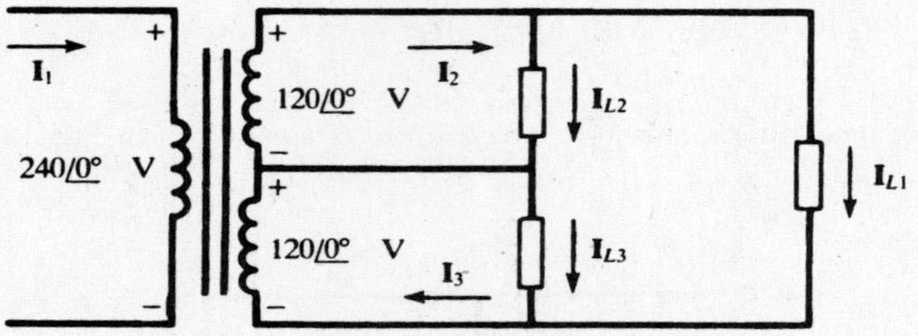

Fig. 14-51

14.42 When the secondary of the linear transformer shown in Fig. 14-52 is open-circulated, the primary current is $\mathbf{I}_1 = 4.0\underline{/-89.69^\circ}$ A. Find the coefficient of coupling *k*. *Ans.* 0.983

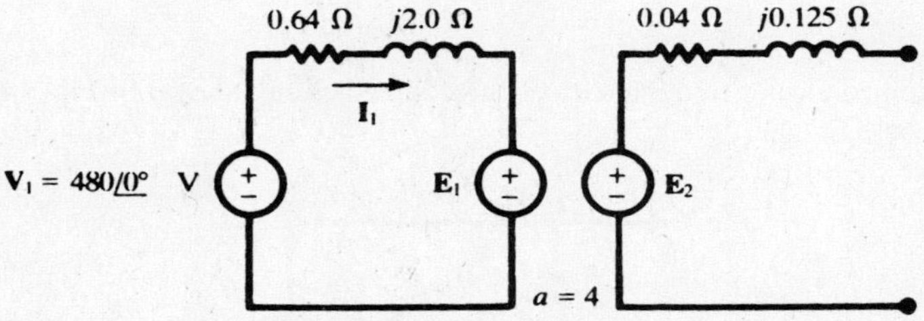

Fig. 14-52

14.43 For the ideal transformer shown in Fig. 14-53, find $\mathbf{I}_1$, given $\mathbf{I}_2 = 50\underline{/-36.87^\circ}$ A and $\mathbf{I}_3 = 16\underline{/0^\circ}$ A.
Ans. $26.6\underline{/-34.29^\circ}$ A

14.44 Considering the autotransformer shown in Fig. 14-54 ideal, obtain the currents $\mathbf{I}_1$, $\mathbf{I}_{cb}$, and $\mathbf{I}_{dc}$.
Ans. $3.70\underline{/22.5^\circ}$ A, $2.12\underline{/86.71^\circ}$ A, $10.34\underline{/11.83^\circ}$ A

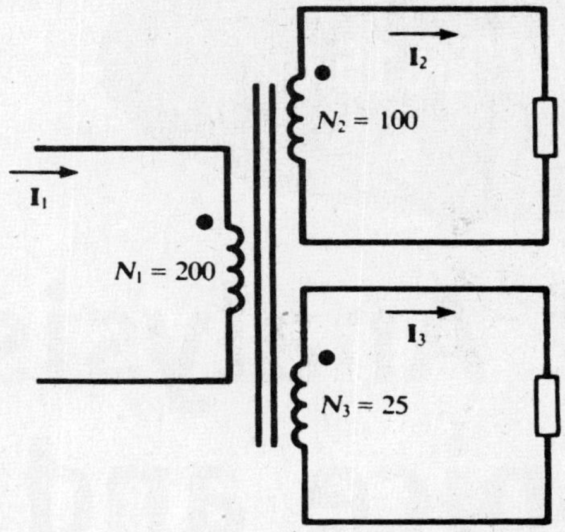

Fig. 14-53

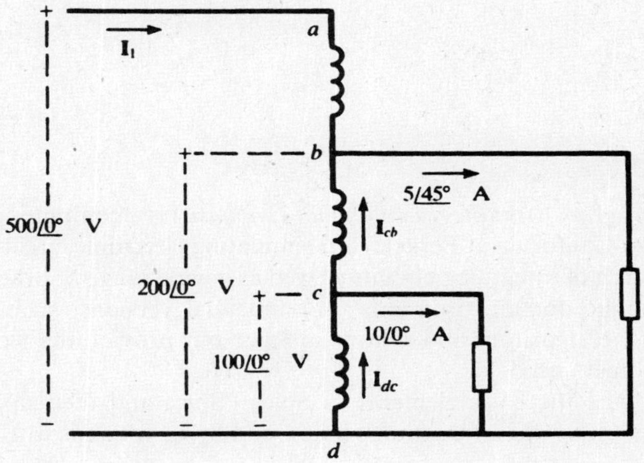

Fig. 14-54

CHAPTER 15

Circuit Analysis Using Spice and PSpice

15.1 SPICE AND PSPICE

Spice (*Simulation Program with Integrated Circuit Emphasis*) is a computer program developed in the 1970s at the University of California at Berkeley for simulating electronic circuits. It is used as a tool for analysis, design, and testing of integrated circuits as well as a wide range of other electronic and electrical circuits. Spice is a public domain program. Commercial versions, such as PSpice by MicroSim Corporation, use the same algorithm and syntax as Spice but provide the technical support and add-ons that industrial customers need.

This chapter introduces the basic elements of Spice/PSpice and their application to some simple circuits. Examples are run on the evaluation version of PSpice which is available free of charge.

15.2 CIRCUIT DESCRIPTION

The circuit description is entered in the computer in the form of a series of statements in a text file prepared by any ASCII text editor and called the *source file*. It may also be entered graphically by constructing the circuit on the computer monitor with the Schematic Capture program from MicroSim. In this chapter, we use the source file with the generic name SOURCE.CIR. To solve the circuit, we run the circuit solver on the source file. The computer puts the solution in a file named SOURCE.OUT.

EXAMPLE 15.1 Use PSpice to find the dc steady-state voltage across the 5-μF capacitor in Fig. 15-1(*a*).

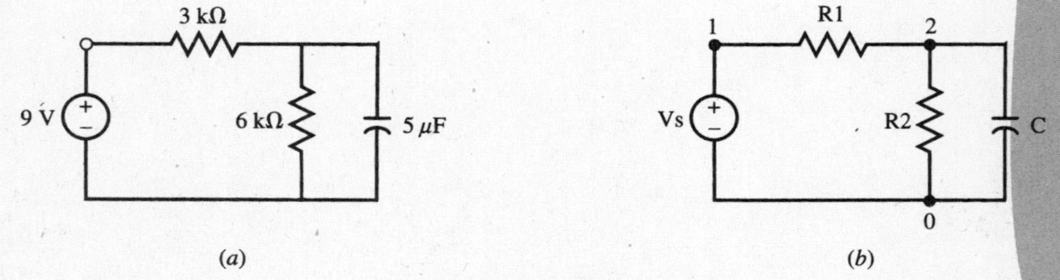

Fig. 15-1

We first label the nodes by the numbers 0, 1, 2 and the elements by the symbols R1, R2, C, and Vs [Fig. 15-1(*b*)]. We then create in ASCII the source file shown below and give it a name, for instance, EXMP1.CIR.

```
DC analysis, Fig. 15-1
    Vs      1 0    DC      9 V
    R1      1 2    3 k
    R2      0 2    6 k
    C       0 2    5 uF
    .END
```

Executing the command PSPICE EXMP1, the computer solves the circuit and writes the following results in the file EXMP1.OUT.

```
    NODE     VOLTAGE      NODE     VOLTAGE
    (1)      9.0000       (2)      6.0000
    VOLTAGE SOURCE CURRENTS
    NAME         CURRENT
    Vs           -1.000E - 03

    TOTAL POWER DISSIPATION     9.00E - 0.3 WATTS
```

The printed output specifies that the voltage at node 2 with reference to node 0 is 6 V, the current entering the voltage source V_s is -10^{-3} A, and the total power dissipated in the circuit is 9×10^{-3} W.

15.3 DISSECTING A SPICE SOURCE FILE

The source file of Example 15.1 is very simple and contains the statements necessary for solving the circuit of Fig. 15-1 by Spice. Each line in the source file is a statement. In general, if a line is too long (over 80 characters), it can be continued onto subsequent lines. The continuation lines must begin with a plus (+) sign in the first column.

PSpice does not differentiate uppercase and lowercase letters and standard units are implied when not specified. We will use both notations.

Title Statement

The first line in the source file of Example 15-1 is called the *title statement*. This line is used by Spice as a label within the output file, and it is not considered in the analysis. Therefore, it is *mandatory* to allocate the first line to the title line, even if it is left blank.

.END Statement

The .END statement is *required* at the end of the source file. Any statement following the .END will be considered a separate source file.

Data Statements

The remaining four data statements in the source file of Example 15.1 completely specify the circuit. The second line states that a voltage source named V_s is connected between node 1 (positive end of the source) and the reference node 0. The source is a dc source with a value of 9 V. The third line states that a resistor named R_1, with the value of 3 kΩ, is connected between nodes 1 and 2. Similarly, the fourth and fifth lines specify the connection of R_2 (6 kΩ) and C (5 μF), respectively, between nodes 0 and 2. In any circuit, one node should be numbered 0 to serve as the reference node. The set of data statements describing the topology of the circuit and element values is called the *netlist*. Data statement syntax is described in Section 15.4.

Control and Output Statements

In the absence of any additional commands, and only based on the netlist, Spice automatically computes the dc steady state of the following variables:

(i) Node voltages with respect to node 0.

(ii) Currents entering each voltage source.
(iii) Power dissipated in the circuit.

However, additional control and output statements may be included in the source file to specify other variables (see Section 15.5).

15.4 DATA STATEMENTS AND DC ANALYSIS

Passive Elements

Data statements for R, L, and C elements contain a minimum of three segments. The first segment gives the name of the element as a string of characters beginning with R, L, or C for resistors, inductors, or capacitors, respectively. The second segment gives the node numbers, separated by a space, between which the element is connected. The third segment gives the element value in ohms, henrys, and farads, optionally using the scale factors given in Table 15-1.

Table 15-1 Scale Factors and Symbols

Name	Symbol	Value
femto	f	$10^{-15} = 1\text{E}-15$
pico	p	$10^{-12} = 1\text{E}-12$
nano	n	$10^{-9} = 1\text{E}-9$
micro	u	$10^{-6} = 1\text{E}-6$
milli	m	$10^{-3} = 1\text{E}-3$
kilo	k	$10^{3} = 1\text{E}3$
mega	meg	$10^{6} = 1\text{E}6$
giga	g	$10^{9} = 1\text{E}9$
tera	t	$10^{12} = 1\text{E}12$

Possible initial conditions can be given in the fourth segment using the form IC = xx. The syntax of the data statement is

⟨name⟩ ⟨nodes⟩ ⟨value⟩ [⟨initial conditions⟩]

The brackets indicate optional segments in the statement.

EXAMPLE 15.2 Write the data statements for R, L, and C given in Fig. 15-2.

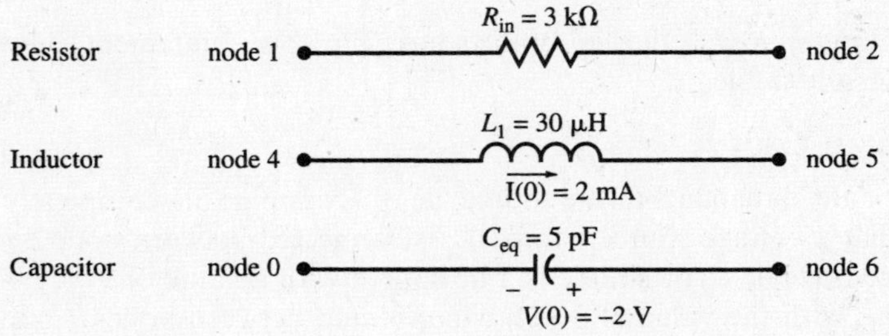

Fig. 15-2

Element	⟨name⟩	⟨nodes⟩	⟨value⟩	[⟨initial condition⟩]
Resistor	Rin	1 2	3 k	
Inductor	L1	5 4	30 uH	IC = −2 mA
Capacitor	Ceq	6	5 pF	IC = −2 V

The third statement for the capacitor connection specifies one node only. The missing node is always taken to be the reference node.

Independent Sources

Independent sources are specified by

⟨name⟩ ⟨nodes⟩ ⟨type⟩ ⟨value⟩

The ⟨type⟩ for dc and ac sources is DC and AC, respectively. Other time-dependent sources will be described in Section 15.12. Names of voltage and current sources begin with V and I, respectively. For voltage sources, the first node indicates the positive terminal. The current in the current source flows from the first node to the second.

EXAMPLE 15.3 Write data statements for the sources of Fig. 15-3.

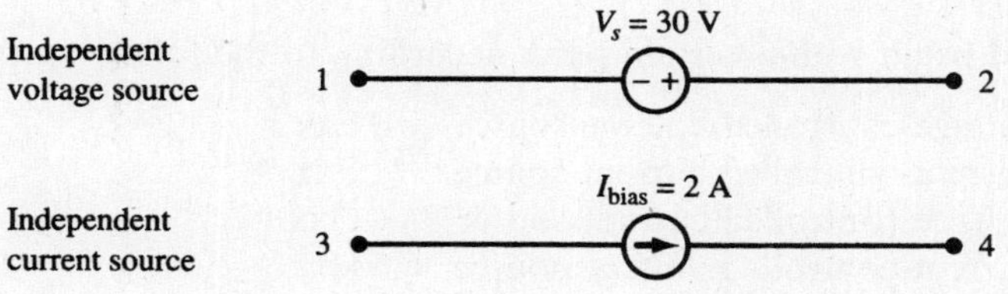

Fig. 15-3

Source	⟨name⟩	⟨nodes⟩	⟨type⟩	⟨value⟩
Independent Voltage Source	Vs	2 1	DC	30 V
Independent Current Source	Ibias	3 4	DC	2 A

EXAMPLE 15.4 Write the netlist for the circuit of Fig. 15-4(*a*) and run PSpice on it for dc analysis.

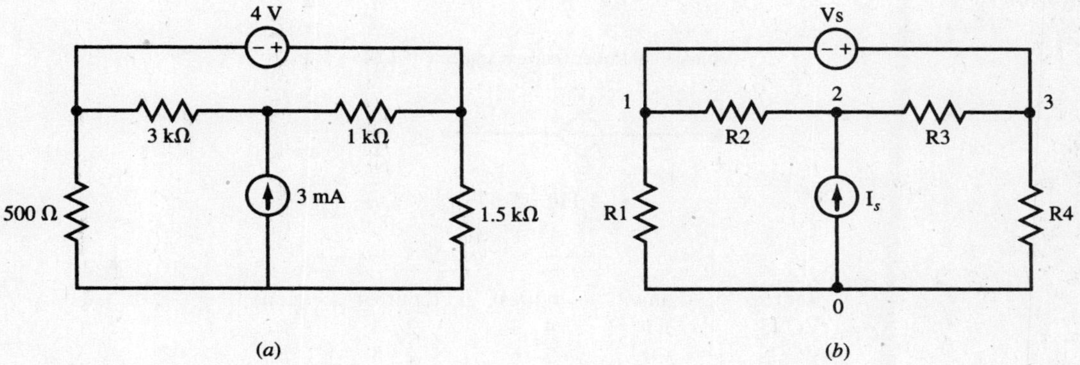

Fig. 15-4

We first number the nodes and name the elements as in Fig. 15-4(*b*). The netlist is

```
DC Analysis, Fig. 15-4
R1     0 1    500
R2     1 2    3 k
R3     2 3    1 k
R4     0 3    1.5 k
Vs     3 1    DC      4 V
Is     0 2    DC      3 mA
.END
```

The results are writte in the output file as follows:

NODE	VOLTAGE	NODE	VOLTAGE	NODE	VOLTAGE
(1)	.1250	(2)	5.3750	(3)	4.1250

VOLTAGE SOURCE CURRENTS

NAME	CURRENT
Vs	−1.500E − 03

TOTAL POWER DISSIPATION 6.00E − 03 WATTS

Dependent Sources

Linearly dependent sources are specified by

⟨name⟩ ⟨nodes⟩ ⟨control⟩ ⟨gain⟩

Each source name should begin with a certain letter according to the following rule:

Voltage-controlled voltage source	Exx
Current-controlled current source	Fxx
Voltage-controlled current source	Gxx
Current-controlled voltage source	Hxx

The order of nodes is similar to that of independent sources. For the voltage-controlled sources, ⟨control⟩ is the pair of nodes whose voltage difference controls the source, with the first node indicating the + terminal. The ⟨gain⟩ is the proportionality factor.

EXAMPLE 15.5 Write the data statements for the voltage-controlled sources of Fig. 15-5.

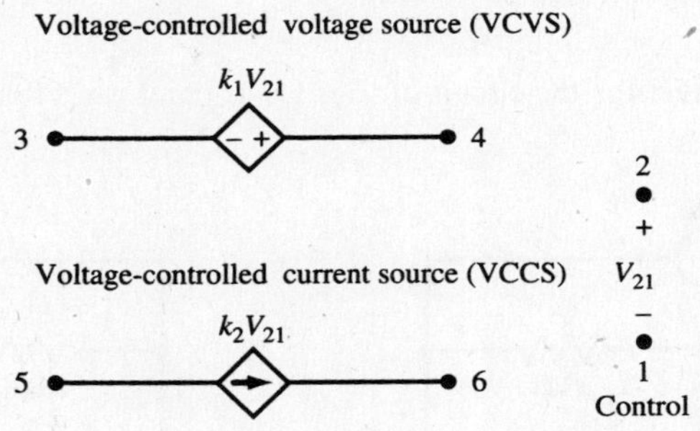

Fig. 15-5

Source	⟨name⟩	⟨nodes⟩	⟨control⟩	⟨gain⟩
VCVS	E1	4 3	2 1	k1
VCCS	G1	5 6	2 1	k2

In the case of current-controlled sources we first introduce a zero-valued voltage source (or dummy voltage V_{dmy}) on the path of the controlling current and use its name as the control variable.

EXAMPLE 15.6 Write data statements for the current-controlled sources in Fig. 15-6.

Introduce V_{dmy} (Vdmy) with current i entering it at node 1.

Vdmy 1 7 DC 0

The data statements for the controlled sources are

Source	⟨name⟩	⟨nodes⟩	⟨control⟩	⟨gain⟩
CCVS	H1	4 3	Vdmy	k3
CCCS	F1	5 6	Vdmy	k4

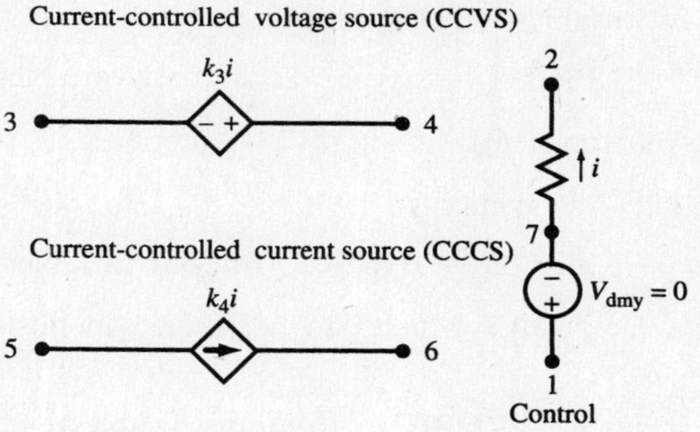

Fig. 15-6

EXAMPLE 15.7 Write the netlist for the circuit of Fig. 15-7(*a*) and run PSpice on it for dc analysis.

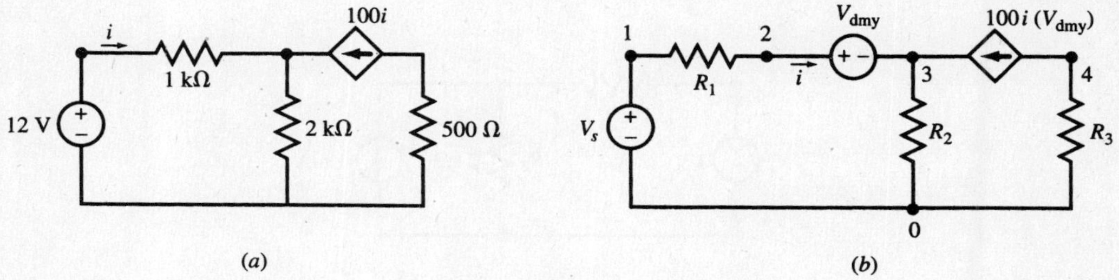

Fig. 15-7

Number the nodes and name the elements as in Fig. 15-7(*b*). Then, the netlist is

```
DC analysis with dependent source, Fig. 15-7
Vs      1 0     DC      12
R1      1 2     1 k
R2      0 3     2 k
R3      0 4     500
Vdmy    2 3     0
F1      4 3     Vdmy    100
.END
```

The results in the output file are

```
NODE  VOLTAGE   NODE  VOLTAGE   NODE  VOLTAGE   NODE  VOLTAGE
(1)   12.0000   (2)   11.9410   (3)   11.9410   (4)   -2.9557
VOLTAGE SOURCE CURRENTS
NAME        CURRENT
Vs          -5.911E-05
Vdmy         5.911E-05
TOTAL POWER DISSIPATION   7.09E-04   WATTS
```

15.5 CONTROL AND OUTPUT STATEMENTS IN DC ANALYSIS

Certain statements control actions or the output format. Examples are:

.OP prints the dc operating point of all independent sources.

.DC sweeps the value of an independent dc source. The syntax is

.DC ⟨name⟩ ⟨initial value⟩ ⟨final values⟩ ⟨step size⟩

.PRINT prints the value of variables. The syntax is

.PRINT ⟨type⟩ ⟨output variables⟩

⟨type⟩ is DC, AC, or TRAN (transient).

.PLOT line-prints variables. The syntax is

.PLOT ⟨type⟩ ⟨output variables⟩

.PROBE generates a data file *.DAT which can be plotted in post-analysis by evoking the Probe program. The syntax is

.PROBE [⟨output variables⟩]

EXAMPLE 15.8 Find the value of V_s in the circuit in Fig. 15-8 such that the power dissipated in the 1-kΩ resistor is zero. Use the .DC command to sweep V_s from 1 to 6 V in steps of 1 V and use .PRINT to show I(Vs), V(1,2), and V(2).

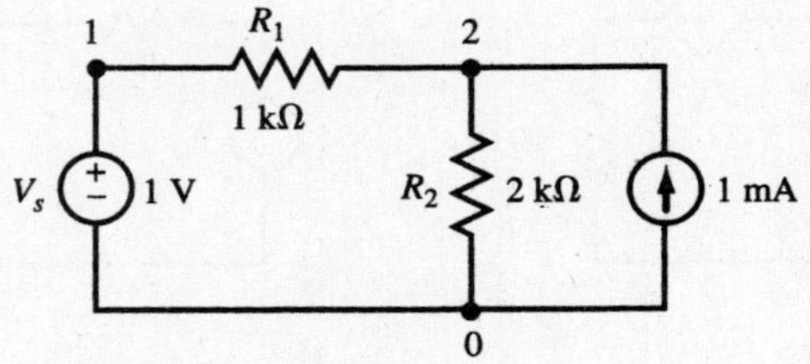

Fig. 15-8

The source file is

```
DC sweep, Fig. 15-8
Vs      1 0     DC      1 V
Is      0 2     DC      1 mA
R1      1 2     1 k
R2      0 2     2 k
.DC     Vs   1 6        1
.PRINT DC I(Vs)  V(1,2)     V(2)
.END
```

The results in the output file are

DC TRANSFER CURVES

Vs	I(Vs)	V(1,2)	V(2)
1.000E + 00	3.333E − 04	−3.333E − 01	1.333E + 00
2.000E + 00	−1.333E − 12	1.333E − 09	2.000E + 00
3.000E + 00	−3.333E − 04	3.333E − 01	2.667E + 00
4.000E + 00	−6.667E − 04	6.667E − 01	3.333E + 00
5.000E + 00	−1.000E − 03	1.000E + 00	4.000E + 00
6.000E + 00	−1.333E − 03	1.333E + 00	4.667E + 00

The answer is $V_s = 2$ V.

EXAMPLE 15.9 Write the source file for the circuit in Fig. 15-9(*a*) using commands .DC, .PLOT, and .PROBE to find the *I*-*V* characteristic equation for *I* varying from 0 to −2 A at the terminal *AB*.

First, we connect a dc current source I_{add} at terminal *AB*, "sweep" its value from 0 to −2 A using the .DC command, and plot *V* versus *I*. Since the circuit is linear, two points are necessary and sufficient. However, for clarity of the plot, ten steps are included in the source file as follows:

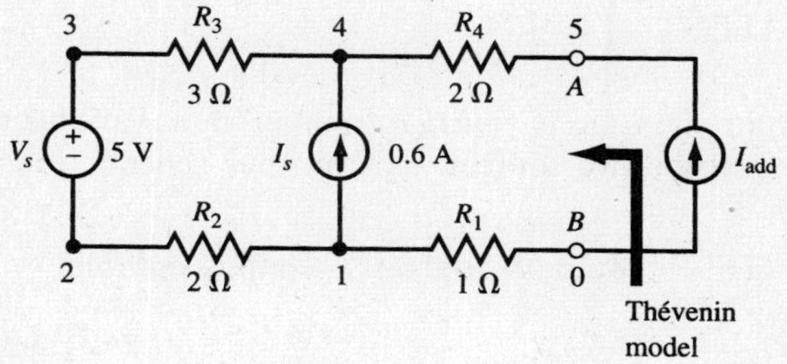

(a)

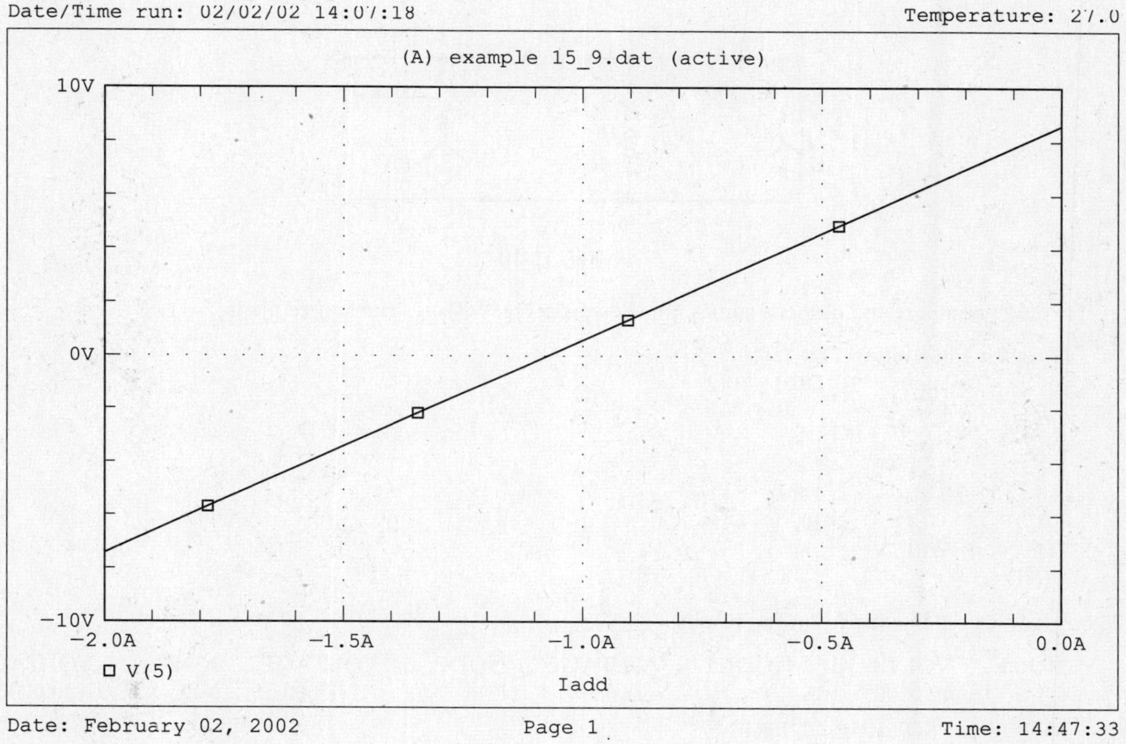

(b)

Fig. 15-9

```
Terminal Characteristic, Fig. 15-9
Iadd     0 5     DC      0
Is       0 4     DC      0.6 A
Vs       3 2     DC      5 V
R1       0 1     1
R2       1 2     2
R3       3 4     3
R4       4 5     2
.DC      Iadd    0    −2 0.2
.PLOT    DC      V(5)
.PROBE
.END
```

The output is shown in Fig. 15-9(*b*). The *I-V* equation is $V = 8I + 8.6$.

15.6 THÉVENIN EQUIVALENT

.TF Statement

The .TF command provides the *transfer function* from an input variable to an output variable and produces the resistances seen by the two sources. It can thus generate the Thévenin equivalent of a resistive circuit. The syntax is

.TF ⟨output variable⟩ ⟨input variable⟩

EXAMPLE 15.10 Use the command .TF to find the Thévenin equivalent of the circuit seen at terminal AB in Fig. 15-10.

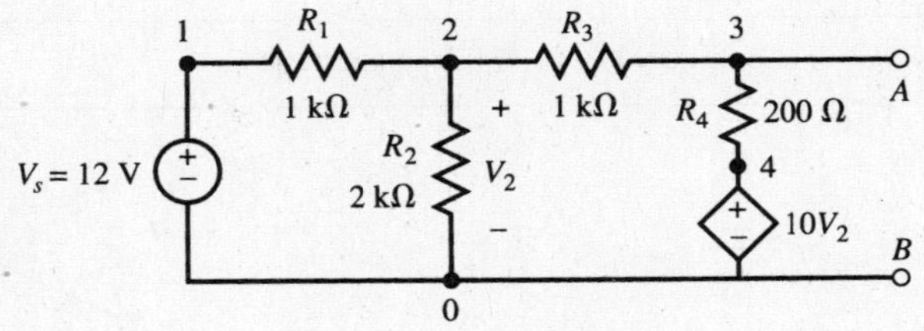

Fig. 15-10

The node numbers and element names are shown on Fig. 15-10. The source file is

```
Transfer Function in Fig. 15-10
Vs      1 0     DC    12
E1      4 0     2 0   10
R1      1 2     1 k
R2      2 0     2 k
R3      2 3     1 k
R4      3 4     200
.TF     V(3) Vs
.END
```

The output file contains the following results:

```
NODE   VOLTAGE   NODE   VOLTAGE   NODE   VOLTAGE    NODE   VOLTAGE
(1)    12.0000   (2)    -2.0000   (3)    -17.0000   (4)    -20.000

VOLTAGE SOURCE CURRENTS
NAME        CURRENT
Vs          -1.400E - 02

TOTAL POWER DISSIPATION     1.68E - 01   WATTS

SMALL-SIGNAL CHARACTERISTICS
V(3)/Vs = -1.417E + 00
INPUT RESISTANCE AT Vs = 8.571E + 02
OUTPUT RESISTANCE AT V(3) = -6.944E + 01
```

Therefore, $V_{\text{Th}} = -1.417(12) = -17$ V and $R_{\text{Th}} = -69.44\ \Omega$.

15.7 OP AMP CIRCUITS

Operational amplifiers may be modeled by high input impedance and high gain voltage-controlled voltage sources. The model may then be used within a net list repeatedly.

EXAMPLE 15.11 Find the transfer function V_3/V_s in the ideal op amp circuit of Fig. 15-11(*a*). The op amp is replaced by a voltage-dependent voltage source with a gain of 10^6 [see Fig. 15-11(*b*)]. The source file is

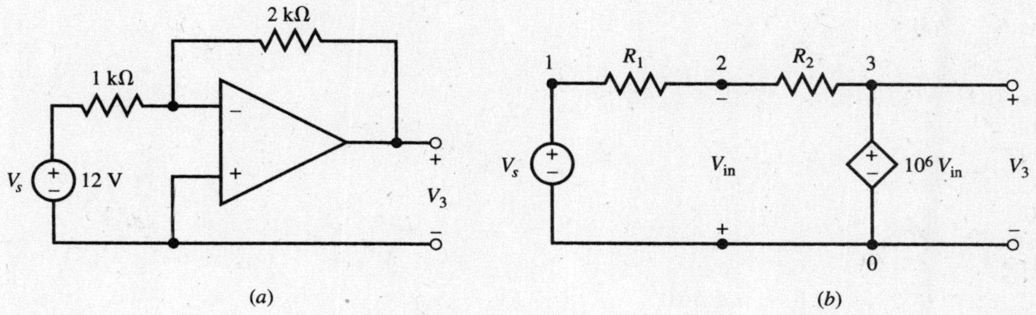

Fig. 15-11

```
Inverting op amp circuit, Fig. 15-11
Vs      1 0     DC     12
E1      3 0     0 2    1E6
R1      1 2     1 k
R2      2 3     2 k
.TF     V(3)  Vs
.END
```

The transfer function is written in the output file:

```
NODE   VOLTAGE    NODE   VOLTAGE       NODE   VOLTAGE
(1)    12.0000    (2)    24.00E – 06   (3)    – 24.0000

VOLTAGE SOURCE CURRENTS
NAME        CURRENT
Vs          – 1.200E – 02

TOTAL POWER DISSIPATION     1.44E – 01 WATTS

SMALL-SIGNAL CHARACTERISTICS
V(3)/Vs = –2.000E + 00
INPUT RESISTANCE AT Vs = 1.000E + 03
OUTPUT RESISTANCE AT V(3) = 0.000E + 00
```

.SUBCKT Statement

A *subcircuit* is defined by a set of statements beginning with

.SUBCKT ⟨name⟩ ⟨external terminals⟩

and terminating with an .ENDS statement. Within a netlist we refer to a subcircuit by

Xaa ⟨name⟩ ⟨nodes⟩

Hence, the .SUBCKT statement can assign a name to the model of an op amp for repeated use.

EXAMPLE 15.12 Given the circuit of Fig. 15-12(*a*), find I_s, I_f, V_2, and V_6 for V_s varying from 0.5 to 2 V in 0.5-V steps. Assume a practical op amp [Fig. 15-12(*b*)], with $R_{in} = 100\,\text{k}\Omega$, $C_{in} = 10$ pF, $R_{out} = 10\,\text{k}\Omega$, and an open loop gain of 10^5.

The source file employs the subcircuit named OPAMP of Fig. 15-12(*b*) whose description begins with .SUBCKT and ends with .ENDS. The X1 and X2 statements describe the two op amps by referring to the OPAMP subcircuit. Note the correspondence of node connections in the X1 and X2 statements with that of the external terminals specified in the .SUBCKT statement. The source file is

```
Op amp circuit of Fig. 15-12 using .SUBCKT
.SUBCKT     OPAMP         1 2 3 4
Rin     1 2       10 E5
Cin     1 2       10 pF
Rout    3 5       10 k
```

```
Eout    5  4          1  2      10 E5
.ENDS

Vs      1  0          DC     .5
Rs      1  2          1 k
R1      2  3          5 k
R2      3  4          9 k
R3      4  5          1.2 k
R4      5  6          6 k
Rf      6  2          40 k
X1      0  3    4  0  OPAMP
X2      0  5    6  0  OPAMP
.DC     Vs      0.5   2      0.5
.PRINT  DC      V(2)  V(6)   I(Vs)   I(R1)   I(Rf)
.TF     V(6)    Vs
.END
```

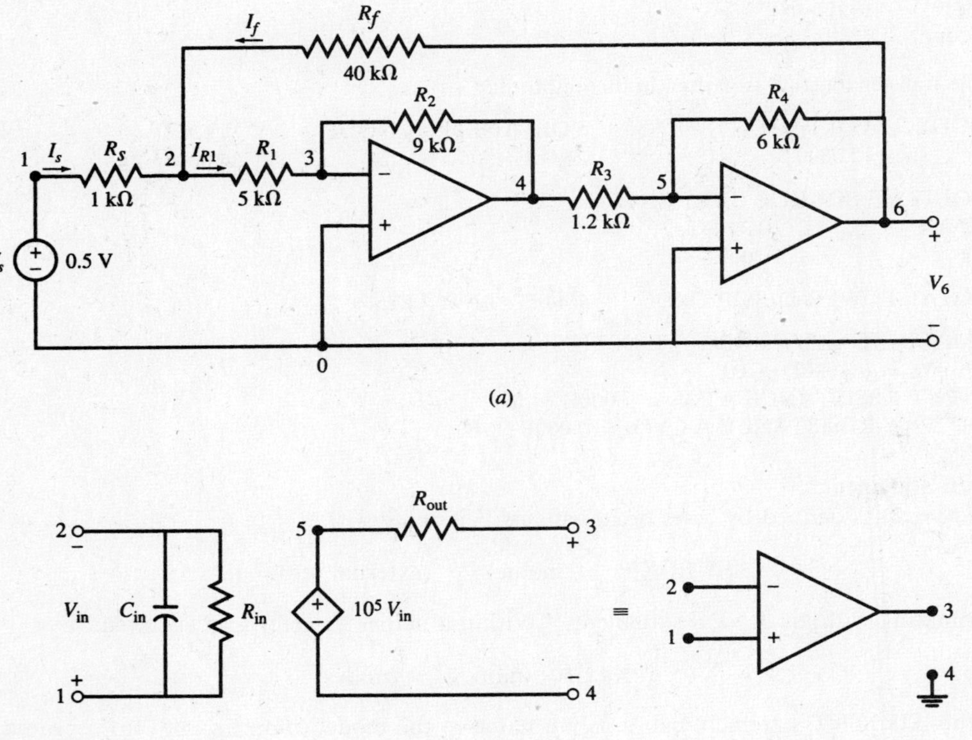

Fig. 15-12

The output file is

DC TRANSFER CURVES

Vs	V(2)	V(6)	I(Vs)	I(R1)	I(Rf)
5.000E – 01	5.000E – 01	4.500E + 00	–3.372E – 09	1.000E – 04	9.999E – 0
1.000E + 00	1.000E + 00	9.000E + 00	–6.745E – 09	2.000E – 04	2.000E – 0
1.500E + 00	1.500E + 00	1.350E + 01	–1.012E – 08	3.000E – 04	3.000E – 0
2.000E + 00	2.000E + 00	1.800E + 01	–1.349E – 08	4.000E – 04	4.000E – 0

NODE	VOLTAGE	NODE	VOLTAGE	NODE	VOLTAGE	NODE	VOLTAGE
(1)	.5000	(2)	.5000	(3)	9.400E − 06	(4)	−.9000
(5)	−13.00E − 06	(6)	4.4998	(X1.5)	−9.3996	(X2.5)	12.9990

VOLTAGE SOURCE CURRENTS

NAME	CURRENT
Vs	−3.372E − 09

TOTAL POWER DISSIPATION 1.69E − 09 WATTS

SMALL-SIGNAL CHARACTERISTICS
V(6)/Vs = 9.000E + 00
INPUT RESISTANCE AT Vs = 1.483E + 08
OUTPUT RESISTANCE AT V(6) = 7.357E − 02

There is no voltage drop across R_s. Therefore, $V(2) = V_s$ and the overall gain is $V(6)/V_s = V(2)/V_s = 9$. The current drawn by R_1 is provided through the feedback resistor R_f.

15.8 AC STEADY STATE AND FREQUENCY RESPONSE

Independent AC Sources

Independent ac sources are described by a statement with the following syntax:

⟨name⟩ ⟨nodes⟩ AC ⟨magnitude⟩ ⟨phase in degrees⟩

Voltage sources begin with V and current sources with I. The convention for direction is the same as that for dc sources.

EXAMPLE 15.13 Write data statements for the sources shown in Fig. 15-13.

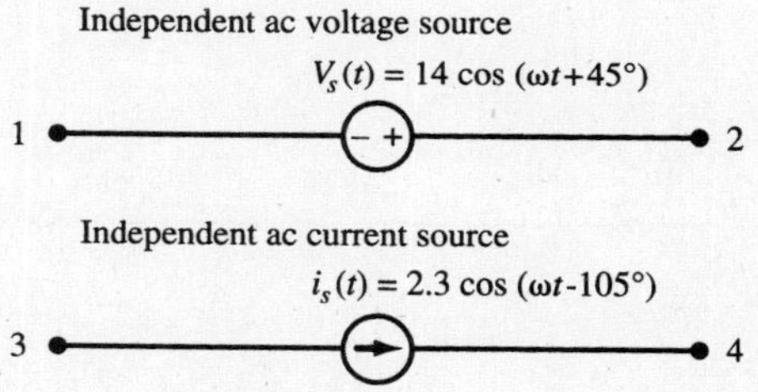

Fig. 15-13

AC Source	⟨name⟩	⟨nodes⟩	⟨type⟩	⟨magnitude⟩	⟨phase⟩
Voltage	Vs	2 1	AC	14	45
Current	Is	3 4	AC	2.3	−105

.AC Statement

The .AC command sweeps the frequency of all ac sources in the circuit through a desired range or sets it at a desired value. The syntax is

.AC ⟨sweep type⟩ ⟨number of points⟩ ⟨starting f⟩ ⟨ending f⟩

For the ac steady state, ⟨sweep type⟩ is LIN. In order to have a single frequency, the starting and ending frequencies are set to the desired value and the number of points is set to one.

.PRINT AC and .PLOT AC Statements

The .PRINT AC statement prints the magnitude and phase of the steady-state output. The syntax is

.PRINT AC ⟨magnitudes⟩ ⟨phases⟩

The magnitudes and phases of voltages are Vm(variable) and Vp(variable), respectively, and the magnitudes and phases of currents are Im(variable) and Ip(variable), respectively. The syntax for .PLOT AC is similar to that for .PRINT AC.

EXAMPLE 15.14 In the series *RLC* circuit of Fig. 15-14(*a*) vary the frequency of the source from 40 to 60 kHz in 200 steps. Find the magnitude and phase of current *I* using .PLOT and .PROBE.

The source file is

```
AC analysis of series RLC, Fig. 15-14
Vs        1 0       AC          1 0
R         1 2       32
L         2 3       2 m
C         3 0       5 n
.AC       LIN       200         40 k  60 k
.PLOT     AC   Im(Vs)   Ip(Vs)
.PROBE    Vm(1, 2)   Vm(2,3)    Vm(3)      Im(Vs)      Ip(Vs)
.END
```

The graph of the frequency response, plotted by Probe, is shown in Fig. 15-14(*b*).

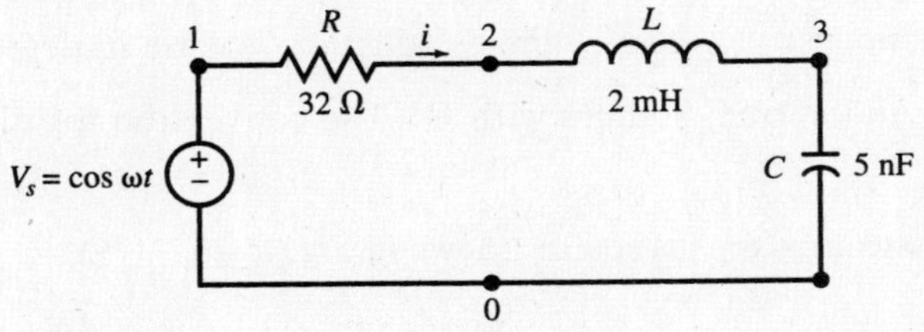

(*a*)

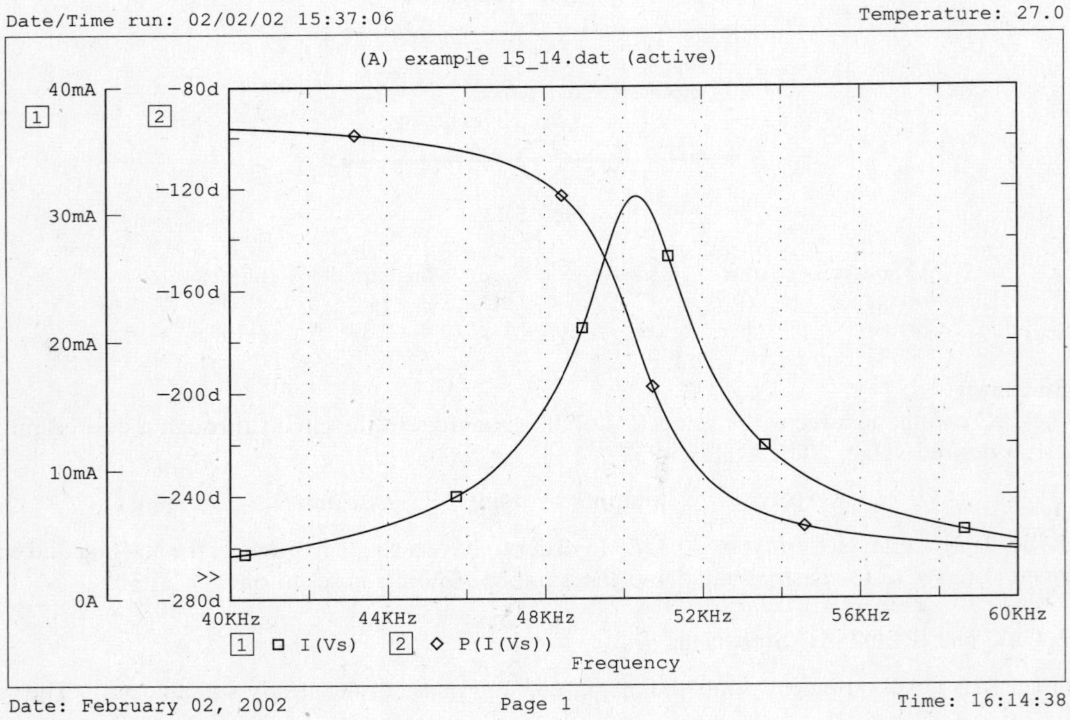

(*b*)

Fig. 15-14

15.9 MUTUAL INDUCTANCE AND TRANSFORMERS

The mutual inductance between inductors is modeled by a device whose name begins with K. The data statement syntax is

⟨name⟩ ⟨inductor 1⟩ ⟨inductor 2⟩ ⟨coupling coefficient⟩

The dot rule, which determines the sign of the mutual inductance term, is observed by making the dotted end of each inductor the first node entered in its data statement.

EXAMPLE 15.15 Write the three data statements which describe the coupled coils of Fig. 15-15.

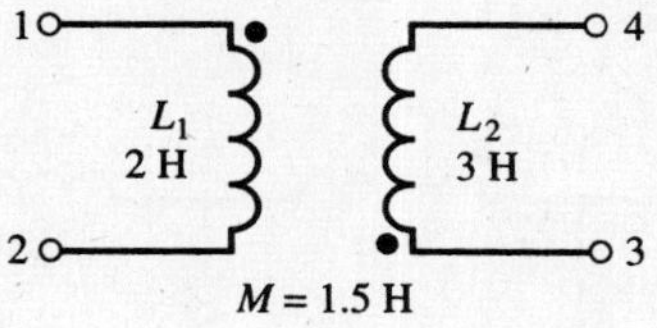

Fig. 15-15

The coupling coefficient is $k12 = 1.5/\sqrt{2(3)} = 0.61$. The netlist contains the following:

```
L1    1    2    2
L2    3    4    3
K12   L1   L2   0.61
```

EXAMPLE 15.16 Plot the input impedance $Z_{in} = V_1/I_1$ in the circuit of Fig. 15-16(*a*) for f varying from 0.01 to 1 Hz.

To find Z_{in}, we connect a 1-A ac current source running from node 0 to node 1 and plot the magnitude and phase of the voltage V(1) across it. The source file is

```
AC analysis of coupled coils, Fig. 15-16
IADD     0     1     AC          1        0
C        0     1     1 000 000 uF
R        0     2     3
L1       1     2     2 H
L2       3     2     5 H
K12      L1    L2    0.6325 H
L3       0     3     1 H
.AC      LIN         20    .01   1
.PRINT   AC        Vm(1)       Vp(1)
.PROBE
.END
```

Vm(1) and Vp(1), which are the magnitude and phase of Z_{in}, are plotted by using Probe and the graph is shown in Fig. 15-16(*b*). Note that the maximum occurs at about 100 mHz.

15.10 MODELING DEVICES WITH VARYING PARAMETERS

.MODEL Statement

The parameters of a passive element can be varied by using .MODEL statement. The syntax is

.MODEL ⟨name⟩ ⟨type⟩ [(⟨parameter⟩ = ⟨value⟩)]

where ⟨name⟩ is the name assigned to the element. For passive linear elements, ⟨type⟩ is

RES for resistor
IND for inductor
CAP for capacitor

We can sweep the parameter of the model though a desired range at desired steps by using the .STEP statement:

.STEP LIN ⟨name⟩ ⟨initial value⟩ ⟨final value⟩ ⟨step size⟩

As an example, the following two statements use .MODEL and .STEP commands to define a resistor called *heater* with the resistance parameter varying from 20 to 40 Ω in 5 steps generating 20, 25, 30, 35, and 40 Ω.

```
.MODEL   heater   RES(R = 20)
.STEP    RES      heater(R)    20   40   5
```

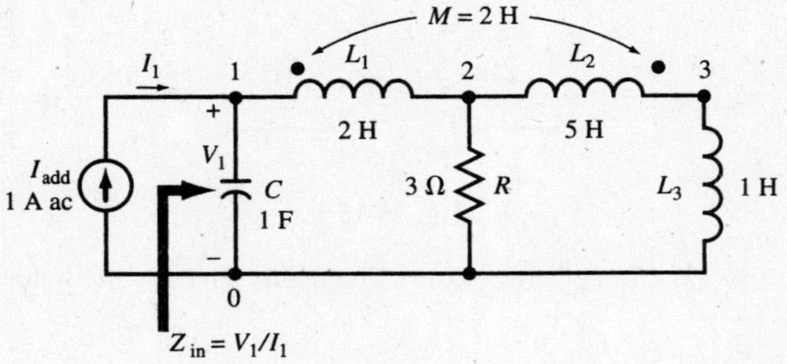

(*a*)

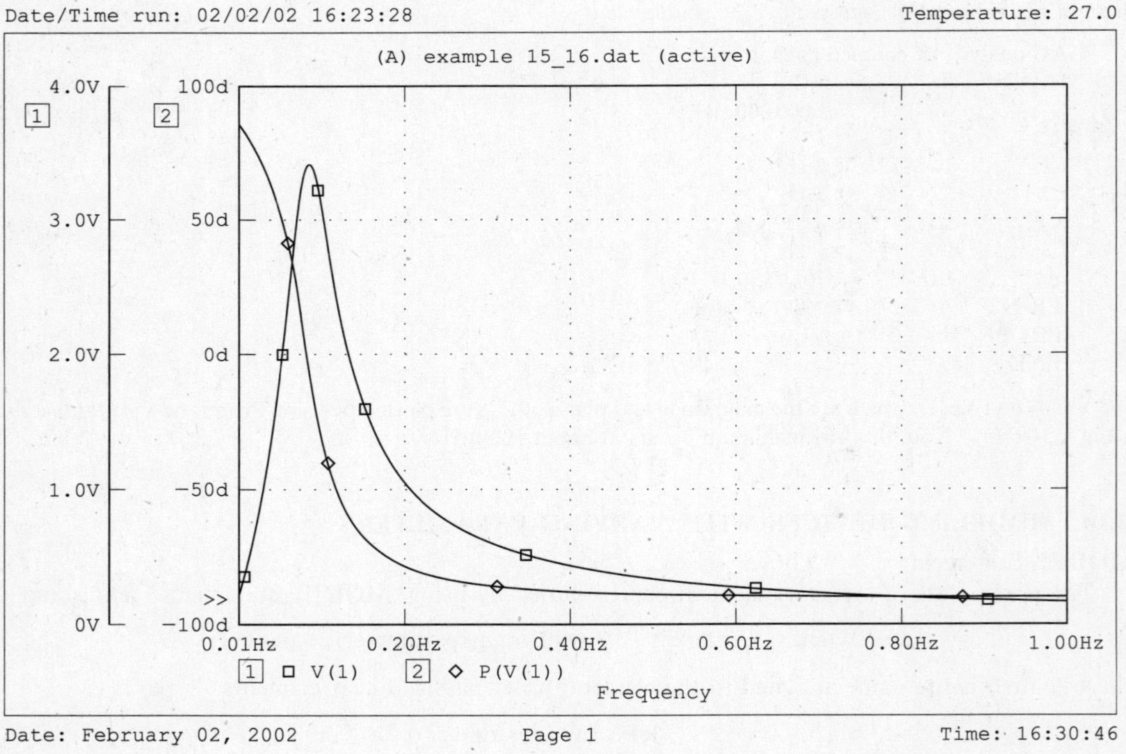

(*b*)

Fig. 15-16

EXAMPLE 15.17 Use Probe to plot V in the circuit in Fig. 15-17(a) for f varying from 1 to 3 kHz in 100 steps. also, R from 500 Ω to 1 kΩ in steps of 100 Ω.

Using .MODEL command we create the resistor RLeak and sweep its value by .STEP in the following source file. The graph of the frequency response V versus f is plotted by using Probe and it is shown in Fig. 15-17(b).

```
Parallel resonance with variable R, Fig. 15-17
I          0 1     AC           1 m  0
R          1 0     RLeak        1
L          1 0     10 m
C          1 0     1 u
.MODEL     RLeak   RES(R = 1)
.STEP      LIN     RES          RLeak(R)    500    1 k    100
.AC        LIN     100    1 k   3 k
.PROBE
.END
```

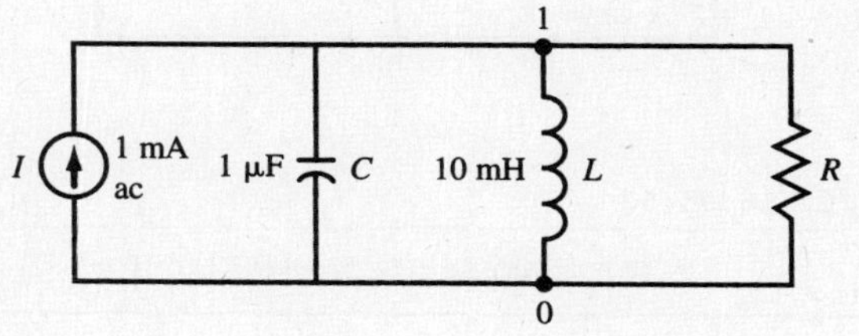

(a)

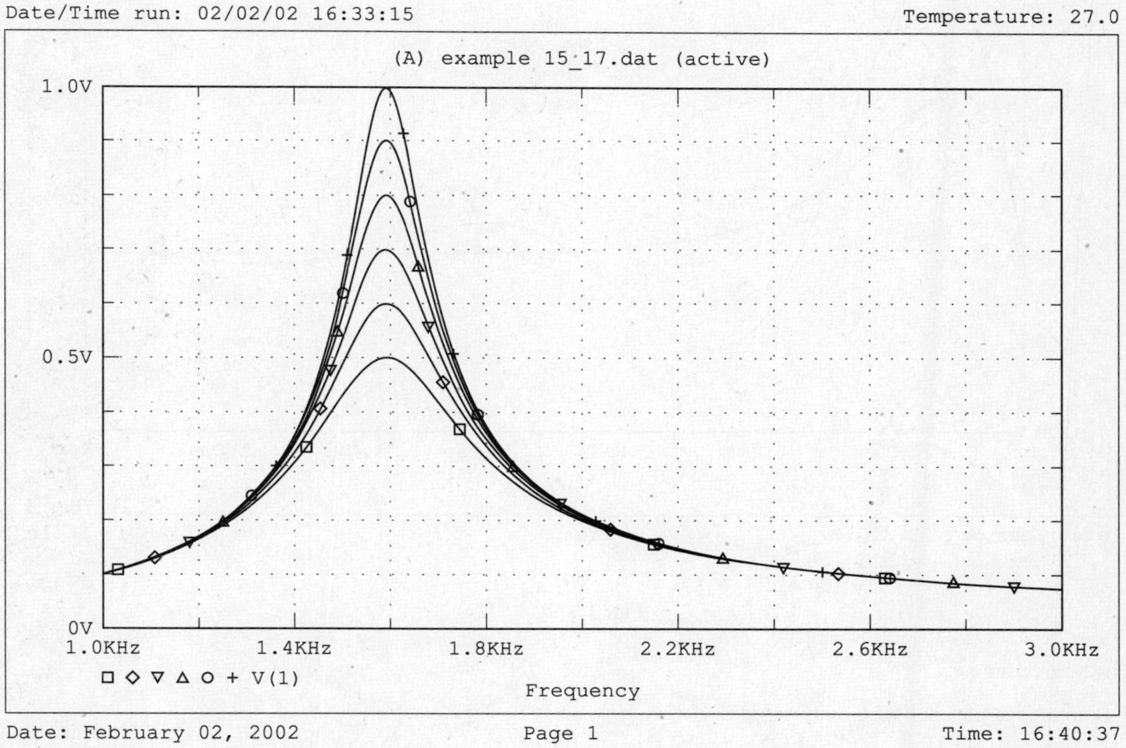

(b)

Fig. 15-17

15.11 TIME RESPONSE AND TRANSIENT ANALYSIS

.TRAN statement

Time responses, such as natural responses to initial conditions in a source-free circuit and responses to step, pulse, exponential, or other time-dependent inputs, are produced by the .TRAN statement. The response begins at $t = 0$. The increment size and final time value are given in the following statement:

.TRAN ⟨increment size⟩ ⟨final time value⟩

EXAMPLE 15.18 Use .TRAN and .PROBE to plot the voltage across the parallel *RLC* combination in Fig. 15-18(*a*) for $R = 50\ \Omega$ and $150\ \Omega$ for $0 < t < 1.4$ ms. The initial conditions are $I(0) = 0.5$ A and $V(0) = 0$.

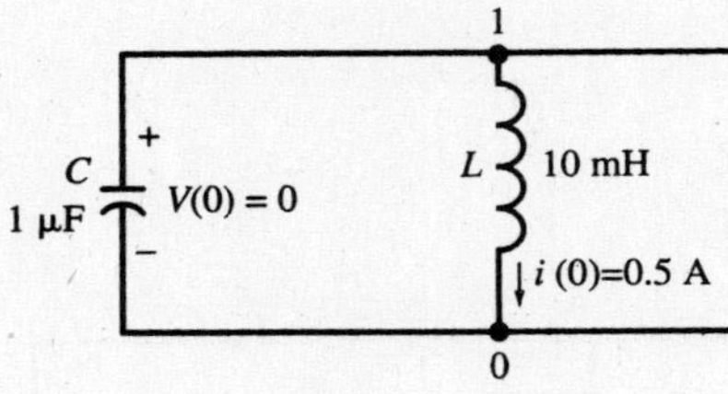

(*a*)

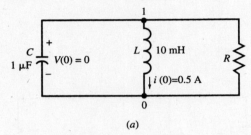

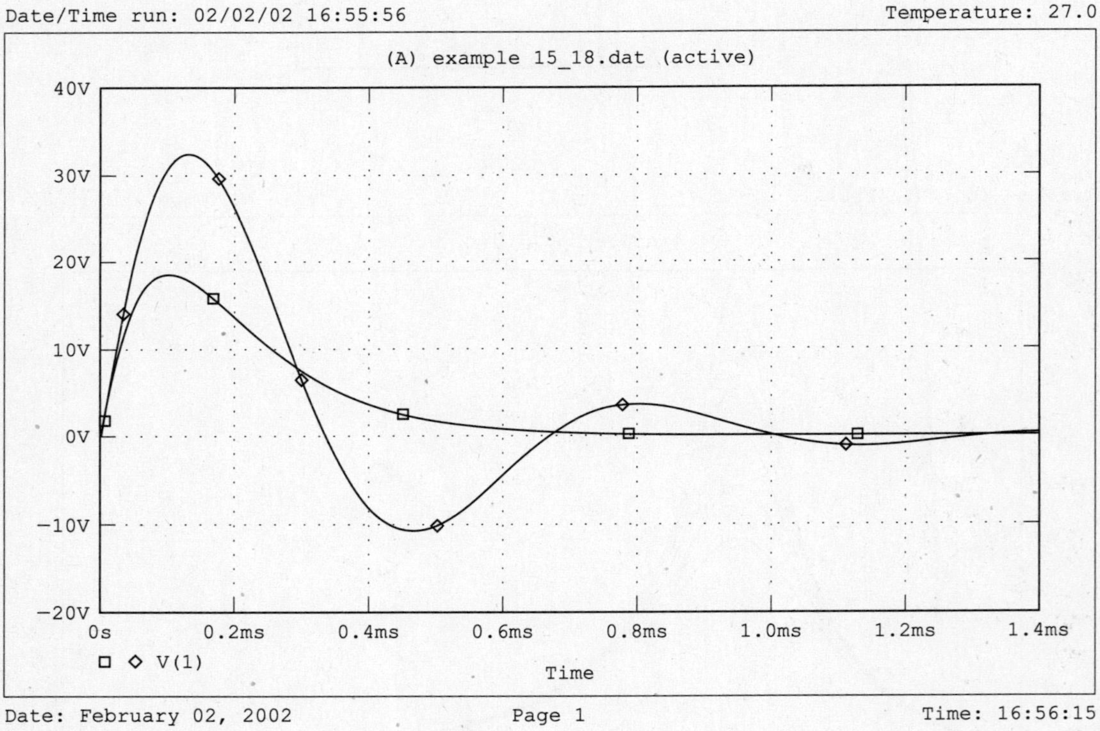

(*b*)

Fig. 15-18

The source file is

```
Source-free parallel RLC with variable R
R        1 0     LOSS        1
L        0 1     10 m        IC = .5
C        1 0     1 u         IC = 0
.MODEL   LOSS    RES(R = 6)
```

```
.STEP    RES        LOSS(R)    50    150    100
.TRAN    2.0E - 6   1.4E - 3   UIC
.PROBE
.END
```

Figure 15-18(*b*) shows the graph of the voltage plotted by Probe. For $R = 50\ \Omega$ there are no oscillations.

15.12 SPECIFYING OTHER TYPES OF SOURCES

Time-dependent sources which include dc, ac, and transient components are expressed by

$$\langle\text{name}\rangle \quad \langle\text{nodes}\rangle \quad \langle\text{dc comp.}\rangle \quad \langle\text{ac comp.}\rangle \quad \langle\text{transient comp.}\rangle$$

The default for the unspecified dc or ac component is zero. The transient component appears for $t > 0$. Several transient components are described below.

Exponential Source

The source starts at a constant initial value V_0. At t_0, it changes exponentially from V_0 to a final value V_1 with a time constant tau1. At $t = T$, it returns exponentially to V_0 with a time constant tau2. Its syntax is

$$\text{EXP}(\text{V}_0 \quad \text{V}_1 \quad \text{t}_0 \quad \text{tau1} \quad \text{T} \quad \text{tau2})$$

EXAMPLE 15.19 A 1-V dc voltage source starts increasing exponentially at $t = 5$ ms, with a time constant of 5 ms and an asymptote of 2 V. After 15 ms, it starts decaying back to 1 V with a time constant of 2 ms. Write the data statement for the source and use Probe to plot the waveform.

The data statement is

```
Vs  1  0  EXP(1  2  5m  5m  20m  2m)
```

The waveform is plotted as shown in Fig. 15-19.

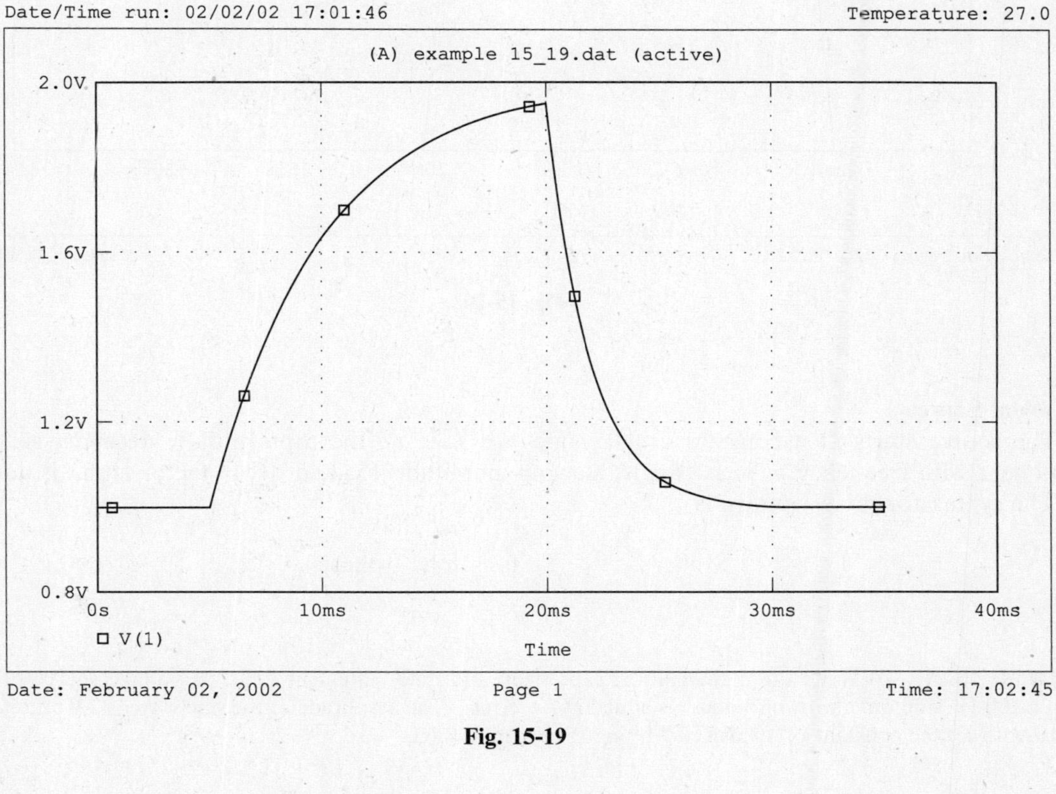

Fig. 15-19

Pulse Source

A periodic pulse waveform which goes from V_0 to V_1 and back can be represented by

$$\text{PULSE}(V_0 \quad V_1 \quad \text{delay} \quad \text{risetime} \quad \text{falltime} \quad \text{duration} \quad \text{period})$$

EXAMPLE 15.20 (*a*) Write the data statement for a pulse waveform which switches 10 times in one second between 1 V and 2 V, with a rise and fall time of 2 ms. The pulse stays at 2 V for 11 ms. The first pulse starts at $t = 5$ ms. (*b*) Using Probe, plot the waveform in (*a*).

(*a*) The data statement is

Vs 1 0 PULSE(1 2 5m 2m 2m 11m 100m)

(*b*) The waveform is plotted as shown in Fig. 15-20.

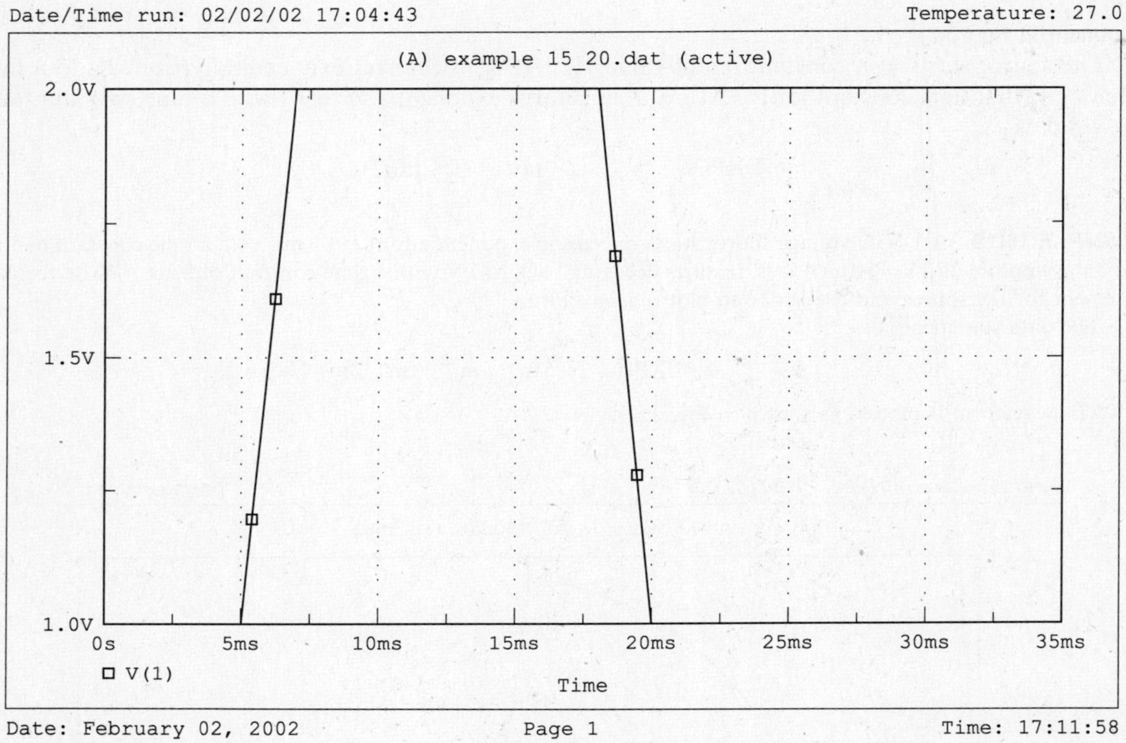

Fig. 15-20

Sinusoidal Source

The source starts at a constant initial value V_0. At t_0, the exponentially decaying sinusoidal component with frequency f, phase angle, starting amplitude V_1, and decay factor alpha is added to it. The syntax for the waveform is

$$\text{SIN}(V_0 \quad V_1 \quad f \quad t_0 \quad \text{alpha} \quad \text{phase})$$

EXAMPLE 15.21 (*a*) Write the mathematical expression and data statement for a dc voltage source of 1 V to which a 100-Hz sine wave with zero phase is added at $t = 5$ ms. The amplitude of the sine wave is 2 V and it decays to zero with a time constant of 10 ms. (*b*) Using Probe, plot $V_s(t)$.

(*a*) The decay factor is the inverse of the time constant and is equal to alpha = 1/0.01 = 100. For $t > 0$, the voltage is expressed by

$$V_s(t) = 1 + 2e^{-100(t-0.005)} \sin 628.32(t - 0.005)u(t - 0.005)$$

The data statement is

```
Vs   1 0   SIN(1  2  100  5m  100)
```

(*b*) The waveform is plotted as shown in Fig. 15-21.

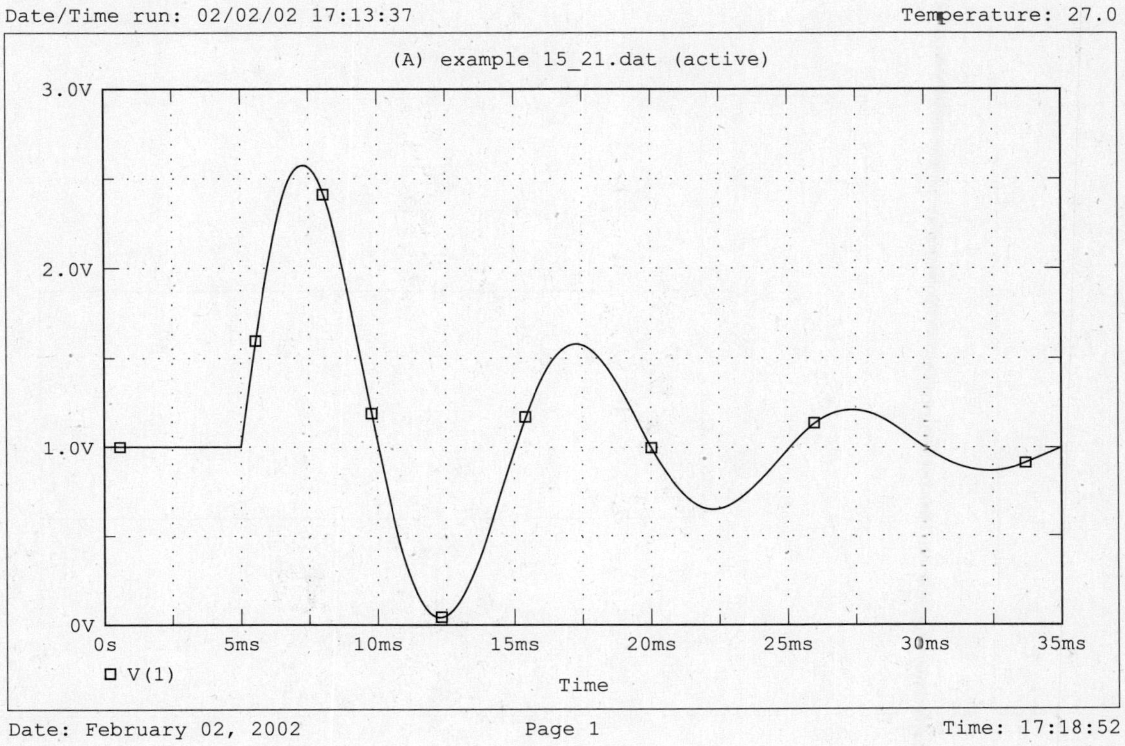

Fig. 15-21

EXAMPLE 15.22 Find the voltage across a 1-μF capacitor, with zero initial charge, which is connected to a voltage source through a 1-kΩ resistor as shown in the circuit in Fig. 15-22(*a*). The voltage source is described by

$$V_s = \begin{cases} 15.819 \text{ V} & \text{for } 0 < t < 1 \text{ ms} \\ 10 \text{ V} & \text{for } t > 1 \text{ ms} \end{cases}$$

We use the exponential waveform to represent V_s. The file is

```
Dead-beat Pulse-Step response of RC
Vs      1 0      EXP(10  15.819  0  1.0E-6  1.0E-3  1.0E-6)
R       1 2      1 k
C       2 0      1 uF
.TRAN   1.0E-6   5.0E-3   UIC
.PROBE
.END
```

The graph of the capacitor voltage is shown in Fig. 15-22(*b*). During $0 < t < 1$ ms, the transient response grows exponentially toward a dc steady-state value of 15.819 V. At $t = 1$ ms, the response reaches the value of 10 V. Also at $t = 1$ ms, the voltage source drops to 10 V. Since the source and capacitor voltages are equal, the current in the resistor becomes zero and the steady state is reached. The transient response lasts only 1 ms.

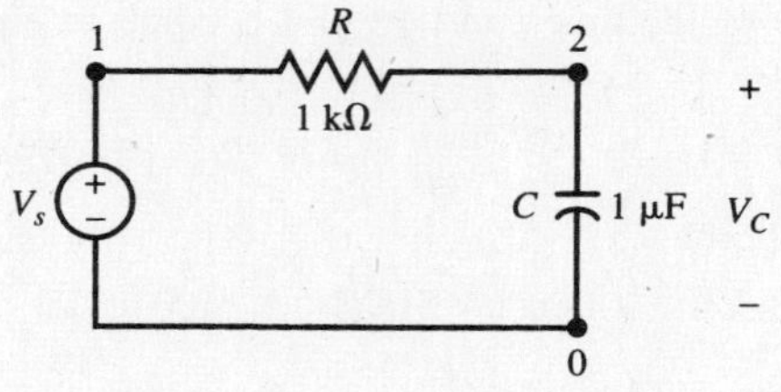

(*a*)

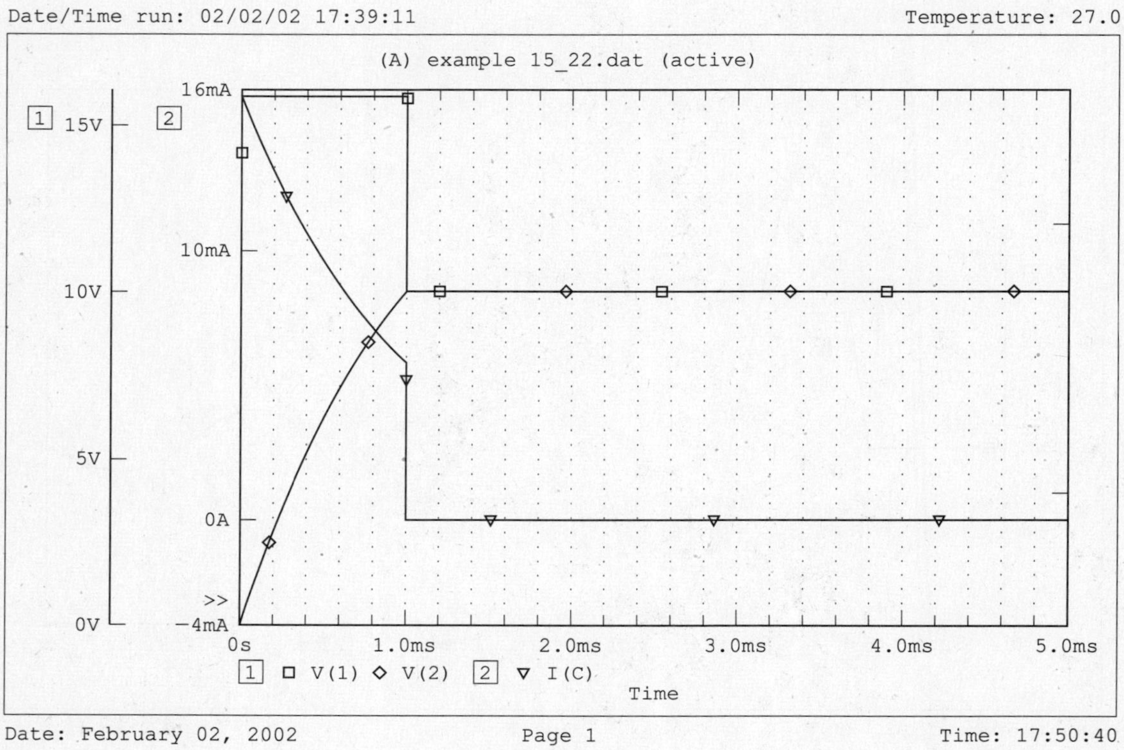

(*b*)

Fig. 15-22

15.13 SUMMARY

In addition to the linear elements and sources used in the preceding sections, nonlinear devices, such as diodes (Dxx), junction field-effect transistors (Jxx), mosfets (Mxx), transmission lines (Txx), voltage controlled switches (Sxx), and current controlled switches (Wxx), may be included in the netlist. Sensitivity analysis is done using the .SENS statement. Fourier analysis is done using the .FOUR statement. These can be found in books or manuals for PSpice or Spice. The following summarizes the statements used in this chapter.

Data Statements:

R, L, C	⟨name⟩	⟨nodes⟩	⟨value⟩	[⟨initial conditions⟩]
Mutual Inductance	kxx	⟨ind.a⟩	⟨ind.b⟩	⟨coupling coefficient⟩
Subcircuit Call	Xxx	⟨name⟩	⟨connection nodes⟩	
DC Voltage source	Vxx	⟨nodes⟩	DC	⟨value⟩
DC Current source	Ixx	⟨nodes⟩	DC	⟨value⟩

AC Voltage source	Vxx	⟨nodes⟩	AC	⟨magnitude⟩	⟨phase⟩
AC Current source	Ixx	⟨nodes⟩	AC	⟨magnitude⟩	⟨phase⟩
VCVS	Exx	⟨nodes⟩	⟨control⟩	⟨gain⟩	
CCCS	Fxx	⟨nodes⟩	⟨control⟩	⟨gain⟩	
VCCS	Gxx	⟨nodes⟩	⟨control⟩	⟨gain⟩	
CCVS	Hxx	⟨nodes⟩	⟨control⟩	⟨gain⟩	

Control Statements:

.AC	⟨sweep type⟩	⟨number of points⟩	⟨starting f⟩	⟨ending f⟩	
.DC	⟨name⟩	⟨initial value⟩	⟨final value⟩	⟨step size⟩	
.END					
.ENDS					
.IC	⟨V(node) = value⟩				
.MODEL	⟨name⟩	⟨type⟩ [(⟨parameter⟩ = ⟨value⟩)]			
		⟨type⟩ is RES for resistor			
		⟨type⟩ is IND for inductor			
		⟨type⟩ is CAP for capacitor			
.LIB	[⟨file name⟩]				
.OP					
.PRINT	DC	⟨output variables⟩			
.PLOT	DC	⟨output variables⟩			
.PRINT	AC	⟨magnitudes⟩	⟨phases⟩		
.PLOT	AC	⟨magnitudes⟩	⟨phases⟩		
.PRINT	TRAN	⟨output variables⟩			
.PROBE	[⟨output variables⟩]				
.STEP LIN	⟨type⟩	⟨name(param.)⟩	⟨initial value⟩	⟨final value⟩	⟨step size⟩
.SUBCKT	⟨name⟩	⟨external terminals⟩			
.TF	⟨output variable⟩	⟨input source⟩			
.TRAN	⟨increment size⟩	⟨final value⟩			

Solved Problems

15.1 Use PSpice to find $V(3, 4)$ in the circuit of Fig. 15-23.

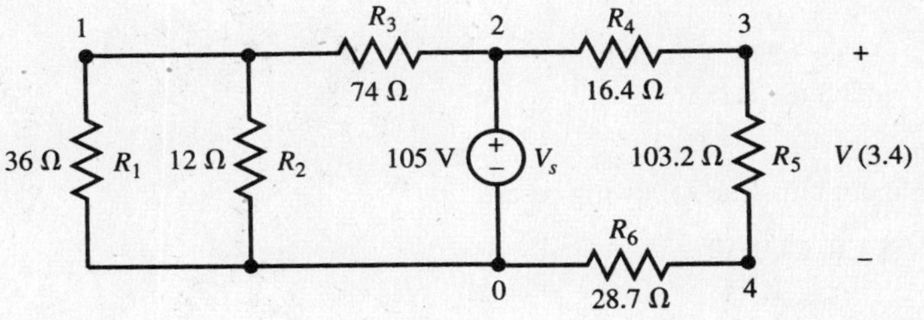

Fig. 15-23

The source file is

```
DC analysis, Fig. 15-23
Vs      2    0      DC        105 V
R1      0    1      36
R2      0    1      12
```

```
R3      1    2     74
R4      2    3     16.4
R5      3    4     103.2
R6      4    0     28.7
.DC     Vs   105   105      1
.PRINT  DC   V(1)  V(3, 4)
.END
```

The output file contains the following:

```
DC TRANSFER CURVES
Vs            V(1)          V(3, 4)
1.050E + 02   1.139E + 01   7.307E + 01
```

Therefore, $V(3, 4) = 73.07$ V.

15.2 Write the source file for the circuit of Fig. 15-24 and find I in R_4.

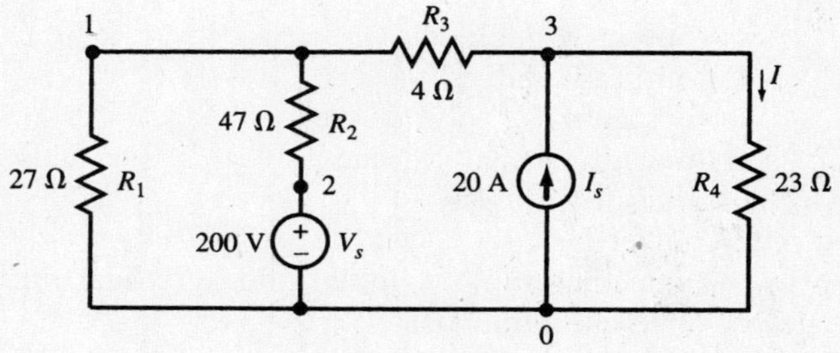

Fig. 15-24

The source file is

```
DC analysis, Fig. 15-24
VS       2    0      DC     200 V
Is       0    3      DC     20 A
R1       0    1      27
R2       1    2      47
R3       1    3      4
R4       3    0      23
.DC      Vs   200    200  1
.PRINT   DC   I(R4)
.END
```

The output file contains the following results:

```
DC TRANSFER CURVE
Vs            I(R4)
2.000E + 02   1.123E + 01
```

Current I(R4) = 11.23 A flows from node 3 to node 0 according to the order of nodes in the data statement for R4.

15.3 Find the three loop currents in the circuit of Fig. 15-25 using PSpice and compare your solution with the analytical approach.

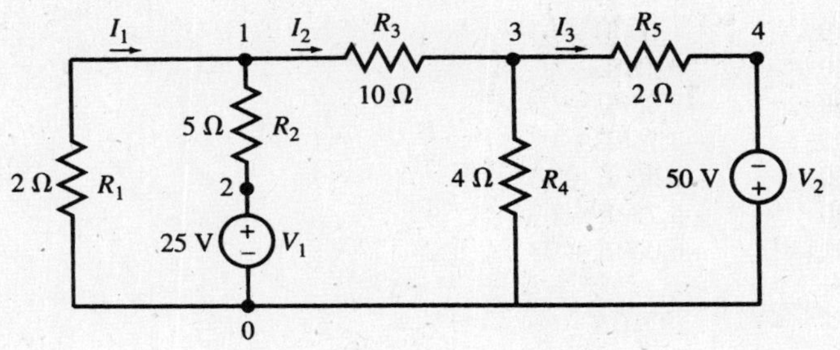

Fig. 15-25

The source file is

```
DC analysis, Fig. 15-25
V1        2    0      DC     25
V2        0    4      DC     50
R1        0    1      2
R2        1    2      5
R3        1    3      10
R4        3    0      4
R5        3    4      2
.DC       V1   25     25     1
.PRINT    DC   I(R1)  I(R3)  I(R5)
.END
```

The output file includes the following results:

```
DC TRANSFER CURVES
V1           I(R1)          I(R3)         I(R5)
2.500E+01    -1.306E+00     3.172E+00     1.045E+01
```

The analytical solution requires solving three simultaneous equations.

15.4 Using PSpice, find the value of Vs in Fig. 15-4 such that the voltage source does not supply any power.

We sweep Vs from 1 to 10 V. The source and output files are

```
DC sweep in the circuit of Fig. 15-4
R1        0    1      500
R2        1    2      3 k
R3        2    3      1 k
R4        0    3      1.5 k
Vs        3    1      DC     4 V
Is        0    2      DC     3 mA
.DC       Vs          1      10     1
.PRINT    DC   I(Vs)
.PROBE
.PLOT     DC   I(Vs)
.END
```

The output file contains the following results:

```
DC TRANSFER CURVES
Vs             I(Vs)
1.000E+00      7.500E-04
2.000E+00      -2.188E-12
3.000E+00      -7.500E-04
```

4.000E + 00	−1.500E − 03
5.000E + 00	−2.250E − 03
6.000E + 00	−3.000E − 03
7.000E + 00	−3.750E − 03
8.000E + 00	−4.500E − 03
9.000E + 00	−5.250E − 03
1.000E + 01	−6.000E − 03

The current in Vs is zero for Vs = 2 V.

15.5 Perform a dc analysis on the circuit of Fig. 15-26 and find its Thévenin equivalent as seen from terminal *AB*.

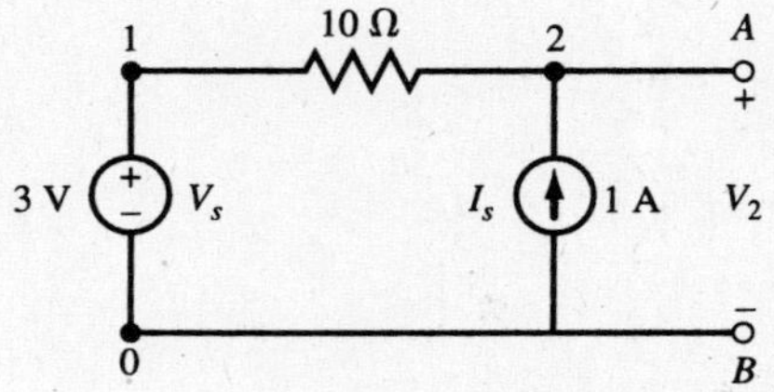

Fig. 15-26

We include a .TF statement in the following netlist:

```
Thévenin equivalent of Fig. 15-26
Vs      1     0     DC     3
R1      1     2     10
Is      0     2     DC     1
.TF     V(2)        Is
.END
```

The output file includes the following results:

```
NODE     VOLTAGE     NODE     VOLTAGE
(1)      3.0000      (2)      13.000

VOLTAGE SOURCE CURRENTS
NAME     CURRENT
Vs       1.000E + 00

TOTAL POWER DISSIPATION     −3.00E + 00     WATTS

SMALL-SIGNAL CHARACTERISTICS
V(2)/Is = 1.000E + 01
INPUT RESISTANCE AT Is = 1.000E + 01
OUTPUT RESISTANCE AT V(2) = 1.000E + 01
```

The Thévenin equivalent is $V_{\text{Th}} = V_2 = 13$ V, $R_{\text{Th}} = 10\ \Omega$.

15.6 Perform an ac analysis on the circuit of Fig. 15-27(*a*). Find the complex magnitude of $\mathbf{V}_2$ for f varying from 100 Hz to 10 kHz in 10 steps.

We add to the netlist an .AC statement to sweep the frequency and obtain V(2) by any of the commands .PRINT, .PLOT, or .PROBE. The source file is

```
AC analysis of Fig. 15-27(a).
Vs      1     0     AC     10     0
R1      1     2     1 k
R2      2     0     2 k
```

```
C          2    0      1 uF
.AC        LIN         10        100      10000
.PRINT     AC          Vm(2)              Vp(2)
.PLOT      AC          Vm(2)              Vp(2)
.PROBE                 Vm(2)              Vp(2)
.END
```

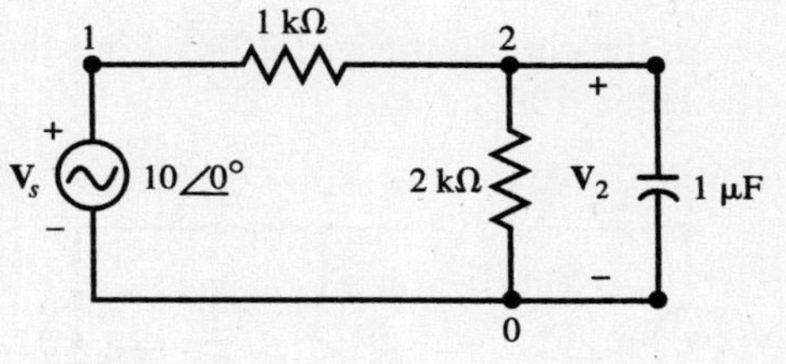

(*a*)

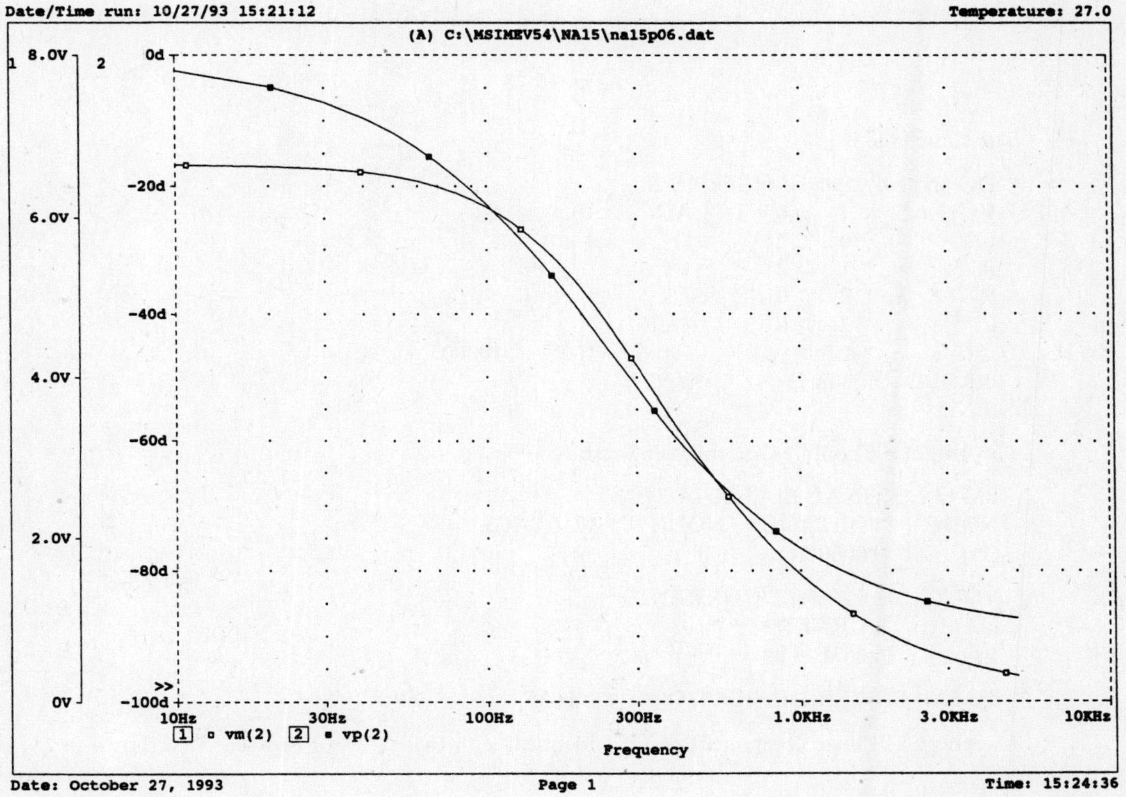

(*b*)

Fig. 15-27

The output file contains the following results:

AC ANALYSIS

FREQ	VM(2)	VP(2)
1.000E + 02	6.149E + 00	−2.273E + 01
1.200E + 03	1.301E + 00	−7.875E + 01
2.300E + 03	6.883E − 01	−8.407E + 01
3.400E + 03	4.670E − 01	−8.598E + 01
4.500E + 03	3.532E − 01	−8.696E + 01

```
5.600E + 03      2.839E - 01      -8.756E + 01
6.700E + 03      2.374E - 01      -8.796E + 01
7.800E + 03      2.039E - 01      -8.825E + 01
8.900E + 03      1.788E - 01      -8.846E + 01
1.000E + 04      1.591E - 01      -8.863E + 01
```

The magnitude and phase of $\mathbf{V}_2$ are plotted with greater detail in Fig. 15-27(*b*).

15.7 Perform dc and ac analysis on the circuit in Fig. 15-28. Find the complex magnitude of $\mathbf{V}_2$ for f varying from 100 Hz to 10 kHz in 100 steps.

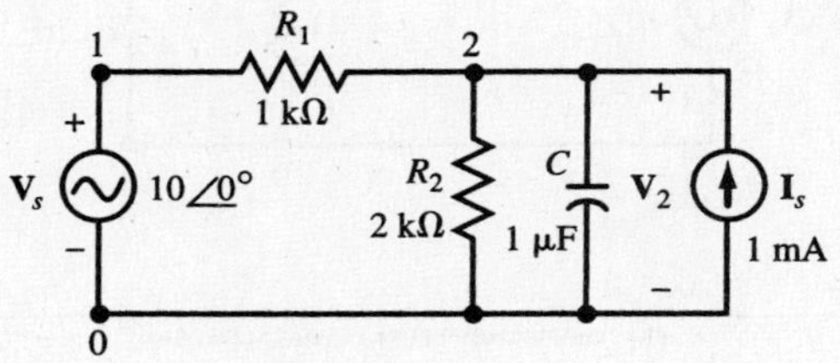

Fig. 15-28

The source file is

```
DC and AC analysis of Fig. 15-28
Vs        1     0     AC      10      0
Is        0     2     DC      1 mA
R1        1     2     1 k
R2        2     0     2 k
C         2     0     1 uF
.AC       LIN         100     100     10000
.PROBE    Vm(2)       Vp(2)
.END
```

The output file contains the following results:

```
SMALL SIGNAL BIAS SOLUTION
NODE     VOLTAGE     NODE     VOLTAGE
(1)      0.0000      (2)      .6667

VOLTAGE SOURCE CURRENTS
NAME     CURRENT
Vs       6.667E - 04

TOTAL POWER DISSIPATION     -0.00E + 00     WATTS
```

The graph of the ac component of $\mathbf{V}_2$ is identical with that of $\mathbf{V}_2$ of Problem 15.6 shown in Fig. 15-27(*b*).

15.8 Plot resonance curves for the circuit of Fig. 15-29(*a*) for $R = 2$, 4, 6, 8, and 10 Ω.

We model the resistor as a single-parameter resistor element with a single-parameter **R** and change the value of its parameter R from 2 to 10 in steps of 2 Ω. We use the .AC command to sweep the frequency from 500 Hz to 3 kHz in 100 steps. The source file is

```
Parallel resonance of practical coil, Fig. 15-29
I         0     2     AC          1 m     0
R         0     2     RLOSS       1
L         1     2     10 m
C         0     2     1 u
.MODEL    RLOSS       RES(R = 1)
```

```
.STEP    RES    RLOSS(R)   2    10   2
.AC      LIN    100      500  3000
.PROBE
.END
```

The resonance curves are shown with greater detail in Fig. 15-29(*b*).

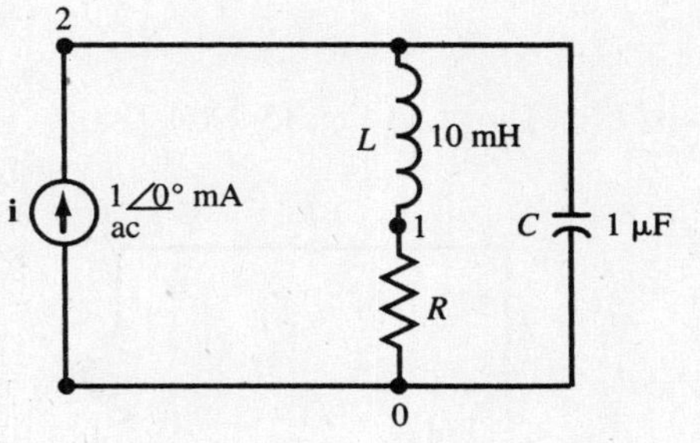

(*a*)

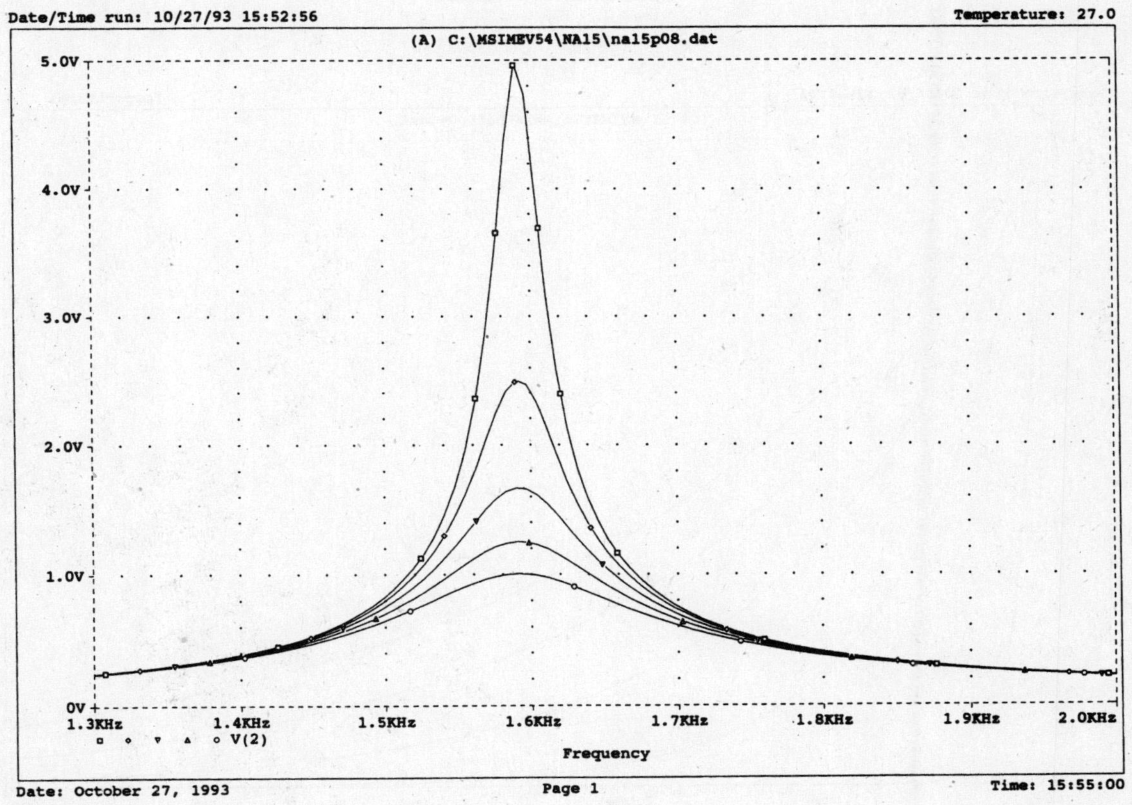

(*b*)

Fig. 15-29

15.9 Use .TRAN and .PROBE to plot V_C across the 1-μF capacitor in the source-free circuit of Fig. 15-30(*a*) for $R = 100$, 600, 1100, 1600, and 2100 Ω. The initial voltage is $V_C(0) = 10$ V.

The values of the resistor R are changed by using .MODEL and .STEP. The source file is

```
Natural response of RC, Fig. 15-30(a)
R        0      1       Rshunt        1
C        1      0       1 uF          IC = 10
.MODEL   Rshunt         RES(R = 1)
.STEP    LIN            RES           Rshunt(R)   100   2.1 k   500
.TRAN    1E - 4         50E - 4       UIC
.PLOT    TRAN           V(1)
.PROBE
.END
```

The graph of the voltage V_C is shown in Fig. 15-30(*b*).

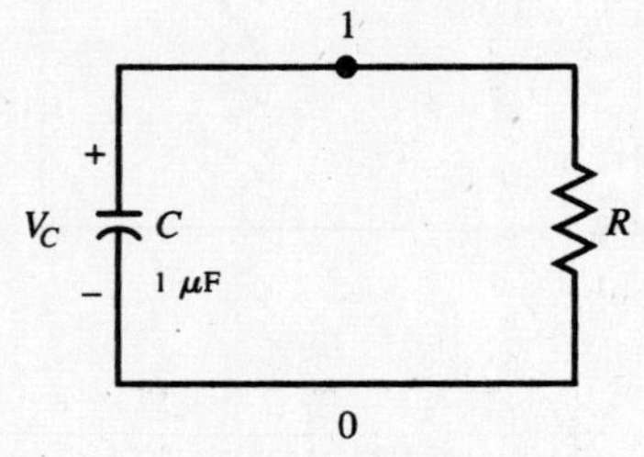

(*a*)

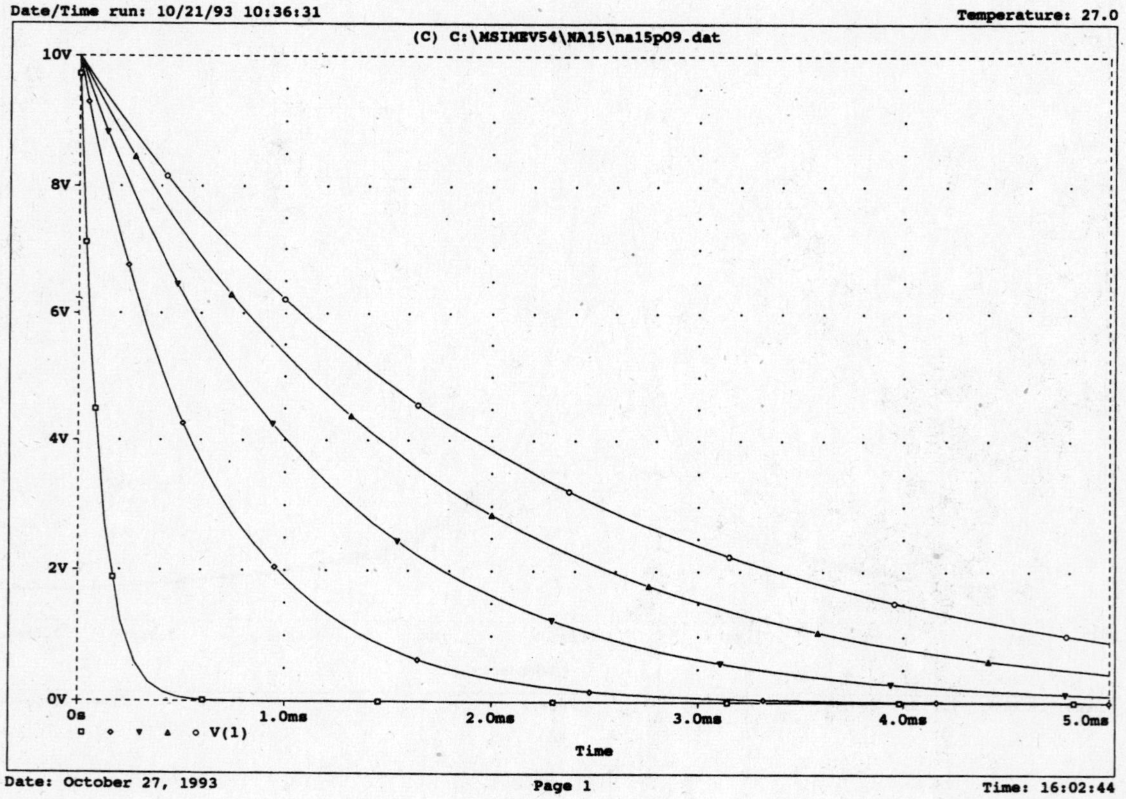

(*b*)

Fig. 15-30

15.10 Plot the voltages between the two nodes of Fig. 15-31(*a*) in response to a 1-mA step current source for $R = 100$, 600, 1100, 1600, and 2100 Ω.

The source file is

```
Step response of RC, Fig. 15-31(a)
I          0        1        1 m
R          0        1        Rshunt       1
C          1        0        1 uF
.MODEL     Rshunt            RES(R = 1)
.STEP      LIN               RES          Rshunt(R)    100    2.1 k    500
.TRAN      1E - 4            50E - 4      UIC
.PLOT      TRAN              V(1)
.PROBE
.END
```

The graphs of the step responses are given in Fig. 15-31(*b*).

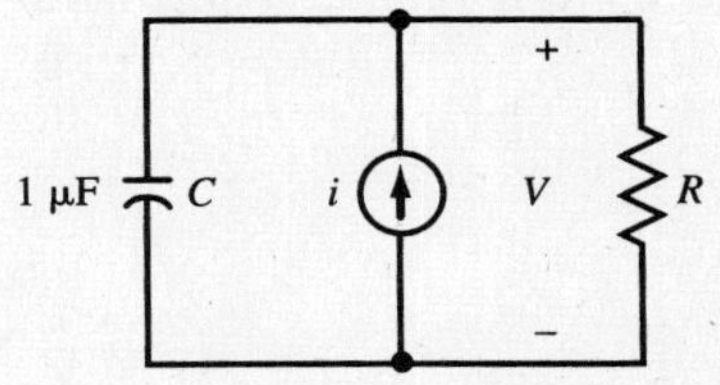

(*a*)

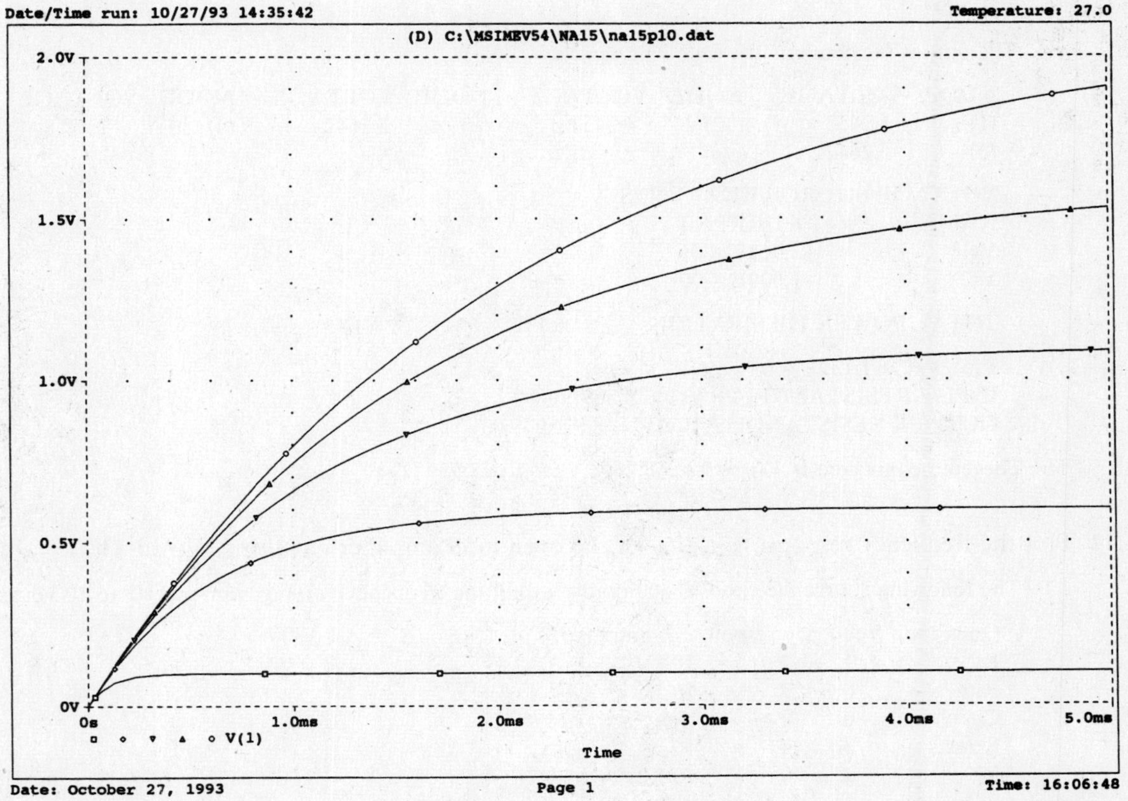

(*b*)

Fig. 15-31

15.11 Find the Thévenin equivalent of Fig. 15-32 seen at the terminal AB.

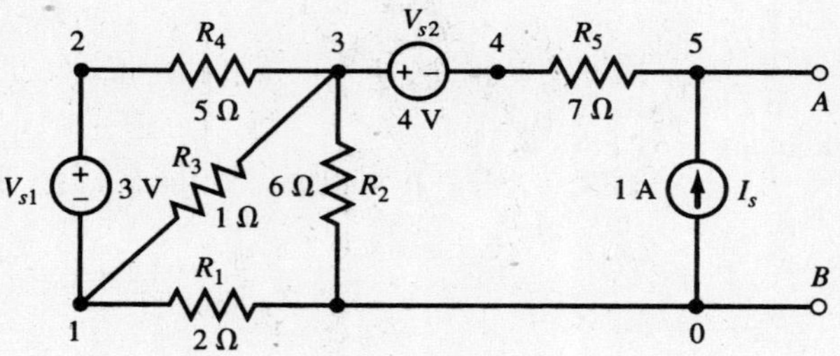

Fig. 15-32

From dc analysis we find the open-circuit voltage at AB. We also use .TF to find the output resistance at AB. The source file and the output files are

```
Solution to Fig. 15-32 and Thévenin equivalent at terminal AB
R1      0     1     2
R2      0     3     6
R3      1     3     1
R4      2     3     5
R5      4     5     7
Vs1     2     1     DC    3
Vs2     3     4     DC    4
Is      0     5     DC    1
.TF     V(5)        Vs1
.END
```

The output file contains the following results:

```
NODE  VOLTAGE    NODE  VOLTAGE    NODE  VOLTAGE    NODE  VOLTAGE
(1)    1.2453    (2)    4.2453    (3)    2.2642    (4)    -1.7358
(5)    5.2642

VOLTAGE SOURCE CURRENTS
NAME        CURRENT
Vs1         -3.962E-01
Vs2         -1.000E+00

TOTAL POWER DISSIPATION         5.19E+00     WATTS

V(5)/Vs1 = 1.132E-01
INPUT RESISTANCE AT Vs1 = 5.889E+00
OUTPUT RESISTANCE AT V(5) = 8.925E+00
```

The Thévenin equivalent is $V_{Th} = V_5 = 5.2642$ V, $R_{Th} = 8.925\ \Omega$.

15.12 Plot the frequency response V_{AB}/V_{ac} of the open-loop amplifier circuit of Fig. 15-33(*a*).

The following source file chooses 500 points within the frequency varying from 100 Hz to 10 Mhz.

```
Open loop frequency response of amplifier, Fig. 15-33
Rs       1     2      10 k
Rin      0     2      10 E5
Cin      0     2      short     1
Rout     3     4      10 k
R1       4     0      10 E9
Eout     3     0      0  2      1 E5
Vac      1     0      AC        10 u     0
.MODEL   short        CAP(C = 1)
```

```
.STEP    LIN    CAP    short(C)    1 pF    101 pF    25 pF
.AC      LIN    500    10    10000 k
.PROBE
.END
```

The frequency response is plotted by Probe for the frequency varying from 10 kHz to 10 MHz as shown in Fig. 15-33(*b*).

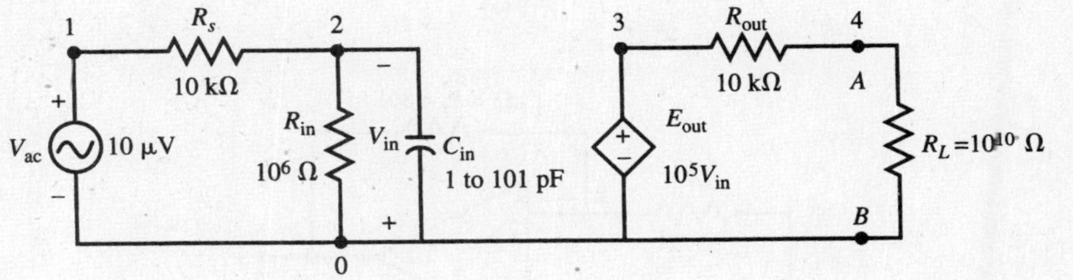

(*a*)

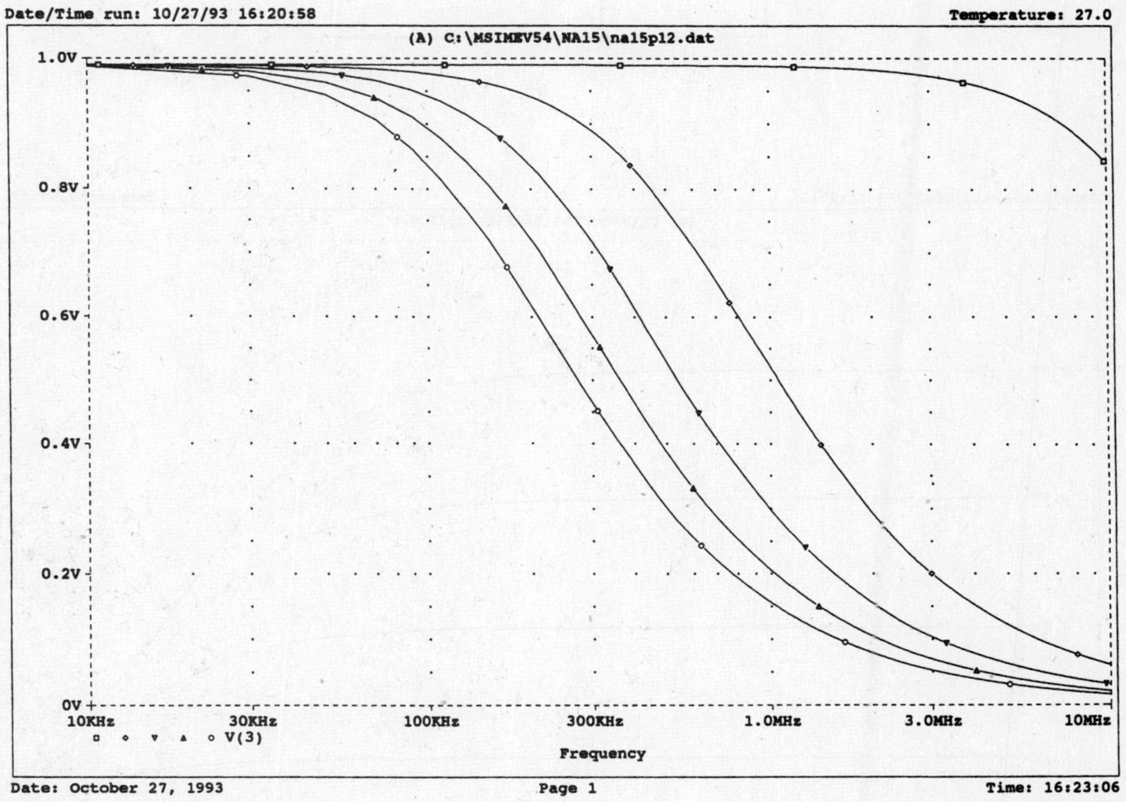

(*b*)

Fig. 15-33

15.13 Model the op amp of Fig. 15-34(*a*) as a subcircuit and use it to find the frequency response of $\mathbf{V}_3/\mathbf{V}_{ac}$ in Fig. 15-34(*b*) for f varying from 1 MHz to 1 GHz.

The source file is

```
Closed loop frequency response of amplifier, Fig. 15-34
.SUBCKT    OPAMP    1  2  3  4
```

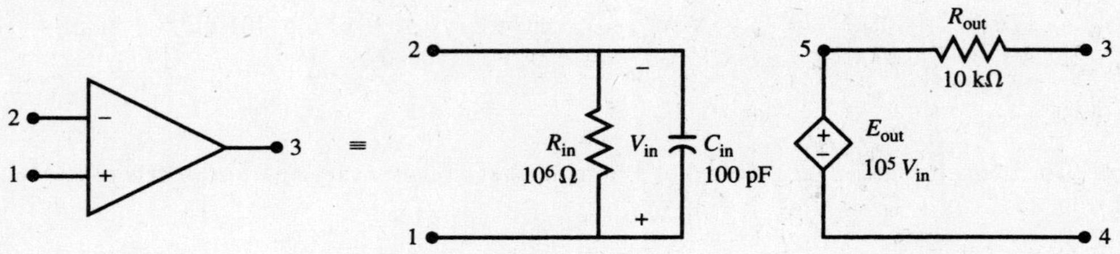
2
1
3
≡
2
1
R_{in}
$10^6\ \Omega$
V_{in}
C_{in}
100 pF
5
R_{out}
10 kΩ
3
E_{out}
$10^5\ V_{in}$
4

(*a*)

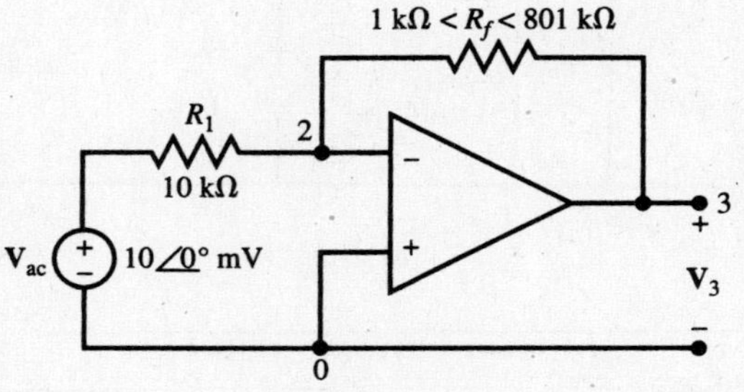
1 kΩ < R_f < 801 kΩ
R_1
10 kΩ
2
$\mathbf{V}_{ac}$
10∠0° mV
3
$\mathbf{V}_3$
0

(*b*)

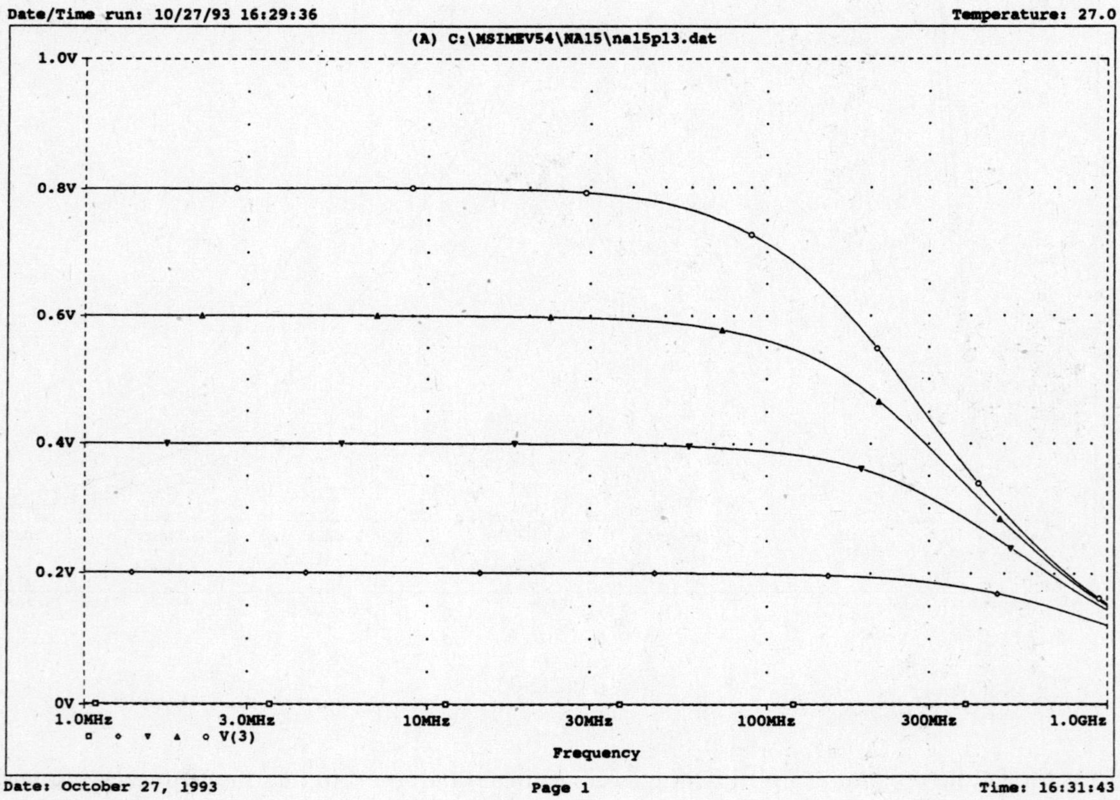
Date/Time run: 10/27/93 16:29:36
Temperature: 27.0
(A) C:\MSIMEV54\NA15\na15p13.dat
1.0V
0.8V
0.6V
0.4V
0.2V
0V
1.0MHz
3.0MHz
10MHz
30MHz
100MHz
300MHz
1.0GHz
V(3)
Frequency
Date: October 27, 1993
Page 1
Time: 16:31:43

(*c*)

Fig. 15-34

```
*       node 1 is the non-inverting input
*       node 2 is the inverting input
*       node 3 is the output
*       node 4 is the output reference (negative end of dependent source)
*       node 5 is the positive end of dependent source
Rin         1 2            10 E5
Cin         1 2            100 pF
Rout        3 5            10 k
Eout        5 4            1 2            1 E5
.ENDS
Vac         1 0            AC             10 m     0
R1          1 2            10 k
Rf          2 3            Rgain          1
X1          0 2 3 0        OPAMP
.MODEL      GAIN           RES(R = 1)
.STEP       LIN            RES     Rgain(R)       1 k    801 k    200 k
.AC         LIN            500     1000 k      1 000 000 k
.PROBE
.END
```

The frequency response is graphed in Fig. 15-34(*c*). Compared with the open-loop circuit of Fig. 15-33(*a*), the dc gain is reduced and the bandwidth is increased.

15.14 Referring to the *RC* circuit of Fig. 15-22, choose the height of the initial pulse such that the voltage across the capacitor reaches 10 V in 0.5 ms. Verify your answer by plotting V_c for $0 < t < 2\,\text{ms}$.

The pulse amplitude A is computed from

$$A(1 - e^{-1/2}) = 10 \qquad \text{from which} \qquad A = 25.415\ \text{V}$$

We describe the voltage source using PULSE syntax. The source file is

```
Pulse-Step response of RC, dead beat in RC/2 seconds
Vs        1 0         PULSE(10   25.415   1.0E - 6   1.0E - 6   0.5 m   3 m)
R         1 2         1 k
C         2 0         1 u
.TRAN     1.0E - 6    2.0E - 3     UIC
.PROBE
.END
```

The response shape is similar to the graph in Fig. 15-22(*b*). During the transition period of $0 < t < 0.5$ ms, the voltage increases exponentially toward a dc steady state value of 25.415 V. However, at $t = 0.5$ ms, when the capacitor voltage reaches 10 V, the source also has 10 V across it. The current in the resistor becomes zero and steady state is reached.

15.15 Plot the voltage across the capacitor in the circuit in Fig. 15-35(*a*) for $R = 0.01\ \Omega$ and $4.01\ \Omega$. The current source is a 1 mA square pulse which lasts 1256.64 μs as shown in the $i - t$ graph.

Model the resistor as a single-parameter resistor element with a single parameter R and change the value of R from 0.01 to 4.01 in step of 4. We use the .AC command to sweep the frequency from 500 Hz to 3 kHz in 100 steps. The source file is

```
Pulse response of RLC with variable R
Is        0 1       Pulse(0   1 m   100 u   0.01 u   0.01 u   1256.64 u   5000 u)
R         1 2       LOSS         1
C         1 0       2000 n       IC = 0
L         2 0       5 m          IC = 0
```

```
.MODEL    LOSS    RES(R = 1)
.STEP     RES     LOSS(R)     .01    4.01     4
.TRAN     10 u    3500 u    0    1 u    UIC
.PROBE
.END
```

The result is shown in Fig. 15-35(*b*).

The transient response is almost zero for $R = 0.01\ \Omega$. This is because pulse width is a multiple of the period of natural oscillations of the circuit.

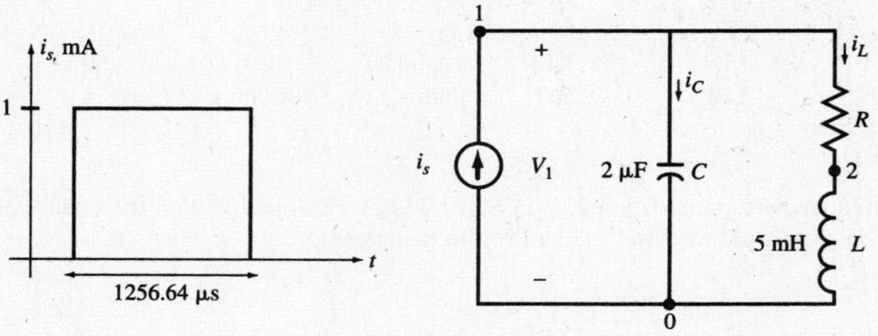

(*a*)

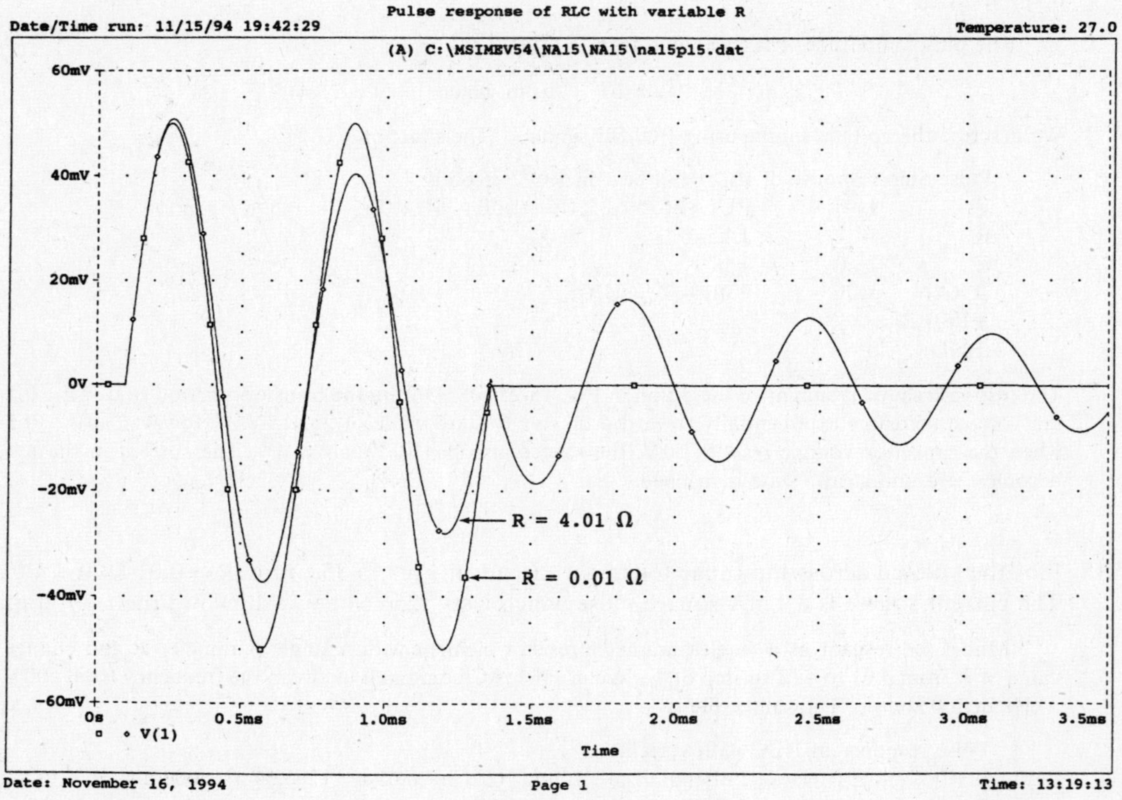

(*b*)

Fig. 15-35

Supplementary Problems

In the following problems, use PSpice to repeat the indicated problems and examples.

15.16 Solve Example 5.9 (Fig. 5-12).

15.17 Solve Example 5.11 (Fig. 5-16).

15.18 Solve Example 5.14 (Fig. 5-20).

15.19 Solve Example 5.15 (Fig. 5-21).

15.20 Solve Example 5.20 (Fig. 5-28) for $x(t) = 1$ V.

15.21 Solve Problem 5.12 (Fig. 5-37).

15.22 Solve Problem 5.16 (Fig. 5-39).

15.23 Solve Problem 5.25 (Fig. 5-48).

15.24 Solve Problem 5.26 (Fig. 5-49).

15.25 Solve Problem 5.48 (Fig. 5-55) for $v_{s1} = v_{s2} = 1$ V.

15.26 Solve Example 7.3.

15.27 Solve Example 7.6 (Fig. 7-12).

15.28 Solve Example 7.7 [Fig. 7-13(*a*)].

15.29 Solve Example 7.11 [Fig. 7-17(*a*)].

15.30 Solve Problem 8.27 (Fig. 8-31).

15.31 Solve Problem 9.11 (Fig. 9-20).

15.32 Solve Problem 9.18 (Fig. 9-28).

15.33 Solve Problem 9.19 (Fig. 9-29).

15.34 Solve Example 11.5 [Fig. 11-15(*a*)].

15.35 Solve Example 11.6 [Fig. 11-16(*a*)].

15.36 Solve Example 11.7 (Fig. 11-17).

15.37 Solve Problem 12.7.

15.38 Solve Problem 12.14 (Fig. 12-40).

15.39 Solve Problem 12.16 (Fig. 12-43).

15.40 Solve Problem 13.28 (Fig. 13-31) for $s = j$.

15.41 Solve Problem 13.31 (Fig. 13-33)

15.42 Solve Problem 14.8 (Fig. 14-24).

15.43 Solve Problem 14.12 (Fig. 14-28).

15.44 Solve Problem 14.13 (Fig. 14-29)

15.45 Solve Problem 14.20 (Fig. 14-35)

15.46 Solve Problem 14.21 (Fig. 14-36) for $s = j$.

CHAPTER 16

The Laplace Transform Method

16.1 INTRODUCTION

The relation between the response $y(t)$ and excitation $x(t)$ in *RLC* circuits is a linear differential equation of the form

$$a_n y^{(n)} + \cdots + a_j y^{(j)} + \cdots + a_1 y^{(1)} + a_0 y = b_m x^{(m)} + \cdots + b_i x^{(i)} + \cdots + b_1 x^{(1)} + b_0 x \qquad (1)$$

where $y^{(j)}$ and $x^{(i)}$ are the jth and ith time derivatives of $y(t)$ and $x(t)$, respectively. If the values of the circuit elements are constant, the corresponding coefficients a_j and b_i of the differential equation will also be constants. In Chapters 7 and 8 we solved the differential equation by finding the natural and forced responses. We employed the complex exponential function $x(t) = \mathbf{X}e^{\mathbf{s}t}$ to extend the solution to the complex frequency **s**-domain.

The Laplace transform method described in this chapter may be viewed as generalizing the concept of the **s**-domain to a mathematical formulation which would include not only exponential excitations but also excitations of many other forms. Through the Laplace transform we represent a large class of excitations as an infinite collection of complex exponentials and use superposition to derive the total response.

16.2 THE LAPLACE TRANSFORM

Let $f(t)$ be a time function which is zero for $t \le 0$ and which is (subject to some mild conditions) arbitrarily defined for $t > 0$. Then the *direct Laplace transform* of $f(t)$, denoted $\mathscr{L}[f(t)]$, is defined by

$$\mathscr{L}[f(t)] = \mathbf{F}(\mathbf{s}) = \int_{0^+}^{\infty} f(t)e^{-st}\,dt \qquad (2)$$

Thus, the operation $\mathscr{L}[\ ]$ transforms $f(t)$, which is in the *time domain*, into $\mathbf{F}(\mathbf{s})$, which is in the *complex frequency domain*, or simply the **s**-*domain*, where **s** is the complex variable $\sigma + j\omega$. While it appears that the integration could prove difficult, it will soon be apparent that application of the Laplace transform method utilizes tables which cover all functions likely to be encountered in elementary circuit theory.

There is a uniqueness in the transform pairs; that is, if $f_1(t)$ and $f_2(t)$ have the same **s**-domain image $\mathbf{F}(\mathbf{s})$, then $f_1(t) = f_2(t)$. This permits going back in the other direction, from the **s**-domain to the time domain, a process called the *inverse Laplace transform*, $\mathscr{L}^{-1}[\mathbf{F(s)}] = f(t)$. The inverse Laplace transform can also be expressed as an integral, the *complex inversion integral*:

$$\mathscr{L}^{-1}[\mathbf{F(s)}] = f(t) = \frac{1}{2\pi j}\int_{\sigma_0 - j\infty}^{\sigma_0 + j\infty} \mathbf{F(s)}e^{\mathbf{s}t}\, d\mathbf{s} \qquad (3)$$

In *(3)* the path of integration is a straight line parallel to the $j\omega$-axis, such that all the poles of $\mathbf{F(s)}$ lie to the left of the line. Here again, the integration need not actually be performed unless it is a question of adding to existing tables of transform pairs.

It should be remarked that taking the direct Laplace transform of a physical quantity introduces an extra time unit in the result. For instance, if $i(t)$ is a current in A, then $\mathbf{I(s)}$ has the units A · s (or C). Because the extra unit s will be removed in taking the inverse Laplace transform, we shall generally omit to cite units in the **s**-domain, shall still call $\mathbf{I(s)}$ a "current," indicate it by an arrow, and so on.

16.3 SELECTED LAPLACE TRANSFORMS

The Laplace transform of the unit step function is easily obtained:

$$\mathscr{L}[u(t)] = \int_0^\infty (1)e^{-\mathbf{s}t}\, dt = -\frac{1}{\mathbf{s}}[e^{-\mathbf{s}t}]_0^\infty = \frac{1}{\mathbf{s}}$$

From the linearity of the Laplace transform, it follows that $v(t) = Vu(t)$ in the time domain has the **s**-domain image $\mathbf{V(s)} = V/\mathbf{s}$.

The exponential decay function, which appeared often in the transients of Chapter 7, is another time function which is readily transformed.

$$\mathscr{L}[Ae^{-at}] = \int_0^\infty Ae^{-at}e^{-\mathbf{s}t}\, dt = \frac{-A}{A+\mathbf{s}}[e^{-(a+\mathbf{s})t}]_0^\infty = \frac{A}{\mathbf{s}+a}$$

or, inversely,

$$\mathscr{L}^{-1}\left[\frac{A}{\mathbf{s}+a}\right] = Ae^{-at}$$

The transform of a sine function is also easily obtained.

$$\mathscr{L}[\sin \omega t] = \int_0^\infty (\sin \omega t)e^{-\mathbf{s}t}\, dt = \left[\frac{-\mathbf{s}(\sin \omega t)e^{-\mathbf{s}t} - e^{-\mathbf{s}t}\omega \cos \omega t}{\mathbf{s}^2 + \omega^2}\right]_0^\infty = \frac{\omega}{\mathbf{s}^2+\omega^2}$$

It will be useful now to obtain the transform of a derivative, $df(t)/dt$.

$$\mathscr{L}\left[\frac{df(t)}{dt}\right] = \int_0^\infty \frac{df(t)}{dt}e^{-\mathbf{s}t}\, dt$$

Integrating by parts,

$$\mathscr{L}\left[\frac{df(t)}{dt}\right] = [e^{-\mathbf{s}t}f(t)]_{0^+}^\infty - \int_0^\infty f(t)(-\mathbf{s}e^{-\mathbf{s}t})\, dt = -f(0^+) + \mathbf{s}\int_0^\infty f(t)e^{-\mathbf{s}t}\, dt = -f(0^+) + \mathbf{sF(s)}$$

A small collection of transform pairs, including those obtained above, is given in Table 16-1. The last five lines of the table present some general properties of the Laplace transform.

EXAMPLE 16.1 Consider a series RL circuit, with $R = 5\ \Omega$ and $L = 2.5$ mH. At $t = 0$, when the current in the circuit is 2 A, a source of 50 V is applied. The time-domain circuit is shown in Fig. 16-1.

Time Domain		*s-Domain*
(i) $Ri + L\dfrac{di}{dt} = v$	→	(ii) $R\mathbf{I}(\mathbf{s}) + L[-i(0^+) + \mathbf{sI}(\mathbf{s})] = \mathbf{V}(\mathbf{s})$
↓		(iii) $5\mathbf{I}(\mathbf{s}) + (2.5\times 10^{-3})[-2+\mathbf{sI}(\mathbf{s})] = \dfrac{50}{\mathbf{s}}$
(classical methods)		↓
↓		(iv) $\mathbf{I}(\mathbf{s}) = \dfrac{10}{\mathbf{s}} + \dfrac{-8}{\mathbf{s}+2000}$
(vii) $i(t) = 10 - 8e^{-2000t}$ (A)	←	(v) $10\mathscr{L}^{-1}\left[\dfrac{1}{\mathbf{s}}\right] = 10$ (vi) $(-8)\mathscr{L}^{-1}\left[\dfrac{1}{\mathbf{s}+2000}\right] = -8e^{-2000t}$

Table 16-1 Laplace Transform Pairs

	$f(t)$	$\mathbf{F}(\mathbf{s})$
1.	1	$\dfrac{1}{\mathbf{s}}$
2.	t	$\dfrac{1}{\mathbf{s}^2}$
3.	e^{-at}	$\dfrac{1}{\mathbf{s}+a}$
4.	te^{-at}	$\dfrac{1}{(\mathbf{s}+a)^2}$
5.	$\sin \omega t$	$\dfrac{\omega}{\mathbf{s}^2+\omega^2}$
6.	$\cos \omega t$	$\dfrac{\mathbf{s}}{\mathbf{s}^2+\omega^2}$
7.	$\sin(\omega t+\theta)$	$\dfrac{\mathbf{s}\sin\theta+\omega\cos\theta}{\mathbf{s}^2+\omega^2}$
8.	$\cos(\omega t+\theta)$	$\dfrac{\mathbf{s}\cos\theta-\omega\sin\theta}{\mathbf{s}^2+\omega^2}$
9.	$e^{-at}\sin\omega t$	$\dfrac{\omega}{(\mathbf{s}+a)^2+\omega^2}$
10.	$e^{-at}\cos\omega t$	$\dfrac{\mathbf{s}+a}{(\mathbf{s}+a)^2+\omega^2}$
11.	$\sinh\omega t$	$\dfrac{\omega}{\mathbf{s}^2-\omega^2}$
12.	$\cosh\omega t$	$\dfrac{\mathbf{s}}{\mathbf{s}^2-\omega^2}$
13.	$\dfrac{df}{dt}$	$\mathbf{sF}(\mathbf{s})-f(0^+)$
14.	$\displaystyle\int_0^t f(\tau)\,d\tau$	$\dfrac{\mathbf{F}(\mathbf{s})}{\mathbf{s}}$
15.	$f(t-t_1)$	$e^{-t_1\mathbf{s}}\mathbf{F}(\mathbf{s})$
16.	$c_1f_1(t)+c_2f_2(t)$	$c_1\mathbf{F}_1(\mathbf{s})+c_2\mathbf{F}_2(\mathbf{s})$
17.	$\displaystyle\int_0^t f_1(\tau)f_2(t-\tau)\,d\tau$	$\mathbf{F}_1(\mathbf{s})\mathbf{F}_2(\mathbf{s})$

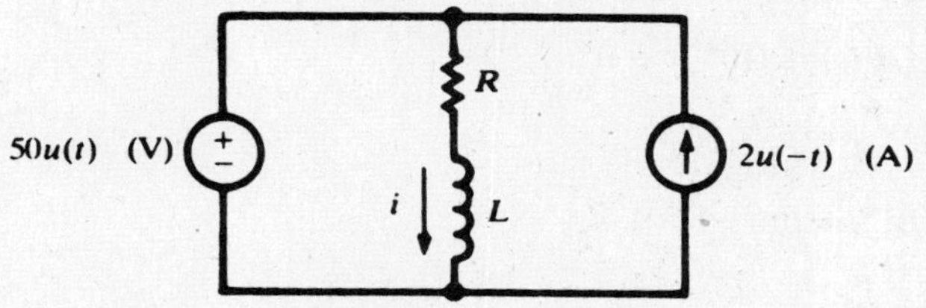

Fig. 16-1

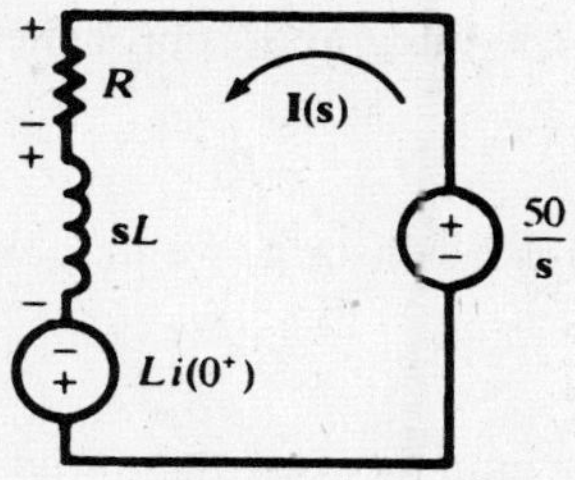

Fig. 16-2

Kirchhoff's voltage law, applied to the circuit for $t > 0$, yields the familiar differential equation (i). This equation is transformed, term by term, into the **s**-domain equation (ii). The unknown current $i(t)$ becomes $\mathbf{I}(\mathbf{s})$, while the known voltage $v = 50u(t)$ is transformed to $50/\mathbf{s}$. Also, di/dt is transformed into $-i(0^+) + \mathbf{sI(s)}$, in which $i(0^+)$ is 2 A. Equation (iii) is solved for $\mathbf{I}(\mathbf{s})$, and the solution is put in the form (iv) by the techniques of Section 16.6. Then lines 1, 3, and 16 of Table 16-1 are applied to obtain the inverse Laplace transform of $\mathbf{I}(\mathbf{s})$, which is $i(t)$.

A circuit can be drawn in the **s**-domain, as shown in Fig. 16-2. The initial current appears in the circuit as a voltage source, $Li(0^+)$. The **s**-domain current establishes the voltage terms $R\mathbf{I}(\mathbf{s})$ and $sL\mathbf{I}(\mathbf{s})$ in (ii) just as a phasor current **I** and an impedance **Z** create a phasor voltage **IZ**.

16.4 CONVERGENCE OF THE INTEGRAL

For the Laplace transform to exist, the integral (*2*) should converge. This limits the variable $\mathbf{s} = \sigma + j\omega$ to a part of the complex plane called the *convergence region*. As an example, the transform of $x(t) = e^{-at}u(t)$ is $1/(\mathbf{s} + a)$, provided Re [**s**] $> -a$, which defines its region of convergence.

EXAMPLE 16.2 Find the Laplace transform of $x(t) = 3e^{2t}u(t)$ and show the region of convergence.

$$\mathbf{X}(\mathbf{s}) = \int_0^\infty 3e^{2t}e^{-\mathbf{s}t}\,dt = \int_0^\infty 3e^{-(\mathbf{s}-2)t}\,dt = \frac{3}{\mathbf{s}-2}\,[e^{-(\mathbf{s}-2)t}]_0^\infty = \frac{3}{\mathbf{s}-2}, \qquad \mathrm{Re}\,[\mathbf{s}] > 2$$

The region of convergence of $\mathbf{X}(\mathbf{s})$ is the right half plane $\sigma > 2$, shown hatched in Fig. 16-3.

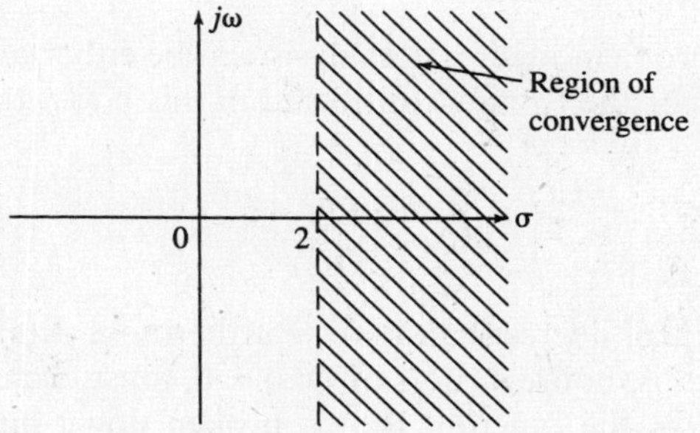

Fig. 16-3

16.5 INITIAL-VALUE AND FINAL-VALUE THEOREMS

Taking the limit as $\mathbf{s} \to \infty$ (through real values) of the direct Laplace transform of the derivative, $df(t)/dt$,

$$\lim_{s\to\infty} \mathscr{L}\left[\frac{df(t)}{dt}\right] = \lim_{s\to\infty} \int_0^\infty \frac{df(t)}{dt}\,e^{-st}\,dt = \lim_{s\to\infty} \{s\mathbf{F}(s) - f(0^+)\}$$

But e^{-st} in the integrand approaches zero as $s \to \infty$. Thus,

$$\lim_{s\to\infty} \{s\mathbf{F}(s) - f(0^+)\} = 0$$

Since $f(0^+)$ is a constant, we may write

$$f(0^+) = \lim_{s\to\infty} \{s\mathbf{F}(s)\}$$

which is the statement of the initial-value theorem.

EXAMPLE 16.3 In Example 16.1,

$$\lim_{s\to\infty} \{s\mathbf{I}(s)\} = \lim_{s\to\infty} \left(10 - \frac{8s}{s+2000}\right) = 10 - 8 = 2$$

which is indeed the initial current, $i(0^+) = 2$ A.

The final-value theorem is also developed from the direct Laplace transform of the derivative, but now the limit is taken as $\mathbf{s} \to 0$ (through real values).

$$\lim_{s\to 0} \mathscr{L}\left[\frac{df(t)}{dt}\right] = \lim_{s\to 0} \int_0^\infty \frac{df(t)}{dt} e^{-st}\, dt = \lim_{s\to 0} \{s\mathbf{F}(s) - f(0^+)\}$$

But

$$\lim_{s\to 0} \int_0^\infty \frac{df(t)}{dt} e^{-st}\, dt = \int_0^\infty df(t) = f(\infty) - f(0^+)$$

and $f(0^+)$ is a constant. Therefore,

$$f(\infty) - f(0^+) = -f(0^+) + \lim_{s\to 0} \{s\mathbf{F}(s)\}$$

or

$$f(\infty) = \lim_{s\to 0} \{s\mathbf{F}(s)\}$$

This is the statement of the final-value theorem. The theorem may be applied only when all poles of **sF(s)** have negative real parts. This excludes the transforms of such functions as e^t and $\cos t$, which become infinite or indeterminate as $t \to \infty$.

16.6 PARTIAL-FRACTIONS EXPANSIONS

The unknown quantity in a problem in circuit analysis can be either a current $i(t)$ or a voltage $v(t)$. In the **s**-domain, it is **I(s)** or **V(s)**; for the circuits considered in this book, this will be a rational function of the form

$$\mathbf{R}(\mathbf{s}) = \frac{\mathbf{P}(\mathbf{s})}{\mathbf{Q}(\mathbf{s})}$$

where the polynomial **Q(s)** is of higher degree than **P(s)**. Furthermore, **R(s)** is real for real values of **s**, so that any nonreal poles of **R(s)**, that is, nonreal roots of **Q(s)** = 0, must occur in complex conjugate pairs.

In a *partial-fractions expansion*, the function **R(s)** is broken down into a sum of simpler rational functions, its so-called *principal parts*, with each pole of **R(s)** contributing a principal part.

***Case 1*: s = a *is a simple pole*.** When **s** = **a** is a nonrepeated root of **Q(s)** = 0, the corresponding principal part of **R(s)** is

$$\frac{\mathbf{A}}{\mathbf{s}-\mathbf{a}} \qquad \text{where} \qquad \mathbf{A} = \lim_{s\to \mathbf{a}} \{(\mathbf{s}-\mathbf{a})\mathbf{R}(\mathbf{s})\}$$

If **a** is real, so will be **A**; if **a** is complex, then $\mathbf{a}^*$ is also a simple pole and the numerator of its principal part is $\mathbf{A}^*$. Notice that if **a** = 0, **A** is the final value of $r(t)$

***Case 2*: s = b *is a double pole*.** When **s** = **b** is a double root of **Q(s)** = 0, the corresponding principal part of **R(s)** is

$$\frac{\mathbf{B}_1}{\mathbf{s}-\mathbf{b}}+\frac{\mathbf{B}_2}{(\mathbf{s}-\mathbf{b})^2}$$

where the constants $\mathbf{B}_2$ and $\mathbf{B}_1$ may be found as

$$\mathbf{B}_2=\lim_{\mathbf{s}\to\mathbf{b}}\{(\mathbf{s}-\mathbf{b})^2\mathbf{R}(\mathbf{s})\} \qquad \text{and} \qquad \mathbf{B}_1=\lim_{\mathbf{s}\to\mathbf{b}}\left\{(\mathbf{s}-\mathbf{b})\left[\mathbf{R}(\mathbf{s})-\frac{\mathbf{B}_2}{(\mathbf{s}-\mathbf{b})^2}\right]\right\}$$

$\mathbf{B}_1$ may be zero. Similar to Case 1, $\mathbf{B}_1$ and $\mathbf{B}_2$ are real if $\mathbf{b}$ is real, and these constants for the double pole $\mathbf{b}^*$ are the conjugates of those for $\mathbf{b}$.

The principal part at a higher-order pole can be obtained by analogy to Case 2; we shall assume, however, that $\mathbf{R}(\mathbf{s})$ has no such poles. Once the partial-functions expansion of $\mathbf{R}(\mathbf{s})$ is known, Table 16-1 can be used to invert each term and thus to obtain the time function $r(t)$.

EXAMPLE 16.4 Find the time-domain current $i(t)$ if its Laplace transform is

$$\mathbf{I}(\mathbf{s})=\frac{\mathbf{s}-10}{\mathbf{s}^4+\mathbf{s}^2}$$

Factoring the denominator,

$$\mathbf{I}(\mathbf{s})=\frac{\mathbf{s}-10}{\mathbf{s}^2(\mathbf{s}-j)(\mathbf{s}+j)}$$

we see that the poles of $\mathbf{I}(\mathbf{s})$ are $\mathbf{s}=0$ (double pole) and $\mathbf{s}=\pm j$ (simple poles).

The principal part at $\mathbf{s}=0$ is

$$\frac{B_1}{\mathbf{s}}+\frac{B_2}{\mathbf{s}^2}=\frac{1}{\mathbf{s}}-\frac{10}{\mathbf{s}^2}$$

since

$$B_2=\lim_{\mathbf{s}\to 0}\left[\frac{\mathbf{s}-10}{(\mathbf{s}-j)(\mathbf{s}+j)}\right]=-10$$

$$B_1=\lim_{\mathbf{s}\to 0}\left\{\mathbf{s}\left[\frac{\mathbf{s}-10}{\mathbf{s}^2(\mathbf{s}^2+1)}+\frac{10}{\mathbf{s}^2}\right]\right\}=\lim_{\mathbf{s}\to 0}\left(\frac{10\mathbf{s}+1}{\mathbf{s}^2+1}\right)=1$$

The principal part at $\mathbf{s}=+j$ is

$$\frac{\mathbf{A}}{\mathbf{s}-j}=-\frac{0.5+j5}{\mathbf{s}-j}$$

since

$$\mathbf{A}=\lim_{\mathbf{s}\to j}\left[\frac{\mathbf{s}-10}{\mathbf{s}^2(\mathbf{s}+j)}\right]=-(0.5+j5)$$

It follows at once that the principal part at $\mathbf{s}=-j$ is

$$-\frac{0.5-j5}{\mathbf{s}+j}$$

The partial-fractions expansion of $\mathbf{I}(\mathbf{s})$ is therefore

$$\mathbf{I}(\mathbf{s})=\frac{1}{\mathbf{s}}-10\,\frac{1}{\mathbf{s}^2}-(0.5+j5)\,\frac{1}{\mathbf{s}-j}-(0.5-j5)\,\frac{1}{\mathbf{s}+j}$$

and term-by-term inversion using Table 16-1 gives

$$i(t)=1-10t-(0.5+j5)e^{jt}-(0.5-j5)e^{-jt}=1-10t-(\cos t-10\sin t)$$

Heaviside Expansion Formula

If all poles of $\mathbf{R}(\mathbf{s})$ *are simple*, the partial-fractions expansion and termwise inversion can be accomplished in a single step:

$$\mathscr{L}^{-1}\left[\frac{\mathbf{P}(\mathbf{s})}{\mathbf{Q}(\mathbf{s})}\right]=\sum_{k=1}^{n}\frac{\mathbf{P}(\mathbf{a}_k)}{\mathbf{Q}'(\mathbf{a}_k)}\,e^{\mathbf{a}_k t} \tag{4}$$

where $\mathbf{a}_1,\mathbf{a}_2,\ldots,\mathbf{a}_n$ are the poles and $\mathbf{Q}'(\mathbf{a}_k)$ is $d\mathbf{Q}(\mathbf{s})/d\mathbf{s}$ evaluated at $\mathbf{s}=\mathbf{a}_k$.

16.7 CIRCUITS IN THE s-DOMAIN

In Chapter 8 we introduced and utilized the concept of generalized impedance, admittance, and transfer functions as functions of the complex frequency **s**. In this section, we extend the use of the complex frequency to transform an *RLC* circuit, containing sources and initial conditions, from the time domain to the **s**-domain.

Table 16-2

Time Domain	s-Domain	s-Domain Voltage Term
		$R\mathbf{I}(\mathbf{s})$
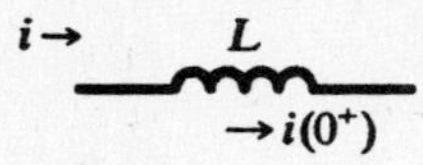	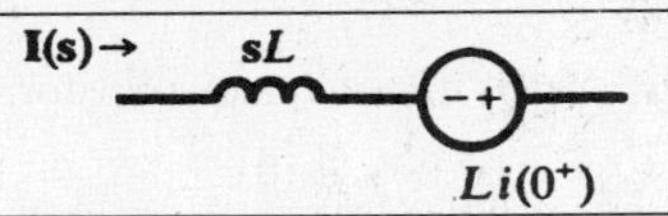	$\mathbf{s}L\mathbf{I}(\mathbf{s}) + Li(0^+)$
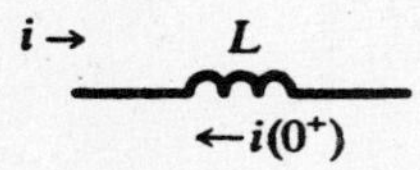	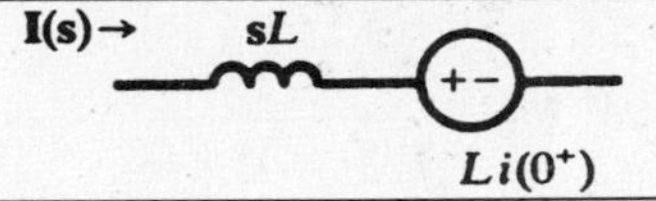	$\mathbf{s}L\mathbf{I}(\mathbf{s}) + Li(0^+)$
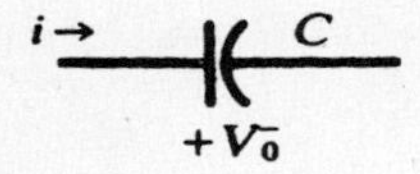	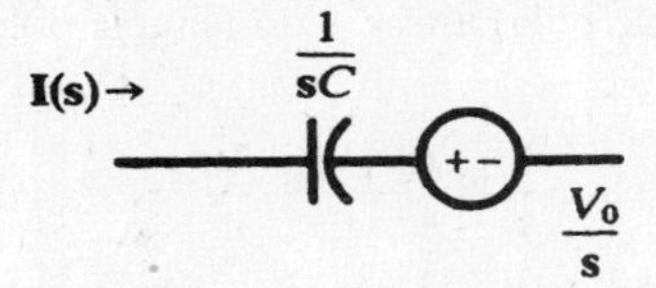	$\frac{\mathbf{I}(\mathbf{s})}{\mathbf{s}C} + \frac{V_0}{\mathbf{s}}$
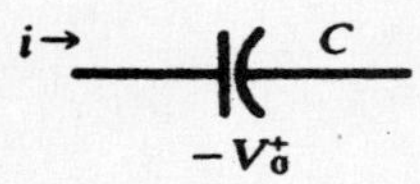	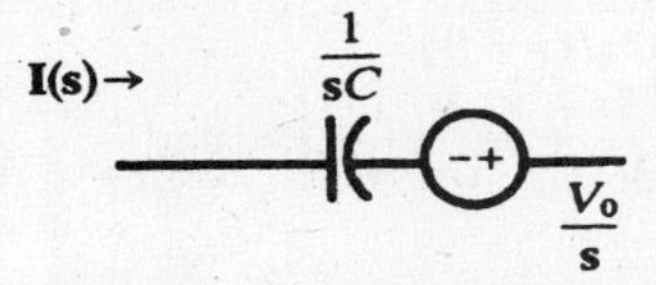	$\frac{\mathbf{I}(\mathbf{s})}{\mathbf{s}c} - \frac{V_0}{\mathbf{s}}$

Table 16-2 exhibits the elements needed to construct the **s**-domain image of a given time-domain circuit. The first three lines of the table were in effect developed in Example 16.1. As for the capacitor, we have, for $t > 0$,

$$v_C(t) = V_0 + \frac{1}{C}\int_0^t i(\tau)\,d\tau$$

so that, from Table 16-1,

$$\mathbf{V}_C(\mathbf{s}) = \frac{V_0}{\mathbf{s}} + \frac{\mathbf{I}(\mathbf{s})}{C\mathbf{s}}$$

EXAMPLE 16.5 In the circuit shown in Fig. 16-4(*a*) an initial current i_1 is established while the switch is in position *1*. At $t = 0$, it is moved to position *2*, introducing both a capacitor with initial charge Q_0 and a constant-voltage source V_2.

The **s**-domain circuit is shown in Fig. 16-4(*b*). The **s**-domain equation is

$$R\mathbf{I}(\mathbf{s}) + \mathbf{s}L\mathbf{I}(\mathbf{s}) - Li(0^+) + \frac{\mathbf{I}(\mathbf{s})}{\mathbf{s}C} + \frac{V_0}{\mathbf{s}C} = \frac{V_2}{\mathbf{s}}$$

in which $V_0 = Q_0/C$ and $i(0^+) = i_1 = V_1/R$.

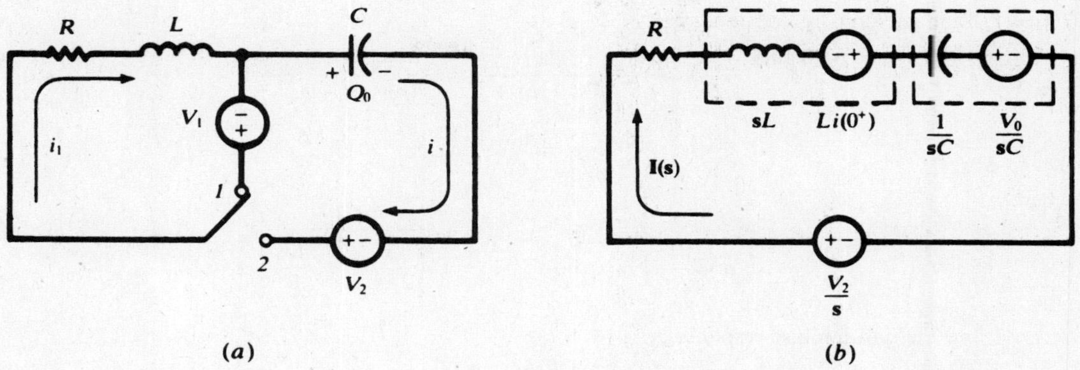

Fig. 16-4

16.8 THE NETWORK FUNCTION AND LAPLACE TRANSFORMS

In Chapter 8 we obtained responses of circuit elements to exponentials e^{st}, based on which we introduced the concept of complex frequency and generalized impedance. We then developed the network function $H(s)$ as the ratio of input-output amplitudes, or equivalently, the input-output differential equation, natural and forced responses, and the frequency response.

In the present chapter we used the Laplace transform as an alternative method for solving differential equations. More importantly, we introduce Laplace transform models of R, L, and C elements which, contrary to generalized impedances, incorporate initial conditions. The input-output relationship is therefore derived directly in the transform domain.

What is the relationship between the complex frequency and the Laplace transform models? A short answer is that the generalized impedance is the special case of the Laplace transform model (i.e., restricted to zero state), and the network function is the Laplace transform of the unit-impulse response.

EXAMPLE 16.17 Find the current developed in a series RLC circuit in response to the following two voltage sources applied to it at $t = 0$: (*a*) a unit-step, (*b*) a unit-impulse.

The inductor and capacitor contain zero energy at $t = 0^-$. Therefore, the Laplace transform of the current is $I(s) = V(s)Y(s)$.

(*a*) $V(s) = 1/s$ and the unit-step response is

$$I(s) = \frac{1}{s}\,\frac{Cs}{LCs^2 + RCs + 1} = \frac{1}{L}\,\frac{1}{(s+\sigma)^2 + \omega_d^2}$$

$$i(t) = \frac{1}{L\omega_d}\,e^{-\sigma t}\sin(\omega_d t)u(t)$$

where

$$\sigma = \frac{R}{2L}, \qquad \text{and} \qquad \omega_d = \sqrt{\left(\frac{R}{2L}\right)^2 - \frac{1}{LC}}$$

(*b*) $V(s) = 1$ and the unit-impulse response is

$$I(s) = \frac{1}{L}\,\frac{s}{(s+\sigma)^2 + \omega_d^2}$$

$$i(t) = \frac{1}{L\omega_d}\,e^{-\sigma t}[\omega_d\cos(\omega_d t) - \sigma\sin(\omega_d t)]u(t)$$

The unit-impulse response may also be found by taking the time-derivative of the unit-step response.

EXAMPLE 16.18 Find the voltage across terminals of a parallel RLC circuit in response to the following two current sources applied at $t = 0$: (*a*) a unit-step, (*b*) a unit-impulse.

Again, the inductor and capacitor contain zero energy at $t = 0^-$. Therefore, the Laplace transform of the current is $V(s) = I(s)Z(s)$.

(*a*) $I(s) = 1/s$ and the unit-step response is

$$V(s) = \frac{1}{s}\frac{RLs}{RLCs^2 + Ls + 1} = \frac{1}{C}\frac{1}{(s+\sigma)^2 + \omega_d^2}$$

$$v(t) = \frac{1}{C\omega_d} e^{-\sigma t} \sin(\omega_d t) u(t)$$

where

$$\sigma = \frac{1}{RC}, \qquad \text{and} \qquad \omega_d = \sqrt{\left(\frac{1}{2RC}\right)^2 - \frac{1}{LC}}$$

(*b*) $I(s) = 1$ and the unit-impulse response is

$$V(s) = \frac{1}{C}\frac{1}{(s+\sigma)^2 + \omega_d^2}$$

$$v(t) = \frac{1}{C\omega_d} e^{-\sigma t}[\omega_d \cos(\omega_d t) - \sigma \sin(\omega_d t)]u(t)$$

Solved Problems

16.1 Find the Laplace transform of $e^{-at} \cos \omega t$, where a is a constant.

Applying the defining equation $\mathscr{L}[f(t)] = \int_0^\infty f(t)e^{-\mathbf{s}t}\,dt$ to the given function, we obtain

$$\mathscr{L}[e^{-at}\cos\omega t] = \int_0^\infty \cos\omega t e^{-(\mathbf{s}+a)t}\,dt$$
$$= \left[\frac{-(\mathbf{s}+a)\cos\omega t e^{-(\mathbf{s}+a)t} + e^{-(\mathbf{s}+a)t}\omega\sin\omega t}{(\mathbf{s}+a)^2 + \omega^2}\right]_0^\infty$$
$$= \frac{\mathbf{s}+a}{(\mathbf{s}+a)^2 + \omega^2}$$

16.2 If $\mathscr{L}[f(t)] = \mathbf{F}(\mathbf{s})$, show that $\mathscr{L}[e^{-at}f(t)] = \mathbf{F}(\mathbf{s}+a)$. Apply this result to Problem 16.1.

By definition, $\mathscr{L}[f(t)] = \int_0^\infty f(t)e^{-\mathbf{s}t}\,dt = \mathbf{F}(\mathbf{s})$. Then,

$$\mathscr{L}[e^{-at}f(t)] = \int_0^\infty [e^{-at}f(t)]e^{-\mathbf{s}t}\,dt = \int_0^\infty f(t)e^{-(\mathbf{s}+a)t}\,dt = \mathbf{F}(\mathbf{s}+a) \tag{5}$$

Applying (*5*) to line 6 of Table 16-1 gives

$$\mathscr{L}[e^{-at}\cos\omega t] = \frac{\mathbf{s}+a}{(\mathbf{s}+a)^2 + \omega^2}$$

as determined in Problem 16.1.

16.3 Find the Laplace transform of $f(t) = 1 - e^{-at}$, where a is a constant.

$$\mathscr{L}[1 - e^{-at}] = \int_0^\infty (1 - e^{-at})e^{-\mathbf{s}t}\,dt = \int_0^\infty e^{-\mathbf{s}t}\,dt - \int_0^\infty e^{-(\mathbf{s}+a)t}\,dt$$
$$= \left[-\frac{1}{\mathbf{s}}e^{-\mathbf{s}t} + \frac{1}{\mathbf{s}+a}e^{-(\mathbf{s}+a)t}\right]_0^\infty = \frac{1}{\mathbf{s}} - \frac{1}{\mathbf{s}+a} = \frac{a}{\mathbf{s}(\mathbf{s}+a)}$$

Another Method

$$\mathscr{L}\left[a\int_0^t e^{-a\tau}\,d\tau\right] = a\,\frac{1/(\mathbf{s}+a)}{\mathbf{s}} = \frac{a}{\mathbf{s}(\mathbf{s}+a)}$$

16.4 Find

$$\mathscr{L}^{-1}\left[\frac{1}{\mathbf{s}(\mathbf{s}^2-a^2)}\right]$$

Using the method of partial fractions,

$$\frac{1}{\mathbf{s}(\mathbf{s}^2-a^2)} = \frac{A}{\mathbf{s}} + \frac{B}{\mathbf{s}+a} + \frac{C}{\mathbf{s}-a}$$

and the coefficients are

$$A = \left.\frac{1}{\mathbf{s}^2-a^2}\right|_{\mathbf{s}=0} = -\frac{1}{a^2} \qquad B = \left.\frac{1}{\mathbf{s}(\mathbf{s}-a)}\right|_{\mathbf{s}=-a} = \frac{1}{2a^2} \qquad C = \left.\frac{1}{\mathbf{s}(\mathbf{s}+a)}\right|_{s=a} = \frac{1}{2a^2}$$

Hence,

$$\mathscr{L}^{-1}\left[\frac{1}{\mathbf{s}(\mathbf{s}^2-a^2)}\right] = \mathscr{L}^{-1}\left[\frac{-1/a^2}{\mathbf{s}}\right] + \mathscr{L}^{-1}\left[\frac{1/2a^2}{\mathbf{s}+a}\right] + \mathscr{L}^{-1}\left[\frac{1/2a^2}{\mathbf{s}-a}\right]$$

The corresponding time functions are found in Table 16-1:

$$\mathscr{L}^{-1}\left[\frac{1}{\mathbf{s}(\mathbf{s}^2-a^2)}\right] = -\frac{1}{a^2} + \frac{1}{2s^2}\,e^{-at} + \frac{1}{2a^2}\,e^{at}$$

$$= -\frac{1}{a^2} + \frac{1}{a^2}\left(\frac{e^{at}+e^{-at}}{2}\right) = \frac{1}{a^2}\,(\cosh at - 1)$$

Another Method

By lines 11 and 14 of Table 16-1,

$$\mathscr{L}^{-1}\left[\frac{1/(\mathbf{s}^2-a^2)}{\mathbf{s}}\right] = \int_0^t \frac{\sinh a\tau}{a}\,d\tau = \left[\frac{\cosh a\tau}{a^2}\right]_0^t = \frac{1}{a^2}\,(\cosh at - 1)$$

16.5 Find

$$\mathscr{L}^{-1}\left[\frac{\mathbf{s}+1}{\mathbf{s}(\mathbf{s}^2+4\mathbf{s}+4)}\right]$$

Using the method of partial fractions, we have

$$\frac{\mathbf{s}+1}{\mathbf{s}(\mathbf{s}+2)^2} = \frac{A}{\mathbf{s}} + \frac{B_1}{\mathbf{s}+2} + \frac{B_2}{(\mathbf{s}+2)^2}$$

Then

$$A = \left.\frac{\mathbf{s}+1}{(\mathbf{s}+2)^2}\right|_{\mathbf{s}=0} = \frac{1}{4} \qquad B_2 = \left.\frac{\mathbf{s}+1}{\mathbf{s}}\right|_{\mathbf{s}=-2} = \frac{1}{2}$$

and

$$B_1 = (\mathbf{s}+2)\left.\frac{\mathbf{s}+2}{2\mathbf{s}(\mathbf{s}+2)^2}\right|_{\mathbf{s}=-2} = -\frac{1}{4}$$

Hence,

$$\mathscr{L}^{-1}\left[\frac{\mathbf{s}+1}{\mathbf{s}(\mathbf{s}^2+4\mathbf{s}+4)}\right] = \mathscr{L}^{-1}\left[\frac{\frac{1}{4}}{\mathbf{s}}\right] + \mathscr{L}^{-1}\left[\frac{-\frac{1}{4}}{\mathbf{s}+2}\right] + \mathscr{L}^{-1}\left[\frac{\frac{1}{2}}{(\mathbf{s}+2)^2}\right]$$

The corresponding time functions are found in Table 16-1:

$$\mathscr{L}^{-1}\left[\frac{\mathbf{s}+1}{\mathbf{s}(\mathbf{s}^2+4\mathbf{s}+4)}\right] = \frac{1}{4} - \frac{1}{4}\,e^{-2t} + \frac{1}{2}\,te^{-2t}$$

16.6 In the series *RC* circuit of Fig. 16-5, the capacitor has an initial charge 2.5 mC. At $t = 0$, the switch is closed and a constant-voltage source $V = 100$ V is applied. Use the Laplace transform method to find the current.

The time-domain equation for the given circuit after the switch is closed is

$$Ri(t) + \frac{1}{C}\left[Q_0 + \int_0^t i(\tau)\,d\tau\right] = V$$

or
$$10i(t) + \frac{1}{50 \times 10^{-6}}\left[(-2.5 \times 10^{-3}) + \int_0^t i(\tau)\,d\tau\right] = V \tag{6}$$

Q_0 is opposite in polarity to the charge which the source will deposit on the capacitor. Taking the Laplace transform of the terms in (*6*), we obtain the **s**-domain equation

$$10\mathbf{I}(\mathbf{s}) - \frac{2.5 \times 10^{-3}}{50 \times 10^{-6}\mathbf{s}} + \frac{\mathbf{I}(\mathbf{s})}{50 \times 10^{-6}\mathbf{s}} = \frac{100}{\mathbf{s}}$$

or
$$\mathbf{I}(\mathbf{s}) = \frac{15}{\mathbf{s} + (2 \times 10^3)} \tag{7}$$

The time function is now obtained by taking the inverse Laplace transform of (*7*):

$$i(t) = \mathscr{L}^{-1}\left[\frac{15}{\mathbf{s} + (2 \times 10^3)}\right] = 15e^{-2\times 10^3 t} \quad \text{(A)} \tag{8}$$

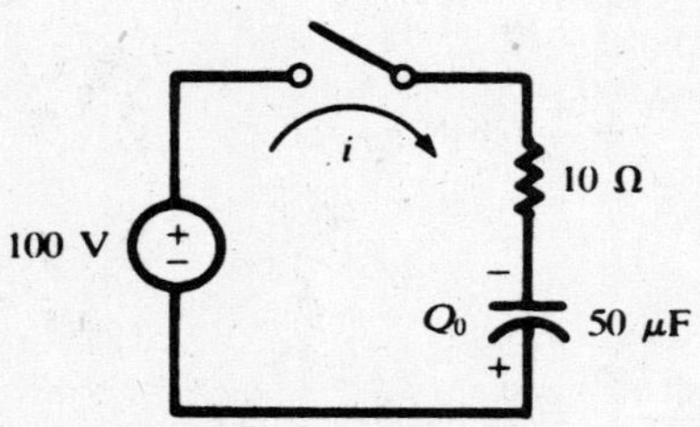

Fig. 16-5

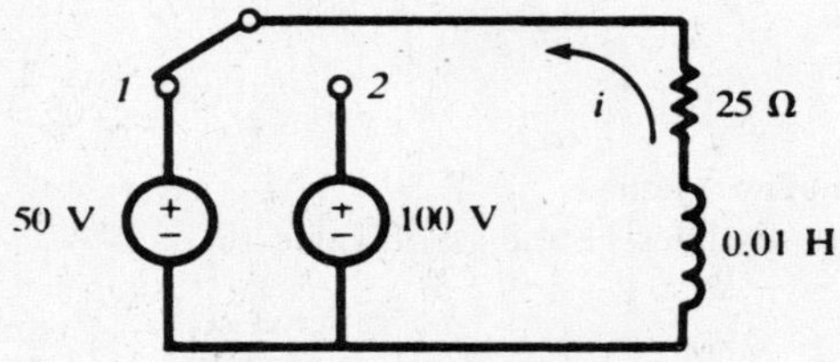

Fig. 16-6

16.7 In the *RL* circuit shown in Fig. 16-6, the switch is in position *1* long enough to establish steady-state conditions, and at $t = 0$ it is switched to position *2*. Find the resulting current.

Assume the direction of the current as shown in the diagram. The initial current is then $i_0 = -50/25 = -2$ A.

The time-domain equation is

$$25i + 0.01\frac{di}{dt} = 100 \tag{9}$$

Taking the Laplace transform of (*9*),

$$25\mathbf{I}(\mathbf{s}) + 0.01\mathbf{s}\mathbf{I}(\mathbf{s}) - 0.01i(0^+) = 100/\mathbf{s} \tag{10}$$

Substituting for $i(0^+)$,

$$25\mathbf{I}(\mathbf{s}) + 0.01\mathbf{s}\mathbf{I}(\mathbf{s}) + 0.01(2) = 100/\mathbf{s} \tag{11}$$

and
$$\mathbf{I}(\mathbf{s}) = \frac{100}{\mathbf{s}(0.01\mathbf{s} + 25)} - \frac{0.02}{0.01\mathbf{s} + 25} = \frac{10^4}{\mathbf{s}(\mathbf{s} + 2500)} - \frac{2}{\mathbf{s} + 2500} \tag{12}$$

Applying the method of partial fractions,

$$\frac{10^4}{\mathbf{s}(\mathbf{s}+2500)} = \frac{A}{\mathbf{s}} + \frac{B}{\mathbf{s}+2500} \tag{13}$$

with
$$A = \left.\frac{10^4}{\mathbf{s}+2500}\right|_{\mathbf{s}=0} = 4 \quad \text{and} \quad B = \left.\frac{10^4}{\mathbf{s}}\right|_{\mathbf{s}=-2500} = -4$$

Then,
$$\mathbf{I}(\mathbf{s}) = \frac{4}{\mathbf{s}} - \frac{4}{\mathbf{s}+2500} - \frac{2}{\mathbf{s}+2500} = \frac{4}{\mathbf{s}} - \frac{6}{\mathbf{s}+2500} \tag{14}$$

Taking the inverse Laplace transform of (*14*), we obtain $i = 4 - 6e^{-2500t}$ (A).

16.8 In the series *RL* circuit of Fig. 16-7, an exponential voltage $v = 50e^{-100t}$ (V) is applied by closing the switch at $t = 0$. Find the resulting current.

The time-domain equation for the given circuit is

$$Ri + L\frac{di}{dt} = v \tag{15}$$

In the s-domain, (*15*) has the form

$$R\mathbf{I}(\mathbf{s}) + sL\mathbf{I}(\mathbf{s}) - Li(0^+) = \mathbf{V}(\mathbf{s}) \tag{16}$$

Substituting the circuit constants and the transform of the source, $\mathbf{V}(\mathbf{s}) = 50/(\mathbf{s}+100)$, in (*16*),

$$10\mathbf{I}(\mathbf{s}) + \mathbf{s}(0.2)\mathbf{I}(\mathbf{s}) = \frac{5}{\mathbf{s}+100} \quad \text{or} \quad \mathbf{I}(\mathbf{s}) = \frac{250}{(\mathbf{s}+100)(\mathbf{s}+50)} \tag{17}$$

By the Heaviside expansion formula,

$$\mathscr{L}^{-1}[\mathbf{I}(\mathbf{s})] = \mathscr{L}^{-1}\left[\frac{\mathbf{P}(\mathbf{s})}{\mathbf{Q}(\mathbf{s})}\right] = \sum_{n=1,2} \frac{\mathbf{P}(\mathbf{a}_n)}{\mathbf{Q}'(\mathbf{a}_n)} e^{\mathbf{a}_n t}$$

Here, $\mathbf{P}(\mathbf{s}) = 250$, $\mathbf{Q}(\mathbf{s}) = \mathbf{s}^2 + 150\mathbf{s} + 5000$, $\mathbf{Q}'(\mathbf{s}) = 2\mathbf{s} + 150$, $\mathbf{a}_1 = -100$, and $\mathbf{a}_2 = -50$. Then,

$$i = \mathscr{L}^{-1}[\mathbf{I}(\mathbf{s})] = \frac{250}{-50}e^{-100t} + \frac{250}{50}e^{-50t} = -5e^{-100t} + 5e^{-50t} \quad \text{(A)}$$

16.9 The series *RC* circuit of Fig. 16-8 has a sinusoidal voltage source $v = 180\sin(2000t + \phi)$ (V) and an initial charge on the capacitor $Q_0 = 1.25$ mC with polarity as shown. Determine the current if the switch is closed at a time corresponding to $\phi = 90°$.

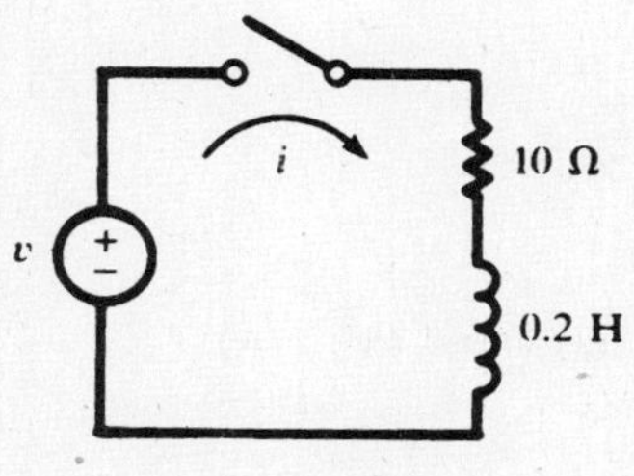

Fig. 16-7

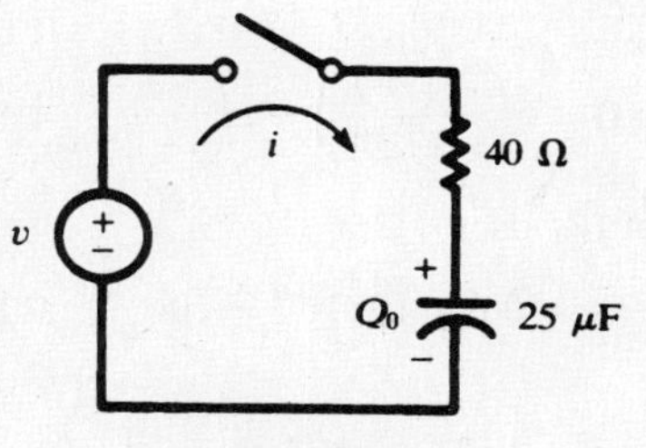

Fig. 16-8

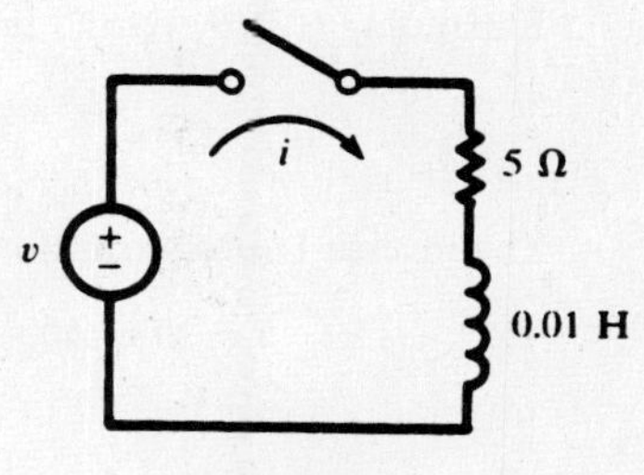

Fig. 16-9

The time-domain equation of the circuit is

$$40i(t) + \frac{1}{25 \times 10^{-6}}\left[(1.25 \times 10^{-3}) + \int_0^t i(\tau)\,d\tau\right] = 180\cos 2000t \tag{18}$$

The Laplace transform of *(18)* gives the **s**-domain equation

$$40\mathbf{I}(\mathbf{s}) + \frac{1.25 \times 10^{-3}}{25 \times 10^{-6}\mathbf{s}} + \frac{4 \times 10^4}{\mathbf{s}}\mathbf{I}(\mathbf{s}) = \frac{180\mathbf{s}}{\mathbf{s}^2 + 4 \times 10^6} \tag{19}$$

or

$$\mathbf{I}(\mathbf{s}) = \frac{4.5\mathbf{s}^2}{(\mathbf{s}^2 + 4 \times 10^6)(\mathbf{s} + 10^3)} - \frac{1.25}{\mathbf{s} + 10^3} \tag{20}$$

Applying the Heaviside expansion formula to the first term on the right in *(20)*, we have $\mathbf{P}(\mathbf{s}) = 4.5\mathbf{s}^2$, $\mathbf{Q}(\mathbf{s}) = \mathbf{s}^3 + 10^3\mathbf{s}^2 + 4 \times 10^6\mathbf{s} + 4 \times 10^9$, $\mathbf{Q}'(\mathbf{s}) = 3\mathbf{s}^2 + 2 \times 10^3\mathbf{s} + 4 \times 10^6$, $\mathbf{a}_1 = -j2 \times 10^3$, $\mathbf{a}_2 = j2 \times 10^3$, and $\mathbf{a}_3 = -10^3$. Then,

$$\begin{aligned} i &= \frac{\mathbf{P}(-j2 \times 10^3)}{\mathbf{Q}'(-j \times 10^3)} e^{-j2\times10^3 t} + \frac{\mathbf{P}(j2 \times 10^3)}{\mathbf{Q}'(j2 \times 10^3)} e^{j2\times10^3 t} + \frac{\mathbf{P}(-10^3)}{\mathbf{Q}'(-10^3)} e^{-10^3 t} - 1.25e^{-10^3 t} \\ &= (1.8 - j0.9)e^{-j2\times10^3 t} + (1.8 + j0.9)e^{j2\times10^3 t} - 0.35e^{-10^3 t} \qquad (21) \\ &= -1.8 \sin 2000t + 3.6 \cos 2000t - 0.35e^{-10^3 t} \\ &= 4.02 \sin(2000t + 116.6^\circ) - 0.35e^{-10^3 t} \quad \text{(A)} \end{aligned}$$

At $t = 0$, the current is given by the instantaneous voltage, consisting of the source voltage and the charged capacitor voltage, divided by the resistance. Thus,

$$i_0 = \left(180 \sin 90^\circ - \frac{1.25 \times 10^{-3}}{25 \times 10^{-6}}\right) \Big/ 40 = 3.25 \text{ A}$$

The same result is obtained if we set $t = 0$ in *(21)*.

16.10 In the series RL circuit of Fig. 16-9, the source is $v = 100 \sin(500t + \phi)$ (V). Determine the resulting current if the switch is closed at a time corresponding to $\phi = 0$.

The **s**-domain equation of a series RL circuit is

$$R\mathbf{I}(\mathbf{s}) + sL\mathbf{I}(\mathbf{s}) - Li(0^+) = \mathbf{V}(\mathbf{s}) \tag{22}$$

The transform of the source with $\phi = 0$ is

$$\mathbf{V}(\mathbf{s}) = \frac{(100)(500)}{\mathbf{s}^2 + (500)^2}$$

Since there is no initial current in the inductance, $Li(0^+) = 0$. Substituting the circuit constants into *(22)*,

$$5\mathbf{I}(\mathbf{s}) + 0.01\mathbf{s}\mathbf{I}(\mathbf{s}) = \frac{5 \times 10^4}{\mathbf{s}^2 + 25 \times 10^4} \quad \text{or} \quad \mathbf{I}(\mathbf{s}) = \frac{5 \times 10^6}{(\mathbf{s}^2 + 25 \times 10^4)(\mathbf{s} + 500)} \tag{23}$$

Expanding *(23)* by partial fractions,

$$\mathbf{I}(\mathbf{s}) = 5\left(\frac{-1 + j}{\mathbf{s} + j500}\right) + 5\left(\frac{-1 - j}{\mathbf{s} - j500}\right) + \frac{10}{\mathbf{s} + 500} \tag{24}$$

The inverse Laplace transform of *(24)* is

$$i = 10 \sin 500t - 10 \cos 500t + 10e^{-500t} = 10e^{-500t} + 14.14 \sin(500t - 45^\circ) \quad \text{(A)}$$

16.11 Rework Problem 16.10 by writing the voltage function as

$$v = 100e^{j500t} \quad \text{(V)} \tag{25}$$

Now $\mathbf{V}(\mathbf{s}) = 100/(\mathbf{s} - j500)$, and the **s**-domain equation is

$$5\mathbf{I}(\mathbf{s}) + 0.01\mathbf{s}\mathbf{I}(\mathbf{s}) = \frac{100}{\mathbf{s} - j500} \quad \text{or} \quad \mathbf{I}(\mathbf{s}) = \frac{10^4}{(\mathbf{s} - j500)(\mathbf{s} + 500)}$$

Using partial fractions,

$$\mathbf{I}(\mathbf{s}) = \frac{10 - j10}{\mathbf{s} - j500} + \frac{-10 + j10}{\mathbf{s} + 500}$$

and inverting,

$$\begin{aligned} i &= (10 - j10)e^{j500t} + (-10 + j10)e^{-500t} \\ &= 14.14e^{j(500t - \pi/4)} + (-10 + j10)e^{-500t} \quad \text{(A)} \end{aligned} \tag{26}$$

The actual voltage is the imaginary part of (*25*); hence the actual current is the imaginary part of (*26*).

$$i = 14.14 \sin(500t - \pi/4) + 10e^{-500t} \quad \text{(A)}$$

16.12 In the series *RLC* circuit shown in Fig. 16-10, there is no initial charge on the capacitor. If the switch is closed at $t = 0$, determine the resulting current.

The time-domain equation of the given circuit is

$$Ri + L\frac{di}{dt} + \frac{1}{C}\int_0^t i(\tau)\,d\tau = V \tag{27}$$

Because $i(0^+) = 0$, the Laplace transform of (*27*) is

$$R\mathbf{I}(\mathbf{s}) + sL\mathbf{I}(\mathbf{s}) + \frac{1}{\mathbf{s}C} = \mathbf{I}(\mathbf{s})\frac{V}{\mathbf{s}} \tag{28}$$

or

$$2\mathbf{I}(\mathbf{s}) + 1\mathbf{s}\mathbf{I}(\mathbf{s}) + \frac{1}{0.5\mathbf{s}}\mathbf{I}(\mathbf{s}) = \frac{50}{\mathbf{s}} \tag{29}$$

Hence,

$$\mathbf{I}(\mathbf{s}) = \frac{50}{\mathbf{s}^2 + 2\mathbf{s} + 2} = \frac{50}{(\mathbf{s} + 1 + j)(\mathbf{s} + 1 - j)} \tag{30}$$

Expanding (*30*) by partial fractions,

$$\mathbf{I}(\mathbf{s}) = \frac{j25}{(\mathbf{s} + 1 + j)} - \frac{j25}{(\mathbf{s} + 1 - j)} \tag{31}$$

and the inverse Laplace transform of (*31*) gives

$$i = j25\{e^{(-1-j)t} - e^{(-1+j)t}\} = 50e^{-t}\sin t \quad \text{(A)}$$

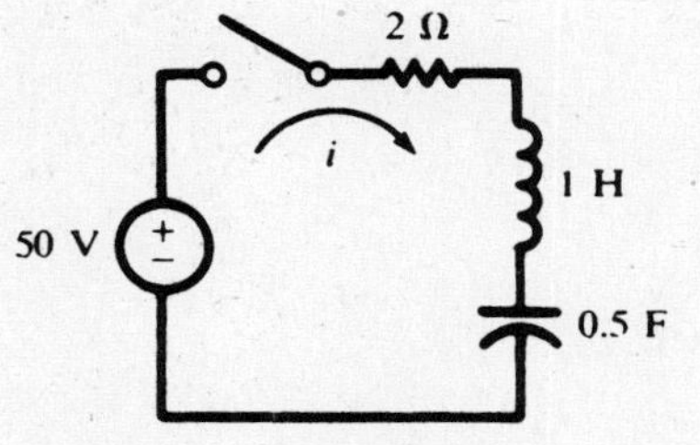

Fig. 16-10

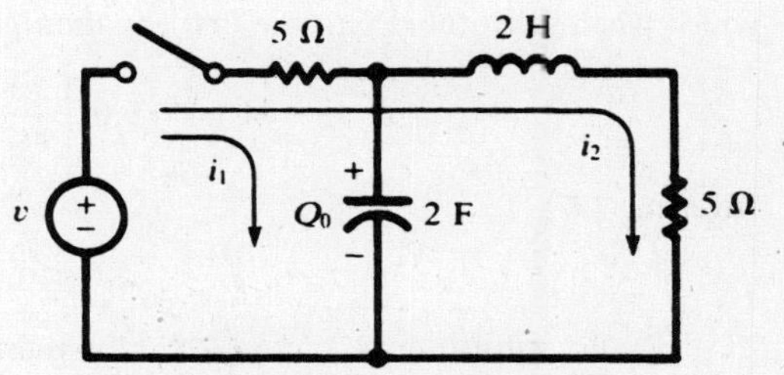

Fig. 16-11

16.13 In the two-mesh network of Fig. 16-11, the two loop currents are selected as shown. Write the **s**-domain equations in matrix form and construct the corresponding circuit.

Writing the set of equations in the time domain,

$$5i_1 + \frac{1}{2}\left[Q_0 + \int_0^t i_1(\tau)d\tau\right] + 5i_2 = v \quad \text{and} \quad 10i_2 + 2\frac{di_2}{dt} + 5i_1 = v \tag{32}$$

Taking the Laplace transform of (*32*) to obtain the corresponding **s**-domain equations,

$$5\mathbf{I}_1(\mathbf{s}) + \frac{Q_0}{2\mathbf{s}} + \frac{1}{2\mathbf{s}}\mathbf{I}_1(\mathbf{s}) + 5\mathbf{I}_2(\mathbf{s}) = \mathbf{V}(\mathbf{s}) \qquad 10\mathbf{I}_2(\mathbf{s}) + 2\mathbf{s}\mathbf{I}_2(\mathbf{s}) - 2i_2(0^+) + 5\mathbf{I}_1(\mathbf{s}) = \mathbf{V}(\mathbf{s}) \tag{33}$$

When this set of s-domain equations is written in matrix form,

$$\begin{bmatrix} 5+(1/2\mathbf{s}) & 5 \\ 5 & 10+2\mathbf{s} \end{bmatrix} \begin{bmatrix} \mathbf{I}_1(\mathbf{s}) \\ \mathbf{I}_2(\mathbf{s}) \end{bmatrix} = \begin{bmatrix} \mathbf{V}(\mathbf{s}) - (Q_0/2\mathbf{s}) \\ \mathbf{V}(\mathbf{s}) + 2i_2(0^+) \end{bmatrix}$$

the required s-domain circuit can be determined by examination of the **Z**(s), **I**(s), and **V**(s) matrices (see Fig. 16-12).

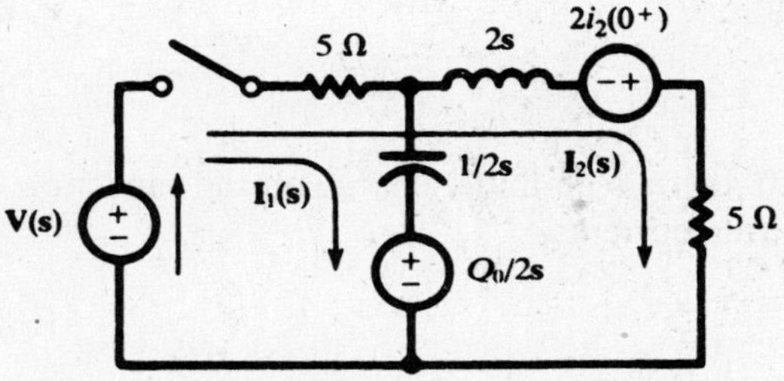

Fig. 16-12

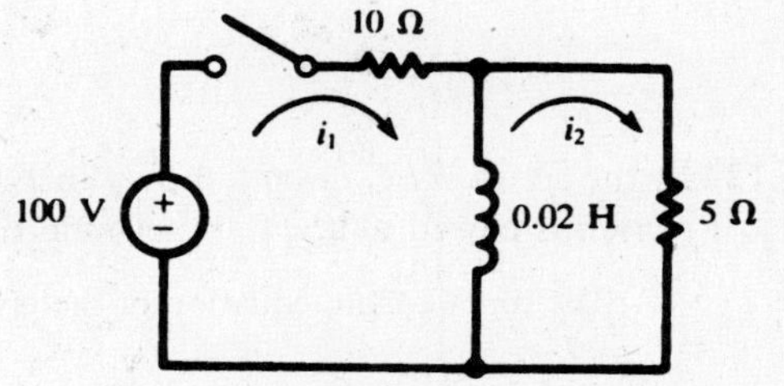

Fig. 16-13

16.14 In the two-mesh network of Fig. 16-13, find the currents which result when the switch is closed.

The time-domain equations for the network are

$$\begin{aligned} 10i_1 + 0.02\frac{di_1}{dt} - 0.02\frac{di_2}{dt} &= 100 \\ 0.02\frac{di_2}{dt} + 5i_2 - 0.02\frac{di_1}{dt} &= 0 \end{aligned} \tag{34}$$

Taking the Laplace transform of set (*34*),

$$(10 + 0.02\mathbf{s})\mathbf{I}_1(\mathbf{s}) - 0.02\mathbf{s}\mathbf{I}_2(\mathbf{s}) = 100/\mathbf{s} \qquad (5 + 0.02\mathbf{s})\mathbf{I}_2(\mathbf{s}) - 0.02\mathbf{s}\mathbf{I}_1(\mathbf{s}) = 0 \tag{35}$$

From the second equation in set (*35*) we find

$$\mathbf{I}_2(\mathbf{s}) = \mathbf{I}_1(\mathbf{s})\left(\frac{\mathbf{s}}{\mathbf{s}+250}\right) \tag{36}$$

which when substituted into the first equation gives

$$\mathbf{I}_1(\mathbf{s}) = 6.67\left[\frac{\mathbf{s}+250}{\mathbf{s}(\mathbf{s}+166.7)}\right] = \frac{10}{\mathbf{s}} - \frac{3.33}{\mathbf{s}+166.7} \tag{37}$$

Inverting (*37*),

$$i_1 = 10 - 3.33e^{-166.7t} \quad \text{(A)}$$

Finally, substitute (*37*) into (*36*) and obtain

$$\mathbf{I}_2(\mathbf{s}) = 6.67\left(\frac{1}{\mathbf{s}+166.7}\right) \qquad \text{whence} \qquad i_2 = 6.67e^{-166.7t} \quad \text{(A)}$$

16.15 Apply the initial- and final-value theorems in Problem 16.14.

The initial value of i_1 is given by

$$i_1(0^+) = \lim_{s\to\infty}[s\mathbf{I}_1(s)] = \lim_{s\to\infty}\left[6.667\left(\frac{s+250}{s+166.7}\right)\right] = 6.67 \text{ A}$$

and the final value is

$$i_1(\infty) = \lim_{s\to 0}[s\mathbf{I}_1(s)] = \lim_{s\to 0}\left[6.67\left(\frac{s+250}{s+166.7}\right)\right] = 10 \text{ A}$$

The initial value of i_2 is given by

$$i_2(0^+) = \lim_{s\to\infty}[s\mathbf{I}_2(s)] = \lim_{s\to\infty}\left[6.667\left(\frac{s}{s+166.7}\right)\right] = 6.67 \text{ A}$$

and the final value is

$$i_2(\infty) = \lim_{s\to 0}[s\mathbf{I}_2(s)] = \lim_{s\to 0}\left[6.67\left(\frac{s}{s+166.7}\right)\right] = 0$$

Examination of Fig. 16-13 verifies each of the preceding initial and final values. At the instant of closing, the inductance presents an infinite impedance and the currents are $i_1 = i_2 = 100/(10+5) = 6.67$ A. Then, in the steady state, the inductance appears as a short circuit; hence, $i_1 = 10$ A, $i_2 = 0$.

16.16 Solve for i_1 in Problem 16.14 by determining an equivalent circuit in the **s**-domain.

In the **s**-domain the 0.02-H inductor has impedance $\mathbf{Z(s)} = 0.02\mathbf{s}$. Therefore, the equivalent impedance of the network as seen from the source is

$$\mathbf{Z(s)} = 10 + \frac{(0.02\mathbf{s})(5)}{0.02\mathbf{s}+5} = 15\left(\frac{\mathbf{s}+166.7}{\mathbf{s}+250}\right)$$

and the **s**-domain equivalent circuit is as shown in Fig. 16-14. The current is then

$$\mathbf{I}_1(\mathbf{s}) = \frac{\mathbf{V(s)}}{\mathbf{Z(s)}} = \frac{100}{\mathbf{s}}\left[\frac{\mathbf{s}+250}{15(\mathbf{s}+166.7)}\right] = 6.67\left[\frac{\mathbf{s}+250}{\mathbf{s}(\mathbf{s}+166.7)}\right]$$

This expression is identical with (*37*) of Problem 16.14, and so the same time function i_1 is obtained.

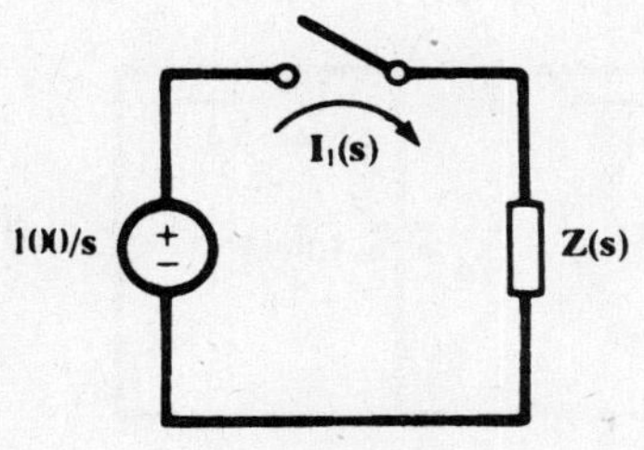

Fig. 16-14

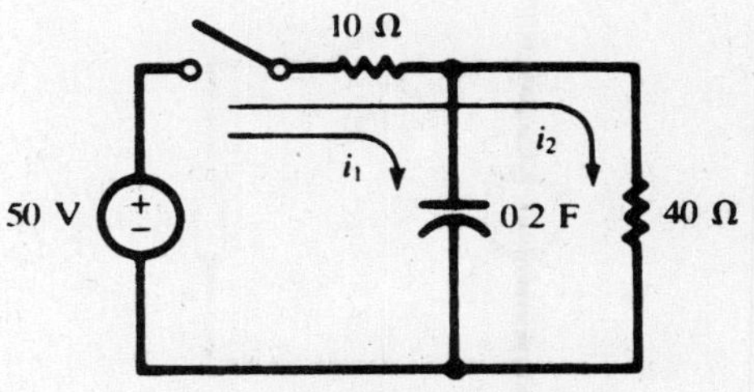

Fig. 16-15

16.17 In the two-mesh network shown in Fig. 16-15 there is no initial charge on the capacitor. Find the loop currents i_1 and i_2 which result when the switch is closed at $t = 0$.

The time-domain equations for the circuit are

$$10i_1 + \frac{1}{0.2}\int_0^t i_1\,d\tau + 10i_2 = 50 \qquad 50i_2 + 10i_1 = 50$$

The corresponding **s**-domain equations are

$$10\mathbf{I}_1(\mathbf{s}) + \frac{1}{0.2\mathbf{s}}\mathbf{I}_1(\mathbf{s}) + 10\mathbf{I}_2(\mathbf{s}) = \frac{50}{\mathbf{s}} \qquad 50\mathbf{I}_2(\mathbf{s}) + 10\mathbf{I}_1(\mathbf{s}) = \frac{50}{\mathbf{s}}$$

Solving,
$$\mathbf{I}_1(\mathbf{s}) = \frac{5}{\mathbf{s}+0.625} \qquad \mathbf{I}_2(\mathbf{s}) = \frac{1}{\mathbf{s}} - \frac{1}{\mathbf{s}+0.625}$$

which invert to

$$i_1 = 5e^{-0.625t} \quad \text{(A)} \qquad i_2 = 1 - e^{-0.625t} \quad \text{(A)}$$

16.18 Referring to Problem 16.17, obtain the equivalent impedance of the s-domain network and determine the total current and the branch currents using the current-division rule.

The s-domain impedance as seen by the voltage source is

$$\mathbf{Z(s)} = 10 + \frac{40(1/0.2\mathbf{s})}{40 + 1/0.2\mathbf{s}} = \frac{80\mathbf{s} + 50}{8\mathbf{s} + 1} = 10\left(\frac{\mathbf{s} + 5/8}{\mathbf{s} + 1/8}\right) \tag{38}$$

The equivalent circuit is shown in Fig. 16-16; the resulting current is

$$\mathbf{I(s)} = \frac{\mathbf{V(s)}}{\mathbf{Z(s)}} = 5\,\frac{\mathbf{s} + 1/8}{\mathbf{s}(\mathbf{s} + 5/8)} \tag{39}$$

Expanding **I(s)** in partial fractions,

$$\mathbf{I(s)} = \frac{1}{\mathbf{s}} + \frac{4}{\mathbf{s} + 5/8} \qquad \text{from which} \qquad i = 1 + 4e^{-5t/8} \quad \text{(A)}$$

Now the branch currents $\mathbf{I}_1(\mathbf{s})$ and $\mathbf{I}_2(\mathbf{s})$ can be obtained by the current-division rule. Referring to Fig. 16-17, we have

$$\mathbf{I}_1(\mathbf{s}) = \mathbf{I(s)}\left(\frac{40}{40 + 1/0.2\mathbf{s}}\right) = \frac{5}{\mathbf{s} + 5/8} \qquad \text{and} \qquad i_1 = 5e^{-0.625t} \quad \text{(A)}$$

$$\mathbf{I}_2(\mathbf{s}) = \mathbf{I(s)}\left(\frac{1/0.2\mathbf{s}}{40 + 1/0.2\mathbf{s}}\right) = \frac{1}{\mathbf{s}} - \frac{1}{\mathbf{s} + 5/8} \qquad \text{and} \qquad i_2 = 1 - e^{-0.625t} \quad \text{(A)}$$

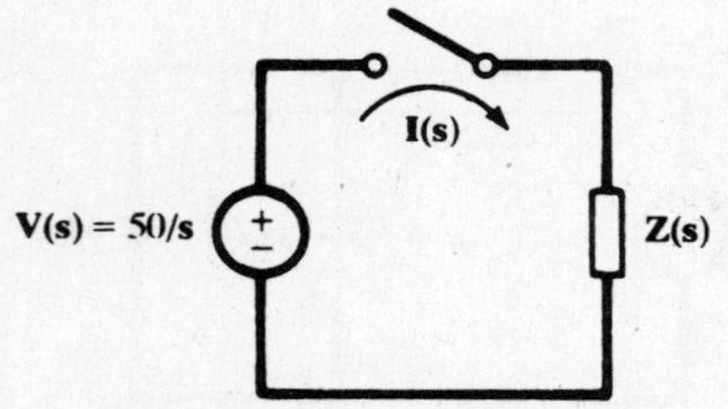

Fig. 16-16

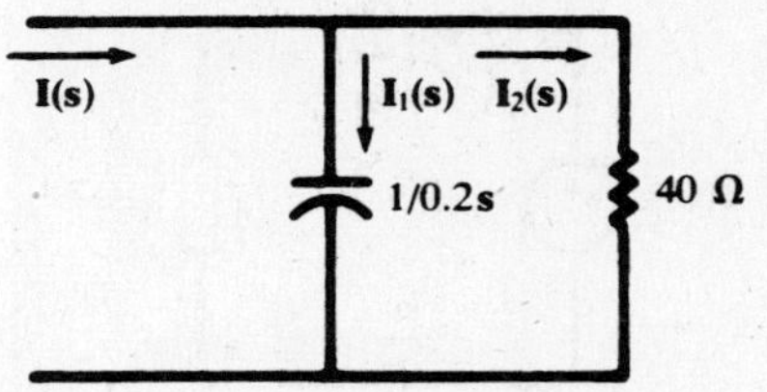

Fig. 16-17

16.19 In the network of Fig. 16-18 the switch is closed at $t = 0$ and there is no initial charge on either of the capacitors. Find the resulting current i.

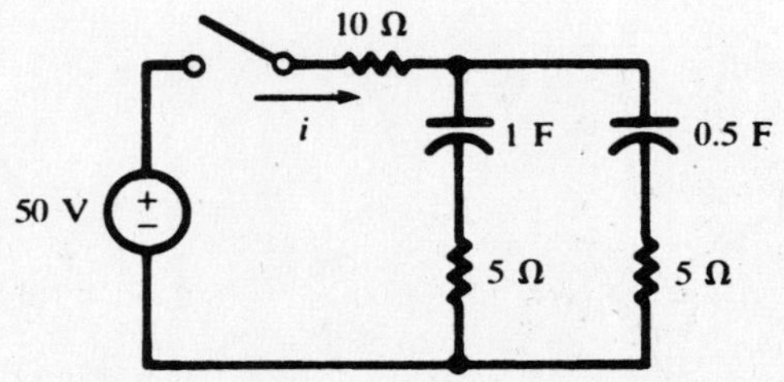

Fig. 16-18

The network has an equivalent impedance in the s-domain

$$\mathbf{Z(s)} = 10 + \frac{(5+1/\mathbf{s})(5+1/0.5\mathbf{s})}{10+1/\mathbf{s}+1/0.5\mathbf{s}} = \frac{125\mathbf{s}^2+45\mathbf{s}+2}{\mathbf{s}(10\mathbf{s}+3)}$$

Hence, the current is

$$\mathbf{I(s)} = \frac{\mathbf{V(s)}}{\mathbf{Z(s)}} = \frac{50}{\mathbf{s}}\,\frac{\mathbf{s}(10\mathbf{s}+3)}{(125\mathbf{s}^2+45\mathbf{s}+2)} = \frac{4(\mathbf{s}+0.3)}{(\mathbf{s}+0.308)(\mathbf{s}+0.052)}$$

Expanding $\mathbf{I(s)}$ in partial fractions,

$$\mathbf{I(s)} = \frac{1/8}{\mathbf{s}+0.308} + \frac{31/8}{\mathbf{s}+0.052} \qquad \text{and} \qquad i = \frac{1}{8}e^{-0.308t} + \frac{31}{8}e^{-0.052t} \qquad \text{(a)}$$

16.20 Apply the initial- and final-value theorems to the s-domain current of Problem 16.19.

$$i(0^+) = \lim_{s\to\infty}[s\mathbf{I}(s)] = \lim_{s\to\infty}\left[\frac{1}{8}\left(\frac{s}{s+0.308}\right) + \frac{31}{8}\left(\frac{s}{s+0.052}\right)\right] = 4\text{ A}$$

$$i(\infty) = \lim_{s\to 0}[s\mathbf{I}(s)] = \lim_{s\to 0}\left[\frac{1}{8}\left(\frac{s}{s+0.308}\right) + \frac{31}{8}\left(\frac{s}{s+0.052}\right)\right] = 0$$

Examination of Fig. 16-18 shows that initially the total circuit resistance is $R = 10 + 5(5)/10 = 12.5\ \Omega$, and thus, $i(0^+) = 50/12.5 = 4$ A. Then, in the steady state, both capacitors are charged to 50 V and the current is zero.

Supplementary Problems

16.21 Find the Laplace transform of each of the following functions.

(a) $f(t) = At$ (c) $f(t) = e^{-at}\sin\omega t$ (e) $f(t) = \cosh\omega t$
(b) $f(t) = te^{-at}$ (d) $f(t) = \sinh\omega t$ (f) $f(t) = e^{-at}\sinh\omega t$

Ans. (a)–(e) See Table 16-1
(f) $\dfrac{\omega}{(\mathbf{s}+a)^2-\omega^2}$

16.22 Find the inverse Laplace transform of each of the following functions.

(a) $\mathbf{F(s)} = \dfrac{\mathbf{s}}{(\mathbf{s}+2)(\mathbf{s}+1)}$ (d) $\mathbf{F(s)} = \dfrac{3}{\mathbf{s}(\mathbf{s}^2+6\mathbf{s}+9)}$ (g) $\mathbf{F(s)} = \dfrac{2\mathbf{s}}{(\mathbf{s}^2+4)(\mathbf{s}+5)}$

(b) $\mathbf{F(s)} = \dfrac{1}{\mathbf{s}^2+7\mathbf{s}+12}$ (e) $\mathbf{F(s)} = \dfrac{\mathbf{s}+5}{\mathbf{s}^2+2\mathbf{s}+5}$

(c) $\mathbf{F(s)} = \dfrac{5\mathbf{s}}{\mathbf{s}^2+3\mathbf{s}+2}$ (f) $\mathbf{F(s)} = \dfrac{2\mathbf{s}+4}{\mathbf{s}^2+4\mathbf{s}+13}$

Ans. (a) $2e^{-2t} - e^{-t}$ (d) $\frac{1}{3} - \frac{1}{3}e^{-3t} - te^{-3t}$ (g) $\frac{10}{29}\cos 2t + \frac{4}{29}\sin 2t - \frac{10}{29}e^{-5t}$
(b) $e^{-3t} - e^{-4t}$ (e) $e^{-t}(\cos 2t + 2\sin 2t)$
(c) $10e^{-2t} - 5e^{-t}$ (f) $2e^{-2t}\cos 3t$

16.23 A series RL circuit, with $R = 10\ \Omega$ and $L = 0.2$ H, has a constant voltage $V = 50$ V applied at $t = 0$. Find the resulting current using the Laplace transform method. *Ans.* $i = 5 - 5e^{-50t}$ (A)

16.24 In the series RL circuit of Fig. 16-19, the switch is in position *1* long enough to establish the steady state and is switched to position *2* at $t = 0$. Find the current. *Ans.* $i = 5e^{-50t}$ (A)

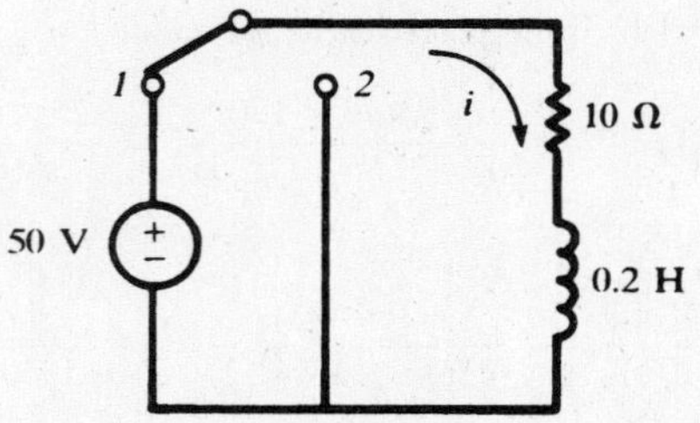

Fig. 16-19

16.25 In the circuit shown in Fig. 16-20, switch *1* is closed at $t = 0$ and then, at $t = t' = 4$ ms, switch *2* is opened. Find the current in the intervals $0 < t < t'$ and $t > t'$.
Ans. $i = 2(1 - e^{-500t})$ A, $i = 1.06e^{-1500(t-t')} + 0.667$ (A)

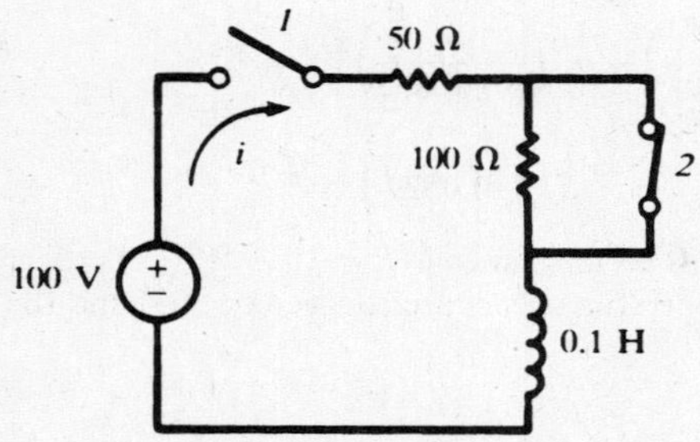

Fig. 16-20

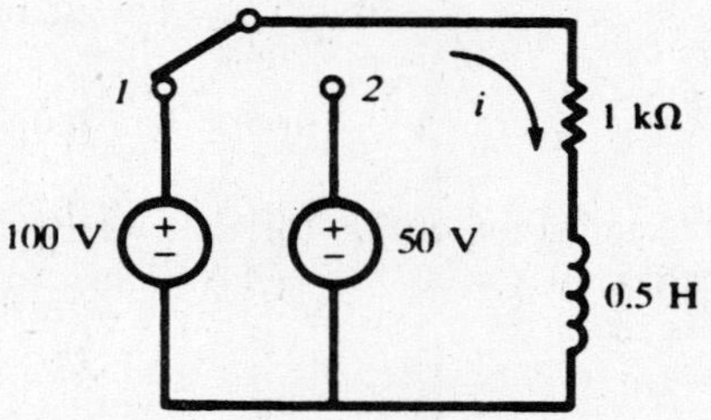

Fig. 16-21

16.26 In the series *RL* circuit shown in Fig. 16-21, the switch is closed on position *1* at $t = 0$ and then, at $t = t' = 50$ μs, it is moved to position *2*. Find the current in the intervals $0 < t < t'$ and $t > t'$.
Ans. $i = 0.1(1 - e^{-2000t})$ (A), $i = 0.06e^{-2000(t-t')} - 0.05$ (A)

16.27 A series *RC* circuit, with $R = 10\ \Omega$ and $C = 4\ \mu$F, has an initial charge $Q_0 = 800\ \mu$C on the capacitor at the time the switch is closed, applying a constant-voltage source $V = 100$ V. Find the resulting current transient if the charge is (*a*) of the same polarity as that deposited by the source, and (*b*) of the opposite polarity.
Ans. (*a*) $i = -10e^{-25\times10^3 t}$ (A); (*b*) $i = 30e^{-25\times10^3 t}$ (A)

16.28 A series *RC* circuit, with $R = 1$ kΩ and $C = 20\ \mu$F, has an initial charge Q_0 on the capacitor at the time the switch is closed, applying a constant-voltage source $V = 50$ V. If the resulting current is $i = 0.075e^{-50t}$ (A), find the charge Q_0 and its polarity.
Ans. 500 μC, opposite polarity to that deposited by source

16.29 In the *RC* circuit shown in Fig. 16-22, the switch is closed on position *1* at $t = 0$ and then, at $t = t' = \tau$ (the time constant) is moved to position *2*. Find the transient current in the intervals $0 < t < t'$ and $t > t'$.
Ans. $i = 0.5e^{-200t}$ (A), $i = -.0516e^{-200(t-t')}$ (A)

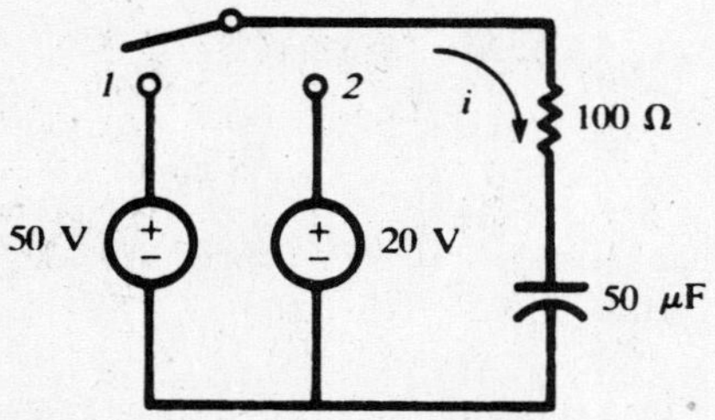

Fig. 16-22

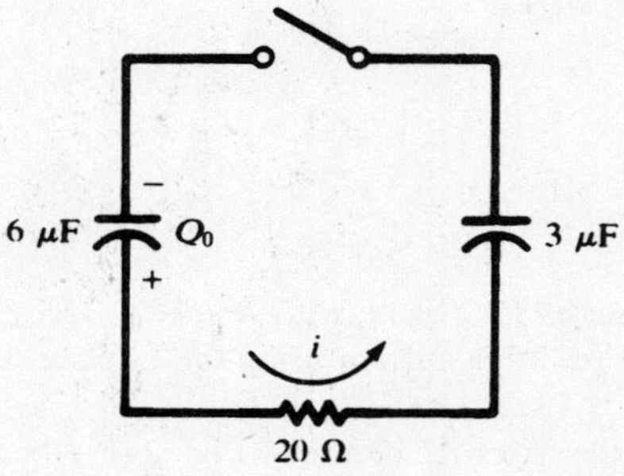

Fig. 16-23

16.30 In the circuit of Fig. 16-23, $Q_0 = 300\ \mu C$ at the time the switch is closed. Find the resulting current transient. *Ans.* $i = 2.5e^{-2.5\times10^4 t}$ (A)

16.31 In the circuit shown in Fig. 16-24, the capacitor has an initial charge $Q_0 = 25\ \mu C$ and the sinusoidal voltage source is $v = 100 \sin(1000t + \phi)$ (V). Find the resulting current if the switch is closed at a time corresponding to $\phi = 30°$. *Ans.* $i = 0.1535e^{-4000t} + 0.0484 \sin(1000t + 106°)$ (A)

16.32 A series *RLC* circuit, with $R = 5\ \Omega$, $L = 0.1$ H, and $C = 500\ \mu F$, has a constant voltage $V = 10$ V applied at $t = 0$. Find the resulting current. *Ans.* $i = 0.72e^{-25t} \sin 139t$ (A)

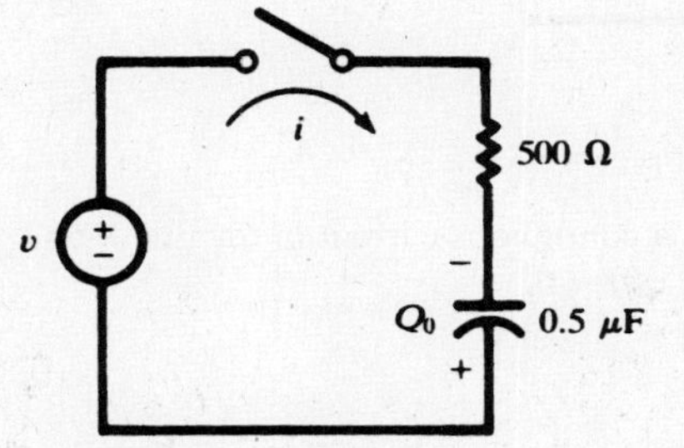

Fig. 16-24

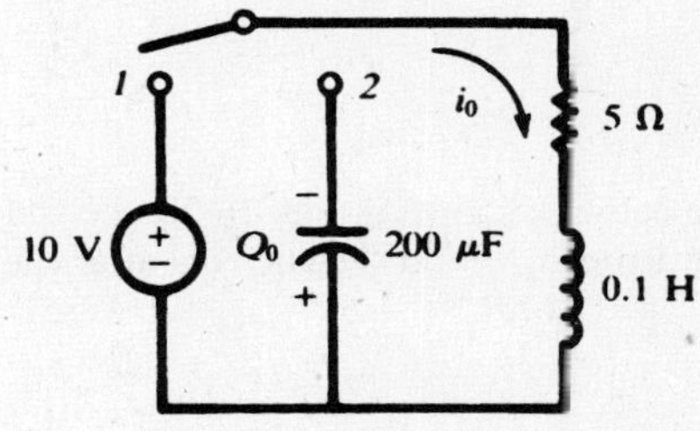

Fig. 16-25

16.33 In the series *RLC* circuit of Fig. 16-25, the capacitor has an initial charge $Q_0 = 1$ mC and the switch is in position *1* long enough to establish the steady state. Find the transient current which results when the switch is moved from position *1* to *2* at $t = 0$. *Ans.* $i = e^{-25t}(2 \cos 222t - 0.45 \sin 222t)$ (A)

16.34 A series *RLC* circuit, with $R = 5\ \Omega$, $L = 0.2$ H, and $C = 1$ F has a voltage source $v = 10e^{-100t}$ (V) applied at $t = 0$. Find the resulting current.
Ans. $i = -0.666e^{-100t} + 0.670e^{-24.8t} - 0.004e^{-0.2t}$ (A)

16.35 A series *RLC* circuit, with $R = 200\ \Omega$, $L = 0.5$ H, and $C = 100\ \mu F$ has a sinusoidal voltage source $v = 300 \sin(500t + \phi)$ (V). Find the resulting current if the switch is closed at a time corresponding to $\phi = 30°$. *Ans.* $i = 0.517e^{-341.4t} - 0.197e^{-58.6t} + 0.983 \sin(500t - 19°)$ (A)

16.36 A series *RLC* circuit, with $R = 5\ \Omega$, $L = 0.1$ H, and $C = 500\ \mu F$ has a sinusoidal voltage source $v = 100 \sin 250t$ (V). Find the resulting current if the switch is closed at $t = 0$.
Ans. $i = e^{-25t}(5.42 \cos 139t + 1.89 \sin 139t) + 5.65 \sin(250t - 73.6°)$ (A)

16.37 In the two-mesh network of Fig. 16-26, the currents are selected as shown in the diagram. Write the time-domain equations, transform them into the corresponding **s**-domain equations, and obtain the currents i_1 and i_2. *Ans.* $i_1 = 2.5(1 + e^{-10^5 t})$ (A), $i_2 = 5e^{-10^5 t}$ (A)

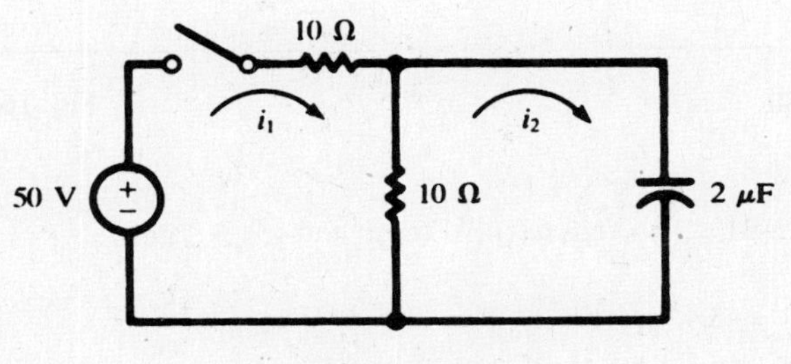

Fig. 16-26

16.38 For the two-mesh network shown in Fig. 16-27, find the currents i_1 and i_2 which result when the switch is closed at $t = 0$. *Ans.* $i_1 = 0.101e^{-100t} + 9.899e^{-9950t}$ (A), $i_2 = -5.05e^{-100t} + 5 + 0.05e^{-9950t}$ (A)

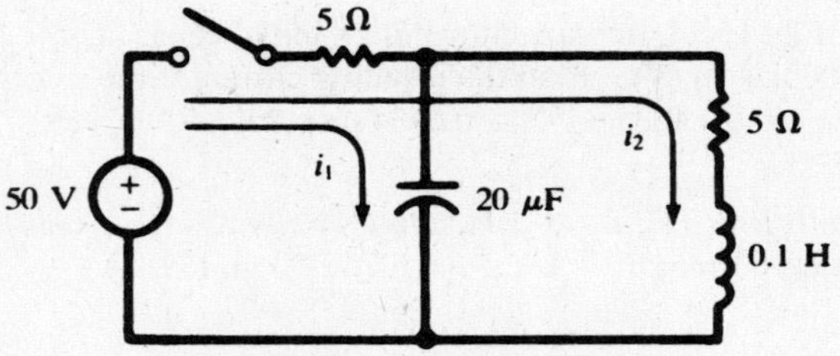

Fig. 16-27

16.39 In the network shown in Fig. 16-28, the 100-V source passes a continuous current in the first loop while the switch is open. Find the currents after the switch is closed at $t = 0$.
Ans. $i_1 = 1.67e^{-6.67t} + 5$ (A), $i_2 = 0.555e^{-6.67t} + 5$ (A)

16.40 The two-mesh network shown in Fig. 16-29 contains a sinusoidal voltage source $v = 100 \sin(200t + \phi)$ (V). The switch is closed at an instant when the voltage is increasing at its maximum rate. Find the resulting mesh currents, with directions as shown in the diagram.
Ans. $i_1 = 3.01e^{-100t} + 8.96 \sin(200t - 63.4°)$ (A), $i_2 = 1.505e^{-100t} + 4.48 \sin(200t - 63.4°)$ (A)

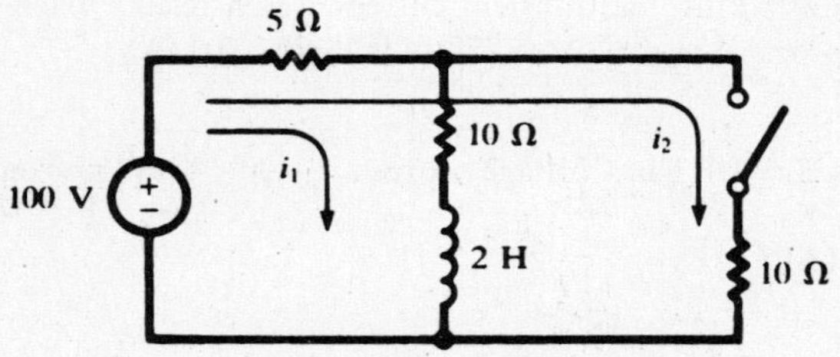

Fig. 16-28

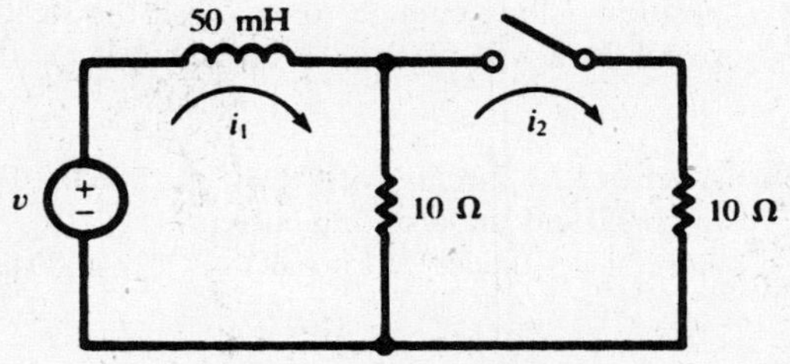

Fig. 16-29

16.41 In the circuit of Fig. 16-30, $v(0) = 1.2$ V and $i(0) = 0.4$ A. Find v and i for $t > 0$.

Ans. $v = 1.3334e^{-t} - 0.1334e^{-2.5t}, t > 0$
$i = 0.66667e^{-t} - 0.2667e^{-2.5t}, t > 0$

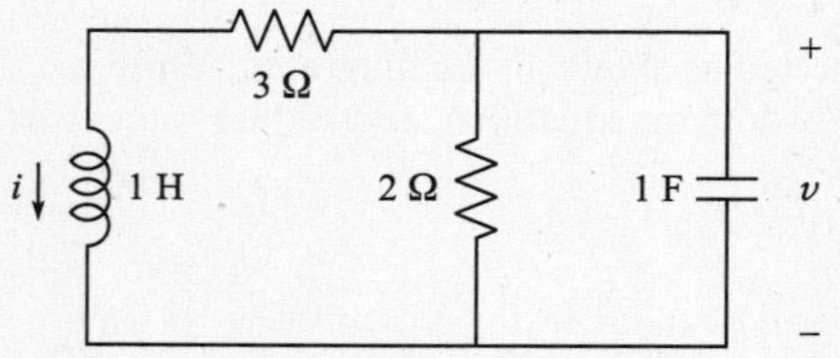

Fig. 16-30

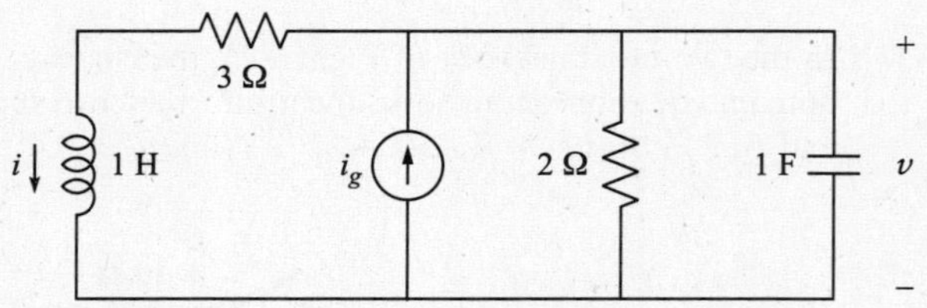

Fig. 16-31

16.42 In the circuit of Fig. 16-31, $i_g(t) = \cos t u(t)$. Find v and i.

Ans. $v = 0.8305 \cos(t - 48.4°), t > 0$
$i = 0.2626 \cos(t - 66.8°), t > 0$

16.43 In the circuit of Fig. 16-31, $i_g = \begin{cases} 1\text{ A} & t < 0 \\ \cos t & t > 0 \end{cases}$. Find v and i for $t > 0$ and compare with results of Problems 16.41 and 16.42.

Ans. $v = 0.6667e^{-t} - 0.0185e^{-2.5t} + 0.8305\cos(t - 48.4°),\ t > 0$

$i = 0.3332e^{-t} - 0.0368e^{-2.5t} + 0.2626\cos(t - 66.8°),\ t > 0$

16.44 Find capacitor voltage $v(t)$ in the circuit shown in Fig. 16-32.
Ans. $v = 20 - 10.21e^{-4t}\cos(4.9t + 11.53°),\ t > 0$

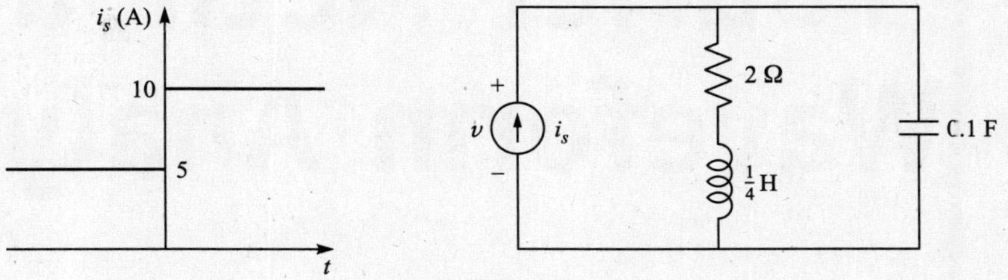

Fig. 16-32

16.45 Find inductor current $i(t)$ in the circuit shown in Fig. 16-32.
Ans. $i = 10 - 6.45e^{-4t}\cos(4.9t - 39.2°),\ t > 0$

CHAPTER 17

Fourier Method of Waveform Analysis

17.1 INTRODUCTION

In the circuits examined previously, the response was obtained for excitations having constant, sinusoidal, or exponential form. In such cases a single expression described the forcing function for all time; for instance, $v =$ constant or $v = V \sin \omega t$, as shown in Fig. 17-1(*a*) and (*b*).

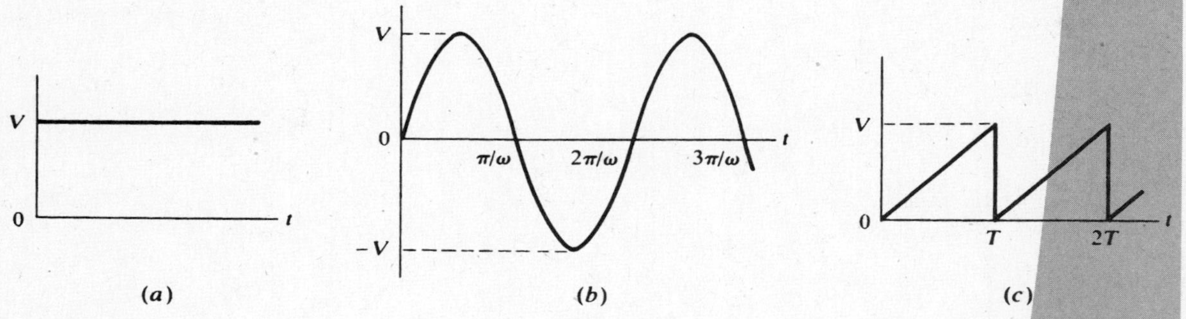

Fig. 17-1

Certain periodic waveforms, of which the sawtooth in Fig. 17-1(*c*) is an example, can be only locally defined by single functions. Thus, the sawtooth is expressed by $f(t) = (V/T)t$ in the interval $0 < t < T$ and by $f(t) = (V/T)(t - T)$ in the interval $T < t < 2T$. While such piecemeal expressions describe the waveform satisfactorily, they do not permit the determination of the circuit response. Now, if a periodic function can be expressed as the sum of a finite or infinite number of sinusoidal functions, the responses of linear networks to nonsinusoidal excitations can be determined by applying the superposition theorem. The Fourier method provides the means for solving this type of problem.

In this chapter we develop tools and conditions for such expansions. Periodic waveforms may be expressed in the form of Fourier series. Nonperiodic waveforms may be expressed by their Fourier transforms. However, a piece of a nonperiodic waveform specified over a finite time period may also be expressed by a Fourier series valid within that time period. Because of this, the Fourier series analysis is the main concern of this chapter.

17.2 TRIGONOMETRIC FOURIER SERIES

Any periodic waveform—that is, one for which $f(t) = f(t+T)$—can be expressed by a Fourier series provided that

(1) If it is discontinuous, there are only a finite number of discontinuities in the period T;

(2) It has a finite average value over the period T;

(3) It has a finite number of positive and negative maxima in the period T.

When these *Dirichlet conditions* are satisfied, the Fourier series exists and can be written in trigonometric form:

$$\begin{aligned} f(t) = \tfrac{1}{2}a_0 + a_1 \cos \omega t + a_2 \cos 2t + a_3 \cos 3\omega t + \cdots \\ + b_1 \sin \omega t + b_2 \sin 2\omega t + b_3 \sin 3\omega t + \cdots \end{aligned} \tag{1}$$

The Fourier coefficients, a's and b's, are determined for a given waveform by the evaluation integrals. We obtain the cosine coefficient evaluation integral by multiplying both sides of (*1*) by $\cos n\omega t$ and integrating over a full period. The period of the fundamental, $2\pi/\omega$, is the period of the series since each term in the series has a frequency which is an integral multiple of the fundamental frequency.

$$\begin{aligned} \int_0^{2\pi/\omega} f(t) \cos n\omega t \, dt = \int_0^{2\pi/\omega} \frac{1}{2} a_0 \cos n\omega t \, dt + \int_0^{2\pi/\omega} a_1 \cos \omega t \cos n\omega t \, dt + \cdots \\ + \int_0^{2\pi/\omega} a_n \cos^2 n\omega t \, dt + \cdots + \int_0^{2\pi/\omega} b_1 \sin \omega t \cos n\omega t \, dt \\ + \int_0^{2\pi/\omega} b_2 \sin 2\omega t \cos n\omega \, dt + \cdots \end{aligned} \tag{2}$$

The definite integrals on the right side of (*2*) are all zero except that involving $\cos^2 n\omega t$, which has the value $(\pi/\omega)a_n$. Then

$$a_n = \frac{\omega}{\pi} \int_0^{2\pi/\omega} f(t) \cos n\omega t \, dt = \frac{2}{T} \int_0^T f(t) \cos \frac{2\pi n t}{T} \, dt \tag{3}$$

Multiplying (*1*) by $\sin n\omega t$ and integrating as above results in the sine coefficient evaluation integral.

$$b_n = \frac{\omega}{\pi} \int_0^{2\pi/\omega} f(t) \sin n\omega t \, dt = \frac{2}{T} \int_0^T f(t) \sin \frac{2\pi n t}{T} \, dt \tag{4}$$

An alternate form of the evaluation integrals with the variable $\psi = \omega t$ and the corresponding period 2π radians is

$$a_n = \frac{1}{\pi} \int_0^{2\pi} F(\psi) \cos n\psi \, d\psi \tag{5}$$

$$b_n = \frac{1}{\pi} \int_0^{2\pi} F(\psi) \sin n\psi \, d\psi \tag{6}$$

where $F(\psi) = f(\psi/\omega)$. The integrations can be carried out from $-T/2$ to $T/2$, $-\pi$ to $+\pi$, or over any other full period that might simplify the calculation. The constant a_0 is obtained from (*3*) or (*5*) with $n = 0$; however, since $\frac{1}{2}a_0$ is the average value of the function, it can frequently be determined by inspection of the waveform. The series with coefficients obtained from the above evaluation integrals converges uniformly to the function at all points of continuity and converges to the mean value at points of discontinuity.

EXAMPLE 17.1 Find the Fourier series for the waveform shown in Fig. 17-2.

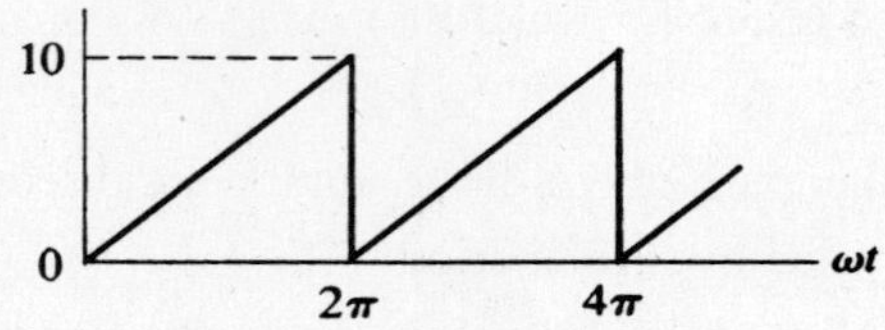

Fig. 17-2

The waveform is periodic, of period $2\pi/\omega$ in t or 2π in ωt. It is continuous for $0 < \omega t < 2\pi$ and given therein by $f(t) = (10/2\pi)\omega t$, with discontinuities at $\omega t = n2\pi$ where $n = 0, 1, 2, \ldots$. The Dirichlet conditions are satisfied. The average value of the function is 5, by inspection, and thus, $\frac{1}{2}a_0 = 5$. For $n > 0$, (5) gives

$$a_n = \frac{1}{\pi}\int_0^{2\pi}\left(\frac{10}{2\pi}\right)\omega t \cos n\omega t \, d(\omega t) = \frac{10}{2\pi^2}\left[\frac{\omega t}{n}\sin n\omega t + \frac{1}{n^2}\cos n\omega t\right]_0^{2\pi}$$
$$= \frac{10}{2\pi^2 n^2}(\cos n2\pi - \cos 0) = 0$$

Thus, the series contains no cosine terms. Using (6), we obtain

$$b_n = \frac{1}{\pi}\int_0^{2\pi}\left(\frac{10}{2\pi}\right)\omega t \sin n\omega t \, d(\omega t) = \frac{10}{2\pi^2}\left[-\frac{\omega t}{n}\cos n\omega t + \frac{1}{n^2}\sin n\omega t\right]_0^{2\pi} = -\frac{10}{\pi n}$$

Using these sine-term coefficients and the average term, the series is

$$f(t) = 5 - \frac{10}{\pi}\sin \omega t - \frac{10}{2\pi}\sin 2\omega t - \frac{10}{3\pi}\sin 3\omega t - \cdots = 5 - \frac{10}{\pi}\sum_{n=1}^{\infty}\frac{\sin n\omega t}{n}$$

The sine and cosine terms of like frequency can be combined as a single sine or cosine term with a phase angle. Two alternate forms of the trigonometric series result.

$$f(t) = \tfrac{1}{2}a_0 + \sum c_n \cos(n\omega t - \theta_n) \tag{7}$$

and

$$f(t) = \tfrac{1}{2}a_0 + \sum c_n \sin(n\omega t + \phi_n) \tag{8}$$

where $c_n = \sqrt{a_n^2 + b_n^2}$, $\theta_n = \tan^{-1}(b_n/a_n)$, and $\phi_n = \tan^{-1}(a_n/b_n)$. In (7) and (8), c_n is the harmonic amplitude, and the harmonic phase angles are θ_n or ϕ_n.

17.3 EXPONENTIAL FOURIER SERIES

A periodic waveform $f(t)$ satisfying the Dirichlet conditions can also be written as an exponential Fourier series, which is a variation of the trigonometric series. The exponential series is

$$f(t) = \sum_{n=-\infty}^{\infty} \mathbf{A}_n e^{jn\omega t} \tag{9}$$

To obtain the evaluation integral for the $\mathbf{A}_n$ coefficients, we multiply (9) on both sides by $e^{-jn\omega t}$ and integrate over the full period:

$$\int_0^{2\pi} f(t)e^{-jn\omega t}\,d(\omega t) = \cdots + \int_0^{2\pi} \mathbf{A}_{-2}e^{-j2\omega t}e^{-jn\omega t}\,d(\omega t) + \int_0^{2\pi} \mathbf{A}_{-1}e^{-j\omega t}e^{-jn\omega t}\,d(\omega t)$$
$$+ \int_0^{2\pi} \mathbf{A}_0 e^{-jn\omega t}\,d(\omega t) + \int_0^{2\pi} \mathbf{A}_1 e^{j\omega t}e^{-jn\omega t}\,d(\omega t) + \cdots$$
$$+ \int_0^{2\pi} \mathbf{A}_n e^{jn\omega t}e^{-jn\omega t}\,d(\omega t) + \cdots \tag{10}$$

The definite integrals on the right side of (*10*) are all zero except $\int_0^{2\pi} \mathbf{A}_n\,d(\omega t)$, which has the value $2\pi\mathbf{A}_n$. Then

$$\mathbf{A}_n = \frac{1}{2\pi}\int_0^{2\pi} f(t)e^{-jn\omega t}\,d(\omega t) \qquad \text{or} \qquad \mathbf{A}_n = \frac{1}{T}\int_0^{T} f(t)e^{-j2\pi nt/T}\,dt \tag{11}$$

Just as with the a_n and b_n evaluation integrals, the limits of integration in (*11*) may be the endpoints of any convenient full period and not necessarily 0 to 2π or 0 to T. Note that, $f(t)$ being real, $\mathbf{A}_{-n} = \mathbf{A}_n^*$, so that only positive n needed to be considered in (*11*). Furthermore, we have

$$a_n = 2\,\mathrm{Re}\,\mathbf{A}_n \qquad b_n = -2\,\mathrm{Im}\,\mathbf{A}_n \tag{12}$$

EXAMPLE 17.2 Derive the exponential series (*9*) from the trigonometric series (*1*).

Replace the sine and cosine terms in (*1*) by their complex exponential equivalents.

$$\sin n\omega t = \frac{e^{jn\omega t} - e^{-jn\omega t}}{2j} \qquad \cos n\omega t = \frac{e^{jn\omega t} + e^{-jn\omega t}}{2}$$

Arranging the exponential terms in order of increasing n from $-\infty$ to $+\infty$, we obtain the infinite sum (*9*) where $A_0 = a_0/2$ and

$$\mathbf{A}_n = \tfrac{1}{2}(a_n - jb_n) \qquad \mathbf{A}_{-n} = \tfrac{1}{2}(a_n + jb_n) \qquad \text{for } n = 1, 2, 3, \ldots$$

EXAMPLE 17.3 Find the exponential Fourier series for the waveform shown in Fig. 17-2. Using the coefficients of this exponential series, obtain a_n and b_n of the trigonometric series and compare with Example 17.1.

In the interval $0 < \omega t < 2\pi$ the function is given by $f(t) = (10/2\pi)\omega t$. By inspection, the average value of the function is $A_0 = 5$. Substituting $f(t)$ in (*11*), we obtain the coefficients $\mathbf{A}_n$.

$$\mathbf{A}_n = \frac{1}{2\pi}\int_0^{2\pi}\left(\frac{10}{2\pi}\right)\omega t e^{-jn\omega t}\,d(\omega t) = \frac{10}{(2\pi)^2}\left[\frac{e^{-jn\omega t}}{(-jn)^2}(-jn\omega t - 1)\right]_0^{2\pi} = j\,\frac{10}{2\pi n}$$

Inserting the coefficients $\mathbf{A}_n$ in (*12*), the exponential form of the Fourier series for the given waveform is

$$f(t) = \cdots - j\,\frac{10}{4\pi}\,e^{-j2\omega t} - j\,\frac{10}{2\pi}\,e^{-j\omega t} + 5 + j\,\frac{10}{2\pi}\,e^{j\omega t} + j\,\frac{10}{4\pi}\,e^{j2\omega t} + \cdots \tag{13}$$

The trigonometric series coefficients are, by (*12*),

$$a_n = 0 \qquad b_n = -\frac{10}{\pi n}$$

and so
$$f(t) = 5 - \frac{10}{\pi}\sin\omega t - \frac{10}{2\pi}\sin 2\omega t - \frac{10}{3\pi}\sin 3\omega t - \cdots$$

which is the same as in Example 17.1.

17.4 WAVEFORM SYMMETRY

The series obtained in Example 17.1 contained only sine terms in addition to a constant term. Other waveforms will have only cosine terms; and sometimes only odd harmonics are present in the series, whether the series contains sine, cosine, or both types of terms. This is the result of certain types of symmetry exhibited by the waveform. Knowledge of such symmetry results in reduced calculations in determining the Fourier series. For this reason the following definitions are important.

1. A function $f(x)$ is said to be *even* if $f(x) = f(-x)$.

The function $f(x) = 2 + x^2 + x^4$ is an example of even functions since the functional values for x and $-x$ are equal. The cosine is an even function, since it can be expressed as the power series

$$\cos x = 1 - \frac{x^2}{2!} + \frac{x^4}{4!} - \frac{x^6}{6!} + \frac{x^8}{8!} - \cdots$$

The sum or product of two or more even functions is an even function, and with the addition of a constant the even nature of the function is still preserved.

In Fig. 17-3, the waveforms shown represent even functions of x. They are symmetrical with respect to the vertical axis, as indicated by the construction in Fig. 17-3(*a*).

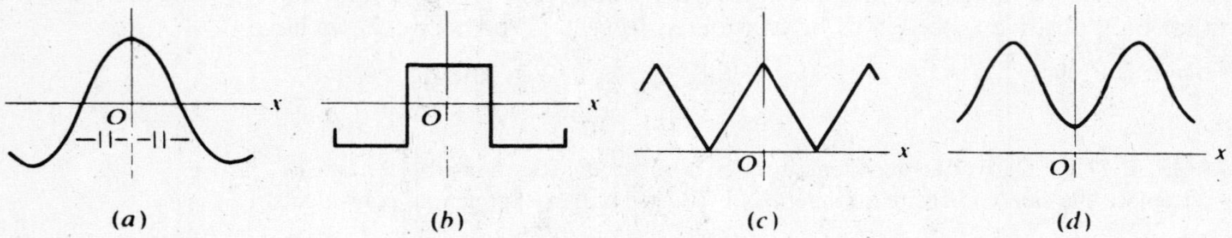

Fig. 17-3

2. A function $f(x)$ is said to be *odd* if $f(x) = -f(-x)$.

The function $f(x) = x + x^3 + x^5$ is an example of odd functions since the values of the function for x and $-x$ are of opposite sign. The sine is an odd function, since it can be expressed as the power series

$$\sin x = x - \frac{x^3}{3!} + \frac{x^5}{5!} - \frac{x^7}{7!} + \frac{x^9}{9!} - \cdots$$

The sum of two or more odd functions is an odd function, but the addition of a constant removes the odd nature of the function. The product of two odd functions is an even function.

The waveforms shown in Fig. 17-4 represent odd functions of x. They are symmetrical with respect to the origin, as indicated by the construction in Fig. 17-4(*a*).

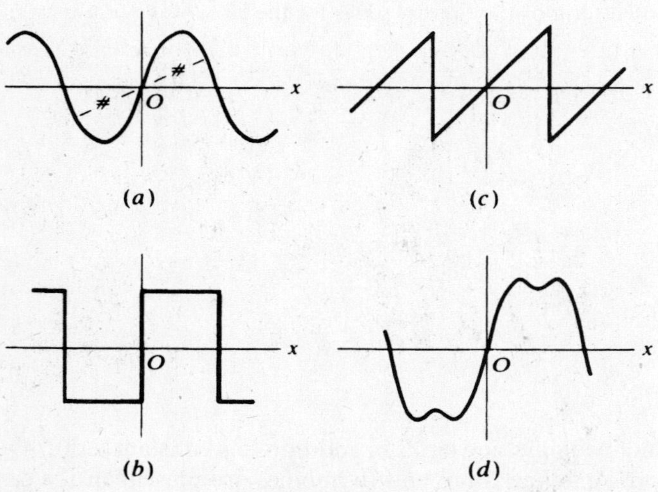

Fig. 17-4

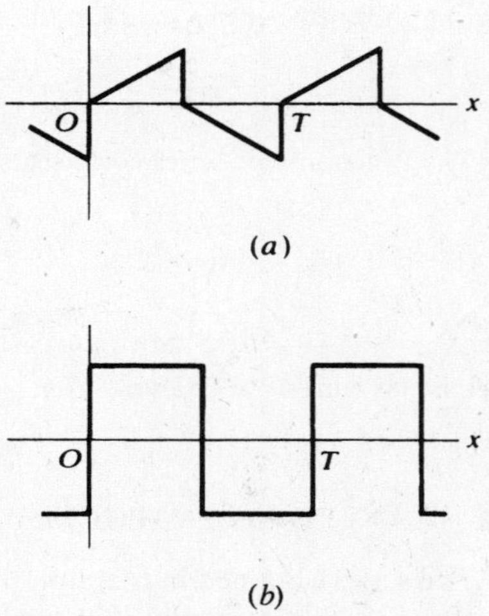

Fig. 17-5

3. A periodic function $f(x)$ is said to have *half-wave symmetry* if $f(x) = -f(x + T/2)$ where T is the period. Two waveforms with half-wave symmetry are shown in Fig. 17-5.

When the type of symmetry of a waveform is established, the following conclusions are reached. If the waveform is even, all terms of its Fourier series are cosine terms, including a constant if the waveform has a nonzero average value. Hence, there is no need of evaluating the integral for the coefficients b_n, since no sine terms can be present. If the waveform is odd, the series contains only sine terms. The wave may be odd only after its average value is subtracted, in which case its Fourier representation will simply contain that constant and a series of sine terms. If the waveform has half-wave symmetry, only odd harmonics are present in the series. This series will contain both sine and cosine terms unless the function is also odd or even. In any case, a_n and b_n are equal to zero for $n = 2, 4, 6, \ldots$ for any waveform with half-wave symmetry. Half-wave symmetry, too, may be present only after subtraction of the average value.

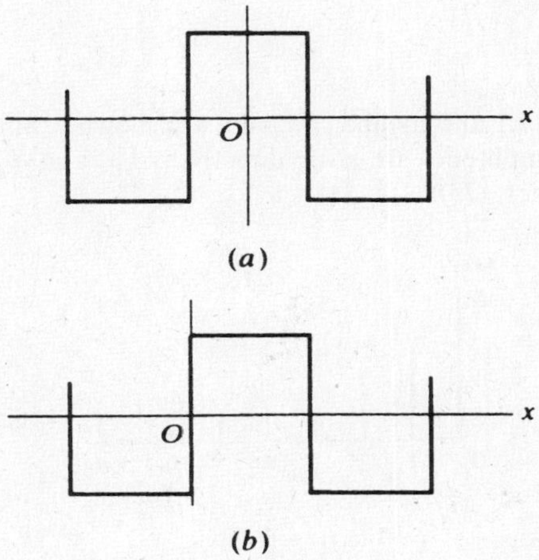

Fig. 17-6

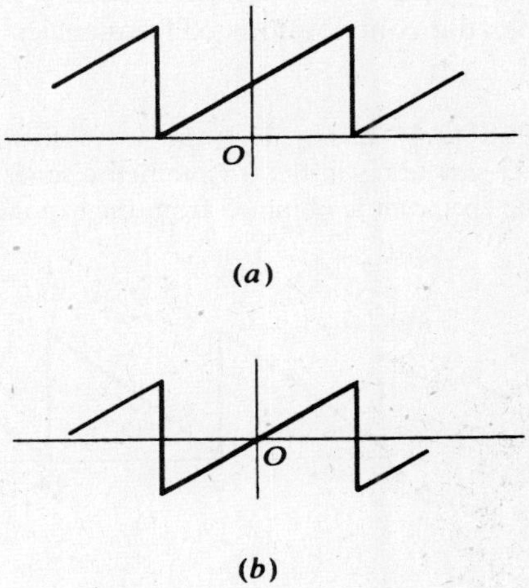

Fig. 17-7

Certain waveforms can be odd or even, depending upon the location of the vertical axis. The square wave of Fig. 17-6(a) meets the condition of an even function: $f(x) = f(-x)$. A shift of the vertical axis to the position shown in Fig. 17-6(b) produces an odd function $f(x) = -f(-x)$. With the vertical axis placed at any points other than those shown in Fig. 17-6, the square wave is neither even nor odd, and its series contains both sine and cosine terms. Thus, in the analysis of periodic functions, the vertical axis should be conveniently chosen to result in either an even or odd function, if the type of waveform makes this possible.

The shifting of the horizontal axis may simplify the series representation of the function. As an example, the waveform of Fig. 17-7(a) does not meet the requirements of an odd function until the average value is removed as shown in Fig. 17-7(b). Thus, its series will contain a constant term and only sine terms.

The preceding symmetry considerations can be used to check the coefficients of the exponential Fourier series. An even waveform contains only cosine terms in its trigonometric series, and therefore the exponential Fourier coefficients must be pure real numbers. Similarly, an odd function whose trigonometric series consists of sine terms has pure imaginary coefficients in its exponential series.

17.5 LINE SPECTRUM

A plot showing each of the harmonic amplitudes in the wave is called the *line spectrum*. The lines decrease rapidly for waves with rapidly convergent series. Waves with discontinuities, such as the sawtooth and square wave, have spectra with slowly decreasing amplitudes, since their series have strong high harmonics. Their 10th harmonics will often have amplitudes of significant value as compared to the fundamental. In contrast, the series for waveforms without discontinuities and with a generally smooth appearance will converge rapidly, and only a few terms are required to generate the wave. Such rapid convergence will be evident from the line spectrum where the harmonic amplitudes decrease rapidly, so that any above the 5th or 6th are insignificant.

The harmonic content and the line spectrum of a wave are part of the very nature of that wave and never change, regardless of the method of analysis. Shifting the origin gives the trigonometric series a completely different appearance, and the exponential series coefficients also change greatly. However, the same harmonics always appear in the series, and their amplitudes,

$$c_0 = |\tfrac{1}{2}a_0| \quad \text{and} \quad c_n = \sqrt{a_n^2 + b_n^2} \quad (n \geq 1) \tag{14}$$

or

$$c_0 = |A_0| \quad \text{and} \quad c_n = |\mathbf{A}_n| + |\mathbf{A}_{-n}| = 2|\mathbf{A}_n| \quad (n \geq 1) \tag{15}$$

remain the same. Note that when the exponential form is used, the amplitude of the nth harmonic combines the contributions of frequencies $+n\omega$ and $-n\omega$.

EXAMPLE 17.4 In Fig. 17-8, the sawtooth wave of Example 17.1 and its line spectrum are shown. Since there were only sine terms in the trigonometric series, the harmonic amplitudes are given directly by $\frac{1}{2}a_0$ and $|b_n|$. The same line spectrum is obtained from the exponential Fourier series, (*13*).

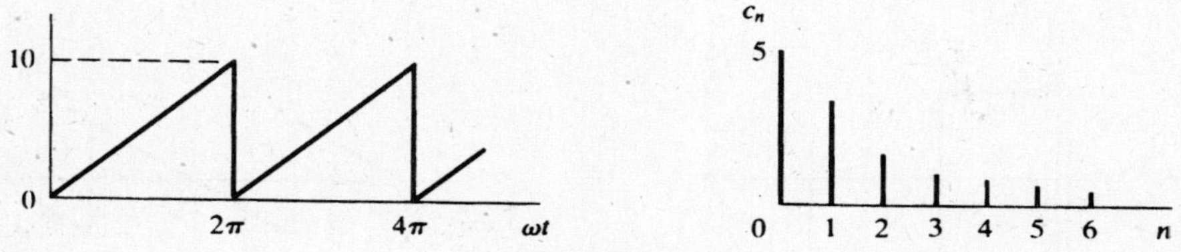

Fig. 17-8

17.6 WAVEFORM SYNTHESIS

Synthesis is a combination of parts so as to form a whole. Fourier synthesis is the recombination of the terms of the trigonometric series, usually the first four or five, to produce the original wave. Often it is only after synthesizing a wave that the student is convinced that the Fourier series does in fact represent the periodic wave for which it was obtained.

The trigonometric series for the sawtooth wave of Fig. 17-8 is

$$f(t) = 5 - \frac{10}{\pi}\sin\omega t - \frac{10}{2\pi}\sin 2\omega t - \frac{10}{3\pi}\sin 3\omega t - \cdots$$

These four terms are plotted and added in Fig. 17-9. Although the result is not a perfect sawtooth wave, it appears that with more terms included the sketch will more nearly resemble a sawtooth. Since this wave has discontinuities, its series is not rapidly convergent, and consequently, the synthesis using only four terms does not produce a very good result. The next term, at the frequency 4ω, has amplitude $10/4\pi$, which is certainly significant compared to the fundamental amplitude, $10/\pi$. As each term is added in the synthesis, the irregularities of the resultant are reduced and the approximation to the original wave is improved. This is what was meant when we said earlier that *the series converges to the function at all points of continuity and to the mean value at points of discontinuity*. In Fig. 17-9, at 0 and 2π it is clear that a value of 5 will remain, since all sine terms are zero at these points. These are the points of discontinuity; and the value of the function when they are approached from the left is 10, and from the right 0, with the mean value 5.

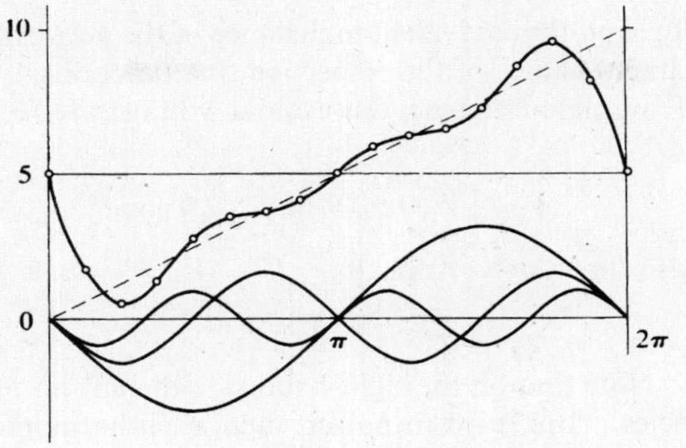

Fig. 17-9

17.7 EFFECTIVE VALUES AND POWER

The effective or rms value of the function

$$f(t) = \tfrac{1}{2}a_0 + a_1 \cos \omega t + a_2 \cos 2\omega t + \cdots + b_1 \sin \omega t + b_2 \sin 2\omega t + \cdots$$

is
$$F_{\text{rms}} = \sqrt{(\tfrac{1}{2}a_0)^2 + \tfrac{1}{2}a_1^2 + \tfrac{1}{2}a_2^2 + \cdots + \tfrac{1}{2}b_1^2 + \tfrac{1}{2}b_2^2 + \cdots} = \sqrt{c_0^2 + \tfrac{1}{2}c_1^2 + \tfrac{1}{2}c_2^2 + \tfrac{1}{2}c_3^3 + \cdots} \qquad (16)$$

where *(14)* has been used.

Considering a linear network with an applied voltage which is periodic, we would expect that the resulting current would contain the same harmonic terms as the voltage, but with harmonic amplitudes of different relative magnitude, since the impedance varies with $n\omega$. It is possible that some harmonics would not appear in the current; for example, in a pure LC parallel circuit, one of the harmonic frequencies might coincide with the resonant frequency, making the impedance at that frequency infinite. In general, we may write

$$v = V_0 + \sum V_n \sin(n\omega t + \phi_n) \quad \text{and} \quad i = I_0 + \sum I_n \sin(n\omega t + \psi_n) \qquad (17)$$

with corresponding effective values of

$$V_{\text{rms}} = \sqrt{V_0^2 + \tfrac{1}{2}V_1^2 + \tfrac{1}{2}V_2^2 + \cdots} \quad \text{and} \quad I_{\text{rms}} = \sqrt{I_0^2 + \tfrac{1}{2}I_1^2 + \tfrac{1}{2}I_2^2 + \cdots} \qquad (18)$$

The average power P follows from integration of the instantaneous power, which is given by the product of v and i:

$$p = vi = \left[V_0 + \sum V_n \sin(n\omega t + \phi_n)\right]\left[I_0 + \sum I_n \sin(n\omega t + \psi_n)\right] \qquad (19)$$

Since v and i both have period T, their product must have an integral number of its periods in T. (Recall that for a single sine wave of applied voltage, the product vi has a period half that of the voltage wave.) The average may therefore be calculated over one period of the voltage wave:

$$P = \frac{1}{T}\int_0^T \left[V_0 + \sum V_n \sin(n\omega t + \phi_n)\right]\left[I_0 + \sum I_n \sin(n\omega t + \psi_n)\right] dt \qquad (20)$$

Examination of the possible terms in the product of the two infinite series shows them to be of the following types: the product of two constants, the product of a constant and a sine function, the product of two sine functions of different frequencies, and sine functions squared. After integration, the product of the two constants is still $V_0 I_0$ and the sine functions squared with the limits applied appear as $(V_n I_n/2)\cos(\phi_n - \psi_n)$; all other products upon integration over the period T are zero. Then the average power is

$$P = V_0 I_0 + \tfrac{1}{2}V_1 I_1 \cos\theta_1 + \tfrac{1}{2}V_2 I_2 \cos\theta_2 + \tfrac{1}{2}V_3 I_3 \cos\theta_3 + \cdots \qquad (21)$$

where $\theta_n = \phi_n - \psi_n$ is the angle on the equivalent impedance of the network at the angular frequency $n\omega$, and V_n and I_n are the maximum values of the respective sine functions.

In the special case of a single-frequency sinusoidal voltage, $V_0 = V_2 = V_3 = \cdots = 0$, and (*21*) reduces to the familiar

$$P = \tfrac{1}{2} V_1 I_1 \cos\theta_1 = V_{\text{eff}} I_{\text{eff}} \cos\theta$$

Compare Section 10.2. Also, for a dc voltage, $V_1 = V_2 = V_3 = \cdots = 0$, and (*21*) becomes

$$P = V_0 I_0 = VI$$

Thus, (*21*) is quite general. Note that on the right-hand side there is no term that involves voltage and current of different frequencies. In regard to power, then, each harmonic acts independently, and

$$P = P_0 + P_1 + P_2 + \cdots$$

17.8 APPLICATIONS IN CIRCUIT ANALYSIS

It has already been suggested above that we could apply the terms of a voltage series to a linear network and obtain the corresponding harmonic terms of the current series. This result is obtained by superposition. Thus we consider each term of the Fourier series representing the voltage as a single source, as shown in Fig. 17.10. Now the equivalent impedance of the network at each harmonic frequency $n\omega$ is used to compute the current at that harmonic. The sum of these individual responses is the total response i, in series form, to the applied voltage.

EXAMPLE 17.5 A series *RL* circuit in which $R = 5\ \Omega$ and $L = 20$ mH (Fig. 17-11) has an applied voltage $v = 100 + 50\sin\omega t + 25\sin 3\omega t$ (V), with $\omega = 500$ rad/s. Find the current and the average power.

Compute the equivalent impedance of the circuit at each frequency found in the voltage function. Then obtain the respective currents.

At $\omega = 0$, $Z_0 = R = 5\ \Omega$ and

$$I_0 = \frac{V_0}{R} = \frac{100}{5} = 20 \text{ A}$$

At $\omega = 500$ rad/s, $\mathbf{Z}_1 = 5 + j(500)(20 \times 10^{-3}) = 5 + j10 = 11.15\underline{/63.4^\circ}\ \Omega$ and

$$i_1 = \frac{V_{1,\max}}{Z_1}\sin(\omega t - \theta_1) = \frac{50}{11.15}\sin(\omega t - 63.4^\circ) = 4.48\sin(\omega t - 63.4^\circ) \quad \text{(A)}$$

At $3\omega = 1500$ rad/s, $\mathbf{Z}_3 = 5 + j30 = 30.4\underline{/80.54^\circ}\ \Omega$ and

$$i_3 = \frac{V_{3,\max}}{Z_3}\sin(3\omega t - \theta_3) = \frac{25}{30.4}\sin(3\omega t - 80.54^\circ) = 0.823\sin(3\omega t - 80.54^\circ) \quad \text{(A)}$$

The sum of the harmonic currents is the required total response; it is a Fourier series of the type (*8*).

$$i = 20 + 4.48\sin(\omega t - 63.4^\circ) + 0.823\sin(3\omega t - 80.54^\circ) \quad \text{(A)}$$

This current has the effective value

$$I_{\text{eff}} = \sqrt{20^2 + (4.48^2/2) + (0.823^2/2)} = \sqrt{410.6} = 20.25 \text{ A}$$

which results in a power in the 5-Ω resistor of

$$P = I_{\text{eff}}^2 R = (410.6)5 = 2053 \text{ W}$$

As a check, we compute the total average power by calculating first the power contributed by each harmonic and then adding the results.

At $\omega = 0$: $P_0 = V_0 I_0 = 100(20) = 2000$ W

At $\omega = 500$ rad/s: $P_1 = \tfrac{1}{2} V_1 I_1 \cos\theta_1 = \tfrac{1}{2}(50)(4.48)\cos 63.4^\circ = 50.1$ W

At $3\omega = 1500$ rad/s: $P_3 = \tfrac{1}{2} V_3 I_3 \cos\theta_3 = \tfrac{1}{2}(25)(0.823)\cos 80.54^\circ = 1.69$ W

Then, $P = 2000 + 50.1 + 1.69 = 2052$ W

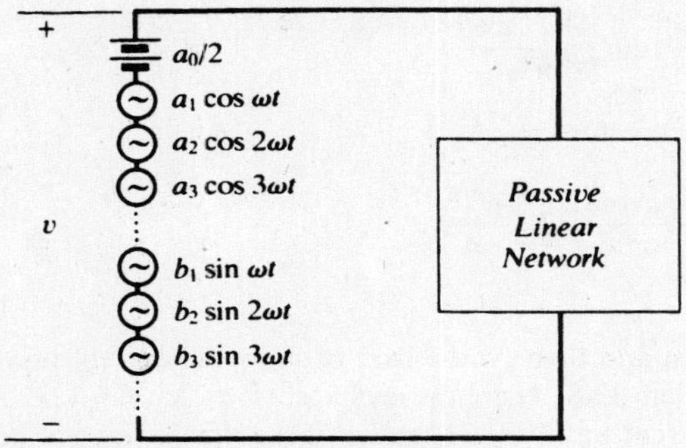

Fig. 17-10

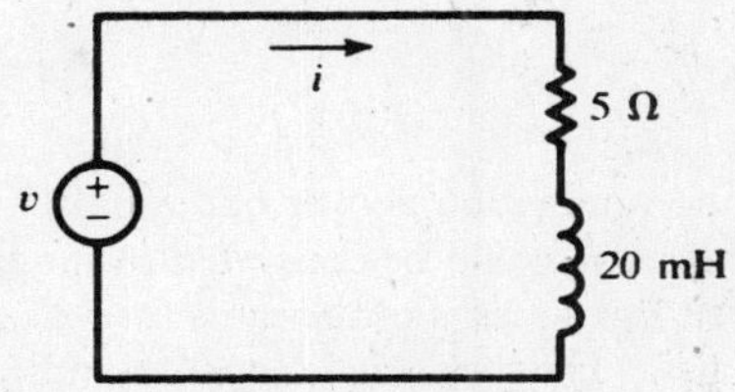

Fig. 17-11

Another Method

The Fourier series expression for the voltage across the resistor is

$$v_R = Ri = 100 + 22.4 \sin(\omega t - 63.4^\circ) + 4.11 \sin(3\omega t - 80.54^\circ) \quad \text{(V)}$$

and

$$V_{R\text{eff}} = \sqrt{100^2 + \frac{1}{2}(22.4)^2 + \frac{1}{2}(4.11)^2} = \sqrt{10\,259} = 101.3 \text{ V}$$

Then the power delivered by the source is $P = V_{R\text{eff}}^2/R = (10\,259)/5 = 2052$ W.

In Example 17.5 the driving voltage was given as a trigonometric Fourier series in t, and the computations were in the time domain. (The complex impedance was used only as a shortcut; Z_n and θ_n could have been obtained directly from R, L, and $n\omega$). If, instead, the voltage is represented by an exponential Fourier series,

$$v(t) = \sum_{-\infty}^{+\infty} \mathbf{V}_n e^{jn\omega t}$$

then we have to do with a superposition of *phasors* $\mathbf{V}_n$ (rotating counterclockwise if $n > 0$, clockwise if $n < 0$), and so frequency-domain methods are called for. This is illustrated in Example 17.6.

EXAMPLE 17.6 A voltage represented by the triangular wave shown in Fig. 17-12 is applied to a pure capacitor C. Determine the resulting current.

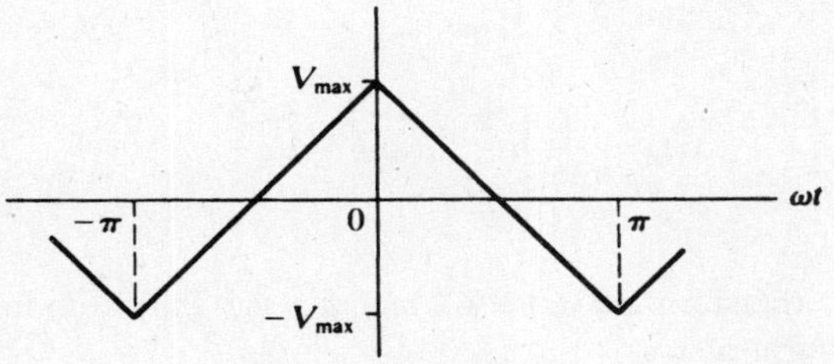

Fig. 17-12

In the interval $-\pi < \omega t < 0$ the voltage function is $v = V_{max} + (27V_{max}/\pi)\omega t$; and for $0 < \omega t < \pi$, $v = V_{max} - (2V_{max}/\pi)\omega t$. Then the coefficients of the exponential series are determined by the evaluation integral

$$\mathbf{V}_n = \frac{1}{2\pi}\int_{-\pi}^{0} [V_{max} + (2V_{max}/\pi)\omega t] e^{-jn\omega t}\, d(\omega t) + \frac{1}{2\pi}\int_{0}^{\pi} [V_{max} - (2V_{max}/\pi)\omega t] e^{-jn\omega t}\, d(\omega t)$$

from which $\mathbf{V}_n = 4V_{max}/\pi^2 n^2$ for odd n, and $\mathbf{V}_n = 0$ for even n.

The phasor current produced by $\mathbf{V}_n$ (n odd) is

$$\mathbf{I}_n = \frac{\mathbf{V}_n}{\mathbf{Z}_n} = \frac{4V_{\max}/\pi^2 n^2}{1/jn\omega C} = j\frac{4V_{\max}\omega C}{\pi^2 n}$$

with an implicit time factor $e^{jn\omega t}$. The resultant current is therefore

$$i(t) = \sum_{-\infty}^{+\infty} \mathbf{I}_n e^{jn\omega t} = j\frac{4V_{\max}\omega C}{\pi^2}\sum_{-\infty}^{+\infty}\frac{e^{jn\omega t}}{n}$$

where the summation is over odd n only.

The series could be converted to the trigonometric form and then synthesized to show the current waveform. However, this series is of the same form as the result in Problem 17.8, where the coefficients are $\mathbf{A}_n = -j(2V/n\pi)$ for odd n only. The sign here is negative, indicating that our current wave is the negative of the square wave of Problem 17.8 and has a peak value $2V_{\max}\omega C/\pi$.

17.9 FOURIER TRANSFORM OF NONPERIODIC WAVEFORMS

A nonperiod waveform $x(t)$ is said to satisfy the Dirichlet conditions if

(*a*) $x(t)$ is absolutely integrable, $\int_{-\infty}^{+\infty} |x(t)|\, dt < \infty$, and

(*b*) the number of maxima and minima and the number of discontinuities of $x(t)$ in every finite interval is finite.

For such a waveform, we can define the Fourier transform $\mathbf{X}(f)$ by

$$\mathbf{X}(f) = \int_{-\infty}^{\infty} x(t)e^{-j2\pi ft}\, dt \tag{22a}$$

where f is the frequency. The above integral is called the *Fourier integral*. The time function $x(t)$ is called the *inverse Fourier transform* of $\mathbf{X}(f)$ and is obtained from it by

$$x(t) = \int_{-\infty}^{\infty} \mathbf{X}(f)e^{j2\pi ft}\, df \tag{22b}$$

$x(t)$ and $\mathbf{X}(f)$ form a Fourier transform pair. Instead of f, the angular velocity $\omega = 2\pi f$ may also be used, in which case, (*22a*) and (*22b*) become, respectively,

$$\mathbf{X}(\omega) = \int_{-\infty}^{\infty} x(t)e^{-j\omega t}\, dt \tag{23a}$$

and

$$x(t) = \frac{1}{2\pi}\int_{-\infty}^{\infty} X(\omega)e^{j\omega t}\, d\omega \tag{23b}$$

EXAMPLE 17.7 Find the Fourier transform of $x(t) = e^{-at}u(t)$, $a > 0$. Plot $\mathbf{X}(f)$ for $-\infty < f < +\infty$.

From (*22a*), the Fourier transform of $x(t)$ is

$$\mathbf{X}(f) = \int_0^{\infty} e^{-at}e^{-j2\pi ft}\, dt = \frac{1}{a + j2\pi f} \tag{24}$$

$\mathbf{X}(f)$ is a complex function of a real variable. Its magnitude and phase angle, $|\mathbf{X}(f)|$ and $\angle\mathbf{X}(f)$, respectively, shown in Figs. 17-13(*a*) and (*b*), are given by

$$|\mathbf{X}(f)| = \frac{1}{\sqrt{a^2 + 4\pi^2 f^2}} \tag{25a}$$

and

$$\angle\mathbf{X}(f) = -\tan^{-1}(2\pi f/a) \tag{25b}$$

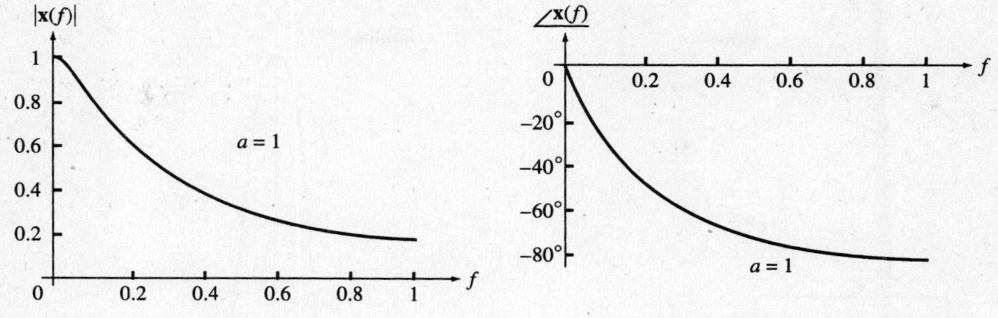

(*a*) (*b*)

Fig. 17-13

Alternatively, $\mathbf{X}(f)$ may be shown by its real and imaginary parts, Re $[\mathbf{X}(f)]$ and Im $[\mathbf{X}(f)]$, as in Figs. 17-14(*a*) and (*b*).

$$\operatorname{Re}[\mathbf{X}(f)] = \frac{a}{a^2 + 4\pi^2 f^2} \tag{26a}$$

$$\operatorname{Im}[\mathbf{X}(f)] = \frac{-2\pi f}{a^2 + 4\pi^2 f^2} \tag{26b}$$

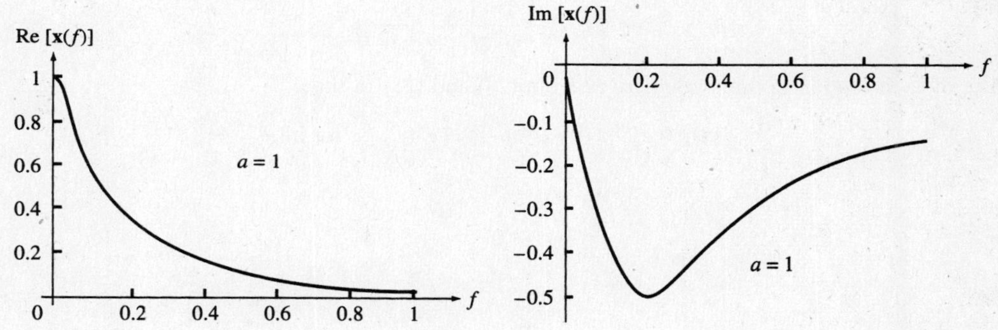

(*a*) (*b*)

Fig. 17-14

EXAMPLE 17.8 Find the Fourier transform of the square pulse

$$x(t) = \begin{cases} 1 & \text{for } -T < t < T \\ 0 & \text{otherwise} \end{cases}$$

From (*22a*),

$$\mathbf{X}(f) = \int_{-T}^{T} e^{-j2\pi ft}\, dt = \frac{1}{-j2\pi f}\left[e^{j2\pi f}\right]_{-T}^{T} = \frac{\sin 2\pi fT}{\pi f} \tag{27}$$

Because $x(t)$ is even, $\mathbf{X}(f)$ is real. The transform pairs are plotted in Figs. 17-15(*a*) and (*b*) for $T = \frac{1}{2}$ s.

EXAMPLE 17.9 Find the Fourier transform of $x(t) = e^{at}u(-t)$, $a > 0$.

$$\mathbf{X}(f) = \int_{-\infty}^{0} e^{at} e^{-j2\pi ft}\, dt = \frac{1}{a - j2\pi f} \tag{28}$$

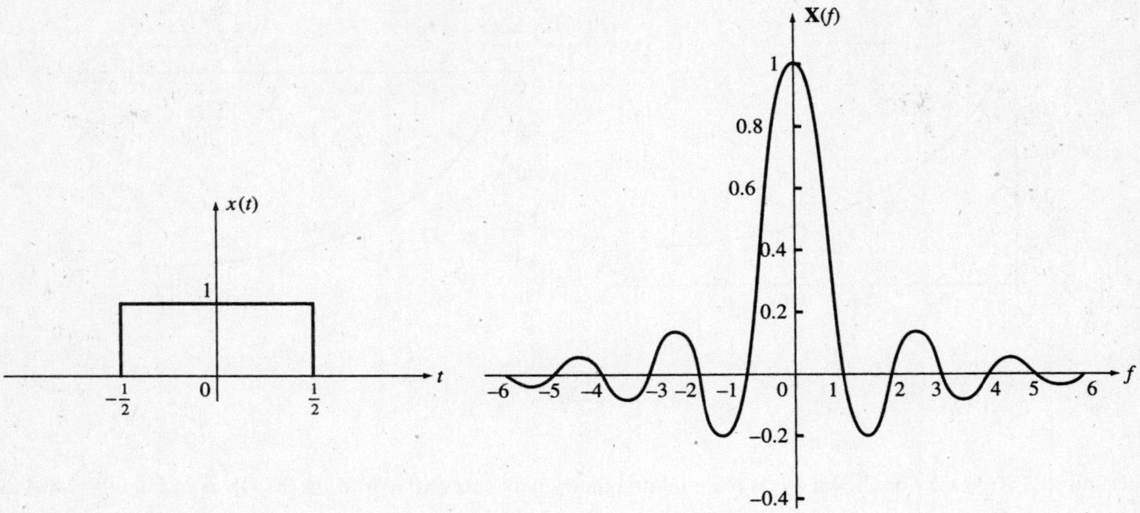

Fig. 17-15

EXAMPLE 17.10 Find the inverse Fourier transform of $\mathbf{X}(f) = 2a/(a^2 + 4\pi^2 f^2)$, $a > 0$.

By partial fraction expansion we have

$$\mathbf{X}(f) = \frac{1}{a + j2\pi f} + \frac{1}{a - j2\pi f} \tag{29}$$

The inverse of each term in (29) may be derived from (24) and (28) so that

$$x(t) = e^{-at}u(t) + e^{at}u(-t) = e^{-a|t|} \qquad \text{for all } t$$

See Fig. 17-16.

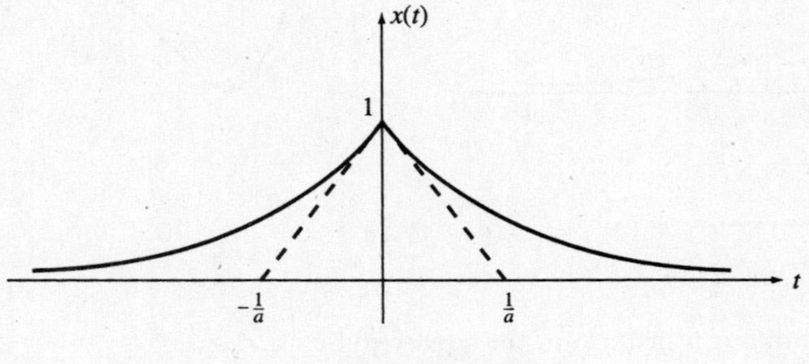

Fig. 17-16

17.10 PROPERTIES OF THE FOURIER TRANSFORM

Some properties of the Fourier transform are listed in Table 17-1. Several commonly used transform pairs are given in Table 17-2.

17.11 CONTINUOUS SPECTRUM

$|\mathbf{X}(f)|^2$, as defined in Section 17.9, is called the *energy density* or the *spectrum* of the waveform $x(t)$. Unlike the periodic functions, the energy content of a nonperiodic waveform $x(t)$ at each frequency is zero. However, the energy content within a frequency band from f_1 to f_2 is

Table 17-1 Fourier Transform Properties

	Time Domain $x(t) = \int_{-\infty}^{\infty} X(f) e^{j2\pi ft}\, dt$	Frequency Domain $\mathbf{X}(f) = \int_{-\infty}^{\infty} x(t) e^{-j2\pi ft}\, dt$
1.	$x(t)$ real	$X(f) = X^*(-f)$
2.	$x(t)$ even, $x(t) = x(-t)$	$X(f) = X(-f)$
3.	$x(t)$, odd, $x(t) = -x(-t)$	$X(f) = -X(-f)$
4.	$X(t)$	$x(-f)$
5.	$x(0) = \int_{-\infty}^{\infty} X(f)\, df$	$X(0) = \int_{-\infty}^{\infty} x(t)\, dt$
6.	$y(t) = x(at)$	$Y(f) = \frac{1}{\lvert a \rvert} X(f/a)$
7.	$y(t) = tx(t)$	$Y(f) = -\frac{1}{j2\pi} \frac{dX(f)}{df}$
8.	$y(t) = x(-t)$	$Y(f) = X(-f)$
9.	$y(t) = x(t - t_0)$	$Y(f) = e^{-j2\pi f t_0} X(f)$

Table 17-2 Fourier Transform Pairs

	$x(t)$	$X(f)$
1.	$e^{-at}u(t), a > 0$	$\frac{1}{a + j2\pi f}$
2.	$e^{-a\lvert t \rvert}, a > 0$	$\frac{2a}{a^2 + 4\pi^2 f^2}$
3.	$te^{-at}u(t), a > 0$	$\frac{1}{(a + j2\pi f)^2}$
4.	$\exp(-\pi t^2/\tau^2)$	$\tau \exp(-\pi f^2 \tau^2)$
5.	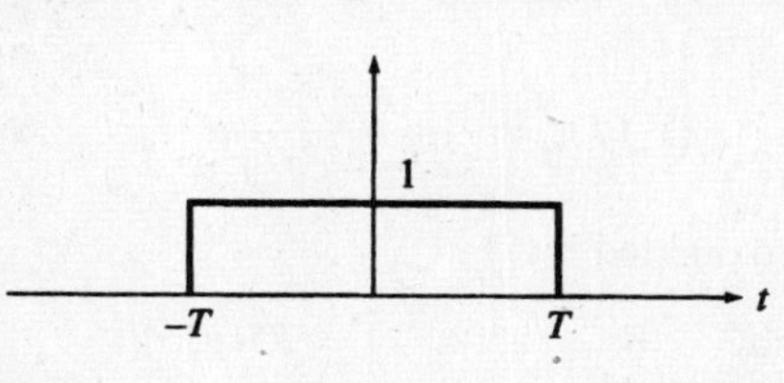	$\frac{\sin 2\pi fT}{\pi f}$ 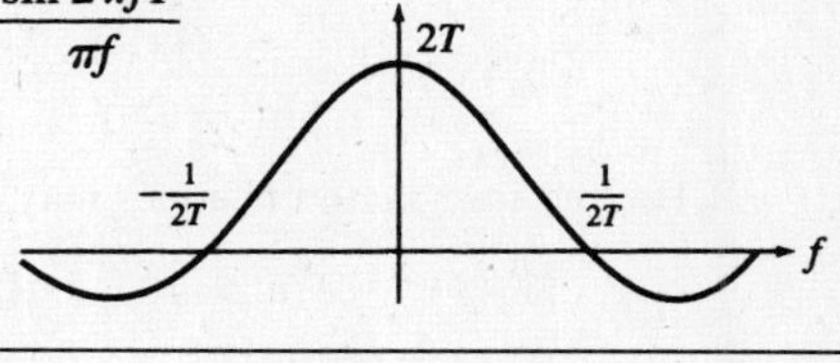
6.	$\frac{\sin 2\pi f_0 t}{\pi t}$ 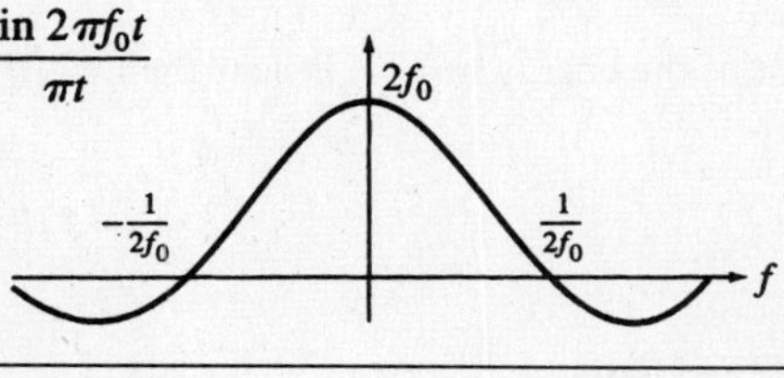	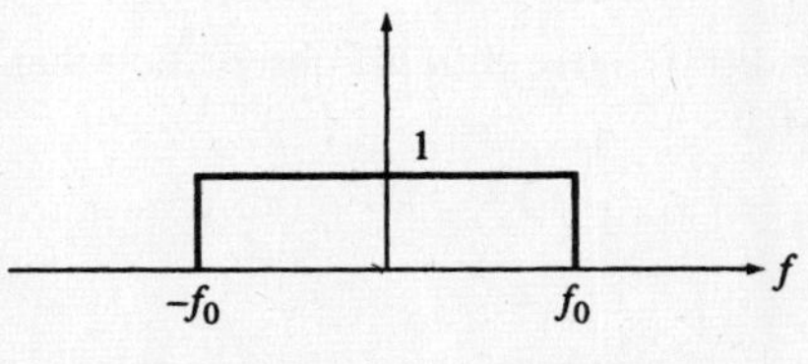
7.	1	$\delta(f)$
8.	$\delta(t)$	1
9.	$\sin 2\pi f_0 t$	$\frac{\delta(f - f_0) - \delta(f + f_0)}{2j}$
10.	$\cos 2\pi f_0 t$	$\frac{\delta(f - f_0) + \delta(f + f_0)}{2}$

$$W = 2\int_{f_1}^{f_2} |\mathbf{x}(f)|^2\, df \tag{30}$$

EXAMPLE 17.11 Find the spectrum of $x(t) = e^{-at}u(t) - e^{at}u(-t)$, $a > 0$, shown in Fig. 17-17.

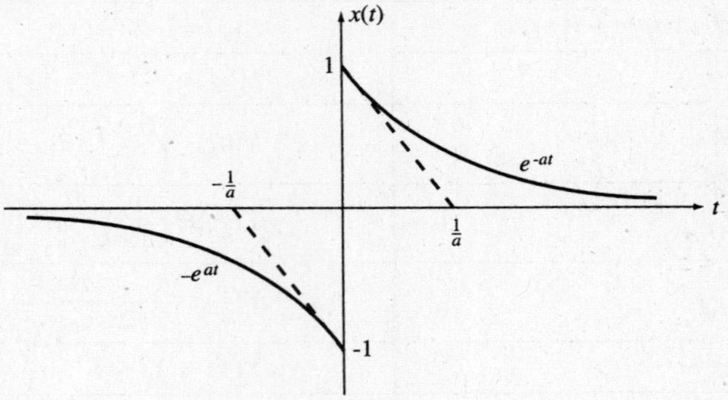

Fig. 17-17

We have $x(t) = x_1(t) - x_2(t)$. Since $x_1(t) = e^{-at}u(t)$ and $x_2(t) = e^{at}u(-t)$,

$$\mathbf{X}_1(f) = \frac{1}{a + j2\pi f} \qquad \mathbf{X}_2(f) = \frac{1}{a - j2\pi f}$$

Then
$$\mathbf{X}(f) = \mathbf{X}_1(f) - \mathbf{X}_2(f) = \frac{-j4\pi f}{a^2 + 4\pi^2 f^2}$$

from which
$$|\mathbf{X}(f)|^2 = \frac{16\pi^2 f^2}{(a^2 + 4\pi^2 f^2)^2}$$

EXAMPLE 17.12 Find and compare the energy contents W_1 and W_2 of $y_1(t) = e^{-|at|}$ and $y_2(t) = e^{-at}u(t) - e^{at}u(-t)$, $a > 0$, within the band 0 to 1 Hz. Let $a = 200$.

From Examples 17.10 and 17.11,

$$|\mathbf{Y}_1(f)|^2 = \frac{4a^2}{(a^2 + 4\pi^2 f^2)^2} \qquad \text{and} \qquad |\mathbf{Y}_2(f)|^2 = \frac{16\pi^2 f^2}{(a^2 + 4\pi^2 f^2)^2}$$

Within $0 < f < 1$ Hz, the spectra and energies may be approximated by

$$|\mathbf{Y}_1(f)|^2 \approx 4/a^2 = 10^{-4}\ \text{J/Hz} \qquad \text{and} \qquad W_1 = 2(10^{-4})\ \text{J} = 200\ \mu\text{J}$$
$$|\mathbf{Y}_2(f)^2| \approx 10^{-7} f^2 \qquad \text{and} \qquad W_2 \approx 0$$

The preceding results agree with the observation that most of the energy in $y_1(t)$ is near the low-frequency region in contrast to $y_2(t)$.

Solved Problems

17.1 Find the trigonometric Fourier series for the square wave shown in Fig. 17-18 and plot the line spectrum.

In the interval $0 < \omega t < \pi$, $f(t) = V$; and for $\pi < \omega t < 2\pi$, $f(t) = -V$. The average value of the wave is zero; hence, $a_0/2 = 0$. The cosine coefficients are obtained by writing the evaluation integral with the functions inserted as follows:

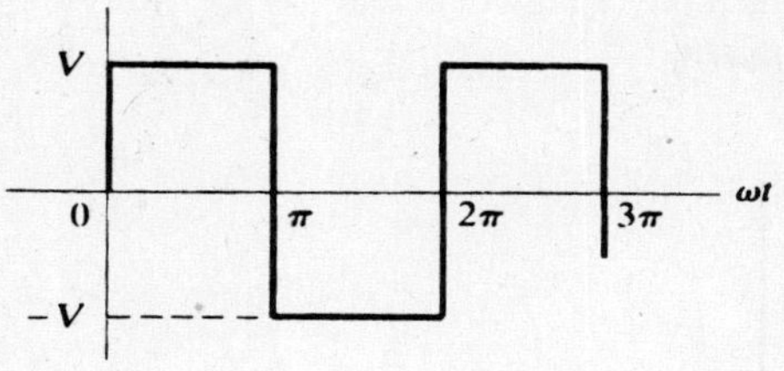

Fig. 17-18

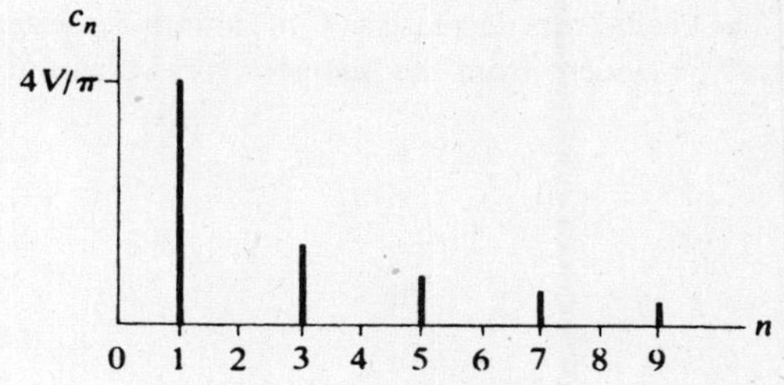

Fig. 17-19

$$a_n = \frac{1}{\pi}\left[\int_0^{\pi} V\cos n\omega t\, d(\omega t) + \int_{\pi}^{2\pi}(-V)\cos n\omega t\, d(\omega t)\right] = \frac{V}{\pi}\left\{\left[\frac{1}{n}\sin n\omega t\right]_0^{\pi} - \left[\frac{1}{n}\sin n\omega t\right]_{\pi}^{2\pi}\right\}$$
$$= 0 \qquad \text{for all } n$$

Thus, the series contains no cosine terms. Proceeding with the evaluation integral for the sine terms,

$$b_n = \frac{1}{\pi}\left[\int_0^{\pi} V\sin n\omega t\, d(\omega t) + \int_{\pi}^{2\pi}(-V)\sin n\omega t\, d(\omega t)\right]$$
$$= \frac{V}{\pi}\left\{\left[-\frac{1}{n}\cos n\omega t\right]_0^{\pi} + \left[\frac{1}{n}\cos n\omega t\right]_{\pi}^{2\pi}\right\}$$
$$= \frac{V}{\pi n}(-\cos n\pi + \cos 0 + \cos n2\pi - \cos n\pi) = \frac{2V}{\pi n}(1 - \cos n\pi)$$

Then $b_n = 4V/\pi n$ for $n = 1, 3, 5, \ldots$, and $b_n = 0$ for $n = 2, 4, 6, \ldots$. The series for the square wave is

$$f(t) = \frac{4V}{\pi}\sin\omega t + \frac{4V}{3\pi}\sin 3\omega t + \frac{4V}{5\pi}\sin 5\omega t + \cdots$$

The line spectrum for this series is shown in Fig. 17-19. This series contains only odd-harmonic sine terms, as could have been anticipated by examination of the waveform for symmetry. Since the wave in Fig. 17-18 is odd, its series contains only sine terms; and since it also has half-wave symmetry, only odd harmonics are present.

17.2 Find the trigonometric Fourier series for the triangular wave shown in Fig. 17-20 and plot the line spectrum.

The wave is an even function, since $f(t) = f(-t)$, and if its average value, $V/2$, is subtracted, it also has half-wave symmetry, that is, $f(t) = -f(t+\pi)$. For $-\pi < \omega t < 0$, $f(t) = V + (V/\pi)\omega t$; and for $0 < \omega t < \pi$, $f(t) = V - (V/\pi)\omega t$. Since even waveforms have only cosine terms, all $b_n = 0$. For $n \geq 1$,

$$a_n = \frac{1}{\pi}\int_{-\pi}^{0}[V + (V/\pi)\omega t]\cos n\omega t\, d(\omega t) + \frac{1}{\pi}\int_0^{\pi}[V - (V/\pi)\omega t]\cos n\omega t\, d(\omega t)$$
$$= \frac{V}{\pi}\left[\int_{-\pi}^{\pi}\cos n\omega t\, d(\omega t) + \int_{-\pi}^{0}\frac{\omega t}{\pi}\cos n\omega t\, d(\omega t) - \int_0^{\pi}\frac{\omega t}{\pi}\cos n\omega t\, d(\omega t)\right]$$
$$= \frac{V}{\pi^2}\left\{\left[\frac{1}{n^2}\cos n\omega t + \frac{\omega t}{\pi}\sin n\omega t\right]_{-\pi}^{0} - \left[\frac{1}{n^2}\cos n\omega t + \frac{\omega t}{n}\sin n\omega t\right]_0^{\pi}\right\}$$
$$= \frac{V}{\pi^2 n^2}[\cos 0 - \cos(-n\pi) - \cos n\pi + \cos 0] = \frac{2V}{\pi^2 n^2}(1 - \cos n\pi)$$

As predicted from half-wave symmetry, the series contains only odd terms, since $a_n = 0$ for $n = 2, 4, 6, \ldots$. For $n = 1, 3, 5, \ldots$, $a_n = 4V/\pi^2 n^2$. Then the required Fourier series is

$$f(t) = \frac{V}{2} + \frac{4V}{-\pi^2}\cos\omega t + \frac{4V}{(3\pi)^2}\cos 3\omega t + \frac{4V}{(5\pi)^2}\cos 5\omega t + \cdots$$

The coefficients decrease as $1/n^2$, and thus the series converges more rapidly than that of Problem 17.1. This fact is evident from the line spectrum shown in Fig. 17-21.

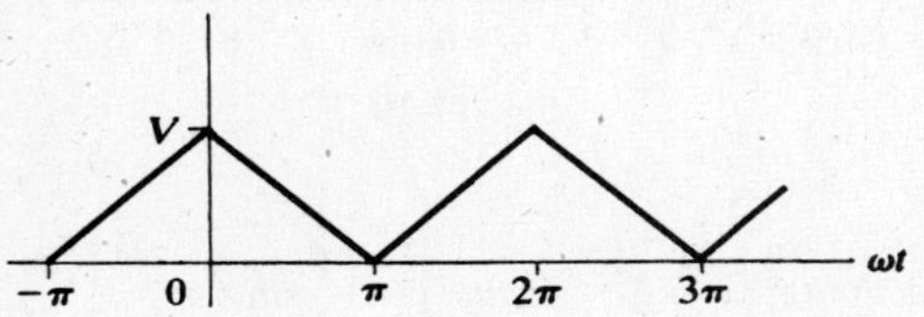

Fig. 17-20

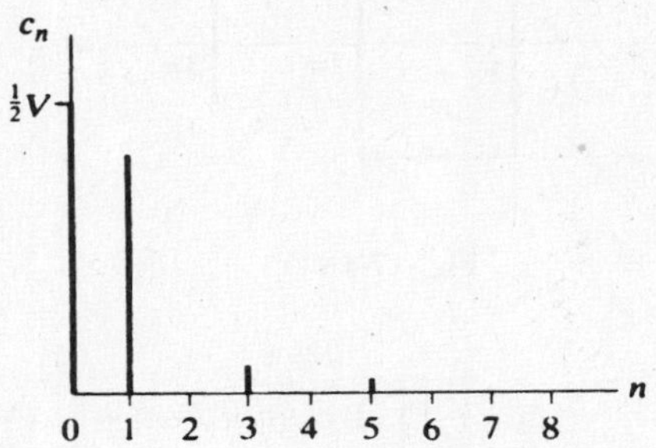

Fig. 17-21

17.3 Find the trigonometric Fourier series for the sawtooth wave shown in Fig. 17-22 and plot the line spectrum.

By inspection, the waveform is odd (and therefore has average value zero). Consequently the series will contain only sine terms. A single expression, $f(t) = (V/\pi)\omega t$, describes the wave over the period from $-\pi$ to $+\pi$, and we will use these limits on our evaluation integral for b_n.

$$b_n = \frac{1}{\pi}\int_{-\pi}^{\pi} (V/\pi)\omega t \sin n\omega t \, d(\omega t) = \frac{V}{\pi^2}\left[\frac{1}{n^2}\sin n\omega t - \frac{\omega t}{n}\cos n\omega t\right]_{-\pi}^{\pi} = -\frac{2V}{n\pi}(\cos n\pi)$$

As $\cos n\pi$ is $+1$ for even n and -1 for odd n, the signs of the coefficients alternate. The required series is

$$f(t) = \frac{2V}{\pi}\{\sin \omega t - \tfrac{1}{2}\sin 2\omega t + \tfrac{1}{3}\sin 3\omega t - \tfrac{1}{4}\sin 4\omega t + \cdots\}$$

The coefficients decrease as $1/n$, and thus the series converges slowly, as shown by the spectrum in Fig. 17-23. Except for the shift in the origin and the average term, this waveform is the same as in Fig. 17-8; compare the two spectra.

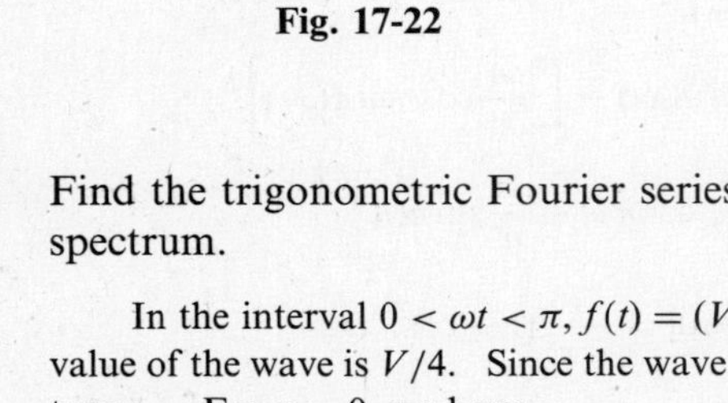

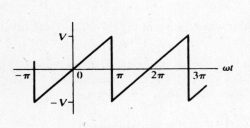

Fig. 17-22

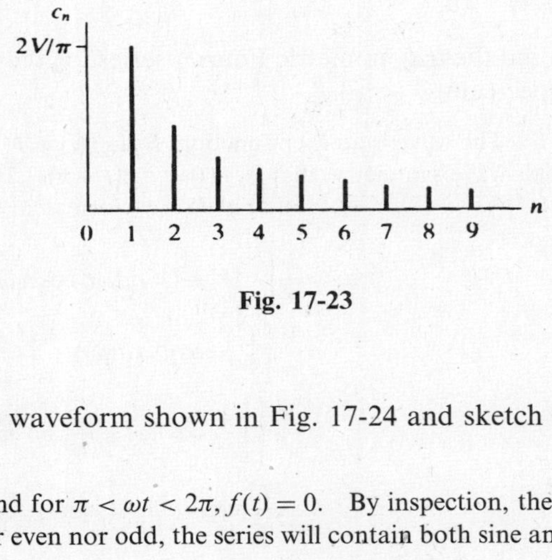

Fig. 17-23

17.4 Find the trigonometric Fourier series for the waveform shown in Fig. 17-24 and sketch the line spectrum.

In the interval $0 < \omega t < \pi$, $f(t) = (V/\pi)\omega t$; and for $\pi < \omega t < 2\pi$, $f(t) = 0$. By inspection, the average value of the wave is $V/4$. Since the wave is neither even nor odd, the series will contain both sine and cosine terms. For $n > 0$, we have

$$a_n = \frac{1}{\pi}\int_0^{\pi} (V/\pi)\omega t \cos n\omega t \, d(\omega t) = \frac{V}{\pi^2}\left[\frac{1}{n^2}\cos n\omega t + \frac{\omega t}{n}\sin n\omega t\right]_0^{\pi} = \frac{V}{\pi^2 n^2}(\cos n\pi - 1)$$

When n is even, $\cos n\pi - 1 = 0$ and $a_n = 0$. When n is odd, $a_n = -2V/(\pi^2 n^2)$. The b_n coefficients are

$$b_n = \frac{1}{\pi}\int_0^{\pi}(V/\pi)\omega t \sin n\omega t\, d(\omega t) = \frac{V}{\pi^2}\left[\frac{1}{n^2}\sin n\omega t - \frac{\omega t}{n}\cos n\omega t\right]_0^{\pi} = -\frac{V}{\pi n}(\cos n\pi) = (-1)^{n+1}\frac{V}{\pi n}$$

Then the required Fourier series is

$$f(t) = \frac{V}{4} - \frac{2V}{\pi^2}\cos\omega t - \frac{2V}{(3\pi)^2}\cos 3\omega t - \frac{2V}{(5\pi)^2}\cos 5\omega t - \cdots$$
$$+\frac{V}{\pi}\sin\omega t - \frac{V}{2\pi}\sin 2\omega t + \frac{V}{3\pi}\sin 3\omega t - \cdots$$

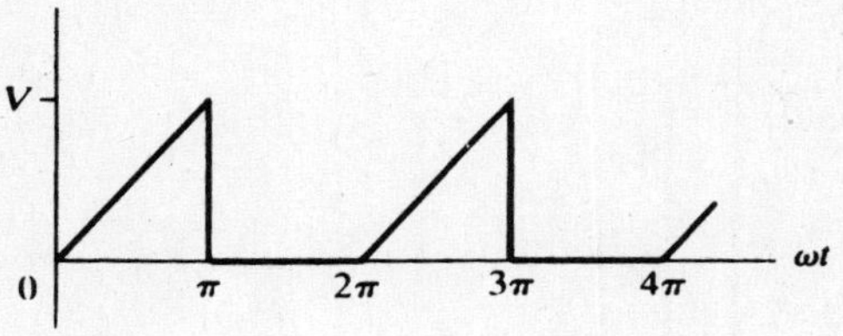

Fig. 17-24

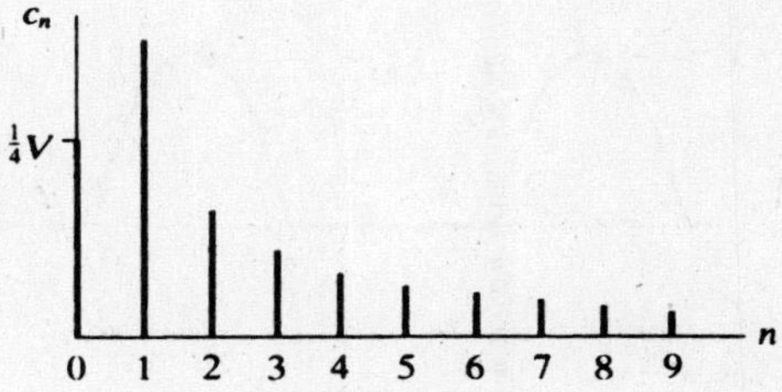

Fig. 17-25

The even-harmonic amplitudes are given directly by $|b_n|$, since there are no even-harmonic cosine terms. However, the odd-harmonic amplitudes must be computed using $c_n = \sqrt{a_n^2 + b_n^2}$. Thus,

$$c_1 = \sqrt{(2V/\pi^2)^2 + (V/\pi)^2} = V(0.377) \qquad c_3 = V(0.109) \qquad c_5 = V(0.064)$$

The line spectrum is shown in Fig. 17-25.

17.5 Find the trigonometric Fourier series for the half-wave-rectified sine wave shown in Fig. 17-26 and sketch the line spectrum.

The wave shows no symmetry, and we therefore expect the series to contain both sine and cosine terms. Since the average value is not obtainable by inspection, we evaluate a_0 for use in the term $a_0/2$.

$$a_0 = \frac{1}{\pi}\int_0^{\pi} V\sin\omega t\, d(\omega t) = \frac{V}{\pi}[-\cos\omega t]_0^{\pi} = \frac{2V}{\pi}$$

Next we determine a_n:

$$a_n = \frac{1}{\pi}\int_0^{\pi} V\sin\omega t\cos n\omega t\, d(\omega t)$$
$$= \frac{V}{\pi}\left[\frac{-n\sin\omega t\sin n\omega t - \cos n\omega t\cos\omega t}{-n^2+1}\right]_0^{\pi} = \frac{V}{\pi(1-n^2)}(\cos n\pi + 1)$$

With n even, $a_n = 2V/\pi(1-n^2)$; and with n odd, $a_n = 0$. However, this expression is indeterminate for $n = 1$, and therefore we must integrate separately for a_1.

$$a_1 = \frac{1}{\pi}\int_0^{\pi} V\sin\omega t\cos\omega t\, d(\omega t) = \frac{V}{\pi}\int_0^{\pi}\tfrac{1}{2}\sin 2\omega t\, d(\omega t) = 0$$

Now we evaluate b_n:

$$b_n = \frac{1}{\pi}\int_0^{\pi} V\sin\omega t\sin n\omega t\, d(\omega t) = \frac{V}{\pi}\left[\frac{n\sin\omega t\cos n\omega t - \sin n\omega t\cos\omega t}{-n^2+1}\right]_0^{\pi} = 0$$

Here again the expression is indeterminate for $n = 1$, and b_1 is evaluated separately.

$$b_1 = \frac{1}{\pi}\int_0^{\pi} V \sin^2 \omega t \, d(\omega t) = \frac{V}{\pi}\left[\frac{\omega t}{2} - \frac{\sin 2\omega t}{4}\right]_0^{\pi} = \frac{V}{2}$$

Then the required Fourier series is

$$f(t) = \frac{V}{\pi}\left(1 + \frac{\pi}{2}\sin \omega t - \frac{2}{3}\cos 2\omega t - \frac{2}{15}\cos 4\omega t - \frac{2}{35}\cos 6\omega t - \cdots\right)$$

The spectrum, Fig. 17-27, shows the strong fundamental term in the series and the rapidly decreasing amplitudes of the higher harmonics.

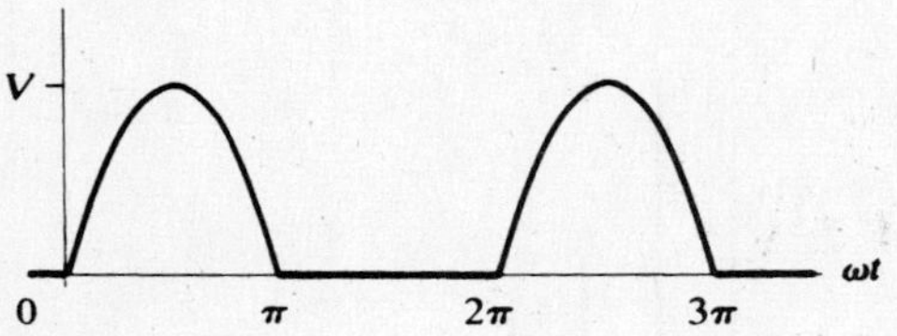

Fig. 17-26

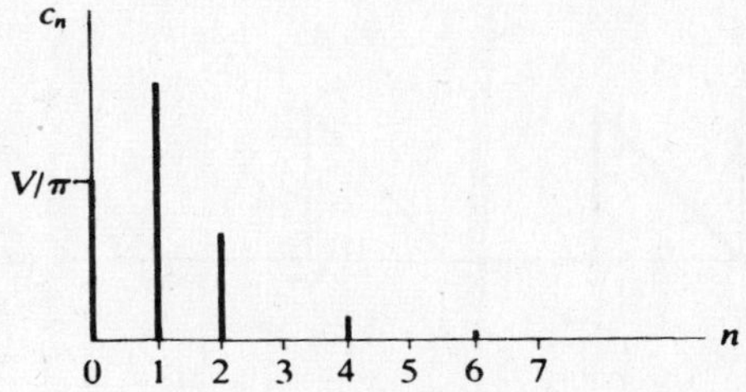

Fig. 17-27

17.6 Find the trigonometric Fourier series for the half-wave-rectified sine wave shown in Fig. 17-28, where the vertical axis is shifted from its position in Fig. 17-26.

The function is described in the interval $-\pi < \omega t < 0$ by $f(t) = -V \sin \omega t$. The average value is the same as that in Problem 17.5, that is, $\frac{1}{2}a_0 = V/\pi$. For the coefficients a_n, we have

$$a_n = \frac{1}{\pi}\int_{-\pi}^{0} (-V \sin \omega t)\cos n\omega t \, d(\omega t) = \frac{V}{\pi(1 - n^2)}(1 + \cos n\pi)$$

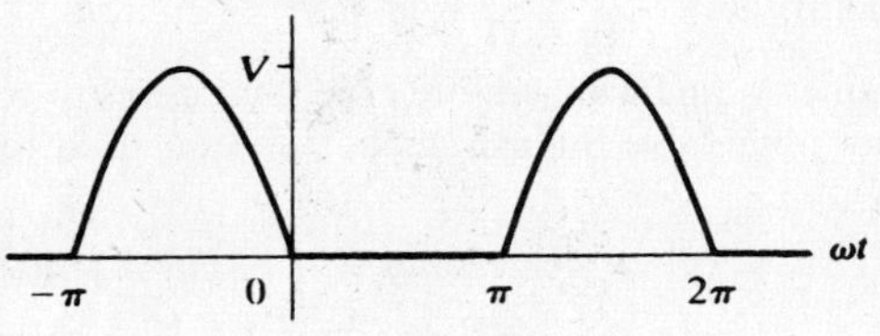

Fig. 17-28

For n even, $a_n = 2V/\pi(1 - n^2)$; and for n odd, $a_n = 0$, except that $n = 1$ must be examined separately.

$$a_1 = \frac{1}{\pi}\int_{-\pi}^{0} (-V \sin \omega t)\cos \omega t \, d(\omega t) = 0$$

For the coefficients b_n, we obtain

$$b_n = \frac{1}{\pi}\int_{-\pi}^{0} (-V \sin \omega t)\sin n\omega t \, d(\omega t) = 0$$

except for $n = 1$.

$$b_1 = \frac{1}{\pi}\int_{-\pi}^{0} (-V)\sin^2 \omega t \, d(\omega t) = -\frac{V}{2}$$

Thus, the series is

$$f(t) = \frac{V}{\pi}\left(1 - \frac{\pi}{2}\sin \omega t - \frac{2}{3}\cos 2\omega t - \frac{2}{15}\cos 4\omega t - \frac{2}{35}\cos 6\omega t - \cdots\right)$$

This series is identical to that of Problem 17.5, except for the fundamental term, which has a negative coefficient in this series. The spectrum would obviously be identical to that of Fig. 17-27.

Another Method

When the sine wave $V \sin \omega t$ is subtracted from the graph of Fig. 17.26, the graph of Fig. 17-28 results.

17.7 Obtain the trigonometric Fourier series for the repeating rectangular pulse shown in Fig. 17-29 and plot the line spectrum.

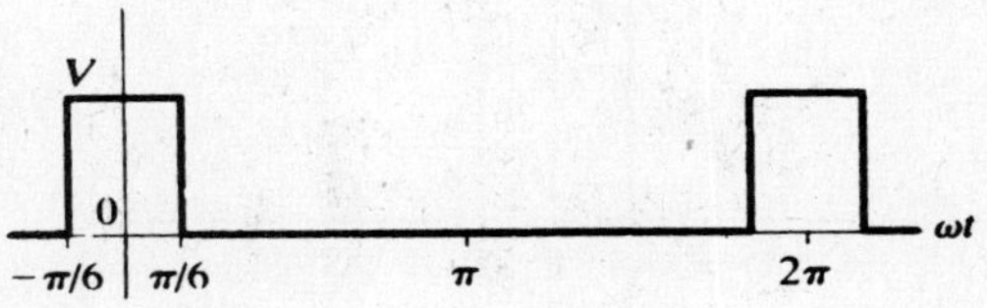

Fig. 17-29

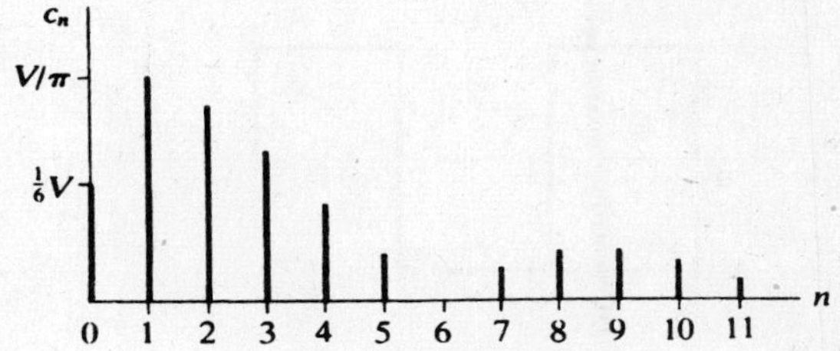

Fig. 17-30

With the vertical axis positioned as shown, the wave is even and the series will contain only cosine terms and a constant term. In the period from $-\pi$ to $+\pi$ used for the evaluation integrals, the function is zero except from $-\pi/6$ to $+\pi/6$.

$$a_0 = \frac{1}{\pi}\int_{-\pi/6}^{\pi/6} V\,d(\omega t) = \frac{V}{3} \qquad a_n = \frac{1}{\pi}\int_{-\pi/6}^{\pi/6} V\cos n\omega t\,d(\omega t) = \frac{2V}{n\pi}\sin\frac{n\pi}{6}$$

Since $\sin n\pi/6 = 1/2, \sqrt{3}/2, 1, \sqrt{3}/2, 1/2, 0, -1/2, \ldots$ for $n = 1, 2, 3, 4, 5, 6, 7, \ldots$, respectively, the series is

$$f(t) = \frac{V}{6} + \frac{2V}{\pi}\left[\frac{1}{2}\cos\omega t + \frac{\sqrt{3}}{2}\left(\frac{1}{2}\right)\cos 2\omega t + 1\left(\frac{1}{3}\right)\cos 3\omega t + \frac{\sqrt{3}}{2}\left(\frac{1}{4}\right)\cos 4\omega t + \frac{1}{2}\left(\frac{1}{5}\right)\cos 5\omega t - \frac{1}{2}\left(\frac{1}{7}\right)\cos 7\omega t - \cdots\right]$$

or

$$f(t) = \frac{V}{6} + \frac{2V}{\pi}\sum_{n=1}^{\infty}\frac{1}{n}\sin(n\pi/6)\cos n\omega t$$

The line spectrum, shown in Fig. 17-30, decreases very slowly for this wave, since the series converges very slowly to the function. Of particular interest is the fact that the 8th, 9th, and 10th harmonic amplitudes exceed the 7th. With the simple waves considered previously, the higher-harmonic amplitudes were progressively lower.

17.8 Find the exponential Fourier series for the square wave shown in Figs. 17-18 and 17-31, and sketch the line spectrum. Obtain the trigonometric series coefficients from those of the exponential series and compare with Problem 17.1.

In the interval $-\pi < \omega t < 0$, $f(t) = -V$; and for $0 < \omega t < \pi$, $f(t) = V$. The wave is odd; therefore, $A_0 = 0$ and the $\mathbf{A}_n$ will be pure imaginaries.

$$\mathbf{A}_n = \frac{1}{2\pi}\left[\int_{-\pi}^{0}(-V)e^{-jn\omega t}\,d(\omega t) + \int_{0}^{\pi} Ve^{-jn\omega t}\,d(\omega t)\right]$$

$$= \frac{V}{2\pi}\left\{-\left[\frac{1}{(-jn)}e^{-jn\omega t}\right]_{-\pi}^{0} + \left[\frac{1}{(-jn)}e^{-jn\omega t}\right]_{0}^{\pi}\right\}$$

$$= \frac{V}{-j2\pi n}(-e^0 + e^{jn\pi} + e^{-jn\pi} - e^0) = j\frac{V}{n\pi}(e^{jn\pi} - 1)$$

For n even, $e^{jn\pi} = +1$ and $\mathbf{A}_n = 0$; for n odd, $e^{jn\pi} = -1$ and $\mathbf{A}_n = -j(2V/n\pi)$ (half-wave symmetry). The required Fourier series is

$$f(t) = \cdots + j\frac{2V}{3\pi}e^{-j3\omega t} + j\frac{2V}{\pi}e^{-j\omega t} - j\frac{2V}{\pi}e^{j\omega t} - j\frac{2V}{3\pi}e^{j3\omega t} - \cdots$$

The graph in Fig. 17-32 shows amplitudes for both positive and negative frequencies. Combining the values at $+n$ and $-n$ yields the same line spectrum as plotted in Fig. 17-19.

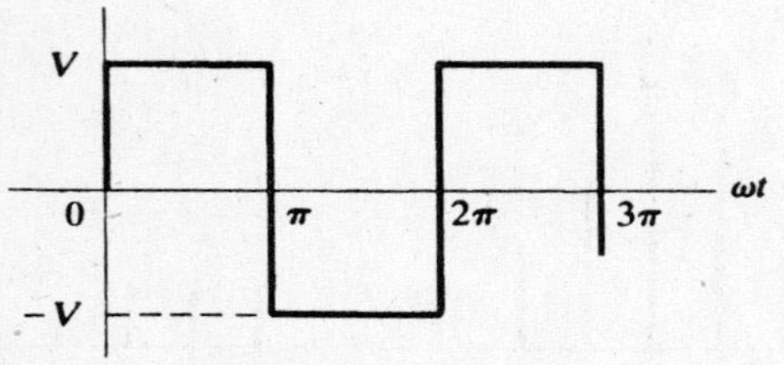

Fig. 17-31

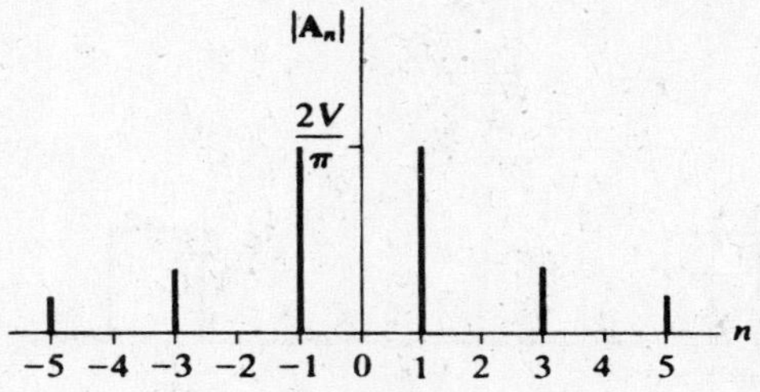

Fig. 17-32

The trigonometric-series cosine coefficients are

$$a_n = 2\,\text{Re}\,\mathbf{A}_n = 0$$

and $$b_n = -2\,\text{Im}\,\mathbf{A}_n = \frac{4V}{n\pi} \qquad \text{for odd } n \text{ only}$$

These agree with the coefficients obtained in Problem 17.1.

17.9 Find the exponential Fourier series for the triangular wave shown in Figs. 17-20 and 17-33 and sketch the line spectrum.

In the interval $-\pi < \omega t < 0$, $f(t) = V + (V/\pi)\omega t$; and for $0 < \omega t < \pi$, $f(t) = V - (V/\pi)\omega t$. The wave is even and therefore the $\mathbf{A}_n$ coefficients will be pure real. By inspection the average value is $V/2$.

$$\mathbf{A}_n = \frac{1}{2\pi}\left\{\int_{-\pi}^{0}[V + (V/\pi)\omega t]e^{-jn\omega t}\,d(\omega t) + \int_{0}^{\pi}[V - (V/\pi)\omega t]e^{-jn\omega t}\,d(\omega t)\right\}$$
$$= \frac{V}{2\pi^2}\left[\int_{-\pi}^{0}\omega t e^{-jn\omega t}\,d(\omega t) + \int_{0}^{\pi}(-\omega t)e^{-jn\omega t}\,d(\omega t) + \int_{-\pi}^{\pi}\pi e^{-jn\omega t}\,d(\omega t)\right]$$
$$= \frac{V}{2\pi^2}\left\{\left[\frac{e^{-jn\omega t}}{(-jn)^2}(-jn\omega t - 1)\right]_{-\pi}^{0} - \left[\frac{e^{-jn\omega t}}{(-jn)^2}(-jn\omega t - 1)\right]_{0}^{\pi}\right\} = \frac{V}{\pi^2 n^2}(1 - e^{jn\pi})$$

For even n, $e^{jn\pi} = +1$ and $\mathbf{A}_n = 0$; for odd n, $\mathbf{A}_n = 2V/\pi^2 n^2$. Thus the series is

$$f(t) = \cdots + \frac{2V}{(-3\pi)^2}e^{-j3\omega t} + \frac{2V}{(-\pi)^2}e^{-j\omega t} + \frac{V}{2} + \frac{2V}{(\pi)^2}e^{j\omega t} + \frac{2V}{(3\pi)^2}e^{j3\omega t} + \cdots$$

The harmonic amplitudes

$$c_0 = \frac{V}{2} \qquad c_n = 2|\mathbf{A}_n| = \begin{cases} 0 & (n = 2, 4, 6, \ldots) \\ 4V/\pi^2 n^2 & (n = 1, 3, 5, \ldots) \end{cases}$$

are exactly as plotted in Fig. 17-21.

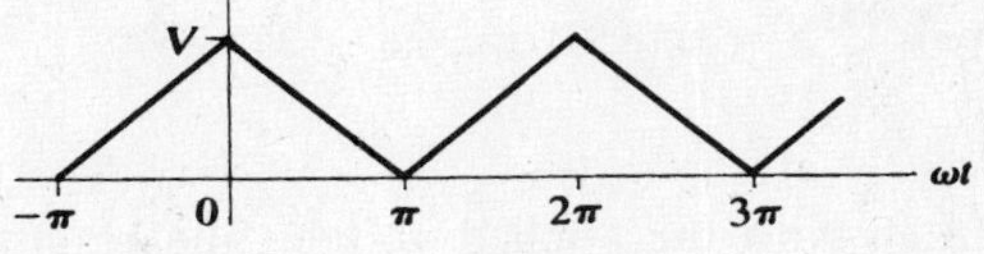

Fig. 17-33

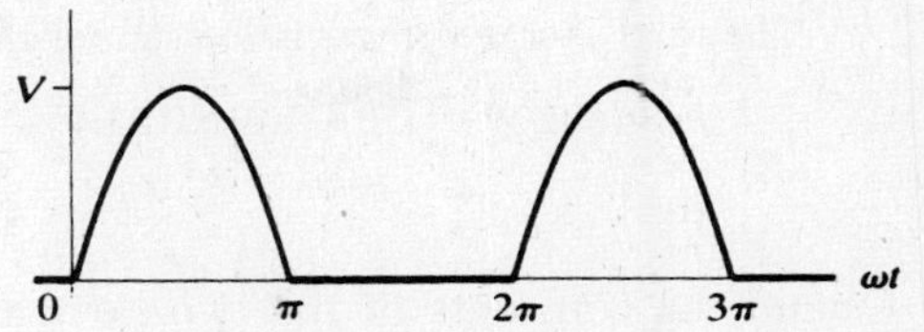

Fig. 17-34

17.10 Find the exponential Fourier series for the half-wave rectified sine wave shown in Figs. 17-26 and 17-34, and sketch the line spectrum.

In the interval $0 < \omega t < \pi$, $f(t) = V \sin \omega t$; and from π to 2π, $f(t) = 0$. Then

$$\mathbf{A}_n = \frac{1}{2\pi}\int_0^{\pi} V \sin \omega t \, e^{-jn\omega t} \, d(\omega t)$$
$$= \frac{V}{2\pi}\left[\frac{e^{-jn\omega t}}{(1-n^2)}(-jn \sin \omega t - \cos \omega t)\right]_0^{\pi} = \frac{V(e^{-jn\pi}+1)}{2\pi(1-n^2)}$$

For even n, $\mathbf{A}_n = V/\pi(1-n^2)$; for odd n, $\mathbf{A}_n = 0$. However, for $n = 1$, the expression for $\mathbf{A}_n$ becomes indeterminate. L'Hôpital's rule may be applied; in other words, the numerator and denominator are separately differentiated with respect to n, after which n is allowed to approach 1, with the result that $\mathbf{A}_1 = -j(V/4)$.

The average value is

$$A_0 = \frac{1}{2\pi}\int_0^{\pi} V \sin \omega t \, d(\omega t) = \frac{V}{2\pi}\Big[-\cos \omega t\Big]_0^{\pi} = \frac{V}{\pi}$$

Then the exponential Fourier series is

$$f(t) = \cdots - \frac{V}{15\pi}e^{-j4\omega t} - \frac{V}{3\pi}e^{-j2\omega t} + j\frac{V}{4}e^{-j\omega t} + \frac{V}{\pi} - j\frac{V}{4}e^{j\omega t} - \frac{V}{3\pi}e^{j2\omega t} - \frac{V}{15\pi}e^{j4\omega t} - \cdots$$

The harmonic amplitudes,

$$c_0 = A_0 = \frac{V}{\pi} \qquad c_n = 2|\mathbf{A}_n| = \begin{cases} 2V/\pi(n^2-1) & (n = 2, 4, 6, \ldots) \\ V/2 & (n = 1) \\ 0 & (n = 3, 5, 7, \ldots) \end{cases}$$

are exactly as plotted in Fig. 17-27.

17.11 Find the average power in a resistance $R = 10\ \Omega$, if the current in Fourier series form is $i = 10 \sin \omega t + 5 \sin 3\omega t + 2 \sin 5\omega t$ (A).

The current has an effective value $I_{\text{eff}} = \sqrt{\frac{1}{2}(10)^2 + \frac{1}{2}(5)^2 + \frac{1}{2}(2)^2} = \sqrt{64.5} = 8.03$ A. Then the average power is $P = I_{\text{eff}}^2 R = (64.5)10 = 645$ W.

Another Method

The total power is the sum of the harmonic powers, which are given by $\frac{1}{2}V_{\max}I_{\max}\cos\theta$. But the voltage across the resistor and the current are in phase for all harmonics, and $\theta_n = 0$. Then,

$$v_R = Ri = 100 \sin \omega t + 50 \sin 3\omega t + 20 \sin 5\omega t$$

and $P = \frac{1}{2}(100)(10) + \frac{1}{2}(50)(5) + \frac{1}{2}(20)(2) = 645$ W.

17.12 Find the average power supplied to a network if the applied voltage and resulting current are

$$v = 50 + 50 \sin 5 \times 10^3 t + 30 \sin 10^4 t + 20 \sin 2 \times 10^4 t \quad \text{(V)}$$
$$i = 11.2 \sin (5 \times 10^3 t + 63.4^\circ) + 10.6 \sin (10^4 t + 45^\circ) + 8.97 \sin (2 \times 10^4 t + 26.6^\circ) \quad \text{(A)}$$

The total average power is the sum of the harmonic powers:

$$P = (50)(0) + \tfrac{1}{2}(50)(11.2)\cos 63.4^\circ + \tfrac{1}{2}(30)(10.6)\cos 45^\circ + \tfrac{1}{2}(20)(8.97)\cos 26.6^\circ = 317.7 \text{ W}$$

17.13 Obtain the constants of the two-element series circuit with the applied voltage and resultant current given in Problem 17.12.

The voltage series contains a constant term 50, but there is no corresponding term in the current series, thus indicating that one of the elements is a capacitor. Since power is delivered to the circuit, the other element must be a resistor.

$$I_{\text{eff}} = \sqrt{\tfrac{1}{2}(11.2)^2 + \tfrac{1}{2}(10.6)^2 + \tfrac{1}{2}(8.97)^2} = 12.6 \text{ A}$$

The average power is $P = I_{\text{eff}}^2 R$, from which $R = P/I_{\text{eff}}^2 = 317.7/159.2 = 2\ \Omega$.

At $\omega = 10^4$ rad/s, the current leads the voltage by 45°. Hence,

$$1 = \tan 45^\circ = \frac{1}{\omega CR} \qquad \text{or} \qquad C = \frac{1}{(10^4)(2)} = 50\ \mu\text{F}$$

Therefore, the two-element series circuit consists of a resistor of 2 Ω and a capacitor of 50 μF.

17.14 The voltage wave shown in Fig. 17-35 is applied to a series circuit of $R = 2\,\text{k}\Omega$ and $L = 10$ H. Use the trigonometric Fourier series to obtain the voltage across the resistor. Plot the line spectra of the applied voltage and v_R to show the effect of the inductance on the harmonics. $\omega = 377$ rad/s.

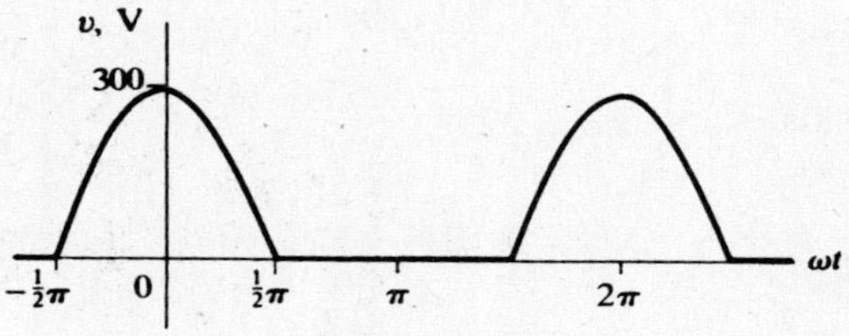

Fig. 17-35

The applied voltage has average value $V_{\max}/\pi$, as in Problem 17.5. The wave function is even and hence the series contains only cosine terms, with coefficients obtained by the following evaluation integral:

$$a_n = \frac{1}{\pi}\int_{-\pi/2}^{\pi/2} 300\cos\omega t \cos n\omega t\, d(\omega t) = \frac{600}{\pi(1-n^2)}\cos n\pi/2 \quad \text{V}$$

Here, $\cos n\pi/2$ has the value -1 for $n = 2, 6, 10, \ldots$, and $+1$ for $n = 4, 8, 12, \ldots$. For n odd, $\cos n\pi/2 = 0$. However, for $n = 1$, the expression is indeterminate and must be evaluated separately.

$$a_1 = \frac{1}{\pi}\int_{-\pi/2}^{\pi/2} 300\cos^2\omega t\, d(\omega t) = \frac{300}{\pi}\left[\frac{\omega t}{2} + \frac{\sin 2\omega t}{4}\right]_{-\pi/2}^{\pi/2} = \frac{300}{2} \text{ V}$$

Thus,
$$v = \frac{300}{\pi}\left(1 + \frac{\pi}{2}\cos\omega t + \frac{2}{3}\cos 2\omega t - \frac{2}{15}\cos 4\omega t + \frac{2}{35}\cos 6\omega t - \cdots\right) \quad \text{(V)}$$

In Table 17-3, the total impedance of the series circuit is computed for each harmonic in the voltage expression. The Fourier coefficients of the current series are the voltage series coefficients divided by the Z_n; the current terms lag the voltage terms by the phase angles θ_n.

Table 17-3

n	$n\omega$, rad/s	R, kΩ	$n\omega L$, kΩ	Z_n, kΩ	θ_n
0	0	2	0	2	0°
1	377	2	3.77	4.26	62°
2	754	2	7.54	7.78	75.1°
4	1508	2	15.08	15.2	82.45°
6	2262	2	22.62	22.6	84.92°

$$I_0 = \frac{300/\pi}{2} \text{ mA}$$

$$i_1 = \frac{300/2}{4.26} \cos(\omega t - 62°) \quad \text{(mA)}$$

$$i_2 = \frac{600/3\pi}{7.78} \cos(2\omega t - 75.1°) \quad \text{(mA)}$$

$$\cdots\cdots\cdots\cdots\cdots\cdots\cdots\cdots\cdots\cdots$$

Then the current series is

$$i = \frac{300}{2\pi} + \frac{300}{(2)(4.26)} \cos(\omega t - 62°) + \frac{600}{3\pi(7.78)} \cos(2\omega t - 75.1°)$$
$$- \frac{600}{15\pi(15.2)} \cos(4\omega t - 82.45°) + \frac{600}{35\pi(22.6)} \cos(6\omega t - 84.92°) - \cdots \quad \text{(mA)}$$

and the voltage across the resistor is

$$v_R = Ri = 95.5 + 70.4 \cos(\omega t - 62°) + 16.4 \cos(2\omega t - 75.1°)$$
$$- 1.67 \cos(4\omega t - 82.45°) + 0.483 \cos(6\omega t - 84.92°) - \cdots \quad \text{(V)}$$

Figure 17-36 shows clearly how the harmonic amplitudes of the applied voltage have been reduced by the 10-H series inductance.

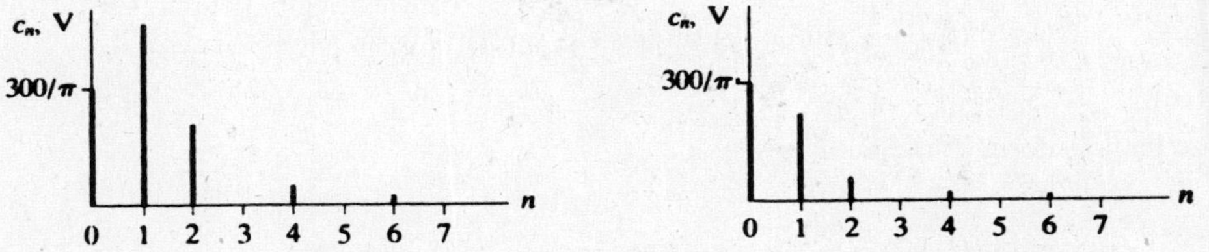

(*a*) **Spectrum of v** (*b*) **Spectrum of v_R**

Fig. 17-36

17.15 The current in a 10-mH inductance has the waveform shown in Fig. 17-37. Obtain the trigonometric series for the voltage across the inductance, given that $\omega = 500$ rad/s.

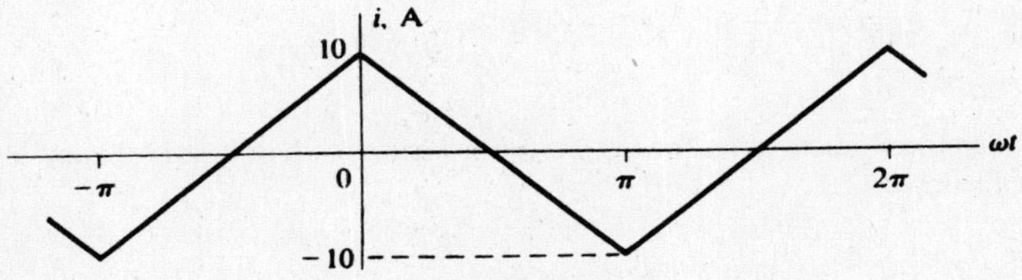

Fig. 17-37

The derivative of the waveform of Fig. 17-37 is graphed in Fig. 17-38. This is just Fig. 17-18 with $V = -20/\pi$. Hence, from Problem 17.1,

$$\frac{di}{d(\omega t)} = -\frac{80}{\pi^2}\left(\sin \omega t + \tfrac{1}{3}\sin 3\omega t + \tfrac{1}{5}\sin 5\omega t + \cdots\right) \quad \text{(A)}$$

and so

$$v_L = L\omega \frac{di}{d(\omega t)} = -\frac{400}{\pi^2}\left(\sin \omega t + \tfrac{1}{3}\sin 3\omega t + \tfrac{1}{5}\sin 5\omega t + \cdots\right) \quad \text{(V)}$$

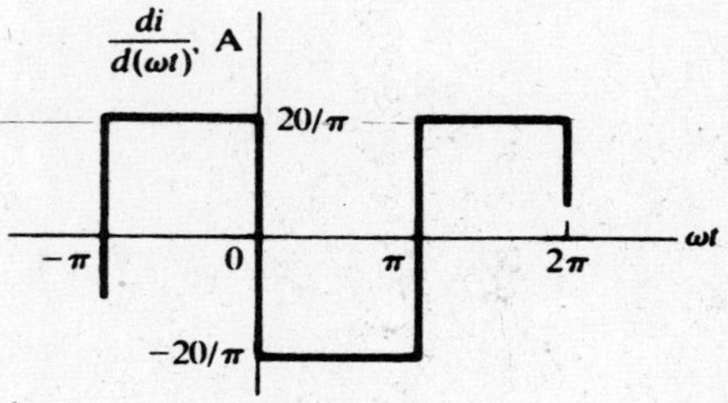

Fig. 17-38

Supplementary Problems

17.16 Synthesize the waveform for which the trigonometric Fourier series is

$$f(t) = \frac{8V}{\pi^2}\left\{\sin \omega t - \tfrac{1}{9}\sin 3\omega t + \tfrac{1}{25}\sin 5\omega t - \tfrac{1}{49}\sin 7\omega t + \cdots\right\}$$

17.17 Synthesize the waveform if its Fourier series is

$$f(t) = 5 - \frac{40}{\pi^2}\left(\cos \omega t + \tfrac{1}{9}\cos 3\omega t + \tfrac{1}{25}\cos 5\omega t + \cdots\right) + \frac{20}{\pi}\left(\sin \omega t - \tfrac{1}{2}\sin 2\omega t + \tfrac{1}{3}\sin 3\omega t - \tfrac{1}{4}\sin 4\omega t + \cdots\right)$$

17.18 Synthesize the waveform for the given Fourier series.

$$f(t) = V\left(\frac{1}{2\pi} - \frac{1}{\pi}\cos \omega t - \frac{1}{3\pi}\cos 2\omega t + \frac{1}{2\pi}\cos 3\omega t - \frac{1}{15\pi}\cos 4\omega t - \frac{1}{6\pi}\cos 6\omega t + \cdots + \frac{1}{4}\sin \omega t - \frac{2}{3\pi}\sin 2\omega t + \frac{4}{15\pi}\sin 4\omega t - \cdots\right)$$

17.19 Find the trigonometric Fourier series for the sawtooth wave shown in Fig. 17-39 and plot the line spectrum. Compare with Example 17.1.

Ans. $f(t) = \dfrac{V}{2} + \dfrac{V}{\pi}\left(\sin \omega t + \tfrac{1}{2}\sin 2\omega t + \tfrac{1}{3}\sin 3\omega t + \cdots\right)$

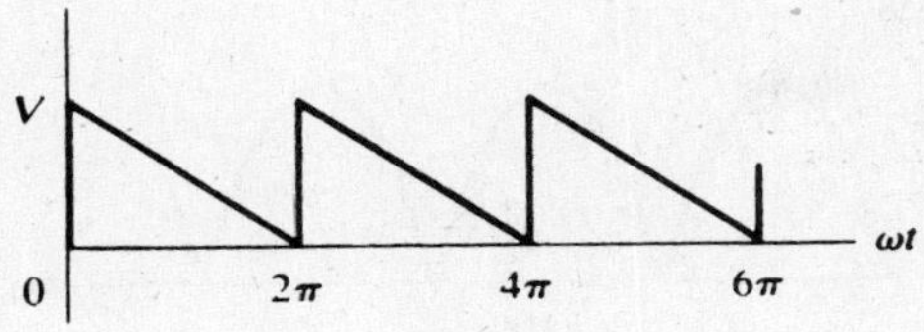

Fig. 17-39

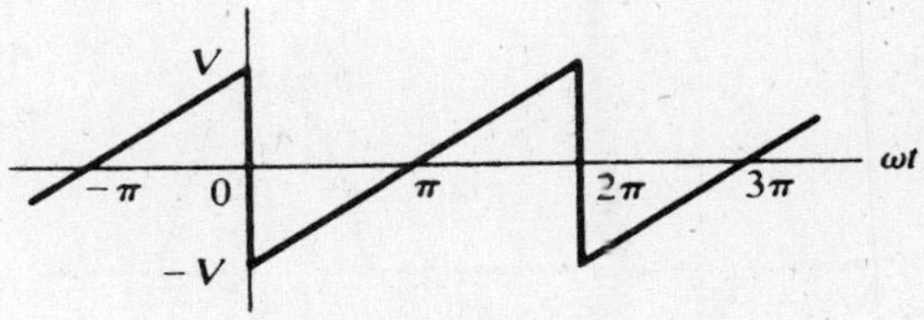

Fig. 17-40

17.20 Find the trigonometric Fourier series for the sawtooth wave shown in Fig. 17-40 and plot the spectrum. Compare with the result of Problem 17.3.

Ans. $f(t) = \dfrac{-2V}{\pi}\{\sin\omega t + \frac{1}{2}\sin 2\omega t + \frac{1}{3}\sin 3\omega t + \frac{1}{4}\sin 4\omega t + \cdots\}$

17.21 Find the trigonometric Fourier series for the waveform shown in Fig. 17-41 and plot the line spectrum.

Ans. $f(t) = \dfrac{4V}{\pi^2}\{\cos\omega t + \frac{1}{9}\cos 3\omega t + \frac{1}{25}\cos 5\omega t + \cdots\} - \dfrac{2V}{\pi}\{\sin\omega t + \frac{1}{3}\sin 3\omega t + \frac{1}{5}\sin 5\omega t + \cdots\}$

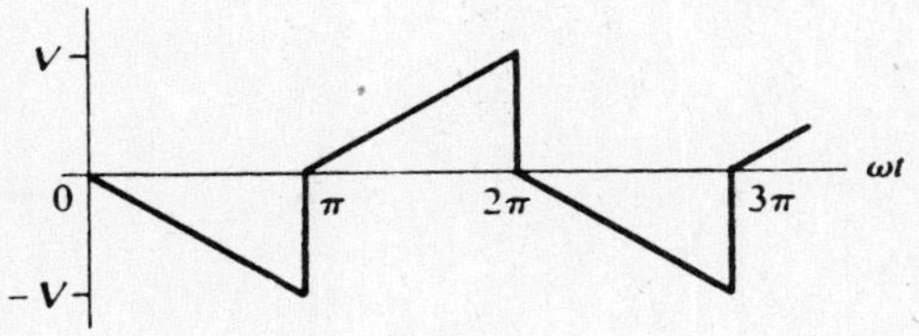

Fig. 17-41

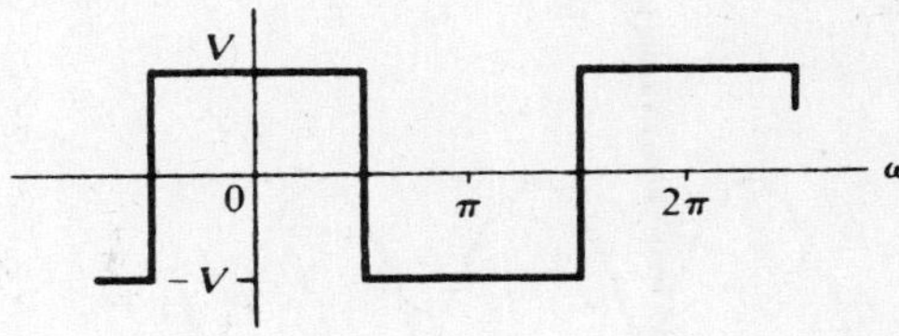

Fig. 17-42

17.22 Find the trigonometric Fourier series of the square wave shown in Fig. 17-42 and plot the line spectrum. Compare with the result of Problem 17.1.

Ans. $f(t) = \dfrac{4V}{\pi}\{\cos\omega t - \frac{1}{3}\cos 3\omega t + \frac{1}{5}\cos 5\omega t - \frac{1}{7}\cos 7\omega t + \cdots\}$

17.23 Find the trigonometric Fourier series for the waveforms shown in Fig. 17-43. Plot the line spectrum of each and compare.

Ans. (*a*) $f(t) = \dfrac{5}{12} + \displaystyle\sum_{n=1}^{\infty}\left[\frac{10}{n\pi}\left(\sin\frac{n\pi}{12}\right)\cos n\omega t + \frac{10}{n\pi}\left(1 - \cos\frac{n\pi}{12}\right)\sin n\omega t\right]$

(*b*) $f(t) = \dfrac{50}{6} + \displaystyle\sum_{n=1}^{\infty}\left[\frac{10}{n\pi}\left(\sin\frac{n5\pi}{3}\right)\cos n\omega t + \frac{10}{n\pi}\left(1 - \cos\frac{n5\pi}{3}\right)\sin n\omega t\right]$

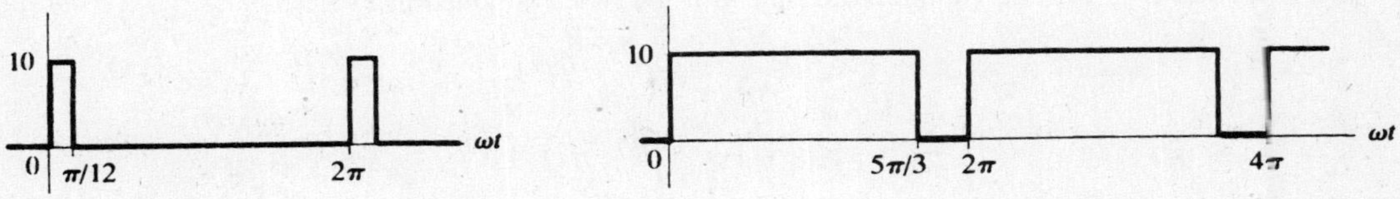

(*a*) (*b*)

Fig. 17-43

17.24 Find the trigonometric Fourier series for the half-wave-rectified sine wave shown in Fig. 17-44 and plot the line spectrum. Compare the answer with the results of Problems 17.5 and 17.6.

Ans. $f(t) = \dfrac{V}{\pi}\left(1 + \dfrac{\pi}{2}\cos\omega t + \dfrac{2}{3}\cos 2\omega t - \dfrac{2}{15}\cos 4\omega t + \dfrac{2}{35}\cos 6\omega t - \cdots\right)$

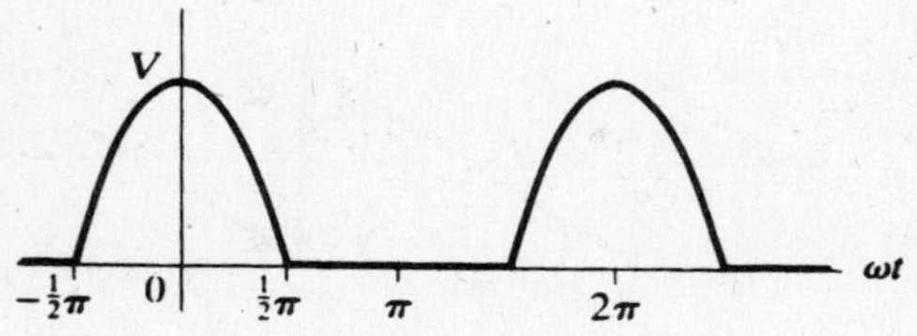

Fig. 17-44

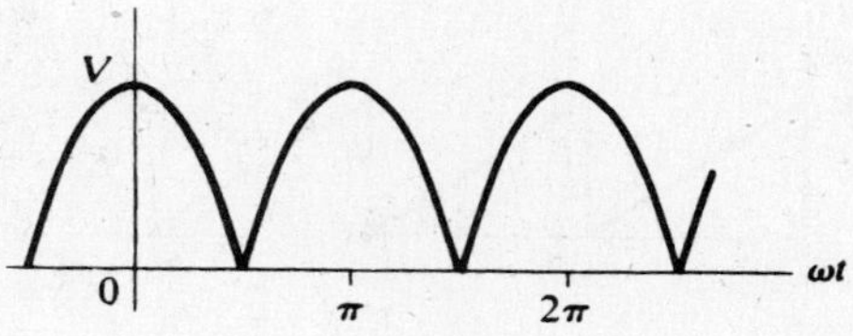

Fig. 17-45

17.25 Find the trigonometric Fourier series for the full-wave-rectified sine wave shown in Fig. 17-45 and plot the spectrum.

Ans. $f(t) = \dfrac{2V}{\pi}\left(1 + \frac{2}{3}\cos 2\omega t - \frac{2}{15}\cos 4\omega t + \frac{2}{35}\cos 6\omega t - \cdots\right)$

17.26 The waveform in Fig. 17-46 is that of Fig. 17-45 with the origin shifted. Find the Fourier series and show that the two spectra are identical.

Ans. $f(t) = \dfrac{2V}{\pi}\left(1 - \frac{2}{3}\cos 2\omega t - \frac{2}{15}\cos 4\omega t - \frac{2}{35}\cos 6\omega t - \cdots\right)$

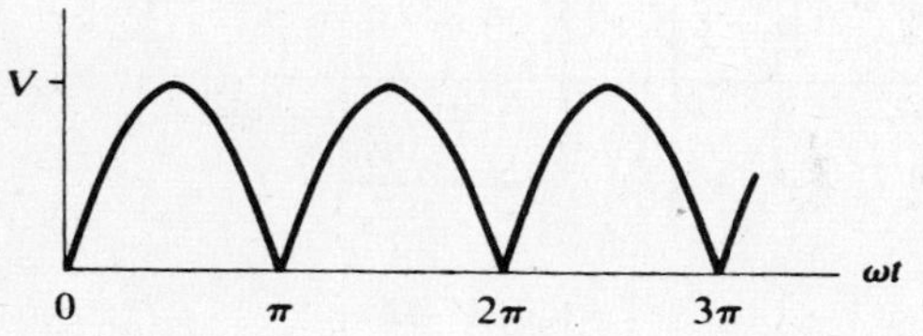

Fig. 17-46

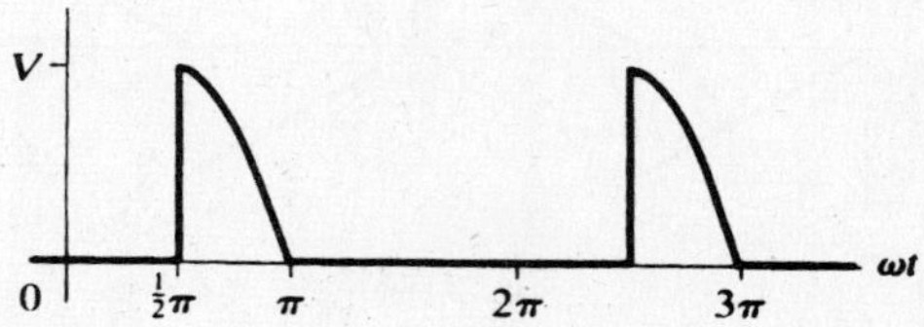

Fig. 17-47

17.27 Find the trigonometric Fourier series for the waveform shown in Fig. 17-47.

Ans.
$$f(t) = \frac{V}{2\pi} - \frac{V}{2\pi}\cos\omega t + \sum_{n=2}^{\infty}\frac{V}{\pi(1-n^2)}(\cos n\pi + n\sin n\pi/2)\cos n\omega t$$
$$+\frac{V}{4}\sin\omega t + \sum_{n=2}^{\infty}\left[\frac{-nV\cos n\pi/2}{\pi(1-n^2)}\right]\sin n\omega t$$

17.28 Find the trigonometric Fourier series for the waveform shown in Fig. 17-48. Add this series termwise to that of Problem 17.27, and compare the sum with the series obtained in Problem 17.5.

Ans.
$$f(t) = \frac{V}{2\pi} + \frac{V}{2\pi}\cos\omega t + \sum_{n=2}^{\infty}\frac{V(n\sin n\pi/2 - 1)}{\pi(n^2-1)}\cos n\omega t + \frac{V}{4}\sin\omega t + \sum_{n=2}^{\infty}\frac{nV\cos n\pi/2}{\pi(1-n^2)}\sin n\omega t$$

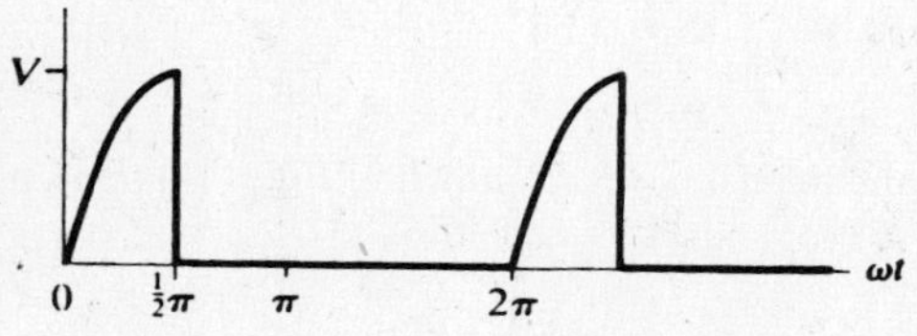

Fig. 17-48

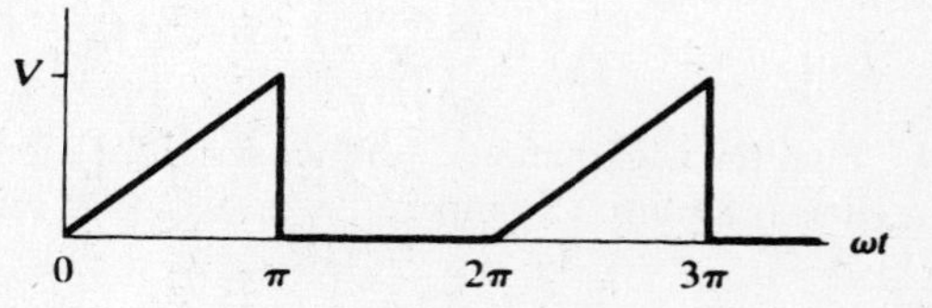

Fig. 17-49

17.29 Find the exponential Fourier series for the waveform shown in Fig. 17-49 and plot the line spectrum. Convert the coefficients obtained here into the trigonometric series coefficients, write the trigonometric series, and compare it with the result of Problem 17.4.

$$\textit{Ans.}\quad f(t) = V\left[\cdots + \left(\frac{1}{9\pi^2} - j\frac{1}{6\pi}\right)e^{-j3\omega t} - j\frac{1}{4\pi}e^{-j2\omega t} - \left(\frac{1}{\pi^2} - j\frac{1}{2\pi}\right)e^{-j\omega t} + \frac{1}{4}\right.$$
$$\left. - \left(\frac{1}{\pi^2} + j\frac{1}{2\pi}\right)e^{j\omega t} + j\frac{1}{4\pi}e^{j2\omega t} - \left(\frac{1}{9\pi^2} + j\frac{1}{6\pi}\right)e^{j3\omega t} - \cdots\right]$$

17.30 Find the exponential Fourier series for the waveform shown in Fig. 17-50 and plot the line spectrum.

$$\textit{Ans.}\quad f(t) = V\left[\cdots + \left(\frac{1}{9\pi^2} + j\frac{1}{6\pi}\right)e^{-j3\omega t} + j\frac{1}{4\pi}e^{-j2\omega t} + \left(\frac{1}{\pi^2} + j\frac{1}{2\pi}\right)e^{-j\omega t} + \frac{1}{4}\right.$$
$$\left. + \left(\frac{1}{\pi^2} - j\frac{1}{2\pi}\right)e^{j\omega t} - j\frac{1}{4\pi}e^{j2\omega t} + \left(\frac{1}{9\pi^2} - j\frac{1}{6\pi}\right)e^{j3\omega t} + \cdots\right]$$

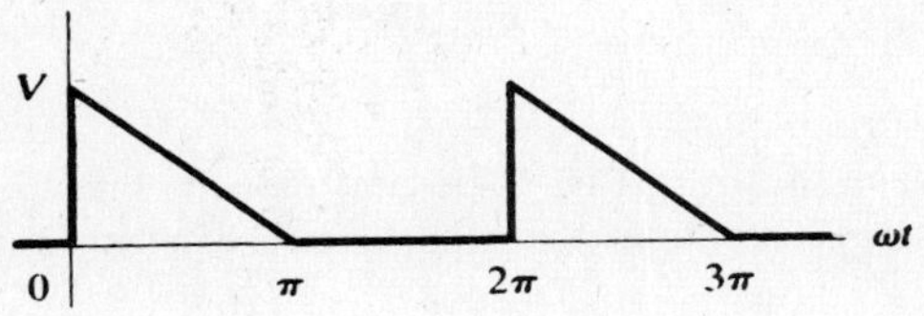

Fig. 17-50

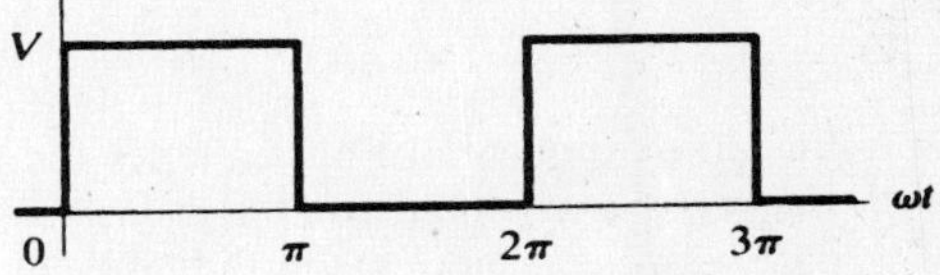

Fig. 17-51

17.31 Find the exponential Fourier series for the square wave shown in Fig. 17-51 and plot the line spectrum. Add the exponential series of Problems 17.29 and 17.30 and compare the sum to the series obtained here.

$$\textit{Ans.}\quad f(t) = V\left(\cdots + j\frac{1}{3\pi}e^{-j3\omega t} + j\frac{1}{\pi}e^{-j\omega t} + \frac{1}{2} - j\frac{1}{\pi}e^{j\omega t} - j\frac{1}{3\pi}e^{j3\omega t} - \cdots\right)$$

17.32 Find the exponential Fourier series for the sawtooth waveform shown in Fig. 17-52 and plot the spectrum. Convert the coefficients obtained here into the trigonometric series coefficients, write the trigonometric series, and compare the results with the series obtained in Problem 17.19.

$$\textit{Ans.}\quad f(t) = V\left(\cdots + j\frac{1}{4\pi}e^{-j2\omega t} + j\frac{1}{2\pi}e^{-j\omega t} + \frac{1}{2} - j\frac{1}{2\pi}e^{j\omega t} - j\frac{1}{4\pi}e^{j2\omega t} - \cdots\right)$$

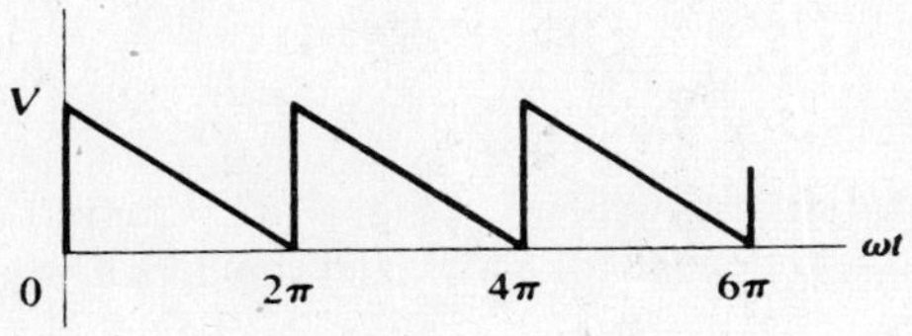

Fig. 17-52

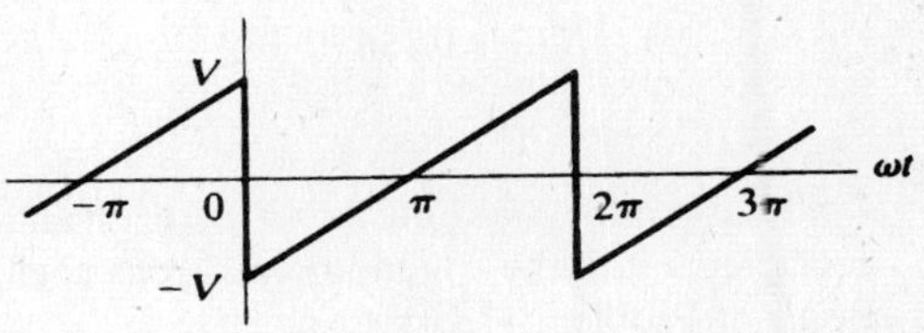

Fig. 17-53

17.33 Find the exponential Fourier series for the waveform shown in Fig. 17-53 and plot the spectrum. Convert the trigonometric series coefficients found in Problem 17.20 into exponential series coefficients and compare them with the coefficients of the series obtained here.

$$\textit{Ans.}\quad f(t) = V\left(\cdots - j\frac{1}{2\pi}e^{-j2\omega t} - j\frac{1}{\pi}e^{-j\omega t} + j\frac{1}{\pi}e^{j\omega t} + j\frac{1}{2\pi}e^{j2\omega t} + \cdots\right)$$

17.34 Find the exponential Fourier series for the waveform shown in Fig. 17-54 and plot the spectrum. Convert the coefficients to trigonometric series coefficients, write the trigonometric series, and compare it with that obtained in Problem 17.21.

$$Ans.\quad f(t) = V\left[\cdots + \left(\frac{2}{9\pi^2} - j\frac{1}{3\pi}\right)e^{-j3\omega t} + \left(\frac{2}{\pi^2} - j\frac{1}{\pi}\right)e^{-j\omega t} + \left(\frac{2}{\pi^2} + j\frac{1}{\pi}\right)e^{j\omega t} + \left(\frac{2}{9\pi^2} + j\frac{1}{3\pi}\right)e^{j3\omega t} + \cdots\right]$$

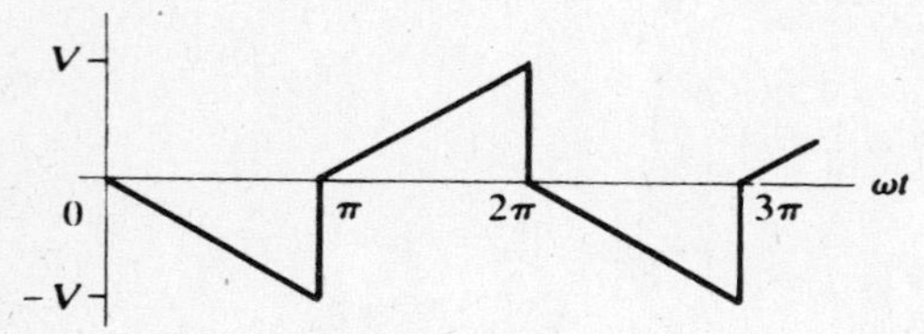

Fig. 17-54

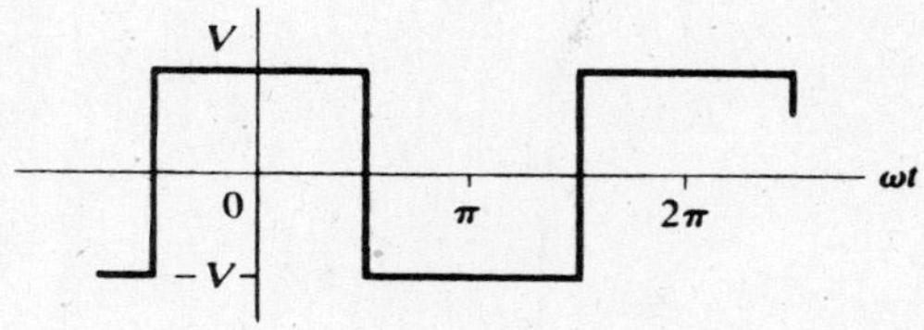

Fig. 17-55

17.35 Find the exponential Fourier series for the square wave shown in Fig. 17-55 and plot the line spectrum. Convert the trigonometric series coefficients of Problem 17.22 into exponential series coefficients and compare with the coefficients in the result obtained here.

$$Ans.\quad f(t) = \frac{2V}{\pi}\left(\cdots + \tfrac{1}{5}e^{-j5\omega t} - \tfrac{1}{3}e^{-j3\omega t} + e^{-j\omega t} + e^{j\omega t} - \tfrac{1}{3}e^{-j3\omega t} + \tfrac{1}{5}e^{j5\omega t} - \cdots\right)$$

17.36 Find the exponential Fourier series for the waveform shown in Fig. 17-56 and plot the line spectrum.

$$Ans.\quad f(t) = \cdots + \frac{V}{2\pi}\sin\left(\frac{2\pi}{6}\right)e^{-j2\omega t} + \frac{V}{\pi}\sin\left(\frac{\pi}{6}\right)e^{-j\omega t} + \frac{V}{6} + \frac{V}{\pi}\sin\left(\frac{\pi}{6}\right)e^{j\omega t} + \frac{V}{2\pi}\sin\left(\frac{2\pi}{6}\right)e^{j2\omega t} + \cdots$$

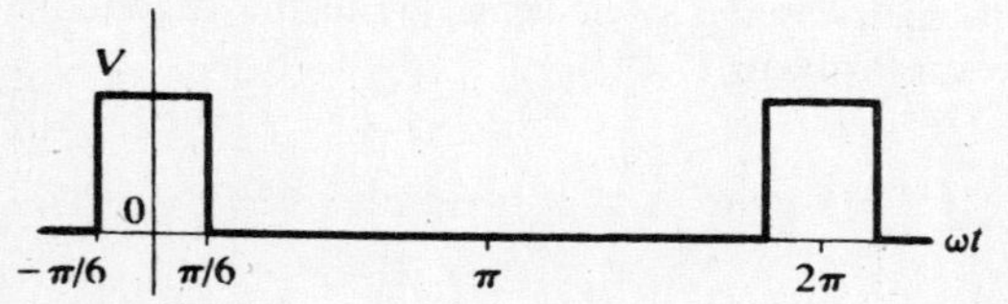

Fig. 17-56

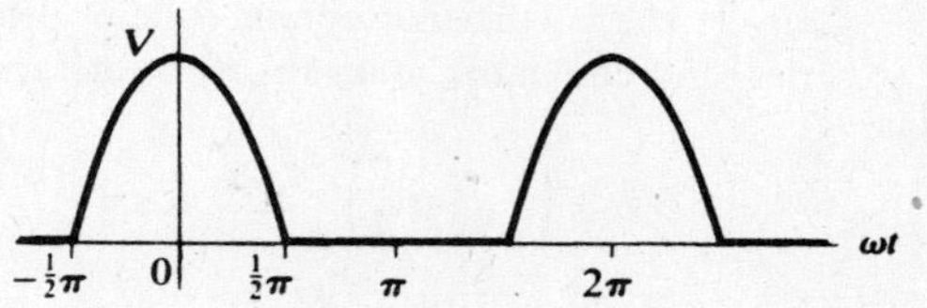

Fig. 17-57

17.37 Find the exponential Fourier series for the half-wave-rectified sine wave shown in Fig. 17-57. Convert these coefficients into the trigonometric series coefficients, write the trigonometric series, and compare it with the result of Problem 17.24.

$$Ans.\quad f(t) = \cdots - \frac{V}{15\pi}e^{-j4\omega t} + \frac{V}{3\pi}e^{-j2\omega t} + \frac{V}{4}e^{-j\omega t} + \frac{V}{\pi} + \frac{V}{4}e^{j\omega t} + \frac{V}{3\pi}e^{j2\omega t} - \frac{V}{15\pi}e^{j4\omega t} + \cdots$$

17.38 Find the exponential Fourier series for the full-wave rectified sine wave shown in Fig. 17-58 and plot the line spectrum.

$$Ans.\quad f(t) = \cdots - \frac{2V}{15\pi}e^{-j4\omega t} + \frac{2V}{3\pi}e^{-j2\omega t} + \frac{2V}{\pi} + \frac{2V}{3\pi}e^{j2\omega t} - \frac{2V}{15\pi}e^{j4\omega t} + \cdots$$

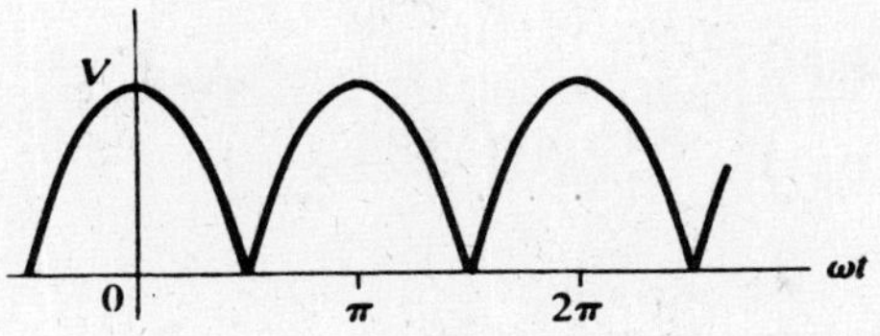

Fig. 17-58

17.39 Find the effective voltage, effective current, and average power supplied to a passive network if the applied voltage is $v = 200 + 100\cos(500t + 30°) + 75\cos(1500t + 60°)$ (V) and the resulting current is $i = 3.53\cos(500t + 75°) + 3.55\cos(1500t + 78.45°)$ (A). *Ans.* 218.5 V, 3.54 A, 250.8 W

17.40 A voltage $v = 50 + 25\sin 500t + 10\sin 1500t + 5\sin 2500t$ (V) is applied to the terminals of a passive network and the resulting current is

$$i = 5 + 2.23\sin(500t - 26.6°) + 0.556\sin(1500t - 56.3°) + 0.186\sin(2500t - 68.2°) \quad \text{(A)}$$

Find the effective voltage, effective current, and the average power. *Ans.* 53.6 V, 5.25 A, 276.5 W

17.41 A three-element series circuit, with $R = 5\ \Omega$, $L = 5$ mH, and $C = 50\ \mu$F, has an applied voltage $v = 150\sin 1000t + 100\sin 2000t + 75\sin 3000t$ (V). Find the effective current and the average power for the circuit. Sketch the line spectrum of the voltage and the current, and note the effect of series resonance. *Ans.* 16.58 A, 1374 W

17.42 A two-element series circuit, with $R = 10\ \Omega$ and $L = 20$ mH, has current

$$i = 5\sin 100t + 3\sin 300t + 2\sin 500t \quad \text{(A)}$$

Find the effective applied voltage and the average power. *Ans.* 48 V, 190 W

17.43 A pure inductance, $L = 10$ mH, has the triangular current wave shown in Fig. 17-59, where $\omega = 500$ rad/s. Obtain the exponential Fourier series for the voltage across the inductance. Compare the answer with the result of Problem 17.8.

Ans. $v_L = \dfrac{200}{\pi^2}(\cdots - j\tfrac{1}{3}e^{-j3\omega t} - je^{-j\omega t} + je^{j\omega t} + j\tfrac{1}{3}e^{j\omega t} + \cdots)$ (V)

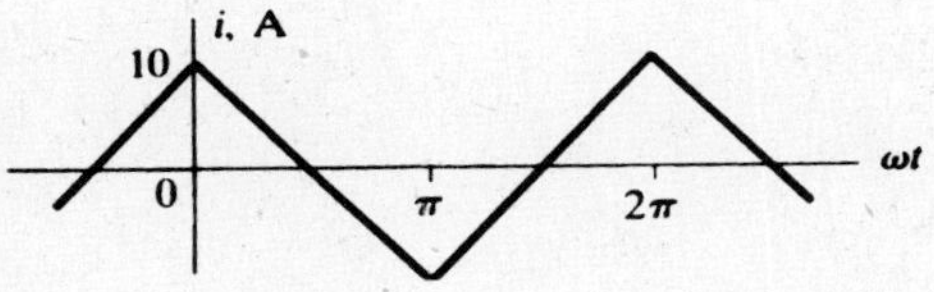

Fig. 17-59

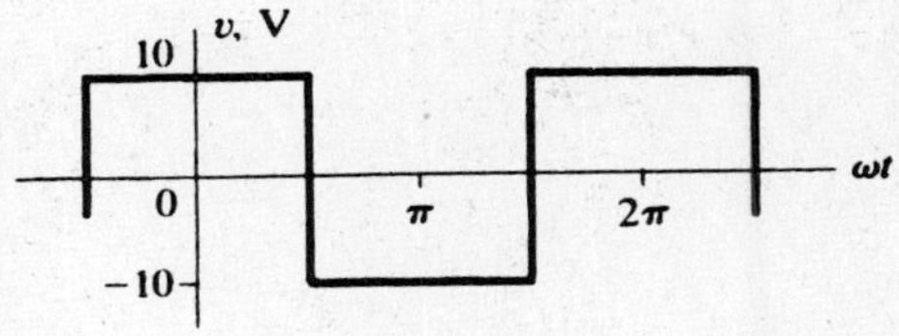

Fig. 17-60

17.44 A pure inductance, $L = 10$ mH, has an applied voltage with the waveform shown in Fig. 17-60, where $\omega = 200$ rad/s. Obtain the current series in trigonometric form and identify the current waveform.

Ans. $i = \dfrac{20}{\pi}(\sin\omega t - \tfrac{1}{9}\sin 3\omega t + \tfrac{1}{25}\sin 5\omega t - \tfrac{1}{49}\sin 7\omega t + \cdots)$ (A); triangular

17.45 Figure 17-61 shows a full-wave-rectified sine wave representing the voltage applied to the terminals of an *LC* series circuit. Use the trigonometric Fourier series to find the voltages across the inductor and the capacitor.

$$\textit{Ans.} \quad v_L = \frac{4V_m}{\pi}\left[\frac{2\omega L}{3\left(2\omega L - \dfrac{1}{2\omega C}\right)}\cos 2\omega t - \frac{4\omega L}{15\left(4\omega L - \dfrac{1}{4\omega C}\right)}\cos 4\omega t + \cdots\right]$$

$$v_C = \frac{4V_m}{\pi}\left[\frac{1}{2} - \frac{1}{3(2\omega C)\left(2\omega L - \dfrac{1}{2\omega C}\right)}\cos 2\omega t + \frac{1}{15(4\omega C)\left(4\omega L - \dfrac{1}{4\omega C}\right)}\cos 4\omega t - \cdots\right]$$

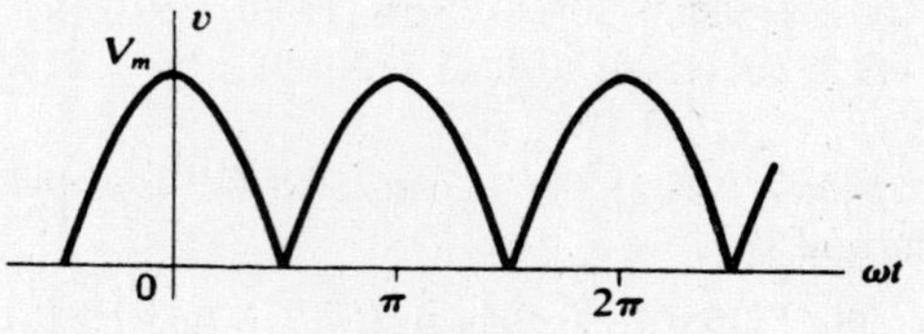

Fig. 17-61

17.46 A three-element circuit consists of $R = 5\ \Omega$ in series with a parallel combination of L and C. At $\omega = 500$ rad/s, $X_L = 2\ \Omega$, $X_C = 8\ \Omega$. Find the total current if the applied voltage is given by $v = 50 + 20 \sin 500t + 10 \sin 1000t$ (V). *Ans.* $i = 10 + 3.53 \sin (500t - 28.1^\circ)$ (A)

APPENDIX A

Complex Number System

A1 COMPLEX NUMBERS

A *complex number* $\mathbf{z}$ is a number of the form $x + jy$, where x and y are real numbers and $j = \sqrt{-1}$. We write $x = \text{Re}\,\mathbf{z}$, the *real part of* $\mathbf{z}$; $y = \text{Im}\,\mathbf{z}$, the *imaginary part of* $\mathbf{z}$. Two complex numbers are equal if and only if their real parts are equal and their imaginary parts are equal.

A2 COMPLEX PLANE

A pair of orthogonal axes, with the horizontal axis displaying Re $\mathbf{z}$ and the vertical axis $j\,\text{Im}\,\mathbf{z}$, determine a complex plane in which each complex number is a unique point. Refer to Fig. A-1, on which six complex numbers are shown. Equivalently, each complex number is represented by a unique vector from the origin of the complex plane, as illustrated for the complex number $\mathbf{z}_6$ in Fig. A-1.

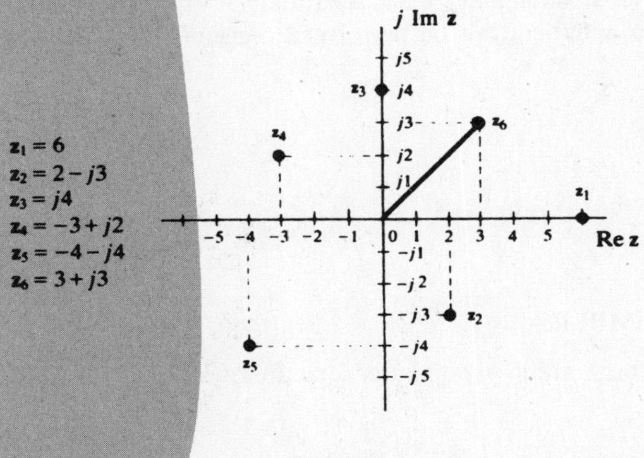

Fig. A-1

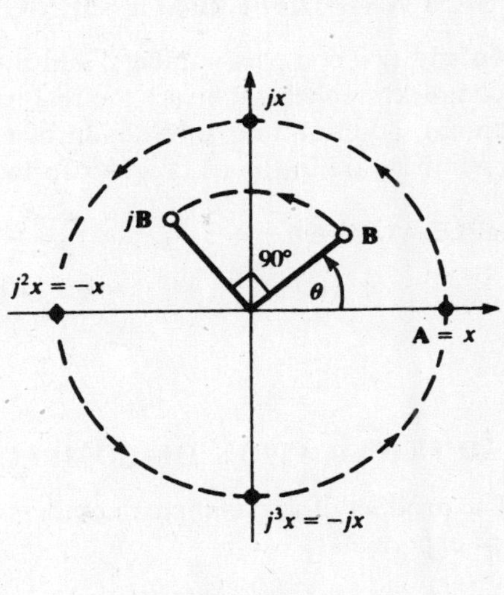

Fig. A-2

A3 VECTOR OPERATOR *j*

In addition to the definition of j given in Section A1, it may be viewed as an operator which rotates any complex number (vector) **A** 90° in the counterclockwise direction. The case where **A** is a pure real number, x, is illustrated in Fig. A-2. The rotation sends **A** into jx, on the positive imaginary axis. Continuing, j^2 advances **A** 180°; j^3, 270°; and j^4, 360°. Also shown in Fig. A-2 is a complex number **B** in the first quadrant, at angle θ. Note that $j\mathbf{B}$ is in the second quadrant, at angle $\theta + 90°$.

A4 OTHER REPRESENTATIONS OF COMPLEX NUMBERS

In Section A1 complex numbers were defined in *rectangular form*. In Fig. A-3, $x = r\cos\theta$, $y = r\sin\theta$, and the complex number **z** can be written in *trigonometric form* as

$$\mathbf{z} = x + jy = r(\cos\theta + j\sin\theta)$$

where r is the *modulus* or *absolute value* (the notation $r = |\mathbf{z}|$ is common), given by $r = \sqrt{x^2 + y^2}$, and the angle $\theta = \tan^{-1}(y/x)$ is the *argument* of **z**.

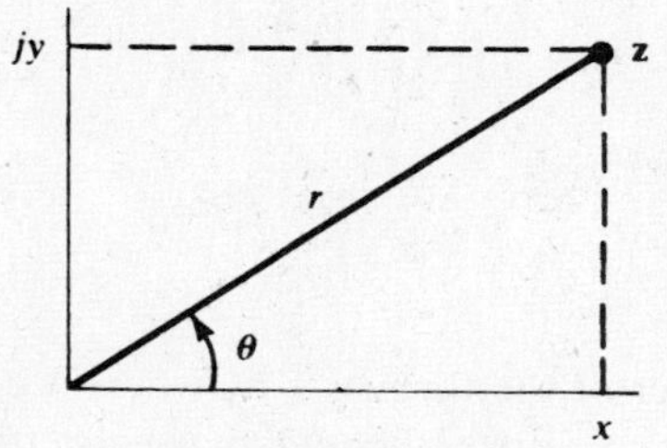

Fig. A-3

Euler's formula, $e^{j\theta} = \cos\theta + j\sin\theta$, permits another representation of a complex number, called the *exponential form*:

$$\mathbf{z} = r\cos\theta + jr\sin\theta = re^{j\theta}$$

A third form, widely used in circuit analysis, is the *polar* or *Steinmetz form*, $\mathbf{z} = r\angle\theta$, where θ is usually in degrees.

A5 SUM AND DIFFERENCE OF COMPLEX NUMBERS

To add two complex numbers, add the real parts and the imaginary parts separately. To subtract two complex numbers, subtract the real parts and the imaginary parts separately. From the practical standpoint, addition and subtraction of complex numbers can be performed conveniently only when both numbers are in the rectangular form.

EXAMPLE A1 Given $\mathbf{z}_1 = 5 - j2$ and $\mathbf{z}_2 = -3 - j8$,

$$\mathbf{z}_1 + \mathbf{z}_2 = (5 - 3) + j(-2 - 8) = 2 - j10$$
$$\mathbf{z}_2 - \mathbf{z}_1 = (-3 - 5) + j(-8 + 2) = -8 - j6$$

A6 MULTIPLICATION OF COMPLEX NUMBERS

The product of two complex numbers when both are in exponential form follows directly from the laws of exponents.

$$\mathbf{z}_1\mathbf{z}_2 = (r_1e^{j\theta_1})(r_2e^{j\theta_2}) = r_1r_2e^{j(\theta_1+\theta_2)}$$

The polar or Steinmetz product is evident from reference to the exponential form.

$$\mathbf{z}_1\mathbf{z}_2 = (r_1\,\underline{/\theta_1})(r_2\,\underline{/\theta_2}) = r_1r_2\,\underline{/\theta_1+\theta_2}$$

The rectangular product can be found by treating the two complex numbers as binomials.

$$\begin{aligned}\mathbf{z}_1\mathbf{z}_2 &= (x_1+jy_1)(x_2+jy_2) = x_1x_2 + jx_1y_2 + jy_1x_2 + j^2y_1y_2\\ &= (x_1x_2 - y_1y_2) + j(x_1y_2 + y_1x_2)\end{aligned}$$

EXAMPLE A2 If $\mathbf{z}_1 = 5e^{j\pi/3}$ and $\mathbf{z}_2 = 2e^{-j\pi/6}$, then $\mathbf{z}_1\mathbf{z}_2 = (5e^{j\pi/3})(2e^{-j\pi/6}) = 10e^{j\pi/6}$.

EXAMPLE A3 If $\mathbf{z}_1 = 2\,\underline{/30^\circ}$ and $\mathbf{z}_2 = 5\,\underline{/-45^\circ}$, then $\mathbf{z}_1\mathbf{z}_2 = (2\,\underline{/30^\circ})(5\,\underline{/-45^\circ}) = 10\,\underline{/-15^\circ}$.

EXAMPLE A4 If $\mathbf{z}_1 = 2+j3$ and $\mathbf{z}_2 = -1-j3$, then $\mathbf{z}_1\mathbf{z}_2 = (2+j3)(-1-j3) = 7-j9$.

A7 DIVISION OF COMPLEX NUMBERS

For two complex numbers in exponential form, the quotient follows directly from the laws of exponents.

$$\frac{\mathbf{z}_1}{\mathbf{z}} = \frac{r_1e^{j\theta_1}}{r_2e^{j\theta_2}} = \frac{r_1}{r_2}\,e^{j(\theta_1-\theta_2)}$$

Again, the polar or Steinmetz form of division is evident from reference to the exponential form.

$$\frac{\mathbf{z}_1}{\mathbf{z}_2} = \frac{r_1\,\underline{/\theta_1}}{r_2\,\underline{/\theta_2}} = \frac{r_1}{r_2}\,\underline{/\theta_1-\theta_2}$$

Division of two complex numbers in the rectangular form is performed by multiplying the numerator and denominator by the conjugate of the denominator (see Section A8).

$$\frac{\mathbf{z}_1}{\mathbf{z}_2} = \frac{x_1+jy_1}{x_2+jy_2}\left(\frac{x_2-jy_2}{x_2-jy_2}\right) = \frac{(x_1x_2+y_1y_2)+j(y_1x_2-y_2x_1)}{x_2^2+y_2^2} = \frac{x_1x_2+y_1y_2}{x_2^2+y_2^2} + j\,\frac{y_1x_2-y_2x_1}{x_2^2+y_2^2}$$

EXAMPLE A5 Given $\mathbf{z}_1 = 4e^{j\pi/3}$ and $\mathbf{z}_2 = 2e^{j\pi/6}$,

$$\frac{\mathbf{z}_1}{\mathbf{z}_2} = \frac{4e^{j\pi/3}}{2e^{j\pi/6}} = 2e^{j\pi/6}$$

EXAMPLE A6 Given $\mathbf{z}_1 = 8\,\underline{/-30^\circ}$ and $\mathbf{z}_2 = 2\,\underline{/-60^\circ}$,

$$\frac{\mathbf{z}_1}{\mathbf{z}_2} = \frac{8\,\underline{/-30^\circ}}{2\,\underline{/-60^\circ}} = 4\,\underline{/30^\circ}$$

EXAMPLE A7 Given $\mathbf{z}_1 = 4-j5$ and $\mathbf{z}_2 = 1+j2$,

$$\frac{\mathbf{z}_1}{\mathbf{z}_2} = \frac{4-j5}{1+j2}\left(\frac{1-j2}{1-j2}\right) = -\frac{6}{5} - j\,\frac{13}{5}$$

A8 CONJUGATE OF A COMPLEX NUMBER

The *conjugate* of the complex number $\mathbf{z} = x + jy$ is the complex number $\mathbf{z}^* = x - jy$. Thus,

$$\operatorname{Re}\mathbf{z} = \frac{\mathbf{z}+\mathbf{z}^*}{2} \qquad \operatorname{Im}\mathbf{z} = \frac{\mathbf{z}-\mathbf{z}^*}{2j} \qquad |\mathbf{z}| = \sqrt{\mathbf{z}\mathbf{z}^*}$$

In the complex plane, the points $\mathbf{z}$ and $\mathbf{z}^*$ are mirror images in the axis of reals.

In exponential form: $\mathbf{z} = re^{j\theta}$, $\mathbf{z}^* = re^{-j\theta}$.

In polar form: $\mathbf{z} = r\underline{/\theta}$, $\mathbf{z}^* = r\underline{/-\theta}$.

In trigonometric form: $\mathbf{z} = r(\cos\theta + j\sin\theta)$, $\mathbf{z}^* = r(\cos\theta - j\sin\theta)$.

Conjugation has the following useful properties:

(i) $(\mathbf{z}^*)^* = \mathbf{z}$

(ii) $(\mathbf{z}_1 \pm \mathbf{z}_2)^* = \mathbf{z}_1^* \pm \mathbf{z}_2^*$

(iii) $(\mathbf{z}_1\mathbf{z}_2)^* = \mathbf{z}_1^*\mathbf{z}_2^*$

(iv) $\left(\dfrac{\mathbf{z}_1}{\mathbf{z}_2}\right)^* = \dfrac{\mathbf{z}_1^*}{\mathbf{z}_2^*}$

APPENDIX B

Matrices and Determinants

B1 SIMULTANEOUS EQUATIONS AND THE CHARACTERISTIC MATRIX

Many engineering systems are described by a set of linearly independent simultaneous equations of the form

$$
\begin{aligned}
y_1 &= a_{11}x_1 + a_{12}x_2 + a_{13}x_3 + \cdots + a_{1n}x_n \\
y_2 &= a_{21}x_1 + a_{22}x_2 + a_{23}x_3 + \cdots + a_{2n}x_n \\
&\cdots\cdots\cdots\cdots\cdots\cdots\cdots\cdots \\
y_m &= a_{m1}x_1 + a_{m2}x_2 + a_{m3}x_3 + \cdots + a_{mn}x_n
\end{aligned}
$$

where the x_j are the independent variables, the y_i the dependent variables, and the a_{ij} are the coefficients of the independent variables. The a_{ij} may be constants or functions of some parameter.

A more convenient form may be obtained for the above equations by expressing them in matrix form.

$$
\begin{bmatrix} y_1 \\ y_2 \\ \cdots \\ y_m \end{bmatrix} = \begin{bmatrix} a_{11} & a_{12} & a_{13} & \cdots & a_{1n} \\ a_{21} & a_{22} & a_{23} & \cdots & a_{2n} \\ \cdots & \cdots & \cdots & \cdots & \cdots \\ a_{m1} & a_{m2} & a_{m3} & \cdots & a_{mn} \end{bmatrix} \begin{bmatrix} x_1 \\ x_2 \\ \cdots \\ x_n \end{bmatrix}
$$

or $\mathbf{Y} = \mathbf{AX}$, by a suitable definition of the product $\mathbf{AX}$ (see Section B3). Matrix $\mathbf{A} \equiv [a_{ij}]$ is called the *characteristic matrix* of the system; its *order* or *dimension* is denoted as

$$d(\mathbf{A}) \equiv m \times n$$

where m is the number of rows and n is the number of columns.

B2 TYPES OF MATRICES

Row matrix. A matrix which may contain any number of columns but only one row; $d(\mathbf{A}) = 1 \times n$. Also called a *row vector*.

Column matrix. A matrix which may contain any number of rows but only one column; $d(\mathbf{A}) = m \times 1$. Also called a *column vector*.

Diagonal matrix. A matrix whose nonzero elements are all on the principal diagonal.

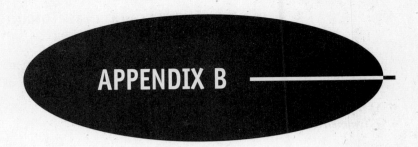

Unit matrix. A diagonal matrix having every diagonal element unity.

Null matrix. A matrix in which every element is zero.

Square matrix. A matrix in which the number of rows is equal to the number of columns; $d(\mathbf{A}) = n \times n$.

Symmetric matrix. Given

$$\mathbf{A} \equiv \begin{bmatrix} a_{11} & a_{12} & a_{13} & \dots & a_{1n} \\ a_{21} & a_{22} & a_{23} & \dots & a_{2n} \\ \dots & \dots & \dots & \dots & \dots \\ a_{m1} & a_{m2} & a_{m3} & \dots & a_{mn} \end{bmatrix} \qquad d(\mathbf{A}) = m \times n$$

the *transpose of* **A** is

$$\mathbf{A}^T \equiv \begin{bmatrix} a_{11} & a_{21} & a_{31} & \dots & a_{m1} \\ a_{12} & a_{22} & a_{32} & \dots & a_{m2} \\ a_{13} & a_{23} & a_{33} & \dots & a_{m3} \\ \dots & \dots & \dots & \dots & \dots \\ a_{1n} & a_{2n} & a_{3n} & \dots & a_{mn} \end{bmatrix} \qquad d(\mathbf{A}^T) = n \times m$$

Thus, the rows of **A** are the columns of $\mathbf{A}^T$, and vice versa. Matrix **A** is *symmetric* if $\mathbf{A} = \mathbf{A}^T$; a symmetric matrix must then be square.

Hermitian matrix. Given

$$\mathbf{A} \equiv \begin{bmatrix} a_{11} & a_{12} & a_{13} & \dots & a_{1n} \\ a_{21} & a_{22} & a_{23} & \dots & a_{2n} \\ \dots & \dots & \dots & \dots & \dots \\ a_{m1} & a_{m2} & a_{m3} & \dots & a_{mn} \end{bmatrix}$$

the *conjugate* of **A** is

$$\mathbf{A}^* \equiv \begin{bmatrix} a^*_{11} & a^*_{12} & a^*_{13} & \dots & a^*_{1n} \\ a^*_{21} & a^*_{22} & a^*_{23} & \dots & a^*_{2n} \\ \dots & \dots & \dots & \dots & \dots \\ a^*_{m1} & a^*_{m2} & a^*_{m3} & \dots & a^*_{mn} \end{bmatrix}$$

Matrix **A** is *hermitian* if $\mathbf{A} = (\mathbf{A}^*)^T$; that is, a hermitian matrix is a square matrix with real elements on the main diagonal and complex conjugate elements occupying positions that are mirror images in the main diagonal. Note that $(\mathbf{A}^*)^T = (\mathbf{A}^T)^*$.

Nonsingular matrix. An $n \times n$ square matrix **A** is *nonsingular* (or *invertible*) if there exists an $n \times n$ square matrix **B** such that

$$\mathbf{AB} = \mathbf{BA} = \mathbf{I}$$

where **I** is the $n \times n$ unit matrix. The matrix **B** is called the *inverse* of the nonsingular matrix **A**, and we write $\mathbf{B} = \mathbf{A}^{-1}$. If **A** is nonsingular, the matrix equation $\mathbf{Y} = \mathbf{AX}$ of Section B1 has, for any **Y**, the unique solution

$$\mathbf{X} = \mathbf{A}^{-1}\mathbf{Y}$$

B3 MATRIX ARITHMETIC

Addition and Subtraction of Matrices

Two matrices of the same order are conformable for addition or subtraction; two matrices of different orders cannot be added or subtracted.

The sum (difference) of two $m \times n$ matrices, $\mathbf{A} = [a_{ij}]$ and $\mathbf{B} = [b_{ij}]$, is the $m \times n$ matrix **C** of which each element is the sum (difference) of the corresponding elements of **A** and **B**. Thus, $\mathbf{A} \pm \mathbf{B} = [a_{ij} \pm b_{ij}]$.

EXAMPLE B1 If

$$\mathbf{A} = \begin{bmatrix} 1 & 4 & 0 \\ 2 & 7 & 3 \end{bmatrix} \qquad \mathbf{B} = \begin{bmatrix} 5 & 2 & 6 \\ 0 & 1 & 1 \end{bmatrix}$$

then

$$\mathbf{A}+\mathbf{B} = \begin{bmatrix} 1+5 & 4+2 & 0+6 \\ 2+0 & 7+1 & 3+1 \end{bmatrix} = \begin{bmatrix} 6 & 6 & 6 \\ 2 & 8 & 4 \end{bmatrix}$$

$$\mathbf{A}-\mathbf{B} = \begin{bmatrix} -4 & 2 & -6 \\ 2 & 6 & 2 \end{bmatrix}$$

The transpose of the sum (difference) of two matrices is the sum (difference) of the two transposes:

$$(\mathbf{A} \pm \mathbf{B})^T = \mathbf{A}^T \pm \mathbf{B}^T$$

Multiplication of Matrices

The product **AB**, in that order, of a $1 \times m$ matrix **A** and an $m \times 1$ matrix **B** is a 1×1 matrix $\mathbf{C} \equiv [c_{11}]$, where

$$\mathbf{C} = [a_{11} \quad a_{12} \quad a_{13} \quad \dots \quad a_{1m}] \begin{bmatrix} b_{11} \\ b_{21} \\ b_{31} \\ \dots \\ b_{m1} \end{bmatrix}$$

$$= [a_{11}b_{11} + a_{12}b_{21} + \dots + a_{1m}b_{m1}] = \left[\sum_{k=1}^{m} a_{1k}b_{k1}\right]$$

Note that each element of the row matrix is multiplied into the corresponding element of the column matrix and then the products are summed. Usually, we identify **C** with the *scalar* c_{11}, treating it as an ordinary number drawn from the number field to which the elements of **A** and **B** belong.

The product **AB**, in that order, of the $m \times s$ matrix $\mathbf{A} = [a_{ij}]$ and the $s \times n$ matrix $\mathbf{B} = [b_{ij}]$ is the $m \times n$ matrix $\mathbf{C} = [c_{ij}]$, where

$$c_{ij} = \sum_{k=1}^{s} a_{ik}b_{kj} \qquad (i = 1, 2, \dots, m, \quad j = 1, 2, \dots, n)$$

EXAMPLE B2

$$\begin{bmatrix} a_{11} & a_{12} \\ a_{21} & a_{22} \\ a_{31} & a_{32} \end{bmatrix} \begin{bmatrix} b_{11} & b_{12} \\ b_{21} & b_{22} \end{bmatrix} = \begin{bmatrix} a_{11}b_{11} + a_{12}b_{21} & a_{11}b_{12} + a_{12}b_{22} \\ a_{21}b_{11} + a_{22}b_{21} & a_{21}b_{12} + a_{22}b_{22} \\ a_{31}b_{11} + a_{32}b_{21} & a_{31}b_{12} + a_{32}b_{22} \end{bmatrix}$$

$$\begin{bmatrix} 3 & 5 & -8 \\ 2 & 1 & 6 \\ 4 & -6 & 7 \end{bmatrix} \begin{bmatrix} I_1 \\ I_2 \\ I_3 \end{bmatrix} = \begin{bmatrix} 3I_1 + 5I_2 - 8I_3 \\ 2I_1 + 1I_2 + 6I_3 \\ 4I_1 - 6I_2 + 7I_3 \end{bmatrix}$$

$$\begin{bmatrix} 5 & -3 \\ 4 & 2 \end{bmatrix} \begin{bmatrix} 8 & -2 & 6 \\ 7 & 0 & 9 \end{bmatrix} = \begin{bmatrix} 5(8)+(-3)(7) & 5(-2)+(-3)(0) & 5(6)+(-3)(9) \\ 4(8)+2(7) & 4(-2)+2(0) & 4(6)+2(9) \end{bmatrix} = \begin{bmatrix} 19 & -10 & 3 \\ 46 & -8 & 42 \end{bmatrix}$$

Matrix **A** is conformable to matrix **B** for multiplication. In other words, the product **AB** is defined, only when the number of columns of **A** is equal to the number of rows of **B**. Thus, if **A** is a 3×2 matrix and **B** is a 2×5 matrix, then the product **AB** is defined, but the product **BA** is not defined. If **D** and **E** are 3×3 matrices, both products **DE** and **ED** are defined. However, it is not necessarily true that **DE** = **ED**.

The transpose of the product of two matrices is the product of the two transposes *taken in reverse order*:

$$(\mathbf{AB})^T = \mathbf{B}^T\mathbf{A}^T$$

If **A** and **B** are nonsingular matrices of the same dimension, then **AB** is also nonsingular, with

$$(\mathbf{AB})^{-1} = \mathbf{B}^{-1}\mathbf{A}^{-1}$$

Multiplication of a Matrix by a Scalar

The product of a matrix $\mathbf{A} \equiv [a_{ij}]$ by a scalar k is defined by

$$k\mathbf{A} = \mathbf{A}k \equiv [ka_{ij}]$$

that is, each element of **A** is multiplied by k. Note the properties

$$k(\mathbf{A}+\mathbf{B}) = k\mathbf{A} + k\mathbf{B} \qquad k(\mathbf{AB}) = (k\mathbf{A})\mathbf{B} = \mathbf{A}(k\mathbf{B}) \qquad (k\mathbf{A})^T = k\mathbf{A}^T$$

B4 DETERMINANT OF A SQUARE MATRIX

Attached to any $n \times n$ matrix $\mathbf{A} \equiv [a_{ij}]$ is a certain scalar function of the a_{ij}, called the *determinant of* **A**. This number is denoted as

$$\det \mathbf{A} \quad \text{or} \quad |\mathbf{A}| \quad \text{or} \quad \Delta_{\mathbf{A}} \quad \text{or} \quad \begin{vmatrix} a_{11} & a_{12} & \dots & a_{1n} \\ a_{21} & a_{22} & \dots & a_{2n} \\ \dots & \dots & \dots & \dots \\ a_{n1} & a_{n2} & \dots & a_{nn} \end{vmatrix}$$

where the last form puts into evidence the elements of **A**, upon which the number depends. For determinants of order $n = 1$ and $n = 2$, we have explicitly

$$|a_{11}| = a_{11} \qquad \begin{vmatrix} a_{11} & a_{12} \\ a_{21} & a_{22} \end{vmatrix} = a_{11}a_{22} - a_{12}a_{21}$$

For larger n, the analogous expressions become very cumbersome, and they are usually avoided by use of Laplace's expansion theorem (see below). What is important is that the determinant is defined in such a way that

$$\det \mathbf{AB} = (\det \mathbf{A})(\det \mathbf{B})$$

for any two $n \times n$ matrices **A** and **B**. Two other basic properties are:

$$\det \mathbf{A}^T = \det \mathbf{A} \qquad \det k\mathbf{A} = k^n \det \mathbf{A}$$

Finally, $\det \mathbf{A} \neq 0$ if and only if **A** is nonsingular.

EXAMPLE B3 Verify the deteminant multiplication rule for

$$\mathbf{A} = \begin{bmatrix} 1 & 4 \\ 3 & 2 \end{bmatrix} \qquad \mathbf{B} = \begin{bmatrix} -2 & 9 \\ 1 & \pi \end{bmatrix}$$

We have

$$\mathbf{AB} = \begin{bmatrix} 1 & 4 \\ 3 & 2 \end{bmatrix}\begin{bmatrix} -2 & 9 \\ 1 & \pi \end{bmatrix} = \begin{bmatrix} 2 & 9+4\pi \\ -4 & 27+2\pi \end{bmatrix}$$

and

$$\begin{vmatrix} 2 & 9+4\pi \\ -4 & 27+2\pi \end{vmatrix} = 2(27+2\pi) - (9+4\pi)(-4) = 90 + 20\pi$$

But

$$\begin{vmatrix} 1 & 4 \\ 3 & 2 \end{vmatrix} = 1(2) - 4(3) = -10$$

$$\begin{vmatrix} -2 & 9 \\ 1 & \pi \end{vmatrix} = -2(\pi) - 9(1) = -9 - 2\pi$$

and indeed $90 + 20\pi = (-10)(-9 - 2\pi)$.

Laplace's Expansion Theorem

The *minor*, M_{ij}, of the element a_{ij} of a determinant of order n is the determinant of order $n-1$ obtained by deleting the row and column containing a_{ij}. The *cofactor*, Δ_{ij}, of the element a_{ij} is defined as

$$\Delta_{ij} = (-1)^{i+j} M_{ij}$$

Laplace's theorem states: In the determinant of a square matrix $\mathbf{A}$, multiply each element in the pth row (column) by the cofactor of the corresponding element in the qth row (column), and sum the products. Then the result is 0, for $p \neq q$; and $\det \mathbf{A}$, for $p = q$.

It follows at once from Laplace's theorem that if $\mathbf{A}$ has two rows or two columns the same, then $\det \mathbf{A} = 0$ (and $\mathbf{A}$ must be a singular matrix).

Matrix Inversion by Determinants; Cramer's rule

Laplace's expansion theorem can be exhibited as a matrix multiplication, as follows:

$$\begin{bmatrix} a_{11} & a_{12} & a_{13} & \cdots & a_{1n} \\ a_{21} & a_{22} & a_{23} & \cdots & a_{2n} \\ \cdots & \cdots & \cdots & \cdots & \cdots \\ a_{n1} & a_{n2} & a_{n3} & \cdots & a_{nn} \end{bmatrix} \begin{bmatrix} \Delta_{11} & \Delta_{21} & \Delta_{31} & \cdots & \Delta_{n1} \\ \Delta_{12} & \Delta_{22} & \Delta_{32} & \cdots & \Delta_{n2} \\ \cdots & \cdots & \cdots & \cdots & \cdots \\ \Delta_{1n} & \Delta_{2n} & \Delta_{3n} & \cdots & \Delta_{nn} \end{bmatrix}$$
$$= \begin{bmatrix} \Delta_{11} & \Delta_{21} & \Delta_{31} & \cdots & \Delta_{n1} \\ \Delta_{12} & \Delta_{22} & \Delta_{32} & \cdots & \Delta_{n2} \\ \cdots & \cdots & \cdots & \cdots & \cdots \\ \Delta_{1n} & \Delta_{2n} & \Delta_{3n} & \cdots & \Delta_{nn} \end{bmatrix} \begin{bmatrix} a_{11} & a_{12} & a_{13} & \cdots & a_{1n} \\ a_{21} & a_{22} & a_{23} & \cdots & a_{2n} \\ \cdots & \cdots & \cdots & \cdots & \cdots \\ a_{n1} & a_{n2} & a_{n3} & \cdots & a_{nn} \end{bmatrix}$$
$$= \begin{bmatrix} \det \mathbf{A} & 0 & 0 & \cdots & 0 \\ 0 & \det \mathbf{A} & 0 & \cdots & 0 \\ \cdots & \cdots & \cdots & \cdots & \cdots \\ 0 & 0 & 0 & \cdots & \det \mathbf{A} \end{bmatrix}$$

or

$$\mathbf{A}(\operatorname{adj} \mathbf{A}) = (\operatorname{adj} \mathbf{A})\mathbf{A} = (\det \mathbf{A})\mathbf{I}$$

where $\operatorname{adj} \mathbf{A} \equiv [\Delta_{ji}]$ is the transposed matrix of the cofactors of the a_{ij} in the determinant of $\mathbf{A}$, and $\mathbf{I}$ is the $n \times n$ unit matrix.

If $\mathbf{A}$ is nonsingular, one may divide through by $\det \mathbf{A} \neq 0$, and infer that

$$\mathbf{A}^{-1} = \frac{1}{\det \mathbf{A}} \operatorname{adj} \mathbf{A}$$

This means that the unique solution of the linear system $\mathbf{Y} = \mathbf{AX}$ is

$$\mathbf{X} = \left(\frac{1}{\det \mathbf{A}} \operatorname{adj} \mathbf{A} \right) \mathbf{Y}$$

which is Cramer's rule in matrix form. The ordinary, determinant form is obtained by considering the rth row ($r = 1, 2, \ldots, n$) of the matrix solution. Since the rth row of $\operatorname{adj} \mathbf{A}$ is

$$[\Delta_{1r} \quad \Delta_{2r} \quad \Delta_{3r} \quad \cdots \quad \Delta_{nr}]$$

we have:

$$x_r = \left(\frac{1}{\det \mathbf{A}}\right)[\,\Delta_{1r} \quad \Delta_{2r} \quad \Delta_{3r} \quad \dots \quad \Delta_{nr}\,] \begin{bmatrix} y_1 \\ y_2 \\ y_3 \\ \dots \\ y_n \end{bmatrix}$$

$$= \left(\frac{1}{\det \mathbf{A}}\right)(y_1\Delta_{1r} + y_2\Delta_{2r} + y_3\Delta_{3r} + \cdots + y_n\Delta_{nr})$$

$$= \left(\frac{1}{\det \mathbf{A}}\right) \begin{vmatrix} a_{11} & \cdots & a_{1(r-1)} & y_1 & a_{1(r+1)} & \cdots & a_{1n} \\ a_{21} & \cdots & a_{2(r-1)} & y_2 & a_{2(r+1)} & \cdots & a_{2n} \\ \dots & \dots & \dots & \dots & \dots & \dots & \dots \\ a_{n1} & \cdots & a_{nr-1)} & y_n & a_{n(r+1)} & \cdots & a_{nn} \end{vmatrix}$$

The last equality may be verified by applying Laplace's theorem to the rth column of the given determinant.

B5 EIGENVALUES OF A SQUARE MATRIX

For a linear system $\mathbf{Y} = \mathbf{AX}$, with $n \times n$ characteristic matrix $\mathbf{A}$, it is of particular importance to investigate the "excitations" $\mathbf{X}$ that produce a proportionate "response" $\mathbf{Y}$. Thus, letting $\mathbf{Y} = \lambda\mathbf{X}$, where λ is a scalar,

$$\lambda\mathbf{X} = \mathbf{AX} \qquad \text{or} \qquad (\lambda\mathbf{I} - \mathbf{A})\mathbf{X} = \mathbf{O}$$

where $\mathbf{O}$ is the $n \times 1$ null matrix. Now, if the matrix $\lambda\mathbf{I} - \mathbf{A}$ were nonsingular, only the trival solution $\mathbf{X} = \mathbf{Y} = \mathbf{O}$ would exist. Hence, for a nontrivial solution, the value of λ must be such as to make $\lambda\mathbf{I} - \mathbf{A}$ a singular matrix; that is, we must have

$$\det(\lambda\mathbf{I} - \mathbf{A}) = \begin{vmatrix} \lambda - a_{11} & -a_{12} & -a_{13} & \cdots & -a_{1n} \\ -a_{21} & \lambda - a_{22} & -a_{23} & \cdots & -a_{2n} \\ \dots & \dots & \dots & \dots & \dots \\ -a_{n1} & -a_{n2} & -a_{n3} & \cdots & \lambda - a_{nn} \end{vmatrix} = 0$$

The n roots of this polynomial equation in λ are the *eigenvalues* of matrix $\mathbf{A}$; the corresponding nontrivial solutions $\mathbf{X}$ are known as the *eigenvectors* of $\mathbf{A}$.

Setting $\lambda = 0$ in the left side of the above *characteristic equation*, we see that the constant term in the equation must be

$$\det(-\mathbf{A}) = \det[(-1)\mathbf{A}] = (-1)^n(\det \mathbf{A})$$

Since the coefficient of λ^n in the equation is obviously unity, the constant term is also equal to $(-1)^n$ times the product of all the roots. *The determinant of a square matrix is the product of all its eigenvalues*—an alternate, and very useful, definition of the determinant.

INDEX